Solid State Physics

Solid State Physics

Gerald Burns
IBM Thomas J. Watson Research Center
Yorktown Heights, New York

Academic Press, Inc.
Harcourt Brace Jovanovich, Publishers
Orlando   San Diego   San Francisco   New York
London   Toronto   Montreal   Sydney   Tokyo   São Paulo

*To Neil, Andrew, and Tracie
who, like many others, are just
beginning their quest. I wish them well.*

Copyright © 1985 by Academic Press, Inc.

All rights reserved.
No part of this publication may be reproduced or transmitted in any form or by any means, electronic or mechanical, including photocopy, recording, or any information storage and retrieval system, without permission in writing from the publisher.

Academic Press, Inc.
Orlando, Florida 32887

United Kingdom edition published by
Academic Press, Inc. (London) Ltd.
24/28 Oval Road, London NW1 7DX

ISBN: 0-12-146070-3
Library of Congress Catalog Card Number: 85-70104
Printed in the United States of America

Preface

The objective of SOLID STATE PHYSICS is to introduce college seniors and first-year graduate students in physics, electrical engineering, materials science, chemistry, and related areas to this diverse and fascinating field. I have attempted to present this complex subject matter in a coherent, integrated manner, emphasizing fundamental scientific ideas to give the student a strong understanding and "feel" for the physics and the orders of magnitude involved. The subject is varied, covering many important, sophisticated, and practical areas, which, at first, may appear unrelated but which are actually built on the same foundation: the bonding between atoms, the periodic translational symmetry, and the resulting electron energy levels. The text is comprehensive enough so that the basics of broad areas of present research are covered, yet flexible enough so that courses of varying lengths can be satisfied. The exercises at the end of each chapter serve to reinforce and extend the text.

This textbook is the outgrowth of notes for an introductory solid state physics course that I have taught several times. Parts of it have been used by several other instructors as well. To make the book more versatile, certain sections have been designated "A" or "S". "A" means the subject matter of that section is more advanced and might be saved for later study. "S" means the subject matter of the section can be saved for a second reading, not because it is difficult, but because it is less crucial in an elementary course. By using or not using the various sections and chapters, the level of the course can be changed easily. Instructors of short courses can eliminate some or all of the "A" and "S" parts and still maintain continuity.

There is often disagreement as to the most desirable order of the topics in a course such as this. Some instructors favor the traditional approach of first discussing unit cells, crystal structures, and so forth, while others claim it "turns the student off" and prefer, instead, that either bonding of solids or electron theory in metals be considered first. Naturally, each of these approaches has merit, and I have tried to satisfy instructors of both persuasions. By starting at Chapter 1 the student follows the traditional approach. However, one can get right into the physics, and considerably shorten the course, by starting at Chapter 6, where bonding is discussed. Further, one may start at Chapter 9 with the free electron theory. The details of the subject ordering for a course are discussed after a brief overview of the subject matter.

Subject Matter

Broadly speaking, the subjects are organized as follows:

(a) Chapters 1 through 3 discuss the symmetry aspects of crystalline solids. Chapter 4 covers some aspects of the use of X-rays in solid state science. Then, in Chapter 5, the anisotropic character of crystals is discussed.

(b) Chapters 6 through 8 discuss the five common types of bonding in solids and show how the underlying ideas are similar.

(c) Chapters 9 and 10 are on free electron theory and band theory. The latter covers the ramifications of translational symmetry, which forms the fundamental basis of the modern theory of solids.

(d) Chapters 11 and 12 consider the effects caused by the movement of atoms; specific heat, diffusion, and phonons are discussed.

(e) Chapter 13 brings together these various topics in a discussion of the optical properties of solids. Thus, Chapters 6 through 13 or 9 through 13 form the backbone of a traditional solid state course.

(f) Chapters 14 through 18 can be covered in any order, with no loss of continuity, depending on the interests of the readers. "A" and "S" designations are included less often in these chapters as the reader and instructor can make their own decisions.

Chapters 1 through 5 provide a firm foundation in crystal structures and crystal physics. However, if time is short, the discussions on space groups can be skipped and all of Chapter 5 can be omitted. Or, as an alternative, students may read this material on their own, followed by some class discussion. The subjects are simple enough for this to be accomplished, especially among more mature students.

I have tried to keep the chapters focused on different fundamental areas of solid state physics to help unify this field, which sometimes may seem diverse and even fragmented. This results in some subjects being covered in several different chapters. For example, the properties of semiconductors are covered in Chapter 8 (covalent bonds and the forbidden gap), Chapter 10 (bands and effects due to bands), Chapter 13 (optical properties), and Chapter 18 (superlattice and inversion layer properties). Similarly, various aspects of metals are discussed in Chapters 9, 10, 13, and 18. In this manner the interrelationships of the different types of solids are better understood.

Where to Start

The book can be started conveniently in Chapter 1, Chapter 6 (where bonding is discussed), or Chapter 9 (where the free electron theory of metals is presented). Each of these chapters is denoted by an asterisk in the table of

(1) If the book is started in Chapter 6, Section 6-1 ("Summary of Chapters 1–3") provides a review of the elements of translational symmetry. Also covered in Section 6-1 is a brief discussion of the 7 crystal systems, 14 Bravais lattices, the meaning of a crystal structure, and related matters. However, these topics can be put off until required in later studies.

(2) If the book is started in Chapter 9, no preliminary subjects are required. However, before going on to Chapter 10, Section 6-1 should be covered, as well as Section 4-3 and Section 8-6.

(3) If the book is started in Chapter 1, after covering Chapters 1 through 3, you may proceed to either of the above starting points.

Units

For most of the equations, any consistent system of units can be used. However, when equations would appear differently in SI units than in cgs-Gaussian units, an explicit statement is given of how to convert the latter to SI units. Important equations, such as Maxwell's equations, are rewritten in SI units. In the tables and text, experimental results are given in the units used by most practicing solid state physicists. Their usage was developed for practical reasons and although unsatisfactory to the purist, there is little likelihood of change. Thus, energy magnitudes tend to be measured in electron volts (eV), degrees Kelvin (°K), and in the common spectroscopic term wave numbers (cm^{-1}). Formally, given an energy $\mathscr{E}$, the energy expressed in °K is $\mathscr{E}/k_B$, and expressed in cm^{-1} is $\mathscr{E}/hc$. The following equivalent values for energies should help (10^{-3} eV = 1 meV).

$$1 \text{ meV} = 8.07 \text{ cm}^{-1} = 11.6°K$$
$$1 \text{ cm}^{-1} = 1.44°K$$
$$1 \text{ cm}^{-1} = 3 \times 10^{10} \text{ cycles/sec} = 3 \times 10^{10} \text{ Hz}$$

Approximate room temperature values of often used quantities are $k_B T = 0.025$ eV and $k_B T/e = 0.025$ V. The appendix in the back of the book gives conversion tables for the different units and values for many fundamental constants. Last, abbreviations for seconds, degrees Kelvin, and degrees Centigrade are sec, °K, and °C, respectively. In a book that covers such a wide range of subjects using so many letters of the alphabet, these abbreviations seem more desirable and easier for the student than the naked letters s, K, and C. Undoubtedly, there will be other inconsistencies, unclear sections, or problems with the text. I would appreciate hearing from the readers about any shortcomings that can be corrected in future printings.

Acknowledgements

In writing this book I have constantly appreciated my debt for the huge amount of help that I have had from so many people. To all of you who have helped in large and small ways, I would like to say, thank you very much.

The book could not possibly have been completed without the immense help of Frank Dacol and Janis Riznychok. Not only did they take care of all aspects of manuscript preparation—figures, word processing, organizing, retyping, correcting many of my mistakes, making helpful suggestions—but they also kept me going when I needed encouragement. I truly owe them much, and I thank them both.

Major portions of the book were read and, consequently, many improvements suggested by Hans Hoffmann, Ann Carol Hohl, and Arthur Zingher. Also, Marshall Nathan and Ted Schultz helped a great deal on many important and difficult points. The input from these five people has improved the book considerably.

Various parts of the book were used in courses at several universities. As a result, Elias Lopez-Cruz, Roy Clark, Mel Lax, and Gertrude Neumark made many suggestions that helped improve the presentation of the material.

Extensive portions of the text were read by Gail Bunt, Eli Burstein, Mike Heaney, John Slonczewski, and Frank Stern. I am very grateful to these people for the help that they have given me.

John Axe, Norm Brauslau, Andrew Burns, Annette Bussman-Holder, G. V. Chandrashekhar, Linda Geppert, Linda Hadel, Bob Keyes, Jim Janak, Paul Marcus, Farrokh Mehran, Arthur Nethercot, Art Nowick, and Theo Siegrist have read sizeable portions of the text. Their work resulted in many important improvements in the text.

Due to the efforts of Slade Cargill, Cheng Chung Chi, Wan Yee Cheug, Bill Dumke, Alan Fowler, Franz Himpsel, Franco Jona, Roger Koch, Bob Laibowitz, Norton Lang, Pam Leary, Jim Misewich, Pat Mooney, Fred Morehead, Tom Morgan, Tom Penney, Mel Pomerantz, Roger Pollak, Erling Pytte, Bruce Scott, Claudia Tesche, Bob von Gutfeld, and Henry Weinberg many other important improvements resulted. The cartoons were done by Claire Albahae, and I thank her very much.

Also, I would like to thank the Academic Press staff for the help they have given me.

The text is directly photo offset from output of our word processing programs here at IBM, Yorktown Heights. I would like to thank the people who have made the system possible and answered the many questions that arose in the course of this work.

Contents

Symmetry Operations   *1

- 1-1   A Symmetry Operation   4
- 1-2   Point Symmetry Operations   5
- 1-3   The Point Groups of a Molecule   9
- 1-4   Other Symmetry Operations of Crystals   17
- Notes   20
- Problems   20

Symmetry Description of Crystals   2

- 2-1   Lattice   25
- 2-2   Primitive Unit Cell   26
- 2-3   The 7 Crystal Systems   26
- 2-4   The 14 Bravais Lattices   29
- 2-5   The 32 Crystallographic Point Groups   35
- 2-6   Space Groups   38
- 2-7   Definitions of Directions, Coordinates, and Planes   43
- Appendix to Chapter 2   46
- Notes   47
- Problems   48

Simple Crystal Structures   3

- 3-1   Introduction   51
- 3-2   Several Cubic Symmorphic Structures   51

---

*The book may be started in Chapter 1, Chapter 6, or Chapter 9.

- 3-3 Diamond and Zinc Blende Structures   56
- 3-4 Point Group of a Space Group (S)   58
- 3-5 Examples of Defect Structures   60
- 3-6 Different Points of View of a Structure   61
- 3-7 Close Packing (and the Hexagonal Close-Packed Structure)   62
- 3-8 Volume Effects for Simple Structures   65
- 3-9 Wurtzite Structure   66
- 3-10 Site Symmetry (S)   67
  - Notes   68
  - Problems   69

4 X-Ray Diffraction

- 4-1 Electron, Neutron, and X-ray Diffraction   73
- 4-2 Bragg's Law   75
- 4-3 The Laue Formulation   77
- 4-4 Experimental X-ray Diffraction Methods (S)   81
  - Notes   83
  - Problems   83

5 Crystal Symmetry and Physical Properties (S)

- 5-1 Introduction   87
- 5-2 Neumann's Principle   88
- 5-3 Tensors   88
- 5-4 Crystal Symmetry and Physical Properties   90
- 5-5 Nonlinear Optics   96
  - Notes   98
  - Problems   98

*6 Classification of Solids

- 6-1 Summary of Chapters 1–3   103
- 6-2 Introduction to Classification of Solids   112
- 6-3 Five Types of Bonds   112
- 6-4 Repulsive Potential Energy   115
- 6-5 Molecular Bond   118
- 6-6 Hydrogen Bond (S)   124
  - Notes   127
  - Problems   128

CONTENTS   xi

7 The Ionic Bond

7-1   Transfer of Electrons   131
7-2   Ionic Radii   133
7-3   Typical Structures   134
7-4   Cohesive Energies of Ionic Crystals   138
      Notes   143
      Problems   144

8 The Covalent Bond

8-1   Introduction   149
8-2   Bonding and Antibonding   150
8-3   The Hydrogen Molecule   154
8-4   Maximum Overlap   157
8-5   The Formation of a Crystal   164
8-6   "Classical" Semiconductors   168
8-7   Continuous Range of Bonding (S)   175
      Appendix   183
      Notes   184
      Problems   185

*9 Metals

PART A   DRUDE'S MODEL   191

9-1   Drude's Free Electron Theory   191
9-2   Drude's Assumptions   195
9-3   DC Conductivity   196
9-4   Wiedemann-Franz Law   197
9-5   Frequency-Dependent Conductivity (S, A)   198
9-6   Problems of Drude's Model   201

PART B   QUANTUM MECHANICS APPLIED   203

9-7   Eigenfunctions of Free Electrons in a Metal   203
9-8   Fermi Energy, Density of States, and Fermi
       Surface   208
9-9   Soft X-rays, Heat Capacities   213
9-10   Fermi-Dirac Statistics   215
9-11   Low Temperature Expansion Using F-D Statistics   216
9-12   Thermal Properties of the Electron Gas   217
9-13   DC Conductivity (with F-D Statistics)   223
9-14   Electron–Electron Collisions (S)   225
9-15   Hall Effect (and Other Magnetic Field Effects) (S)   228

9-16 Landau Levels (S, A) 233
Notes 235
Problems 236

10 Band Theory

PART A QUALITATIVE DISCUSSION 243

10-1 Nearly Free Electrons 243
10-2 Classifications of Solids 247
10-3 Effective Mass 248

PART B WAVE FUNCTIONS AND ENERGY LEVELS 252

10-4 Bloch Functions 252
10-5 Nearly Free Electrons 257
10-6 Brillouin Zones 260
10-7 Examples of Brillouin Zones 263
10-8 Wigner-Seitz Approximation — The Binding Energy (S) 273
10-9 The Tight Binding Approximation (S) 276
10-10 Crystal Momentum 280

PART C SEMICONDUCTORS, REAL BANDS, AND RELATED CONCEPTS 281

10-11 Holes 281
10-12 Band Preliminaries (A) 289
10-13 $\mathscr{E}_{(k)}$ for a Two-Dimensional Square Lattice 293
10-14 Body-Centered Cubic Lattice — Sodium (S, A) 302
10-15 Si, Ge, GaAs, and GaP 304
10-16 Carrier Concentration at Thermal Equilibrium 313
10-17 p-n Junctions 323
10-18 Metal-Semiconductor Junctions 334
10-19 The Gunn Effect (S) 337
10-20 Other Topcis (S) 339
10-21 Summary 345
Notes 348
Problems 349

11 Some Thermal Effects in Solids

PART A HEAT CAPACITY 355

11-1 Specific Heat at Constant Volume and Pressure 355
11-2 Energy and C_V from Statistical Mechanics 357
11-3 Classical Results for C_V 360
11-4 Einstein's Model 362
11-5 Debye's Calculation of C_V 365

CONTENTS   xiii

PART B   EFFECTS ASSOCIATED WITH DISORDER   371

11-6   Orientational Disorder in Molecular and Ionic
       Crystals   371
11-7   Polarization by Orientation (S)   379
11-8   Point Imperfections in Crystals   385
11-9   Diffusion (S)   389
11-10 Color Centers in Ionic Crystals (S)   396
11-11 Localized Vibrational Modes (S)   398
       Notes   399
       Problems   401

Lattice Vibrations   12

12-1   Introduction   407
12-2   Vibrations of a One-Dimensional Monatomic
       Chain   408
12-3   Vibrations of a One-Dimensional Diatomic Chain   412
12-4   Real Crystal Systems   419
12-5   Phonons (A)   425
12-6   Crystal Momentum (A)   428
12-7   Neutron Diffraction from Phonons   430
12-8   Thermal Conductivity (S)   433
       Notes   441
       Problems   443

Optical Properties of Crystals   13

PART A   MACROSCOPIC THEORY   450

13-1   Dielectric Polarization   450
13-2   Oscillating Fields   452
13-3   Electromagnetic Waves in Solids   454
13-4   Reflectivity at an Interface   457
13-5   Kramers-Kronig Relations (S, A)   458
13-6   Damped Harmonic Oscillator   461
13-7   Dielectric Response of a Quantum System   464

PART B   LATTICE VIBRATIONS   465

13-8   Introduction   465
13-9   Long Wavelength Optical Vibrations   466
13-10 Measurements and Results   471
13-11 Polaritons (S)   476
13-12 A Microscopic Model (S)   480
13-13 Clausius-Mossotti (Lorenz-Lorentz) Equations (S)   484

PART C FREE CARRIER ABSORPTION 486

13-14 Introduction 486
13-15 Oscillator Model 487
13-16 Experimental Results 490
13-17 Transverse and Longitudinal Free Electron Modes (S) 495

PART D INTERBAND TRANSITIONS 498

13-18 Introduction 498
13-19 Fundamental Absorption Near $\mathscr{E}_g$ 500
13-20 Excitons (Mostly Weakly Bound Excitons) 509
13-21 Fundamental Absorption Above $\mathscr{E}_g$ 520
13-22 Urbach Edge (S) 521
Notes 524
Problems 526

14 Ferroelectricity and Structural Phase Transitions

14-1 Introduction 531
14-2 The Free Energy 536
14-3 Soft Modes 542
14-4 Microscopic Model of Soft Modes 550
14-5 Renormalization Group 552
14-6 Optical Properties of Ferroelectrics (S) 554
14-7 Other Related Properties
Notes 559
Problems 562

15 Magnetism

PART A DIAMAGNETISM AND PARAMAGNETISM 565

15-1 Introduction 565
15-2 Diamagnetism 567
15-3 Paramagnetism 569

PART B FERROMAGNETISM, ANTIFERRO-MAGNETISM, AND RELATED TOPICS 584

15-4 Introduction 584
15-5 Molecular Field Theory 584
15-6 The Heisenberg Exchange Interaction 588
15-7 Magnetic Structures 590
15-8 Special Techniques Used to Study Magnetic Structures 596

CONTENTS   xv

PART C OTHER TOPICS 604

15-9   Spin Waves (S, A)   604
15-10 Anisotropy, Hysteresis, Domains, and Bloch Walls   612
15-11 Metals and Magnetism (S, A)   619
15-12 Spin Glasses (S)   625
       Notes   627
       Problems   629

Superconductivity   16

16-1   Introduction (dc Conductivity)   633
16-2   The Occurrence of Superconductivity   634
16-3   Effects that Destroy Superconductivity   635
16-4   Magnetic Properties   637
16-5   The BCS Theory   642
16-6   BCS Predictions   648
16-7   BCS Related Measurements   653
16-8   The Josephson Effect   659
       Notes   666
       Problems   668

Surface Science   17

17-1   Introduction — The Need for UHV   674
17-2   Crystal Shape   675
17-3   Preparation of Clean Surfaces and LEED   677
17-4   The Structure of Surfaces   679
17-5   Interaction of Gases with Surfaces   686
17-6   Surface Related Techniques   692
17-7   Electronic Surface Structure   702
       Notes   710
       Problems   712
       Appendix to Chapter 17   712

Artificial Structures   18

PART A SEMICONDUCTORS 716

18-1   Introduction   716
18-2   A Particle in a 1-D Rectangular Well   717
18-3   3-D Motion with a 1-D Rectangular Well   719
18-4   Experimental Aspects   724
18-5   Semiconductor Superlattices   726
18-6   Inversion Layers   737

PART B METALS 747
18-7 Introduction 747
18-8 Sample Preparation 748
18-9 Properties of Layered Metal Structures 749
18-10 Other Artificial Structures (S) 752
      Notes 753
      Problems 755

Appendix — Units 757

Bibliography 761

Index 765

1

Symmetry Operations

1-1 A Symmetry Operation
1-2 Point Symmetry Operations
    a Principal axis
    b Point symmetry operations
    c Inverse of symmetry operations
    d Examples
    e Stereographic projection
1-3 The Point Groups of a Molecule
    a Development of point groups
    b Point group flowchart
1-4 Other Symmetry Operations of Crystals
    a Translation
    b Screw operations (S)
    c Glide planes (S)
    d Point group of a crystal

Notes
Problems

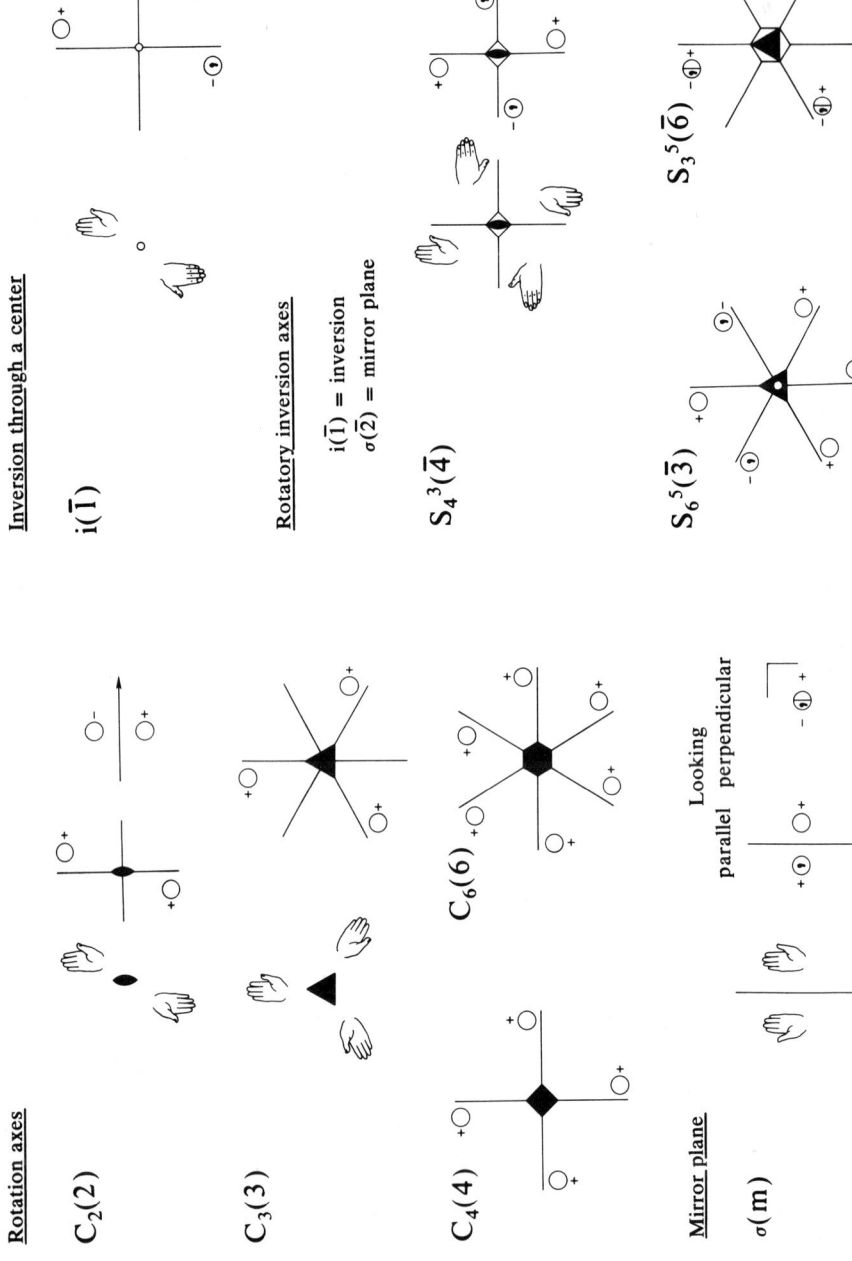

Fig. 1-2 The point symmetry operations and the symbols used to designate them. Hands are used to show the effect for several symmetry operations.

SYMMETRY OPERATIONS

All that will come when it will; but we, meanwhile,
Have much to do. Leave the future to itself.

Sophocles, *"Antigone"*

One of the most noticeable things about molecules is their "symmetry," just as in looking at people one notices that the left side is (approximately) the mirror image of the right side. (Try it with your hands and a mirror.) Similarly, crystals have a fundamental symmetry in that they are formed from a "unit cell" repeated throughout space, much as identical bricks repeated in three dimensions form a wall. A closer look also reveals symmetries within this "unit cell" similar to those in an isolated molecule. The terms "symmetry" and "unit cell" will be defined shortly. This paragraph is intended only to point out that they are not new concepts and are used intuitively by all of us. Even though the concepts are simple and intuitive they are extremely important. In fact, the idea that a crystal is made up of a unit cell repeated throughout space forms the foundation of the modern theory of solids. For this reason the first three chapters of this book are devoted to a discussion and amplification of these ideas. However, as mentioned in the preface, one also may start this book with Chapter 6, where there is a very brief overview of the important concepts needed in the rest of the text. Nevertheless, when the first three chapters are studied in detail, the result will be a good background in crystal structures and their symmetries. The reader will find this basic to all studies in solid state science.

In this chapter, the point group operations for molecules are discussed first. Next, the important symmetry operations for crystals are introduced which, along with the point group operations, include translation, glide operations, and screw operations. Then, given any molecule, we show how the point group, and a symbol for the point group, can be determined. This symbol alone describes all the symmetry operations possessed by the molecule. A similar procedure is used to determine the point group of a crystal. In fact, in Chapter 5 it is shown how the point group of the crystal determines the physical behavior of crystals in different directions (the anisotropy).

Warning: There are many new concepts and a considerable amount of new language introduced in this chapter. Neither the concepts nor the language are difficult, but it is important to understand them as they will be used often in the future.

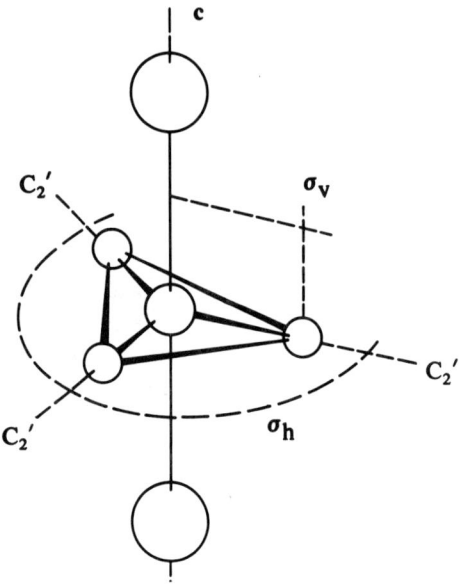

Fig. 1-1 A diagram of a PF_3Cl_2 molecule showing some of the symmetry operations. Naturally, an operation that takes a F atom into a Cl atom cannot be a symmetry operation, so it might be helpful to label the atoms.

1-1 A Symmetry Operation

A **symmetry operation** for a molecule or crystal is an operation that interchanges the positions of the various atoms in such a way that the molecule or crystal appears exactly as before the operation. (The molecule or crystal is said to be in an *equivalent position*.) Furthermore, such an operation, when repeatedly applied, must also produce an equivalent position. Symmetry operations are sometimes called **covering operations**.

As an example, consider the PF_3Cl_2 molecule shown in Fig. 1-1. Twelve symmetry operations can be found. Some of the easier ones to see are: three rotations by 180° about each of the P-F axes, labeled C_2' in the figure; similarly, three mirror planes, each plane containing the Cl atoms, the P atom, and a different F atom, one of which is labeled σ_v in the figure; and rotations by 120° and 240° about the c-axis. One very important symmetry operation is the **identity** operation, which leaves the molecule unchanged. This is possessed by all objects.

In general, many different types of symmetry operations are used, such as the interchange of electrons, time reversal, and the interchange of quarks in particle physics. However, in the following chapters we shall consider only the spatial interchange of atoms. In the next section these types of symmetry operations are examined systematically and in many of the latter chapters the symmetry is used implicitly if not explicitly.

CHAPTER 1   SYMMETRY OPERATIONS   5

1-2 Point Symmetry Operations

A **point symmetry operation** is a symmetry operation carried out with respect to a fixed point in space, that is, the point does not move during the operation. For example, the P-atom in Fig. 1-1 would be such a point for the operations just discussed. In order to discuss point symmetry operations it is useful to define a principal axis.

1-2a Principal axis   The axis of highest symmetry of a molecule or crystal is called the **principal axis** or **c-axis** or **z-axis**. In Fig. 1-1 the Cl-P-Cl axis is the principal axis because there are more symmetry operations about this axis than any other. Axial molecules or crystals will always have a principal axis. For example, a fourfold axis (rotations by 0°, 90°, 180°, and 270° is higher than a twofold axis (rotations by 0° and 180°). However, this concept is not always appropriate since there are many molecules and crystals for which no one axis has higher rotation symmetry than any of the others. For these cases there is no principal axis. (For example, an orthorhombic crystal may have three twofold axes; in CH_4 or other high symmetry molecules there is no principal axis.)

1-2b Point symmetry operations   Table 1-1 summarizes the meaning of the point symmetry operations using the **Schoenflies notation**. Figure 1-2 (page 2) shows diagrams of the effects of these symmetry operations, repeated to bring the object back to the starting position. Also shown are the diagrammatic symbols, such as the triangle, diamond, and hexagon, that are used to denote the symmetry operations. The **notation** is simple. The circles with +, −, and commas are convenient ways to represent any *general object* such as a collection of atoms. Hands are also used for the simpler operations because they are a convenient general object. The + and − signs signify a general object either above or below the plane of the paper. A circle with a comma in it is the **enantiomorphic** or mirror image of a circle without a comma. The left and right hands are the enantiomorphic images of each other. Mirror planes and centers of inversion will produce enantiomorphic images of molecules. A split circle means one circle is above the plane and one below. These symbols, which are used particularly in space group diagrams, conveniently show the meanings of the point symmetry operations listed in Table 1-1. Warning: A circle with a "+" next to it does not mean a circle is above the plane of the paper. It means a general arbitrarily shaped object is above the paper. Hands are ideal to use but they tend to clutter the diagrams.

Although the Schoenflies notation is used by most solid state chemists and physicists, crystallographers always use the **International**

Table 1-1 Symmetry operations.

E   **Identity** – The molecule or crystal is not rotated at all (or rotated by 2π about any axis). All objects possess this symmetry operation. Taking **r** as a vector from the origin to any point. E**r** → **r**.

i   **Inversion** – The molecule is inverted through some origin, which is called the center of inversion. (An equilateral triangle and a five-pointed star do not have i as a symmetry operation, but a square and six-pointed star do.) The operation is $i(x, y, z) \to (-x, -y, -z)$.

C_n **Rotation** – Sometimes called a **proper rotation**. A rotation of the molecule by $360°/n$ or $2\pi/n$ about an axis, in the sense of a right-hand screw by **convention**. If the axis is not the principal axis (i.e, c-axis or z-axis), often, but not always, there will be a prime or other superscript: C_n' or C_n^x where C_n^x means a C_n operation about the x-axis. This is different from a numerical superscript, for example C_n^2 means apply C_n two times, that is, C_3^2 is the same as a rotation by $240°$. (PF_3Cl_2 in Fig. 1-1 has C_3, C_3^2 and three C_2'.) For example, taking the z-axis as the principal axis, $C_4(x, y, z) \to (y, -x, z)$, and $C_2^x(x, y, z) \to (x, -y, -z)$.

σ **Reflection** – Reflection of the molecule in a plane. The particular plane of reflection will sometimes be specified by a subscript. Reflection in a plane means the transfer of all of the points to the other side of the plane an equal distance along perpendiculars to the plane. For example, for a reflection across the x-plane, which contains the yz-axis, $\sigma(x, y, z) \to (-x, y, z)$.

σ_h **Reflection in the horizontal plane** – The plane of reflection is perpendicular to the principal axis and contains the origin. (PF_3Cl_2 has one σ_h.)

σ_v **Reflection in a vertical plane** – The plane contains the principal axis. (PF_3Cl_2 has three σ_v through the z-axis.)

σ_d **Reflection in a diagonal plane** – Like σ_v this plane also contains the principal axis but bisects the angle between the twofold axes perpendicular to the principal axis.

S_n **Improper rotation** – A rotation by $2\pi/n$ followed by a reflection in a horizontal plane. For example, $S_n = \sigma_h C_n$. Again $S_n^m \equiv (S_n)^m \equiv (\sigma_h C_n)^m$. (This is a subtle operation so please <u>note</u>: $S_2 = i$; $S_3^2 S_3 \neq E$ rather $S_3^3 = \sigma_h$ and $S_3^6 = E$; try it. An object with S_n symmetry need not have σ_h and C_n as symmetry operations. Methane CH_4 is one such example.) $S_4(x, y, z) \to (y, -x, -z)$.

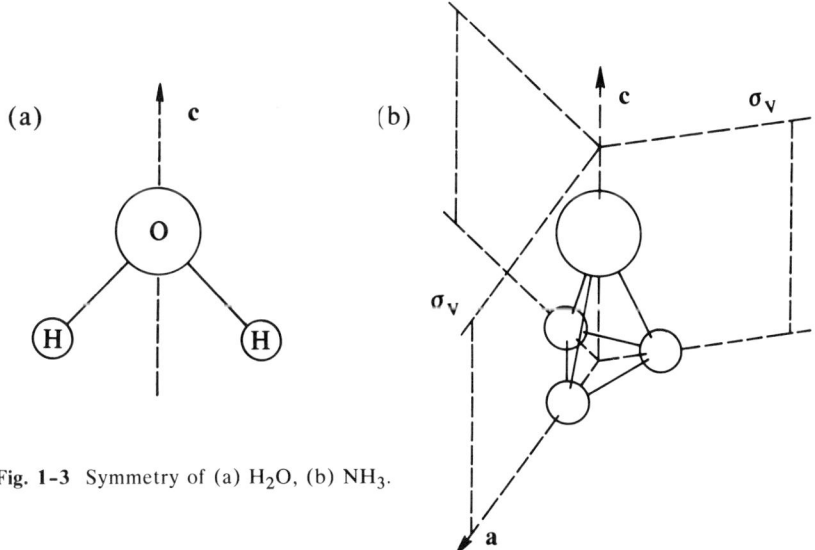

Fig. 1-3 Symmetry of (a) H_2O, (b) NH_3.

(or **Hermann-Mauguin**) **notation**. The correspondence of the symbols 1, $\bar{1}$, n, and m to the Schoenflies symbols is

$$1 \equiv E, \quad \bar{1} \equiv i, \quad n \equiv C_n, \quad m \equiv \sigma, \quad \bar{n} \equiv iC_n \tag{1-1}$$

The use of the rotatory inversion axis, $\bar{n}$, where a rotation is followed by an inversion rather than a reflection as in the Schoenflies approach, sometimes makes going from one notation to another a bit of a nuisance, as can be seen in Fig. 1-2. Nevertheless, it is necessary, for a solid state scientist at least, not to be intimidated by the International notation. Thus, for pedagogical purposes we will follow an unusual **convention**. *When not inconvenient, we will write the symmetry operation symbols (and later point group and space group symbols) in the Schoenflies notation immediately followed, in parentheses, by the International symbol.* This is done in Fig. 1-2. Problem 2 helps us to understand how to go between these two notations and to see why $\bar{6} = S_3^5$ and $\bar{3} = S_6^5$. Don't be intimidated!

1-2c Inverse of symmetry operations Note that the inverse of every symmetry operation must also be a symmetry operation. The **inverse** of a symmetry operation A is a symmetry operation B such that the application of A followed by B will return the molecule or crystal to a position identical (not just equivalent) to the starting position, that is, $BA = E$. Note the **convention** in the order of writing B to the left of A. Several symmetry operations, such as $E(1)$, $i(\bar{1})$, and $\sigma(m)$ are their own inverses. Other situations are more complicated, that is, $C_3^2 C_3 = E$ or $S_3^5 S_3 = E$; the latter must be tried to be believed. In general

Symmetry Operation	Inverse	
C_n^m	C_n^{n-m}	
S_n^m	S_n^{n-m}	m, n even
S_n^m	S_n^{2n-m}	m, n odd

$S_n^m = C_n^m$ for n odd and m even.

1-2d Examples We give several examples of the symmetry operations of molecules shown in Fig's. 1-1 and 1-3.

Example 1. PF_3Cl_2 has E, C_3 and C_3^2 about the principal axis. There are three σ_v containing the principal axis. Three C_2 are perpendicular to the principal axis as is a σ_h-plane. Then there are S_3 and S_3^5. The 12 operations would usually be written as E, C_3, C_3^2, $3C_2'$, $3\sigma_v$, σ_h, S_3, S_3^5.

Example 2. The four symmetry operations of H_2O are E, C_2, $2\sigma_v$. Both planes contain the c-axis and one contains the molecule.

Example 3. NH_3 has E, C_3, C_3^2, $3\sigma_v$. Note that in PF_3Cl_2 there is one point that must be considered as the origin for these point symmetry operations, while for H_2O and NH_3 any point along the principal axis can be considered the origin.

1-2e Stereographic projection A **stereographic projection** is a very useful way to visualize the effects of point symmetry operations. It is usually constructed by assuming that a unit sphere surrounds the object. A point in the +z hemisphere is projected onto the xy-plane by determining the intersection, with that plane, of the line connecting the point to the south pole of the sphere. If the point is in the −z hemisphere the north pole is used. We observe the projections by looking down the +z-axis. The points in the +z or −z hemisphere are respectively labeled by a circle (o) or a dot (•) on the xy-plane. It is important to realize that **general points**, also called **general equivalent positions** (perhaps best called arbitrary points), are pictured in the stereographic projection. That is, the points must not be on a symmetry plane, line, or point. Several examples should clarify this sometimes confusing subject. Refer to Fig. 1-4.

Example a. H_2O has four symmetry operations E, C_2, and $2\sigma_v$. Start with a circle at position 1 and apply E(1), naturally resulting in the same circle at position 1; apply $C_2(2)$ to the starting circle at 1, resulting in 2; apply $\sigma_v(m)$ to the starting circle 1, resulting in 3; apply the other $\sigma_v(m)$ to the starting circle at 1, resulting in 4. Note that the order of application of the symmetry operations and the starting position are irrelevant. Also note the **convention** that thick lines representing the $2\sigma_v$ and the symbol for C_2 in the center, as in Fig. 1-2. *Since stereograms are for general points there always are as many points as there are symmetry operations.*

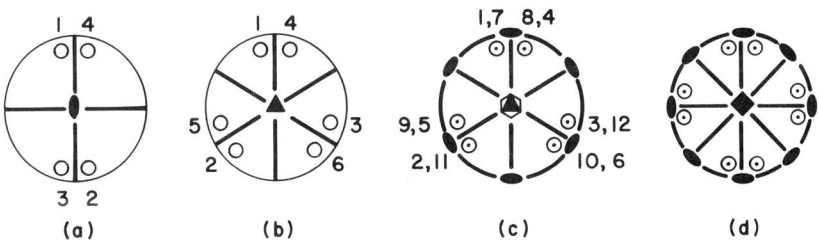

Fig. 1-4 Stereograms for some of the examples.

Example b. The molecule NH_3 has E, C_3, C_3^2, $3\sigma_v$. E(1) applied to 1 gives 1; $C_3(3)$ applied to 1 gives 2; $C_3^2(3^2)$ applied to 1 gives 3; and each of the $3\sigma_v$ applied to 1, 2, and 3, respectively, gives 4, 5, and 6. Again notice the thick lines in the figure for the $3\sigma_v$ and the symbol in the center for C_3.

Example c. PF_3Cl_2 has E, C_3, C_3^2, $3C_2'$, σ_h, $3\sigma_v$, S_3, S_3^5. The results are labeled in the same order as the symmetry operations where the first entry refers to the circle and the second to the dot. Notice that the thick lines again represent the $3\sigma_v$ and the line for the large circle, which represents σ_h, is also thick. The symbol in the center refers to $S_3(\overline{6}^5)$, while the $3C_2'$ in the plane of the paper, the xy-plane, are also shown.

Example d. The stereogram for an object with the following symmetry operations is given: E, C_4, C_2, C_4^3, $2C_2'$, $2C_2''$, i, S_4, S_4^3, σ_h, $2\sigma_v$, $2\sigma_d$. (C_2' is in the σ_v plane and C_2'' is in the σ_d plane.) Note the thick lines for the $2\sigma_d$ as well as $2\sigma_v$ and the symbols for the four twofold axes in the xy-plane. The line for the large circle also is thick, representing σ_h.

Warning: The notations in Fig. 1-2 and 1-4 are different. In Fig. 1-2 an object below the plane of the paper is a circle with a $-$ sign, while in Fig. 1-4 it is a dot. Commas are used in Fig. 1-2, while no such comparable symbol appears in Fig. 1-4. The notation of circles with $\pm$ signs and commas in Fig. 1-2 is used for space groups while the circle and dots in Fig. 1-4 are used for point groups. Although it may be confusing at first, this is common usage and we shall go along with this notation.

1-3 The Point Groups of a Molecule

Now we can write down all the symmetry operations for any molecule. Further, given all the symmetry operations, its stereogram may be drawn. We might ask if there is some shorthand notation so that this information can be passed succinctly to others. The answer is yes, and we now discuss how to determine the *point group of a mole-*

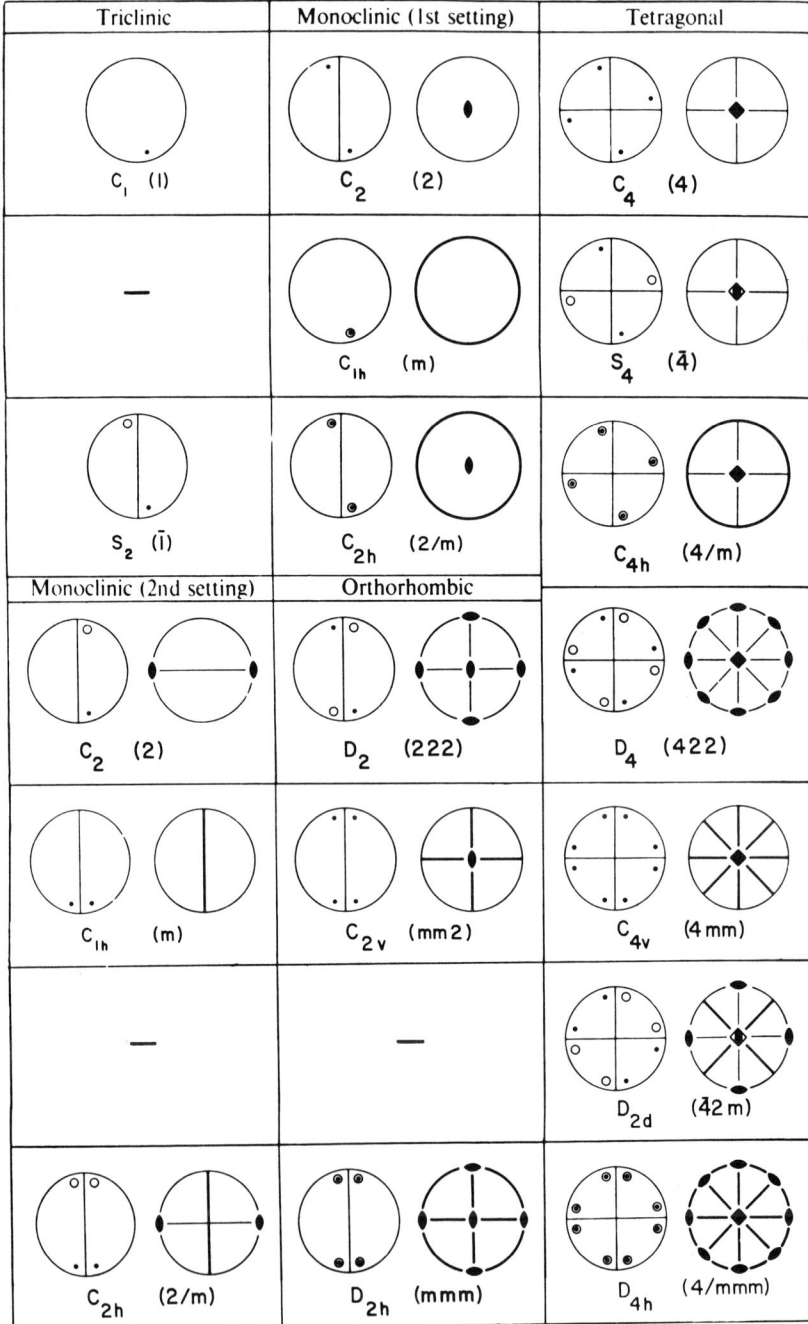

Fig. 1-5 Stereograms of the 32 crystallographic point groups.

CHAPTER 1 SYMMETRY OPERATIONS

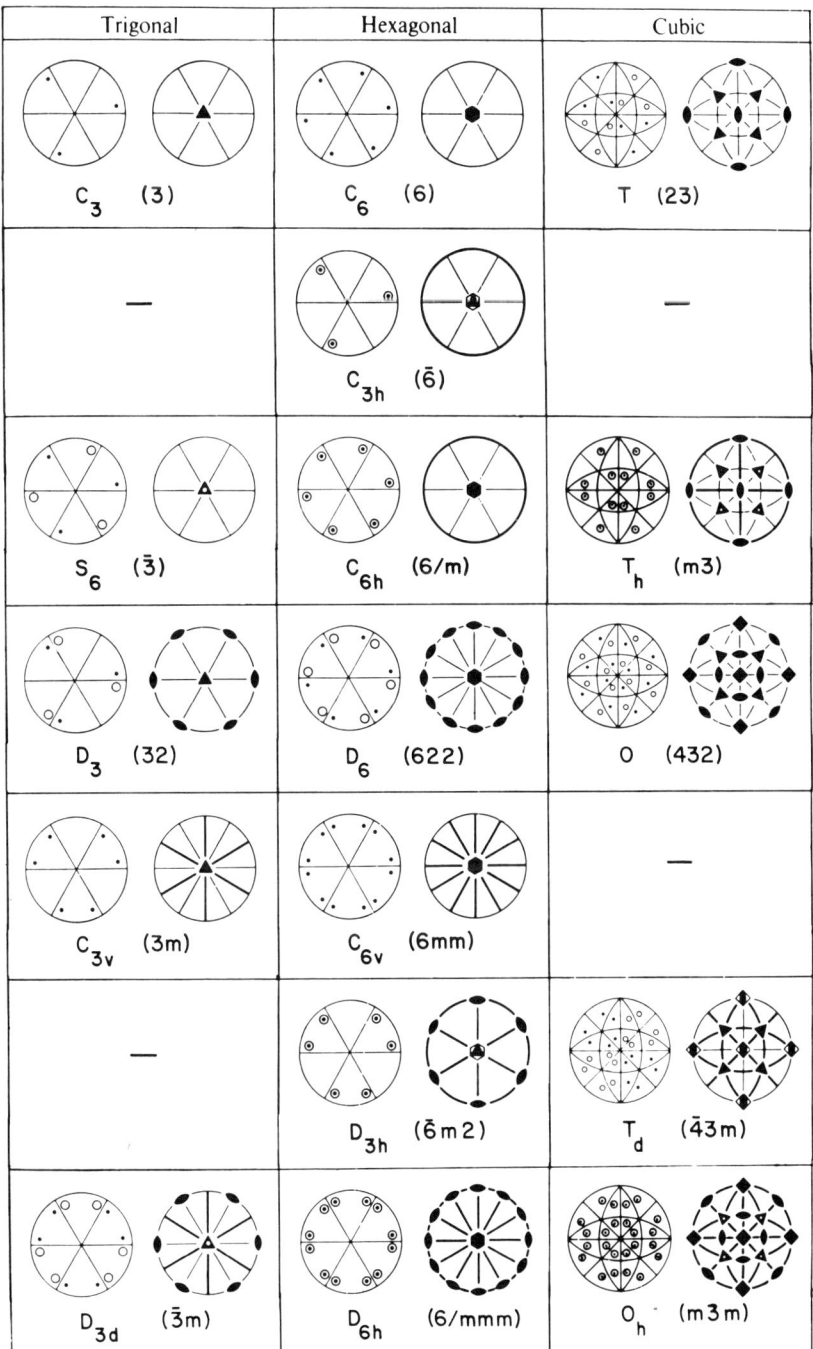

Fig. 1-5 Stereograms of the 32 crystallographic point groups.

cule and the associated symbol. Given just this symbol we can determine the symmetry operations or vice versa. (The concept of a point group has a deeper meaning and important uses. The point group of a crystal determines the symmetry of the macroscopic physical properties, (Chapter 5). Actually, the collection of symmetry operations is a group in the mathematical sense, but this is not discussed here. Thus, the wave functions associated with the electronic and vibrational properties of the crystal can be classified according to the irreducible representations of the point group or its subgroups.)

1-3a Development of point groups We will develop the point groups and their symbols. For molecules there is no restriction on n for the rotational symmetry operations C_n or S_n. However, for crystals, due to translational symmetry the values of n are restricted to n = 1, 2, 3, 4, and 6, as shown in the appendix in Chapter 2. Therefore, we continue this discussion using only these values, while understanding that for molecules there are no such restrictions. The discussion is appropriate for either case. However, using these values of n, we will have the well-known 32 crystallographic point groups.

First consider molecules that can have just rotational, C_n, symmetries. We can visualize a molecule with just one symmetry operation, E(1); or a molecule with two symmetry operations, E(1) and C_2(2); or a molecule with three symmetry operations, E(1), C_3(3), and $C_3^2(3^2)$; etc., right up to a molecule with six symmetry operations, which would be E, C_6, C_3, C_2, C_3^2, and C_6^5. Each of these five collections of symmetry operations, which just includes the identity and proper rotations, is given a name: the **point group** C_1, C_2, C_3, C_4, and C_6. (Note the confusing nomenclature. For example, the point group C_6 also has a symmetry operation called C_6. Naturally, the point group also has other symmetry operations, for example, C_6^2, C_6^3, and so on. For most of the point groups discussed below this confusion does not occur.) We list these point groups including the International symbol.

$$C_1(1), \quad C_2(2), \quad C_3(3), \quad C_4(4), \quad C_6(6) \qquad (1-2)$$

Molecules or crystals that possess a particular set of symmetry operations are said to have the **point symmetry** C_n(n) and Fig. 1-2 shows generalized arrangements for all of these point groups except the first. The number of symmetry operations in each of these five point groups is 1, 2, 3, 4, and 6, respectively.

Next we consider molecules that contain symmetry operations other than just proper rotations. Five more point groups can be developed by adding a σ_h-plane to each of the five point groups in Eq. 1-2. For example, the point group C_3 has symmetry operations {E, C_3, C_3^2}; by adding a σ_h-plane we obtain the point group C_{3h}, which has {E, C_3, C_3^2, $\sigma_h E = \sigma_h$, $\sigma_h C_3 = S_3$, $\sigma_h C_3^2 = S_3^5$}. The shorthand, group theoretic, way to write this is $C_{nh} = C_n \times \{E, \sigma_h\}$, which means the collection of operations obtained by multiplying every operation in the point group C_n by E plus every one in C_n by σ_h. Thus, from the five point groups in Eq. 1-2, five new point groups are obtained, each with twice as many symmetry operations. We list these using both Schoenflies and International notation, except for the last, which is not one of the 32 crystallographic point groups and thus has no International symbol. (It could be called 6/m.)

$$C_{1h}(m), \quad C_{2h}(2/m), \quad C_{3h}(\overline{6}), \quad C_{4h}(4/m), \quad C_{6h} \tag{1-3}$$

Stereograms for these and the point groups in Eq. 1-3 can be found in Fig. 1-5. In the International notation the n/m symbol means that the m-mirror plane is perpendicular to the n-rotation axis.

Next, new point groups are obtained by allowing still more symmetry operations. By adding a vertical plane to the original five point groups in Eq. 1-2 we obtain five other point groups called $C_{nv} = C_n \times \{E, \sigma_v\}$. However, only four of these are new since n = 1 gives C_{1h}, which contains $\{E, \sigma_h\}$, and already has the symmetry operations of one that would be called C_{1v}. Stereograms for these as well as the rest of the point groups discussed below are in Fig. 1-5; the figure should be consulted to help you understand the large amount of information that is being presented here. The point groups are listed below in the two notations. Refer to the H_2O and NH_3 examples for clarification of the symmetry operations in C_{2v} or C_{3v}.

$$C_{2v}(2mm), \quad C_{3v}(3m), \quad C_{4v}(4mm), \quad C_{6v}(6mm) \tag{1-4}$$

In the International notation no slash between the mirror planes and the n-axis means that the mirror plane contains the axis.

By considering improper as well as proper rotations about the principal axis, we can find three new point groups from Eq. 1-2 that have not been generated before. These are S_2 with symmetry operations $\{E, i\}$; $S_4\{E, C_2, S_4, S_4^3\}$; and $S_6\{E, C_3, C_3^2, S_6, i, S_6^5\}$, that is, $S_6 = C_3 \times \{E, S_6\}$. Note that the point group that would be called $S_3(\overline{6})$ is identical to the one called $C_{3h}(\overline{6})$ and is already accounted for. Also note that the symmetry operation $S_2 = i$ and the latter is normally used. The point group S_2 is conventionally called C_i. So we have

$$C_i(\overline{1}), \quad S_4(\overline{4}), \quad S_6(\overline{3}) \tag{1-5}$$

By going back to the original five point groups in Eq. 1-2 and adding a twofold axis perpendicular to the principal axis, four new, D_n, point groups with twice as many symmetry operations are obtained, that is, $D_n = C_n \times \{E, C_2'\}$. Clearly $D_1 \equiv C_2$ so it is not repeated here. Note that the symbol D for point groups means there is a twofold axis perpendicular to the principal axis. The new point groups are

$$D_2(222), \quad D_3(32), \quad D_4(422), \quad D_6(622) \qquad (1\text{-}6)$$

By adding a C_2' axis to the C_{nh} point groups four new point groups $D_{nh} = C_{nh} \times \{E, C_2'\}$ are obtained. These have twice as many symmetry operations as the C_{nh} and four times as many as the C_n point groups. The new point groups are

$$D_{2h}(mmm), \quad D_{3h}(\overline{6}m2), \quad D_{4h}(4/mmm), \quad D_{6h}(6/mmm) \quad (1\text{-}7)$$

Last, by adding a C_2' axis to the S_n point groups, two new point groups $D_{nd} = S_n \times \{E, C_2'\}$ are obtained.

$$D_{2d}(\overline{4}2m), \quad D_{3d}(\overline{3}m) \qquad (1\text{-}8)$$

All of the above point groups have a principal axis. In addition, there are five cubic point groups that are distinguished by possessing no unique axis but by having four threefold axes. The smallest point group, T, has only twelve, purely rotational, symmetry operations $\{E, 4C_3, 4C_3^2, 3C_2\}$, where the C_2 are along the a-, b-, and c-axes. Then T_h and T_d can be obtained from T and have twice as many symmetry operations, $T_h = T \times \{E, \sigma_h\}$ and $T_d = T \times \{E, \sigma_d\}$. The point group O has six fourfold axes and a total of 24 symmetry operations. Last, $O_h = O \times \{E, \sigma_h\}$. Thus, in the two notations we have

$$T(23), \quad T_h(m3), \quad T_d(\overline{4}3m), \quad O(432), \quad O_h(m3m) \quad (1\text{-}9)$$

Table 1-2 lists the 32 crystallographic point groups in both notations and in the full International notation, which is quite descriptive. The symmetry operations are included in a succinct form, along with other information that will be discussed later. The cubic point groups often give people considerable trouble, so we discuss them at greater length with the help of Table 1-2 and Fig. 1-5. (a) As can be seen for T(23), the twelve symmetry operations include E, $C_3(3)$ about the four body diagonals of a cube and four $C_3^2(3^2)$ about the same body diagonals. These latter eight symmetry operations are lumped together under $8C_3$ in Table 1-2. Similarly the three $C_2(2)$ perpendicular to the faces of a cube (along the a-, b-, and c-axes) are lumped together as $3C_2$. (b) Then $T_h(m3)$ has these same twelve symmetry operations plus twelve more that are listed. The 3σ are mirror planes perpendicular to the $3C_2$. (c) $T_d(\overline{4}3m)$ has the same twelve as T(23) plus six diagonal mirror planes across the six face diagonals of a cube and S_4 and S_4^3 along the same axes as the $3C_2$. <u>Note</u> that these three point groups do not have C_4 axes as symmetry operations! (d) O(432) has the same operations as T(23) but it also possesses C_4 and C_4^3 along the a-, b-, and c-axes. These are lumped together as $6C_4$. This point group also has twofold axes along the six face diagonals, which are written as $6C_2$. (e) $O_h(m3m)$ has all of the above operations for a total of 48 symmetry operations.

CHAPTER 1   SYMMETRY OPERATIONS   15

Table 1-2 The 32 crystallographic point groups.

Schoenflies	International	Full Int.	Symmetry Elements	Generating Elements	Space Groups
Triclinic					
C_1	1	1	E	E	1
$S_2(C_i)$	$\bar{1}$	$\bar{1}$	E i	i	2
Monoclinic					
C_2	2	2	E C_2	C_2	3-5
$C_{1h}(C_s)$	m	m	E σ_h	σ_h	6-9
C_{2h}	2/m	$\frac{2}{m}$	E C_2 i σ_h	i C_2	10-17
Orthorhombic					
$D_2(V)$	222	222	E C_2 C_2' C_2'	C_2 C_2^y	16-24
C_{2v}	mm2	mm2	E C_2 σ_v σ_v	C_2 σ_v^y	25-46
$D_{2h}(V_h)$	mmm	$\frac{2}{m}\frac{2}{m}\frac{2}{m}$	E C_2 C_2' C_2' i σ_h σ_v σ_v	i σ_v^y C_2	47-74
Tetragonal					
C_4	4	4	E $2C_4$ C_2	C_4	75-80
S_4	$\bar{4}$	$\bar{4}$	E $2S_4$ C_2	S_4^3	81-82
C_{4h}	4/m	$\frac{4}{m}$	E $2C_4$ C_2 i $2S_4$ σ_h	i C_4	83-88
D_4	422	422	E $2C_4$ C_2 $2C_2'$ $2C_2''$	C_2^y C_4	89-98
C_{4v}	4mm	4mm	E $2C_4$ C_2 $2\sigma_v$ $2\sigma_d$	σ_v^y C_4	99-110
$D_{2d}(V_d)$	$\bar{4}$2m	$\bar{4}$2m	E C_2 $2C_2'$ $2\sigma_d$ $2S_4$	C_2^y S_4^3	111-122
D_{4h}	4/mmm	$\frac{4}{m}\frac{2}{m}\frac{2}{m}$	E $2C_4$ C_2 $2C_2'$ $2C_2''$ i $2S_4$ σ_h $2\sigma_v$ $2\sigma_d$	i C_2^y C_4	123-142
Trigonal (Rhombohedral)					
C_3	3	3	E $2C_3$	C_3	143-146
$S_6(C_{3i})$	$\bar{3}$	$\bar{3}$	E $2C_3$ i $2S_6$	i C_3	147-148
D_3	32	32	E $2C_3$ $3C_2'$	C_2^y C_3	149-155
C_{3v}	3m	3m	E $2C_3$ $3\sigma_v$	σ_v^y C_3	156-161
D_{3d}	$\bar{3}$m	$\bar{3}\frac{2}{m}$	E $2C_3$ $3C_2'$ i $2S_6$ $3\sigma_v$	iC_2^y C_3	162-176
Hexagonal					
C_6	6	6	E $2C_6$ $2C_3$ C_2	C_2 C_3	168-173
C_{3h}	$\bar{6}$	$\bar{6}$	E $2C_3$ σ_h $2S_3$	σ_h C_3	174
C_{6h}	6/m	$\frac{6}{m}$	E $2C_6$ $2C_3$ C_2 i $2S_3$ $2S_6$ σ_h	i C_2C_3	175-176
D_6	622	622	E $2C_6$ $2C_3$ C_2 $3C_2'$ $3C_2''$	C_2 C_2^y C_3	177-182
C_{6v}	6mm	6mm	E $2C_6$ $2C_3$ C_2 $3\sigma_v$ $3\sigma_d$	C_2 σ_v^y C_3	183-186
D_{3h}	$\bar{6}$m2	$\bar{6}$m2	E $2C_3$ $3C_2'$ σ_h $2S_3$ $3\sigma_v$	C_2^y σ_h C_3	187-190
D_{6h}	6/mmm	$\frac{6}{m}\frac{2}{m}\frac{2}{m}$	E $2C_6$ $2C_3$ C_2 $3C_2'$ $3C_2''$ i $2S_3$ $2S_6$ σ_h $3\sigma_v$ $3\sigma_d$	i C_2^y C_2 C_3	191-194
Cubic					
T	23	23	E $8C_3$ $3C_2$	C_2 C_3[111]	195-199
T_h	m3	$\frac{2}{m}\bar{3}$	E $8C_3$ $3C_2$ i $8S_6$ $3\sigma_h$	i C_2 C_3[111]	200-206
O	432	432	E $8C_3$ $3C_2$ $6C_2$ $6C_4$	C_4 C_3[111]	207-214
T_d	$\bar{4}$3m	$\bar{4}$3m	E $8C_3$ $3C_2$ $6\sigma_d$ $6S_4$	S_4^3 C_3[111]	215-220
O_h	m3m	$\frac{4}{m}\bar{3}\frac{2}{m}$	E $8C_3$ $3C_2$ $6C_2$ $6C_4$ i $8S_6$ $3\sigma_h$ $6\sigma_d$ $6S_4$	i C_4 C_3[111]	221-230

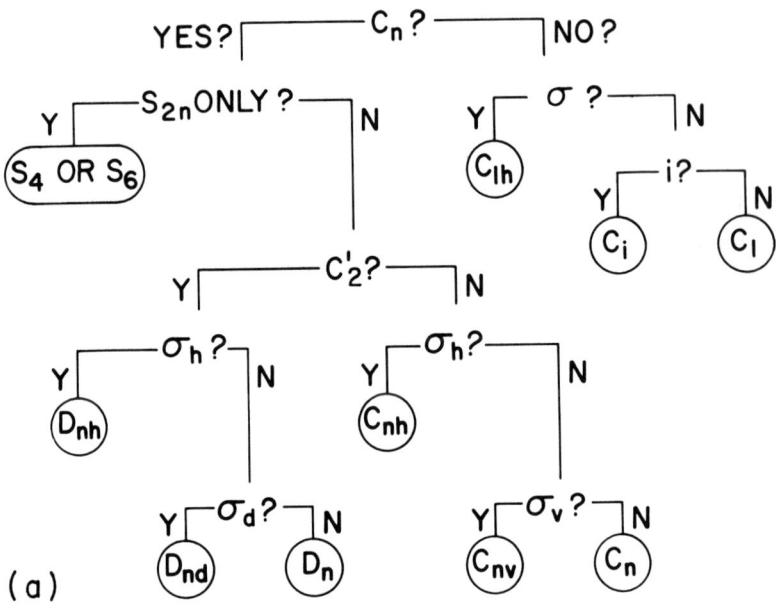

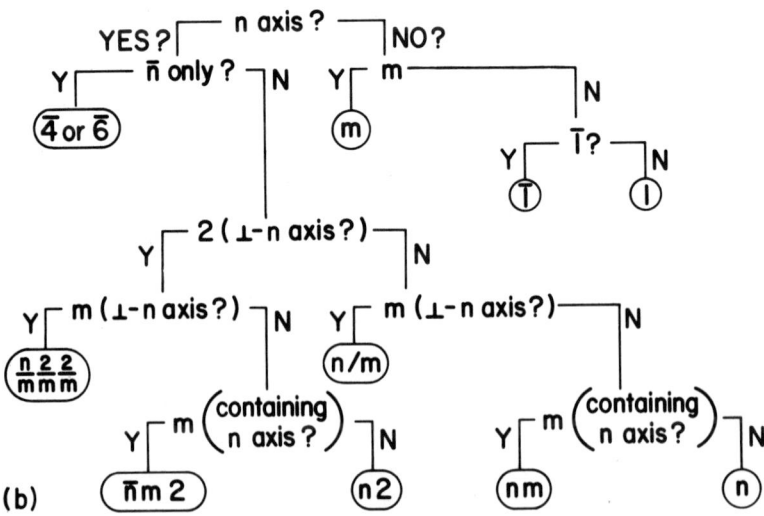

Fig. 1-6 A flowchart showing a systematic way to determine the appropriate point group in the (a) Schoenflies notation and (b) International notation.

CHAPTER 1   SYMMETRY OPERATIONS   17

1-3b Point group flowchart   Figure 1-6 is a flowchart to help label the point groups of a molecule. First, look for the special point groups such as the cubic group or various linear groups such as for Cl_2 or CO molecules. These are usually fairly clear. If there is a principal axis, the chart will help to label the point groups. We give below several examples using the flowchart, and there are many problems to practice with. It is important to appreciate that the flowchart follows the same lines as the development of the point groups.

Consider the H_2O example and follow the flowchart.
  (a) Are there any special groups involved (i.e., cubic ones or $C_{\infty v}$ or $D_{\infty v}$ not discussed here but appropriate for linear molecules)? No.
  (b) Is there a C_n? Yes, there is.
  (c) Are there S_{2n} only? No.
  (d) C_2'? No.
  (e) σ_h? No.
  (f) σ_v? Yes, there is a σ_v.
Thus, we see that the point group is C_{2v}. Drawing a stereogram for C_{2v} will immediately reveal the four symmetry operations.

In a similar manner the point group of NH_3 is found to be C_{3v} and that of PF_3Cl_2 is D_{3h}. Use the flowchart to make sure that you agree with these results.

1-4 Other Symmetry Operations of Crystals

Crystals possess other symmetry operations in addition to the point symmetry operations discussed above for molecules. The most obvious, and the most fundamental of these is associated with the repetitiveness of the crystal, called translational symmetry. However, there also are the screw symmetry operations and the glide planes.

1-4a Translation   Crystals are different from liquids and glasses in that there is a smallest entity, defined as a unit cell, that when moved in three directions parallel to itself reproduces the entire crystal. This is written mathematically in terms of a translation vector

$$t_m = m_1 a + m_2 b + m_3 c \qquad (1\text{-}10a)$$

where m_i can take all integer values and **a**, **b**, and **c** are the edges of the unit cell, which is a parallelepiped. (This will be discussed more clearly in the next chapter, but it is fairly evident at any rate.) The unit cell and **translational symmetry** are fundamental to crystals and distinguishes them from other condensed matter such as liquids and glasses.

18   CHAPTER 1   SYMMETRY OPERATIONS

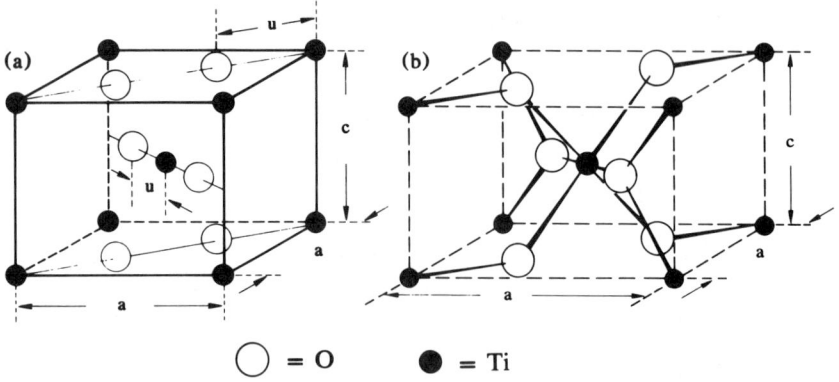

Fig. 1-7 Two diagrams showing a unit cell of TiO$_2$.

The same idea of translational symmetry is expressed by saying that when the atomic arrangement in the crystal is viewed from any point **r**, it is identical when viewed from the point

$$\mathbf{r}' = \mathbf{r} + \mathbf{t}_m \qquad (1\text{-}10\text{b})$$

1-4b Screw operations Figure 1-7 shows one unit cell of crystal TiO$_2$. One must imagine that the crystal is made up of these cells repeated again and again in the sense of Eqs. 1-10a and 1-10b to fill all space. Now examine the symmetry operations. Taking the origin at the center of the unit cell, there are clearly the following eight symmetry operations: E, C_2, $2C_2''$, i, σ_h, $2\sigma_d$. However, as is now shown, there are eight more. Consider a C_4 operation about the c-axis followed by a translation $c/2$ along the c-axis and $a/2$ along the a- and b-axes. This is written, in the **Seitz notation**, as

$$\{C_4 | \tau\} \qquad \text{where } \tau = \mathbf{a}/2 + \mathbf{b}/2 + \mathbf{c}/2 \qquad (1\text{-}11)$$

Note that before and after this operation the crystal, *not one unit cell, but the entire crystal*, is in an equivalent position. Thus, this is a symmetry operation of the crystal and is called a screw operation. The important point to note is that the translation τ is a *fraction of the unit cell size*. (In general, for screw operations the fractions are always 1/2, 1/3, 1/4, 1/6, or integral multiplies of the latter three). Altogether there are eight symmetry operations of the crystal involving τ. These are

$$\{C_4 | \tau\}, \quad \{C_4^3 | \tau\}, \quad 2\{C_2' | \tau\}, \quad 2\{S_4 | \tau\}, \quad 2\{\sigma_v | \tau\}$$

Thus, this crystal has the 16 symmetry operations discussed here and, of course, the infinite number of translations of the unit cell.

We should note that a crystallographer defines a **screw operation** a bit differently. Namely, it is a rotation followed by a translation a fraction of a unit cell *parallel* to the rotation axis. In our example τ

CHAPTER 1 SYMMETRY OPERATIONS   19

has a component perpendicular to the rotation axis. This brings up a complicated point on which we will not dwell. The difference arises because the crystallographer will take screw symmetry operations about different points in the unit cell, i.e., about different origins. Most other scientists prefer one origin for all the symmetry operations, and thus τ will often have a perpendicular component.

1-4c Glide planes A screw operation, like an improper rotation, is a composite operation, i.e., a rotation followed by a translation a fraction of a unit cell. A glide plane is also a composite operation. A **glide plane** involves a reflection across a plane followed by a translation of a fraction of a unit cell.

The concept and execution of glide planes is simple. It is, however, lengthy to describe these planes in detail because there are three different kinds of glide planes, and these behave differently in different crystal systems. (They will be discussed further in Chapter 2.) Two types of glide planes are sketched in Fig. 1-8. An **axial glide** is a reflection followed by a translation parallel to the mirror plane. Figure 1-8a shows a b-glide where the circle in the upper left is reflected across the plane (perpendicular to the a-direction) and then translated half a unit cell in the b-direction to appear as a circle with a comma. In the Seitz notation this is $\{\sigma[100]\,|\,\tau\}$ where $\tau = \mathbf{b}/2$. The operation is repeated to show that the original circle appears but translated one unit cell to the right. Figure 1-8b shows an **n-glide** where the reflection is across the ab-plane and the translation $(\mathbf{a}/2 + \mathbf{b}/2)$ gives a circle with a comma below the plane or $\{\sigma[001]\,|\,\tau\}$, where $\tau = \mathbf{a}/2 + \mathbf{b}/2$. In a **diamond or d-glide** the translations, in cubic crystals, are $(\mathbf{a} \pm \mathbf{b} \pm \mathbf{c})/4$. An example will be shown when the crystal structure of diamond is discussed in Chapter 3.

1-4d Point group of a crystal In Section 1-3 we showed how to determine the point group of a molecule. Here we describe how to determine it for a crystal. One must know all the symmetry operations of the crystal. These include the infinite number of unit cell translation as in Eq. 1-10a, the point group operations, and, if they exist, the screw operations and the glide planes. This set of symmetry operations of the crystal is called the **space group** of the crystal. *The point group of a crystal is obtained by setting all translations equal to zero.* This eliminates the unit cell translations and the fraction of unit cell translations in the screw and glide operations. One is left only with point operations. These point operations will be identical to the collection of operations in one of the 32 point groups, and that is the point group of the crystal. Note that the operations in the point group of the space group are not necessarily symmetry operations of the crystal. For example, in the TiO_2 problem, although $\{C_4\,|\,\tau\}$ is a

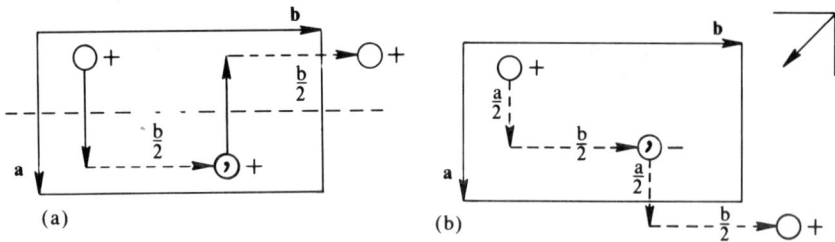

Fig. 1-8 (a) A b-glide operation. (b) An n-glide operation. In each figure, each symmetry operation is applied twice.

symmetry operation of the crystal, $\{C_4|0\}$ is not a symmetry operation of the crystal, yet it is an operation in the point group of the space group. By the **notation** $\{C_4|0\}$ we mean a C_4 rotation followed by a translation of amount zero. Thus, $\{C_4|0\} \equiv C_4$.

To prove and understand the ramifications of the definition of the point group of a crystal, we must understand group theory. However, it should be noted that it is much easier to determine the point group of most crystals than it may appear. In fact, given the symbol of the space group of the crystal the determination is trivial. The examples in Chapter 3 will clarify this.

Notes

A number of the group theory books and several of the solid state books listed in the Bibliography have parts that cover material similar to this chapter. Looking at one or two to see the different approaches will prove interesting. In particular, see Burns, Cotton, and/or Tinkham.

Problems

1. The symmetry operations for H_2O are E, C_2, σ_v, σ_v'. Many students will claim that E and any one other will do for this molecule. However, the point is that you want the maximum number of independent point symmetry operations to describe the molecule. Draw a molecule similar to H_2O, but having only $\{E, C_2\}$ symmetry operations; $\{E, \sigma_v\}$ symmetry operations; $\{E, \sigma_v'\}$ symmetry operations.

2. For the International symbol 3^i, $\overline{3}^i$, 6^i, and $\overline{6}^i$ for i = 1 to 6 show the corresponding Schoenflies symbol.

CHAPTER 1 SYMMETRY OPERATIONS   21

3. (a) What are the symmetry operations of the pyramids of Egypt? A tennis ball including the seam? A boat? A twin-bladed propeller? A maple leaf? (b) What are the point groups of these objects?

4. Problem 3.9 in the second edition of Cotton is excellent. (a) Work it out. I believe the answers are: C_1, C_{2v}, C_{1h}, C_2, C_{2v}, S_2, C_1, C_{4v}, D_{2h}, C_{2h}, C_{2h}, S_4, S_2. (b) Work out the point groups in the International notation, using Fig. 1-6b.

5. For the linear molecules OCO and HCN, besides the rotations about the molecular axis, what are the other symmetry operations and how many are there? The point groups are called $D_{\infty h}$ and $C_{\infty v}$, respectively.

6. What are the point groups for the molecules pictured below? Draw a stereogram for each of these point groups.

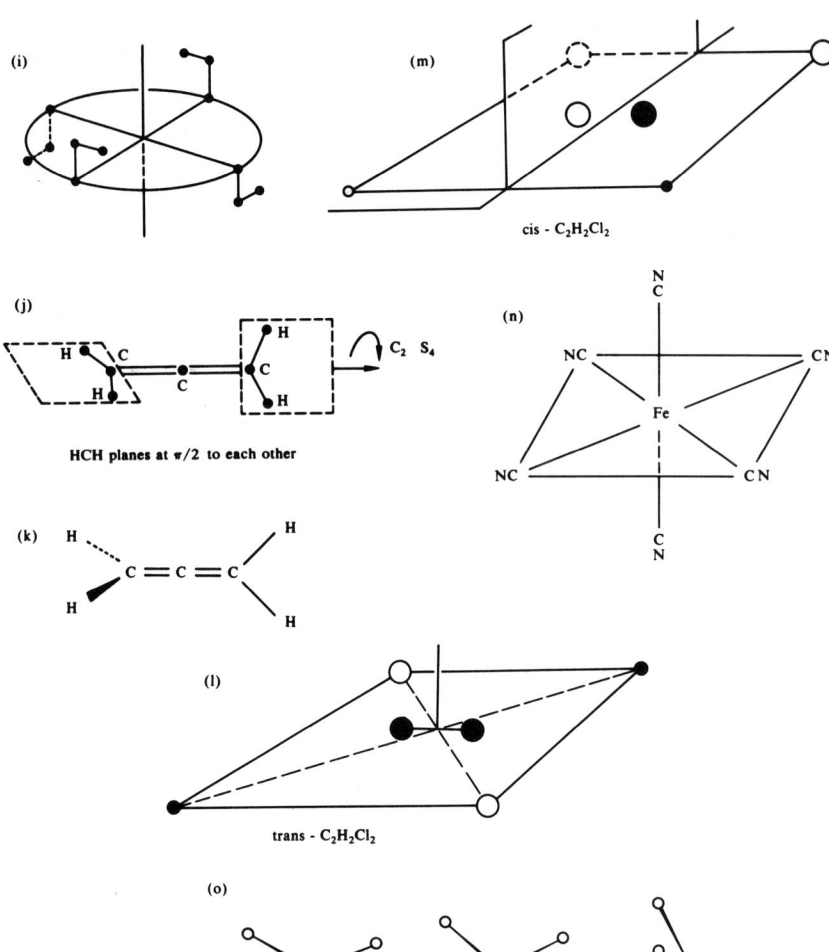

2

Symmetry Description of Crystals

2-1 Lattice
2-2 Primitive Unit Cell
2-3 The 7 Crystal Systems
    a Some clarifications
2-4 The 14 Bravais Lattices
    a Centering of lattices
    b Unit cells
2-5 The 32 Crystallographic Point Groups
2-6 Space Groups
    a Types of space groups
    b Symmorphic space groups
    c Nonsymmorphic space groups (S)
    d International notation for planes and axes (S)
2-7 Definitions of Directions, Coordinates, and Planes
    a Directions
    b Coordinates
    c Planes
    Appendix to Chapter 2

    Notes
    Problems

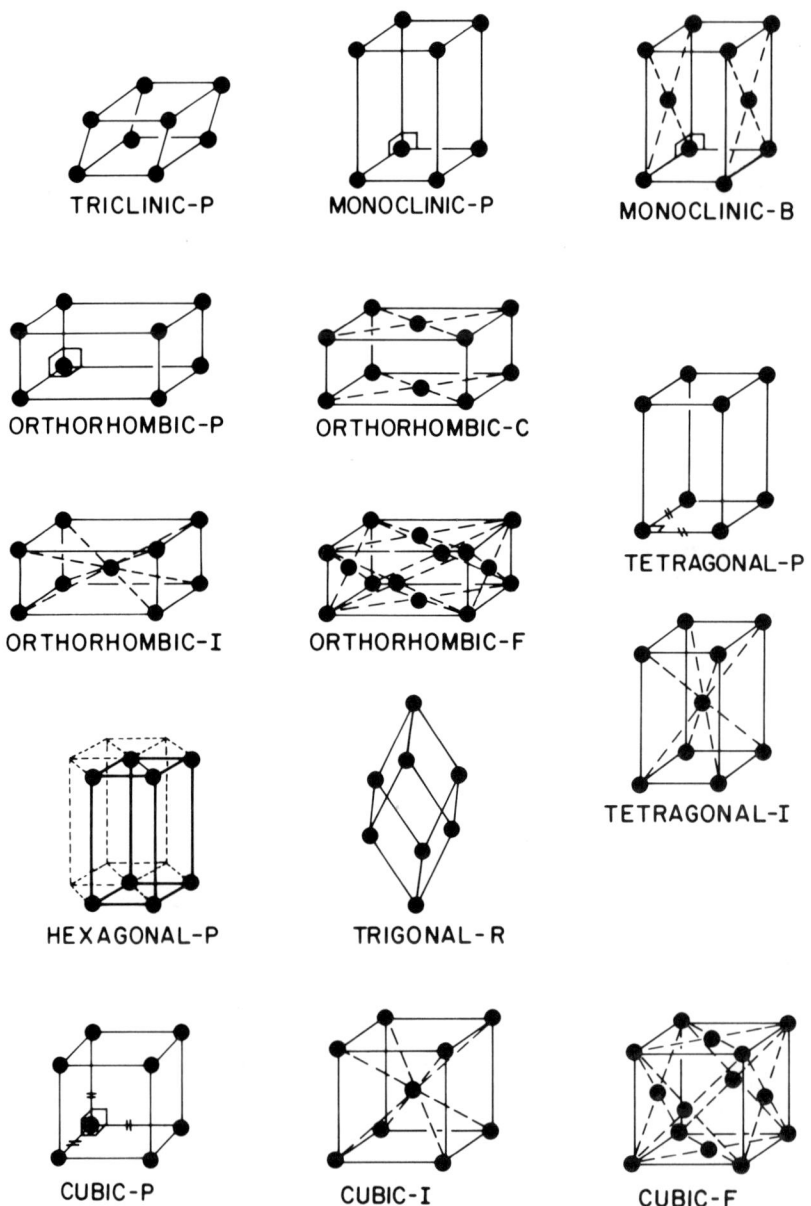

Fig. 2-7 The conventional unit cell for each of the 14 Bravais lattices. (For the hexagonal lattice the unit cell is outlined in solid lines, but a hexagonal prism is also shown, dotted, to help indicate the angles of the unit cell.)

SYMMETRY DESCRIPTION OF CRYSTALS

To see a World in a Grain of Sand
And a Heaven in a Wild Flower,
Hold Infinity in the palm of your hand
And Eternity in an hour.

W. Blake, "Auguries of Innocence"

In this chapter crystals are classified according to their symmetry. This is a rigorous and fundamental description. The topics to be covered include the 7 crystal systems, the 14 Bravais lattices, the 32 crystallographic point groups, and the 230 space groups. In a certain sense each of these successive classifications describes the crystal structure in more detail. However, before beginning this program we define the fundamental terms **lattice** and **unit cell.**

A point of view quite different from a symmetry standpoint could be taken. For example, to describe crystalline solids we might want to ask: how are the atoms bonded to each other; what kind of coordination polyhedra surround each kind of atom; are the electrons strongly bound or free to move as in metals? These questions, falling under the general heading of "bonding in solids," cannot always be answered rigorously but are of fundamental importance and will be discussed in later chapters.

2-1 Lattice

A **lattice** is defined as an infinite array of points in which each point has surroundings identical to those of all the other points. (The distances and angles to all the other points from *any* one point are the same as from *any* other point.) A lattice is a mathematical construction. Lattices are generated by a **primitive lattice translation** vector

$$t_m = m_1 \mathbf{a} + m_2 \mathbf{b} + m_3 \mathbf{c} \qquad (2\text{-}1)$$

where m_i is any integer (positive, zero, or negative) and **a**, **b**, and **c** are three arbitrary, independent vectors. The infinite number of end points of this vector are the points of the lattice. Figure 2-1a shows a lattice (in projection), and Fig. 2-1b shows an array of points that is not a lattice because the angular surroundings of each point are not the same. Notice that the point symmetry at a lattice point of an arbitrary lattice is $C_i(\bar{1})$, since every lattice point is at a center of inversion, that is, for every lattice point at $+m_i$ there is one at $-m_i$.

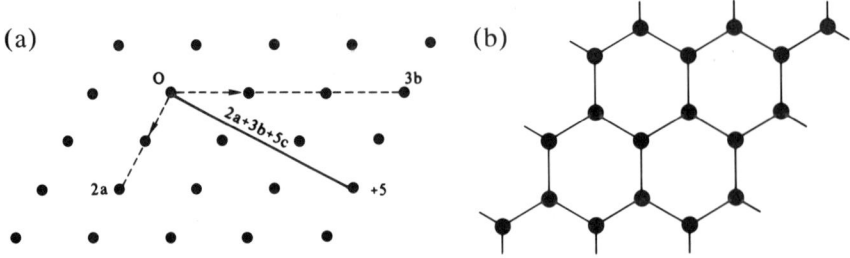

Fig. 2-1 (a) A projection of a lattice of points. (b) An example of a periodic arrangement of points that do not form a lattice.

2-2 Primitive Unit Cell

By completing the parallelepiped formed by **a**, **b**, and **c**, a volume **a**·(**b** × **c**) is enclosed which, if continuously translated parallel to itself by t_m (Eq. 2-1), will fill all space. This volume, which contains *one lattice point*, is called a **primitive unit cell**. When saying "contains" one lattice point, we must remember that for a square lattice, for example, a lattice point at the intersection of four primitive unit cells is considered to be only 1/4th in each cell.

We define the term **unit cell** as a cell also formed by three vectors, in a manner similar to that as above, which will fill all space when translated by multiples of these vectors. The difference is that while a primitive unit cell contains one lattice point, a unit cell may contain more. Cells such as these, containing more than one lattice point, are also called **nonprimitive** or **multiply primitive unit cells**.

Figure 2-2 shows examples of primitive and nonprimitive unit cells. The primitive cell in the lower left is slightly displaced and thus more clearly shows that it only contains one lattice point. Note that there are an infinite number of possible primitive unit cells, but they all must have the same volume (area in this figure). As we shall see, however, there are conventions as to the choices.

2-3 The 7 Crystal Systems

In this section we discuss the symmetry restrictions that can be applied to the unit cell lengths and interaxial angles of a lattice. These restrictions give rise to the 7 crystal systems. Figure 2-3 shows the usual definition of the interaxial angles for a right handed set of axes.

The general idea is simple. Apply a certain rotational symmetry operation to an arbitrary lattice and consider whether the operation puts any restrictions on the values **a**, **b**, **c**, α, β, and/or γ. Recall from

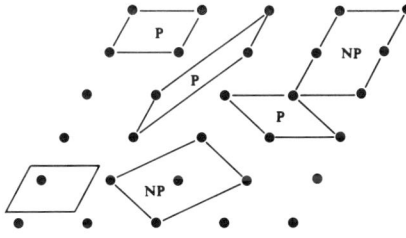

Fig. 2-2 Examples of primitive and nonprimitive unit cells.

P = primitive unit cell
NP = nonprimitive unit cell

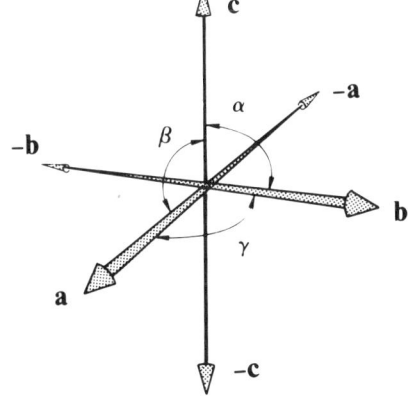

Fig. 2-3 The angles α, β, and γ, used to define the lattices.

Chapter 1 that we need only consider C_n rotations with n = 1, 2, 3, 4, and 6, since only these rotations are consistent with the requirement that the unit cells fill all space. (The proof is in the Appendix.)

We shall discuss the symmetry conditions leading to the seven crystal systems. These conditions are shown in Table 2-1.

(a) It is clear that the symmetry operation E(1) requires no special relationships among the axes and angles. This is true for i($\bar{1}$) since an arbitrary lattice has inversion symmetry. Thus, either one or both of these two symmetry operations define what is called the triclinic crystal system in which there are no symmetry restrictions on the axes or angles, hence the inequality signs in Table 2-1.

(b) Consider the consequences of requiring that $C_2(2)$ be a symmetry operation, so that <u>before and after the operation we must have an equivalent lattice</u>. Figure 2-4a shows the consequences of such a requirement. The $C_2(2)$ is along the c-axis. Observing the lattice points at +**a** and −**a** in the ac-plane, we can see that if $C_2(2)$ operation is to be a symmetry operation the lattice point due to C_2**a** must coincide with that at −**a**. The only way for this to occur is if **a** and **c** are perpendicular to each other. Similarly, **b** and **c** must be perpendicular to each other, i.e., $\alpha = \beta = \pi/2$. In a similar manner a mirror plane perpendicular to the c-axis will impose the same conditions on the angles, as can be seen in Fig. 2-4b. These two symmetry operations, therefore, lead to a new condition on the unit cell, which

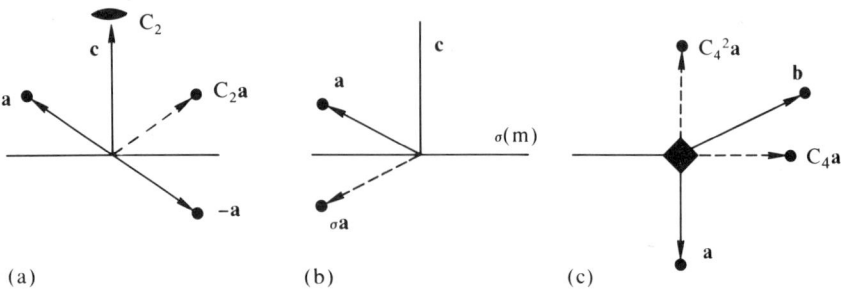

Fig. 2-4 The effects of symmetry operations on a general lattice.

gives what is called the monoclinic crystal system; the conditions on the angles and axes are listed in Table 2-1.

(c) Consider the result of demanding that a C_2 rotation about the b-axis, $C_2[010]$, as well as a C_2 rotation about the c-axis, $C_2[001]$, be a symmetry operation of the lattice. For the same reasons, as shown in Fig. 2-4a, we find the $\alpha = \beta = \gamma = \pi/2$. This is called the orthorhombic crystal system (See Table 2-1). It should be clear that two σ-planes will impose the same conditions as two C_2 operations.

(d) Figure 2-4c is a view along the c-axis of a lattice when we require that this axis have fourfold symmetry, $C_4[001]$. It should now be clear that in order for this to be a symmetry operation rather severe conditions are demanded of the angles and axes. This leads to the tetragonal crystal system.

(e) In this manner the 7 crystal systems are obtained (Table 2-1). We quickly add some clarifications and cautions to this discussion.

2-3a   Some clarifications   When discussing the monoclinic crystal system one should take the unique axis as the c-axis. However, many crystallographers have gotten into a bad habit of taking the unique, twofold axis as the **b**-axis. This usage is called the "second setting." The "first setting" has the unique axis as the **c**-axis.

For the orthorhombic crystal system none of the axes are unique so it is arbitrary as to which is picked as the **c**-axis.

The distinction between the hexagonal and trigonal crystal system is controversial. (We ignore the rhombohedral type for the moment since this lattice cannot be obtained until lattice centering is considered in the next section.) The application of $C_6(6)$ or $S_3{}^5(\bar{6})$ as well as $C_3(3)$ or $S_5{}^5(\bar{3})$ gives the same conditions on the angles and axes. This is why some crystallographers claim there are only six crystal systems. However, there is an advantage to keeping to seven since, as we shall see, the trigonal system can be centered to give the rhombohedral lattice, while the hexagonal system cannot be centered.

CHAPTER 2   SYMMETRY DESCRIPTION OF CRYSTALS

Table 2-1 Conditions that determine the 7 crystal systems. (For each crystal system the top line on the left indicates the conditions in the Schoenflies notation, the next line in the International notation.)

E or i 1 or $\bar{1}$	Triclinic	$a \neq b \neq c$ $\alpha \neq \beta \neq \gamma$
C_2 or σ 2 or $\bar{2}$	Monoclinic	$a \neq b \neq c$ $\alpha = \beta = 90° \neq \gamma$ (1st setting) $\alpha = \gamma = 90° \neq \beta$ (2nd setting)
two C_2 or σ two 2 or $\bar{2}$	Orthorhombic	$a \neq b \neq c$ $\alpha = \beta = \gamma = 90°$
C_4 or S_4 4 or $\bar{4}$	Tetragonal	$a = b \neq c$ $\alpha = \beta = \gamma = 90°$
four 3-fold axes	Cubic	$a = b = c$ $\alpha = \beta = \gamma = 90°$
C_6 or S_3 6 or $\bar{6}$	Hexagonal	$a = b \neq c$ $\alpha = \beta = 90°; \gamma = 120°$
C_3 or S_6 3 or $\bar{3}$	Trigonal (Rhombohedral)	same as hexagonal ($a = b = c; \alpha = \beta = \gamma$)

The meaning of $\neq$ should be reemphasized. It means that symmetry does not demand an equality. Occasionally we find, within experimental accuracy, that two values are found to be equal even though not required by symmetry. This is called an **accidental degeneracy**.

2-4 The 14 Bravais Lattices

We now have seven different lattices each with special conditions on the angles and axes as summarized in Table 2-1. We ask whether we can add more points at judicious locations to any of these lattices and still have a lattice? We shall do this and find seven more lattices, and then have all of the 14 Bravais lattices.

2-4a Centering of lattices The procedure is straightforward. Add points in each crystal system and then ask (a) is the arrangement still a lattice and (b) is it a new lattice in the same crystal system or just another lattice perhaps in a different orientation?

To each lattice, for each crystal system, points can be added at the body center of the unit cell, or at the face centers, etc. In terms of the axes, these points are located at

(I)   body center $(\mathbf{a}/2 + \mathbf{b}/2 + \mathbf{c}/2)$
(F)   face centers $(\mathbf{a}/2 + \mathbf{b}/2)$, $(\mathbf{a}/2 + \mathbf{c}/2)$, and $(\mathbf{b}/2 + \mathbf{c}/2)$
(C)   C-face center $(\mathbf{a}/2 + \mathbf{b}/2)$
(R)   rhombohedral $(2\mathbf{a}/3 + \mathbf{b}/3 + \mathbf{c}/3)$ and

$$(\mathbf{a}/3 + 2\mathbf{b}/3 + 2\mathbf{c}/3) \qquad (2\text{-}2)$$

For a given crystal system some of these points will result in a new unique lattice and some will not. Each case must be considered separately. Figure 2-5 shows these points located in a unit cell. The letters I, F, C, and R are the conventional letters for each of these lattices. Naturally, other possibilities could be tried. However, lattices will not be formed, and we will give one example of such a failure.

First, note that I-, F-, or C-centering can in principle form lattices since the environment of every point can be the same. Whether each of these possibilities actually will form a lattice depends on the relationship of the axes and angles in each crystal system. On the other hand, two-face centering cannot form a lattice, as indicated in Fig. 2-5e. Remember, to form a lattice the environment of every point must be the same. In Fig. 2-5e we can see that the A and B types of points have different neighbors. It should also be mentioned that we have listed C-face centering above (a lattice point in the ab-plane), but could have listed A- or B-face centering just as well (bc- or ac-plane), and these are collectively called one-face centering.

Let us consider several examples of the procedure. (a) The triclinic system can always be centered and the lattice condition will indeed remain, but the lattice will not be new. A smaller primitive cell can always be drawn with the same arbitrariness of the cell edges and angles. Thus, there is only one type of lattice consistent with the triclinic crystal system, and this is the primitive P-lattice. (b) Figure 2-6 shows several aspects of centering a monoclinic lattice. A lattice is obtained in any one-face centering, but, as shown in Fig. 2-6b, C-centering produces nothing new. Another primitive unit cell can be drawn (dotted in the figure) that still has the fundamental monoclinic condition that $\mathbf{a} \neq \mathbf{b} \neq \mathbf{c}$ and $\alpha = \beta = \pi/2 \neq \gamma$ (in the 1st setting nomenclature mentioned above). On the other hand, B-face centering $(\mathbf{a}/2 + \mathbf{c}/2)$, as in Fig. 2-6c, gives a new lattice because it is not possible to draw another primitive unit cell that still retains the fundamental monoclinic conditions. A dotted primitive unit cell is shown, and it can be seen that the angles no longer can be $\pi/2$. Thus, for the monoclinic crystal system there are two possible lattices: the P- and B-lattices. (Instead of B we could say A-centering.) (c) Figures 2-6d

CHAPTER 2   SYMMETRY DESCRIPTION OF CRYSTALS   31

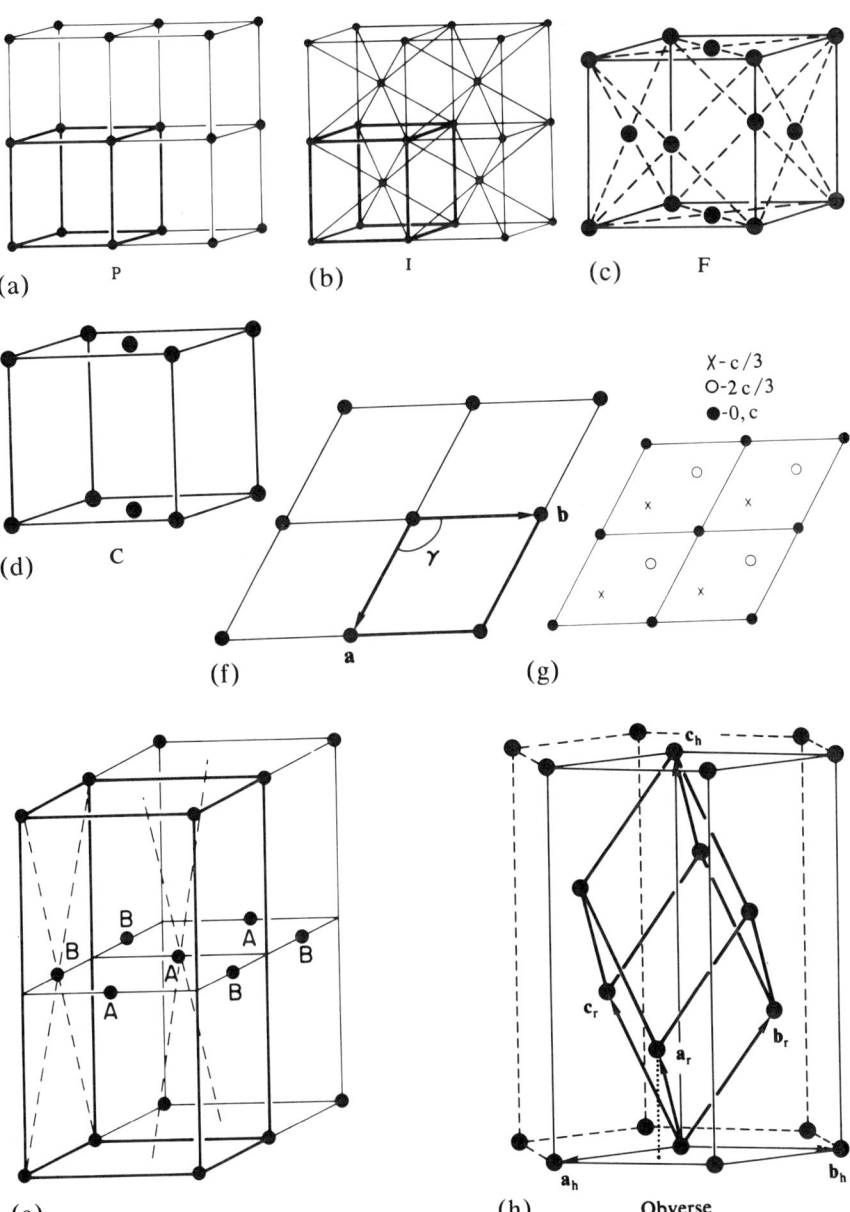

Fig. 2-5 Types of centering of lattices. (a) Four-unit cells (not necessarily cubic). (b) Body centering, showing all four unit cells. (c) All-face centering, showing only one unit cell for clarity. (d) One-face centering, showing one unit cell. (e) An impossible way to center a lattice. (f) A projection of four trigonal unit cells. (g) Rhombohedral centering of these cells. (h) A rhombohedral unit cell in a rhombohedral lattice. The corresponding hexagonal cell, which is three times larger, is also shown.

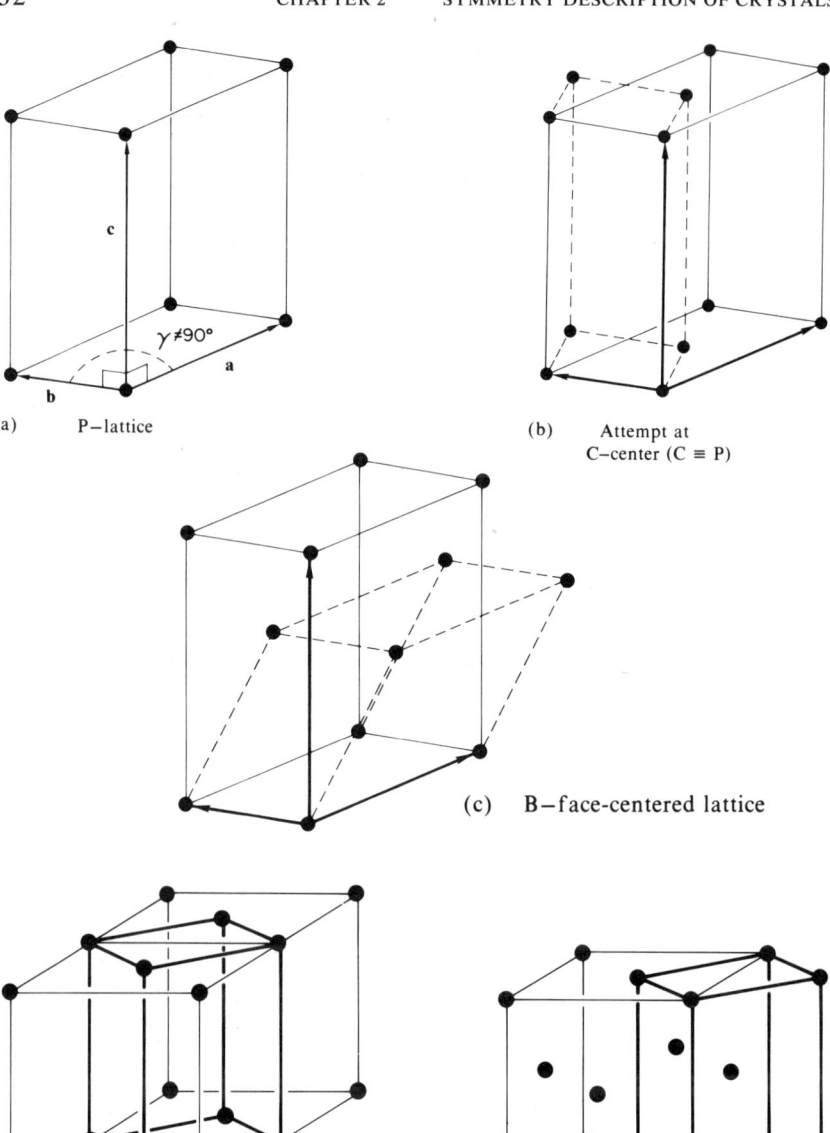

Fig. 2-6 (a) through (c) Several aspects of centering a monoclinic lattice. (d) and (e) Several aspects of centering a tetragonal lattice.

and 2-6e show the results of centering a tetragonal lattice. C-centering gives nothing new since the lattice can be described by a smaller tetragonal primitive cell rotated by $\pi/4$ to the centered cell. Similarly, F-centering can be described in terms of a body-centered cell, as can be seen. Centering the A- or B-face will destroy the fundamental tetragonal condition. Thus, for the tetragonal system there are only two lattices, and these are called a P-lattice and an I-lattice. (These are conventionally chosen in preference to the C- and F-lattice since the former have fewer lattice points per unit cell.)

(d) Similar considerations can be applied to the other crystal systems, and the results can be found in the first two columns of Table 2-2, and are pictured in Fig. 2-7. Only the hexagonal and trigonal crystal classes need further discussion. Figure 2-5f shows the hexagonal or trigonal system in projection, and Fig. 2-5g shows these centered in a special manner as described in Eq. 2-2. The result is indeed a lattice, but it no longer has $C_6(6)$ or $S_3^5(\bar{6})$ symmetry. It does, however, still have $C_3(3)$ and $S_6^5(\bar{3})$ symmetry. The result is a new lattice, and a primitive cell is shown in Fig. 2-5h. This is called a rhombohedral cell of the rhombohedral or R-lattice. Thus, we see that it is possible to center the trigonal crystal class in this special manner, still keep the threefold symmetry, and obtain a new lattice, the rhombohedral or R-lattice. However in this manner it is not possible to center the hexagonal crystal system because the sixfold symmetry is not maintained. There is an extra note of confusion in the trigonal-hexagonal system complication. The rhombohedral R-cell can be, and often is, referred to hexagonal axes and a hexagonal cell with three times as many lattice points as the primitive R-cell. That this is possible is clear from the construction of the R-lattice in Fig. 2-5g and 2-5h. This is usually done because hexagonal axes are much more convenient to deal with than rhombohedral axes. (In the problems it is shown that F- or I-centering of the R-lattice gives nothing new in that the resulting R-lattice can still be described by a primitive R-cell with different sides and angles.)

Figure 2-7 (page 24) shows the resulting conventional unit cells of the 14 Bravais lattices that are obtained from the 7 crystal systems. The P, I, F, C, or R symbols are also shown, indicating whether the lattice is primitive (P), etc., as in Eq. 2-2. The rhombohedral R-cell is primitive, but conventionally R is used instead of P.

2-4b Unit cells We should emphasize that the unit cells shown in Fig. 2-7 are the conventional unit cells of the 14 Bravais lattices. They have the properties required of unit cells, and have been chosen because they easily display the symmetry of the crystal system. They are not, however, the only unit cells that are used.

Figure 2-8 shows the conventional primitive cells that can be drawn for any one I-, F-, and C-lattice in any crystal system. Taken separately from the lattice these primitive cells do not easily display the symmetry as do the conventional multiply primitive cells of the I-, F-, and C-lattices shown in Fig. 2-7. The primitive cell shown in Fig.

Table 2-2 The 73 symmorphic space groups. (The * denotes each of the seven space groups for which the point group operations have more than one possible orientation with respect to the Bravais lattice.)

Crystal system	Bravais lattice	Space group
Triclinic	P	P1, P$\bar{1}$
Monoclinic	P	P2, Pm, P2/m
	B or A	B2, Bm, B2/m (1st setting)
Orthorhombic	P	P222, Pmm2, Pmmm
	C, A or B	C222, Cmm2, Amm2*, Cmmm
	I	I222, Imm2, Immm
	F	F222, Fmm2, Fmmm
Tetragonal	P	P4, P$\bar{4}$, P4/m, P422, P4mm, P$\bar{4}$2m, P$\bar{4}$m2*, P4/mmm.
	I	I4, I$\bar{4}$, I4/m, I422, I4mm, I$\bar{4}$2m, I$\bar{4}$m2*, I4/mmm.
Cubic	P	P23, Pm3, P432, P$\bar{4}$3m, Pm3m
	I	I23, Im3, I432, I$\bar{4}$3m, Im3m
	F	F23, Fm3, F432, F$\bar{4}$3m, Fm3m
Trigonal	P	P3, P$\bar{3}$, P312, P321*, P3m1, P31m*, P$\bar{3}$1m, P$\bar{3}$m1*
(Rhombohedral)	R	R3, R$\bar{3}$, R32, R3m, R$\bar{3}$m
Hexagonal	P	P6, P$\bar{6}$, P6/m, P622, P6mm, P$\bar{6}$m2, P$\bar{6}$2m*, P6/mmm

2-8, for example, for a cubic F-lattice, looks, and is, rhombohedral with $\alpha = 60°$ and the primitive cell of a cubic I-lattice is rhombohedral with $\alpha = 109° 28'$. Nevertheless, these primitive cells properly generate the lattice, which is the only requirement of a unit cell, and they are very useful since they contain only one lattice point in contrast to the I-, F-, and C-cells, which respectively contain 2, 4, and 2 lattice points.

Besides the conventional unit cells shown in Figs. 2-7 and 2-8, there is one more important unit cell that is used particularly in electronic band theory. This is the **Wigner-Seitz cell**, which is primitive and displays the symmetry of the crystal system. To obtain this cell we start at any lattice point, the origin, and draw vectors to all neigh-

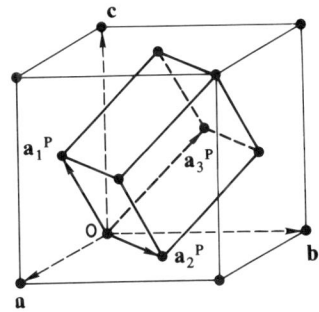

(a) F – Lattice

$a_1^P = (a + c)/2$
$a_2^P = (a + b)/2$
$a_3^P = (b + c)/2$

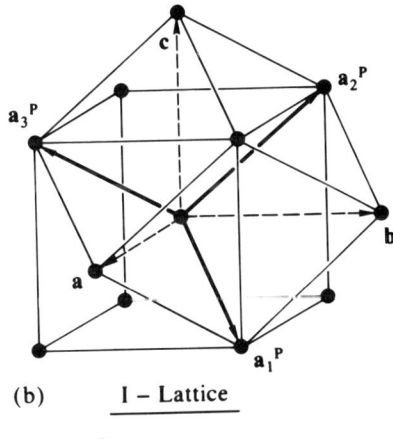

(b) I – Lattice

$a_1^P = (a + b - c)/2$
$a_2^P = (-a + b + c)/2$
$a_3^P = (a - b + c)/2$

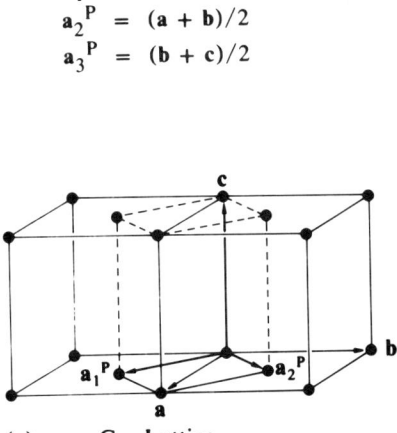

(c) C – Lattice

$a_1^P = (a - b)/2$
$a_2^P = (a + b)/2$
$a_3^P = c$

Fig. 2-8 The conventional multiply primitive and primitive cells of the F-, I-, and C-lattices. The relationships between these two cells can be seen. (Figures 10-13a and 10-14a show another view of the primitive cells for the I- and F-lattices.)

boring lattice points. Planes perpendicular to and passing through the midpoints of these vectors are constructed. The Wigner-Seitz cell is the cell with the smallest volume about the origin bounded by these planes. Figure 2-9a to 2-9d show such a construction for a cubic I-lattice, and 2-9e shows the stacking of such cells that fills all space. In Fig. 2-9f the Wigner-Seitz cell of a cubic F-lattice is shown.

2-5 The 32 Crystallographic Point Groups

There are a number of different ways to develop point groups and in particular the 32 crystallographic point groups. We have already

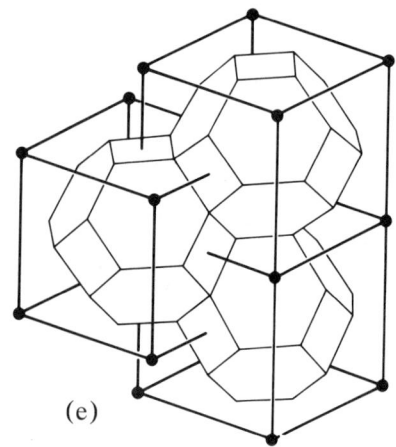

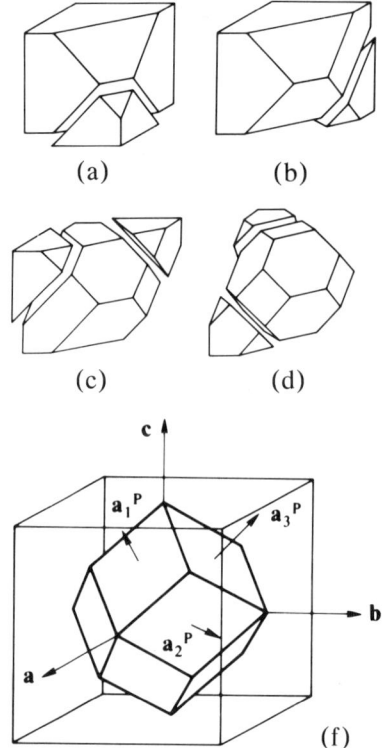

Fig. 2-9 (a)-(d) Construction of the Wigner-Seitz cell of a cubic I-lattice. (e) The stacking of Wigner-Seitz cells of a cubic I-lattice. (f) The Wigner-Seitz cell of a cubic F-lattice.

shown one method in Section 1-3a. The shortcoming of the method developed there is that the results seem totally independent of the 7 crystal systems. Thus, after the point groups are developed, we must determine to which crystal system each belongs. Although this is not difficult, it causes the two types of classification to appear unrelated. Here we shall sketch the development of the 32 point groups within the context of the 7 crystal systems.

The approach is straightforward. Consider each crystal system, one at a time. Starting with the symmetry operations that define the system, as listed in Table 2-1, we add symmetry operations to the lattice and <u>determine which symmetry operations keep the lattice within the starting crystal system</u>. This will determine the point groups allowed within each crystal system.

(a) First, consider the triclinic crystal system. This crystal system is determined by the symmetry operations $E(1)$ or $i(\bar{1})$, as shown in Table 2-1. What happens if a $C_2(2)$ or $\sigma(\bar{2})$ is added? Clearly this will put more conditions on the interaxial angles. In fact, these conditions lead to the monoclinic system. The same is true for any other $C_n(n)$ operations. Thus, there are only two point groups allowed in the triclinic crystal system, and these are the point groups C_1, with just the operation $\{E\}$, and S_2, with the operations $\{E, i\}$. These are appropriately listed in Table 1-2. An *important point* should be made here. A triclinic lattice, by itself, has point symmetry operations E(1) and i($\bar{1}$), so the point group of this lattice is $S_2(\bar{1})$. Nevertheless, the two point groups $C_1(1)$ and $S_2(\bar{1})$ are *consistent* with the triclinic crystal system in that they do not put any *further* conditions on the lattice. It should be remembered that we will want to attach a collection of atoms to each lattice point. Thus, the statement that these two point groups are allowed in the triclinic crystal system means the collection of atoms, associated with each lattice point, may have either of these point groups. Such a collection of atoms will keep the lattice within the triclinic crystal system. (The collection of atoms that will be "attached" to each lattice point is called a **basis** and is discussed in Chapter 3. However, it is useful to bear in mind that so far we have been talking only about mathematical points, lattice points. Eventually we want to give the "skeleton" some "flesh.")

(b) The results for the monoclinic system are also very easy to determine. Table 2-1 shows that $C_2(2)$ or $\sigma(\bar{2})$ determines this crystal system. One can have the point groups $C_2(2)$, with symmetry operations {E, C_2}, and $C_{1h}(m)$, {E, σ_h}. Consider, however, a σ_h plane perpendicular to the C_2-axis. Will that put any new conditions on the axes or interaxial angles? No, since the c-axis is already perpendicular to the ab-plane. Thus, this symmetry operation will keep us within the monoclinic crystal system, and the point group $C_{2h}(2/m)$, {E, C_2, i, σ_h} is a point group in the monoclinic crystal system. (Note that i is generated by the product of the two symmetry operations $C_2\sigma_h$.) On the other hand, a symmetry operation of a mirror plane in the ac-plane will cause the b-axis to be perpendicular to the ac-plane. This will take us out of the monoclinic crystal system. Thus, the monoclinic crystal system has three point groups $C_2(2)$, $C_{1h}(m)$, and $C_{2h}(2/m)$. Again, note that the latter is the point group of a lattice in this crystal class. These point groups are listed in Table 1-2. We should note that in this table the point group listed last for each crystal system is that of a lattice of points in that crystal system. (Each of these groups is called the **holosymmetric point group** for that crystal system.)

(c) Two twofold axes or $\bar{2}$-axes determine the orthorhombic crystal system (Table 2-1). This leads to the point groups $D_2(222)$, {E, C_2, C_2' C_2''}, and $C_{2v}(mm2)$, {E, C_2, σ_v, σ_v'}. (Note that for both point groups a C_2 operation is generated by the product of two other symmetry operations.) A third point group is obtained by having a σ_h perpendicular to a C_2. Such a symmetry operation will keep the system orthorhombic (and other symmetry operations will be generated). The resulting point group is $D_{2h}(mmm)$ and the symmetry operations are listed in Table 1-2. Also note, in Table 1-2, how the full International symbol describes the operations rather nicely.

(d) The point groups for the other crystal systems can be obtained with equal ease and the results are summarized in Table 1-2. See Problem 2 in this chapter.

2-6 Space Groups

The 7 crystal systems, 14 Bravais lattices, and 32 crystallographic point groups are ways of describing, respectively, the axial lengths and angles, the axial lengths and angles and the type of lattice, the axial lengths and angles and the point symmetry conditions (but not the type of lattice). The first two describe what might be called the "geometry" of the lattice. Thus, each of these three approaches describes the symmetry of a crystal in successively more detail. However, to understand thoroughly a crystal structure we must examine the spatial distribution of the electron density (atoms). This distribution need only be described within one unit cell, since Eq. 2-1 will generate the entire crystal. A description of the electron distribution, from a symmetry point of view, requires what are called **space groups.** We shall see how they are developed and thus understand how space groups include all the symmetry of any arrangement of atoms associated with any lattice. (Formally, a space group is defined as the group that contains all of the spatial symmetry operations of the atoms in the crystal.) In the next chapter, many simple crystal structures are described in terms of their appropriate space groups. The development here also enables us to understand better the symbols used to describe them. Even these symbols possess important information.

2-6a Types of space groups There are 230 (nonmagnetic) space groups, and they can be classified into two types, symmorphic and nonsymmorphic. Recall from Section 1-3 that by τ we mean a translation that is a *fraction* of a unit cell length. A **symmorphic space** group is defined as a space group that may be specified entirely by symmetry operations acting at a common point (the operations need not involve τ) as well as the unit cell translations of Eq. 2-1. On the other hand, to specify a **nonsymmorphic space** group at least one operation involving a translation τ is *required*. As we shall see, it is easy to develop the 73 symmorphic space groups, while the others are more difficult.

A cautionary note. We will find many glide planes and screw axes, in symmorphic space groups, but these will be generated from products of unit cell translations, Eq. 2-1, with the point operations. Thus, while they are symmetry operations, they are not *necessary* to describe the symmetry of the crystal. It is helpful to know that if there is no symbol for a glide plane (a, b, c, n, d) or screw axis (n_m) in the standard International symbol for the space group, then the space group is symmorphic. On the other hand, the standard International symbol for a nonsymmorphic space groups always contains a glide plane or screw axis symbol.

2-6b Symmorphic space groups   In order to obtain these space groups we must combine the 32 point groups with the 14 Bravais lattices; that is, each point group in the crystal system (Table 1-2) is combined with the allowed Bravais lattices (P, I, F, and/or C) of that crystal system. With this procedure 73 space groups are obtained.

As an example of the procedure consider an orthorhombic P-lattice. Take an object (a general collection of atoms and later this will be called the most general basis) that belongs to one of the point groups under the orthorhombic crystal system; we start with C_{2v}(mm2). Combine the lattice with the object as shown in Fig. 2-10a. Clearly, we must be a little careful about the meaning of "combining"; in this case the mirror planes of the point group must coincide with those of the lattice. It is apparent that the four circles in the upper left have C_{2v}(mm2) symmetry. The figure on the right of Fig. 2-10a shows the twofold axes, and the heavy lines indicate the vertical mirror planes perpendicular to the paper. The other three collections of atoms are obtained from the one in the upper left by full translations of the unit cell (Eq. 2-1). (The **convention** in this figure and in the International Tables is: origin in the upper left; a-axis down the page; b-axis to the right; c-axis out of the page, that is, a right-hand convention). Figure 2-10b shows the same type of space group construction but starting with an orthorhombic C-lattice; the centering condition puts the four circles at the base-centered position.

Two space groups have been generated with ease. Since this has been accomplished so quickly, a review of what was done might be appropriate. We have taken an orthorhombic primitive P-lattice and a one-face centered C-lattice and, about each (and every) lattice point, we have put a distribution of material (we might as well think in terms of collections of atoms) that has an orthorhombic point group. This simple procedure has generated the space groups, which, in the International notation, are labeled Pmm2 and Cmm2. Clearly the first letter in the symbol refers to the type of lattice, and then the rest of the letters refer to the point symmetry of the collection of atoms. The only word of caution that should be added is that the collection of atoms is not necessarily extremely close to the lattice points as sketched in Fig. 2-10. Nevertheless the collection does (and must) have the mm2 or C_{2v} point symmetry. In Fig. 2-10 the symbols for these two space groups are given in the notation that we are using throughout. Namely, the Schoenflies symbol is given immediately followed, in parentheses, by the International symbol. The Interna-

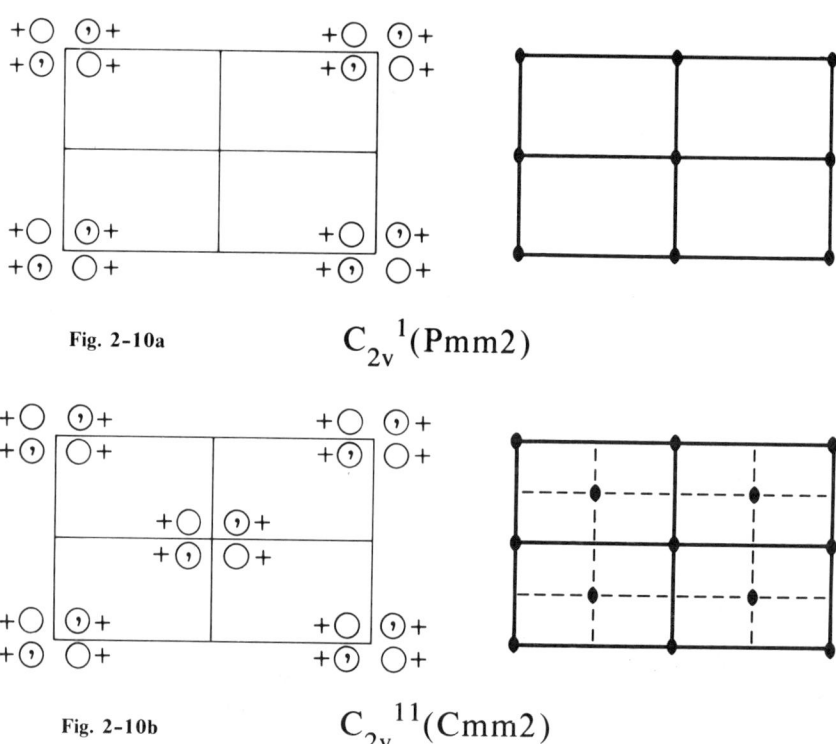

Fig. 2-10a $\quad C_{2v}^{1}(Pmm2)$

Fig. 2-10b $\quad C_{2v}^{11}(Cmm2)$

tional space group symbol conveys more information than the corresponding Schoenflies symbol. As a last comment about these space groups notice how, for both space groups, other symmetry operations result from combining the point symmetry operations with unit cell lattice translations. In Fig. 2-10a there are twofold axes halfway between the lattice points and mirror planes parallel to and halfway between the starting planes. For the C-centered space group even glide planes are generated. The glide planes are represented by the dashed lines. For example, there is a glide plane at ($\mathbf{a}/4$, 0, 0), which causes the collection of atoms to be reflected across this plane and then translated by $\mathbf{b}/2$. The C-lattice imposes this glide plane. Thus, we see that, for a symmorphic space group, one can generate nonsymmorphic operations, but they are not *necessary* to describe the space group.

Now that these two space groups are determined we might want to write the coordinates of the possible positions in the unit cell. For example, consider $C_{2v}^{1}(Pmm2)$. Take the coordinates of the circle to the lower right of the origin as (x, y, z). Although it is easier to think in terms of small values of x, y, and z so that the points are localized about the origin, in fact x, y, and z are independent and can have any value up to 1. The values of x, y, and z are fractions of the unit cell

Table 2-3 Coordinates of equivalent positions for space group C_{2v}^1(Pmm2).

1a	0, 0, z			2e	x, 0, z	$\bar{x}$, 0, z
1b	0, 1/2, z			2f	x, 1/2, z	$\bar{x}$, 1/2, z
1c	1/2, 0, z			2g	0, y, z	0, $\bar{y}$, z
1d	1/2, 1/2, z			2h	1/2, y, z	1/2, $\bar{y}$, z
4i	x, y, z	$\bar{x}$, $\bar{y}$, z	x, $\bar{y}$, z	$\bar{x}$, y, z

axes. An atom at (x, y, z) is at a vector distance **r** = x**a** + y**b** + z**c** from the origin of the particular unit cell. After a $C_2(2)$ symmetry operation, on the circle at (x, y, z) we have the circle to the upper left of the origin at ($\bar{x}$, $\bar{y}$, z). Similarly, after the two mirror operations we have ($\bar{x}$, y, z) and (x, $\bar{y}$, z). Thus, we say that the coordinates of the general equivalent positions, in this space group, are

(x, y, z), ($\bar{x}$, $\bar{y}$, z), ($\bar{x}$, y, z), and (x, $\bar{y}$, z)

The important fact is that these four points, called the **general equivalent positions**, will transform among themselves under all the symmetry operations of the space group. (Naturally by translation symmetry, Eq. 2-1, these points are associated with every lattice point.) There are four different places, within the unit cell, where a twofold axis intersects two vertical planes. One such place is x = 0 and y = 0. In this space group, the coordinates of a point along the twofold axis are (0, 0, z). In Table 2-3 we list these values, calling this position 1a where the "1" means that there is only one position of this type in the unit cell after all the space group operations are applied and the "a" is an arbitrary letter to separate this position from other positions. For example, in Table 2-3 we also list the 1b, 1c, 1d positions, which also lie on twofold axes at the intersection of two vertical planes. The positions 1a, 1b, 1c, and 1d are called **special positions** because under some of the symmetry operations of the space group they transform into themselves. As shown in Figure 2-10a other special positions are those positions that lie on one mirror plane. The other mirror plane will transform the starting position to a symmetry related position. For example, the position at (x, 0, z) will go to ($\bar{x}$, 0, z) since these are two symmetry-related positions. These are called 2e where the "e" is again arbitrary. There are four such mirror planes in the unit cell; this can be seen clearly in the right-hand figure of Figure 2-10a. These positions are labeled 2e, 2f, 2g, and 2h in Table 2-3. Last, the general equivalent positions are labeled 4i and are included in the table.

In the same manner we could generate the other orthorhombic space groups from these two lattices, namely D_2^1(P222), D_2^6(C222), D_{2h}^1(Pmmm), D_{2h}^{19}(Cmmm). Still other orthorhombic space groups can be generated similarly from the I- and F-lattices. Table 2-2 lists

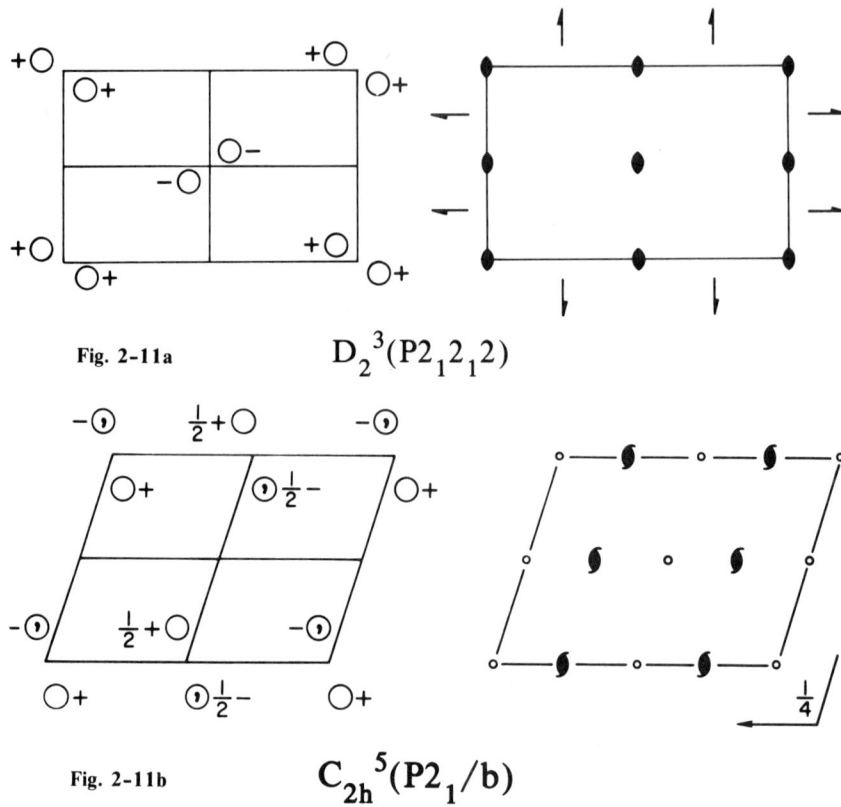

Fig. 2-11a $D_2^3 (P2_1 2_1 2)$

Fig. 2-11b $C_{2h}^5 (P2_1/b)$

the 73 symmorphic space groups that can be generated in this manner for all crystal systems. The International notation displays rather clearly the symmetry information, while the Schoenflies symbol is just the point group symbol with a fairly arbitrary numerical superscript to index each of the space groups with a given point group. Notice that the International symbols for the symmorphic space groups contain no glide plane or screw axis symbols since these are not *essential* to describe the space group; nevertheless, they may be *generated* when the point operations are multiplied (combined) by the unit cell translations or the centering conditions.

2-6c Nonsymmorphic space groups The development of the nonsymmorphic space groups follows in the same general manner as symmorphic space groups, but the addition of glide planes and screw axes makes the problem much more complex. We shall not show how they may be developed (see the notes) but we give two examples of the results.

Figure 2-11a shows one of the orthorhombic nonsymmorphic space groups. It has similarities to P222 but it also has differences.

CHAPTER 2 SYMMETRY DESCRIPTION OF CRYSTALS 43

The half arrows pointing in the a- and b-directions indicate 2_1 screw axes. These operations imply a $C_2(2)$ rotation about each of the screw axes indicated and then a translation by **a**/2 or **b**/2, as indicated by the direction of the half arrow. By following the effects of these operations on the circles we can see that all the screw axes and twofold axes are, indeed, symmetry operations. The 2_1 screw axes do not intersect the twofold axes; thus, the arrangement of circles, which indicates general equivalent positions, is rather different from that in the P222 space group. Nevertheless, both of these space groups have four general equivalent positions.

In Fig. 2-11b a monoclinic, nonsymmorphic space group is shown that has 2_1 screw axes in the c-direction as well as a glide plane parallel to the page at c/4. The combination of the b-glide and 2_1-axes produces centers of inversion, which are indicated by the very small circles. The notation that is used for space group diagrams is summarized in Table 2-4.

2-6d International notation for planes and axes   Table 2-4 shows the International space group symbols for symmetry planes, and Table 2-5 shows the corresponding symbols for symmetry axes.

Figure 2-10 has many vertical planes, indicated by heavy lines. Figures 2-10b has several glide planes, as already discussed. For example, using the Seitz notation, there is a glide plane, at $\{\sigma(a/4, 0, 0) \mid \tau(0, b/2, 0)\}$, as well as three others. Figure 2-11b has a horizontal glide plane at $\{\sigma(0, 0, c/4) \mid \tau(0, b/2, 0)\}$ and screw axes of the type $\{C_2(0, 0, 1) \mid \tau(0, 0, c/2)\}$. Screw axes in the ab-plane can be found in Figure 2-11a.

As can be seen from Table 2-5, using the International notation, an n_α screw axis can be written as $\{n[0, 0, 1] \mid \tau(0, 0, \alpha c/n)\}$. For example, successive operations of a 6_5 screw give points at c = 0, 5c/6, 10c/6, 15c/6, 20c/6, 25c/6. To bring these back into one unit cell 6c/6 = c can always be subtracted. Thus, we have c = 0, 5c/6, 4c/6 = 2c/3, 3c/6 = c/2, 2c/6 = c/3, c/6. Figure 2-12 shows all the possible screw axes operations. Notice how the 6_5 operation is enantiomorphic to the 6_1, operation.

2-7 Definitions of Directions, Coordinates, and Planes

Although some of the nomenclature defined here has already been used, it is summarized here for easy reference. In what follows, **a**, **b**, and **c** refer to the axes of a lattice in which a primitive or nonprimitive unit cell is used.

Table 2-4 Symbols of the symmetry planes (from the International Tables).

Symbol	Symmetry plane	Graphical symbol		Nature of glide translation
		Normal to plane of projection	Parallel to plane of projection	
m	Reflection plane (mirror)	───────	⌐ ∕	None (NOTE. If the plane is at $z=\frac{1}{4}$ this is shown by printing $\frac{1}{4}$ beside the symbol.)
a, b	Axial glide plane	- - - - - -	⌐ ⌐ ⌐	$a/2$ along [100] or $b/2$ along [010]; or along $\langle 100 \rangle$.
c		· · · · · · · · · ·	None	$c/2$ along z-axis; or $(a+b+c)/2$ along [111] on rhombohedral axes.
n	Diagonal glide plane (net)	— · — · — · —	⌐	$(a+b)/2$ or $(b+c)/2$ or $(c+a)/2$; or $(a+b+c)/2$ (tetragonal and cubic).
d	"Diamond" glide plane	— · ← — · — · — · → · — ·	⌐↘	$(a \pm b)/4$ or $(b \pm c)/4$ or $(c \pm a)/4$; or $(a \pm b \pm c)/4$ (tetragonal and cubic). See note below.

NOTE. In the "diamond" glide plane the glide translation is half of the resultant of the two possible axial glide translations. The arrows in the first diagram show the direction of the horizontal component of the translation when the z-component is positive. In the second diagram the arrow shows the actual direction of the glide translation; there is always another diamond-glide reflection plane parallel to the first with a height difference of $\frac{1}{4}$ and with the arrow pointing along the other diagonal of the cell face.

2-7a Directions A lattice point $m_1\mathbf{a} + m_2\mathbf{b} + m_3\mathbf{c}$ lies in the direction $[m_1, m_2, m_3]$; the commas are often left out, but the square bracket is always used. Thus, this number triple defines a direction in the lattice. Usually, these values of m_i are divided by their greatest common divisor yielding the smallest integer set, that is, the body diagonal of a unit cell is the [111] direction, instead of [222].

The $\langle \ \rangle$ bracket represents a family of directions related by symmetry. Thus, $\langle 100 \rangle$ for a cubic lattice refers to the six directions [100], [$\bar{1}$00], [010], [0$\bar{1}$0], [001], and [00$\bar{1}$], where $\bar{1}$ means -1 and is read as "one bar" or "bar one."

2-7b Coordinates The coordinates of points (or atoms) in a unit cell are specified in terms of a fraction of the axial lengths of $\mathbf{a}$, $\mathbf{b}$, and $\mathbf{c}$. Thus, the coordinates of the central point of a unit cell is given by (1/2, 1/2, 1/2). A position along the body diagonal halfway to the center is given by (1/4, 1/4, 1/4). The coordinates of the face center positions are (1/2, 1/2, 0); (1/2, 0, 1/2); (0, 1/2, 1/2).

2-7c Planes The terminology for planes is a little more complicated. A plane is defined in terms of **Miller indices** (hkℓ). If a plane intersects the axes at $m_1\mathbf{a}$, $m_2\mathbf{b}$, $m_3\mathbf{c}$, then the Miller indices of the plane are the set of integers, with no common factors, that are inversely proportional to the intercepts of the plane with the axes, or

CHAPTER 2   SYMMETRY DESCRIPTION OF CRYSTALS   45

Table 2-5 Symbols of the symmetry axes (from the International Tables).

Symbol	Symmetry axis	Graphical symbol	Nature of right-handed screw translation along the axis	Symbol	Symmetry axis	Graphical symbol (normal to plane of paper)	Nature of right-handed screw translation along the axis
1	Rotation monad	None	None	4	Rotation tetrad	◆	None
$\bar{1}$	Inversion monad	○	None	4_1	Screw tetrads	◆	$c/4$
				4_2		◆	$2c/4$
2	Rotation diad	● (normal to paper)	None	4_3		◆	$3c/4$
		→ (parallel to paper)		$\bar{4}$	Inversion tetrad	◆	None
2_1	Screw diad	● (normal to paper)	$c/2$	6	Rotation hexad	⬢	None
		→ (parallel to paper)	Either $a/2$ or $b/2$	6_1	Screw hexads	⬢	$c/6$
		Normal to paper		6_2		⬢	$2c/6$
3	Rotation triad	▲	None	6_3		⬢	$3c/6$
3_1	Screw triads	▲	$c/3$	6_4		⬢	$4c/6$
3_2		▲	$2c/3$	6_5		⬢	$5c/6$
$\bar{3}$	Inversion triad	△	None	$\bar{6}$	Inversion hexad	⬢	None

$$h : k : \ell = m_1^{-1} : m_2^{-1} : m_3^{-1} \tag{2-3}$$

Figure 2-13 shows a few planes and their Miller indices. As can be seen, if $m_i = \infty$ the corresponding Miller index is 0. Often a given (h k ℓ) is used to define an infinite number of parallel planes. (See the problems for the distance between them.) This is indicated in Fig. 2-13 for the (100) planes. Then the (100) planes intersect the axes at $m_1 = 0, 1, 2,..., -1, -2,...$ Similarly the (200) plane intersects at $m_1 = 1/2$, but the (200) planes intersect at $m_1 = 0, 1/2, 1, 3/2, 2$, etc.

In cubic crystals the direction [hkℓ] is perpendicular to the plane with the same numbers, i.e., (hkℓ).

A family of planes consists of planes that are equivalent by symmetry and are denoted by a curly bracket, { }. Thus, the six faces of a cube are {100}. This notation is used extensively in Chapter 17.

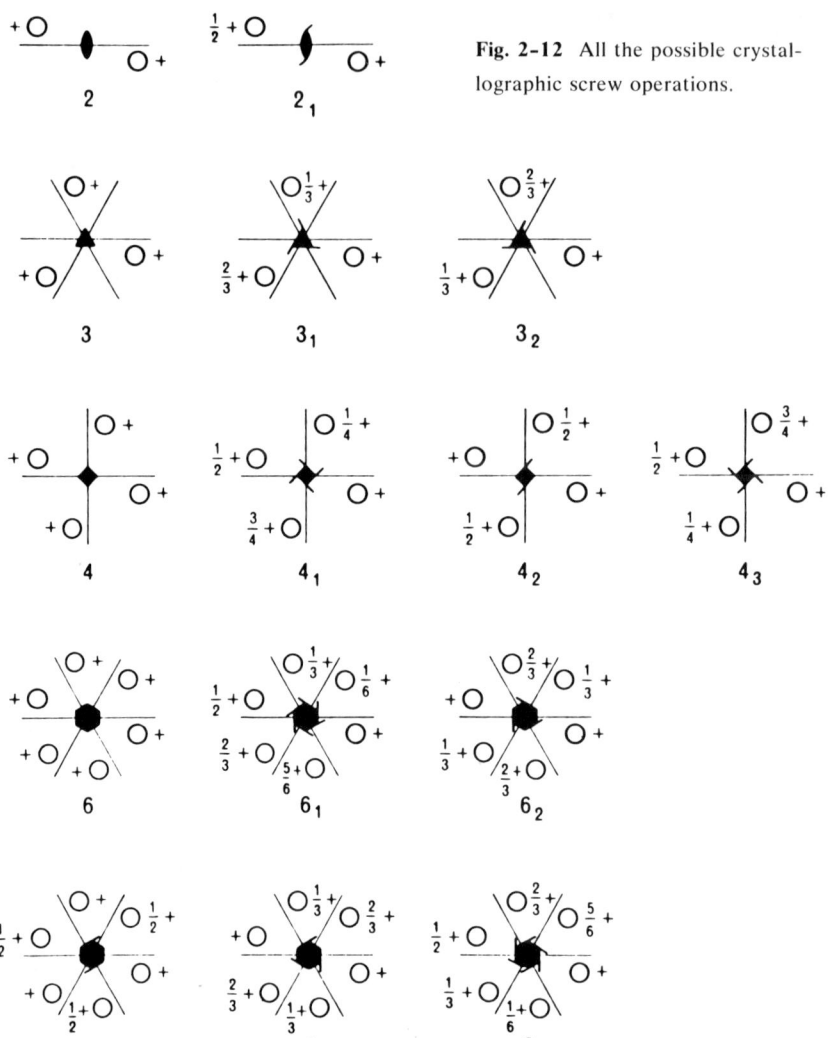

Fig. 2-12 All the possible crystallographic screw operations.

Appendix to Chapter 2 We show that, in two dimensions, only $C_n(n)$ where n = 1, 2, 3, 4, and 6 is compatible with the translation symmetry of a lattice.

With reference to Fig. 2-14 consider two lattice points A and A′, which are separated by a unit lattice translation t. Let R be a rotational symmetry operation. Apply R or its inverse R^{-1}, which is also a symmetry operation as in Section 1-2c, to the lattice at each of these points. This gives the points B and B′, a distance t′ apart, and each point is a distance t from A and A′, respectively, as shown. Since R is a symmetry operation, B and B′ must also be lattice points, so t′ must be an integral multiple of t. Thus,

CHAPTER 2 SYMMETRY DESCRIPTION OF CRYSTALS

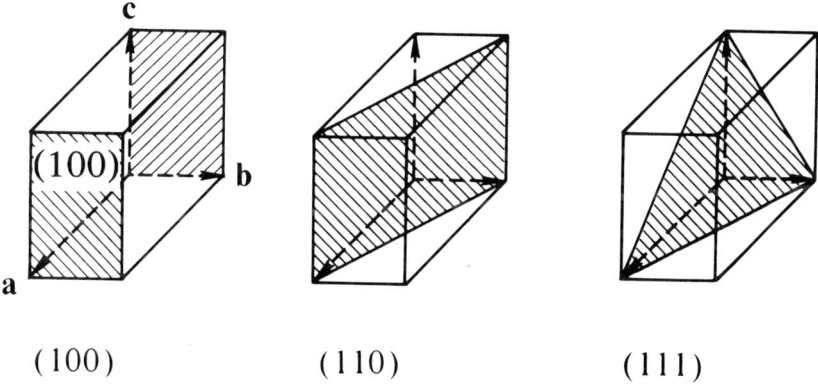

Fig. 2-13 Some important planes and their Miller indices.

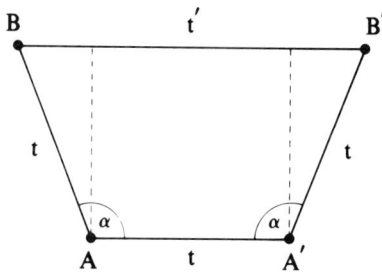

Fig. 2-14 Points in a lattice.

$$t' = mt \qquad \text{m is an integer} \qquad (2\text{-}4a)$$
$$t' = -2t\cos\alpha + t \qquad (2\text{-}4b)$$
$$\cos\alpha = (1-m)/2 \qquad (2\text{-}4c)$$

where the second equation comes from the diagram and the third is obtained by combining the first two equations. Since m is an integer, then $1-m$ is also an integer. Furthermore, α must lie between 0 and 180° in order to obtain closure for the symmetry operation R, that is, $\cos\alpha$ lies between $+1$ and -1. This leads to

$$|\cos\alpha| \le 1$$

so $\qquad m = -1, 0, 1, 2, 3$
or $\qquad \alpha = 0, \pi/3, \pi/2, 2\pi/3, \pi$

Thus, the only allowed rotations are $2\pi/n$ where $n = 1, 6, 4, 3$, or 2, respectively. (Or $n = 1, 2, 3, 4, 6$.)

Notes

The book by Burns and Glazer is written for solid state scientists and gives many details on space groups that are left out here. Kittel,

Chapter 1, deals with the same concepts discussed here but emphasizes the two-dimensional space groups. The interested reader should certainly look at the International Tables, the "Bible" for crystallographers. Look at the frontmatter of the International Tables now, and scan the tables themselves after reading the next chapter.

Problems

1. Every lattice point of an arbitrary lattice is a center of inversion. Show that a center of inversion also exists halfway between any two lattice points. Why must all holosymmetric point groups contain the symmetry operation $i(\bar{1})$?

2. (a) From the defining conditions, in Table 2-1, for the tetragonal crystal system the point groups $C_4(4)$ and $S_4(\bar{4})$ are clearly in this crystal system. The other symmetry operations that are compatible with tetragonal symmetry are C_2', σ_v, and σ_h. Using sterograms, starting with these two point groups and adding, one at a time, these symmetry operations, show that the other point groups shown in Table 1-2 are obtained. (b) Using a similar procedure, develop the point groups for the cubic crystal system.

3. Point group development – Develop the point groups first by finding the 11 point groups that are pure rotational point groups (they have only rotations about various axes). Next, by adding a center of symmetry obtain the 11 centrosymmetric point groups. Then find the ten subgroups of these latter 11 point groups that are different from the purely rotational ones. (Hint: see Bhagavantam.)

4. Space group diagrams – Draw the space group diagrams, showing all of the generated symmetry operations for C_4^1(P4), C_{4v}^1(P4mm), and C_{4h}^1(P4/m). Check your results in the International Tables.

5. Enantiomorphous space groups – Draw the space group diagrams for P4, $P4_1$, $P4_2$, and $P4_3$. Define the term **enantiomorphous operation**. Which of these four space groups are enantiomorphically related to each other? (Hint: see Burns and Glazer. Check your results in the International Tables.)

6. For a cubic crystal with unit cell length a, show that the distance between adjacent planes with Miller indices (h k ℓ) is

$$d = a/(h^2 + k^2 + \ell^2)^{1/2}$$

3

Simple Crystal Structures

3-1 Introduction
3-2 Several Cubic Symmorphic Structures
    a   Space group Pm3m
    b   Space group Im3m
    c   Space group Fm3m
3-3 Diamond and Zinc Blende Structures
3-4 Point Group of a Space Group (S)
3-5 Examples of Defect Structures
3-6 Different Points of View of a Structure
3-7 Close Packing (and the Hexagonal Close-Packed Structure)
3-8 Volume Effects for Simple Structures
    a   Coordination number
    b   High pressure/temperature effects
3-9 Wurtzite Structure
3-10 Site Symmetry (S)
    Notes
    Problems

Table 3-1 Elements with the body-centered cubic O_h^9(Im3m), face-centered cubic O_h^5(Fm3m), and/or hexagonal close-packed D_{6h}^4(P6$_3$/mmc) crystal structures. (Adapted from Galasso.)

bcc	fcc	hcp	c/a for hcp
Ba	Ac	Be	1.566
γ-Ca(to m.p.)	Ag	β-Ca(450°C)	1.638
δ-Ce(730° to m.p.)	Al	Cd	1.885
α-Cr	Am	α-Co	1.623
Cs	Ar	γ-Cr	1.626
Eu	Au	Dy	1.574
α-Fe	α-Ca(to 250°C)	Er	1.571
β-Fe(800°C)	Ce(−10° to 730°)	Gd	1.590
δ-Fe(1425°C)	Co	He	1.633
β-Hf(above 1950°C)	β-Cr	α-Hf	1.582
K	Cu	Ho	1.570
γ-La(868°C to m.p.)	γ-Fe	Li(78°K)	1.637
Li	Ir	Lu	1.583
Mo	Kr	Mg	1.623
Na	β-La(310 to 868°C)	Na(5°K)	1.613
Nb	Ne	Ni	1.634
β-Nd(868°C to m.p.)	Ni	Os	1.579
γ-Np(~600°C)	Pb	Re	1.614
β-Pr(798° to m.p.)	Pd	Ru	1.583
ε-Pu(500°C)	Pt	α-Sc(to 1000°C)	1.591
Rb	δ-Pu(320°C)	β-Sr(248°C)	1.634
β-Sm(917°C to m.p.)	Rh	Tb	1.582
γ-Sr(614°C)	β-Sc(1000°C to m.p.)	Te	1.604
Ta	α-Sr	α-Ti	1.588
β-Th(1450°C)	α-Th	α-Tl	1.598
β-Ti(900°C)	Xe	Tm	1.572
β-Tl	α-Yb(to 798°)	α-Y(to 1490°C)	1.572
γ-U		Zn	1.856
V		α-Zr	1.592
W			
β-Y(1490°C to m.p.)			
β-Yb(798°C to m.p.)			
β-Zr(850°C)			

SIMPLE CRYSTAL STRUCTURES

Nothing happens in nature which can be attributed to the vice of nature, for she is always the same and everywhere one. Her virtue is the same, and her powers of acting; that is to say her laws and rules, according to which all things are and are changed from form to form, are everywhere and always the same; ...

Spinoza, "Ethics"

3-1 Introduction

In this chapter we discuss the basic, simple crystal structures. These structures can be, and often are, discussed without reference to any of the information that we acquired in the first two chapters. However, a discussion within the framework of this information gives a much deeper meaning to the structures and lays the groundwork for all crystal structures no matter how complicated.

A crystal structure is a periodic arrangement of atoms. Focusing for a moment on a lattice point, there is some ordered arrangement of atoms about the lattice point (or lattice points in a multiply primitive unit cell). This arrangement repeats throughout all space by the translational symmetry of the lattice. This arrangement of atoms about a lattice point is called a **basis** or **lattice complex**. We can define a **crystal structure** by the particular lattice (which means the axial lengths, angles between them, and the centering, i.e., P, I, F, C, or R) and the basis. Thus, we say

$$\text{crystal structure} = \text{lattice} + \text{basis} \qquad (3\text{-}1)$$

(This is not a mathematical equation, but just a concise way to write the above definition.)

3-2 Several Cubic Symmorphic Structures

In this section we discuss a number of quite simple, cubic crystal structures. These structures are discussed in different subsections, listed by space group, in order to emphasize the variety of crystal structures that can have the same symmetry operations.

3-2a Space group Pm3m Figure 3-1 shows three different crystal structures, all with the same O_h^1(Pm3m) space group. The structure with atoms only at the lattice points of a cubic P-lattice (Fig.

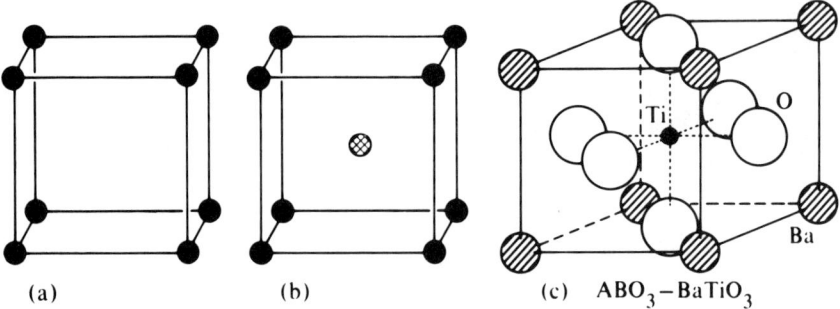

Fig. 3-1 Crystal structures with the O_h^1(Pm3m) space group.

3-1a) is the simplest crystal structure imaginable and is not commonly found in nature. In fact, the only element with this structure is polonium. In this structure each atom has six-nearest neighbors. To describe the structure of Po we give the space group, the length of the unit cell, and the basis, which in this case is Po at (0, 0, 0). Actually, black phosphores above 100 kbars of pressure transforms to this simple structure. However, Po has this **simple cubic (sc) structure** at atmospheric pressure.

The **CsCl structure** (Fig. 3-1b) is slightly more complicated; the basis is Cs at (0, 0, 0) and Cl at (1/2, 1/2, 1/2), or vice versa. About 25% of the alkali halide compounds have this structure. Both the Cs and the Cl-ions have eight-nearest neighbors. This is sometimes called the **coordination number**. For CsCl it would be written as C.N. 8:8; the first 8 indicates that the atom first mentioned in the chemical formula has eight-neighbors, and the second 8 indicates that the second atom has eight-neighbors. (This crystal structure is sometimes incorrectly described as having a body-centered lattice. The lattice is clearly a cubic primitive one!)

The **cubic perovskite structure** is more complicated (Fig. 3-1c). There are a large number of such ABO_3 compounds with this structure. Many of these compounds are slightly distorted from the ideal cubic unit cell shown, in which case they have fewer symmetry operations and thus are classified by different space groups. ($CaTiO_3$ is actually the prototype perovskite material and has symmetry lower than cubic.) The perovskite structure shown in the figure is cubic with a primitive lattice, the O_h(m3m) point group, and the O_h^1(Pm3m) space group. We take cubic $BaTiO_3$ as an example; then the basis is

Ba: (0, 0, 0)
Ti: (1/2, 1/2, 1/2)
3O: (0, 1/2, 1/2); (1/2, 0, 1/2); (1/2, 1/2, 0)   (3-2)

CHAPTER 3   SIMPLE CRYSTAL STRUCTURES   53

$Pm3m$ No. 221 $P\,4/m\,\bar{3}\,2/m$ $m\,3\,m$ Cubic
O_h^1
                              Origin at centre ($m3m$)

Number of positions, Wyckoff notation, and point symmetry			Co-ordinates of equivalent positions						Conditions limiting possible reflections

General:

48 n   1 $x,y,z;\;\; z,x,y;\;\; y,z,x;\;\; x,z,y;\;\; y,x,z;\;\; z,y,x;$ $hkl:$ ⎱
            $x,\bar{y},\bar{z};\;\; z,\bar{x},\bar{y};\;\; y,\bar{z},\bar{x};\;\; x,\bar{z},\bar{y};\;\; y,\bar{x},\bar{z};\;\; z,\bar{y},\bar{x};$ $hhl:$ ⎰ No conditions
            $\bar{x},y,\bar{z};\;\; \bar{z},x,\bar{y};\;\; \bar{y},z,\bar{x};\;\; \bar{x},z,\bar{y};\;\; \bar{y},x,\bar{z};\;\; \bar{z},y,\bar{x};$ $0kl:$ ⎱
            $\bar{x},\bar{y},z;\;\; \bar{z},\bar{x},y;\;\; \bar{y},\bar{z},x;\;\; \bar{x},\bar{z},y;\;\; \bar{y},\bar{x},z;\;\; \bar{z},\bar{y},x;$
            $\bar{x},\bar{y},\bar{z};\;\; \bar{z},\bar{x},\bar{y};\;\; \bar{y},\bar{z},\bar{x};\;\; \bar{x},\bar{z},\bar{y};\;\; \bar{y},\bar{x},\bar{z};\;\; \bar{z},\bar{y},\bar{x};$
            $x,y,\bar{z};\;\; \bar{z},x,y;\;\; \bar{y},z,x;\;\; \bar{x},z,y;\;\; \bar{y},x,z;\;\; \bar{z},y,x;$
            $x,\bar{y},z;\;\; z,\bar{x},y;\;\; y,\bar{z},x;\;\; x,\bar{z},y;\;\; y,\bar{x},z;\;\; z,\bar{y},x;$
            $x,y,\bar{z};\;\; z,x,\bar{y};\;\; y,z,\bar{x};\;\; x,z,\bar{y};\;\; y,x,\bar{z};\;\; z,y,\bar{x}.$

Special:

24 m   m $x,x,z;\;\; z,x,x;\;\; x,z,x;\;\; \bar{x},\bar{x},\bar{z};\;\; \bar{z},\bar{x},\bar{x};\;\; \bar{x},\bar{z},\bar{x};$ No conditions
            $x,\bar{x},\bar{z};\;\; z,\bar{x},\bar{x};\;\; x,\bar{z},\bar{x};\;\; \bar{x},x,z;\;\; \bar{z},x,x;\;\; \bar{x},z,x;$
            $\bar{x},x,\bar{z};\;\; \bar{z},x,\bar{x};\;\; \bar{x},z,\bar{x};\;\; x,\bar{x},z;\;\; z,\bar{x},x;\;\; x,\bar{z},x;$
            $\bar{x},\bar{x},z;\;\; \bar{z},\bar{x},x;\;\; \bar{x},\bar{z},x;\;\; x,x,\bar{z};\;\; z,x,\bar{x};\;\; x,z,\bar{x}.$

24 l   m $\tfrac{1}{2},y,z;\;\; z,\tfrac{1}{2},y;\;\; y,z,\tfrac{1}{2};\;\; \tfrac{1}{2},z,y;\;\; y,\tfrac{1}{2},z;\;\; z,y,\tfrac{1}{2};$
            $\tfrac{1}{2},\bar{y},\bar{z};\;\; \bar{z},\tfrac{1}{2},\bar{y};\;\; \bar{y},\bar{z},\tfrac{1}{2};\;\; \tfrac{1}{2},\bar{z},\bar{y};\;\; \bar{y},\tfrac{1}{2},\bar{z};\;\; \bar{z},\bar{y},\tfrac{1}{2};$
            $\tfrac{1}{2},y,\bar{z};\;\; \bar{z},\tfrac{1}{2},y;\;\; \bar{y},z,\tfrac{1}{2};\;\; \tfrac{1}{2},\bar{z},y;\;\; y,\tfrac{1}{2},\bar{z};\;\; \bar{z},y,\tfrac{1}{2};$
            $\tfrac{1}{2},\bar{y},z;\;\; z,\tfrac{1}{2},\bar{y};\;\; \bar{y},z,\tfrac{1}{2};\;\; \tfrac{1}{2},z,\bar{y};\;\; \bar{y},\tfrac{1}{2},z;\;\; z,\bar{y},\tfrac{1}{2}.$

24 k   m $0,y,z;\;\; z,0,y;\;\; y,z,0;\;\; 0,z,y;\;\; y,0,z;\;\; z,y,0;$
            $0,\bar{y},\bar{z};\;\; \bar{z},0,\bar{y};\;\; \bar{y},\bar{z},0;\;\; 0,\bar{z},\bar{y};\;\; \bar{y},0,\bar{z};\;\; \bar{z},\bar{y},0;$
            $0,y,\bar{z};\;\; \bar{z},0,y;\;\; y,\bar{z},0;\;\; 0,\bar{z},y;\;\; y,0,\bar{z};\;\; \bar{z},y,0;$
            $0,\bar{y},z;\;\; z,0,\bar{y};\;\; \bar{y},z,0;\;\; 0,z,\bar{y};\;\; \bar{y},0,z;\;\; z,\bar{y},0.$

12 j   mm $\tfrac{1}{2},x,x;\;\; x,\tfrac{1}{2},x;\;\; x,x,\tfrac{1}{2};\;\; \tfrac{1}{2},x,\bar{x};\;\; \bar{x},\tfrac{1}{2},x;\;\; x,\bar{x},\tfrac{1}{2};$
            $\tfrac{1}{2},\bar{x},\bar{x};\;\; \bar{x},\tfrac{1}{2},\bar{x};\;\; \bar{x},\bar{x},\tfrac{1}{2};\;\; \tfrac{1}{2},\bar{x},x;\;\; x,\tfrac{1}{2},\bar{x};\;\; \bar{x},x,\tfrac{1}{2}.$

12 i   mm $0,x,x;\;\; x,0,x;\;\; x,x,0;\;\; 0,x,\bar{x};\;\; \bar{x},0,x;\;\; x,\bar{x},0;$
            $0,\bar{x},\bar{x};\;\; \bar{x},0,\bar{x};\;\; \bar{x},\bar{x},0;\;\; 0,\bar{x},x;\;\; x,0,\bar{x};\;\; \bar{x},x,0.$

12 h   mm $x,\tfrac{1}{2},0;\;\; 0,x,\tfrac{1}{2};\;\; \tfrac{1}{2},0,x;\;\; x,0,\tfrac{1}{2};\;\; \tfrac{1}{2},x,0;\;\; 0,\tfrac{1}{2},x;$
            $\bar{x},\tfrac{1}{2},0;\;\; 0,\bar{x},\tfrac{1}{2};\;\; \tfrac{1}{2},0,\bar{x};\;\; \bar{x},0,\tfrac{1}{2};\;\; \tfrac{1}{2},\bar{x},0;\;\; 0,\tfrac{1}{2},\bar{x}.$

8   g   3m $x,x,x;\;\; x,\bar{x},\bar{x};\;\; \bar{x},x,\bar{x};\;\; \bar{x},\bar{x},x;$
            $\bar{x},\bar{x},\bar{x};\;\; \bar{x},x,x;\;\; x,\bar{x},x;\;\; x,x,\bar{x}.$

6   f   4mm $x,\tfrac{1}{2},\tfrac{1}{2};\;\; \tfrac{1}{2},x,\tfrac{1}{2};\;\; \tfrac{1}{2},\tfrac{1}{2},x;\;\; \bar{x},\tfrac{1}{2},\tfrac{1}{2};\;\; \tfrac{1}{2},\bar{x},\tfrac{1}{2};\;\; \tfrac{1}{2},\tfrac{1}{2},\bar{x}.$

6   e   4mm $x,0,0;\;\; 0,x,0;\;\; 0,0,x;\;\; \bar{x},0,0;\;\; 0,\bar{x},0;\;\; 0,0,\bar{x}.$

3   d   4/mmm $\tfrac{1}{2},0,0;\;\; 0,\tfrac{1}{2},0;\;\; 0,0,\tfrac{1}{2}.$

3   c   4/mmm $0,\tfrac{1}{2},\tfrac{1}{2};\;\; \tfrac{1}{2},0,\tfrac{1}{2};\;\; \tfrac{1}{2},\tfrac{1}{2},0.$

1   b   m3m $\tfrac{1}{2},\tfrac{1}{2},\tfrac{1}{2}.$

1   a   m3m $0,0,0.$

Fig. 3-2 $O_h^1(Pm3m)$.

The C.N. 12:6:2 indicates the 12 O-ions about the Ba-ions, the six O-ions about the Ti-ions, and the two Ti-ions about the O-ions.

There are many, much more complicated, crystal structures with this space group. It would be convenient to have a generally accepted way to describe the basis. The **International Tables for X-ray Crystallography**, Volume 1, usually referred to as International Tables, provides the proper description. It has all the symmetry information we would usually like to know for each of the 230 space groups. Figure 3-2 shows the page for O_h^1(Pm3m) from these tables. We briefly describe the information that appears, starting at the top and going across. First, the symbol for the space group is given in the two notations. The space groups are arbitrarily numbered from 1 (triclinic) to 230 (cubic). This space group is number 221. The full International space group symbol is then given. It contains a considerable amount of information. For example, here it shows that there are mirror planes perpendicular to the four- and twofold axes. Then the point group of the space group is given, which is m3m or O_h in the Schoenflies notation. Then the crystal system is given, which for this case is cubic. The next line states that the origin is at a center of symmetry (English, not American, spelling of center) and has point symmetry O_h(m3m). The next line labels the columns. The number of positions, Wyckoff notation, point symmetry, and coordinates of equivalent positions describe the *possible* atom position(s) in this space group. For example, to describe the crystal structure of α-Po we would say that the space group is O_h^1(Pm3m) with Po at the 1a position. For CsCl, the same space group is given with Cs at 1a and Cl at 1b. For $BaTiO_3$: Ba at 1a, Ti at 1b, and 3O at 3c.

There probably is some crystal structure with 48 atoms in the unit cell, all at the 48n sites. Under all the symmetry operations of this space group these 48 sites transform among themselves just as 1a transforms into itself. The same statement applies separately to each of the possible sites from 1a to 48n.

The point symmetry, sometimes called **site symmetry**, for each of the possible sites is also listed in the International notation for each possible position, as can be seen.

As we mentioned there are many crystals, some very complicated, with this space group. For any one of them the position of the atoms will be listed, 1a to 48n, and if the complete structure has been done, the values of (x), (x, y), or (x, y, z) will be given as required for the 6e to 48n sites. Remember what is guaranteed by saying that a particular crystal structure has this space group. All the symmetry operations of this space group O_h^1(Pm3m) can be obtained by combining the lattice translations (Eq. 2-1) of this cubic primitive lattice and the 48 point operations of the O_h(m3m) point group about any lattice site. *Then it is guaranteed that atoms that may be on any one of the types of positions, 1a, or 1b, ... or 48n, will transform among themselves under all the symmetry operations of this space group.*

The last column, which lists conditions limiting possible reflections, is of great importance to x-ray crystallographers in establishing the space group of an unknown crystal, but we shall not concern ourselves with this topic.

CHAPTER 3   SIMPLE CRYSTAL STRUCTURES   55

We make a trivial point. Metals, ionic crystals, and covalently bonded crystals can all have this or any other space group. Later we will classify crystals by types of bonding, but here we are classifying them by symmetry. (Note these and other crystal structures can be discussed in quite a different manner. One can take the picture of the unit cell and its atomic contents, as in Fig. 3-1, naturally repeated throughout all space by the lattice translation, and say that is the structure. However, with the approach that we are using we can generalize and interrelate structure, as well as better appreciate the symmetries.)

3-2b Space group Im3m   Figure 3-3, which looks complicated but really is not, shows some of the possible positions for the space group O_h^9(Im3m). Figure 3-4 shows the corresponding page from the International Tables for this space group. From the International space group symbol we can see that the lattice is a body-centered one, that is, an I-lattice. The large circles in Fig. 3-3 are at the 2a sites of this space group. In Fig. 3-4, at the bottom, the coordinates of these sites are listed as (0, 0, 0) but for an I-lattice we must always add the centering condition as noted in the upper middle of this figure. Thus, the 2a site has atoms at (0, 0, 0) and (1/2, 1/2, 1/2). Crystals with just the 2a site of this space group occupied are said to have a **body-centered cubic**, or **bcc, structure.** A number of elements have this structure (see Table 3-1, page 50).

Figure 3-3 shows the location of several other positions in this space group, including the 24h site for x = 3/8. Note how the 6b sites in the middle of the faces, at first glance, look different from those in the middle of the edges, but more careful consideration, when the rest of the structure is imagined, reveals that they have equivalent surroundings (as they must).

For the space group O_h^1(Pm3m) there were 48 general positions, 48n, while here there are twice as many general equivalent positions because of the centering condition. Thus we could imagine for small values of x, y, and z, 48 sites around the (0, 0, 0) position and another 48 around the (1/2, 1/2, 1/2) position. Under all the symmetry operations of the space group all 96 sites transfer among themselves. There may be some crystal structure with 96 atoms of the same kind with space group O_h^9(Im3m) where the atoms are on the 96ℓ sites. Many compounds have crystal structures with this space group and atoms at various of the possible sites 2a to 96ℓ.

3-2c Space group Fm3m   Figure 3-5 shows three crystal structures with the O_h^5(Fm3m) space group. The lattice is an F-lattice, as is clear from the symbol and the figure. Figure 3-6 shows the page of the International Tables for this space group. The basis for copper

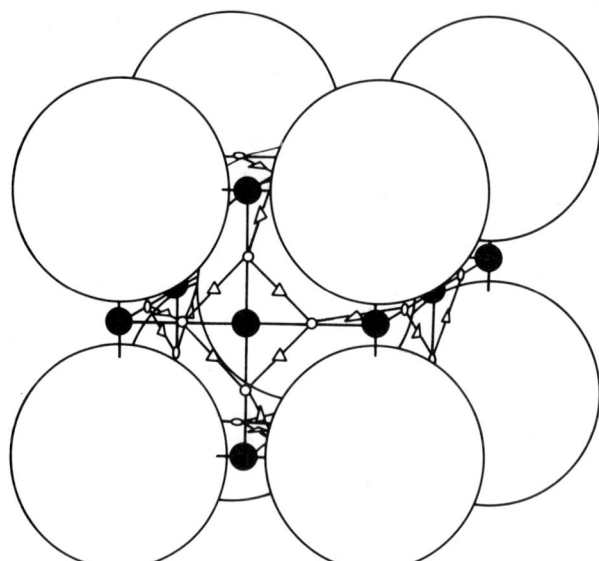

Im3m-O_h^9

● 6b sites
○ 12d sites
△ 24h sites

Fig. 3-3 Some of the positions of the O_h^9(Im3m) space groups. The very large circles represent spheres at the 2a site at (0, 0, 0) and (1/2, 1/2, 1/2).

consists of Cu at the 4a sites or just at the lattice points. For NaCl it consists of Na at 4a and Cl at 4b, or vice versa. For CaF_2 the basis is Ca at 4a and F at 8c. Again we note that the general equivalent point is 192ℓ, and there can be a crystal structure with this space group with atoms just at 192ℓ (4 × 48 = 192). Of course, there are only 48 such atoms in the primitive unit cell. For the structures shown in Fig. 3-5 the coordination numbers are: Cu-12; NaCl-6:6; CaF_2-8:4.

A number of elements have the **face-centered cubic (fcc) structure,** and they are listed in Table 3-1. This means only the 4a sites are occupied as in the Cu example. The fcc structure is one of the two so-called close-packed structures. These are discussed in Section 3-7, but suffice it to say this structure can be obtained by the packing of hard spheres (billiard balls).

3-3 Diamond and Zinc Blende Structures

In this section we discuss the last two simple, important, cubic crystal structures. Figures 3-7a and 3-7b show the diamond structure, space group O_h^7(Fd3m). This is the first nonsymmorphic structure discussed in this chapter. It has an F-lattice with atoms at the 8a positions (0, 0, 0), (1/4, 1/4, 1/4) and the face-centering positions, i.e., the coordinates of equivalent positions are the same for all F-lattices as in Fig. 3-6. The full space group symbol is $F4_1/d\bar{3}2/m$ and Fig. 3-7b shows the 4_1 axis at $x = 1/2$ and $y = 1/4$. Note that there is a center of inversion at (1/8, 1/8, 1/8) and (1/8, 3/8, 3/8). Be-

CHAPTER 3   SIMPLE CRYSTAL STRUCTURES

$Im3m$ No. 229 $I\,4/m\,\bar{3}\,2/m$ $m\,3\,m$ Cubic
O_h^9

Origin at centre ($m3m$)

Number of positions, Wyckoff notation, and point symmetry			Co-ordinates of equivalent positions $(0,0,0;\ \tfrac{1}{2},\tfrac{1}{2},\tfrac{1}{2})+$	Conditions limiting possible reflections

General:

| 96 | l | 1 | $x,y,z;\ z,x,y;\ y,z,x;\ x,z,\bar{y};\ y,x,\bar{z};\ z,y,\bar{x};$
 $x,\bar{y},\bar{z};\ z,\bar{x},\bar{y};\ y,\bar{z},\bar{x};\ x,\bar{z},\bar{y};\ y,\bar{x},\bar{z};\ z,\bar{y},\bar{x};$
 $\bar{x},y,\bar{z};\ \bar{z},x,\bar{y};\ \bar{y},z,\bar{x};\ \bar{x},z,\bar{y};\ \bar{y},x,\bar{z};\ \bar{z},y,\bar{x};$
 $\bar{x},\bar{y},z;\ \bar{z},\bar{x},y;\ \bar{y},\bar{z},x;\ \bar{x},\bar{z},y;\ \bar{y},\bar{x},z;\ \bar{z},\bar{y},x;$
 $\bar{x},\bar{y},\bar{z};\ \bar{z},\bar{x},\bar{y};\ \bar{y},\bar{z},\bar{x};\ \bar{x},\bar{z},\bar{y};\ \bar{y},\bar{x},\bar{z};\ \bar{z},\bar{y},\bar{x};$
 $\bar{x},y,z;\ \bar{z},x,y;\ \bar{y},z,x;\ \bar{x},z,y;\ \bar{y},x,z;\ \bar{z},y,x;$
 $x,\bar{y},z;\ z,\bar{x},y;\ y,\bar{z},x;\ x,\bar{z},y;\ y,\bar{x},z;\ z,\bar{y},x;$
 $x,y,\bar{z};\ z,x,\bar{y};\ y,z,\bar{x};\ x,z,\bar{y};\ y,x,\bar{z};\ z,y,\bar{x}.$ | $hkl:\ h+k+l=2n$
 $hhl:\ (l=2n);\ \circlearrowright$
 $0kl:\ (k+l=2n);\ \circlearrowright$ |

Special: as above, plus

48	k	m	$x,x,z;\ z,x,x;\ x,z,x;\ \bar{x},\bar{x},\bar{z};\ \bar{z},\bar{x},\bar{x};\ \bar{x},\bar{z},\bar{x};$ $x,\bar{x},\bar{z};\ z,\bar{x},\bar{x};\ x,\bar{z},\bar{x};\ \bar{x},x,z;\ \bar{z},x,x;\ \bar{x},z,x;$ $\bar{x},x,\bar{z};\ \bar{z},x,\bar{x};\ \bar{x},z,\bar{x};\ x,\bar{x},z;\ z,\bar{x},x;\ x,\bar{z},x;$ $\bar{x},\bar{x},z;\ \bar{z},\bar{x},x;\ \bar{x},\bar{z},x;\ x,x,\bar{z};\ z,x,\bar{x};\ x,\bar{z},\bar{x}.$	
48	j	m	$0,y,z;\ z,0,y;\ y,z,0;\ 0,z,y;\ y,0,z;\ z,y,0;$ $0,\bar{y},\bar{z};\ \bar{z},0,\bar{y};\ \bar{y},\bar{z},0;\ 0,\bar{z},\bar{y};\ \bar{y},0,\bar{z};\ \bar{z},\bar{y},0;$ $0,y,\bar{z};\ \bar{z},0,y;\ y,\bar{z},0;\ 0,\bar{z},y;\ y,0,\bar{z};\ \bar{z},y,0;$ $0,\bar{y},z;\ z,0,\bar{y};\ \bar{y},z,0;\ 0,z,\bar{y};\ \bar{y},0,z;\ z,\bar{y},0.$	
48	i	2	$\tfrac{1}{4},x,\tfrac{1}{2}-x;\ \tfrac{1}{4},\bar{x},\tfrac{1}{2}+x;\ \tfrac{1}{4},x,\tfrac{1}{2}+x;\ \tfrac{1}{4},\bar{x},\tfrac{1}{2}-x;$ $\tfrac{1}{2}-x,\tfrac{1}{4},x;\ \tfrac{1}{2}+x,\tfrac{1}{4},\bar{x};\ \tfrac{1}{2}+x,\tfrac{1}{4},x;\ \tfrac{1}{2}-x,\tfrac{1}{4},\bar{x};$ $x,\tfrac{1}{2}-x,\tfrac{1}{4};\ \bar{x},\tfrac{1}{2}+x,\tfrac{1}{4};\ x,\tfrac{1}{2}+x,\tfrac{1}{4};\ \bar{x},\tfrac{1}{2}-x,\tfrac{1}{4};$ $\tfrac{1}{4},\tfrac{1}{2}-x,x;\ \tfrac{1}{4},\tfrac{1}{2}+x,\bar{x};\ \tfrac{1}{4},\tfrac{1}{2}+x,x;\ \tfrac{1}{4},\tfrac{1}{2}-x,\bar{x};$ $x,\tfrac{1}{4},\tfrac{1}{2}-x;\ \bar{x},\tfrac{1}{4},\tfrac{1}{2}+x;\ x,\tfrac{1}{4},\tfrac{1}{2}+x;\ \bar{x},\tfrac{1}{4},\tfrac{1}{2}-x;$ $\tfrac{1}{2}-x,x,\tfrac{1}{4};\ \tfrac{1}{2}+x,\bar{x},\tfrac{1}{4};\ \tfrac{1}{2}+x,x,\tfrac{1}{4};\ \tfrac{1}{2}-x,\bar{x},\tfrac{1}{4}.$	no extra conditions
24	h	mm	$0,x,x;\ x,0,x;\ x,x,0;\ 0,x,\bar{x};\ \bar{x},0,x;\ x,\bar{x},0;$ $0,\bar{x},\bar{x};\ \bar{x},0,\bar{x};\ \bar{x},\bar{x},0;\ 0,\bar{x},x;\ x,0,\bar{x};\ \bar{x},x,0.$	
24	g	mm	$x,0,\tfrac{1}{2};\ \tfrac{1}{2},x,0;\ 0,\tfrac{1}{2},x;\ x,\tfrac{1}{2},0;\ 0,x,\tfrac{1}{2};\ \tfrac{1}{2},0,x;$ $\bar{x},0,\tfrac{1}{2};\ \tfrac{1}{2},\bar{x},0;\ 0,\tfrac{1}{2},\bar{x};\ \bar{x},\tfrac{1}{2},0;\ 0,\bar{x},\tfrac{1}{2};\ \tfrac{1}{2},0,\bar{x}.$	
16	f	$3m$	$x,x,x;\ x,\bar{x},\bar{x};\ \bar{x},x,\bar{x};\ \bar{x},\bar{x},x;$ $\bar{x},\bar{x},\bar{x};\ \bar{x},x,x;\ x,\bar{x},x;\ x,x,\bar{x}.$	
12	e	$4mm$	$x,0,0;\ 0,x,0;\ 0,0,x;\ \bar{x},0,0;\ 0,\bar{x},0;\ 0,0,\bar{x}.$	
12	d	$\bar{4}2m$	$\tfrac{1}{4},0,\tfrac{1}{2};\ \tfrac{1}{2},\tfrac{1}{4},0;\ 0,\tfrac{1}{2},\tfrac{1}{4};\ \tfrac{3}{4},0,\tfrac{1}{2};\ \tfrac{1}{2},\tfrac{3}{4},0;\ 0,\tfrac{1}{2},\tfrac{3}{4}.$	
8	c	$3m$	$\tfrac{1}{4},\tfrac{1}{4},\tfrac{1}{4};\ \tfrac{1}{4},\tfrac{3}{4},\tfrac{3}{4};\ \tfrac{3}{4},\tfrac{1}{4},\tfrac{3}{4};\ \tfrac{3}{4},\tfrac{3}{4},\tfrac{1}{4}.$	$hkl:\ h,k,(l)=2n$
6	b	$4/mmm$	$0,\tfrac{1}{2},\tfrac{1}{2};\ \tfrac{1}{2},0,\tfrac{1}{2};\ \tfrac{1}{2},\tfrac{1}{2},0.$	no extra conditions
2	a	$m3m$	$0,0,0.$	

Fig. 3-4 O_h^9(Im3m).

sides diamond and gray tin, the technologically important elements silicon (Si) and germanium (Ge) have this crystal structure.

The zinc blende structure is closely related to that of diamond. ZnS, GaAs, and many other binary compounds have the zinc blende structure. The symmorphic space group is T_d^2(F$\bar{4}$3m) and the basis is

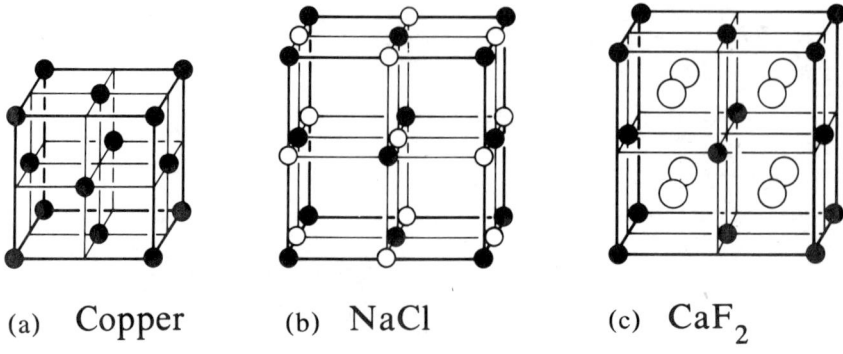

Fig. 3-5 Several structures with the O_h^5(Fm3m) space group.

Zn at the 4a site at (0, 0, 0) and S at the 4c site at (1/4, 1/4, 1/4) plus the equivalent positions due to the F-lattice. Figure 3-7c shows the structure. Unlike diamond, this structure does not have a screw axis or center of symmetry owing to the dissimilarity of the atoms. The lack of an i($\bar{1}$) symmetry operation allows materials with this crystal structure to be piezoelectric (Chapter 5), while diamond is not.

3-4 Point Group of a Space Group

So far in this chapter we have avoided mentioning the meaning of the term the point group of a space group and how it can be determined, although it was mentioned briefly in Chapter 1. Let us define it now. Consider all the symmetry operations of a space group. These operations include the lattice translations as in Eq. 2-1, point symmetry operations like $\{R\,|\,\mathbf{0}\}$, and glide and screw operations $\{R\,|\,\tau\}$. All three types may be written, in the Seitz notation, as $\{R\,|\,\mathbf{t}\}$ where $\mathbf{t}$ is a lattice translation (Eq. 2-1) or a fraction of a unit cell translation τ as in Eq. 1-11. The **point group of a space group** is defined as the group of operations obtained if we take the space group symmetry operations and set *all* translations to zero. We are then left with a set of operations $\{R\,|\,\mathbf{0}\}$. Note that for nonsymmorphic space groups some of these operations will *not* be symmetry operations of the crystal (as discussed in Section 1-3), so we might ask what is the use of the definition. The answer is that it is extremely useful and important. There is a one-to-one correspondence (an isomorphism) between the set of operations of the point group of the space group and the set of operations of one of the 32 crystallographic point groups; this leads to ways to classify the wave functions of the crystal. (This point will not be discussed here.) Also the *macroscopic* properties of crystal are determined by the operations $\{R\,|\,\mathbf{0}\}$ because small translations such as

CHAPTER 3   SIMPLE CRYSTAL STRUCTURES   59

$Fm3m$
O_h^5 No. 225 $F\,4/m\,\bar{3}\,2/m$ $m\,3\,m$ Cubic

Origin at centre ($m3m$)

Number of positions, Wyckoff notation, and point symmetry			Co-ordinates of equivalent positions $(0,0,0;\ 0,\tfrac{1}{2},\tfrac{1}{2};\ \tfrac{1}{2},0,\tfrac{1}{2};\ \tfrac{1}{2},\tfrac{1}{2},0)+$	Conditions limiting possible reflections
				General:
192	l	1	$x,y,z;\ z,x,y;\ y,z,x;\ x,z,y;\ y,x,z;\ z,y,x;$ $x,\bar{y},\bar{z};\ z,\bar{x},\bar{y};\ y,\bar{z},\bar{x};\ x,\bar{z},\bar{y};\ y,\bar{x},\bar{z};\ z,\bar{y},\bar{x};$ $\bar{x},y,\bar{z};\ \bar{z},x,\bar{y};\ \bar{y},z,\bar{x};\ \bar{x},z,\bar{y};\ \bar{y},x,\bar{z};\ \bar{z},y,\bar{x};$ $\bar{x},\bar{y},z;\ \bar{z},\bar{x},y;\ \bar{y},\bar{z},x;\ \bar{x},\bar{z},y;\ \bar{y},\bar{x},z;\ \bar{z},\bar{y},x;$ $\bar{x},\bar{y},\bar{z};\ \bar{z},\bar{x},\bar{y};\ \bar{y},\bar{z},\bar{x};\ \bar{x},\bar{z},\bar{y};\ \bar{y},\bar{x},\bar{z};\ \bar{z},\bar{y},\bar{x};$ $\bar{x},y,z;\ \bar{z},x,y;\ \bar{y},z,x;\ \bar{x},z,y;\ \bar{y},x,z;\ \bar{z},y,x;$ $x,\bar{y},z;\ z,\bar{x},y;\ y,\bar{z},x;\ x,\bar{z},y;\ y,\bar{x},z;\ z,\bar{y},x;$ $x,y,\bar{z};\ z,x,\bar{y};\ y,z,\bar{x};\ x,z,\bar{y};\ y,x,\bar{z};\ z,y,\bar{x}.$	$hkl:\ h+k, k+l, (l+h)=2n$ $hhl:\ (l+h=2n);\ \circlearrowleft$ $0kl:\ (k,l=2n);\ \circlearrowleft$
				Special: as above, plus
96	k	m	$x,x,z;\ z,x,x;\ x,z,x;\ \bar{x},\bar{x},z;\ \bar{z},\bar{x},\bar{x};\ \bar{x},\bar{z},\bar{x};$ $x,\bar{x},\bar{z};\ z,\bar{x},\bar{x};\ x,\bar{z},\bar{x};\ \bar{x},x,\bar{z};\ \bar{z},x,x;\ \bar{x},z,x;$ $\bar{x},x,\bar{z};\ \bar{z},x,\bar{x};\ \bar{x},z,\bar{x};\ x,\bar{x},z;\ z,\bar{x},x;\ x,\bar{z},x;$ $\bar{x},\bar{x},z;\ \bar{z},\bar{x},x;\ \bar{x},\bar{z},x;\ x,x,\bar{z};\ z,x,\bar{x};\ x,z,\bar{x}.$	no extra conditions
96	j	m	$0,y,z;\ z,0,y;\ y,z,0;\ 0,z,y;\ y,0,z;\ z,y,0;$ $0,\bar{y},\bar{z};\ \bar{z},0,\bar{y};\ \bar{y},z,0;\ 0,z,\bar{y};\ \bar{y},0,\bar{z};\ \bar{z},\bar{y},0;$ $0,y,\bar{z};\ \bar{z},0,y;\ y,\bar{z},0;\ 0,\bar{z},y;\ y,0,\bar{z};\ \bar{z},y,0;$ $0,\bar{y},z;\ z,0,\bar{y};\ \bar{y},z,0;\ 0,z,\bar{y};\ \bar{y},0,z;\ z,\bar{y},0.$	
48	i	mm	$\tfrac{1}{2},x,x;\ x,\tfrac{1}{2},x;\ x,x,\tfrac{1}{2};\ \tfrac{1}{2},x,\bar{x};\ \bar{x},\tfrac{1}{2},x;\ x,\bar{x},\tfrac{1}{2};$ $\tfrac{1}{2},\bar{x},\bar{x};\ \bar{x},\tfrac{1}{2},\bar{x};\ \bar{x},\bar{x},\tfrac{1}{2};\ \tfrac{1}{2},\bar{x},x;\ x,\tfrac{1}{2},\bar{x};\ \bar{x},x,\tfrac{1}{2}.$	
48	h	mm	$0,x,x;\ x,0,x;\ x,x,0;\ 0,x,\bar{x};\ \bar{x},0,x;\ x,\bar{x},0;$ $0,\bar{x},\bar{x};\ \bar{x},0,\bar{x};\ \bar{x},\bar{x},0;\ 0,\bar{x},x;\ x,0,\bar{x};\ \bar{x},x,0.$	
48	g	mm	$x,\tfrac{1}{4},\tfrac{1}{4};\ \tfrac{1}{4},x,\tfrac{1}{4};\ \tfrac{1}{4},\tfrac{1}{4},x;\ x,\tfrac{1}{4},\tfrac{3}{4};\ \tfrac{3}{4},x,\tfrac{1}{4};\ \tfrac{1}{4},\tfrac{3}{4},x;$ $\bar{x},\tfrac{1}{4},\tfrac{1}{4};\ \tfrac{1}{4},\bar{x},\tfrac{1}{4};\ \tfrac{1}{4},\tfrac{1}{4},\bar{x};\ \bar{x},\tfrac{1}{4},\tfrac{3}{4};\ \tfrac{3}{4},\bar{x},\tfrac{1}{4};\ \tfrac{1}{4},\tfrac{3}{4},\bar{x}.$	$hkl:\ h,(k,l)=2n$
32	f	$3m$	$x,x,x;\ x,\bar{x},\bar{x};\ \bar{x},x,\bar{x};\ \bar{x},\bar{x},x;$ $\bar{x},\bar{x},\bar{x};\ \bar{x},x,x;\ x,\bar{x},x;\ x,x,\bar{x}.$	no extra conditions
24	e	$4mm$	$x,0,0;\ 0,x,0;\ 0,0,x;\ \bar{x},0,0;\ 0,\bar{x},0;\ 0,0,\bar{x}.$	
24	d	mmm	$0,\tfrac{1}{4},\tfrac{1}{4};\ \tfrac{1}{4},0,\tfrac{1}{4};\ \tfrac{1}{4},\tfrac{1}{4},0;\ 0,\tfrac{1}{4},\tfrac{3}{4};\ \tfrac{3}{4},0,\tfrac{1}{4};\ \tfrac{1}{4},\tfrac{3}{4},0.$	$hkl:\ h,(k,l)=2n$
8	c	$\bar{4}3m$	$\tfrac{1}{4},\tfrac{1}{4},\tfrac{1}{4};\ \tfrac{3}{4},\tfrac{3}{4},\tfrac{3}{4}.$	
4	b	$m3m$	$\tfrac{1}{2},\tfrac{1}{2},\tfrac{1}{2}.$	no extra conditions
4	a	$m3m$	$0,0,0.$	

Fig. 3-6 O_h^5(Fm3m).

τ will not be noticed on a macroscopic scale (Neumann's principle, discussed in Chapter 5).

Given a space group symbol it is very easy to determine the point group of the space group. In the Schoenflies notation the space group symbol is just a point group symbol with a superscript. If the superscript is ignored then we have the point group of the space group. In the International notation the space group symbol is a lattice type followed by a description of the point symmetry operations as well as

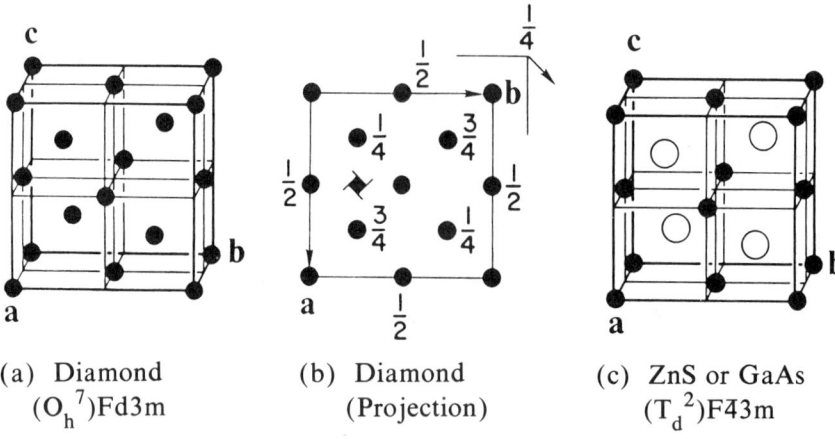

(a) Diamond (O_h^7)Fd3m   (b) Diamond (Projection)   (c) ZnS or GaAs (T_d^2)F$\bar{4}$3m

Fig. 3-7 On page 148 these structures are shown with an emphasis on the tetrahedral bonding.

the glide and screw operations. To obtain the point group we ignore the lattice symbol, since that involves only lattice translations, and set all the $\tau = 0$ in the point group operations, for example, d → m or 4_1 → 4. Here are some examples to show how easy it is. The first few are space groups discussed in this chapter.

Space group	Point group	Space group	Point group	
O_h^1(Pm3m)	O_h(m3m)	T_d^2(F$\bar{4}$3m)	T_d($\bar{4}$3m)	
O_h^9(Im3m)	O_h(m3m)	C_{6v}^4(P6_3mc)	C_{6v}(6mm)	
O_h^5(Fm3m)	O_h(m3m)	D_{6h}^4(P6_3/mmc)	D_{6h}(6/mmm)	
O_h^7(Fd3m)	O_h(m3m)	C_{2h}^5(P2_1/b)	C_{2h}(2/m)	(3-3)

3-5 Examples of Defect Structures

So far we have discussed ordered crystal structures. Now two examples of disordered structures are presented to show that such structures are possible and that these materials may have interesting physical properties (such as very high diffusion constants). The materials are the α-phase of AgI and the α'-phase of CuI. Both of these phases exist at high temperatures, and while in these phases the solid materials are **fast ion conductors**. That is, they are **solid electrolytes** and can conduct ions as fast as liquid electrolytes, the solution in a Pb-acid battery. By studying the crystal structures we can understand how a solid can be such a good conductor of ions.

Up to 147°C AgI has the wurtzite crystal structure, which will be discussed in Section 3-9. Above this temperature the material is in the

α-phase, and the structure can be classified according to the O_h^9(Im3m) space group with the I-ions at the 2a site as in Fig. 3-3. Then where are the Ag-ions? By examining Fig. 3-4 we can see that there are no other sites that permit two ions per unit cell. The original x-ray structure paper in 1935 reported that the two Ag-ions were uniformly and randomly distributed over the 6b, 12d, and 24h sites (see Fig. 3-4). More recent work shows that the two Ag-ions are randomly distributed over only the 12d sites. In either case, (1) there are many more possible sites than there are Ag-ions, and (2) by looking at the structure we can appreciate that there could be a very low potential energy maximum between one 12d site and an unoccupied neighboring 12d site. (It is these two conditions that seem to be required for fast ion diffusion.) It is also of interest to note that Ag-ions in a 12d site are tetrahedrally surrounded by four I-ions in the α-phase, the same as the coordination in the low temperature wurtzite phase.

Our other example of a disordered structure is the α'-phase of CuI. At room temperature this material has the zinc blende structure as shown in Fig. 3-7c; it is a fully ordered structure and a poor ionic conductor. However, at 407°C there is a phase change to the α'-phase and the crystal has been described according to the O_h^5(Fm3m) space group with the I-ions at the 4a sites (see Figs. 3-5 and 3-6). The four Cu-ions per unit cell appear to be distributed among the 8c sites of this space group. These are, of course, the same sites occupied by the eight F-ions per unit cell in CaF_2 (Fig. 3-5c). Notice the great similarity between the zinc blende structure and the CaF_2 structure. In both cases the ions at (1/4, 1/4, 1/4) and related positions are tetrahedrally surrounded by the I- or Ca-ions. Thus, for CuI the Cu-ions are tetrahedrally bound to four I-ions in the room temperature as well as in the fast ion conducting phase. For the α'-phase it is not obvious that the potential energy barrier to get from one 8c site to another should be very low, but apparently this is so.

The above two structures are examples of a large and scientifically interesting group of crystals that have defect structures. As can imagine, the defect nature of the materials allows a new degree of freedom to the materials, which often leads to interesting physical and chemical properties.

3-6 Different Points of View of a Structure

Until now in this chapter, the diagrams of the structures have been what are called **clinographic projections** of the conventional Bravais unit cell. This is a simple and neat way to show the crystal structure. However, there are other, perhaps more realistic, ways to look at a crystal structure, and sometimes these other approaches show interesting features.

Figure 3-8a shows the fcc crystal structure of Cu. The rest of the figure shows different ways of looking at the same crystal structure. For example, in (c) and (e) we see more clearly how the space is used. In (g) and (i) we have a view of the (111) face, which is a higher density face than the (100) face shown in Fig. 3-8c.

For some of the other structures discussed in this chapter the ions have different sizes, and thus pictorial representations such as Fig. 3-8c are useful and sometimes give good insight into the amount of space available for the various small ions when the larger ions are touching. (This will be used in Chapter 7.)

3-7 Close Packing (and the Hexagonal Close-packed Structure)

In Fig. 3-8i we can see that in two dimensions this type of packing of hard spheres is the densest that can be obtained. All spheres are of equal diameter; each sphere has six spheres touching it and each other. Compare this packing to the density obtained on the (100) plane in this structure in Fig. 3-8c or similar packing on the (100) plane of a simple cubic structure.

Figure 3-9a shows this same close-packed plane but with different labeling. There are a series of depressions on the upper surface of the close-packed plane. These depressions are divided into a B-series and a C-series, although for a single layer of spheres there is no fundamental distinction between these two series. Figure 3-9b shows this plane with the same lettering but with the spheres labeled A reduced in size. Now consider a second identical layer of spheres to be superimposed, as closely as possible, on top of this first A layer. Place the bottoms of the spheres of this second layer into the B-depressions of the first layer. (They could be set in the C-depressions and for the two infinite planes there is no difference so far.) A third identical layer of spheres is to be added, but now we have a real and important choice. The third layer can be added to fit on the depressions of the second layer in two ways. First, it can be added so that it is over the A-sites of the first layer or it can be added so that it is over the C-depressions of the first layer. If we choose the first possibility, and then continue this type of stacking forever, we have an ABABAB... sequence. If we choose the second possibility and then continue the stacking to infinity we have an ABCABCAB... sequence. Note first, that in both cases each sphere has 12-nearest neighbors: six in the plane, three above and three below. Second, both sequences represent the closest packing of identical spheres that is possible.

CHAPTER 3 SIMPLE CRYSTAL STRUCTURES

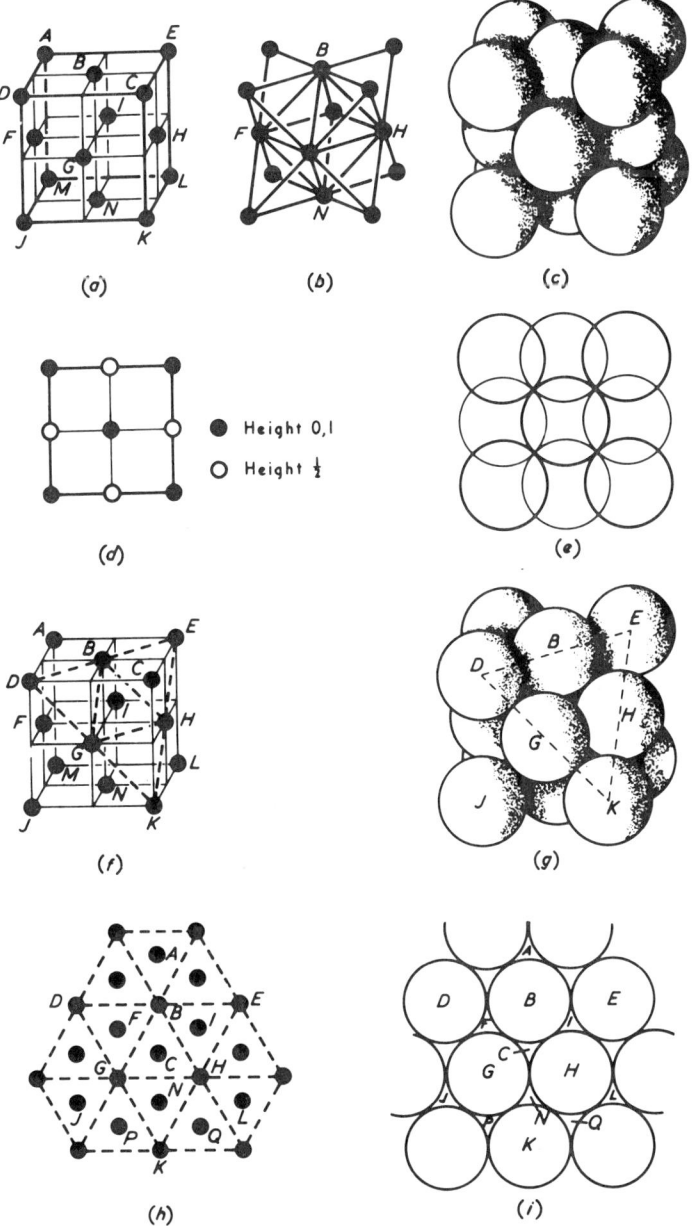

Fig. 3-8 Various ways of looking at the crystal structure of Cu. (After Megaw.)

The first sequence, ABABAB..., gives the **hexagonal close-packed (hcp) structure** for which Fig. 3-9c and 3-9d shows a projection of the structure (with the unit cell outlined) and a clinographic projection of the structure. (Note that we would not quickly guess from the latter

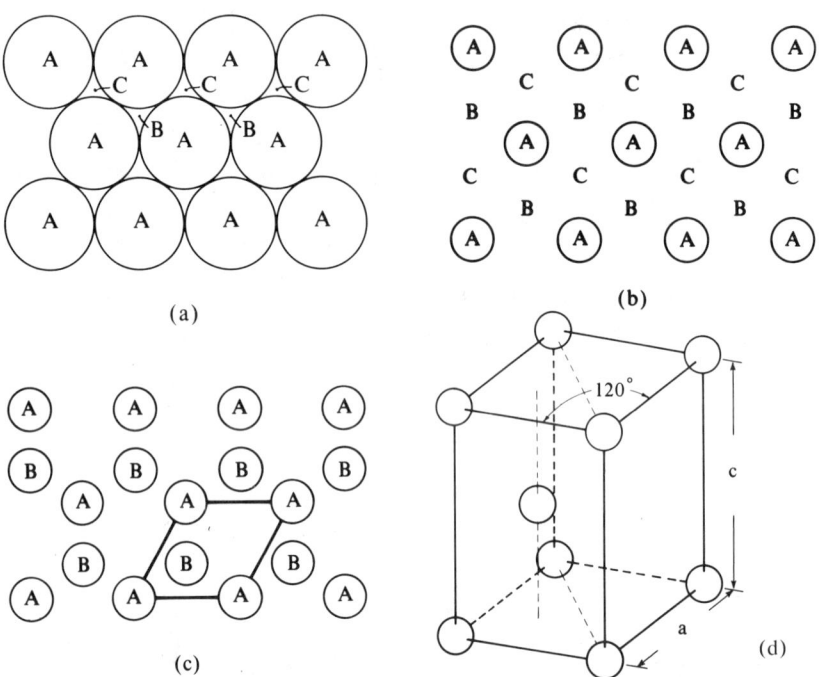

Fig. 3-9 (a) and (b) A close packing of spheres. (c)-(d) Several ways to look at a hexagonal close-packed structure.

that each sphere had 12-nearest neighbors.) The lattice is clearly a hexagonal P-lattice with a basis at (0, 0, 0) and (2/3, 1/3, 1/2). The space group is D_{6h}^4(P6$_3$/mmc), and for packing of identical spheres the c/a ratio is $(8/3)^{1/2} = 1.633$, which is called the **ideal c/a ratio**. (Figure 3-9d defines c and a.) A number of elements crystallize with this structure (Table 3-1) with c/a varying from 1.56 for Be to 1.89 for Cd. Clearly for c/a ratios that differ from the ideal value, the 12-nearest neighbors are not all equidistant.

For the second sequence, ABCABC..., one obtains a **cubic close-packed structure**, which is just the face-centered cubic structure shown in Fig. 3-8, space group O_h^5(Fm3m) with a basis at (0, 0, 0) and the face-centering condition. Many elements have this structure.

Note that while the hexagonal and cubic close-packed crystal structures are certainly the most usual that are found in nature, there are an infinite number of possibilities for close packing, but these rarely occur. Nevertheless, several rare earth elements take on a close-packing structure with the sequence ABACABAC.... We use the word **polytypism** to describe a stacking sequence with a long repeat distance. For example, SiC can be grown in many **polytypes,** that is, many different stacking sequences. See the Notes in Chapter 12.

CHAPTER 3   SIMPLE CRYSTAL STRUCTURES   65

3-8 Volume Effects for Simple Structures

Consider some of the simple structures and fill them with hard spheres until they touch, as was done for the close-packed structures. Then the maximum proportion of the available volume that is filled is

face-centered cubic	$\pi(2)^{1/2}/6$	$= 0.74$
hexagonal close-packed	$\pi(2)^{1/2}/6$	$= 0.74$
body-centered cubic	$\pi(3)^{1/2}/8$	$= 0.68$
simple cubic	$\pi/6$	$= 0.52$
diamond	$\pi(3)^{1/2}/16$	$= 0.34$ (3-4)

The two close-packed structures give the same values, as they must, and are the most efficient for packing hard spheres in the sense that the largest fraction of space is filled by the spheres. On the other hand, the diamond structure has a great deal of space between the spheres. Nevertheless, four elements (diamond, Si, Ge, and gray tin) crystallize in the diamond structure and many simple, binary, very stable compounds, crystallize in the closely related zinc blende crystal structure. This shows that atoms are not hard spheres and that other considerations are involved in determining how they crystallize. (We shall see, in Chapters 6 and 8, that the strong covalent bond leads in a natural way to the diamond structure.)

3-8a Coordination number   As might be expected, the maximum proportion of filled volume for the different simple structures is closely related to the coordination number. The coordination number for the two close-packed structures is 12 for the bcc-8, for the simple cubic-6, for diamond-4. (However, note that for the bcc structure the six second-nearest neighbors are at a distance of only 1.15 times that of the first-nearest neighbors.)

3-8b High pressure/temperature effects   Consider what kinds of changes of structure might be expected as a crystal is subjected to increasing pressure. For simple structures we would expect that as the pressure is increased the crystal will tend to change to structures that fill space more efficiently (and at the same time have a higher coordination number). This is the case for many elements and for complex compounds. For example, many AB compounds with a zinc blende or wurtzite crystal structure at atmospheric pressure (CN 4:4) have the NaCl structure at higher pressure (CN 6:6). At room temperature AgI has the wurtzite crystal structure, but only three kbars are needed to cause a phase change to the NaCl structure.

Temperature can have an effect similar to that of pressure in simple structures. The amplitude of the lattice vibrations (the oscillatory motion of the atoms about their equilibrium positions) increases

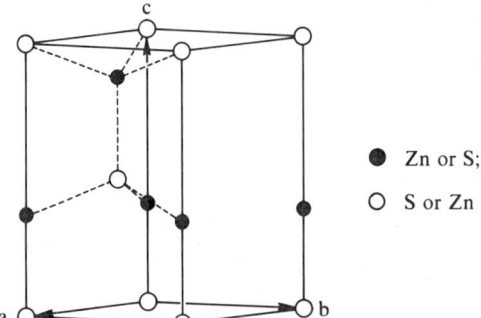

● Zn or S;
○ S or Zn

Fig. 3-10 The wurtzite structure. Note $|a| = |b|$.

with increasing temperature. This increased "banging of the atoms into each other" would be reduced if a transformation occurs to a structure with a lower density and coordination number. This transformation is found in a number of elements; for example Y, Yb, and Zr transform from a close-packed structure at room temperature (CN 12) to a bcc structure at higher temperature (CN 8), as can be seen in Table 3-1. Thus, in simple systems, increasing pressure or decreasing the temperature often have the same effect in causing phase transitions.

3-9 Wurtzite Structure

The last simple structure that we discuss is the wurtzite crystal structure. Figure 3-10 shows a clinographic projection of this hexagonal structure using ZnS as an example. (ZnS actually can be found in this structure as well as in the cubic zinc blende structure.) The space group is C_{6v}^4(P6$_3$mc), Zn at (0, 0, 0) and (2/3, 1/3, 1/2) and S at (0, 0, u) and (2/3, 1/3, 1/2 + u) with u ≈ 3/8. These are the 2b sites of this space group. As can be seen, each Zn-atom is tetrahedrally surrounded by four S-atoms and vice versa, as shown by the dotted lines in the figure. There is a close relation between this structure and the zinc blende structure. In fact for u = 3/8 and the ideal c/a the tetrahedron formed from Zn and its four S neighbors is exactly the same as that found in the zinc blende structure. However, there is a difference in the positions of the second-nearest neighbors.

There are many simple binary II-VI, III-V, or even IV-IV compounds that crystallize in the closely related zinc blende or wurtzite structure. (By II-VI we mean an element from column II and one from column VI in the periodic table, Fig. 3-11, e.g., ZnS or CdTe.) A few of the compounds that crystallize in the zinc blende phase are ZnS, CdTe, AlSb, GaAs, InP, and SiC. Some of these compounds can be made to crystallize with the wurtzite structure if grown by different

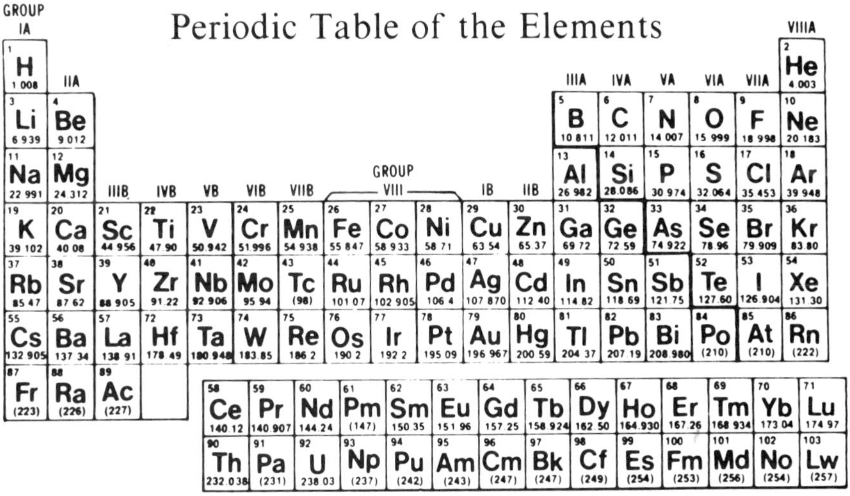

Fig. 3-11

techniques, for example ZnS and SiC. This shows the very small amount of binding energy difference between these two crystal structures. Several compounds that have the wurtzite structure are GaN, CdS, CdSe, and ZnO.

3-10 Site Symmetry

We have mentioned site symmetry but we have not emphasized the meaning and use of the site symmetry of the various positions in a space group or crystal structure. The **site symmetry** of a position is nothing more than its point symmetry, that is, the symmetry of the crystal with the site held fixed.

The site symmetries of all possible positions for each space group are listed in the International Tables. For example, for the space group O_h^1(Pm3m) in Fig. 3-2 we see that the 1a and 1b positions have site symmetry O_h(m3m). This is the same as the point group of the space group. Thus, under all 48 symmetry operations of the point group O_h(m3m), with the 1a or 1b site taken as an origin, a crystal with this space group will transform into itself. The 6e or 6f positions have site symmetry C_{4v}(4mm), and so on, for the other positions. Note that the site symmetry of the general equivalent positions is E(1) for all space groups. Further, there must be as many general equivalent positions as there are symmetry operations of the point group of the space group. Of course, for nonprimitive F-, I-, and C-lattices there will be four, two, and two times as many general equivalent positions because the unit cells used in the International Tables are multiply primitive.

For the space group O_h^9(Im3m) we see (Fig. 3-4) that the site symmetry of the 2a site is O_h(m3m) and all the other sites have lower point symmetry (and hence more positions) as noted in our discussion of defect structures (Section 3-5). Last the 96ℓ positions have E(1) point symmetry. The situation for O_h^5(Fm3m) should be clear.

The space groups discussed above are all symmorphic ones. This is why the position with highest symmetry has the same point symmetry as the point group of the space group. That this must occur is clear if one recalls the development of the symmorphic space groups (Section 2-6). For the same reason, the site of highest symmetry for a nonsymmorphic space group must have a point symmetry that is lower (fewer symmetry operations) than the point symmetry of the space group. That is, there must be some symmetry operation of the space group, of the form $\{R \mid \tau\}$, that takes one highest symmetry point into an equivalent nonidentical one. For example, in the nonsymmorphic space group O_h^2(Pn3m) the highest symmetry site, 2a, has point symmetry O(432) or only half as many symmetry operations as the point group of the space group.

There are many times when we want to know the site symmetry. For example we can add a small amount of impurity atoms having unpaired electron spins into a crystal and by the electron spin resonance technique, transitions between the energy levels of this unpaired spin in a magnetic field can be observed. The position and number of these energy levels is determined by symmetry seen by these impurity atoms rather than the crystal symmetry.

Notes

There are many good books on structures. Wyckoff and Structure Reports are the standard references for the crystal structures of all compounds of most compounds. Megaw, Galasso, Bloss and Wells should be consulted for pictures and an explanation of many more simple structures than are discussed here. The book by Taylor and Kagle lists the space groups, crystal structure and unit cell size of thousands of metals and alloys as well as the elements.

Compounds that have tetrahedral structures usually have four valence electrons per atom on the average. There are many compounds that are tetrahedrally coordinated where this average number of valence electrons is maintained by the substitution of an atom pair by two different atoms. For example, the minerals ZnS (zinc blende) and $CuFeS_4$ (chalcopyrite) are related this way. The latter has the same type of tetrahedral structure as the former. However, since the

CHAPTER 3   SIMPLE CRYSTAL STRUCTURES   69

Cu and Fe atoms are ordered the space groups are different. There are many closely related ordered, tetrahedrally coordinated compounds. For a reference see the review article by Miller, MacKinnon and Weaire mentioned in the Notes of Chapter 12.

There are many review articles on high-pressure results. A recent, good, readily available one is A. Jayaraman, Rev. Mod. Phys. **55**, 65 (1983). This article also has a useful discussion on the diamond anvil cell. The **units of pressure** are given in the notes of Chapter 14.

S. Chandra, "Superionic Solids" (North-Holland, 1981) is a good general reference to this field and discusses applications.

Problems

1. Cu_3Ag has space group Pm3m, unit cell size = 3.74 Å with Ag at 1a and 3Cu at 3c. What is the nearest neighbor and second-nearest neighbor Cu-Cu distance, Cu-Ag distance, and Ag-Ag distance?

2. For the α'-phase of CuI, space group O_h^5(Fm3m), we find that the Cu-ions are slightly displaced (in a random manner) from the 8c sites away from any one of the nearest neighbor I-ions. How would you describe the positions of the Cu-ions? (Hint: look at Figs. 3-5 and 3-6.)

3. Determine the ideal c/a ratio for the hcp structure.

4. Show that the values in Eq. 3-4 are correct.

5. For the **wurtzite structure** with u = 3/8 and the ideal c/a ratio, show that the tetrahedra are the same as those found in the zinc blende structure. Discuss the difference of the next-nearest neighbor positions in these two crystal structures.

6. **Site symmetry** – What is the site symmetry of the occupied positions in diamond? Of the position halfway between the closest carbon atoms? Answer the same questions for the wurtzite structure.

7. Start with cubic $BaTiO_3$. (a) What is the space group of the crystal if it is uniaxially stressed in a ⟨100⟩ direction? (b) What is the space group if the Ti-ions are moved along a ⟨100⟩ direction? (c) What is the space group if both of these effects occur simultaneously?

8. **Symmetry operations of a space group** – Of all 230 space groups there are probably more crystals that have the C_{2h}^5($P2_1/b$) space group than any other. Figure 2-11b comes from the International

Tables. Your task is to find the symmetry operations of the space group. How many symmetry operations must be written? Of course there are the infinite unit cell translations that are trivial. The figure shows that in the (primitive) unit cell there are four screw axes, four centers of inversion, a glide plane parallel to the plane of the paper (c/4 above it), and the ubiquitous E(1). Although all these indeed exist we need to find only four symmetry operations of the type $\{R|\tau\}$, since there are only four general equivalent positions. Equivalently, there are four symmetry operations in the point group of the space group, which clearly is $C_{2h}(2/m)$. We know the four operations for R since they are those of $C_{2h}(2/m)$. Looking at the four general equivalent points *within* the primitive unit cell shown by circles in the figure, taking the origin at the center of the cell at z = 0, $\{E|0\}$ takes the starting circle in the upper left into itself, $\{i|0\}$ takes the starting circle into the lower right. Find the other two symmetry operations of this form $\{R|\tau\}$. (Hint: $\tau \neq 0$ for these two and R = C_2 and σ_h.) (Note that all the symmetry operations of the space group are just the product of these four operations with the unit cell translations, Eq. 2-1.)

9. Symmetry operations of a space group – Just to prove to yourself how easy it is now to write the symmetry operations of a space group, do it for $C_{2v}^2(Pmc2_1)$ and $D_{4h}^{14}(P4_2/mnm)$. (The latter was encountered in Chapter 1.)

10. Cu_2O is cubic, a = 4.26 Å, with a density 6.0 gm/cm³. The following three space groups are consistent with the diffraction pattern: Pn3; $P4_232$; Pn3m. How many formular weights does one unit cell contain? Determine the correct space group (hence the structure).

11. Cu_2O has the **cuprite structure**, space group O_h^4(Pn3m). The basis is oxygen at 2a (0, 0, 0) and (1/2, 1/2, 1/2) and copper at 4b (1/4, 1/4, 1/4), (1/4, 3/4, 3/4), (3/4, 1/4, 3/4), and (3/4, 3/4, 1/4). (a) What are the coordinations for the two atoms? (b) What are the site symmetries of the atoms? (c) Using the Seitz notation, describe one of the diagonal glides. (It helps to know that the full space group symbol is $P4_2\,\bar{3}\,2/m$.)

4

X-Ray Diffraction

4-1 Electron, Neutron, and X-ray Diffraction
    a Neutron diffraction
    b Electron diffraction
    c X-Ray diffraction
4-2 Bragg's Law
    a Missing reflections
4-3 The Laue Formulation
    a Assumptions
    b Reciprocal lattice
    c Laue formulation
4-4 Experimental X-ray Diffraction Methods (S)

Notes
Problems

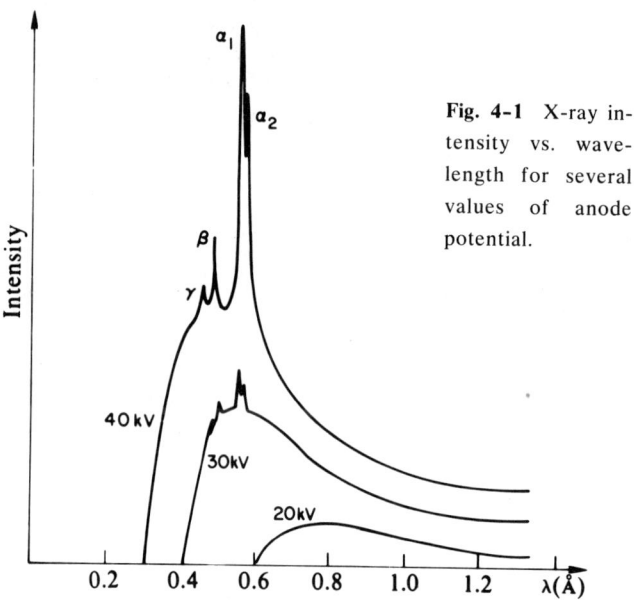

Fig. 4-1 X-ray intensity vs. wavelength for several values of anode potential.

neutrons $\quad\lambda(\text{Å}) = \dfrac{0.28}{[\mathscr{E}(\text{eV})]^{1/2}}$

electrons $\quad\lambda(\text{Å}) = \dfrac{12}{[\mathscr{E}(\text{eV})]^{1/2}}$

x-rays $\quad\lambda(\text{Å}) = \dfrac{12.4}{\mathscr{E}(\text{keV})}$

X-RAY DIFFRACTION

*All I really want to do
Is, baby, be friends with you.*

B. Dylan, "All I really want to do"

The diffraction of waves from the periodic arrangement of atoms in solids to determine the crystal structure was first suggested by von Laue in 1912, developed by Bragg and is now a well developed science. Most of the diffraction work is done using x-rays. Thus, besides a brief introduction that includes diffraction by electrons and neutrons, we specialize the discussion to x-rays. Even then, only the barest of detail is presented. References in the Notes can be found to the many specialized books on diffraction. Also note that in the last ten years there has been a huge expansion in the use of synchrotron radiation. Although this radiation can be of typical x-ray wave lengths, it is not often used for classical structure determinations. Thus, we discuss it in Chapter 17 in the context of surface science, although the radiation is also used for various kinds of studies in the bulk.

4-1 Electron, Neutron, and X-ray Diffraction

The distances between atoms in a crystal are ≈ 1 Å, so waves with approximately this wavelength are required to explore this structure. If the wavelength is much larger, structural details cannot be resolved; rather some averaged interaction occurs, as is found for visible light ($\lambda \sim 5000$ Å). If the wavelengths are much smaller the beam is diffracted by very small angles, making detection difficult.

4-1a Neutron diffraction According to the **de Broglie law**, any free particle with velocity v, momentum p, mass M, and energy $\mathscr{E}$, has a wavelength λ given by

$$\lambda = h/p = h/Mv = h/(2M\mathscr{E})^{1/2} \qquad (4-1)$$

Using the mass of a neutron for M (1.675×10^{-24}g) and the value of Planck's constant h, $\mathscr{E} \approx 0.08$ eV is required for $\lambda = 1$ Å. These are rather slow neutrons since room temperature $k_B 300°K \approx 1/40$ eV = 0.025 eV. Neutrons with these wavelengths are obtained from reactors. The neutrons thermalize by repeatedly scattering from a graphite moderator that is cooled to room temperature. Such 300°K free neutrons will have a mean thermal kinetic energy $\mathscr{E} = 3k_BT/2 \approx 0.04$ eV, which, as mentioned, corresponds to a useful wavelength.

Neutrons scatter appreciably only from the atomic nuclei (electrons and x-rays scatter from the electron change density of the atoms). Thus, neutron diffraction complements x-ray diffraction in the sense that light atoms such as hydrogen through carbon appreciably scatter neutrons while, for example, hydrogen can hardly be observed by x-ray diffraction because of its very low charge density. Consequently, neutron diffraction is extremely useful for structural studies of organic crystals as well as ordinary inorganic materials.

4-1b Electron diffraction The difference in the masses between the electron (0.911×10^{-28}g) and neutron when used in Eq. 4-1 results in 150 eV electrons for the desired $\lambda = 1$ Å.

As might be expected, there is a very strong interaction of a 150 eV beam of electrons with the electrons in solids. Consequently, the electron beam penetrates only a few interatomic distances and is primarily of use for surface studies. Bulk electron diffraction work requires $\approx 50 \times 10^3$ eV electron beams, which penetrate several hundred interatomic distances. The wavelength of the electrons of such an energetic beam is much smaller than the internuclear spacing and the diffraction angle is very small, which causes experimental difficulties. Thus, electron diffraction work is used mostly to study surfaces, goes under the name LEED (low energy electron diffraction) and is discussed in Chapter 17.

4-1c X-ray diffraction X-rays are electromagnetic waves of high frequency, hence small wavelength, and naturally have no rest mass. They are produced in standard laboratory equipment by letting a beam of high energy electrons impinge on a metal anode. These high energy electrons are accelerated from a cathode to the anode by a potential V_{ca}. The deceleration of the electron beam by collisions in the anode causes electromagnetic radiation to be emitted. The maximum amount of energy emitted occurs when an electron is stopped in one collision, but most of the electrons lose their energy in a series of collisions. Thus, there is a maximum energy and a minimum λ.

$$h\nu = \frac{hc}{\lambda} \leq e V_{ca} \qquad (4\text{-}2)$$

Figure 4-1 shows an x-ray spectrum for 20×10^3 eV electrons impinging on a metal anode. The smallest x-ray wavelength is 0.6 Å. As V_{ca} is increased the minimum wavelength decreases, as can be seen. However, another phenomenon occurs. The electron beam begins to have sufficient energy to eject electrons from the inner shells of the metal atoms. Then electrons from the outer shells drop down to fill the

CHAPTER 4   X-RAY DIFFRACTION

inner shells and emit characteristic x-rays with well-defined energies as shown in Fig. 4-1.

If the 1s shell is empty the wavelength of the radiation emitted when an electron drops down from the 2p, j = 3/2 and 2p, j = 1/2 shell is called the $K_{\alpha 1}$ and $K_{\alpha 2}$ radiation. The values of the wavelengths depend on the metal that is used as an anode. Some of the **characteristic wavelengths** for various metals are listed here.

Cu: $K_{\alpha 1}$ = 1.54056 Å   Mo: $K_{\alpha 1}$ = 0.70930 Å
     $K_{\alpha 2}$ = 1.54439 Å $K_{\alpha 2}$ = 0.71359 Å
     $K_{\beta 1}$ = 1.39222 Å $K_{\beta 1}$ = 0.63229 Å

Thus, for x-rays there is a continuous spectrum as well as a choice of very sharply defined peaks available for experiments. Both of these types of radiation are used in experimental situations.

4-2 Bragg's Law

In 1913 Bragg gave a simple explanation for the observed diffraction of an x-ray beam by a crystal. He assumed that the incident x-ray waves were **specularly reflected** (i.e., angle of incidence equals the angle of reflection) from parallel planes of atoms, Fig. 4-2a. Further, each plane of atoms reflects only a small fraction of the incident radiation. (This is consistent with the large penetration depth of the x-ray beam.) Then, diffracted beams (spots or lines on a film) are observed when the rays from adjoining planes add constructively. Figure 4-2a shows that the path difference between rays specularly reflected between two planes is 2d sin θ. When this path difference is an integral number, n, of wavelengths then constructive interference is observed. This is Bragg's law

$$2d \sin \theta = n\lambda \tag{4-3}$$

where d is the spacing between the planes under consideration and n is called the order of the corresponding reflection. Since $\sin \theta \leq 1$, then $\lambda \leq 2d$, gives a maximum value for the wavelength of the x-rays that can be used. From this we see why visible light cannot be used. (See the Notes for how x-rays interact with atoms in crystals. This leads to an understanding of the assumptions in the beginning of this paragraph.)

Figure 4-2b shows how the Bragg angle is determined experimentally as just half the angle between the incident and reflected beam. From Eq. 4-3 for n = 2 we would measure a larger value of θ than for the values obtained from n = 1. It is not too difficult to show that the angle corresponding to this n = 2 reflection corresponds to the angles from other prominent crystallographic planes. Such a possibility can

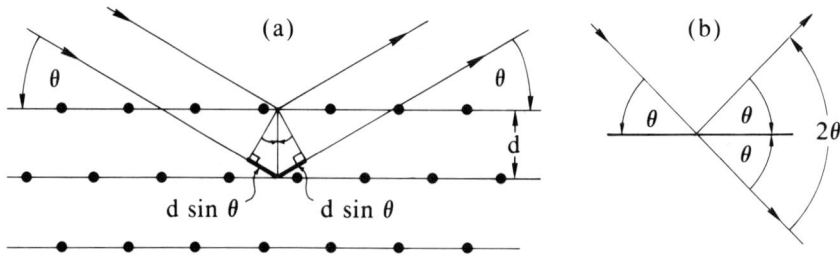

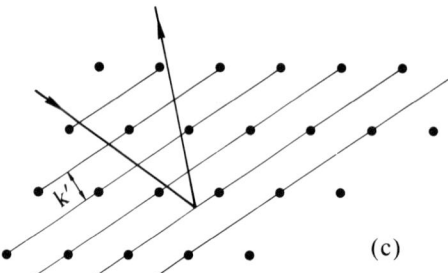

Fig. 4-2 (a) Bragg reflection from a family of lattice planes. (b) The Bragg angle θ, which is half the angle between the incident and scattered beam. (c) The same lattice as in part a, but emphasizing a different set of planes. Clearly there are an infinite set of planes.

be seen in Fig. 4-2c for the identical structure as in Fig. 4-2a. Then Bragg's law can be reinterpreted to be

$$2d' \sin \theta = \lambda \qquad (4\text{-}4)$$

where $d' = d/n$ is the spacing between planes of the crystal. It is clear from Fig. 4-2c that there are an infinite number of such planes, although the ones for very large n have a low density of atoms and hence low scattering amplitude.

4-2a Missing reflections We sketch how centering of lattices can cause certain reflections to be missing. This gives a general idea as how the "conditions limiting possible reflections" come about. These are the conditions in the International Tables, for example, on the right-hand side of Figs. 3-2, 3-4, and 3-6.

Figure 4-3a shows a lattice (and a primitive cell) with a set of planes from which Bragg reflections can be observed. P_1 and P_2 are typical planes. We assume a simple structure with a single atom at each lattice position. The reflected waves from such planes differ by a complete path difference of the wavelength. By knowing λ and by measuring the Bragg angle, d' from Eq. 4-4 can be determined. Now consider the effect of centering the primitive cell. Figure 4-3b shows the lattice. A new set of planes, labeled Q_1 and Q_2 (dashed in the figure), are at a position halfway between the P-set. The P and Q planes contain the same kind of atoms at the same density, so have the

CHAPTER 4 X-RAY DIFFRACTION

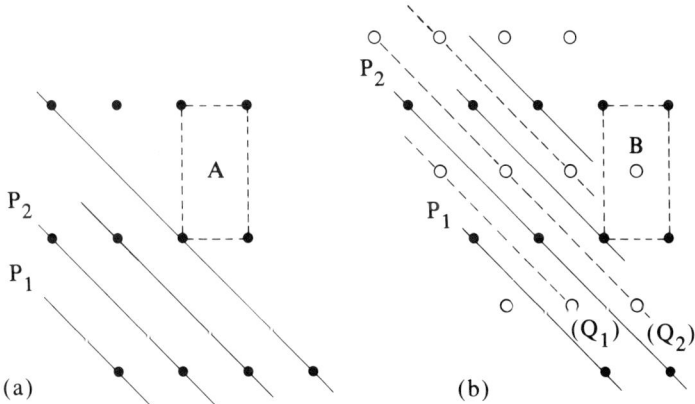

Fig. 4-3 (a) Reflection planes for lattice. (b) Reflection planes for the same lattice but now it is centered.

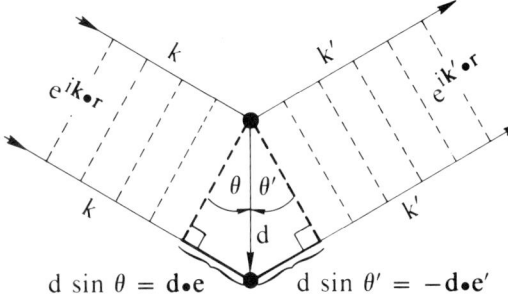

$d \sin \theta = \mathbf{d} \cdot \mathbf{e}$ $d \sin \theta' = -\mathbf{d} \cdot \mathbf{e}'$

Fig. 4-4 The path difference for two arbitrary scattering centers.

same scattering power. Thus, waves reflected from the new set of planes are exactly out of phase with those reflected from the P-set, and cancel the Bragg peak observed in the situation shown in Fig. 4-3a. This shows how centering a lattice will put new conditions on the Bragg peaks.

4-3 The Laue Formulation

Bragg's assumption that planes of atoms specularly reflect x-rays, as does a very lightly silvered mirror, is ad hoc. The approach taken by von Laue is much more satisfying in this fundamental respect.

4-3a Assumptions Von Laue took the point of view that each of the sites (sets of atoms) of the Bravais lattice could radiate the incident radiation, at the same frequency, in all directions. Then, wherever in space the radiated radiation interfered constructively, sharp peaks would be observed.

Figure 4-4 shows the geometry of two scattering centers separated by **d** where the incident and scattered radiation have wave vectors **k** and **k′**, respectively, in the directions of the unit vectors **e** and **e′**, respectively (**e** ≡ **k**/|**k**|, etc.). Thus, $k = 2\pi e/\lambda$; similarly $k' = 2\pi e'/\lambda$. The path difference is

$$d \sin \theta + d \sin \theta' = \mathbf{d} \cdot (\mathbf{e} - \mathbf{e}') \tag{4-5}$$

"Spots" are experimentally observed when **k′** is such that constructive interference occurs, just as for the Bragg law

$$\mathbf{d} \cdot (\mathbf{e} - \mathbf{e}') = n\lambda \qquad \text{or} \qquad \mathbf{d} \cdot (\mathbf{k} - \mathbf{k}') = 2\pi n \tag{4-6}$$

where n is any integer.

Instead of just two scattering centers we inquire as to the conditions on **k′** for constructive interference from an entire structure with atoms located at the positions given by $\mathbf{t}_m$, Eq. 2-1, which form a Bravais lattice. The condition is the same as Eq. 4-6, except now the scattered waves from all the lattice points must simultaneously, constructively interfere. Then

$$\mathbf{t}_m \cdot (\mathbf{k} - \mathbf{k}') = 2\pi n \tag{4-7a}$$

This expression can be written in an equivalent form, which is

$$e^{i(\mathbf{k} - \mathbf{k}') \cdot \mathbf{t}_m} = 1 \tag{4-7b}$$

4-3b Reciprocal lattice The concept of the reciprocal lattice plays a very important role in the fields of x-ray crystallography, electronic band structure, lattice vibrational spectra, and, in fact, all of solid state physics. We discuss it only briefly at this point, in connection with x-ray diffraction, but will come back to this subject in Chapter 10, when electronic band theory is presented. At that time the reader will be able to attach a more immediate physical significance to the reciprocal lattice.

The end points of the Bravais lattice translation vector, Eq. 2-1, represent an infinite lattice of points. Rewriting this vector here

$$\mathbf{t}_m = m_1 \mathbf{a} + m_2 \mathbf{b} + m_3 \mathbf{c} \tag{4-8}$$

for all integer m_i. Then we may define a set of vectors in wave vector space, k-space, as the set of vectors **K** that satisfy

$$e^{i\mathbf{K} \cdot \mathbf{t}_m} = 1 \tag{4-9}$$

for all $\mathbf{t}_m$. The lattice of points determined by the set of **K** vectors is called the **reciprocal lattice**, while the lattice of points determined by $\mathbf{t}_m$ can be called the **direct lattice**. (We leave as a problem to show that the reciprocal lattice of the reciprocal lattice is the direct lattice.) <u>Note each of the 14 Bravais lattices have a different reciprocal lattice.</u>

CHAPTER 4 X-RAY DIFFRACTION

A reciprocal lattice may be generated by three primitive reciprocal lattice vectors **a***, **b***, **c*** as follows

$$\mathbf{a}^* = 2\pi \frac{\mathbf{b} \times \mathbf{c}}{\mathbf{a} \cdot (\mathbf{b} \times \mathbf{c})} \qquad \mathbf{b}^* = 2\pi \frac{\mathbf{c} \times \mathbf{a}}{\mathbf{a} \cdot (\mathbf{b} \times \mathbf{c})}$$

$$\mathbf{c}^* = 2\pi \frac{\mathbf{a} \times \mathbf{b}}{\mathbf{a} \cdot (\mathbf{b} \times \mathbf{c})} \tag{4-10a}$$

where **a**, **b**, and **c** are primitive lattice vectors for any one of the 14 Bravais lattices. Then $\mathbf{a} \cdot (\mathbf{b} \times \mathbf{c}) = V_p$, the volume of the primitive unit cell. The reciprocal lattice is then given by

$$\mathbf{K}_{m'} = m_1' \mathbf{a}^* + m_2' \mathbf{b}^* + m_3' \mathbf{c}^* \tag{4-10b}$$

where m_1' is any integer. That Eqs. 4-10 satisfy the requirement of the reciprocal lattice in Eq. 4-9 is easy to see since

$$\mathbf{a}^* \cdot \mathbf{a} = \mathbf{b}^* \cdot \mathbf{b} = \mathbf{c}^* \cdot \mathbf{c} = 2\pi$$

$$\mathbf{a}^* \cdot \mathbf{b} = 0, \text{ etc.}$$

$$\mathbf{K}_{m'} \cdot \mathbf{t}_m = 2\pi (m_1' m_1 + m_2' m_2 + m_3' m_3) \tag{4-11}$$

Clearly from the construction of $\mathbf{K}_{m'}$, we can write any wave vector defined in this space in terms of the basis vectors of the reciprocal lattice, for example,

$$\mathbf{k} = k_1 \mathbf{a}^* + k_2 \mathbf{b}^* + k_3 \mathbf{c}^* \tag{4-12}$$

for any values of k_i. Then $\exp(i\mathbf{k} \cdot \mathbf{t}_n)$ is not generally 1. However, whenever all the k_i in Eq. 4-12 are integers, **k** in this expression is a reciprocal lattice vector, and $\exp(i\mathbf{K}_{m'} \cdot \mathbf{t}_m) = 1$. The space determined by the reciprocal lattice vectors is called **k-space** or **reciprocal space**.

Important details of the construction of the reciprocal lattices are left to Chapter 10. However, we note that for primitive lattices with orthogonal basis vectors, such as the cubic, tetragonal, and orthorhombic primitive lattices, the reciprocal lattice takes a particularly simple form. The reciprocal lattice also has an orthogonal basis with the magnitudes given by

$$a^* = 2\pi/a, \quad b^* = 2\pi/b, \quad c^* = 2\pi/c \tag{4-10a}$$

a* is perpendicular to the **bc**-plane and thus parallel to **a**; similarly **b*** is parallel to **b**, and **c*** is parallel to **c**. Since typical values of a, b, and c are 6 Å, typical values of a*, b*, and c* are 1 Å^{-1} = 10^{+8} cm^{-1}.

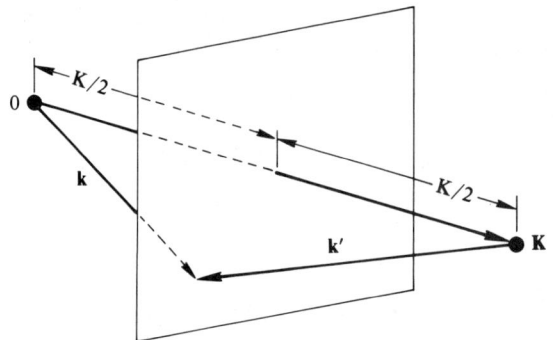

Fig. 4-5 A plane that is a perpendicular bisector of a reciprocal lattice vector, showing the Laue condition.

4-3c Laue formulation The Laue condition for diffraction spots, Eq. 4-7, can now be reformulated in a simple, geometrical manner. Constructive interference will occur whenever the change in wave vector between the incident and scattered wave vectors is a vector of the reciprocal lattice, that is, whenever $\mathbf{k} - \mathbf{k}' = \mathbf{K}_m$. One can now sense how this formulation of diffraction spots begins to reduce the problem to geometric considerations in three dimensions.

Write the condition for constructive interference in terms of only $\mathbf{k}$ and $\mathbf{K}$. First, note that $|\mathbf{k}| = |\mathbf{k}'|$, because we are considering elastic scattering of photons, so the frequency and wavelength of the radiation is the same before and after scattering, only the directions of $\mathbf{k}$ and $\mathbf{k}'$ differ. Thus, $k^2 = (k')^2 = |\mathbf{k} - \mathbf{K}_m|^2$, leading to

$$\mathbf{k} \cdot \mathbf{K}_m = \frac{1}{2} K_m^{\,2} \tag{4-13}$$

This shows that for constructive interference the component of the incident wave vector $\mathbf{k}$ along the reciprocal lattice vector $\mathbf{K}_m$ must be half the length of $\mathbf{K}_m$. Figure 4-5 shows a perpendicular bisector plane of a reciprocal lattice vector $\mathbf{K}_m$. The Laue condition for an incident wave vector $\mathbf{k}$ will be met if the tip of $\mathbf{k}$ lies in this plane. Planes in reciprocal space determined by Eq. 4-13 are called **Bragg planes.**

We can show that the Bragg and Laue formulations of the diffraction problem are equivalent (see the Problems). We find that for a given $\mathbf{K}_m$, the reciprocal space Bragg plane defined above in the Laue formulation (Fig. 4-5) is parallel to a family of direct space lattice planes in the Bragg formulation.

A deeper understanding of the Bragg planes is given in Chapter 10. It will be seen that not only are x-rays scattered at such planes, but electrons with $\mathbf{k}$ vectors that satisfy Eq. 4-13 also are scattered. This leads to forbidden bands of electron energies.

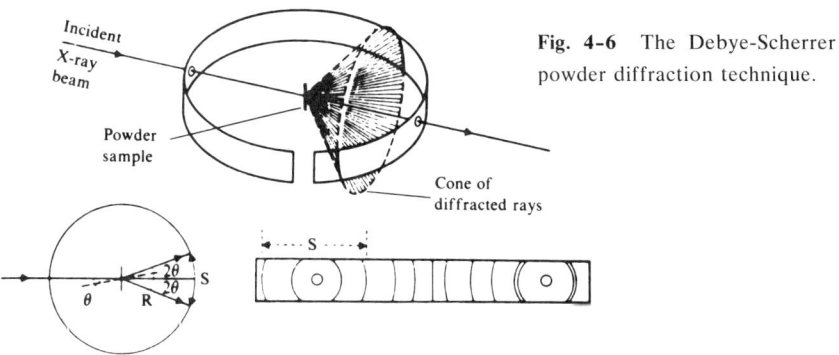

Fig. 4-6 The Debye-Scherrer powder diffraction technique.

4-4 Experimental X-ray Diffraction Methods

We discuss briefly some of the basic x-ray diffraction techniques commonly used.

The **Debye-Scherrer powder method** is illustrated in Fig. 4-6. A monochromatic x-ray beam is incident on a powder sample. (The sample is often contained in a very fine glass tube, which in turn is usually rotated about its axis to reduce graininess of the pattern on the film.) Since the sample is polycrystalline, the Bragg condition will be satisfied for each set of suitably oriented planes. As shown in the figure, the diffracted beams are cones coaxial with the incident beam direction. These cones intersect the film in arcs as shown. From the figure it is clear that the Bragg angle is $\theta = S/4R$, where S is the separation of the reflections on the film and R is the radius of the film cylinder. The most accurate values of the d-spacing between planes can be obtained by back-reflection, or large values of θ. This can be seen by taking the derivative of both sides of the Bragg equation, Eq. 4-3, for constant λ

$$\Delta(d \sin \theta) = (\Delta d) \sin \theta + d \cos \theta (\Delta \theta) = 0$$
$$\Delta\theta/\Delta d = -\tan \theta/d \qquad (4\text{-}14)$$

For a given variation of the distance between Bragg planes, $\Delta\theta$ is largest when θ approaches $90°$, which is just backscattering. For accurate measurements of lattice parameters as a function of temperature, diffracted rays measured in backscattering would be used.

Another important application of the Debye-Scherrer powder technique is in the field of sample identification. For example, some material might have been obtained from a high temperature growth technique, and it is important to determine if all of the starting material reacted and if a single or multiple phase product resulted. The resulting powder is used as the sample and the resulting lines measured. The lines can be compared to more than 20,000 results published, in useful card form, in the JCPDS powder diffraction file.

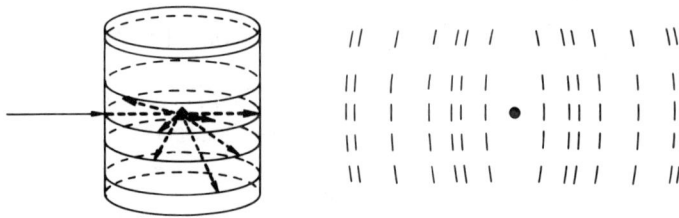

Fig. 4-7 The rotating crystal method of x-ray measurement.

The **Laue method** is used for routine orientation of single crystal samples. In this method a stationary single crystal is irradiated with a continuous wavelength x-ray source (Fig. 4-1). The various sets of planes in the crystal select, from the white radiation, the appropriate λ for the diffraction condition to hold. The particular angle θ then determines the exposure on the film. The resulting pattern shows the symmetry properties of the crystal. In this manner crystals can be orientated. For example, if the crystal is mounted so that it is q degrees from a sixfold axis, the resulting pattern shows this very clearly. Then the goniometer can be rotated by −q degrees to obtain the proper crystal direction. Typically such oriented single crystals are then cut by a saw and later used in experiments, or the crystals could be used while mounted on the goniometer.

The **rotating crystal method** is used by crystallographers to determine actual crystal structures. In this method a monochromatic x-ray beam and a single crystal, rotating about the axis, are used (see Fig. 4-7). As in the Debye-Scherrer method, crystal planes containing the rotation axis give a series of spots on both sides of the x-ray beam in a horizontal plane. Other sets of planes give reflections that are vertically displaced as shown. This diffraction pattern is essentially a map of the reciprocal lattice, and for this reason this method is used for crystal structure determinations.

There are several variations of this method, such as the oscillating crystal method (the crystal being rotated through a limited angular range) and the use of a Weissenberg goniometer (film shifted in synchronism with crystal rotation). Both of these methods eliminate overlapping reflections. However, the most important variation is to collect the data with scintillation or proportional counters, allowing automatic, computer-controlled data collection and accurate intensity measurements. Rather complex crystal structures can thus be solved and simple structures can be done routinely.

CHAPTER 4   X-RAY DIFFRACTION   83

Notes

X-ray diffraction can be studied at many levels and there are many books written for crystallographers. Here we note only Kittel, Chapter 2, which goes into the subject in more detail than we do here, and includes a discussion of atomic form factors and structure factors. Ashcroft and Mermin, Chapter 6, might also be read. The original paper by W. H. Bragg and W. L. Bragg [Proc. Roy. Soc. (London) **A88**, 428 (1913)] is worth reading. Bloss, Chapter 13, has a more detailed account, including examples of the use of the Powder Diffraction File (JCPDS) and B. E. Warren, "X-Ray Diffraction" (Addison-Wesley Pub. Co., 1969) discusses the interactions in a deeper way. Accurate values of the characteristic x-ray lines can be found in J.A. Bearden, Rev. Mod. Phys. **39**, 78 1967.

Problems

1. Nickel has a fcc crystal structure with a = 3.52 Å. For λ = 1.54 Å x-ray radiation, use Bragg's law to determine the angles of diffraction for the (100), (111), (200) and (220) planes. Why is the (100) reflection forbidden?

2. Show that the Bragg and von Laue formulations for constructive interference of x-rays are equivalent. (Hint: see Ashcroft and Mermin, Chapter 6.)

3. In two dimensions show that the reciprocal lattice of a reciprocal lattice is the direct lattice. See Problem 4 for the definition of a reciprocal lattice in two dimensions. Prove that this is also true in three dimensions. (Hint: see Ashcroft and Mermin, Chapter 5.)

4. **Reciprocal lattice in two dimensions** – In two dimensions a direct space lattice is given by

$$t_m = m_1 \mathbf{a} + m_2 \mathbf{b}$$

where m_i is any integer. Then the reciprocal lattice is given by

$$K_{m'} = m_1' \mathbf{a}^* + m_2' \mathbf{b}^*$$
$$\mathbf{a}^* \cdot \mathbf{a} = 2\pi \qquad \mathbf{b}^* \cdot \mathbf{a} = 0$$
$$\mathbf{a}^* \cdot \mathbf{b} = 0 \qquad \mathbf{b}^* \cdot \mathbf{b} = 2\pi$$

where m_i' also takes all integer values. For the direct space lattice shown below, draw the reciprocal lattice and the Wigner-Seitz cell for the reciprocal lattice. (The Wigner-Seitz cell in reciprocal space is called the **first Brillouin zone.**) Compare your results to Fig. 10-6.

5. Draw the reciprocal lattice and first Brillouin zone for the two-dimensional, direct space, rectangular lattice shown below. Compare your results to those in Fig. 10-7. What is the area of the first Brillouin zone?

6. **Reciprocal lattice in three dimensions** – (a) For a simple cubic (sc) lattice the unit cell is given by $\mathbf{a} = a\hat{x}$, $\mathbf{b} = a\hat{y}$, and $\mathbf{c} = a\hat{z}$. Then show that the reciprocal lattice is given by $\mathbf{a}^* = (2\pi/a)\hat{x}$, $\mathbf{b}^* = (2\pi/a)\hat{y}$, and $\mathbf{c}^* = (2\pi/a)\hat{z}$. What is the volume of the primitive cell of the reciprocal lattice and first Brillouin zone in terms of the sc lattice unit cell? (b) For a face-centered cubic lattice the conventional primitive unit cell vectors are given in Fig. 2-8. Show that the reciprocal lattice vectors are given by

$$\mathbf{a}^* = \frac{2\pi}{a}(\hat{x} - \hat{y} + \hat{z}), \quad \mathbf{b}^* = \frac{2\pi}{a}(\hat{x} + \hat{y} - \hat{z}), \quad \mathbf{c}^* = \frac{2\pi}{a}(-\hat{x} + \hat{y} + \hat{z})$$

Notice that these vectors are those of the conventional primitive bcc cell if the side of the conventional unit cell is taken as $4\pi/a$ (Fig. 2-8). Thus, the primitive lattice of an fcc lattice is a bcc lattice (and vice versa). The first Brillouin zones of both of these reciprocal lattices are shown in Figs. 10-13 and 10-14 (and also see Fig. 2-9).

7. **Volume of the first Brillouin zone** – (a) In two dimensions, if the area of the primitive unit cell in direct space is A_p, then show that the area of the first Brillouin zone is $(2\pi)^2/A_p$. (b) If the volume of the primitive cell in direct space is V_p [$=\mathbf{a}\cdot(\mathbf{b}\times\mathbf{c})$], then show that the volume of the first Brillouin zone [$\mathbf{a}^*\cdot(\mathbf{b}^*\times\mathbf{c}^*)$] is $(2\pi)^3/V_p$. (Hint: write $\mathbf{a}^*$ in terms of $\mathbf{a}$, $\mathbf{b}$, and $\mathbf{c}$ using the orthogonality conditions and the vector identity in Section 13-3.)

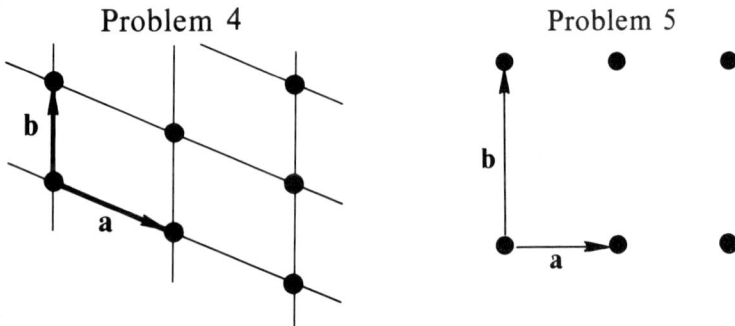

Problem 4   Problem 5

5

Crystal Symmetry and Physical Properties (S)

5-1 Introduction
5-2 Neumann's Principle
5-3 Tensors
5-4 Crystal Symmetry and Physical Properties
    a Pyroelectricity
    b Polarizability
    c Piezoelectricity
    d Elastic coefficients
5-5 Nonlinear Optics

Notes
Problems

CRYSTAL SYMMETRY AND PHYSICAL PROPERTIES

All things intertwine one with another, in a holy bond: scarce one thing is disconnected from another. In due coordination they combine for one and the same order.

M. Aurelius, "To Himself"

One of the ways in which crystalline solids differ from gases, liquids, and glasses is that in crystals the physical properties can vary a great deal depending on the direction of measurement with respect to the crystalline axes. By rather elementary use of symmetry, relations between the values in the various directions can be determined. Further, in certain directions, for certain physical properties, zero must be obtained; this too can be determined by symmetry. In this chapter we discuss these aspects of crystals.

This chapter also serves as a general introduction to certain concerns of solid state science. However, it is of specialized interest and need not be covered on a first reading of this book.

5-1 Introduction

Most physical properties of crystals, the "constants" of the crystal, can be defined in terms of a relationship between two measurable quantities. For example, the density of a crystal is defined as the mass divided by the volume of the crystal. Since both mass and volume are scalars (zero-rank tensors) their magnitudes do not depend on directions in the crystal, so density ($\equiv$ mass/volume) is independent of direction, or **isotropic**. On the other hand, a quantity such as the polarization, P, of a crystal can depend on the direction in which the electric field, E, is applied. In fact, most properties of single crystals will depend on the direction in which they are measured; such a property is said to be **anisotropic**. Thus, if E is applied in a certain direction in a crystal there will be a component of **P** in that direction but there may also be components of **P** in directions perpendicular to **E**. The existence of these perpendicular components depends on the symmetry of the crystal; this chapter is devoted to determining which components are allowable for the different crystal symmetries.

P and **E** are vector quantities (first-rank tensors). The polarizability of a crystal, which is defined in terms of the relation between **P** and **E**, is therefore a second-rank tensor, as we shall see. The "constants" that are properties of solids are tensors. That is, the relation of the constants in different crystallographic directions is just that of different components of a tensor. Neumann's principle can be used to find the relationship of the different tensor components of measurable properties of crystals.

5-2 Neumann's Principle

Neumann's principle states that any macroscopic physical property of a crystal has, at least, the symmetry of the point group of the space group. Restating this principle in the language of symmetry: any macroscopic physical property of a crystal, that is, each separate tensor component, transforms into $+1$ times itself under a symmetry operation of the point group of the crystal.

This is a principle and there is no proof of it; one can only make the idea plausible. The physical argument is this: when measuring a macroscopic property we do not expect to be able to detect the effect of a translation that is a fraction of a primitive unit cell. Let R be a rotation, mirror plane, center of inversion, etc., as in Chapter 1, and τ be a translation a fraction of a unit cell length associated with a screw or glide symmetry operation, as in Chapter 1. Then we expect that from macroscopic properties we cannot tell the difference between the symmetry operation $\{R|\tau\}$ and the operation $\{R|0\}$. On the other hand, the rotational part of the symmetry operation will relate points within the crystal that are separated by macroscopic distances and hence can affect the macroscopic properties. (See Burns and Glazer as well as Bhagavantam.)

After a brief review of the properties of tensors we apply this principle to several important physical properties. The ease of the approach will be appreciated as the examples are worked out.

5-3 Tensors

When measuring any physical property, we must refer to some coordinate system. Yet the physical property must be independent of the system of coordinates. Most macroscopic properties of crystals are described by tensors because they have the proper behavior under coordinate transformation.

If, under a rotation of coordinates, a fixed point in a fixed body in space can be described in terms of the new (x_i' where $i = 1, 2,$ or 3) and the original (x_i) coordinate system, then the relations between x_i' and x_i are

CHAPTER 5 CRYSTAL SYMMETRY AND PHYSICAL PROPERTIES

$$x_i' = \Sigma_j \, \Gamma_{ij} \, x_j \tag{5-1a}$$

where the direction cosines of a particular axis x_i' with respect to x_1, x_2, x_3, are Γ_{i1}, Γ_{i2}, Γ_{i3}, which have the usual property of direction cosines, namely

$$\Sigma_j \, \Gamma_{ij} \, \Gamma_{kj} = \delta_{ik} = \Sigma_j \, \Gamma_{ji} \, \Gamma_{jk} \tag{5-1b}$$

The relations in Eq. 5-1b are called the **orthogonal relations**. The transformations in Eq. 5-1a, in which the coefficients satisfy the orthogonality relations, are called **linear orthogonal transformations**. The inverse transformation is

$$x_i = \Sigma_j \, \Gamma_{ji} \, x_j' \tag{5-1c}$$

Assume that we have a coordinate transformation as in Eq. 5-1a. Then a quantity is an r-rank tensor if it transforms as

$$T_{ijk\ldots}' = \Sigma_{mno\ldots} [\Gamma_{im} \, \Gamma_{jn} \, \Gamma_{ko} \ldots] \, T_{mno\ldots} \tag{5-2}$$

where there are r subscripts on the tensor. An r-rank tensor has 3^r components since each of the subscripts has three values. Then a second-rank tensor is a nine-component object that transforms as

$$T_{ij}' = \Sigma_{mn} \, \Gamma_{im} \, \Gamma_{jn} \, T_{mn} \tag{5-3a}$$

and the inverse transformation of second- and general-rank tensors are

$$T_{ij} = \Sigma_{mn} \, \Gamma_{mi} \, \Gamma_{nj} \, T_{mn}' \tag{5-3b}$$

$$T_{ijk\ldots} = \Sigma_{mno\ldots}[\Gamma_{mi} \, \Gamma_{nj} \, \Gamma_{ok} \ldots] T_{mno}' \tag{5-3c}$$

Examples of tensors that we discuss are: zero-rank, temperature (a scalar); first-rank, electric polarization P_i (three components and is called a vector); second-rank, polarizability a_{ij} (relates external polarization, P_i, obtained by applying an external electric field E_j, i.e., $P_i = \Sigma_j \, a_{ij} E_j$); three-rank, piezoelectricity d_{ijk} (which relates strain, e_{ij}, caused by an external electric field, i.e., $e_{jk} = \Sigma_i \, d_{ijk} \, E_i$); fourth-rank, elastic constant c_{ijmn} (which relates applied stress t_{ij}, to strain, i.e., $t_{ij} = \Sigma_{mn} \, c_{ijmn} \, e_{mn}$).

A general property of tensors is that a quantity that relates an m-rank tensor to an n-rank tensor is an $(m + n)$-rank tensor. For example if **P** and **E** are both first-rank tensors and $\mathbf{P} = a\mathbf{E}$, then a is a second-rank tensor. To prove this, recall that if $E_i' = \Sigma_j \, \Gamma_{ij} \, E_j$, then $E_j = \Sigma_i \, \Gamma_{ij} \, E_i'$, which is just the inverse transformation. Thus,

$$P_i' = \Sigma_j \Gamma_{ij} P_j = \Sigma_{jk} \Gamma_{ij} a_{jk} E_k = \Sigma_{jkm} \Gamma_{ij} a_{jk} \Gamma_{mk} E_m' \quad (5\text{-}4a)$$

but $\quad P_i' = \Sigma_m a_{im}' E_m'$ \hfill (5-4b)

so $\quad a_{im}' = \Sigma_{ij} \Gamma_{ij} \Gamma_{mk} a_{jk}$ \hfill (5-4c)

which is just the condition required for a second-rank tensor, Eq. 5-3. This shows that "constants" mentioned in the above paragraph (a_{ij}, d_{ijk}, c_{ijmn}) indeed are tensors of the rank quoted. The general proof follows along the same lines as Eqs. 5-4.

5-4 Crystal Symmetry and Physical Properties

The **direct inspection method** will be used to find the components that are zero and the relationships between the nonzero components of the tensors that describe the physical properties of crystals. This method uses the fact that an r-rank tensor transforms as the cartesian coordinates r at a time. For example, a second-rank tensor, a_{ij}, transforms in the same way as $x_i x_j$. This can easily be seen:

$$a_{ij}' = \Sigma_{km} \Gamma_{ik} \Gamma_{jm} a_{km}$$

$$x_i' x_j' = \Sigma_{km} (\Gamma_{ik} x_k)(\Gamma_{jm} x_m) = \Sigma_{km} \Gamma_{ik} \Gamma_{jm} x_k x_m \quad (5\text{-}5)$$

We are now in a position to apply Neumann's principle to various rank tensors.

5-4a Pyroelectricity The properties of a first-rank tensor are discussed first, although it is easier to appreciate the method for second-rank tensors. A crystal is **pyroelectric** if the primitive cell possesses a dipole moment. A unit cell has a nonzero dipole moment if the center of mass of the positive charge is at a different position than that of the negative charge. The dipole moment $\mu \equiv \Sigma q_i r_i$ where the sum is over all the charges in the cell of charge q and at position **r**. Polarization, $\mathbf{P} \equiv \mu/\text{volume}$. Since **P**, or μ, transform as **r**, a vector or first-rank tensor, it is easy to see which components transform into +1 times themselves under the symmetry operations of any point group.

Let us see how it is done. First, remember that each component of r(x, y, and z) must be treated separately. Second, the problem is so simple that we can treat it in general. For example, a twofold rotation symmetry operation perpendicular to any component will transform it into minus itself, and thus **P** along that direction must be zero. If the point group has an i($\bar{1}$) as a symmetry operation, then this transforms **r** into minus itself, so all components of **P** must be zero.

Take as a specific example the point group C_{2v}(mm2). Besides E(1), the symmetry operations are two mirror planes containing the z-axis and a twofold rotation $C_2(2)$ along the z-axis (Table 1-2). The two mirror planes transform x into minus itself and y into minus itself, so the components of **P** in these directions must be zero. However, under these two symmetry operations, z is transferred into plus itself. Similarly, under $C_2(2)$, z is transformed into plus itself. Thus, symmetry allows crystals with this point group to be pyroelectric with a polarization along only the z-axis. Symmetry makes no restrictions as to the magnitude of **P**. The ease of the procedure is clear.

In general we find that crystals with the point groups D_n, D_{nd}, D_{nh}, S_n, the cubic ones, and C_{nh} for n > 1 cannot be pyroelectric. The physical reason is straightforward. A component of a vector, which is a macroscopic property of the crystal, cannot be perpendicular to a symmetry plane, because the symmetry plane causes it to transform into minus itself and thus it must be zero. Similarly, a component of a vector cannot be perpendicular to a twofold axis. An inversion center or four threefold axes also causes the vector to be zero.

Only crystals with a C_n, C_{nv}, or C_{1h} point group can be pyroelectric. For most of these point groups, only z transforms into itself under all the symmetry operations; so the polarization may be nonzero only along the c-axis (as in the example for C_{2v} point symmetry). However, for the monoclinic point groups C_{1h}(m), x and *separately* y transform into themselves under the symmetry operations. Therefore, the polarization has a zero z-component but arbitrary x- and y-components, and thus the vector lies anyplace in the ab-plane. For the triclinic point group C_1(1), **P** can be in any direction because x, y, and *separately* z transform into themselves.

The order of magnitude for the maximum dipole moment in a unit cell is that of one electron charge separated by 1 Å. Thus, $\mu \approx$ $(4.8 \times 10^{-10}$ esu$)(10^{-8}$ cm$)$ = 4.8 Debye, where 1 Debye = 10^{-18} esu cm. For a unit cell approximately 4 Å on a side, we thus obtain a polarization value P = $(4.8 \times 10^{-18}$ esu cm$)(4 \times 10^{-8}$cm$)^3 \approx$ $1.5 \times 10^{+6}$ esu/cm^2 = 0.25 Coulomb meter^{-2}.

Table 5-1 lists values of polarization for a few materials. Some of these are ferroelectric and have a transition temperature above which the crystal has a different structure than it has at lower temperatures. The point group of the high temperature structures requires that **P** be identically zero. However, several materials do not have a phase transition and are pyroelectric to their melting temperatures.

Some of these ferroelectric materials with large values of dP/dT, which occurs near the transition temperature, are useful devices as **pyroelectric detectors**. That is, they have an output voltage proportional to a temperature change and this can be large.

Table 5-1 Values of the polarization along the unique crystal axis in several crystals. Those on the right are ferroelectric and room temperature values are given. The point groups are given in parentheses and the units are Coulombs/meter^2.

ZnS	(C_{6v})	0.02	BaTiO$_3$	(C_{4v})	0.26
CdS	(C_{6v})	0.03	LiTaO$_3$	(C_{3v})	0.50
ZnO	(C_{6v})	0.06	LiNbO$_3$	(C_{3v})	0.71
			PbTiO$_3$	(C_{4v})	0.81

5-4b Polarizability This second-rank tensor is defined via Eqs. 5-4. It is perhaps a better example of the application of Neumann's principle than a first-rank tensor, the latter being so elementary it is easy not to realize what is being done.

Not from symmetry, but from energy considerations (see Nye p. 74), one can show that $a_{ij} = a_{ji}$, which leaves just six independent components. Thus, in general, polarizability, which relates an applied electric field to a resultant polarization, can be written in matrix form.

$$\begin{bmatrix} P_1 \\ P_2 \\ P_3 \end{bmatrix} = \begin{bmatrix} a_{11} & a_{12} & a_{13} \\ a_{12} & a_{22} & a_{23} \\ a_{13} & a_{23} & a_{33} \end{bmatrix} \begin{bmatrix} E_1 \\ E_2 \\ E_3 \end{bmatrix} \quad (5\text{-}6)$$

Writing the expression this way shows very clearly that an electric field applied in the 1-direction, E_1, produces a polarization parallel to it, P_1, governed by $a_{11}E_1$. However, there may also be a polarization in the 2-direction $P_2 = a_{12}E_1$ as well as one in the 3-direction, $P_3 = a_{13}E_1$. The polarizability referred to here is often called the optic polarizability when the frequency of the applied electric field is in the optic region of the spectrum ($\approx 10^{14}$ cycles/sec).

For a given point group, we use the symmetry operations to determine which tensor components are zero and which are related to each other. The approach, as outlined above, is simple. If a symmetry operation shows that $x_i\, x_j = -x_i'x_j'$ then the a_{ij} component is zero because Neumann's principle demands that the component transform into $+1$ times itself. Similarly, if a symmetry operation results in $x_1\, x_2 = x_1'x_3'$ then $a_{12} = a_{13}$. We can save work by appreciating that the only symmetry operations of each point group that need be used in these considerations are the generators of the point groups (Table 1-2). The **generators of a group** are defined as a set of elements of the group such that multiplications among them along with the identity will give every element of the group. This set is not always unique but Table 1-2 lists a convenient set for each point group.

Apply the method to the monoclinic point group $C_{2h}(2/m)$ where the generators are $C_2(2)$ and $i(\bar{1})$. We want to determine how x^2, y^2, z^2, xy, xz, and yz transform under these symmetry operations. For

CHAPTER 5   CRYSTAL SYMMETRY AND PHYSICAL PROPERTIES   93

$C_2(2)$ we know that x transforms into $-x$, y into $-y$, z into $+z$. However, for $i(\bar{1})$ they all transform into -1 times themselves.

Thus, C_2: $x^2 \rightarrow x^2$ $xy \rightarrow +xy$
              $y^2 \rightarrow y^2$ $xz \rightarrow -xz$ therefore $a_{13} = 0$
              $z^2 \rightarrow z^2$ $yz \rightarrow -yz$ therefore $a_{23} = 0$

   i: $x^2 \rightarrow x^2$ $xy \rightarrow xy$
              $y^2 \rightarrow y^2$
              $z^2 \rightarrow z^2$

Notice that many of the relations such as $x^2 \rightarrow x^2$ give no information; all that it implies is that $a_{11} = a_{11}$. It is only relations such as $xz \rightarrow -xz$ that provide a nontrivial relation. The result is that for crystals with this point group there are four independent tensor components and the polarizability of Eq. 5-6 has the form

$$\begin{bmatrix} a_{11} & a_{12} & 0 \\ a_{12} & a_{22} & 0 \\ 0 & 0 & a_{33} \end{bmatrix}$$

Applying this method to a cubic crystal, $C_3(3)$ will always give $x^2 \rightarrow y^2 \rightarrow z^2$ so $a_{11} = a_{22} = a_{33}$. Similarly it is very easy to see that $a_{12} = a_{13} = a_{23} = 0$. Thus, for cubic crystals there is only one independent polarizability component and the tensor can be written as

$$\begin{bmatrix} a_{11} & 0 & 0 \\ 0 & a_{11} & 0 \\ 0 & 0 & a_{11} \end{bmatrix}$$

Besides polarizability there are a number of other important physical properties that are symmetric second-rank tensors. The tensors describing all of these properties have the same zero terms and interrelationships. Examples of these physical properties are dielectric polarizability, diamagnetic and paramagnetic polarizability, thermal expansion, and thermal and electrical conductivities. Several of these tensors are discussed in the Problems.

The dielectric constant ε, which relates the displacement **D** to the electric field **E**, i.e., $D_i = \Sigma_j \varepsilon_{ij} E_j$, is discussed in Chapter 13. Table 5-2 lists typical room temperature values for some common materials. For the cubic materials only one value need be listed because $\varepsilon_{11} = \varepsilon_{22} = \varepsilon_{33}$. For noncubic materials there are at least two different values. Near ferroelectric phase transitions, for example, there may be huge anisotropies ($>10^4$). This is discussed in Chapter 14.

A discussion and numerical values for other symmetric second-rank tensors can be found in the books referred to in the Notes.

Table 5-2 Values of the room temperature dielectric constants for some materials. (Other values are listed in Table 8-3.)

		ε_{11}			ε_{11}	ε_{33}
SrTiO$_3$	(O$_h$)	450	CdS	(C$_{6v}$)	8.8	8.0
CuCl	(T$_d$)	9	ZnO	(C$_{6v}$)	9.3	11.0
CuBr	(T$_d$)	8	LiNbO$_3$	(C$_{3v}$)	78	27
CuI	(T$_d$)	6.5				
Si	(O$_h$)	11.9				
Ge	(O$_h$)	16				

5-4c Piezoelectricity The piezoelectricity constant is an example of a third-rank tensor that relates a strain of a crystal to an externally applied electric field, $e_{jk} = \Sigma_i d_{ijk} E_i$. Since strain is a symmetric second-rank tensor, that is, $e_{jk} = e_{kj}$, instead of 27 components for this third-rank tensor there are only 18 because $d_{ijk} = d_{ikj}$. (See Bhagavantam, p. 108, for a discussion of strain.)

To determine which components of d_{ijk} are zero and which are related to each other consider how $x_i x_j x_k$ transforms. For any point group with a center of inversion, $x_i x_j x_k$ transforms into minus itself so that every component in the piezoelectric tensor is zero; *crystals with a center of symmetry are not piezoelectric*.

As an example of crystals with nonzero coefficients, consider the orthorhombic point group $D_2(222)$ with symmetry operation {E, C_2, C_2^x, C_2^y} where C_2 and C_2^y are the generators. (C_2 and C_2^y refer to twofold rotations about the z- and y-axis, respectively.)

C_2 : $x^3 \to -x^3$; $y^3 \to -y^3$ so $d_{111} = d_{222} = 0$
$yx^2 \to -yx^2$; so $d_{211} = d_{133} = d_{233} = d_{122} = 0$
$xz^2 \to -xz^2$; etc.
$xxy \to -xxy$; etc. so $d_{112} = d_{323} = d_{212} = d_{313} = 0$

C_2^y: $z^3 \to -z^3$ so $d_{333} = 0$
$zx^2 \to -zx^2$; so $d_{311} = d_{322} = 0$
$zy^2 \to -zy^2$ $d_{113} = d_{223} = 0$

We are left with three nonzero independent tensor components, d_{123}, d_{213}, and d_{312}, where d_{123} means that for an electric field along the a-axis one obtains a bc-shear, etc., for the other components.

A contracted notation is often used so that tensor components may be written more easily. For any set of *symmetric* $x_i x_j$ components there are only six, not nine, independent indices. Thus, a single index that goes from 1 to 6 is used. The correspondence is

ij:	11	22	33	23	13	12
$x_i x_j$:	x^2	y^2	z^2	yz	xz	xy
contracted index:	1	2	3	4	5	6

(5-7)

CHAPTER 5   CRYSTAL SYMMETRY AND PHYSICAL PROPERTIES   95

Table 5-3 Values of the piezoelectric constants in several representative crystals. The crystals on the left have the zinc blende structure, $T_d(\bar{4}3m)$, while those on the right have the wurtzite structure, $C_{6v}(6mm)$ point symmetry. The values are listed in units of 10^{-12} Coulombs/Newton = 10^{-12} meter/volt.

	d_{14}		d_{31}	d_{33}	d_{15}
CuCl	27.2	CdS	−5.18	10.32	−13.98
CuBr	16.0	ZnO	−5.2	10.6	−13.9
CuI	7.0				
GaAs	−2.7				
GaSb	−2.9				
InAs	−1.1				
InSb	2.4				

Please remember that the contraction of indices is used only when two indices are symmetric.

For the $D_2(222)$ point group example, using the contracted notation, the nonzero components are d_{14}, d_{25}, and d_{36}. The result can, and usually is, written in matrix form, consistent with the defining expression $e_{jk} = \Sigma_i E_i d_{ijk}$, thus

$$\begin{bmatrix} 0 & 0 & 0 & d_{14} & 0 & 0 \\ 0 & 0 & 0 & 0 & d_{25} & 0 \\ 0 & 0 & 0 & 0 & 0 & d_{36} \end{bmatrix}$$

For the cubic crystals with point group $T_d(\bar{4}3m)$ there is only one independent constant since $d_{14} = d_{25} = d_{36}$. Values of d_{14} are given for several materials with this point group in Table 5-3.

For the hexagonal point group $C_{6v}(6mm)$ we determine that $d_{31} = d_{32}$ and that $d_{15} = d_{24}$ so the piezoelectric coefficients can be written as

$$\begin{bmatrix} 0 & 0 & 0 & 0 & d_{15} & 0 \\ 0 & 0 & 0 & d_{15} & 0 & 0 \\ d_{31} & d_{31} & d_{33} & 0 & 0 & 0 \end{bmatrix}$$

and the values for two useful materials are listed in Table 5-3.

5-4d Elastic coefficients The defining relation for the elastic coefficients is $t_{ij} = \Sigma_{mn} c_{ijmn} e_{mn}$, which relates an applied stress t_{ij} to a resultant strain e_{mn}. Being a fourth-rank tensor the elastic coefficient has 81 components, but this is reduced to 21 because both stress and strain are symmetric second-rank tensors; terms in c_{ijmn} are symmetric with respect to interchanges in i and j and separately in m and n; also from energy consideration $c_{ijmn} = c_{mnij}$. Thus, in general,

$$c_{ijmn} = c_{jimn} = c_{ijnm} = c_{jinm} = c_{mnij} = c_{nmij} = c_{mnji} = c_{nmji} \tag{5-8}$$

For the $D_2(222)$ example we find the following nonzero, independent coefficients: c_{1111}, c_{2222}, c_{3333}, c_{1122}, c_{1133}, c_{2233}, c_{2323}, c_{1313}, c_{1212}; which in contracted notation are: c_{11}, c_{22}, c_{33}, c_{12}, c_{13}, c_{23}, c_{44}, c_{55}, c_{66}, respectively. In matrix notation, consistent with the defining equation we can write for the elastic coefficients

$$\begin{bmatrix} c_{11} & c_{12} & c_{13} & 0 & 0 & 0 \\ & c_{22} & c_{23} & 0 & 0 & 0 \\ & & c_{33} & 0 & 0 & 0 \\ & & & c_{44} & 0 & 0 \\ & & & & c_{55} & 0 \\ & & & & & c_{66} \end{bmatrix}$$

The lower left part of the matrix is usually not written because it follows immediately from $c_{ijmn} = c_{mnij}$. This is a matrix, written for convenience in terms of the contracted indices, and is not related to a second-rank tensor. If the transformation properties, under rotation of the crystal, of the elastic coefficients are desired, remember that they are components of a fourth-rank tensor and it is safest to use the full tensor notation when applying the transformation.

For example, for CdS the elastic constants are $c_{11} = 8.4$, $c_{12} = 5.2$, $c_{44} = 1.46$, $c_{13} = 4.6$, and $c_{33} = 9.2$ in the same units used in Table 12-1.

Another fourth-rank tensor, the electrooptic effect, is discussed in Chapter 14.

5-5 Nonlinear Optics

Consider the polarizability tensor discussed in Section 5-4b, $P_i = \Sigma a_{ij}E_j$, i.e. Eq. 5-6. What happens if very, very large electric fields are applied? Is there a linear relation between P and E for all E? The answer is no; rather this expression is just the first term of a series expansion

$$P_i = \Sigma a_{ij}E_j + \Sigma \beta_{ijk}E_jE_k + \Sigma \gamma_{ijkm}E_jE_kE_m + \ldots \tag{5-9}$$

where the series converges rapidly. All of the macroscopic properties discussed in this chapter are just the first terms of a series expansion similar to Eq. 5-9; however, symmetry often allows only the odd (or even) powers in the expansion.

The a, β, and γ coefficients in Eq. 5-9 are all tensors and we could handle them as we handled the other tensors discussed so far. In fact, β_{ijk} is a third-rank symmetric (in j and k) tensor and hence has the same symmetry properties as the piezoelectric tensor, Section 5-4c. It is zero if the point group of the crystal has a center of symmetry. The fourth-rank γ_{ijkm} tensor is symmetric in the last three indices and thus it is different from the elastic coefficients.

Let us not be concerned with the details of the components; rather let us consider some of the consequences of the higher order terms. In many fields of science the higher order terms (those past the first term in Eq. 5-9) lead to interesting nonlinear effects. For example, deviations from Ohm's Law in semiconductors (deviations from the linear current vs. voltage behavior) play an important role not only in understanding the physics of the processes but in their use as devices (see the Notes).

However, the largest area of science concerned with the higher order terms is the field of **nonlinear optics** (NLO). In most laboratory experiments, crystals break down if electric fields much larger than 10^4 V/cm are applied. Although, as noted above, in semiconductors nonlinear effects often can be observed for such fields (usually applied in the form of short pulses so that the current density, and thus the heating, can be kept to a minimum). In nonlinear optics one exploits the very, very large electric field available in focused laser beams ($\approx 10^8$ V/cm). These electric fields are oscillating in the 10^{15} cycle/sec range and thus electrical breakdown does not have time to occur. Many books have been written on this subject and the Notes should be consulted for references. We just give an indication of some of the effects that are found.

Assume that two laser beams are "mixed" in a crystal (the two short intense pulses of laser light arrive at the same place at the same time in a crystal). The two lasers have frequencies ω_1 and ω_2 (in radians/sec) so the electric field applied to the crystal is of the form

$$E = E_1 \sin \omega_1 t + E_2 \sin \omega_2 t \tag{5-10}$$

The polarization from βE^2 in Eq. 5-9 contains the following terms

$$\beta E_1^2 \cos 2\omega_1 t, \qquad \beta E_2^2 \cos 2\omega_2 t,$$
$$\beta E_1 E_2 \cos (\omega_1 + \omega_2)t, \qquad \beta E_1 E_2 \cos (\omega_1 - \omega_2)t \tag{5-11}$$

as well as a d.c. term (i.e. $\omega = 0$, optical rectification). The terms are just second harmonics ($2\omega_1$ and $2\omega_2$) as well as the sum and difference frequency terms caused by the mixing of the laser beams via βE^2. (These polarizations are the source terms that are put into Maxwell's equations and yield electromagnetic radiation at the frequencies shown in Eq. 5-11.) Clearly, the intensity of these second harmonics depends on the magnitude of the nonlinearity, and on the square of the electric field. We could also anticipate that the output waves depend on energy and momentum (of the light) conservation laws. The details of these latter effects can be complicated but have been well studied. See the references in the Notes.

In an analogous manner the γE^3 term leads to third harmonics and related sum and difference terms. Since the electric fields associated with lasers can be so large, effects from this and higher order terms are clearly observable.

As indicated, for point groups that have a center of symmetry all the components of β_{ijk} are zero. However, even for these crystals nonzero components of γ_{ijkm} are allowed so third harmonic generation is observable.

Notes

For fuller discussions of the tensor properties of crystals see Nye, Bhagavantam, Wooster, and Fumi [Acta. Cryst. **5**, 44 (1952)]. The direct inspection method is directly applicable only to crystal point groups in which we can find cartesian orthogonal coordinates that do not transform into linear combinations of themselves under the symmetry operations. This includes 22 of the 32 point groups. For a discussion of the transformation properties of axial vectors see Bhagavantam, Section 2-5, and Briss, Sections 1-3 and 1-4.

Extensive numerical tables can be found in the Landolt-Börnstein Tables, Ed. K. H. Hellwege, Group III: Crystal and Solid State Physics, Volume 11. These tables include values for the elastic, piezoelectric, pyroelectric, piezooptic, and electrooptic constants as well as some of the nonlinear optic constants that have been determined via the use of high-power lasers.

For a review of the nonlinear current versus voltage behavior in Si see C. Jacoboni, C. Canali, G. Ottaviani, and A. A. Quaranta, Solid State Electronics **20**, 77 (1977). There are many books on nonlinear optics. You might start with A. Yariv, "Optical Electronics" (Holt, Rinehart and Winston, 1976) or N. Bloembergen, "Nonlinear Optics" (Benjamin, 1965).

Problems

1. Tensor components – For the point group $C_{4h}(4/m)$ determine the tensor components for the polarizability, piezoelectric, and elastic constant tensors. Also, write the results in matrix form using the contracted notation and check your results with the books in the Notes.

CHAPTER 5   CRYSTAL SYMMETRY AND PHYSICAL PROPERTIES   99

2. Do the same as in problem 1 for the cubic point groups. Notice that the second- and fourth-rank tensors are the same for all five cubic point groups.

3. **Axial tensors** — Our discussion in Section 5-1 refers to what are called true or polar tensors. An **axial tensor** is defined as a quantity that transforms as follows:

$$S_{ijh\ldots}' = \pm \Sigma\, \Gamma_{im}\, \Gamma_{jn}\, \Gamma_{ho} \cdots S_{mno\ldots}$$

where the negative sign is used for a symmetry operation that changes handedness, (i, σ, S_n) and the positive sign for the other symmetry operations (E, C_n). Determine how an axial vector (first-rank tensor) transforms for the C_n point groups. Orbital angular momentum, which is the cross product of two true vectors, is a first-rank axial tensor. (Check the Notes for references.)

4. **Thermal expansion** — If there is a small temperature change ΔT the lengths and angles of the unit cell change. This can be described in terms of the components of a strain e_{ij}. For this small ΔT the change of the components of e is proportional to ΔT, that is, $\Delta e_{ij} = y_{ij}\Delta T$. y_{ij} is the **thermal expansion**, which is a symmetric two-rank tensor. Show that $y_{11} + y_{22} + y_{33}$ is the coefficient of volume expansion. (Hint: see Bhagavantam, Chapter 13.) Find the values for calcite, where one of the coefficients is negative, and for gypsum and graphite, where the anisotropy is large. How do these values compare to those for diamond, Si, and Ge?

5. **Electrical conductivity** — The electrical conductivity tensor σ_{ij} is determined by the relation between the electric field E_k and the current density j_i, i.e., $j_i = \Sigma_k \sigma_{ik} E_k$. The resistivity ρ_{ik} is given by $E_k = \Sigma_k \rho_{ik} j_k$. Both σ_{ik} and ρ_{ik} are symmetric second-rank tensors that have the same symmetry properties as those of the polarizability tensor. If C is the 3 × 3 matrix of the conductivity tensor and R that of the resistivity tensor, show that $R = C^{-1}$.

Claire Albahae 85

6

Classification of Solids

6-1 Summary of Chapters 1–3
    a Translational symmetry
    b Unit cells
    c Seven crystal systems
    d Fourteen Bravais lattices
    e Definitions of directions, coordinates and planes
    f 32 crystallographic point groups and 230 space groups
    g Crystal structure
    h Problems
    i Point groups, space groups, and other terms
6-2 Introduction to Classification of Solids
6-3 Five Types of Bonds
6-4 Repulsive Potential Energy
6-5 Molecular Bond
    a Simple solids
    b Simple structures (S)
    c Packing of molecules (S)
    d Liquid crystals–plastic crystals (S)
    e Lennard-Jones potential (S)
6-6 Hydrogen Bond (S)
    Notes
    Problems

Table 6-1 Examples and characteristics of the five types of bonds. The **cohesive energy** is the energy necessary to disassociate the solid into separated atoms. However, for ionic crystals the term is defined as the energy to separate the solid into isolated ions. For the hydrogen-bonded materials values are the **sublimation energy,** the energy required to separate the solid into separated molecules. (See Evans, Kittel, 2nd Ed., and NBS, Technical Note 270).

Bond Type	Examples	Cohesive Energy (kcal/mole)	Cohesive Energy (eV/molecule)	Striking Characteristics
Ionic	LiCl	199	8.63	Nondirected bonding, giving structures of high coordination; no electrical conductivity at low temperatures; good ionic conductivity at high temperatures.
	NaCl	183	7.94	
	KCl	166	7.20	
	RbCl	159	6.90	
Covalent	Diamond	170	7.37	Spatially directed bonds, giving structures of low coordination; low conductivity at low temperatures for pure crystals.
	Si	108	4.68	
	Ge	89	3.87	
	Sn	72	3.14	
Metallic	Li	37.7	1.63	Nondirected bond, giving structures of very high coordination and density; high electrical conductivity; ductile.
	Na	25.7	1.11	
	K	21.5	0.934	
	Rb	19.6	0.852	
Molecular	Ne	0.46	0.020	Low melting and boiling points; very compressible; the properties of the molecules are retained.
	Ar	1.79	0.078	
	Kr	2.67	0.116	
	Xe	3.92	0.170	
Hydrogen-bonded	H_2O(ice)	12	0.52	Increase in bonding energy over similar molecules without hydrogen bonds.
	HF	7	0.30	

CLASSIFICATION OF SOLIDS

> *—that to you differences mean little, while to me they are the most important things. Mine is the nature of a scholar, and my branch of scholarship is science. And science, to quote your own words, is nothing than a "strange hankering after differences." Her essence could not be better defined. For men of science nothing is so important as the clear definition of differences.*
>
> H. Hesse, *"Narziss & Goldmund"*

6-1 Summary of Chapters 1 – 3

Readers may desire to start their study of solid state science with less emphasis on crystal structures and the formalities associated with them. If so, this chapter on the types of bonding in solids is a good place to begin. However, knowledge of some of the terminology that is covered in Chapters 1-3 will prove useful. Various terms are defined in this first section; more details can be found in the earlier chapters (via the index).

The fundamental property that sets crystalline solids apart from gases, liquids, and amorphous solids is **translational symmetry.** In other words, there is a basic building block in the crystal, called a **unit cell,** which, when displaced parallel to itself in three dimensions, will form a crystal, much as identical bricks would form a thick wall. These two concepts, and related ones, are included in the following discussion.

6-1a Translational symmetry The most important fundamental aspect of the modern theory of solids is translational symmetry. That is, in crystals there is a small basic fundamental unit that is repeated over and over in three dimensions and this makes up a crystal.

It is easiest to define translational symmetry via a lattice. A **lattice** is an infinite array of points in which the surroundings of each point are identical, in every way, to those of all of the other points. A lattice can be generated by three noncoplanar (not necessarily perpendicular to each other) vectors **a**, **b**, and **c**, which are called **primitive lattice translation vectors.** Then the lattice points are the end points of the vector t_m given by

$$t_m = m_1 a + m_2 b + m_3 c \qquad (1\text{-}10a)$$
where $$m_i = 0, \pm 1, \pm 2, ... \qquad (2\text{-}1)$$

Thus, a lattice is a *array of points.* (See Fig. 2-1 and note that we may also define a lattice in one and two dimensions in the same way.) Atoms are not involved in defining a lattice. In fact, we shall continue to discuss unit cells, Bravais lattices, and so forth, without mentioning atoms. It is not until we talk about a crystal structure that we will consider the presence of atoms. Thus, repeating, the word lattice means we are talking about points in space (and crystal structures have atoms located at or near lattice points). This is often confused in the solid state literature.

6-1b Unit cells By completing the parallelepiped formed by **a**, **b**, and **c**, we obtain a **unit cell**. This unit cell has a volume $\mathbf{a} \cdot (\mathbf{b} \times \mathbf{c})$ and if translated parallel to itself by $\mathbf{t}_m$ will fill all space. As defined here, this unit cell contains just one lattice point (Figs. 2-1 and 2-2 and Section 2-2) and is therefore called a **primitive unit cell**. We can also construct unit cells that contain more than one lattice point (Fig. 2-2) called **nonprimitive** (or **multiply primitive**) **unit cells**. These still satisfy the fundamental property of unit cells, namely, that when translated parallel to themselves they fill all space. Given a lattice, the choice of a unit cell is not unique; Fig. 2-2 shows a few of the infinite choices of unit cells for a given lattice. However there are conventions, and these will be discussed. One other type of unit cell that is important in solid state science is the **Wigner-Seitz cell**, which is primitive. To obtain this cell we start at any lattice point, the origin, and draw vectors to all neighboring lattice points. Planes perpendicular to and passing through the midpoints of these vectors are constructed (Section 2-4b and Fig. 2-9). The Wigner-Seitz cell is the cell with the smallest volume about the origin bounded by these planes. Clearly this cell must be primitive; this cell, important in electronic band theory, will be covered in the following discussion.

6-1c Seven crystal systems Consider an arbitrary lattice. An arbitrary lattice is one in which there are no conditions on the values of **a**, **b**, and **c**, and the angles α, β, and γ between the axes; Fig. 2-3 defines the angles in terms of the axes in the conventional way. To this arbitrary lattice apply certain symmetry operations and determine if the application of these symmetry operations put any new restrictions on the values of **a**, **b**, **c**, α, β, and/or γ.

A **symmetry operation** of a molecule or crystal is an operation that moves the positions of the various atoms in such a way that the molecule or crystal appears exactly as before the operation (that is, it appears in an equivalent position). Furthermore, when repeatedly applied, the operation also must be a symmetry operation. Symmetry operations include rotations, inversions, reflections, and translations, as well as others that will be discussed.

The simplest symmetry operation to consider is the inversion operation, $i(\bar{1})$. When applied to an arbitrary lattice it imposes no new restrictions because every lattice already has inversion symmetry, that is, for every lattice point at (m_1 m_2 m_3) there also is a lattice point at ($-m_1$ $-m_2$ $-m_3$). A lattice with no restrictions on the angles and unit cell lengths is said to belong to the **triclinic crystal system.**

The first nontrivial effect occurs when we demand that a lattice have a twofold operation, $C_2(2)$, that is, a rotation by 180°, as a symmetry operation. Taking this operation along the c-axis, then it is straightforward to see (Fig. 2-4a) that this symmetry requirement puts no conditions on the lengths of **a**, **b**, or **c** and none on γ; however, if this operation is to be a symmetry operation, then $\alpha = \beta = 90°$. A lattice with these restrictions is said to belong to the **monoclinic crystal system.**

In a similar way if we demand that an arbitrary lattice have not one but two twofold operations that are symmetry operations, then there are still no restrictions on **a**, **b**, or **c**, but all of the angles must be 90°. (Repeat the operation in Fig. 2-4a along two axes.) A lattice with these restrictions is said to belong to the **orthorhombic crystal system.**

In a similar way we arrive at the 7 crystal systems listed in Table 2-1, where the imposed symmetry is shown and conditions are given on **a**, **b**, and **c**, as well as α, β, and γ.

6-1d Fourteen Bravais lattices We have seven different lattices each having different special conditions on the axes and angles due to different symmetry requirements (Table 2-1). We now ask if to any one of these lattices we can add more lattice points at "judicious locations" and (a) still have a lattice (see the preceding definition) (b) and if it is a lattice, is it a new one in the same crystal system or just one of the known ones in the crystal system in some other orientation?

When these two points are considered for each of the seven crystal systems, one system at a time, it turns out that for several of the crystal systems nothing new can be found and for others more lattices can be found keeping within the same crystal system. For example, for the orthorhombic crystal system three new lattices can be found, for a total of four.

Let us consider the general approach and results; the details are in Section 2-4. First, the "judicious locations" that we consider trying to put lattice points for each of the lattices in the seven crystal systems are: the body-center position (I); the face-center positions (F); the one-face centered position (C); and the rhombohedral position (R). These positions, in an arbitrary lattice, are shown in Figs. 2-5b, 2-5c, 2-5d and 2-5g; the equations for the positions are in Eq. 2-2. Remember, for each crystal system we want to try adding points to the initial lattice and determine the answers to the preceding two questions. If, with the addition of these new points we still have a lattice and it is a new one within the starting crystal system then that crystal system, indeed, has a new lattice.

To understand this better, consider what happens if we start with an arbitrary lattice and consider centering not one nor three faces, but two faces, as shown in Fig. 2-5e. By studying this figure it is clear we no longer have a lattice because the surroundings of all of the points are not identical as indicated in Fig. 2-5e. As a second example, consider C-centering (one-face centering) in the cubic crystal system. We have a lattice but it is no longer in the same crystal system since these no longer are four threefold axes. Convince yourself that the lattice is in the tetragonal crystal system, and see Fig. 2-6d.

Summarizing – We started with a lattice for each of the 7 crystal systems. Each of these lattices is a primitive lattice since the unit cell contains just one lattice point and the lattices are called the cubic-P lattice, tetragonal-P lattice, etc. (the P meaning primitive), for the seven systems (Table 2-1). Then by the above centering approach, for each of the 7 crystal systems we find that some centering does not work and some do work, giving new lattices within the same crystal system. In all, seven new lattices are found, giving 14 in all. These are called the **14 Bravais lattices**, and they are pictured in Fig. 2-7. The face-centered ones are indicated by the letter F, body centered by I, one face centered by C, and rhombohedral centering by R as indicated. Notice that for the orthorhombic crystal system one can have a P-, F-, I-, and C-lattice.

Unit cells of the 14 Bravais lattices – The unit cells shown in Fig. 2-7 for the seven new F-, I-, and C-lattices, all are multiply primitive having four, two, and two lattice points per unit cell, respectively. Naturally these unit cells are not unique but the ones shown for all 14 Bravais lattices are the conventional ones. For the face-centered, body-centered, and one-face centered lattices there are also conventions on how to draw the corresponding primitive cells. These are shown in Fig. 2-6. The conventional primitive and multiply primitive unit cells shown in Fig. 2-7 very clearly display the symmetry of the lattice; pictorially the symmetry is less obvious for the primitive cells of the face-centered, body-centered, and one-face centered lattice. It is this difference that causes us to define and often show the multiply primitive unit cells.

Wigner-Seitz cells – The general method to construct a Wigner-Seitz cell was discussed (Section 2-4b). For the cubic I-lattice (body-centered cubic lattice) the details are indicated in Figs. 2-9a to 2-9d. Then in Fig. 2-9e we show how these stack together in real space and indeed fill all of space, something not obvious from the Wigner-Seitz cell for this lattice as shown in Fig. 2-9d. The resulting Wigner-Seitz cell for cubic F-lattice (face-centered cubic lattice) is shown in Fig. 2-9f. Looking at Fig. 2-8a will convince you that the result shown in

CHAPTER 6   CLASSIFICATION OF SOLIDS   107

Fig. 2-9f is correct. These, as all Wigner-Seitz cells, only contain one lattice point so they are primitive.

6-1e Definitions of directions, coordinates, and planes The terminology of **directions**, [m_1 m_2 m_3] and ⟨m_1 m_2 m_3⟩, in lattices and unit cells is discussed in Section 2-7 very briefly so it need not be reviewed here. The terminology for **coordinates** and **planes**, (hkℓ) and {hkℓ}, also is discussed in that section.

6-1f 32 crystallographic point groups and 230 space groups So far we have not considered the addition of atoms to our unit cells and lattices. We could continue without atoms and discuss the 32 crystallographic point groups and then the 230 space groups. However, we shall not do this but rather we will introduce atoms and thus discuss crystal structures. Then, after crystal structures, the point groups and space groups will be mentioned in the context of crystal structures.

6-1g Crystal structure The best way to think of a crystal structure is in terms of the expression

$$\text{crystal structure} \equiv \text{lattice} + \text{basis} \tag{3-1}$$

This is not a mathematical equation but just a concise way to write a definition. Let us see what it means. For any lattice, P, F, I, C, or R, about every lattice point there is an identical arrangement of atoms. This arrangement of atoms about one lattice point is called a **basis** (or **lattice complex**). The expression Eq. 3-1 implies that if the basis is positioned with respect to one lattice point, then, by translational symmetry, Eq. 2-1 it is about every lattice point. Certainly this means about all the lattice points within a multiply primitive unit cell as well as all of the lattice points in all of the unit cells.

Before discussing examples of some of the simple crystal structures encountered by a solid state scientist, first note that Eq. 3-1 defines all possible crystals and also gives another way to appreciate translational symmetry. If a point in the unit cell at the origin ($t_m = 0$) is given by **r**, then the atomic arrangement and the properties of the crystal are identical, in every way, at the point **r**′ given by

$$\mathbf{r}' = \mathbf{r} + \mathbf{t}_m \tag{1-10b}$$

sc, bcc, and fcc crystal structures – Many people have difficulties with these structures because they are so simple. By using the definition of a crystal structure (Eq. 3-1) we will take each one in turn and discuss them. For example, it will become clear that only elements can have these structures because the basis consists of only one atom.

The element polonium, Po, has the **simple cubic (sc) crystal structure**. That means that the lattice is a primitive (P) lattice with the cubic crystal system and the basis is one Po atom at the (0, 0, 0) position, that is, on the lattice point. Figure 3-1a shows a unit cell of the structure, which looks just like the cubic P-lattice (Fig. 2-7) but it is not because in Fig. 3-1a there are atoms and in Fig. 2-7 there are just mathematical points. In this structure every atom has six nearest neighbors (nn) and 12 next-nearest neighbors (nnn) (see Fig. 2-5a to check these statements).

The element Na has the **body-centered cubic (bcc) crystal structure.** That means that the lattice is a body-centered (I) lattice with the cubic crystal system and the basis is one Na atom at the (0, 0, 0) position. Naturally, if we are using the conventional multiply primitive unit cell for the cubic I-lattice (Fig. 2-7), then another Na atom goes at the (½, ½, ½) or body-centered position. This crystal structure is shown in Fig. 3-3 if we focus only on the very large spheres. In this structure each Na atom has 8 nn atoms; see Fig. 2-5b.

The element Cu has the **face-centered cubic (fcc) crystal structure.** This means that the lattice is a face-centered (F) lattice with the cubic crystal structure and the basis is one Cu atom at (0, 0, 0). Using the conventional multiply primitive unit cell for the cubic F-lattice from Fig. 2-7, this structure is shown in Fig. 3-5a. Then in Fig. 3-8 the same crystal structure is shown from many different points of view. (Each Cu atom has 12 nn atoms.)

At atmospheric pressure only one element has the sc crystal structure. Most have the bcc, fcc or hcp crystal structure. (The latter is discussed later but also see Fig. 3-9.) Table 3-1 lists the elements that have these latter three structures.

CsCl and BaTiO$_3$ crystal structure – Both of these materials are in the cubic crystal system with a primitive (P) lattice. This is the same as the element Po discussed previously but for these materials the basis is more complicated.

For CsCl the basis is Cs at (0, 0, 0) and Cl at (½, ½, ½), or vice versa. The crystal structure is shown in Fig. 3-1b and is called the **CsCl structure.** Many other compounds have this structure, including CsBr, CsI, NH$_4$Cl (if one ignores the hydrogens), AgCe, etc. (Each Cs has 8 nn Cl atoms and each Cl has 8 nn Cs atoms.) Note the fundamental difference between this crystal structure and the bcc crystal structure. Both are cubic, but in the former there are two different kinds of atoms in one primitive unit cell, while in the latter there are two of the same kind of atom in the multiply primitive cell.

For BaTiO$_3$ the basis consists of five atoms, one Ba, one Ti and three O atoms; their positions are listed in Eq. 3-2, and the structure is shown in Fig. 3-1c. There are many compounds of this ABO$_3$ type that have this so-called **perovskite crystal structure,** for example, PbTiO$_3$, KNbO$_3$, KTaO$_3$, LaAlO$_3$, etc. (Each Ti has six nearest-neighbor O atoms, and the Ba atoms have 12 nearest-neighbor O atoms.)

Naturally, for all of these crystal structures we are just showing one primitive or multiply primitive unit cell. We must take this unit cell and by translation (Eq. 1-10a or Eq. 2-1) form the entire structure.

NaCl and CaF$_2$ crystal structures – Both of these structures have a face-centered cubic lattice and are simple to understand if the basic definition Eq. 3-1 is kept in mind.

For NaCl we have a cubic crystal system and a face centered (F) lattice, as in Fig. 2-7, and the basis is an Na atom at (0, 0, 0) and a Cl atom at (½, 0, 0) or vice versa. By remembering that the basis gets translated to every lattice point, it is clear that we obtain the crystal structure shown in Fig. 3-5b, where the conventional multiply primitive unit cell is shown. (Every Na has 6 nn Cl atoms and vice versa. We say that the Na atoms are **octrahedrally coordinated** with Cl atoms.) Many alkali halides have this NaCl structure as do MgO, EuO, AgBr, HfC, ZrB, KH, CeN, BaS, AsCe, and many other compounds with this general formula.

CaF$_2$ has the same crystal system and lattice as the NaCl crystal structure but the basis is

$$\text{Ca: } (0, 0, 0) \qquad 2\text{F: } \pm(\tfrac{1}{4}, \tfrac{1}{4}, \tfrac{1}{4})$$

The structure is shown in Fig. 3-5c and the similarities and differences between the CaF$_2$ and NaCl structures can be seen. A representative few compounds that have this structure are EuF$_2$, CeO$_2$, CeH$_2$, CeH$_2$, K$_2$S, Be$_2$B, and Al$_2$Au. (Each Ca has 8 nn F atoms and each F has 4 nn Ca atoms.)

Zinc blende and diamond crystal structures – As for the NaCl and CaF$_2$ structures, the two structures discussed here also have a cubic face-centered lattice, that is, a cubic F-lattice. However, for zinc blende (typified by GaAs) and diamond (typified by diamond or Si) the basis is

$$\text{zinc blende: } \text{Ga: } (0, 0, 0) \qquad \text{As: } (\tfrac{1}{4}, \tfrac{1}{4}, \tfrac{1}{4})$$
$$\text{diamond: } \quad 2\text{C: } (0, 0, 0); \ (\tfrac{1}{4}, \tfrac{1}{4}, \tfrac{1}{4})$$

The structures are shown in Fig. 3-7. They are similar to each other in many ways except the one is made up with different atoms in the two sites. This has fundamental consequences; for example, the (⅛, ⅛, ⅛) point is a center of symmetry in the diamond structure, while it is not in the zinc blende structure, since an interchange of Ga and As atom does not leave the structure in an equivalent position. (Thus, crystals with the zinc blende structure are piezoelectric, while crystals with the diamond structure are not, Section 5-4e.) However, in both structures every atom has 4 nn and are said to be **tetrahedrally coordinated**; in the zinc blende structure they are different atoms, while in the diamond structure they are the same atoms.

Four elements have the diamond structure: C, Si, Ge, and Sn (in the so-called α-Sn phase). The **tetrahedral bonding** to the four-nearest neighbors is of great importance in determining the properties of these crystals; this will be discussed in Chapter 8.

Tetrahedral bonding also is fundamental to compounds with the zinc blende structure. There are a great many binary compounds that have this structure. A representative list is GaAs, InSb, CdSe, BeS, ZnO, AgI, AsB, etc.

Hexagonal close-packed (hcp) crystal structure — We can treat the hcp structure as a hexagonal lattice with a basis at (0, 0, 0) and (2/3, 1/3, 1/2). The structure is pictured in Fig. 3-9d, and Table 3-1 lists the elements with this structure.

From a fast look at this structure in Fig. 3-9 we might conclude that each atom has six-nearest neighbors. However, this is not true because the atoms in (0 0 ½) plane are not shown. For the **ideal c/a ratio** of $(8/3)^{1/2} = 1.633$ the six nn atoms in the plane are equidistant from the three above and three below, giving a total of 12 nn. This is the same number as in the fcc structure. In fact, there is a very close relationship between these two structures in the sense that either can be obtained for the closest packing of hard spheres (billiard balls). How we arrive at these and related polytype structures from the point of view of close packing is discussed in Section 3-7.

In a consideration related to close packing, we can fill some of the simple structures with hard spheres that touch each other and determine the proportion of space taken up by the spheres. The result is given in Eq. 3-4. The fcc and hcp structures are the most efficient for filling space and are called **close packed**, while the diamond structure is very open. These results are related to the 12- and 4 nn found respectively, in these structures. If we look for phase transitions under high pressures in materials with simple structures then typically materials transform to structures with more nearest neighbors and higher packing densities. For example, RbI at $\approx$5 kbars transforms from NaCl to the CsCl structure. See Section 3-8.

Wurtzite crystal structure — This last simple structure that we discuss is shown in Fig. 3-10; the basis in this hexagonal lattice is given in Section 3-9.

As in the zinc blende structure, in the wurtzite structure each atom is tetrahedrally bonded to four dissimilar atoms, and the close relationship between these structures is discussed in Section 3-9.

CHAPTER 6   CLASSIFICATION OF SOLIDS   111

6-1h Problems To help sharpen your understanding of the material discussed here, the following problems are recommended: Chapter 2, Problems 1 and 6; Chapter 3, Problems 3, 4, and 11a.

6-1i Point groups, space groups, and other terms To give the interested reader some knowledge of these closely related topics we summarize the meaning of these terms.

The **point symmetry** or **point group** of a molecule is the set of symmetry operations taken with respect to a fixed point. The point is chosen to maximize the number of symmetry operations. If only point symmetry operations that are compatible with translational symmetry (Eq. 2-1) are allowed (rotations by 60°, 90°, 120°, 180°, and multiples of these as discussed in the Appendix to Chapter 2), then one is led to the **32 crystallographic point groups**.

The complete way to describe the symmetry operations of a crystal structure is in terms of a space group. The space group symmetry operations include the translations (Eq. 2-1) the point symmetry operations discussed previously, as well as screw operations and glide plane operations mentioned in the following discussion. There are **230 space groups** that describe the symmetry operations of all possible crystal structures (in three dimensions). These space groups can be made up, and thus classified, in two ways. An arrangement of atoms that has the symmetry of the 32 crystallographic point groups can be taken as a basis along with each of the 14 Bravais lattices. This basis plus the lattice form a crystal structure, provided that the symmetry operations of the basis are compatible with those of the lattice (see Section 2-6). This leads to 73 **symmorphic space groups**, i.e., space groups in which only translation symmetry (Eq. 2-1) and the point operations are *necessary* to describe the symmetry operations of the space group. (They are listed in Table 2-2.) There are also 157 **nonsymmorphic space groups**. These are made up of translation symmetry, usually some point group operations, but require *at least* one screw operation or glide plane to describe the space group. As the name implies, a **screw operation** is a rotation followed by a translation a *fraction* of a unit cell (for example, a rotation about the c-axis followed by a translation $c/2$, or $c/4$, etc.). A **glide plane** operation is a mirror plane operation followed by a translation of a *fraction* of a unit cell (typically parallel to the mirror plane). See the discussion in Section 1-4.

For space groups it is useful to define the **point group of a space group**, which is usually just referred to as "the point group" (of the crystal structure). If all of the translations of the space group are set equal to zero (t_m as well as the fractions of the unit cell lengths that are used in screw operations or glide planes), then the operations are identical to one of the 32 point groups. This point group is said to be the point group of the space group. (This concept is important for several reasons. The most obvious is that the point group of the space group determines the symmetry conditions of the tensors that describe the macroscope properties of crystals, that is, the anisotropy of these properties. See Chapter 5.)

The last point in this review mentions the **convention** used in this book to label the symmetry operations, point groups, and space groups. There are two popular notations: the **Schoenflies** and the **International** (or **Hermann-Mauguin**) **notation**. Rather than pick one or the other, we have written, for pedagogical purposes, the Schoenflies symbol, immediately followed by the International symbol in parentheses. Both the metal Cu and for the CaF_2 crystal structures, discussed previously, have the O_h^5 or Fm3m space group in these two notations, respectively. We write the space group as O_h^5(Fm3m), and the point group as O_h(m3m). Thus, you may use whichever you prefer or you may become familiar with both; each has its strong points. However, throughout the text both symbols can be ignored with little loss.

6-2 Introduction to Classification of Solids

In the early chapters we showed how crystalline solids can be classified according to their symmetry. This is a rigorous classification and it determines, for example, the interrelationship of the properties of some physical characteristics in different directions (Neumann's principle) and other fundamental aspects of their behavior. In this chapter we introduce a different, very important way to classify crystals. This classification is according to the type of bonding. Simple crystalline solids can be classified according to five types of bonding: ionic, covalent, metallic, molecular, and hydrogen bonding. There are enormous differences in physical and chemical properties of solids that have different types of bonding, and these differences will be discussed.

In this chapter we cover broadly some of the basic and simple aspects of the classification according to bonding and also discuss the molecular and hydrogen bond. In three subsequent chapters we go into more detail for the ionic, covalent, and metallic bond and properties of materials that are bonded in these ways.

6-3 Five Types of Bonds

Table 6-1 (page 102) lists the five types of bonds, gives examples, lists typical binding energies, and shows other characteristics. These characteristics are determined by the distribution of electrons. Essen-

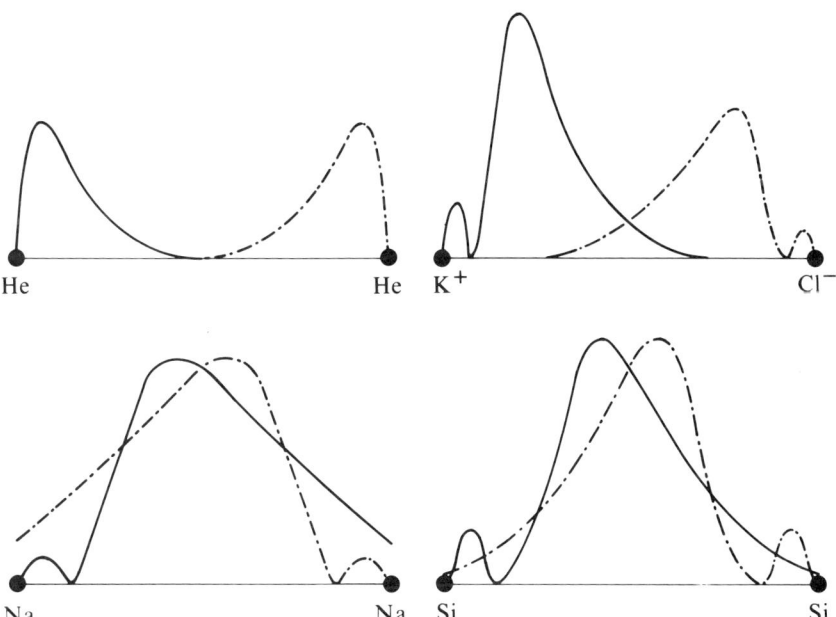

Fig. 6-1 The approximate charge densities along the internuclear axes for crystals that have different types of bonding. The various atoms are placed at the nearest-neighbor distance in the corresponding crystal structures. The charge densities are calculated from atomic wave functions for all the electrons in He, the outer shells for K^+Cl^-, the 3s-electron for Na, and the four valence electrons for Si.

tially all of the static forces that hold together atoms in solids arise from the electrostatic interactions (i.e., Coulomb force law, $e_1 e_2/r^2$). The Pauli principle, by allowing only two electrons in each spatial state, has important implications as to where the electrons may be, but the forces are electrostatic. (See Problem 1).

On a microscopic scale the different bonding types can be characterized by different distributions of electrons around the atoms. For example, in both ionically bonded potassium chloride (KCl) and molecularly bonded helium (He) the electron distribution around the atoms is *approximately spherical*. However, in KCl the atoms are charged ions (i.e., K^+Cl^-), while in He they are neutral atoms. In fact, the interaction between charged, spherical closed-shell ions is just the characteristic of the ionic bond. Molecular bonds are found in solids composed of noble atoms (He, Ne, Ar, Kr, Xe), simple molecules (CH_4, C_6H_6, ferrocene, etc.), as well as in solids composed of more complicated molecules. For both of these types of bonds there is very little electron charge in the region between the atoms or molecules. Figure 6-1 shows the charge distribution for molecularly bonded He and ionically bonded KCl. The model that we use to calculate the binding energy for ionic solids will be just the interaction, via Coulomb's law, of charged balls. For molecular bonds, as we shall see, the effects of Coulomb's law enter in a more subtle manner.

On the other hand, in the metallic bond there is considerable overlap of the electron distribution of the neighboring atoms. Figure 6-1 shows the overlap for metallically bonded Na. For the metallic bond we often say "the ions (i.e., the atomic cores, which for Na are the $1s^2 2s^2 p^6$ electrons) are imbedded in a uniform sea of free electrons." In fact, the simple theory of metals treats the electrons as totally free electrons and often ignores the cores. This approximation enables some of the most striking characteristics of metals to be understood. However, consideration of the translational symmetry of the crystal has profound effects on many properties, as we shall see.

For covalent bonding there also is a large amount of overlap of the electronic charge between the atoms (Fig. 6-1), but it is *directed* in specific directions in which there are neighboring atoms. By directed bonds we mean that there are important angular-dependent forces as well as the usual radially-dependent forces. This can be understood in terms of what are called hybridized orbitals. With this concept we may talk about a *chemical bond*, due to electron sharing, between neighboring atoms. However, the distinction between covalent and metallic bonding is unclear if we only consider the electron distribution in real space since there is considerable overlap for both of these bonds. The distinction can be made rigorous in reciprocal space. This will be discussed in connection with the band theory of solids in Chapter 10.

Figure 6-2 shows schematically the electronic charge distribution for four of the five types of bonds and summarizes what has been said. From careful x-ray diffraction data it has been determined that the electron density gets below 0.1 electron/$Å^3$ along the nearest-neighbor line in NaCl [G. Schoknecht, Z. Naturforschung **12**, 983 (1957)] while it remains above 5 electrons/$Å^3$ in diamond. So, in these two exceptionally good examples, the enormous differences in the electron density for the different types of bonds can be seen. The last type, the hydrogen bond, is rather special. It is best described as a sharing of a hydrogen ion between two neighboring atoms. This is possible because of the special property of hydrogen. A hydrogen ion, being a bare proton, is extremely small on the atomic scale.

We treat the ionic, covalent, and metal bonds in the next three chapters. In this chapter some of basics of the molecular and hydrogen bonds are discussed. However, first we treat the repulsive energy, which occurs between all charged or uncharged closed-shell atoms or ions.

CHAPTER 6   CLASSIFICATION OF SOLIDS   115

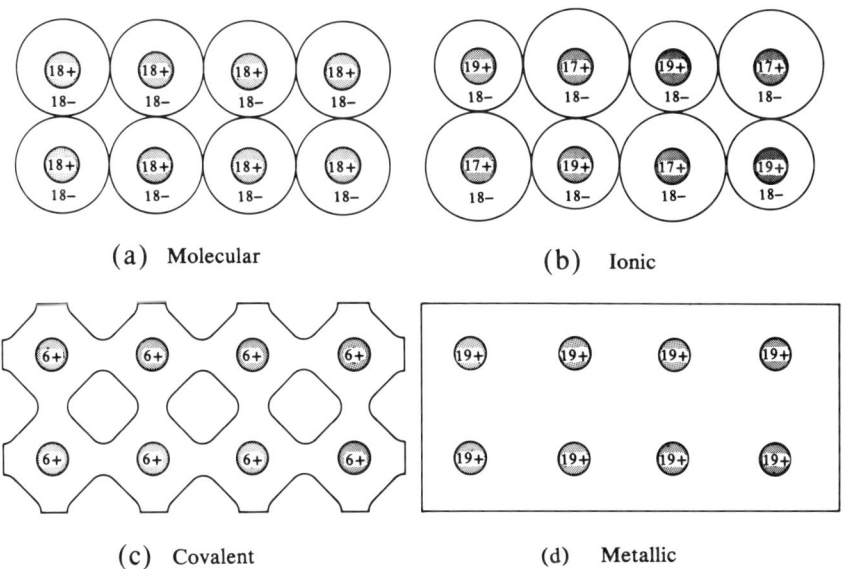

Fig. 6-2 Schematic two-dimensional representations of the electronic charge distributions in the basic solid types. The small circles represent the positively charged nuclei. The white, enclosed areas represent regions in which the electronic density is appreciable (but not uniform). The types are: (a) molecular (represented by two-dimensional argon); (b) ionic (potassium chloride); (c) covalent (carbon); (d) metallic (potassium). (After Ashcroft and Mermin.)

6-4 Repulsive Potential Energy

When two closed-shell atoms, ions, or molecules are brought together, a strong repulsion occurs. The forms of attraction between atoms, ions, or molecules are different (Coulomb's law for ions, van der Waal's forces for molecules, covalent bonding for atoms, although all these are fundamentally electrostatic in origin), but the repulsion is similar. In this section we perform a simple calculation to determine the form of the repulsive potential energy.

As the simplest type of calculation that enables us to determine some of the essentials, consider the repulsion of a proton from a hydrogen atom. Figure 6-3 shows the electron at a distance r_e from its nucleus. A proton is at a distance d from the nucleus, and r_{ep} is the distance between the electron and the proton. We treat the extra energy due to the proton as a perturbation, that is, assume that the hydrogen atom wave function is not affected to first order.

First, recall an important basic result from electrostatics. Take the origin at the nucleus of the hydrogen atom (which is on the left). Expand the potential of the proton of charge +e about this origin.

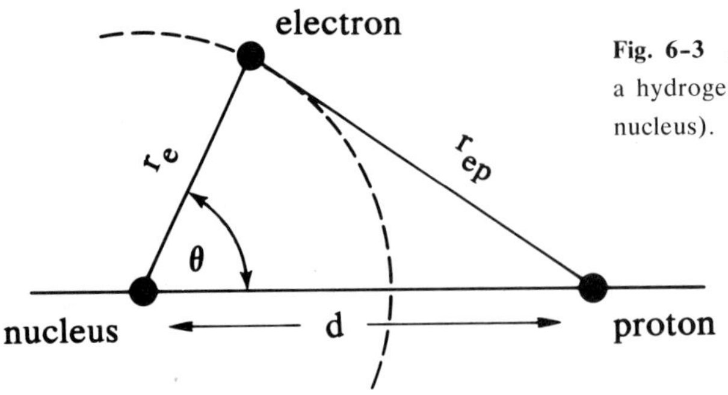

Fig. 6-3 A proton penetrating a hydrogen atom (electron and nucleus).

(We use this origin because the wave function of the electron is known with respect to this origin.) The result, evaluated at a distance r_e from the origin where θ is the angle between the vector r_e and d, is

$$\text{Pot.} = e\left[\frac{1}{d} + \frac{r_e \cos\theta}{d^2} + \frac{r_e^2}{d^3}\left(\frac{3\cos^2\theta - 1}{2}\right)\right.$$

$$\left. + \frac{r_e^3}{d^4}P_3(\cos\theta) + \ldots\right]_{r_e < d}$$

$$\text{Pot.} = e\left[\frac{1}{r_e} + \frac{d\cos\theta}{r_e^2} + \frac{d^2}{r_e^3}P_2(\cos\theta) + \ldots\right]_{r_e > d} \quad (6\text{-}1)$$

where $P_n(\cos\theta)$ is the nth Legendre polynomial. There are two expansions, one when $r_e < d$ and the other when $r_e > d$. These are called the outside and inside expansions, and they give identical results when $r_e = d$. These expansions consist of monopole, dipole, quadrupole, and other terms.

Now return to our original repulsive potential energy problem. The potential energy of a hydrogen atom interacting with the proton is

$$U = \frac{e^2}{d} - \frac{e^2}{r_{ep}} = e^2\left[\frac{1}{d} - \frac{1}{\sqrt{d^2 + r_e^2 - 2dr_e\cos\theta}}\right] \quad (6\text{-}2)$$

The first term comes from the interaction of the external proton with the nucleus of the hydrogen atom (which is also a proton), and the second term is the interaction of the external proton with the electron. We introduce more convenient variables, r and R, then expand U in powers of r/R, which is equivalent to an "inside expansion" (r < R) and an "outside expansion" (r > R). The variables are defined as

$$r \equiv 2r_e/a_0, \quad R \equiv 2d/a_0 \tag{6-3}$$

$$U = -\frac{2e^2}{a_0}\left[\frac{r\cos\theta}{R^2} + \frac{r^2}{2R^3}(3\cos^2\theta - 1) + \ldots\right] \text{ for } r < R \tag{6-4a}$$

$$U = -\frac{2e^2}{a_0}\left[(\frac{1}{r} - \frac{1}{R}) + \frac{R\cos\theta}{r^2} \right.$$

$$\left. + \frac{R^2}{2r^3}(3\cos^2\theta - 1) + \ldots\right] \text{ for } r > R \tag{6-4b}$$

Notice that in Eq. 6-4a the monopole term from the potential energy interaction of the electron and proton is just cancelled by that of the nucleus and proton, while in Eq. 6-4b it is not. The Bohr radius a_0 (=0.529 Å) is used because it is convenient in the perturbation calculation. The two expressions in Eq. 6-4 are identical at $r = R$.

The perturbation energy is given by $U_{rep} = \langle\psi|U|\psi\rangle$ where we take the 1s state of hydrogen as the unperturbed wave function, that is, $\psi = (1/\pi a_0^3)^{1/2}\exp(-r/2)$. We use Eq. 6-4a for the integration from zero to $r_e = d$ and Eq. 6-4b for the integration from r_e equals d to ∞. Upon integration, terms containing angular parts ($\cos\theta$, $3\cos^2\theta - 1$, etc.) give zero; the only nonzero term is

$$U_{rep} = -\frac{e^2}{a_0}\int_{r=R}^{\infty}(\frac{1}{r} - \frac{1}{R})\exp(-r)\, r^2\, dr$$

$$= 2\mathscr{E}_0(1 + a_0/d)\exp(-2d/a_0) \tag{6-5}$$

where $\mathscr{E}_0$ is the energy of the ground state of hydrogen. This result, for the repulsive potential energy, varies very rapidly with internuclear distance and this is a fundamental general characteristic of repulsion. Other, more sophisticated calculations give the same exponential form *because of the exponential form of the charge density of the electron distribution on the atom.*

To model the exponential repulsion of the type in Eq. 6-5, Born and Mayer proposed $U_{rep} = b\exp(-d/\rho)$, where $\rho = 0.345$ Å was found to give good fits for alkali halide crystals as discussed in the next chapter. The constant b is usually obtained by empirically fitting compressibility data.

In the calculations that follow, instead of an exponential form for the repulsive potential energy, we use a power law $U_{rep} = bd^{-n}$ with $n \sim 10$. Like the exponential form, this form varies very rapidly with d (for large n) and it is slightly easier to use. (The important point to model is the very rapid dependence on the internuclear distance.) For ionic bonding we shall see that the repulsive energy contributes only about 10% of the total binding energy of the system, and thus the details of its form are not important.

The physical reason for the repulsion of two closed-shell atoms at distances where their charge clouds overlap is the **Pauli exclusion principle**. This principle states that two electrons cannot have all the same quantum numbers. Thus, when two atoms come close enough so that their electron clouds begin to overlap, the electrons from the first atom start to occupy some of the states of the second atom and vice versa. However, since we have closed-shell atoms, the ground states are already occupied and the Pauli principle prevents further occupancy. For the charge distribution to overlap, some of the electrons must be promoted to higher lying states that are unoccupied. Thus, the overlap of two closed-shell atoms rapidly increases the total energy of the system, which is a repulsive term in the interaction energy.

6-5 Molecular Bond

6-5a Simple solids All of the rare gas elements (column VIII of the periodic table), except He, have a face-centered cubic (fcc) crystal structure. The melting and boiling points of these elements are very low, which implies very weak attractive forces. All of these atoms have closed-shell electron configurations, and in the solid state there is very little overlap between the neighboring atoms.

The type of repulsion for these closed-shell atoms was discussed previously. The very weak attractive forces that hold the solid together are the so-called **van der Waals, London–dispersion**, or **fluctuating dipole forces**. Qualitatively it is easy to understand these forces. Consider two atoms, 1 and 2, separated by a distance d. Both atoms have a spherical, rare gas configuration and thus have no average dipole moment. However, there is an *instantaneous* dipole moment **p**, which causes an electric field of the order of p_1/d^3. This electric field will instantaneously induce in atom 2 a dipole moment given by

$$p_2 = \alpha E \approx \alpha p_1/d^3 \qquad (6\text{-}6)$$

where α is the polarizability of atom 2. The dipole moment on atom 1 interacts with the induced dipole moment on atom 2 to give an attractive interaction energy U_{att} proportional to the products of the dipole moments and inversely proportional to the distance cubed. Thus,

$$U_{att} = -\mathbf{p}_2 \cdot \mathbf{E} \approx -p_1 p_2/d^3 \approx -\alpha p_1^2/d^6 \qquad (6\text{-}7)$$

which is the famous inverse sixth power van der Waals potential energy of attraction. Note that this attraction depends on p_1^2 and although the time average of $\mathbf{p}_1$ vanishes, the time average of the square

CHAPTER 6   CLASSIFICATION OF SOLIDS   119

Table 6-2 The van der Waals radii for some of the elements. (The values are in Å.)

				H	1.2		
N	1.5	O	1.4	F	1.35	Ne	1.6
P	1.9	S	1.85	Cl	1.8	Ar	1.9
As	2.0	Se	2.0	Br	1.95	Kr	2.0
Sb	2.2	Te	2.2	I	2.15	Xe	2.2

does not. However, an evaluation of p_1^2 is difficult. For a simple calculation see Problem 5, where this quantity is related to the zero point vibrational frequency. (This was first done quantum mechanically by London.) For hydrogen atoms calculations yield

$$U_{att} = \left(\frac{6e^2}{a_0}\right)\left[\left(\frac{a_0}{d}\right)^6 + 22.5\left(\frac{a_0}{d}\right)^8 + 236\left(\frac{a_0}{d}\right)^{10} + ...\right] \quad (6\text{-}8)$$

where the second and third terms come from higher order terms in the multipole expansion, (i.e., since the extent of the dipole moment is of the order of internuclear distance, higher order terms need be considered). However, the second term in Eq. 6-8 is only 10% of the first term at the appropriate values of d.

The equilibrium internuclear distance in molecular solids is determined by the distance for which the van der Waals attractive force is just equal to the repulsive force. Values of van der Waals radii can be deduced and are given in Table 6-2. For the rare gases these values are determined from their spacing in the crystalline state. For many of the other atoms the radii are found from the appropriate molecular structure by determining the distance of closest approach between adjacent molecularly bonded molecules. It will be seen later that these radii are much larger than the corresponding covalent radii for the same element. This is expected since covalent bonding, which involves a sharing of electrons between two adjacent atoms, is a very strong bond and the atoms are pulled together.

The form of the van der Waals attraction in Eq. 6-7 is the same as that used to correct the equation of state of a perfect gas, $PV = RT$, where R is the gas constant. The more accurate equation of state, $[P + (\alpha/V^2)](V - \beta) = RT$, is called the **van der Waals equation.** The V term in the perfect gas equation is replaced by the "free-space volume," $V - \beta$, to correct for the fact that the externally measured volume is too large because the molecules take up some space. From straightforward geometrical considerations $\beta \approx (8/2)N\, v_m$ where N is the number of molecules and v_m is the volume of a single molecule. P is replaced by $[P + (\alpha/V^2)]$ because the pressure that is measured at the wall, P, is just the average pressure as seen by the wall due to the bombardment of walls by the molecules. The pressure seen by two molecules as they race toward one another to collide is augmented by the van der Waals attractive force, Eq. 6-7. Thus, the pressure that the molecules see is larger than P by an amount $\propto 1/\bar{d}^6$, where $\bar{d}$ is an average distance between molecules. Since between the molecules $V \propto \bar{d}^3$ the pressure term is $(P + \alpha/V^2)$ where α is a constant.

6-5b Simple structures We have already mentioned that the rare gases that solidify at atmospheric pressure form a cubic close-packed structure, that is, the fcc structure.

The structures of some simple diatomic molecules such as N_2, O_2, CO, Cl_2, Br_2, and I_2 provide interesting examples of several effects. First, in the vapor phase these are all diatomic molecules bound together by a strong covalent bond. Second, this covalent bond remains in the solid state and the bonding between the molecular entities is of the weak van der Waals type; hence we observe low boiling and melting temperatures of the solid. Last, just below the melting points the weak intermolecular bonding in the solid can allow the molecules to rotate freely. At lower temperatures there is a phase transition to a lower symmetry structure in which the molecules are "frozen in" in a more normal crystal structure.

For example, upon solidification we observe the β-N_2 form in which the centers of mass of the N_2 molecules form a hexagonal close-packed structure, but the molecules are freely rotating. Then at 35°K a phase transition occurs to the γ-phase, with the nitrogens at the 8c positions of the cubic T_h^6(Pa3) space group. (See the International Tables.) In this phase there is no free rotation; however, the centers of mass of each N_2-molecular entity are still at very highly symmetric positions. CO is found to be isomorphic to N_2; the γ to β transition occurs at 61°K. O_2 is somewhat different; its high temperature γ-form is isomorphic to the low temperature α-form of N_2. At lower temperatures there are two other phase transitions to orthorhombic crystal structures.

The crystal structure of Cl_2, Br_2, and I_2 is pictured in Fig. 6-4. The atoms are at the 8f position of the orthorhombic D_{2h}^{18}(Bmab) space group. However, as can be seen in the figure, the centers of mass of the molecule are at the higher symmetry face-centered positions of this orthorhombic crystal system. Again we point out that there is a strong covalent bond in the I_2 molecule and that these molecules are weakly bonded to each other by van der Waals forces. It is from the distances between molecules in such structures that one obtains van der Waals radii for Cl, Br, and I as given in Table 6-2. At rather high pressures, a very unusual phenomenon is observed in crystalline iodine. **Molecular dissociation** is observed above 210 kbar

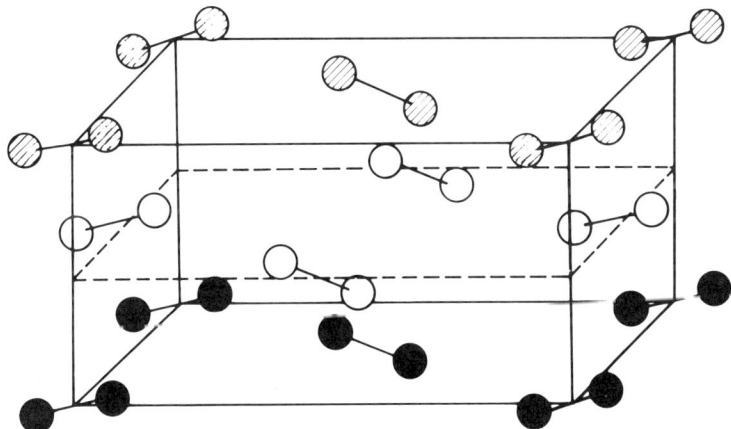

Fig. 6-4 The crystal structure of chlorine, bromine, and iodine. The line between the two atoms represents the strong covalent bond present in the gas as well as solid phase. The diatomic molecules are bonded by van der Waals forces with each other.

when there is a phase transition to a body-centered orthorhombic crystal structure, D_{2h}^{25}(Immm). In this structure all of the I-atoms are equally spaced; thus, the molecular bond is broken.

6-5c Packing of molecules We can consider the quite general problem, What are the structural possibilities for molecular crystals that contain arbitrarily shaped molecules? The basic principle used in these considerations is that the molecular crystals will pack so that the maximum amount of space is used. That is, the protrusions of one molecule will pack into the hollows of the next, or, the packing is determined by the **steric forces,** the strong hard-core closed-shell repulsive forces; constant force contours follow the shape of the molecule. Most complicated, large molecules have rather odd shapes. From these considerations it can be shown that the crystal structures of arbitrarily shaped molecules will adopt one of 13 out of the 230 space groups. This is a most remarkable result and reference to Kitaigorodskii's book, mentioned in the Notes, should be made. We will examine the two-dimensional problems to get some ideas as to why this occurs.

Assume that we are packing two-dimensional, arbitrarily shaped molecules using the preceding principle. Figure 6-5a shows that a rectangular lattice is unsuitable for arbitrarily shaped molecules because the repulsive forces will not allow overlap. Figure 6-5b shows another structure that is unsuitable because the voids are too large. However, Figs. 6-5c and 6-5d show two structures that have a high

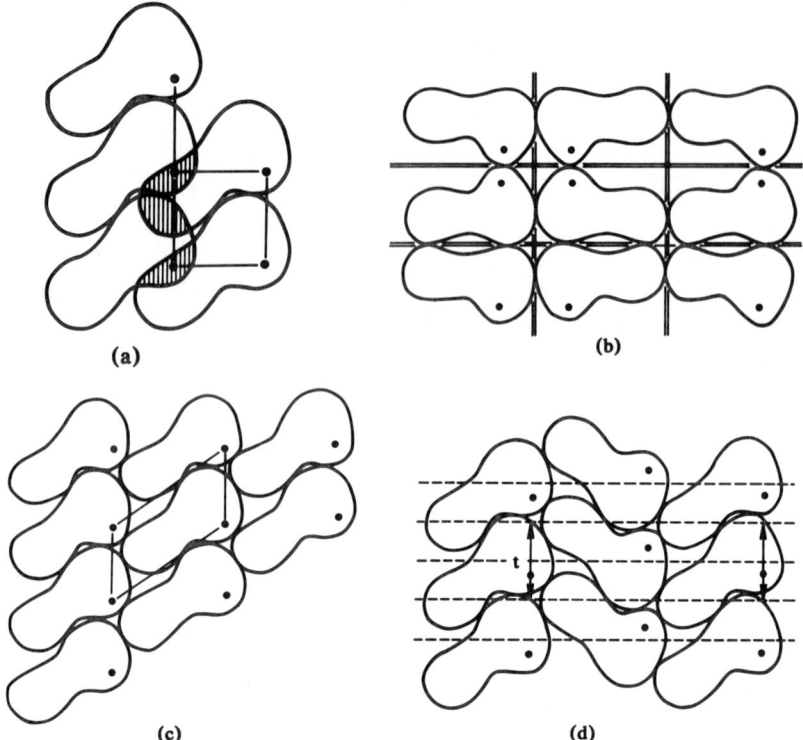

Fig. 6-5 (a) The unsuitability of a tetragonal cell for packing figures of arbitrary shape. (b) Packing of shapes but with too much open space. (c) and (d) Two suitable ways to pack arbitrary shapes. Note the screw axes in the latter.

packing density. In both of these planar structures the mean coordination number of any molecule is six (while in Fig. 6-5b it is four).

For the three-dimensional structures we find a mean coordination number of 12. These two- and three-dimensional results are, of course, just what is found for close packing of spheres, ("billiard balls") Section 3-7. Thus, it appears that when the packing is determined only by steric forces, a coordination number of 12 is obtained (six in two dimensions). For spherical atoms the high symmetry fcc and hcp structures are obtained, while for arbitrarily shaped molecules, crystal structures that are orthorhombic, monoclinic, or triclinic are predicted, depending on the point symmetry of the molecule.

6-5d Liquid crystals – plastic crystals Some molecular crystals exhibit interesting behavior between their normal crystalline phase observed at very low temperatures and their liquid phase. Normally as a molecular crystal is heated up, at some temperature (the melting point, M.P.) the periodic, regular symmetry associated with the crystalline state disappears. That is, one loses the translational order and the rotational order as the crystal melts. However, there are at least two other possibilities that are observed.

For molecular crystals in which the molecules tend toward being spherical, (for example, see Fig. 8-7) we may see a phase transition between 0°K and the M.P. At this phase change the translational symmetry is kept but the molecules either rotate freely or take on one of many possible orientational positions. This phase is known as the **plastic phase** and is characterized by high entropy, high symmetry crystal structures (because of the averaging), and physical softness. (See the Notes.)

For molecular crystals that tend toward irregular cigar shape we may observe a phase transition between 0°K and the M.P. at which the translational symmetry is lost but the orientational symmetry is kept. This phase is known as the **liquid crystal** phase and has some characteristics similar to the plastic crystal phase. Both of these phases represent "intermediate" states of matter.

6-5e Lennard-Jones potential As an example of the use of the idea of a steep repulsion and a van der Waals attraction between atoms, we use the Lennard-Jones potential to calculate the lattice constants and binding energies of the rare gases. The philosophy of this calculation is similar to that for ionic crystals discussed in the next chapter. The Lennard-Jones potential giving the potential energy between two neutral atoms is of the form $(\alpha/d^{12}) - (\beta/d^6)$ and is sometimes called the 6-12 potential. The pair potential energy is

$$U = 4U_0 \left[\left(\frac{\sigma}{r}\right)^{12} - \left(\frac{\sigma}{r}\right)^6 \right] \quad (6\text{-}9)$$

This semiempirical form has a steeply rising repulsive potential energy and a van der Waals attractive term. Table 6-3 gives the values of U_0 and σ for rare gas atoms, which are obtained from measurements of the second virial coefficients.

We can easily apply Eq. 6-9 to solids by neglecting the kinetic energy of the vibrating atoms and assuming that the total energy, U_t, is given by the potential energy in Eq. 6-9. Then for N atoms

$$U_t = \frac{N}{2} 4U_0 \left[A\left(\frac{\sigma}{d}\right)^{12} - B\left(\frac{\sigma}{d}\right)^6 \right] \quad (6\text{-}10)$$

The 1/2 comes from the sum over pairs of atoms. We have used $r = pd$ where d is the nearest-neighbor distance so $A = \Sigma' p_{ij}^{-12}$ and $B = \Sigma' p_{ij}^{-6}$, where the sums are from the ith atom to all the j-atoms (except the ith, which is indicated by the prime). These sums are

Table 6-3 Values of the nearest-neighbor distance d_0, melting and boiling points, U_0/k_B (k_B is the Boltzmann constant, which converts ergs to °K), and σ in the Lennard-Jones potential, Eq. 6-9. (At atmospheric pressure He does not solidify.) See N. Bernards, Phys. Rev. **112**, 1534 (1958) and Hirschfelder, Curtiss, and Bird.

	d_0 (Å)	M.P. (°K)	B.P. (°K)	U_0/k_B (°K)	σ (Å)	$2^{1/6}\sigma$ (Å)
He	—	—	4.2	10.0	2.56	2.79
Ne	3.13	24	27	34.9	2.78	3.03
Ar	3.76	84	87	119.8	3.40	3.71
Kr	4.01	117	120	171	3.60	3.93
Xe	4.35	161	165	221	4.10	4.47

rapidly converging and the values for the face-centered cubic, hexagonal close-packed, and body-centered cubic structures are:

fcc	A = 12.12188	B = 14.45392
hcp	12.13229	14.45489
bcc	9.11418	12.25330

Since the sum for A, in particular, is so very rapidly converging, A almost equals the number of nearest neighbors in the different structures, as can be seen. The equilibrium value of the internuclear distance, d_0, is found by setting the derivative of the total energy with respect to distance equal to zero. Thus,

$$\partial U_t/\partial d = 0 = -2NU_0\left[(12A\sigma^{12}/d^{13}) - (6B\sigma^6/d^7)\right] \quad (6\text{-}11)$$

from which $d_0/\sigma = (2A/B)^{1/6} = 1.090$ is obtained for an fcc structure. This is close to experimental, as can be seen in Table 6-3. By substituting d_0/σ into Eq. 6-10 we obtain

$$U_t = -8.60NU_0 \quad (6\text{-}12)$$

for rare gases with the fcc structure. Kinetic energy and quantum mechanical effects reduce this energy but only by 4% for the heaviest rare gas, although by larger amounts for the lighter ones. (See the Notes for the appropriate references.) Thus, we see how a simple force law between pairs of neutral atoms can be used to calculate the binding energy of atoms in a solid. The bond is found in O-H, F-H, N-H, and to a lesser extent in Cl-H.

6-6 Hydrogen Bond

If a sequence of related substances such as H_2Te, H_2Se, and H_2S are considered, the melting and boiling points decrease, as in Fig. 6-6. This is expected since, owing to the decreasing polarizabilities, the van

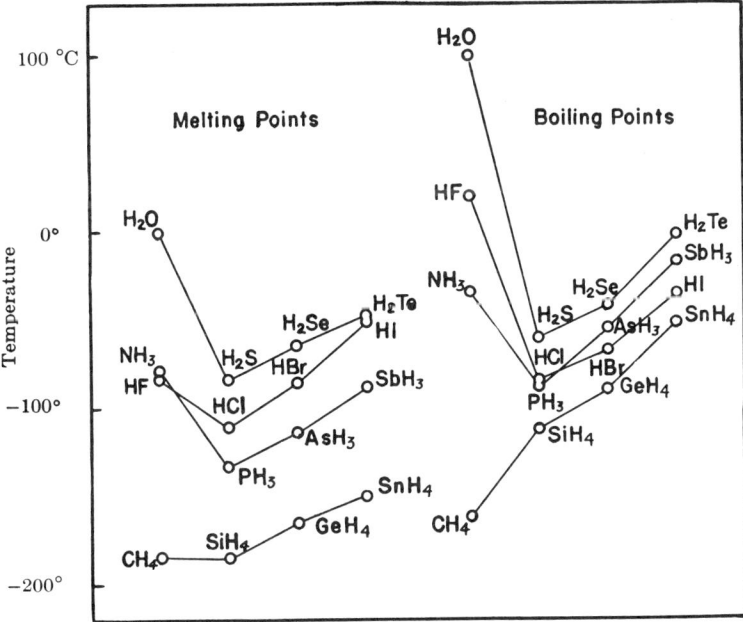

Fig. 6-6 The melting and boiling points of isoelectronic sequences of hydrides. (After Pauling.)

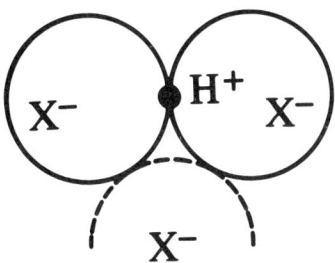

Fig. 6-7 A diagram showing that a bare hydrogen ion cannot be shared by more than two anions.

der Waals attraction becomes smaller going from Te to S. However, as can be seen in the figure, there is an abrupt change when the series is extended to H_2O. Similar effects can be seen for other series, as shown in Fig. 6-6, although this effect is not found in the SnH_4-CH_4 series. These large changes are due to the formation of hydrogen bonding, which takes place only with the most highly electronegative elements.

There are several characteristics that make hydrogen unique among the elements. First, a H^+ ion is just a bare proton (10^{-5} Å radius) and certainly cannot behave like the other alkali metal atoms. (Also the ionization potential of hydrogen, 13.59 eV, is about three times larger than that of the other alkali atoms). Second, the negative ion, H^-, is very weakly bound (0.75 eV) and has a very large radius. Last, hydrogen has one electron, so it can form a covalent bond with only one atom rather than with several, as is the case for most atoms that form covalent bonds. Thus, we expect that hydrogen might be unique in its bonding possibilities. The extremely small size of a bare proton enables it to bond with only two other atoms as indicated in Fig. 6-7.

The energy of the hydrogen bond, ≈ 0.4 eV, is approximately ten times room temperature thermal energies. This fact makes the study of the bond interesting because the bond is fundamental to life processes and also enables us to understand many simple organic materials, including water and ice.

The effects of hydrogen bonding can be seen in the crystal structures that are formed. For example, in crystalline H_2S and H_2Se each S or Se atom (or H_2S or H_2Se molecule) has 12 neighbors. (Each molecule is at the face-centered position of a cubic structure.) However, ice has quite a different structure. Each oxygen atom is tetrahedrally surrounded by four other oxygen atoms, at a distance of 2.76 Å. The structure is similar to that of wurtzite, Fig. 3-10. Is a given hydrogen atom midway between the two oxygen atoms or closer to one than another? The answer is that the hydrogen is closer to one. This is clear because in the gas phase the O-H distance is 0.96 Å, and the changes in properties between the vapor and solid are much too small to allow one to conclude that the distance increased to 1.38 Å.

One rather interesting property of ice is related to the preceding structure. At low temperatures it is found experimentally that appreciable amounts of entropy are retained. This can be understood in terms of the different possible configurations of a crystal of ice. Each H_2O molecule in the crystal can be formed in a number of ways since each hydrogen atom can be closely associated (spatially) with either one of the two oxygen atoms along the O-O axis. The residual entropy is given by $k_B \ln W$, where k_B is the Boltzmann constant and W is the number of possible configurations accessible to the crystal. We can calculate W. In one mole of ice there are 2N hydrogen nuclei and each has a choice of one of two positions. Thus, there are 2^{2N} configurations. However, many of these are ruled out because we demand that only two hydrogen atoms can be associated to any one oxygen atom, the **ice rule**. Consider a particular oxygen atom with four surrounding hydrogen atoms. There are 16 possible arrangements. One arrangement has all four hydrogens close to the oxygen corresponding to $(H_4O)^{2+}$, four arrangements correspond to $(H_3O)^+$, six to (H_2O), four to $(OH)^-$, and one to O^{2-}. We consider only the arrangements that obey the **ice rule**, which means $6/16 = 3/8$, of the total. Then $W = 2^{2N}(3/8)^N = (3/2)^N$. The residual entropy $= k_B \ln(3/2)^N = R \ln(3/2) = 0.806$ cal mole^{-1} °K^{-1} is very close to the experimental value of 0.82 cal mole^{-1} °K^{-1} (the gas constant $R = k_B N_A$, where N_A is Avogadro's number, given in the Notes.) Thus, we see some of the peculiarities and interesting aspects of hydrogen bonding.

Notes

Pauling's book presents many approaches to bonding, and the book should be studied. The chapter on hydrogen bonding is very good, and the effect of the dipole moments of some of the molecules shown in Fig. 6-6 is discussed there. Other detailed discussions can be found in G. C. Pimentel and A. L. McClellan, "The Hydrogen Bond," (W. A. Freeman and Co., 1960) and S. N. Vinogrodov and R. H. Linnell, "Hydrogen Bonding," (Van Nostrand Reinhold, 1971).

Kitaigorodskii's books (see the Bibliography) present the bonding of molecular crystals and the theory of close packing of molecules.

The classic paper on **plastic crystals** by Timmermans [J. Phys. and Chem. of Solids **18**, 1 (1961)] is worth reading, and a good summary of many aspects of the field can be found in "The Plastically Crystalline State—Orientationally Disordered Crystals," Ed. J. N. Sherwood (John Wiley and Sons, 1979). A number of articles on **liquid crystals** can be found in Solid State Physics, Supplement 14; also see P.G. deGennes, "The Physics of Liquid Crystals" (Clarendon Press, Oxford, 1974); and Physics Today, May 1982.

See Ashcroft and Mermin, Chapter 20, as well as Kittel, Chapter 3, for related discussions on the bonding of rare gas crystals. The bulk modulus is calculated in both books. The latter book has some comments on quantum corrections. The excellent book by Hirschfelder, Curtiss, and Bird, has a rather complete discussion on van der Waals forces and the Lennard-Jones potential is discussed and applied to the interactions between molecules in many ways.

Molecular dissociation of crystalline iodine was observed by K. Takemura, S. Minomura, O. Shimomura, and Y. Fujii, Phys. Rev. Letters **45**, 1881 (1980). Also see the high pressure review article by Jayaraman quoted in the Notes of Chapter 3.

The **gas constant** $R = 1.987$ calorie mole^{-1} °K^{-1} = 8.3143 joule mole^{-1} °K^{-1} = 82.057 cm^3 atmosphere mole^{-1} °K^{-1} = $k_B N_A$ where N_A is **Avogadro's number** of atoms per mole = 6.022×10^{23} atoms/mole; instead of atoms/mole, Avogadro's number could be molecules/mole or formular units/mole.

Problems

1. Forces between atoms – Using a typical internuclear distance in a solid, calculate the energy due to the gravitational attraction between two neighboring atoms. If these two atoms have a typical magnetic moment of that of a free electron, calculate the energy due to the magnetic dipole attraction. Last, compare these energies with that from the Coulomb attraction of two oppositely charged ions.

2. Discuss the nature of the bonding in high polymers. (Hint: see Lovell, Avery, and Vernon.)

3. NaCl has a density of 2.167 g/cm^3. What is the distance between adjacent atoms?

4. Lennard-Jones potential – Calculate the values of A and B in Eq. 6-10 by summing up to fourth nearest neighbors for an fcc and a bcc structure. Calculate U_{tot} for Ne in these two structures so that you can compare their relative stabilities. In which structure would you expect the larger quantum effects? (See the Notes).

5. van der Waals interaction – As a model that gives the form of the van der Waals interaction consider two identical harmonic oscillators at x = 0 and x = R constrained to oscillate along the x-axis. Each oscillator has charges $\neq$e. By considering the Coulomb interaction between the charges, show that the energy is lowered by an amount proportional to R^{-6} over what it would be if there were no interaction between the oscillators. (Hint: see Kittel or Zhdanov.)

6. When AgI undergoes a phase transition at 147°C from an ordered wurtzite structure to the disordered α-phase structure (Fig. 3-3) there is a measured entropy increase = 4 cal mole^{-1} °K^{-1}. If, in the high temperature phase, the two Ag ions are statistically distributed over the twelve tetrahedrally coordinated positions (12d-sites), what is the increase in configurational entropy at this transition? Why do you think that this calculated result is smaller than observed?

7. HCN is a linear molecule with a dipole moment of 1.17 x 10^{-18} esu-cm. The crystal structure is such that the carbon atoms are on the lattice points of a body-centered lattice with the dipoles all pointing in the +z-direction. Determine the H-C-N distances from Table 8-2 and the distances between the molecules from Table 6-2. Find the dipole energy of interaction from just the first- and second-neighbor molecules. What is the polarization of this pyroelectric crystal (P $\equiv \mu$/vol)? What is the space group?

7

The Ionic Bond

7-1 Transfer of Electrons
7-2 Ionic Radii
    a Coordination number effect
7-3 Typical Structures
    a Structures (S)
    b Geometric basis of morphotropy (S)
    c Pressure effects (S)
7-4 Cohesive Energies of Ionic Crystals
    a Introduction
    b Interaction Energy
    c Madelung constant
    d Some results

Notes
Problems

Table 7-1 Values of the distance, d, between anions and nearest-neighbor cations, in the alkali halide crystals. Values are given in Å.

	Li	Na	K	Rb	Cs
F	2.01	2.31	2.67	2.82	3.00
Cl	2.57	2.82	3.15	3.29	3.57
Br	2.75	2.99	3.30	3.43	3.71
I	3.00	3.24	3.53	3.67	3.95

Table 7-2 (a) Ionic radii. (b) Other empirical ionic radii. (Both are from Pauling.)

(a)

			H^- 2.08	Li^+ 0.60	Be^{2+} 0.32	B^{3+} 0.20	C^{4+} 0.15	N^{5+} 0.11	O^{6+} 0.09	F^{7+} 0.07
C^{4-} 2.60	N^{3-} 1.71	O^{2-} 1.40	F^- 1.36	Na^+ 0.95	Mg^{2+} 0.65	Al^{3+} 0.50	Si^{4+} 0.41	P^{5+} 0.31	S^{6+} 0.29	Cl^{7+} 0.26
Si^{4-} 2.71	P^{3-} 2.12	S^{2-} 1.84	Cl^- 1.81	K^+ 1.33	Ca^{2+} 0.99	Sc^{3+} 0.81	Ti^{4+} 0.68	V^{5+} 0.59	Cr^{6+} 0.52	Mn^{7+} 0.46
				Cu^+ 0.96	Zn^{2+} 0.74	Ga^{3+} 0.62	Ge^{4+} 0.53	As^{5+} 0.47	Se^{6+} 0.42	Br^{7+} 0.39
Ge^{4-} 2.72	As^{3-} 2.22	Se^{2-} 1.98	Br^- 1.95	Rb^+ 1.48	Sr^{2+} 1.13	Y^{3+} 0.93	Zr^{4+} 0.80	Nb^{5+} 0.70	Mo^{6+} 0.62	
				Ag^+ 1.26	Cd^{2+} 0.97	In^{3+} 0.81	Sn^{4+} 0.71	Sb^{5+} 0.62	Te^{6+} 0.56	I^{7+} 0.50
Sn^{4-} 2.94	Sb^{3-} 2.45	Te^{2-} 2.21	I^- 2.16	Cs^+ 1.69	Ba^{2+} 1.35	La^{3+} 1.15	Ce^{4+} 1.01			
				Au^+ 1.37	Hg^{2+} 1.10	Tl^{3+} 0.95	Pb^{4+} 0.84	Bi^{5+} 0.74		

(b)

NH_4^+ 1.48 Å			
Tl^+ 1.44			
Mn^{++} 0.80 Å	Ti^{3+} 0.69		
Fe^{++} .75	V^{3+} .66		
Co^{++} .72	Cr^{3+} .64		
Ni^{++} .70	Mn^{3+} .62		
	Fe^{3+} .60		

Ionic radii for trivalent rare earth ions 0.90 ± 0.05 Å

THE IONIC BOND

There arises thus a certain insincerity in our philosophic discussions: the potentest of all our premises is never mentioned.

W. James, *"Pragmatism"*

The static behavior of the ionicly bonded crystals is simple to calculate and understand because the attractive forces are basically coulombic and the ions can be treated classically as spherically discrete entities (charged billiard balls).

7-1 Transfer of Electrons

The formation of an ionic bond involves the transfer of electrons from one atom to another such that the resulting ions have closed completed shells. For example, think of forming NaF from the free atoms. The free Na atom has a 3s electron outside a $2s^2p^6$ closed shell, while the F atom is deficient by one electron of having a closed shell, so that the electrons in the atoms and ions are

	atom	ion
Na	$1s^2 2s^2 p^6 3s$	$1s^2 2s^2 p^6$
F	$1s^2 2s^2 p^5$	$1s^2 2s^2 p^6$

When the Na and F atoms are brought together, we may say that the 3s electron of the Na atom is transferred to the F atom. This leaves sodium with one more positive charge than negative charge, that is, Na^+ and similarly F^-. We may write Na^+F^- to emphasize the ionic nature of the compound. In a similar manner two electrons can be transferred in, for example, $Mg^{2+}O^{2-}$.

The energy required to take a single outermost electron from a neutral atom to infinity is the **first ionization energy**. In the opposite sense an extra electron attached to a neutral atom can be bound (have an energy less than zero). The energy of binding is called the **electron affinity**, which is equal to the first ionization energy of the negative ion. For alkali atoms the electron affinity is very small, that is, an extra electron gains very little energy by binding to such an atom. However, for a free halide atom an extra electron is strongly bound. In fact the electron affinity increases as one goes across any row in the periodic table. The binding occurs because the "extra" electron has a reasonably large probability of being fairly far "in" the atom so the other electrons do not completely screen the positive nuclear charge. Hence a net attractive force is found. For the alkali atoms the "extra" outer electron has a much larger effective radius and it is screened better from the positive nucleus by the other electrons, in particular by those electrons with lower principal quantum numbers.

The first ionization energy for the alkalies and the electron affinity for the halides are listed.

	ionization energy		electron affinity
H	13.60 eV	F	3.4 eV
Li	5.39	Cl	3.61
Na	5.14	Br	3.36
K	4.34	I	3.06
Rb	4.18		
Cs	3.89		

For example, formation of an isolated Na^+ ion from a neutral Na atom requires 5.14 eV (to be put into the system). 3.4 eV is obtained (from the system) if the electron attaches to an isolated F atom to form a F^- ion. For an isolated Na^+ and an isolated F^- there is thus a net cost of 1.74 eV. However, if these two isolated ions are brought together so that their separation is 2.31Å (the internuclear separation in a NaF crystal), from Coulomb's law there is a binding energy of $-e^2/R = -6.2$ eV. Thus, assuming this ionic model, we can see why alkali halide molecules and crystals are formed. The **cohesive energy** of a crystal, which is calculated in the discussion that follows, is defined as the energy to disassociate it into separate atoms but for ionic solids it is the energy to disassociate the solid into separate ions.

Consider the question of size for two ions in the same row of the Periodic Table, for example Na^+ and F^-. Both ions have the same electron configuration, but the Na^+ ion has a nuclear charge of $+11e$ with an electron charge of $-10e$; the F^- ion has a nuclear charge of $+9e$ but an electron charge of $-10e$. Thus, for the negative ions the screening of the nuclear charge by the electrons is more complete than for the positive ions. Hence we expect the ionic size of Na^+ to be much smaller than that of F^-; this is found to be so.

We should mention that many complicated salts can be thought of as primarily ionic. An example is the spinel, $MgAl_2O_4$ formally written as $Mg^{2+}Al_2^{3+}O_4^{2-}$. In the normal spinel the Al^{3+} ions are octrahedrally surrounded by six O^{2-} ions, and the Mg^{2+} ions are tetrahedrally surrounded by four O^{2-} ions. In considering garnet crystals, $Ca_3Al_2Si_3O_{12}$ and many related materials, again we think of the crystal as primarily ionic. The Al^{3+} is octahedrally surrounded by O^{2-} ions and the Si^{4+} tetrahedrally surrounded by O^{2-} ions. However, we would guess that the Si-O bond has a good deal of covalent as well as ionic character.

CHAPTER 7   THE IONIC BOND   133

7-2 Ionic Radii

There are 20 alkali halide crystals. As a model we assume that each type of ion is a hard sphere of a definite **ionic radius** R. Then we see whether this model fits the experimental data for the lattice constants. All the alkali halides have either the NaCl structure, O_h^5(Fm3m), or the CsCl structure, O_h^1(Pm3m). See Figs. 3-5b and 3-1b. The distance between the centers of the positive ions (**cations**) and nearest-neighbor negative ions (**anions**) is defined as d and in terms of the unit cell length, a, is

$$\begin{array}{ccc} & \text{NaCl} & \text{CsCl} \\ d = & a/2 & (3)^{1/2}a/2 \end{array}$$

as can be seen from the figures. Table 7-1 (page 130) lists the values of d. Then, for a given alkali halide, AH, we can fit $d_{AH} = R_A + R_H$ very well for nine values of R_X. This would give each ion a definite size. The problem is that the choice of ionic radii is not unique. If a fixed amount is added to all of the alkali radii and subtracted from all of the halogen radii we still obtain the same d values. Table 7-2 (page 130) gives values of ionic radii, due to Pauling, that are often used. As can be seen, ionic radii of many other ions are included but these are less certain. Pauling's values are based on O^{2-} having a radius of 1.40 Å, obtained from calculations and experimental data. (See the Notes.)

7-2a Coordination number effect   We might expect that the ionic radii will depend, to a small extent, on the coordination number, although this is not considered in this simple model. Indeed, it is observed that the interionic distance for an 8-coordinated crystal, for example, CsCl structure, is about 3% larger than what would be calculated using Table 7-2; similarly interionic distances for 4-coordinated crystals, for example, zinc blende structure, are about 5% smaller.

Shannon and Prewitt have considered the ionic radii, with coordination number taken into account. For example, the ionic radius of Rb^+ is 1.52 Å or 1.61 Å, depending on whether it is 6- or 8-coordinated. See the Notes for a reference to this work.

We should always remember that radii come from a model (hard spheres). They are very useful in many ways but are strictly appropriate only to materials that are highly ionic and even then to relatively simple structures where the point symmetry of the ions is high. In low symmetry crystals a 10% variation of the internuclear distances between a central ion and its immediate neighbors is possible because of strong dipole and other multipole forces.

7-3 Typical Structures

7-3a Structures All of the alkali halides are cubic. Most have the NaCl structure, Fig. 3-5b. Three salts, CsCl, CsBr, and CsI have the CsCl structure, Fig. 3-1b. These statements apply at atmospheric pressure (high pressure results will be discussed later). In the next section we examine the possibility of predicting these structures.

There are thousands of crystals that are predominantly ionic. We mentioned spinel, $MgAl_2O_4$, and garnet, $Ca_3Al_2Si_3O_{12}$ in Section 7-1 and the cubic perovskite, $BaTiO_3$, Fig. 3-1c. The first two are important because some very useful magnetic materials have these structures. Yttrium iron garnet, $Y_3Fe_2(FeO_4)_3$ has the garnet structure; we have written the formulas in a way that emphasizes the fact that three Fe^{3+} ions are tetrahedrally surrounded by O^{2-} ions and two are octrahedrally surrounded by O^{2-} ions. The perovskite structure is of interest because there are hundreds of crystals that have this structure, such as $A^+B^{5+}O_3$, $A^{2+}B^{4+}O_3$, $A^{3+}B^{3+}O_3$, as well as ordered and disordered mixed systems of these (see Galasso). Many of these materials have phase transitions to structures that are slightly distorted variations of the ideal cubic perovskite. These lower symmetry structures are ferroelectric, ferroelastic, ferromagnetic, or have combinations of these and other interesting properties.

7-3b Geometric basis of morphotropy The change in crystal structure that is due to systematic chemical substitution is defined as a **morphotropic transition**. Using the model of hard spheres we can calculate the energy of crystal assuming different crystal structures. Thus we can consider questions such as the possibility of a transformation from CsCl to the NaCl structure due to a change of size of the ions. Similarly, other simple structures may be considered.

Figure 7-1a shows the atoms in the (001) plane of the NaCl structure. We can think of the larger ions as Cl^- with an ionic radius r_ℓ and the smaller ions as Na^+ with ionic radius r_s. If r_s is big enough so that the large ions do not touch each other, only the small ions are in contact with the large ions at point A. Consider what happens as r_s decreases. First, the ions get closer together and the binding energy will increase (larger negative number) owing to the Coulomb energy $(-e^2/r)$. Second, at a critical value of the radii, r_s^c and r_ℓ^c, the larger ions touch each other at point B. Last, any decrease of r_s below r_s^c does not change the cohesive energy. This is because the internuclear distances do not decrease since the large ions are in contact with each other (at point B). Remembering that the internuclear distance $d = r_s + r_\ell$, from the figure it can be seen that when the ions are in contact

CHAPTER 7 THE IONIC BOND

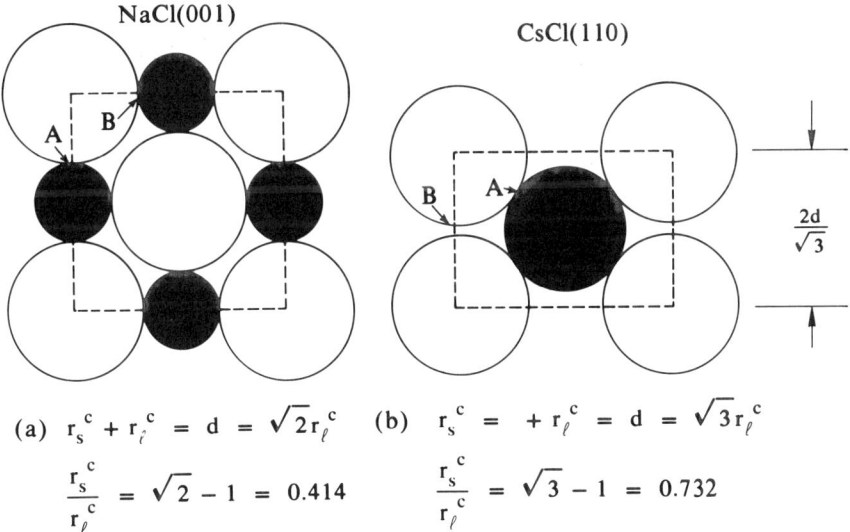

Fig. 7-1 Figures and results to calculate the critical radius ratio where the large ions are just touching.

at point A as well as B one has $\sqrt{2}r_\ell^c = d$. These results are shown in Fig. 7-1a, where the critical radius ratio is given.

Figure 7-1b is similar to Fig. 7-1a but it shows the (110) plane of the CsCl structure. Again it is assumed that r_s is large enough so that the larger ions do not touch each other at the point B. For this structure the unit cell dimension is related to d as $2d/\sqrt{3}$, as can be seen in Fig. 3-1b. As r_s decreases a critical value is reached where the ions touch at points A and B; then $2d/\sqrt{3} = 2r_\ell^c$. The results are shown in Fig. 7-1b.

In a similar manner the result for the zinc blende structure, for large ion contact, is $r_s^c/r_\ell^c = (\sqrt{6}/2) - 1 = 0.225$. As will be shown in the next section, one finds from considerations of the attractive and repulsive forces that at a given, large, interionic distance the CsCl structure has the largest binding energy, followed by the NaCl and then the zinc blende structures. These results, along with the critical radii ratio results, as we have found, enable the curves in Fig. 7-2 to be drawn. The lowest energy for a given r_s/r_ℓ will determine which structure the ions choose. Thus, the figure shows that as the radius ratio decreases we should observe structures in the different materials that vary from the CsCl to the NaCl to the zinc blende structures. The "transitions" should occur for values of r_s/r_ℓ slightly smaller than the critical values given in the preceding discussion, as drawn in the figure.

Table 7-3 gives the values of r_s/r_ℓ for the alkali halides. The CsCl structure would be expected for CsCl, CsBr and CsI. We would

136   CHAPTER 7   THE IONIC BOND

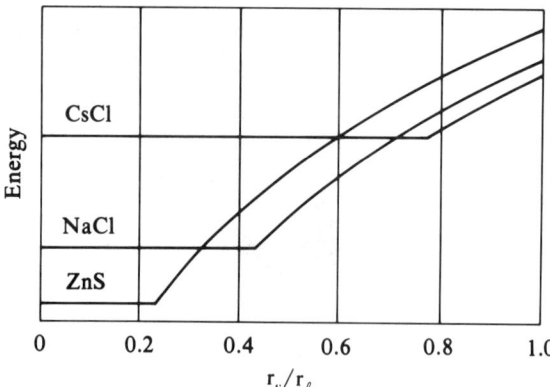

Fig. 7-2 The electrostatic part of the cohesive energy, for the various structures, as a function of r_s/r_ℓ where r_ℓ is assumed constant.

Table 7-3 Values of r_s/r_ℓ for the alkali halides.

	Li	Na	K	Rb	Cs
F	0.44	0.70	0.98	0.92	0.80
Cl	0.33	0.52	0.74	0.82	0.93
Br	0.31	0.49	0.68	0.76	0.87
I	0.28	0.44	0.62	0.69	0.78

expect several more alkali halides to have the CsCl structure and at least LiI to have the zinc blende structure. Yet, all the other alkali halides have the NaCl structure. This shows that while the simple ionic model may be good for overall understanding, it cannot be used to understand subtle differences. Related discrepancies can be found by comparing the sum of the ionic radii (Table 7-2) to the internuclear distance (Table 7-1) in, for example, the Li-halide salts. The agreement gets worse as we go from LiF to LiI. This may be attributed to the fact that the Li$^+$ ions are very much smaller than the calculated geometric space left for them by the halide ions. This causes other effects to come into play that are not taken into account by a charged billiard ball model.

Nevertheless, radius ratios can be helpful in understanding trends, and they are used effectively for other crystal structures. For example, Table 7-4 shows the structures and radius ratio values for compounds of the form AF$_2$ and BO$_2$. The rutile structure (Fig. 1-7) has coordination number 6:3 while fluorite has 8:4 (Fig. 3-5c). As r_+/r_- increases there is a morphic transition from a less coordinated to a more coordinated structure, just as the transition from zinc blende (4:4) to NaCl (6:6) to CsCl (8:8). The oxide ZrO$_2$ has a radius ratio that is smaller than that of TeO$_2$, yet the structures are the reverse of what we would expect. Often such materials have interesting properties. In this case ZrO$_2$ has a low temperature phase that has a distorted fluorite structure.

CHAPTER 7   THE IONIC BOND

Table 7-4 Structural types for some fluorides and oxides for ratios of the +ion/−ion.

Rutile				Fluorite			
r_+/r_-		r_+/r_-		r_+/r_-		r_+/r_-	
MgF_2	0.59	MnO_2	0.39	CdF_2	0.77	ZrO_2	0.66
NiF_2	0.59	RuO_2	0.49	CaF_2	0.80	PrO_2	0.76
FeF_2	0.62	MoO_2	0.52	HgF_2	0.84	CeO_2	0.77
ZnF_2	0.62	PbO_2	0.64	PbF_2	0.99	UO_2	0.80
MnF_2	0.68	TeO_2	0.67	BaF_2	1.08	ThO_2	0.84

7-3c Pressure effects   We have discussed only the structures that are stable at atmospheric pressure. Now we consider some results of high pressure experiments.

For example RbCl has a radius ratio that should put it in the CsCl structure range (Fig. 7-2), but it has the NaCl structure. However, above 5200 atmospheres it transforms and has the CsCl structure. This is a very small pressure for a polymorphic transition. The reasons for the transition are just the arguments that lead to Fig. 7-2, namely, packing. We look at this further.

In Table 7-5 we consider the alkali chlorides. The ionic radius of Cl⁻ is larger than that of all the alkalis and r_A/r_{Cl} varies smoothly with increasing r_A, as shown in the table. However, notice that the **packing fraction** (volume of the ions/volume of the crystal) of the compounds decreases as r_A/r_{Cl} increases up to RbCl. This is really another way of looking at Fig. 7-2. Also notice that the **compressibility**, K, which is the fractional change of volume per unit of pressure $\equiv -(1/V)(dV/dP)$, increases from LiCl to RbCl by almost a factor two. (The **bulk modulus**, B, is equal to K^{-1}.) Thus, at some high pressure, RbCl could become considerably smaller than LiCl if this extrapolation continued. By increasing the coordination number from six to eight in going from RbCl to CsCl there is a large increase in the packing fraction and decrease in the compressibility. Similarly when RbCl undergoes a transition to the CsCl structure under pressure there is about a 15% reduction in volume. That is a huge effect.

As was discussed in Section 3-8b, reducing the temperature is, in general, similar to increasing the pressure; conversely, increasing the temperature is equivalent to reducing the pressure. Indeed CsCl at 470°C is observed to transform and acquire the NaCl structure. (See the Notes for further discussion.)

From these low transition temperatures and pressures we see that in these simple compounds with different structures the energies can be rather close. Thus, the fact that the exact predictions, summarized in Fig. 7-2, of the hard sphere model are not found should not deter us very much. The model seems to be a reasonable one for these ionic materials and we shall proceed to calculate their energies.

Table 7-5 Some results for the alkali chlorides. r_A/r_{Cl} is the ionic radius ratio of alkali/chloride.

	r_A/r_{Cl}	Structure	Packing fraction (%)	Compressibility (10^{-12} cm^2/dyne)
LiCl	0.38	NaCl	78.7	3.36
NaCl	0.54	NaCl	65.7	4.17
KCl	0.73	NaCl	56.0	5.73
RbCl	0.82	NaCl	50.2	6.40
CsCl	0.92	CsCl	68.4	5.55

7-4 Cohesive Energies of Ionic Crystals

7-4a Introduction So far the discussion of the ionic bond has been of a qualitative or semiquantitative nature. Now we proceed to calculate quantitatively the cohesive energy of simple ionic crystals from our charged billiard ball model (the **Born model**). We should really calculate the Gibbs free energy, G, which is the thermodynamic quantity that describes the stability of solids as a function of temperature and pressure. $G = U_0 + PV + (1/2)\omega - TS$, where U_0 is the lattice energy that we will calculate. The pressure volume term will be ignored since at atmospheric pressure this term is very small compared to U_0. The $\omega/2$ term is due to zero point vibrations which can also be ignored for these heavy ions. The entropy term will be neglected because only low temperatures are considered. Moreover, for the calculation of U_0 we consider only the point charge coulomb term and a short-range two-body repulsive term, neglecting covalent bonding, three-body forces, higher electrostatic moments, and kinetic energy. Nevertheless, the results are good; in fact about 90% of U_0 comes from the Coulomb energy and only 10% from the short-range repulsive energy, so all the other terms should not be important.

7-4b Interaction energy The interaction energy u between a pair of ions with charges Z_1 and Z_2, separated by a distance r, is

$$u(r) = (Z_1 Z_2 e^2/r) + (b/r^n) \qquad (7-1)$$

The first term is the electrostatic interaction of point ions and the second term represents a short range ($n \gg 1$) repulsion term. The repulsion term is required, otherwise oppositely charged ions would collapse one into another. It arises because both ions have filled, closed shells and the Pauli principle do not allow another electron in the orbits as discussed in Section 6-3. Figure 7-3 shows the long-range attractive coulomb and short-range repulsive energies as well as the resultant energy. <u>Note</u> that as long as $n \gg 1$, most of the total energy at the equilibrium distance comes from the coulomb term, as

CHAPTER 7   THE IONIC BOND

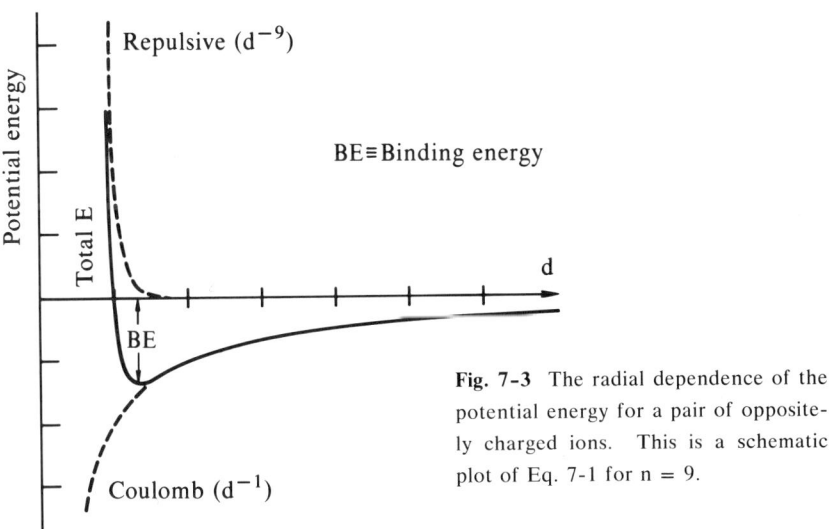

Fig. 7-3 The radial dependence of the potential energy for a pair of oppositely charged ions. This is a schematic plot of Eq. 7-1 for n = 9.

can be seen in the figure. In fact, for n = ∞ all of the energy comes from the coulomb term.

We want to calculate the interaction energy that an ion i for convenience (assumed to be positively charged) has with all the other j ions considering only singly charged ions,

$$U_i = \Sigma_j' u(r)_{ij} = \Sigma_j' (b/r_{ij}^n) \pm (e^2/r_{ij}) \qquad (7\text{-}2)$$

where the prime in the sum means sum over all ions except i = j, + is used for a positively charged ion, and − for a negatively charged ion. The value of U_i does not depend on whether the reference ion is positive or negative, so that we drop the subscript on U_i and note that if there are 2N ions in the crystal the total energy is $U_0 = NU$. (N rather than 2N is used because in the total energy we must count each interaction between a pair of ions only once.) As a convenience let $r_{ij} = p_{ij}d$, where d is the distance between the nearest neighbors. Then Eq. 7-2 becomes

$$U = \frac{bA_n}{d^n} - \frac{\alpha e^2}{d} \qquad \begin{cases} A_n = \Sigma_j' p_{ij}^{-n} \\ \alpha = \Sigma_j'(\pm)p_{ij}^{-1} \end{cases} \qquad (7\text{-}3)$$

This is an important result and a few points should be discussed. First, the sum A_n is rapidly convergent because n is large. This was also seen in Section 6-5e. However, it is not necessary to calculate A_n because the value of b is not known except from a first principles calculation. Rather we eliminate the product bA_n by demanding that $\partial U/\partial d = 0$ at the equilibrium separation, d_0. Second, the value of n is, in a manner similar to bA_n, evaluated in terms of the compressibility of the structure. Compressibility is proportional to the second derivative $\partial^2 U/\partial V^2$, of the interaction energy with respect to volume rather than the first derivative. Third, α is called the **Madelung constant** and like any lattice sum it is a property of the geometry of the crystal structure. Its evaluation will be discussed in the next section. Since the $\alpha e^2/d$ term will turn out to be the dominant term in the energy of an ionic material, this constant, and its variation from structure to structure, is very important in the theory of ionic crystals. We now proceed to carry out this threefold program.

First, eliminate bA_n from Eq. 7-3 by setting $\partial U/\partial d = 0$ at the equilibrium separation.

$$\left(\frac{\partial U}{\partial d}\right)_{d_0} = -\frac{nbA_n}{d_0^{n+1}} + \frac{\alpha e^2}{d_0^2} = 0 \qquad (7\text{-}4a)$$

or

$$U_0 = -\frac{\alpha e^2}{d_0}\left(1 - \frac{1}{n}\right) \qquad (7\text{-}4b)$$

Notice that for $n \gg 1$ most of the energy is given by the Coulomb term $\alpha e^2/d_0$. This also can be seen in Fig. 7-3.

Second, we evaluate n in terms of the **compressibility** $K \equiv -(1/V)(dV/dP)$. At low temperatures, from the first law of thermodynamics, $dU_0 = -PdV$ or $dP/dV = -d^2U_0/dV^2$. Thus, at low temperatures, and equilibrium volume, denoted by zero subscript,

$$1/K = V(\partial^2 U_0/\partial V^2)_0 \qquad (7\text{-}5)$$

To evaluate the right side of Eq. 7-5 we recall that in general

$$\partial U/\partial V = (\partial U/\partial d)(\partial d/\partial V)$$
$$\partial^2 U/\partial V^2 = (\partial U/\partial d)(\partial^2 d/\partial V^2) + (\partial^2 U/\partial d^2)(\partial d/\partial V)^2 \qquad (7\text{-}6)$$

Noting that at equilibrium $(\partial U/\partial d)_0 = 0$ and using $V = 2Nd^3$ for the NaCl structure so that $(\partial d/\partial V)^2 = (36N^2d^4)^{-1}$, the compressibility can be evaluated from Eqs. 7-5 and 7-6.

$$\frac{1}{K} = \frac{1}{18Nd_0}\left(\frac{\partial^2 U_0}{\partial d^2}\right)_{d_0} \qquad (7\text{-}7)$$

(Remember that $U_0 = NU$.) Determining $\partial^2 U_0/\partial d^2$ from Eq. 7-3 and then eliminating bA_n by using Eq. 7-4a, we find

$$\frac{1}{K} = \frac{(n-1)e^2\alpha}{18d_0^4}$$

or

$$n = 1 + (18d_0^4/Ke^2\alpha) \qquad (7\text{-}8)$$

CHAPTER 7   THE IONIC BOND   141

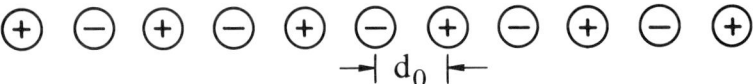

Fig. 7-4 A univalent, one-dimensional ionic crystal.

For NaCl with K = 3.3 × 10⁻¹² cm²/dyne and α = 1.748 (α is calculated in the next section), n = 9.4 is obtained. Thus, n ≫ 1, which shows self consistency of the hard-sphere model.

7-4c Madelung constant   Third, and last, we discuss the Madelung constant, defined in Eq. 7-3. To appreciate the problem more clearly evaluate α for the one-dimensional ionic crystal in Fig. 7-4. Recall in the definition of $\alpha \equiv \Sigma_j' (\pm) p_{ij}^{-1}$ that p_{ij} is defined by $r_{ij} = p_{ij}d$. Then picking the origin on a negative ion, we have

$$\alpha = 2\left[1 - \frac{1}{2} + \frac{1}{3} - \frac{1}{4} + ...\right] \tag{7-9}$$

where the factor 2 occurs because for each term in the sum there are two ions, one to the right and one to the left. Although this series does not converge rapidly, it may be evaluated by using the series expansion.

$$\ln(1 + x) = x - (x^2/2) + (x^3/3) - (x^2/4) + ... \tag{7-10}$$

Thus, for the one-dimensional ionic crystal α = 2 ln 2.

In three dimensions the convergence problem is more difficult. Consider the NaCl structure and take the origin at a negative ion. In the Madelung sum there are six positive-ion nearest neighbors with p = 1, 12 next-nearest neighbor negative ions with p = 2^(1/2), eight positive ions with p = 3^(1/2); and so on. Thus,

$$\alpha = \frac{6}{1} - \frac{12}{2^{1/2}} + \frac{8}{3^{1/2}} - \frac{6}{2} + ... = 6 - 8.485 + 4.620 - 3 + ... \tag{7-11}$$

Convergence is not apparent when the problem is approached in this manner. Other methods must be used.

Much better convergence is obtained in evaluating the sum for α by using units of the crystal that are nearly neutral. For example, consider the unit cell as shown in Fig. 3-5b, and formally break up the ions into charges inside the cell and outside, so that 1/2 of the charge of ions on the cube faces is considered inside the cube and the other 1/2 is outside. Then if an ion is on an edge, 1/4 is inside the cube; if on a corner 1/8 is considered inside. Thus, the cubic symmetry is maintained in the sum. Then, the contribution to α from the ions

Table 7-6 The top numbers, for each alkali halide, are the measured cohesive energies in units of 10^{-11} ergs per ion pair (from Tosi's article referenced in the problems). The numbers in parentheses are the calculated electrostatic energies obtained from $-\alpha e^2/d_0$. Note that all these energies are negative, that is, binding, but we have left out the negative sign.

	Li	Na	K	Rb	Cs
F	1.68	1.49	1.32	1.26	1.20
	(2.01)	(1.75)	(1.51)	(1.43)	(1.34)
Cl	1.38	1.27	1.15	1.11	
	(1.57)	(1.43)	(1.28)	(1.23)	
Br	1.32	1.21	1.10	1.06	
	(1.47)	(1.35)	(1.22)	(1.18)	
I	1.23	1.13	1.04	1.01	
	(1.34)	(1.24)	(1.14)	(1.10)	

inside this cubic cell is

$$\frac{6/2}{1} - \frac{12/4}{2^{1/2}} + \frac{8/8}{3^{1/2}} = 1.45$$

Continuing this process, the contribution from the next larger cube is 0.3. Thus, for just these two terms in the sum, $\alpha = 1.75$, which is close to the accurate value.

Thus, by using more subtle methods than straightforward summing, as in Eq. 7-11, the Madelung constants can be obtained. The Ewald method is the one most often used, and references can be found in the Notes. Some results for α in different structures are

CsCl structure	1.7627	Wurtzite structure	1.641
NaCl structure	1.7476	CaF$_2$ structure	5.0388
Zinc blende structure	1.6381	Cu$_2$O structure	4.4425

Finally, we see that the Coulomb energy of the CsCl structure favors binding slightly more than for the NaCl structure, for the same internuclear distance. (The values of α shown here for the CsCl, NaCl, and zinc blende structures determine the ordering of the curves at large distances in Fig. 7-2.)

7-4d Some results   Table 7-6 shows the experimentally observed cohesive energies and those calculated from just the electrostatic term. As expected, the $\alpha e^2/d$ term accounts for most of the energy. Consistent with this, the exponential in the repulsive term is very large

($n \gg 1$, as can be seen in the problems). Thus, the simple charged billiard ball model of ionic crystals can describe bonding energies in simple salts.

We could use the values in Table 7-6 to obtain n via Eq. 7-4b. This would not be totally consistent since the values given in the table are experimentally determined ones. Rather, n should be obtained from the compressibility and the electrostatic term via Eq. 7-8, as in the problems.

As was mentioned previously, the CsCl structure has a slightly larger electrostatic energy than the NaCl structure for the same value of d. There are some other small contributions that should be included in a better theory before detailed structural predictions can be made. These other contributions include van der Waals forces between ions; the use of a repulsive term of the form $\exp(-r/\rho)$ rather than r^{-n}, since the exponential form is suggested by theoretical work; zero point vibrations; and quantum mechanical many-body effects, which are of the order of 5% to 10% of the cohesive energy when the sizes of the component ions differ significantly. However, errors due to many of these smaller contributions are compensated for in the Born approach by the use of experimental data via the internuclear distance and compressibility in fitting the parameters. (See Tosi's article for a discussion of these points.)

Notes

See Pauling for ionization energies and electron affinities. For more recent tables see National Bureau of Standards, Circular 467 for first and second ionization energies (Kittel 5th Ed. has these values), and H. Hotop and W.C. Lineberger, J. Phys. Chem. Ref. Data **4**, 539 (1975) for electron affinities.

A recent, updated table of the Shannon and Prewitt coordination dependent ionic radii is R. D. Shannon, Acta Cryst. A **32**, 751 (1976).

More details and/or a different emphasis on the material covered in this chapter can be found in the many books listed under solid state science in the bibliography.

A nice discussion of the Ewald method [Ann. Physik **64**, 253 (1921)] can be found in J.C. Stater, Insulators, Semiconductors and Metals, McGraw-Hill, N.Y. 1967. R. Sherman, Chem. Revs. **11**, 93 (1932) lists values of α for many structures.

Tosi's article, referenced in the problems, is a good source of more detailed information on cohesion in ionic solids. Lövdin's classic paper [Phil. Mag. Suppl. **5**, 1 (1956)] deals with the quantum mechanical effect of overlap (not covalency or deformation) of the wave functions of the ions, alluded to in Section 7-4d.

The pressure effects on the alkali halides are basically as noted in Section 7-3c but have some complicating features that are not understood. The transition pressures for the transition from the NaCl to the CsCl structures depend strongly on the cations (Na halides transforming ~300 kbar, K halides ~19 kbar and Rb halides ~4 kbar) and much less on the anions. (There is only a slight increase in transition pressures along the sequence I to Br to Cl.) The situation for several of the fluorides is unclear. A review article by C.W.F.T. Pistorius, Prog. in Sol. State Chem. **11**, 1 (1976), is worth looking at.

The structures of **ammonium halides** have been calculated using the approach discussed here but with a distribution of charge on the NH_4^+-ion. The phase transitions at different temperatures and pressures also have been calculated. [G. Raghurama and R. Narayon, J. Phys. Chem. Solids **44**, 633 (1983).]

Problems

1. Use the simple formulas for hydrogen-like wave functions for the eigenvalues and the average values of the various powers of the radial wave functions given, for example, in L. Pauling and E. B. Wilson, Jr., "Introduction to Quantum Mechanics" (McGraw-Hill Book Co., 1935). From these values determine the approximate binding energies and value of $\langle r \rangle$ for a 3s and 3p electron on Na^-. Compare these values to a 3p electron on Cl^-. Relate these quantities to electron affinity and ionic radius of these ions. [To calculate the effective nuclear charges see, G. Burns, J. Chem. Phys. **41** 1521 (1964).]

2. For which alkali halides do the cation radii correspond to r_ℓ? Why is it unimportant as to which ion is larger? For example, why will this have no effect on the considerations that lead to Fig. 7-2?

3. Madelung constant – (a) How does the Madelung constant vary with the number of nearest neighbors? Would you expect this? (b) Assuming that the repulsive potential is unchanged in NaCl, what is the effect on α, d_0, K, and U_0 of doubling the charges on all of the ions? (c) Determine the total energy for the one-dimensional crystal in Fig. 7-4. Why is the result the same as Eq. 7-4b?

4. Discuss the **Born-Haber cycle** from which we may determine the lattice energy of alkali halide crystals. (See, for example, Evans or Pauling.)

5. n determination – Using the tabulated data in Tosi's article [Solid State Physics, Vol. 16, Ed. F. Seitz and D. Turnbull, Academic Press, N.Y., 1964], determine the values for the exponent n in the repulsive

CHAPTER 7   THE IONIC BOND   145

term for all of the alkali halides. As you will see, most of the values are greater than 8 except for the Li-salts where n ≈ 6 to 7.

6. Define the **reduced Madelung constant** and discuss an empirical way to determine it for any structure. See Tosi's article Section II-3 and the references given there.

7. Do Problem 4 in Chapter 3.

8. BaO has the NaCl structure with the internuclear distance equal to 2.76 Å. Using the Born-Haber cycle estimate the cohesive energies per molecule, assuming Ba^+O^- and $Ba^{2+}O^{2-}$. Which valence state has the lower energy? The first and second ionization potentials of Ba are 5.19 eV and 9.96 eV. The electron affinities of the first and second electrons added to neutral oxygen are 1.5 eV and −9.0 eV.

9. By standard radio frequency techniques one can measure the **nuclear quadrupole coupling** constant, which gives the product of the nuclear quadrupole moment, Q, with the electric field gradient, q, at the nucleus. Assume that Qq is measured as a function of temperature for the Ti-nucleus in $BaTiO_3$ below the phase transition for this material. Below the phase transition temperature, T_c, assume that the material goes from the cubic structure discussed in Section 3-2a to one in which the Ti-atom moves along the c-axis and the coordinates are given by $(1/2, 1/2, 1/2 + \Delta)$ with no other changes in the other atom positions. In general $q = \Sigma_i\, e_i(3\cos^2\theta_i - 1)/r_i^3$ where an ion with charge e_i is at a distance r_i and θ_i is the angle between $\mathbf{r}_i$ and the c-axis. (a) Calculate q as a function of Δ taking only the nearest neighbor Ba^{2+}, Ti^{4+}, and O^{2-} ions into account, and assume that Δ is small compared to nearest-neighbor internuclear distances. (b) For a second-order phase transition we usually find $\Delta^2 \propto T_c - T$. What is the temperature dependence of q close to T_c?

8

The Covalent Bond

8-1 Introduction
8-2 Bonding and Antibonding
    a Hydrogen molecular ion
    b Dissimilar atoms (S)
8-3 The Hydrogen Molecule
    a Molecular orbital approach
    b Heitler-London approach
8-4 Maximum Overlap
    a The basic idea
    b Hybridization (σ-bonding)
    c Double and triple bonding (S)
    d Covalent radii (S)
    e Examples (S)
8-5 The Formation of a Crystal
8-6 "Classical" Semiconductors
    a Introduction (and band gap)
    b Impurity effects
    c The metal–nonmetal transition
    d High pressure phases (S)
8-7 Continuous Range of Bonding (S)
    a Pauling's electronegativity model
    b Electronic charge densities in semiconductors (S)
    c Phillips' and VanVechten's Determinations of f_i (A)
    Appendix

    Notes
    Problems

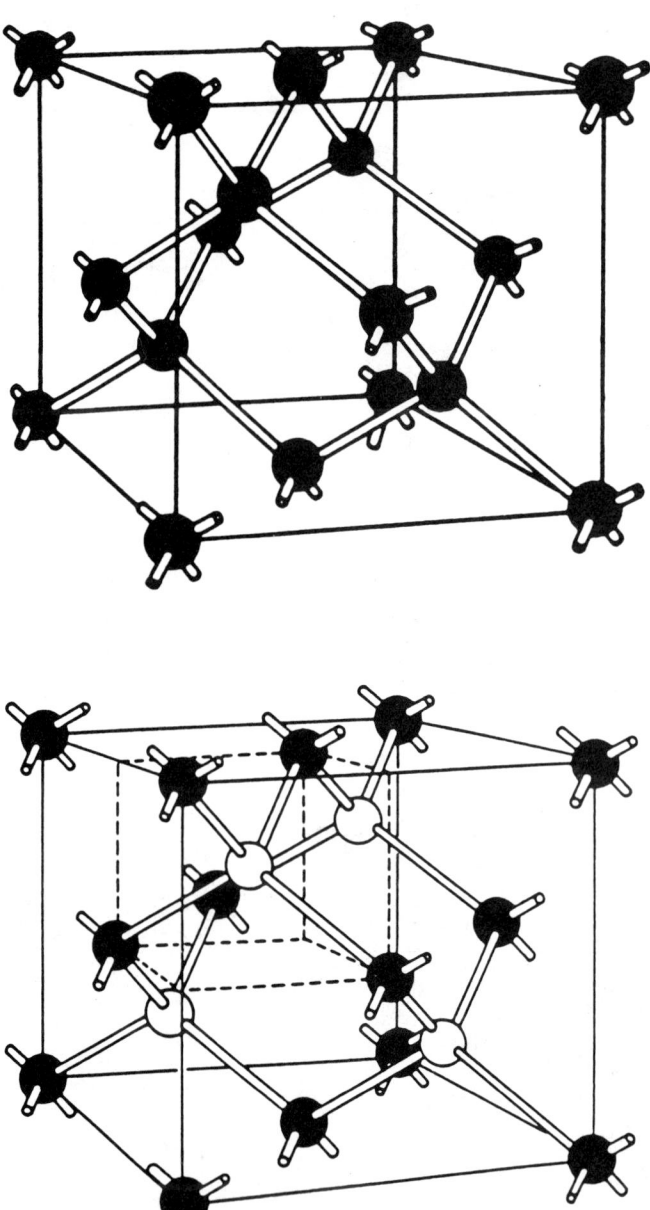

A unit cell of the diamond crystal structure (top) and the zinc blende crystal structure. The rods are meant to emphasize the tetrahedral bonds between each atom and its four nearest neighbors. These are also shown in Fig. 3-7 but here the emphasis is on the tetrahedral bonding.

THE COVALENT BOND

One can philosophize indefinitely about matters of this kind, but in practice one is limited by circumstance ...

Aristotle, "Politics"

8-1 Introduction

There is a large difference between the covalent bond and the ionic and van der Waals bonds. The latter two are characterized by rather small overlap of the ions, neutral atoms, or molecules. If these ions, neutral atoms, or molecules were in solution or free in a gas there would be little difference in their electronic configuration from that found in the solid state. The models that are useful for the elemental materials with ionic or van der Waals bonds are "billiard ball" models where the interactions are between either charged or neutral atoms, as was discussed in the last two chapters. The electronic configuration of the charged or neutral atoms is what is normally expected for that particular atom. However, for atoms that form covalent bonds there usually is a major rearrangement of the electronic configuration when going from the atomic state to the bonding state. The covalent bond can be thought of as due to, or associated with, the formation of an electronic state different from the free atom ground state and the overlap of the electronic wave function of one atom with that of its neighbor. In fact, we shall see that we may think qualitatively in terms of maximizing the overlap of the atoms to give stronger bonds. (For ionic or van der Waals bonds the small overlap is neglected.)

One striking characteristic of simple covalently bonded solids, already noted in Table 6-1, is that they have a low coordination number and hence a low density. The atoms in the classic covalently bonded, elemental solids (the group IV elements, carbon with the diamond structure, silicon, germanium, and gray tin) have four neighbors. At high pressures phase transitions occur to structures with a higher coordination number and density than those at zero pressure.

The plan for this chapter is simple. We start by discussing the hydrogen molecular ion H_2^+, which does not form a covalent bond in the traditional sense, since there is only one electron. Nevertheless, by using very simple mathematics, we arrive at the essence of this type of bond by finding the so-called bonding and antibonding states. The nature of these states is fundamental to covalent bonding in molecules and solids. Then we discuss the hydrogen molecule, which has two electrons and forms a true covalent bond. H_2 will be discussed in the molecular orbital and then in the Heitler-London approximation. These ideas are extended in a qualitative manner, applying the principle of maximum overlap to the concept of hybridization; in this way the covalent diamond structure can be understood. Last, the concept of a continuous range of bond types between covalent and ionic is considered. This topic is of present research concern.

8-2 Bonding and Antibonding

8-2a Hydrogen molecular ion Consider the H_2^+ molecular ion in Fig. 8-1a. The single electron is a distance r_a and r_b from the two nuclei and the nuclei are a distance d apart. The Hamiltonian is

$$H = \frac{\hbar^2}{2m}\nabla^2 - \frac{e^2}{r_a} - \frac{e^2}{r_b} + \frac{e^2}{d} \tag{8-1}$$

We want to determine only the low-lying energy levels; hence we approximate the atomic orbitals as follows: a sum of the 1s-wave functions centered on each of the two nuclei (each wave function is labeled a and b) is used for the total wave function. Then the approximate wave function for our problem is

$$\psi = c_a a + c_b b \tag{8-2}$$

The charge density on the two nuclei must be equivalent, so $c_a^2 = c_b^2$ and $c_a = \pm c_b$; the two possible wave functions, $\psi_\pm = N_\pm(a \pm b)$, are shown in Fig. 8-1b. To determine the wave functions explicitly we normalize them to unity as follows:

$$1 = \langle \psi | \psi \rangle = N_\pm^2 \int (a^* \pm b^*)(a \pm b) d^3 r$$

$$= N_\pm^2 (2 \pm 2S) \tag{8-3}$$

where $S \equiv \langle a | b \rangle$ is the overlap integral between the functions a and b. We have also used the fact that unperturbed 1s-orbitals, a and b, are real and normalized. The two wave functions are

$$\psi_+ = \frac{a+b}{\sqrt{2(1+S)}^{1/2}} \qquad \psi_- = \frac{a-b}{\sqrt{2(1-S)}^{1/2}} \tag{8-3}$$

The energies, $\mathscr{E}_\pm$, associated with these two wave functions can be found easily. We use the usual notation $H_{11} \equiv \langle a|H|a \rangle$, $H_{22} \equiv \langle b|H|b \rangle$, and $\beta \equiv \langle a|H|b \rangle$. (We can fix the phase of the wave functions so that all the matrix elements are real.) We see that $\mathscr{E}_+ =$

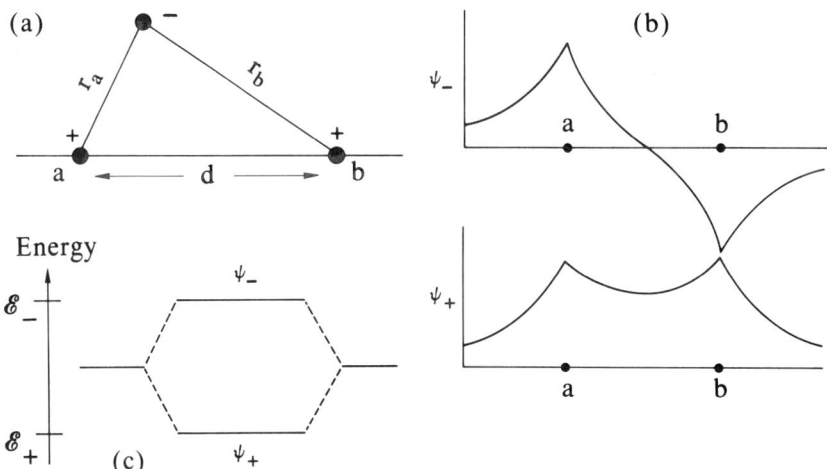

Fig. 8-1 (a) The hydrogen molecular ion. (b) The bonding, ψ_+, and antibonding orbitals, ψ_-, for H_2. (c) The energy level diagram for H_2.

$\langle \psi_+ | H | \psi_+ \rangle = N_+^2 \{H_{11} + 2\beta + H_{22}\}$ and $H_{11} = H_{22}$ for this problem. Similarly, $\mathscr{E}_-$ can be determined. The results are

$$\mathscr{E}_\pm = (H_{11} \pm \beta)/(1 \pm S) \qquad (8\text{-}4)$$

Identical results can be obtained via the secular determinant and secular equations, which are shown here for tutorial purposes. For the wave function shown in Eq. 8-2 the secular determinant immediately can be written and it is

$$\begin{vmatrix} H_{11} - \mathscr{E} & \beta - S\mathscr{E} \\ \beta - S\mathscr{E} & H_{22} - \mathscr{E} \end{vmatrix} = 0, \quad (H_{11} - \mathscr{E})(H_{22} - \mathscr{E}) = (\beta - S\mathscr{E})^2 \quad (8\text{-}5)$$

Since $H_{11} = H_{22}$, the two solutions for the energies are the same as shown in Eq. 8-4. To determine the wave functions associated with these energies, we take the solution for one value of the energy and substitute it back into the secular equations. For example, substituting for $\mathscr{E}_+$ we obtain $c_a(H_{11} - \mathscr{E}_+) + c_b(\beta - S\mathscr{E}_+) = 0$ from which $c_a = +c_b$ is determined. Similarly, using $\mathscr{E}_-$, $c_a = -c_b$ is obtained. The normalization is, of course, the same as the preceding and thus the same wave functions as in Eq. 8-3 are obtained.

$\mathscr{E}_+$ is a lower energy than $\mathscr{E}_-$ because H_{11} and β are negative numbers. (For most simple systems the matrix elements are all negative energies.) Figure 8-1c shows these two energy levels labeled by the wave functions. ψ_+ is called the **bonding state** because a great deal of the charge density is located between the two nuclei, conforming to what we would think of as bonding of the two nuclei. ψ_- is called the

antibonding state because there is a node between the two nuclei. The reason that ψ_+ has a lower energy than ψ_- is that ψ_+ has a great deal of charge in the region where it can be attracted to both nuclei, while ψ_- has much less.

Experimentally we find a dissociation energy D = 2.791 eV and an internuclear distance d = 1.06 Å. Evaluating the integrals H_{11}, β, and S in the preceding equations (see the Notes), we obtain the largest D (the minimum energy) of 1.76 eV at d = 1.32 Å. This is poor agreement, but only the ground state 1s-wave functions were used for a and b in Eq. 8-2. If the same 1s-wave functions are used but with an effective nuclear charge Z_{eff}, instead of 1, and if the total energy is minimized with respect to this parameter, then D = 2.25 eV and d = 1.06 Å are obtained for Z_{eff} = 1.228. (See the Notes for a discussion of this method.) Half of the 1 eV discrepancy is eliminated by this very simple change. By adding more orbitals to the basis set in Eq. 8-2 further improvements can be obtained. For example, if, besides the 1s-orbitals, 2p-orbitals are included, we obtain D = 2.71 eV at 1.06 Å. Thus, large improvements can be obtained by simple changes of the Z_{eff} or expansion of the basis set. However, irrespective of the details of the numerical agreement the qualitative concept of a bonding and antibonding state, as shown in Fig. 8-1, remains. Thus, this elementary example shows the concept of bonding and antibonding states, and we see that the ground and excited state wave functions of a covalent bond are rather different from those of the free atom.

8-2b Dissimilar atoms Consider multielectron atoms. In the previous section we took the two orbitals, a and b, to be the same, but there is no fundamental reason why the two eigenfunctions, a and b in Eq. 8-2, need be the same; they may be associated with dissimilar atoms. The eigenfunctions could be associated with a situation as complicated as the following: the a-state might be a central atom and the b-state a linear combination of atomic orbitals (LCAO) of the atoms surrounding the a-atom. For example, we might be dealing with the $(FeF_6)^{3-}$ complex where Fe^{3+} is the central a-atom and some linear combination of the F^- ions orbitals is the b-state.

The details are not important. It is merely important to remember that we are considering the bonding between an atom (or some combination of atoms) in the a-state with another atom (or some combinations of atoms) in a b-state. For convenience, in Eq. 8-4 we take $H_{11} > H_{22}$ (H_{22} is more negative; for example the F^- ions are more stable than the Fe^{3+} ions). If we assume that the effect of bonding is small, then we can use perturbation theory. Using the matrix elements

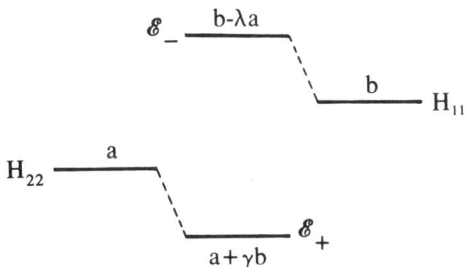

Fig. 8-2 An energy level diagram showing a bonding and an antibonding level in the center. The noninteracting states are on the left and right. The wave functions are indicated.

shown in Eq. 8-5, to second order in energy, the solution is

$$\mathscr{E}_+ = H_{22} - \frac{(\beta - H_{22}S)^2}{H_{11} - H_{22}}, \quad \mathscr{E}_- = H_{11} + \frac{(\beta - H_{11}S)^2}{H_{11} - H_{22}} \quad (8\text{-}6a)$$

$$\psi_+ = N_+ (a + \gamma b), \quad \psi_- = N_- (b - \lambda a) \quad (8\text{-}6b)$$

From $\langle \psi_+ | \psi_- \rangle = 0$ we find $\lambda = (\gamma + S)/(1 + \gamma S) \approx \gamma + S$. Then by inserting a value of the energy from Eqs. 8-6a into the secular equations we may solve for γ and λ. The result is

$$\gamma \approx -\frac{\beta - H_{22}S}{H_{11} - H_{22}} \qquad \lambda \approx -\frac{\beta - H_{11}S}{H_{11} - H_{22}} \quad (8\text{-}7)$$

(Remember that β is negative.) The need for the γ-coefficient in the bonding state wave function, Eq. 8-6b, is clear. Since the a-state and b-state atoms (or complexes) are different, nonequal amounts of them will be required in the bonding state. Similarly the λ-coefficient is required for the antibonding state.

The results of these general considerations are shown in Fig. 8-2 and are quite general. The energy repulsion shown here is really just a manifestation of the famous "no-crossing rule." This rule states that when eigenstates have a nonzero interaction due to a perturbation, they repel each other. On the other hand, if the wave functions, a-state and b-state, have zero interaction the off-diagonal terms β and S are zero and there is no shift of the energy levels.

Consider the large amount of energy that may be gained in the bond indicated in Fig. 8-2. If each atom has one "outer" electron, when the atoms come together both outer electrons can have opposite spin and be in the bonding eigenstate ψ_+. The binding energy is $H_{22} - H_{11}$ plus $2(\mathscr{E}_+ - H_{22})$. For example, if the two atoms are both hydrogen atoms then $H_{11} = H_{22}$. If the overlap term is ignored and the energy of the interacting atoms is compared to the energy of the well separated atoms, it is clear that the former is lower in energy by 2β. (Use Eq. 8-4.) This can be several electron volts.

To obtain actual numbers we usually resort to approximate techniques to calculate the energies. The Wolfsberg-Helmholtz approximation relates the off-diagonal terms to the diagonal ones $H_{ij} = g\,S(H_{ii} + H_{jj})/2$. Then g is usually taken as 2.0 or 1.75, the overlap S is evaluated in a straightforward manner, and the diagonal matrix elements are taken as the experimental ionization energies. Consult the Notes for the extensive literature in this field.

8-3 The Hydrogen Molecule

Since only two electrons are involved the H_2 molecule is the simplest example in which we can treat a true covalent bond. Figure 8-3a shows the coordinates used for the Hamiltonian

$$H = -\frac{\hbar^2}{2m}(\nabla_1^2 + \nabla_2^2) - \frac{e^2}{r_{a1}} - \frac{e^2}{r_{a2}} - \frac{e^2}{r_{b1}} - \frac{e^2}{r_{b2}} + \frac{e^2}{r_{12}} + \frac{e^2}{d}$$

(8-8)

Although we have not discussed it, for H_2^+ an exact solution is possible by transforming to elliptical coordinates. However, for H_2 an exact solution is not possible because of the **electron–electron repulsion**, e^2/r_{12}, term. Two well-known approximate methods of solving this problem will be examined.

8-3a Molecular orbital approach In a manner similar to that described for the H_2^+ ion, we would expect to approximate the lowest molecular orbital by the same type of wave function as used previously, namely $a(1) + b(1)$, where a and b refer to wave functions centered on nucleus a and b, respectively, and the 1 refers to electron 1. For ease of thinking, we can always picture using the 1s wave functions for orbitals. As a starting point take the unnormalized wave function as

$$\psi = [a(1) + b(1)][a(2) + b(2)] \quad (8\text{-}9)$$

which is the product of the wave functions of electrons 1 and 2. (The conforming of this function to the Pauli principle is discussed in the next subsection.) If we ignore the e^2/r_{12} term, the Hamiltonian in Eq. 8-8 is approximately the sum of two Hamiltonians, one for each electron of the type for H_2^+ (Eq. 8-1). In this approximation the total wave function is the product of two wave functions of the Eq. 8-2 type. This is justification of the function in Eq. 8-9. Using 1s-orbitals we know that $a(1) \propto \exp(-r_{a1}/a_0)$, $a(2) \propto \exp(-r_{a2}/a_0)$, and so forth. The experimental values at equilibrium are $D = 4.72$ eV and $d = 1.40a_0$. Using Eq. 8-9 as an approximation we obtain $D = 2.65$ eV and $d = 1.6a_0$. (a_0, the Bohr radius, equals 0.52915 Å.) The lower curve in Fig. 8-3b shows $\mathscr{E}$ vs. d for the H_2 molecule obtained from

CHAPTER 8   THE COVALENT BOND

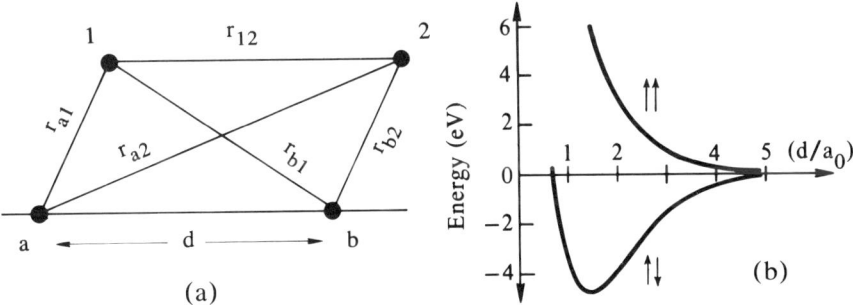

Fig. 8-3 (a) An H_2 molecule. (b) $\mathscr{E}$ vs. internuclear distance (in units of the Bohr radius) for the hydrogen molecule. The lower curve is for oppositely paired spins, and the upper curve is for parallel spins (which is an antibonding state).

very elaborate calculations. About the minimum energy, $\mathscr{E}$ vs. d can be approximated by a harmonic well, and the curvature yields a vibrational frequency. 4400 cm^{-1} is found experimentally for this frequency, which can be computed from the $\mathscr{E}$ vs. d result. The agreement with the results of this simple trail wave function is poor.

The reason for the poor agreement can be seen by multiplying the right side of Eq. 8-9 to obtain the individual terms. We obtain

$$\psi = a(1)\,a(2) + a(1)\,b(2) + b(1)\,a(2) + b(1)\,b(2) \qquad (8\text{-}9)$$

The first and last terms represent ionic states with both electrons on the same nucleus, which corresponds to H^+H^-. Such states are not very stable and are less likely to occur than the middle two terms, which closely represent one electron on each nucleus and correspond to a large amount of charge between the two nuclei, that is, bonding states. For covalent bonding, weighing the ionic terms equally with the two middle states does not make physical sense. This leads to the Heitler–London approximation.

8-3b Heitler–London approach In this approach, sometimes called the **valence bond method**, instead of Eq. 8-9 we take

$$\psi = a(1)\,b(2) + a(2)\,b(1) \qquad (8\text{-}10)$$

as the wave function. This function completely ignores the ionic configurations. The solutions for the problem can be found in Eyring, Walter, and Kimball. The results are D = 3.14 eV and d = $1.64a_0$, an improvement over the molecular orbital approach.

As is always the case, improvements can be made by using more parameters in the starting wave function and then minimizing the energy with respect to these parameters. Of course, we would like to add sensible parameters. If, instead of a nuclear charge of unity, a variable Z_{eff} is used, for example, $a(1) \propto \exp(-Z_{eff} r_{a1}/a_0)$, and so on, then we obtain $Z_{eff} = 1.17$, D = 3.76 eV and d = $1.41 a_0$. If we add some p_z-function, for example, $a(1) \propto (1 + c_1 z) \exp(-Z_{eff} r_{a1}/a_0)$, and so on, then with $Z_{eff} = 1.17$ and $c_1 = 0.10$, D = 4.02 eV and d = $1.40 a_0$ are obtained.

It is interesting to see the result of starting with Eq. 8-10 and adding the ionic terms in Eq. 8-9, but multiplied by a parameter λ,

$$\psi = \psi_{cov} + \lambda \psi_{ion}$$

Solving, by minimizing the energy, we obtain $Z_{eff} = 1.19$, $c_1 = 0.07$, $\lambda = 0.175$, and D = 4.10 eV. The small value found for λ confirms the fact that Eq. 8-10 is a sensible approach. The difference in energy, 4.10 eV − 4.02 eV, is sometimes called the covalent-ionic **resonance energy**, that is, the amount of lowering of the energy by adding the new configuration. However, care must be exercised in its use because it has no absolute meaning but is only an energy relative to the starting orbitals. For example, if a much poorer ψ_{cov} were used to begin with, then the addition of $\lambda \psi_{ion}$ would have a much larger effect.

So far we have not mentioned electron spin. The Hamiltonian, Eq. 8-8, has no spin-dependent terms so that the wave functions are separable into an orbital and a spin part. Until now we have written just the orbital function; the spin part will now be added. Using the usual notation where α is a spin eigenfunction with eigenvalue $+1/2$ and β has eigenvalue $-1/2$, we write the total wave function in the Heitler-London approximation. Remember that the total wave function must be antisymmetric with respect to the interchange of electrons according to the Pauli principle.

$$\psi(^1\Sigma_g^+) = \{a(1) b(2) + a(2) b(1)\}\{\alpha(1) \beta(2) - \alpha(2) \beta(1)\} \quad (8\text{-}11a)$$

$$\psi(^3\Sigma_u^+) = \{a(1) b(2) - a(2) b(1)\} \begin{Bmatrix} \{\alpha(1) \alpha(2)\} \\ \{\alpha(1) \beta(2) + \alpha(2) \beta(1)\} \\ \{\beta(1) \beta(2)\} \end{Bmatrix} \quad (8\text{-}11b)$$

Equation 8-11a has the orbital part that has been discussed, along with an antisymmetric spin function. Under interchange of electron 1 with electron 2 we can see that the orbital part transforms into $+1$ times itself and the spin part into -1 times itself. The total electron spin, S, is zero so that the **spin degeneracy** $2S + 1 = 1$. The three wave functions in Eq. 8-11b all have an antisymmetric orbital state (it transforms into -1 times itself if electrons 1 and 2 are interchanged) and symmetric spin states. For these functions S = 1 and the spin degeneracy is 3. The Σ_g^+ and Σ_u^+ nomenclature in these equations refers to irreducible representation labels for the $D_{\infty v}$ point group and the spin degeneracy appears as an upper left superscript as in atomic

CHAPTER 8   THE COVALENT BOND   157

physics. The $\mathscr{E}$ vs. d for the antibonding $^3\Sigma_u{}^+$ state is the upper curve in Fig. 8-3b.

Summary The Heitler-London approach leads to results like those found in Section 8-2. The bonding state, Eq. 8-11a, has charge concentrated between the nuclei and the two bonding electrons have opposite spin so $S = 0$ (and the charge density is high between the two nuclei). The antibonding state at higher energy, Eq. 8-11b, has $S = 1$ and this state has a node in the charge density between the two nuclei. The ionic configurations are ignored but can be added, if desired, as shown previously. Although we may become much more sophisticated when dealing with larger molecules, these simple ideas are basic to an understanding of the chemistry of molecules and solids.

8-4 Maximum Overlap

We may appreciate, from the last two sections, that for larger molecules or solids the calculations rapidly become more cumbersome. Thus, to understand a broad sweep of materials, some qualitative principles must be developed. The idea used for covalent materials is the principle of **maximum overlap**. This principle states that the binding energy is lowered as neighboring electron clouds overlap each other to a larger extent. This applies to antiparallel spin states. However, the principle cannot be pushed to the extreme or else all the atoms in the world would combine to form one very large atom. Also, the promotion energy must be kept in mind; this will be discussed. We shall see how this principle helps us to understand hybridization.

8-4a The basic idea First, in a hydrogen atom the single s-electron has a spin. However, as we have seen, for H_2 there is a bond and, to a reasonable approximation, the bond can be called an s-s bond since the s-orbitals from both atoms can be used. The electron spins in this bond are paired; that is, one spin is up and one is down to give a total spin $S = 0$, Eq. 8-11a. Second, a fluorine atom has a $1s^2 2s^2 2p_x^2 2p_y^2 2p_z^1$ electron configuration. Note that within the atom all the spins are paired except the $2p_z$. An F_2 molecule is formed by the bond formed by the 2p-orbitals on the two atoms overlapping in such a manner to give $S = 0$, that is, paired spins. This is a p-p bond. Last, the molecule HF is formed by overlap and by pairing the spins of the 1s of hydrogen with the $2p_z$ of fluorine to give an s-p bond. The bond strengths differ and decrease in order

$$p - p > s - p > s - s$$

This is also in the order of decreasing overlap.

Now consider some more complicated situations. An oxygen atom has an electron configuration $1s^2 2s^2 2p_z^2 2p_y^1 2p_x^1$. These two unpaired electrons on the O-atom can bond to two other atoms (oxygen is divalent). Note that the unpaired orbitals are at right angles to one another. This explains the characteristic bonding angles found for oxygen. In H_2O the angle is 105°. The increase over 90° is partly due to the H-H interaction. For H_2S, where the H-atoms are farther inside the more polarizable sulphur atom, the angle is 92°.

Even with this simple H_2O molecule example, a fundamental difference between covalent and ionic bonding can be seen. If the H_2O molecule were ionicly bonded, the forces would be central forces, that is, the force between one H and the O-atom would be independent of where the other H-atom was located and the Coulomb repulsion between H-ions would lead to a linear molecule. However, in the covalently bonded H_2O molecule there is a noncentral or bond bending force, that is, the force between one H and the O-atom depends on where the other H-atom is located. In other words, once one H atom is attached to an oxygen atom the covalent part of the potential energy of the bonding has a minimum at 90°.

Now, getting back to our general problem, a nitrogen atom has the configuration $1s^2 2s^2 2p_x^1 2p_y^1 2p_z^1$ so that there are three unpaired electrons (the element is trivalent). Thus, it should form bonds with three other atoms and the bonds should all be at right angles to each other. This is found in NH_3 with a slight increase in the bond angles, again mainly due to a repulsion among the hydrogen atoms.

The carbon atom has a $1s^2 2s^2 2p_x^1 2p_y^1$ configuration, so that we would expect it to behave like oxygen and be divalent. Actually, it is tetravalent and behaves quite different from oxygen. The concept of hybridization is basic to the understanding of the behavior.

8-4b Hybridization (σ-bonding) Instead of starting with a discussion of carbon, the classical example of hybridization, we first consider the simpler case of an atom with fewer valence electrons.

Consider a beryllium atom, $1s^2 2s^2$. There are no unpaired electrons in the ground state so a covalent bond is not expected. However, there is a low-lying excited state, $1s^2 2s^1 2p_y^1$, with two unpaired electrons. (Instead of p_y we may say p_x or p_z.) We can form two hybrid orbitals ϕ_1 and ϕ_2 from the unpaired s- and p-orbitals.

$$\phi_1 = (s + p)/\sqrt{2} \qquad \phi_2 = (s - p)/\sqrt{2} \qquad (8\text{-}12)$$

The original s- and p-orbitals as well as ϕ_1 and ϕ_2 are shown in Fig. 8-4a. As can be seen, the latter two orbitals have larger charge densities to the right and left, respectively, than a p-function or an s-function alone. Thus, these new orbitals can overlap and form a

CHAPTER 8   THE COVALENT BOND   159

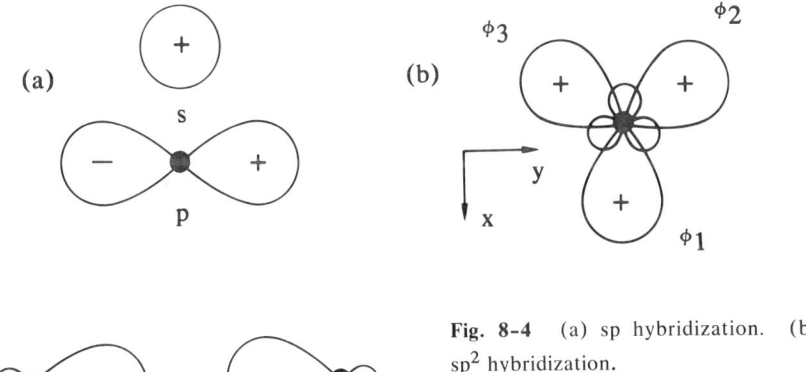

Fig. 8-4 (a) sp hybridization. (b) sp² hybridization.

covalent bond with two atoms. ϕ_1 can bond to the atom on the right, and ϕ_2 to one on the left. The formation of these hybrid orbitals, Eq. 8-12, from the $2s^2$ ground state "costs" an energy called the **promotional energy** (the energy difference between $2s^1 2p^1$ and $2s^2$) but we "get back" the overlap energy. *The balance between these two energies determines to what extent this process actually takes place.*

A boron atom has $1s^2 2s^2 2p_x^1$ for a ground state configuration, so only one unpaired electron is available for bond formation. However, one of the 2s-electrons may be promoted to a 2p-orbital so that the excited state configuration is $1s^2 2s^1 2p_x^1 2p_y^1$. A hybrid bond can then be formed. The details are left as a problem. The results are shown in Fig. 8-4b for the hybrid function ϕ_1, ϕ_2, and ϕ_3 where

$$\begin{bmatrix} \phi_1 \\ \phi_2 \\ \phi_3 \end{bmatrix} = \begin{bmatrix} a & 2b & 0 \\ a & -b & c \\ a & -b & -c \end{bmatrix} \begin{bmatrix} s \\ p_x \\ p_y \end{bmatrix} \qquad (8\text{-}13)$$

There are large charge lobes in the three directions, separated by 120°. This type of hybridization is called an sp² hybrid; it is planar with orbitals pointed toward the corners of an equilateral triangle.

Carbon with an atomic configuration $1s^2 2s^2 2p_x^1 p_y^1$ can be promoted to $1s^2 2s^1 2p_x^1 2p_y^1 2p_z^1$ and sp³ hybrid bonds can be formed. These bonds point to the corners of a regular tetrahedron. This tetrahedral hybridization is, of course, an important example of a hybrid function; it is the basis of the diamond crystal structure that is so important in the semiconductor field.

We shall not discuss in detail any more hybrid orbitals, but Table 8-1 lists some of the common orbitals and Eyring, Walter, and Kimball, Table 12.2, show a much more extensive list.

Table 8-1 Some common hybrid orbitals. (Note for p^3 there is no hybridization.)

Number of bonds	Hybrid	Direction of bonds
2	sp	linear
3	sp^2	trigonal, planar
	sd^2	trigonal, planar
	p^3	trigonal pyramid
4	sp^3	tetrahedral
	dsp^2	tetragonal, planar
5	d^4s	tetragonal pyramid
6	d^2sp^3	octrahedral
	d^4sp	trigonal prism
8	sp^3d^3f	cube vertices

Summary – For many atoms the covalent bond is quite a strong bond. In fact, the ground state electron configuration will be greatly distorted (for example, a carbon 2s electron will be promoted to a 2p) in order to have the proper configuration to form these bonds. Of course, the processes of promotion, hybridization, and then overlap to form the bond actually occur together and not in step-by-step sequence. The step-by-step sequence is discussed only to help a scientist conceptualize a quantum system going into its lowest energy state.

For covalently bonded atoms there is a great deal of charge midway between the atoms, unlike ionicly bonded materials. Further, covalent bonding is highly directional. The spatial distribution can be described by atomic or hybrid orbitals with very specific angles between the orbitals, leading to special structures. These angles are independent of the radius ratios discussed in Section 7-3b for the ionic bond. In fact, the simple tetrahedrally coordinated solids with the diamond and zinc blende structures have very low density, which leads to interesting high pressure effects (Section 8-6d). The charge density and directional effects mentioned here will be discussed again in Section 8-6 for binary compounds. There we will show that as the crystals become less covalent the structures change from ones with tetrahedral bonds into ones that have the NaCl structure (which is associated with ionic bonding).

8-4c Double and triple bonding So far, in Section 8-4b, single bonds between neighboring atoms have been discussed. Now we consider double and after that, triple bonds.

First several definitions are needed. A **nodal surface** is the locus of points for which the wave function is zero. For example, a p_z-wave function has the xy-plane as a nodal surface. A *σ*-**bond** is defined as one for which the bond direction does not contain a nodal plane. A

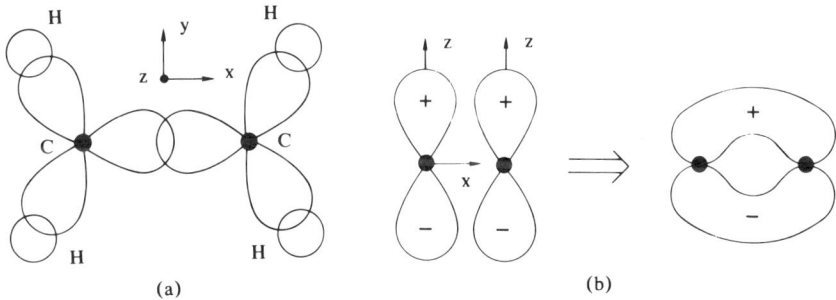

Fig. 8-5 (a) Top view of an ethylene molecule showing the sp² σ-bonding but not the π-bonds. (b) The overlap of two orbitals to form a π-bond.

π-bond is defined as one for which the bond direction contains one nodal plane, and an example will be given.

Consider an ethylene (C_2H_4) molecule, Fig. 8-5a. Each carbon atom is bonded to three neighbors: two hydrogen atoms and one carbon atom. The sp² hybrid orbitals are shown in Fig. 8-4b. Using these three orbitals, σ-bonds with the three neighbors of each carbon atom can be formed; each of these orbitals is filled with two electrons of opposite spin. This bonding scheme leaves an unpaired p_z-electron on each carbon. Figure 8-5b shows how these p_z-orbitals can overlap and form a π-bond. The addition of π-bonding to the σ-bonding will lower the energy of the entire molecule. Atoms connected by σ- as well as π-bonds are said to be **double bonded**, and the formula is usually written as $H_2C = CH_2$. Since the two double bonded carbon atoms are more tightly bonded than two carbon atoms that are just σ-bonded, the potential energy is lower and the internuclear distance of the former is shorter than that of the latter.

These ideas can be extended to the linear molecule acetylene (C_2H_2), which has a triple bond between the carbons. Each carbon can have two σ-bonds, one to a carbon and one to a hydrogen atom. These are formed from $s \pm p_x$ hybrid functions. Then the p_y- and p_z-orbitals can overlap and bond to the same orbitals on the other carbon atom. Each orbital will contain oppositely paired spins. The formula for acetylene can be written as HC ≡ CH. Because the single, double, and triple bonds are progressively stronger (deeper energy well) the internuclear distances are progressively shorter (1.54 Å, 1.33 Å, and 1.20 Å) and vibration frequencies are higher (greater curvature of the energy on internuclear distance curve).

8-4d Covalent radii We have discussed ionic and van der Waals radii so it is natural to inquire into the characteristic radii for atoms that are covalently bonded. Such values have already been mentioned for carbon single, double, and triple bonds and the existence of such values in general would enable us to predict internuclear distances.

Pauling (Chapter 7) discusses such radii at length, and Table 8-2 summarizes the values. **Part a** gives the radii for single, double, and triple bonds as discussed in Section 8-4c. Naturally, atoms like fluorine, with only one unpaired electron, can form only a single bond and atoms like oxygen can form only a single or double bond. **Part b** gives the radii for tetrahedral sp^3 hybridization. For elements in group IV of the periodic table these radii are just the single bond radii. **Part c** shows octahedral radii (six neighbors) appropriate to d^2sp^3 and sp^3d^2 hybridization. From a symmetry point of view these hybrid orbitals are the same and give orbitals pointing to the corners of a regular octahedron (Table 8-1). However sp^3d^2 refers to orbitals obtained by promoting s and/or p electrons to empty d-levels to form the hybrid orbitals, while d^2sp^3 refers to promoting $(n-1)d$ electrons to the ns- and/or np-levels to form the hybrid bonds. **Part d** shows the planar square radii appropriate for $d\,sp^2$ hybrid bonds. **Part e** gives the covalent linear radii for sp hybridization.

Note, because of the different electron configuration between a free atom and one with a covalent bond, the covalent radii *may not* be interpreted as the radii of spherical atoms as may be done for ionic radii; they are radii that can be used to calculate internuclear distances between atoms that are covalently bonded. Figure 8-6 stresses this rather dramatically for Cl_2 where the van der Waals radius is almost twice as large as the single bond covalent radius.

8-4e Examples There are literally thousands of examples of structures that have covalent bonds. Pauling's book has many examples. We consider just three simple examples.

Figure 8-7 shows the structure of the molecule adamantane, $C_{10}H_{16}$. It is a highly symmetric molecule composed of four chair-like rings of six carbon atoms where six carbon atoms are bonded to 2C + 2H and four carbon atoms are bonded to 3C + 1H. The molecule has $T_d(\bar{4}3m)$ point symmetry, as can be seen. All the carbon atoms have four neighbors, as we expect for sp^3 hybridization. The measured C-C and C-H distances are 1.54 Å and 1.09 Å, which are very close to the sum of the covalent radii in Table 8-2. Closely related to this molecule is hexamethylene-tetramine $(C_6H_2)_6N_4$, obtained by replacing the four carbon and hydrogen atoms that are located along the body diagonal in adamantane with four nitrogen atoms. Note that it too has $T_d(\bar{4}3m)$ symmetry and the remaining six carbon atoms still have sp^3 hybridization. However, since adamantane is a somewhat spherical molecule, it has a plastic phase in the solid state as discussed in Section 6-5d.

CHAPTER 8   THE COVALENT BOND

Table 8-2 Covalent radii (in Å).

(a) Normal radii

	H	C	N	O	F
Single bond	0.30	0.77	0.74	0.74	0.72
Double bond	—	0.67	0.62	0.62	—
Triple bond	—	0.60	0.55	—	—

	Si	P	S	Cl
Single bond	1.17	1.10	1.04	0.99
Double bond	1.07	1.00	0.94	—
Triple bond	1.00	0.93	—	—

	Ge	As	Se	Br
Single bond	1.22	1.21	1.17	1.14
Double bond	1.12	1.11	1.07	—

	Sn	Sb	Te	I
Single bond	1.40	1.41	1.37	1.33
Double bond	1.30	1.31	1.27	—

(b) Tetrahedral radii

		Be	B	C	N	O	F
		1.06	0.88	0.77	0.70	0.66	0.64
		Mg	Al	Si	P	S	Cl
		1.40	1.26	1.17	1.10	1.04	0.99
Cu	Zn	Ga	Ge	As	Se	Br	
1.35	1.31	1.26	1.22	1.18	1.14	1.11	
Ag	Cd	In	Sn	Sb	Te	I	
1.52	1.48	1.44	1.40	1.36	1.32	1.28	
Au	Hg	Tl	Pb	Bi			
1.50	1.48	1.47	1.46	1.46			

(c) Octahedral radii, d^2sp^3 hybrids

Fe^{II}	Co^{II}	Ni^{II}	Ru^{II}		Os^{II}		
1.23	1.32	1.39	1.33		1.33		
	Co^{III}	Ni^{III}		Rh^{III}		Ir^{III}	
	1.22	1.30		1.32		1.32	
Fe^{IV}	Ni^{IV}			Pd^{IV}		Pt^{IV}	Au^{IV}
1.20	1.21			1.31		1.31	1.40

sp^3d^2 hybrids

	Ti^{IV}	Zr^{IV}	Sn^{IV}	Te^{IV}	Pb^{IV}
	1.36	1.48	1.45	1.52	1.50

(d) Square radii

Ni^{II}	Pd^{II}	Pt^{II}	Au^{III}
1.39	1.31	1.31	1.40

(e) Linear radii

Cu^{I}	Ag^{I}	Hg^{II}
1.18	1.39	1.29

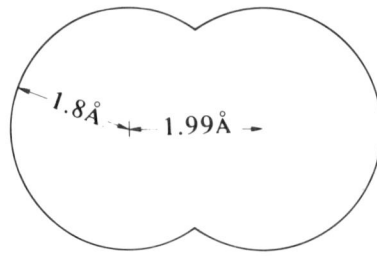

Fig. 8-6 The dimensions of a Cl_2 molecule showing the single bond covalent radius of (1.99/2) Å and the van der Waals radius of 1.8 Å.

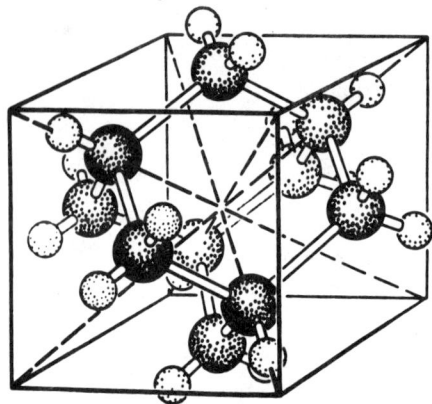

Fig. 8-7 Adamantane.

Figure 8-8 shows a unit cell of the K_2PtCl_4 structure. It has space group D_{4h}^1(P4/mmm) with positions

Pt	1a	(0, 0, 0)		
K	2e	(0, 1/2, 1/2)	(1/2, 0, 1/2)	
Cl	4j	$\pm$(u, u, 0)	$\pm$(u, $\bar{u}$, 0)	u ≈ 0.233

a = 6.99 Å and c = 4.13 Å. (This notation is discussed in Chapter 3.) The computed distances are Pt-Cl = 2.32 Å and K-Cl = 3.28 Å. As can be seen the Pt-atoms are in square planar coordination (dsp^2 hybridization) and the radii in Table 8-2 very closely predict the Pt-Cl distance. The compound $(NH_4)_2PtCl_4$ also has this crystal structure.

Figure 8-9 shows the cubic unit cell of K_2PtCl_6. The space group is O_h^5(Fm3m) and the structure is similar to the CaF_2 structure with the $PtCl_6$ groups at the face-centered positions and the K-atoms at the 8c positions (as are the F-atoms in CaF_2). The Pt-Cl distance is 2.33 Å, in close agreement with what is found from Table 8-2. A very wide variety of compounds have this structure and many phase transitions are found associated with various kinds of rotation of the octrahedra.

8-5 The Formation of a Crystal

In Section 8-2 and Fig. 8-1 we have seen how a bond can be formed between two atoms. Now we discuss from a conceptual point of view, how a covalent crystal is formed.

CHAPTER 8   THE COVALENT BOND   165

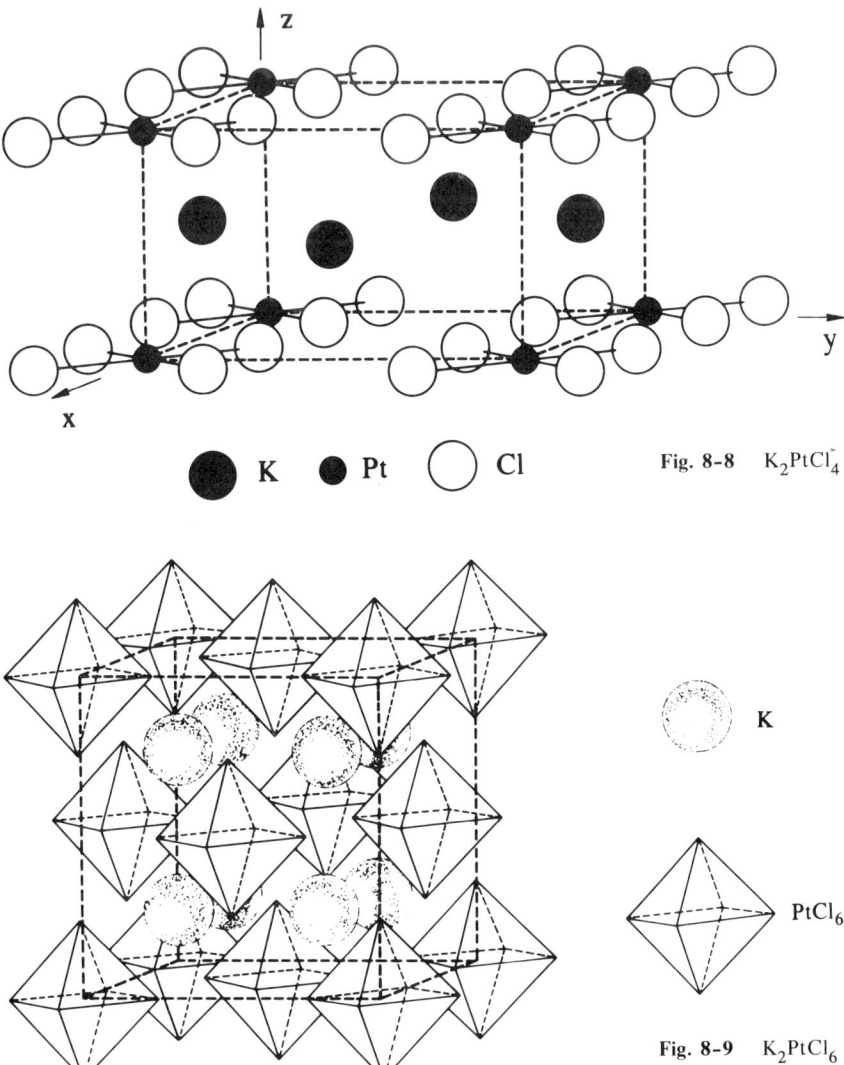

Fig. 8-8 $K_2PtCl_4^-$

Fig. 8-9 K_2PtCl_6

Just as Fig. 8-1 shows the basic effect on the eigenvalues when two similar atoms are brought together, Fig. 8-10a shows this effect in a solid where very many similar atoms ($\approx 10^{22}$) are brought together. For ease of discussion we will consider all of the atoms to be the same kind and discuss the two situations in parallel. When the atoms are very far apart, the eigenfunctions all have the same energy. For two atoms there is a twofold degeneracy and for N atoms there is an N-fold degeneracy. (Actually, the degeneracy will be somewhat less since some of the eigenfunctions may have a two- or threefold essential degeneracy.) When the atoms are close enough so there is a weak

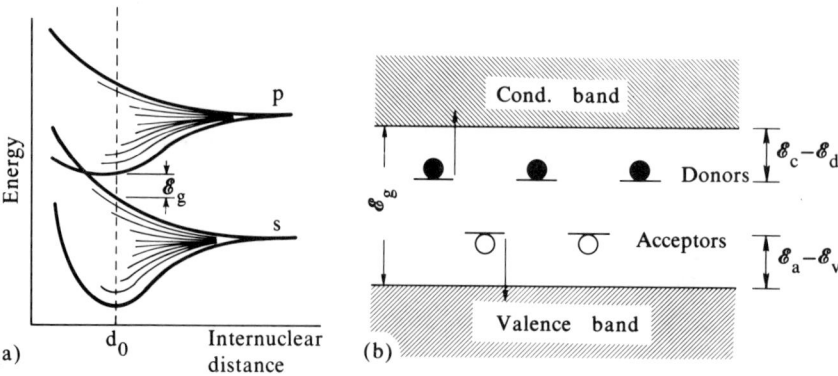

Fig. 8-10 (a) A schematic diagram showing how single-atom energy levels broaden into bands as the atoms are brought together. (b) An energy gap in a semiconductor showing donor and acceptor levels.

interaction (a small amount of overlap of the wave functions), the degeneracy is lifted; for two atoms there are two separate quantum levels (Fig. 8-1), for six atoms there are six separate quantum levels, and for N atoms there are N separate quantum levels. For the N-atom case the light lines in Fig. 8-10a labeled "s" indicate this situation. This is a general effect: as soon as there is interaction between the eigenfunctions on the separated atoms, the N-fold degeneracy is lifted and approximately N quantum levels are obtained. For 10^{22} atoms this results in a large number of separate but very closely spaced states and we say, "the levels broaden into a band."

Let us consider the width of these "bands." For the two-atom case the energy separation is 2β, where $\beta \equiv \langle a|H|b \rangle$. For a six-atom linear chain (or ring) of atoms, or N atoms, the maximum and minimum eigenvalues will be separated by about the same value as for the two-atom case (assuming the same internuclear separation). This is because the lowest eigenvalue corresponds to each atom being in a bonding state with each of its neighbors. The highest eigenvalue corresponds to each atom being in antibonding state with each of its neighbors. Intermediate eigenvalues correspond to a phase relation of the wave functions between neighboring atoms that is in between 0 (bonding) and π (antibonding). The important point is that the maximum spread essentially is independent of the number N, and for ordinary internuclear distances the spread of eigenvalues is a few electron volts. Thus, for very large N ($\approx 10^{22}$) the energy levels are extremely closely spaced and for all practical purposes a quasicontinuous energy band is formed. This is the implication in Fig. 8-10a of the many energy levels (lines) drawn for each of the two bands that are shown. (The two separate bands, labeled s and p, come from entirely different

CHAPTER 8   THE COVALENT BOND   167

energy levels in the atoms and are discussed in the material that follows.)

With reference to Fig. 8-1a, we considered the case of only one eigenvalue for the isolated atom. However, as indicated in Fig. 8-10a, the situation can be complicated because isolated atoms typically have more than one quantum level with energy separations of only a few electron volts. For example, this is the situation for the 3s and 3p valence levels in silicon. The labels s and p in Fig. 8-10a indicate this situation schematically. When N silicon atoms are close enough to begin to interact, the N 3s levels begin to broaden into bands as discussed previously. Independently the N 3p levels also broaden into bands. Both of these effects are shown in Fig. 8-10a. The effects are complex and more difficult to describe when the upper part of the lower band has approximately the same energy as the lower part of the upper band. When this happens for the silicon case we get sp^3 hybridization and a repulsion of the energy levels, all of which depend on the spatial directions. Some of the results are discussed in Chapter 10. However, these complex effects do not alter the basic band situation shown in Fig. 8-10a.

There is one very important point that we discussed in Section 8-2 but not here as yet. Besides having bands of energy levels we want to put electrons into these levels; two electrons, one with spin up and one with spin down, can occupy each of these energy levels. In Chapter 10, the filling of these energy levels will be discussed and the various fundamentally different possibilities will be emphasized. Here, referring to Fig. 8-10a, we just discuss two simplified cases.

(1) If each "s-level" (one on each of the N isolated atoms) has just one electron, then when the levels are broadened into a band only the lower half of the energy levels will be occupied; this is the same situation as in Section 8-2. Then there are unoccupied energy levels extremely close in energy to the occupied ones. Thus, with very small perturbations, electrons can be excited from occupied states into empty quantum states. If the small perturbation is a weak electric field this will lead to electrical conduction; such materials are metals and are discussed in the next chapter.

(2) If each "s-level" (one on each of the N isolated atoms) has two electrons, then when the levels are broadened into a band all of the quantum levels in the band are filled. In Fig. 8-10a, at the equilibrium internuclear distance, d_0, there is an energy separation between the top of the s-band and the bottom of the p-band. Thus, the filled states in the s-band cannot easily be excited into the empty states in the p-band. They must be given enough energy to overcome the **energy gap** $\mathscr{E}_g$ indicated in the figure, and the energy gap can be several electron volts. (The energy gap is defined as the energy separation between the filled and empty electron states.) Materials with an energy

gap are called semiconductors or insulators and are discussed briefly in the rest of this chapter and in Chapter 10. We call the band with the filled states the **valence band** while the empty band is called the **conduction band**. The reasons for the names will become more obvious later. Figure 8-10b shows the situation but emphasizes only the band gap region, and note that the x-coordinate is no longer internuclear distance; it is just ordinary space in the crystal. (Ignore the donors and acceptor levels for now.)

The situation for the classic covalently bonded materials (diamond, silicon, and germanium) (page 148) in principle is the same as described previously. There are complications due to hybridization, but the important point to remember is that there is a band gap and this has some very important implications as we will discuss in the following section.

8-6 "Classical" Semiconductors

8-6a Introduction (and band gap) The well-known and technologically important semiconductors, like silicon, should not really be discussed until band theory (Chapter 10) is introduced. Nevertheless, we mention a few aspects of these materials because they are classic covalently bonded solids. In the simple elemental (Si, Ge) and binary (GaP, GaAs, InAs) materials the bonding is primarily covalent and each atom has four neighbors. Table 8-3 lists many of these simple materials. The four elemental materials have the diamond crystal structure (Fig. 3-7a), while most of the binary compounds have the zinc blende structure (Fig. 3-7c). Both of these crystal structures also display the classic sp^3 hybridization. The tetrahedral bonding to the four-nearest neighbor atoms is emphasized in a sketch of these structures which appears on page 148. The bonding is completely covalent in the four elemental materials, and predominantly covalent in the other materials listed. Considerations of the fraction of covalency of these various materials are discussed later. Binary compounds that are more ionic tend to possess the NaCl structure.

The first question to ask is, What is a semiconductor? As the name implies, it is a solid that has an electrical conductivity intermediate between a metal and an insulator. This is a gross overall statement and we can easily find exceptions. (We are talking about charge carried by electrons, not ions. As mentioned in Section 3-5, some materials have a high ionic conductivity at high temperatures.) A distinctive characteristic of a semiconductor is that the conductivity increases with increasing temperature while in metals it decreases. In metals this effect can be understood because at higher temperatures the amplitude of vibrations of the ions increases and the electrons

CHAPTER 8 THE COVALENT BOND

Table 8-3 Values of the band gap (in eV) at 300°K and the dielectric constant for the elemental and some of the more common III-V, II-VI, and IVA-VI semiconductors. The group IV materials have the diamond crystal structure, most of the III-V and II-VI materials have the zinc blende structure (the rest have the wurtzite structure), and the IVA-VI materials have the rock salt structure. For more extensive tables see Sze as well as H. F. Wolf, "Semiconductors," (Wiley-Interscience, 1971). Also see the appendix to this chapter and the tables in Chapter 10.

Material	$\mathscr{E}_g$	ε	III-V	$\mathscr{E}_g$	ε	IVA-VI	$\mathscr{E}_g$	ε
C(d)	5.47	5.7	AlAs	2.2	8.5	PbS	0.41	17
Si	1.12	11.9	AlP	3.0	11.6	PbSe	0.27	24
Ge	0.66	16.0	AlSb	1.58	14.4	PbTe	0.31	30
α−Sn	0.08	--	GaAs	1.42	13.1			
SiC	3.0	10.0	GaP	2.26	11.1			
			GaSb	0.72	15.7			
II-VI			InP	1.35	12.4			
CdS	2.42	5.4	InAs	0.36	14.6			
CdSe	1.70	10.0	InSb	0.17	17.7			
CdTe	1.56	10.2						
ZnS	3.68	5.2						
ZnSe	2.68	5.9						
ZnTe	2.25	9						

"bump into" the ions more often, that is, the mean free path of the electrons decreases. In semiconductors the conductivity increases with increasing temperature as

$$\sigma = \sigma_0 \exp(-\Delta\mathscr{E}/k_B T) \qquad (8\text{-}14)$$

where σ_0 is a constant, k_B is the Boltzman constant, and $\Delta\mathscr{E}$ is an activation energy. With increasing temperature, carriers are excited across an energy barrier and the increase in conductivity is thus due to the presence of more carriers.

Figure 8-10b shows a simplified picture of the band gap in a semiconductor. (Ignore the donor and acceptor levels for the present.) Electrons are in eigenstates that just fill the valence band; the conduction band is empty. As the temperature is increased from 0°K some electrons from the valence band can be excited across the energy gap, $\mathscr{E}_g$, into the conduction band and conduct electricity. The number of such electrons is given by an exponential law like Eq. 8-14 with $\Delta\mathscr{E}$ equals $\mathscr{E}_g/2$. For every electron excited into the conduction band a hole is left in the valence band. When the number of holes is equal to the number of electrons we have an **intrinsic semiconductor**; the properties are dependent only on the solid itself and not on any impurities. Table 8-3 shows the values of $\mathscr{E}_g$ for the four elemental semiconductors and a number of binary compounds.

Materials like Si and Ge look metallic to the eye; they have a high coefficient of reflection for visible light. However, this large reflectivity occurs only for light that has an energy higher than $\mathscr{E}_g$ (for light $h\nu = \mathscr{E}$). These materials are transparent to light with $h\nu < \mathscr{E}_g$. Absorption of light near and above $\mathscr{E}_g$ leads to the phenomenon called **photoconductivity**, where the conductivity of the semiconductor can be varied with light. For $\nu < (\mathscr{E}_g/h)$ there is no absorption in the materials and the light passes through the solid except for reflection losses from the boundaries due to index of refraction mismatch. However, for $\nu \geq \mathscr{E}_g/h$, the light can excite an electron from the valence band to the conduction band where it (and the hole left in the valence) can conduct electricity. The detailed frequency response of the photoconductivity depends on many material properties of the solid (such as the absorption constant and mobility of the carriers). However, the general idea is simple. In practical photoconductors the conductivity is usually due to excitation from impurity levels.

8-6b Impurity effects If all semiconductors were intrinsic they would be an interesting group of solids but never would have caused a technological revolution. The useful aspects of some of these materials are that they can be grown easily in a state of very high purity and can also be "doped" with other elements. For example, silicon can be grown routinely and can be doped with controlled impurity densities of 10^{14} atoms/cm^3 where the Si density in the crystal is about 5.0×10^{22} atoms/cm^3. Particular impurities, added in a well-controlled manner, can completely dominate the electrical behavior of these semiconductors.

Elements from group V (i.e., phosphorus) of the periodic table can be added to Si. These dopants enter the lattice substitutionally on Si sites rather than interstitially. This can be determined because the covalent radii of Si and P are 1.17 Å and 1.10 Å, respectively (Table 8-2), and the lattice constant of the doped crystal decreases as P is added. If P entered the lattice interstitially the lattice constant would increase. The phosphorus atom has one more electron than Si so that *besides* forming a covalent sp^3 bond, this fifth electron is weakly bound. Relatively little energy is required to set this electron free into the conduction band. Hence elements like P, As, and so forth, are called **donors**; when added to Si or Ge they donate an electron.

The energy required to ionize this extra electron can be estimated by treating the donor as a hydrogen atom. If this extra electron is far away from the P atom impurity, it will see a positively charged P-ion. The two differences between this problem and the hydrogen atom are that these charges are embedded in a medium of high dielectric constant, ($\epsilon = 12$ for Si and 16 for Ge) and the effective mass of the electron m* ≈ m/10. This is due to subtle effects that will be discussed in Section 10-3. The eigenvalue problem can be solved as for the hydrogen atom but with a potential energy

$$\text{PE} = -e^2/\epsilon r \tag{8-15a}$$

For the average radius of a Bohr orbit, $\langle r \rangle$, and the energies of the various levels we obtain results just as in the hydrogen case except for the appearance of ϵ and m^* in the appropriate places.

$$\frac{\langle r_n \rangle}{a_0} = \left(\frac{m}{m^*}\right)(n^2 \epsilon) \tag{8-15b}$$

$$\frac{\mathscr{E}}{13.6 \text{eV}} = -\left(\frac{m^*}{m}\right)\left(\frac{1}{\epsilon^2}\right)\left(\frac{1}{n^2}\right) \tag{8-15c}$$

where the Bohr radius $a_0 = \hbar^2/me^2 \approx 0.529$ Å, $me^4/\hbar^2 = 13.6$ eV, and $n = 1, 2, ...\infty$ which is the principal quantum number. Notice that the entire hydrogen-like energy scale for these donor electrons is reduced by ϵ^2 compared to atomic hydrogen and the first power of the effective mass. For the large values of ϵ (Table 8-3) and small values of m^* the ionization energies for these donors ≈ 10 meV, considerably smaller than the 13.6 eV for atomic hydrogen. The effective radius of the orbital states is increased, compared to atomic hydrogen, by the first powers of ϵ and m^* so $\langle r_1 \rangle \approx 50$ Å. The results in Eq. 8-15 are self-consistent in the sense that $\langle r \rangle$ is large enough so that the approximation that the extra electron sees a uniform macroscopic dielectric constant is sensible. However, the m^* comes about due to band theory, that is, the periodic arrangement of all of the other atoms in the crystal, in this simple hydrogen-like model m^* is added in an ad hoc manner. (To obtain Eqs. 8-15 in **SI units,** replace ϵ by $4\pi\epsilon\epsilon_0$.)

What is the zero energy reference state for this $\mathscr{E}$ value? It is the energy that corresponds to the electron completely ionized from the donor atom to the conduction band. This occurs for $n = \infty$. Thus, $\mathscr{E}$ is measured down from the bottom of the conduction band. Figure 8-11 shows the theoretical levels for the first few values of n, as well as some experimentally determined levels. The experimental values can be determined by far infrared absorption; the transitions can be rather sharp in favorable cases. This hydrogen-like theory predicts that the energy levels in a given material should be independent of the donor. As seen in Fig. 8-11, this is not true especially for the $n = 1$ quantum level. The breakdown of this simple **effective mass theory** in the lowest energy state occurs because in this state the electron has the largest probability of being within the donor cell and Eq. 8-15a is no longer a very good approximation. The agreement is better for $n = 2$, 3, and 4. However, there is another basic theoretical complication. Due to band structure effects the effective mass is anisotropic. This complicates the theoretical picture and is discussed in Chapter 10.

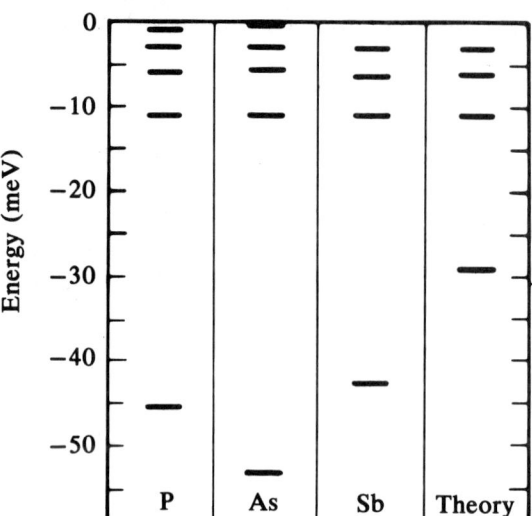

Fig. 8-11 The theoretical and experimental energy levels of several donors in Si. [From W. Kohn, Solid State Phys. 5, 257 (1957).] Zero energy is the bottom of the conduction band.

Table 8-4 (a) Values of $\mathscr{E}_c - \mathscr{E}_d$ for group V donors and (b) $\mathscr{E}_a - \mathscr{E}_v$ for group III acceptors in Si and Ge. (c) Typical donor and acceptor binding energies in several III-V compounds. The values are taken from the Landolt-Börnstein table listed in the Notes in Chapter 10 and they are given in meV where 1 meV = 10^{-3} eV.

		P	As	Sb	Bi	
(a)	Si	45	54	43	71	
	Ge	13	14	10	13	

		B	Al	Ga	In	Tl
(b)	Si	45	67	71	155	246
	Ge	10	11	11	12	13

		GaP	GaAs	InP	InSb
(c)	$\mathscr{E}_c - \mathscr{E}_d$	59	5.7	7.1	0.7
	$\mathscr{E}_a - \mathscr{E}_v$	45	27-40	46-100	8-10

Table 8-4a lists the binding energies ($\mathscr{E}_c - \mathscr{E}_d$ = the difference in energy between the n = 1 and n = ∞ levels) for some common donors in Si and Ge.

In the same manner elements from column III of the periodic table can be used to dope Si or Ge, but now the impurities are one electron short of the four required for sp³ hybridization. Thus, each impurity atom gives rise to a vacant electron level slightly above the valence band (in the same manner as the donor levels are below the conduction band). These are called **acceptor levels** because they may accept an electron from the filled band if the electron is thermally excited. An empty electron state is called a **hole**. In a manner quite similar to the donor electrons discussed previously, one can describe an acceptor level as a hole describing a Bohr orbit about the impurity atom. Thus ionization of an acceptor level is described as the excitation of a valence electron into an acceptor impurity atom leaving a hole in the valence band. Table 8-4b lists the binding energies of some acceptors in Si and Ge.

The III-V semiconductor, such as GaAs, and so forth, also can be doped with donors and acceptors. For example, Si and Ge on a Ga-site act as donors, as do S and Se on an As-site. Similarly, Si and Ge on an As-site act as acceptors, as do Zn and Cd on a Ga-site. Typical donor and acceptor binding energies are listed in Table 8-4c. The binding energy values will be discussed again in Section 10-15.

What we have just described are usually called **shallow donors** and **acceptors**. They are elements from Group V (donors) or Group III (acceptors) of the periodic table and are used to dope Ge or Si which are in Group IV; the word shallow is used because the energy levels are close to the conduction or valence bands and the hydrogen-like theory is appropriate.

Now the donor and acceptor energy levels in Fig. 8-10 can be understood. Note that electrons are excited upwards while holes are excited downwards. The binding energies for the donors and acceptors are usually called $\mathcal{E}_d$ and $\mathcal{E}_a$ (the difference in energy between the n = 1 and n = ∞ levels) and that of the bottom of the conduction band and the top of the valence band $\mathcal{E}_c$ and $\mathcal{E}_v$, respectively. For some donors and acceptors the ionization energies are small and $\mathcal{E}_c - \mathcal{E}_d$ or $\mathcal{E}_a - \mathcal{E}_v$ is of the order of $k_B T$ at room temperature (Table 8-4). For a reasonable number of such impurities the electrical conduction can be completely dominated by the presence of these donors or acceptors. Such semiconductors are called **extrinsic** materials (impurity dominated). Note that for low temperatures where the electrons are in the donor ground state, photoconductivity can be observed at $\nu = (\mathcal{E}_c - \mathcal{E}_d)/h$. In fact, photoconductivity measurements are a good method of studying the energy levels of donors and acceptors. Also the photoconductive response can be used to detect infrared light.

When Si (or Ge) is doped with impurities from groups other than III or V, **deep levels** can be produced. This means that the energy levels are well separated from either the conduction or valence bands. The preceding theory is generally not satisfactory for deep levels, since the orbitals are much more tightly bound and details of the potential of the impurity atom as well as band theory are required for a suitable theory. These will not be discussed here but reference should be made to the Notes in Chapter 13 and this chapter, where several books on semiconductor properties are listed. Typical of the effects that can occur when Si is doped with elements from Group II such as Zn or Cd is that the acceptors can act as double acceptors since they can trap two holes. The first hole is tightly bound but the second is weakly

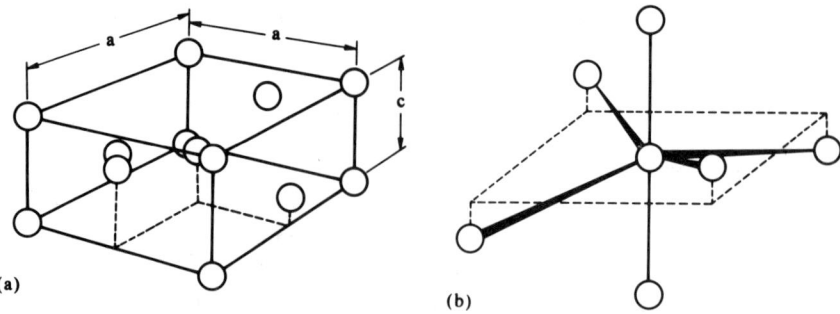

Fig. 8-12 (a) The β-Sn structure. (b) The nearest neighbor coordination.

bound and the preceding theory is appropriate. Elements such as Mn, Co, and Fe act as double acceptors, but both holes are tightly bound.

8-6c The metal-nonmetal transition The hydrogen-like model of shallow donors and acceptors conceptually is simple and useful. The self-consistent prediction of large orbits is in agreement with experiments. However, with such large orbits we can ask what happens when the density of impurities becomes large enough so that the orbits begin to overlap appreciably.

Up until now we have been talking about the extra electron (in the donor case) revolving around the donor atom in hydrogen-like states. Clearly this is a localized picture and at $0°K$ this leads to no conductivity. However, if the donor density becomes large enough for appreciable overlap of the orbits, the electrons no longer will be localized and they can move among the donors with no energy loss even at $0°K$. The basic ideas behind this type of motion are the topics of Chapters 9 and 10. We only point out that as the donor electron density, N/V, increases the electrons increasingly get screened from the charged donor atoms. At some critical density, N_c/V, the electrons are no longer bound to the donor atoms and are free to move in the crystal (they become delocalized). So at N_c/V there is a transition from semiconductor behavior to "metallic behavior." This is called the **metal-insulator** or **metal-nonmetal transition**. Mott has shown that this could occur for a critical density of

$$(N_c/V)\langle r_1 \rangle^3 \simeq 0.2 \qquad (8\text{-}15d)$$

Conceptually, this is a different sort of phase transition in that we have to deal with different samples to vary N/V, while previously we considered the behavior of one sample as the temperature or pressure is changed. Nevertheless, the concept is the same. Actually, in systems other than the idealized semiconductors already discussed, we can observe and study this transition in a sample as a function of pressure,

CHAPTER 8   THE COVALENT BOND   175

temperature, or magnetic field. Also, in Chapter 18, we discuss how by injection methods the density of carriers can be varied and this metal-nonmetal transition can be studied in a single sample.

Mott originally investigated this transition in conjunction with the peculiar behavior of more conventional insulators such as VO and V_2O_3. In these types of materials the density of carriers is larger but their effective radius is smaller, so the reasoning behind the universal scaling law in Eq. 8-15d is the same. For a further discussion see the book by Mott or the other references in the Notes.

8-6d High pressure phases One of the striking aspects of tetrahedrally coordinated structures, and certainly of the diamond and zinc blende structures, is their low density. See, for example, Section 3-8 and the table there. Thus, under high pressures we expect, and find, that materials with these structures transform to structures with higher coordination and higher density.

Under hydrostatic pressures of 80-150 kbar the following materials transform to the NaCl structure: AlSb; InAs; InP; ZnS. However, ZnO, Si, Ge, GaSb, and InSb transform to the β-Sn (white-tin) structure. This tetragonal structure, shown in Fig. 8-12, has space group $D_{4h}^{19}(I4_1/amd)$ with Sn at 4a (0, 0, 0), (0, 1/2, 1/4) plus the body-centering condition. Figure 8-12b shows the nearest-neighbor positions. Even though the coordination is six there is a vestige of the tetrahedral bond. For β-Sn, where c = 3.18 Å and a = 5.83 Å, the two-nearest neighbors along the c-axis are about 5% farther away than the other four neighbors.

Due to the coordination changes at the phase boundary between these low pressure structures and either of these two high pressure structures, there are fractional volume decreases of about 20%, which is quite drastic.

8-7 Continuous Range of Bonding

In the discussion of the difference between the molecular orbital and Heitler-London approaches we have seen how a partial covalent, partial ionic bond may be defined. Here this problem is discussed in greater detail. In Section 8-7a we discuss Pauling's electronegativity model and how it can be used to define the fraction of ionicity for a given bond. This model is most appropriate for an AB diatomic gas molecule. In Section 8-7b calculations of the electron density of simple diatomic solids are discussed. These calculations show graphically that some bonds are more ionic than others. Section 8-7c sketches the Phillips-Van Vechten model of the fraction of ionic character of bonds. From the onset this model is directed toward simple diatomic solids and it gives formulas for the various terms from which related properties can be calculated.

8-7a Pauling's electronegativity model Pauling defined electronegativity as "the power of an atom in a molecule to attract electrons to itself." He assigned a number, X, to each atom as a measure of the electronegativity. Table 8-5 gives values of X; they are taken with reference to 4.0 for fluorine, the most electronegative element. The object of this concept is to understand the occurrence and stability of molecules. Consider how these values were obtained.

Pauling observed that the following reaction is almost always exothermic (emits energy) in molecules

$$AA + BB \rightarrow 2\ AB$$

Define the A-B bond energy as D_{AB}. Since the reaction is exothermic D_{AB} must be lower (larger negative value) than $(D_{AA} + D_{BB})/2$. However, by convention these values are taken as positive; so larger positive values mean stronger bonding. This lower energy must come from the fact that the A-B bond is partially ionic since the A-A and B-B bonds are covalent (homopolar). This extra ionic energy Δ_{AB} is defined as

$$\Delta_{AB} \equiv D_{AB} - (D_{AA} + D_{BB})/2 \tag{8-16}$$

It is clear that Δ_{AB} can depend only on even powers of the electronegativity differences; taking the lowest order term we have $\Delta_{AB} \propto (X_A - X_B)^2$. Pauling found that the equation

$$D_{AB} = (D_{AA} + D_{BB})/2 + 23(\text{kcal/mole})(X_A - X_B)^2 \tag{8-17}$$

reproduces the experimental D_{AB} values fairly well, where the D values are in kcal/mol. Only differences of electronegativity enter Eq. 8-17, so we may arbitrarily set a scale and 4.0 was picked for fluorine.

As can be seen in Table 8-5, fluorine and oxygen are the most electronegative elements with nitrogen and chlorine next. It is just these elements that form hydrogen bonds, as discussed in Chapter 6. Note that besides the obvious increase in the electronegativity in going from the left to right periods, there is also a decrease of electronegativity as we go down the periodic table.

It would be useful to relate electronegativity differences to the fraction of ionic character in a bond, f_i. This was done by comparing the calculated dipole moments to the measured dipole moments in HF, HCl, HBr, and HI. It was concluded that these materials are 60%, 19%, 11%, and 4% ionic, respectively. Then for these data a formula was proposed that relates f_i and $X_A - X_B$. It is

$$f_i = 1 - \exp[-(X_A - X_B)^2/4] \tag{8-18}$$

CHAPTER 8   THE COVALENT BOND   177

Table 8-5 The electronegativities of the elements. (After Pauling.)

H																
2.1																
Li	Be											B	C	N	O	F
1.0	1.5											2.0	2.5	3.0	3.5	4.0
Na	Mg											Al	Si	P	S	Cl
0.9	1.2											1.5	1.8	2.1	2.5	3.0
K	Ca	Sc	Ti	V	Cr	Mn	Fe	Co	Ni	Cu	Zn	Ga	Ge	As	Se	Br
0.8	1.0	1.3	1.5	1.6	1.6	1.5	1.8	1.8	1.8	1.9	1.6	1.6	1.8	2.0	2.4	2.8
Rb	Sr	Y	Zr	Nb	Mo	Tc	Ru	Rh	Pd	Ag	Cd	In	Sn	Sb	Te	I
0.8	1.0	1.2	1.4	1.6	1.8	1.9	2.2	2.2	2.2	1.9	1.7	1.7	1.8	1.9	2.1	2.5
Cs	Ba	La	Hf	Ta	W	Re	Os	Ir	Pt	Au	Hg	Tl	Pb	Bi	Po	At
0.7	0.9	1.1	1.3	1.5	1.7	1.9	2.2	2.2	2.2	2.4	1.9	1.8	1.8	1.9	2.0	2.2
Fr	Ra	Ac														
0.7	0.9	1.1														

The lanthanide and actinide elements

Ce-Lu	Th	Pa	U	Np-No
1.1-1.2	1.3	1.5	1.7	1.3

Thus, we have a formula that relates the fraction of ionic bonding, ionicity, to the electronegativity differences. For $X_A - X_B = 1.7$ the bond is 50% ionic and 50% covalent. Bonds with larger electronegativity differences are more ionic. This includes most, but not all, of the alkali halide molecules. See the Notes for references. This formula for f_i is for a single bond as in an AB molecule. See the problems for its value for molecules where we have multiple bonds.

8-7b Electron charge densities in semiconductors In this section we describe some results that illustrate a change of bonding from purely covalent in Ge to more ionic in going to GaAs and ZnSe.

We can calculate the electron density in solids, at least in simple solids. The methods involve band theory, to be discussed in Chapter 10, and pseudopotential theory, which is outside the scope of this book. The results, nevertheless, are clear and easy to understand. We describe briefly and show the result for the Ge row of the periodic table, which includes Ge, the III-V material GaAs, and the II-VI ZnSe. For all three of these compounds the atoms are tetrahedrally bonded. The crystal structures are, of course, the diamond structure for Ge and zinc blende structure for the other two (Fig. 3-7 and also see page 148). The core electrons, $1s^2 2s^2 p^6 3s^2 p^6 d^{10}$, are ignored as far as the valence charge density, $\rho(\mathbf{r})$, is concerned (although the core is required to determine the pseudopotential). There are two atoms per primitive unit cell and thus eight valence electrons in these crystal structures. For Ge these eight electrons come from the $4s^2p^2$ electrons of the Ge atoms. For GaAs, three come from the Ga atom and five from the As atom, and so forth.

$\rho(\mathbf{r})$ is a three-dimensional function but the results are displayed in a plane. The $(1\bar{1}0)$ plane is a convenient plane since two neighboring atoms lie in it. Figure 8-13 shows the results for these three solids. The values of constant $\rho(\mathbf{r})$ form contour plots, or relief maps, as shown. The units of $\rho(\mathbf{r})$ are $4e/a^3$ where a is the lattice parameter so $a^3/4$ is the volume of the primitive unit cell. For Ge the charge density is a maximum exactly in the middle along the line between the two atoms. Of course, this is the classic covalent bond. For GaAs there is also a great deal of charge between the two atoms but it is shifted toward As. This is just what is expected for a solid with some ionic character, since for a completely ionic material all the charge would be on As, that is, $Ga^{3+}As^{3-}$. ZnSe shows a larger shift of charge; in fact, the material looks more ionic than covalent. Thus, we can see the continuous range of bonding from purely covalent to ionic. Similar results have been found for the series in the next row of the periodic table, Sn, InSb, and CdTe. (In Fig. 8-13 the core electrons are indicated by the dark circles. Naturally, they give a very high concentration of electrons near the atom sites. For example, in Ge the core radius is about one-fifth the Ge-Ge distance.)

We can gain insight into antibonding orbitals and the conduction band by calculating $\rho(\mathbf{r})$ if two more electrons per primitive cell were added to Ge. The results are shown in Fig. 8-14. There is a stronger buildup of charge on the atoms rather than halfway between them. This is just like the antibonding orbitals in Fig. 8-1b.

8-7c Phillips' and Van Vechten's determination of f_i There are other scales of electronegativity that can be used to determine values of f_i. Mulliken and Coulson are names associated with some of this work. (See Coulson's book.) However, the more recent method of Phillips and Van Vechten has met with success and we briefly discuss this approach. Further references can be found in the Notes. Their thought from the start is to understand bonding in solids (as opposed to molecules) relative to a free electron gas with one band gap. (A free electron like gas with one, average, gap is sometimes called the **Penn model**.)

Consider the binary compounds A^NB^{8-N}, where N is the group in the periodic table (the number of the column). The periodic potential of the eight valence electrons per primitive unit cell in the diamond or zinc blende structures can be written in terms of a symmetric and antisymmetric part (with respect to the interchange of atoms A and B). This potential is related to the homopolar (covalent) and heteropolar

CHAPTER 8   THE COVALENT BOND   179

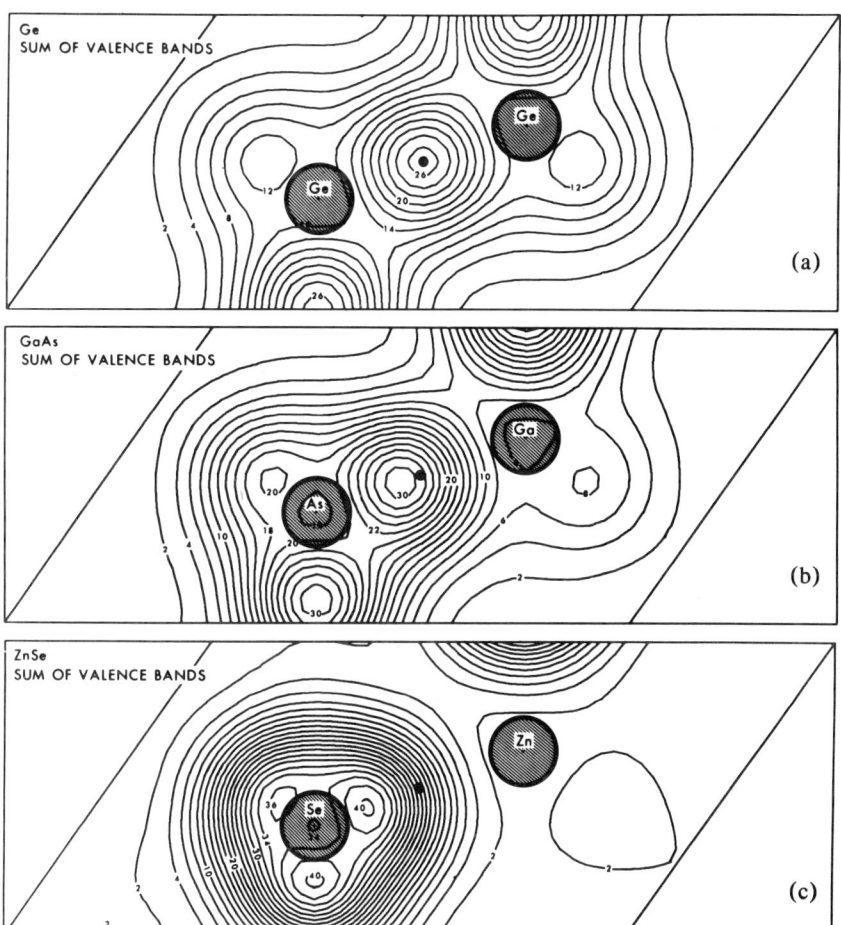

Fig. 8-13 These diagrams show how the valence electron density changes from totally covalent to partially ionic. (a) The total calculated valence charge density for Ge in the ($1\bar{1}0$) plane. This plane cuts through two neighboring Ge atoms as shown. (b) and (c) Same but for GaAs and ZnSe. (From M.L. Cohen; see the Notes.)

(ionic) bonding, respectively. Associated with each of these potentials is an energy, $\mathscr{E}_h$ and C, respectively, related to an energy between a bonding and antibonding state. Using pseudopotential formalism we obtain a secular equation for the energy

$$\begin{vmatrix} 0 & \mathscr{E}_h + iC \\ \mathscr{E}_h - iC & 0 \end{vmatrix} = 0 \qquad (8\text{-}19)$$

Diagonalizing this equation gives the average band gap of the solid, denoted as $\mathscr{E}_p$ referring to the **Penn gap**.

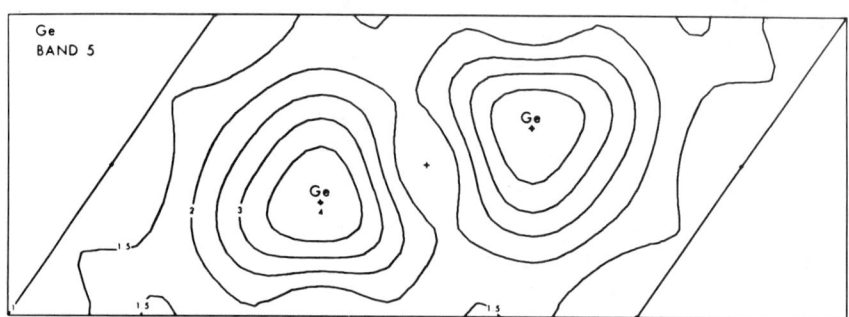

Fig. 8-14 Calculated charge density of Ge if two more electrons were added to the results in Fig. 8-13. This is the first conduction band in Ge. (From M.L. Cohen; see the Notes.)

$$\mathscr{E}_p^2 = \mathscr{E}_h^2 + C^2 \tag{8-20}$$

In order to relate these terms to known quantities we consider first purely covalent materials such as diamond, Si, Ge, and αSn, for which $C = 0$ and $\mathscr{E}_p = \mathscr{E}_h$. The real part of the frequency-dependent dielectric constant, $\varepsilon_1(\omega)$, as for atoms, is related to oscillator strengths and the energy separations of the unoccupied excited states. This is also what is found in semiconductors and we write

$$\varepsilon_1(0) = 1 + (\hbar\omega_p/\mathscr{E}_p)^2 \tag{8-21}$$

where ω_p is the plasma frequency, that is, $\omega_p^2 = 4\pi N e^2 / V m$, where N/V is the electron density and e and m are the charge and mass of the electron. This equation relates the dielectric constant at $\omega = 0$, a known quantity, to $\mathscr{E}_p$ or $\mathscr{E}_h$. Remember $\mathscr{E}_p$ in Eq. 8-21 is not the band gap discussed in Section 8-5; it is an average gap that determines the optical properties of the semiconductor. $\mathscr{E}_p > \mathscr{E}_g$ because the matrix elements connecting the filled states with high lying states are much stronger than those connecting the filled states to the state at $\mathscr{E}_g$. Using experimental values for $\varepsilon_1(0)$ of 5.66 and 12 we obtain $\mathscr{E}_h =$ 13.5 eV and 4.77 eV for diamond and Si, respectively. Assuming that $\mathscr{E}_h$ scales with the lattice constant or nearest-neighbor distance, according to some power s, then $d(\ln \mathscr{E}_h)/d(\ln a) = s$. Using these two values of $\mathscr{E}_h$ as well as the pressure dependence of $\mathscr{E}_h$ for each of these two crystals, we obtain

$$\mathscr{E}_h \propto a^{-2.5} \tag{8-22}$$

This relation is assumed to be universally valid for the $A^N B^{8-N}$ binary compounds. (There are some small corrections for atoms with d and f electrons, which we will ignore.)

The quantity C is obtained using an approach similar to the way that $\mathscr{E}_h$ is obtained. Values of $\mathscr{E}_p$ for many $A^N B^{8-N}$ materials were obtained from $\varepsilon_1(0)$ experimental values via Eq. 8-21. For each of these materials $\mathscr{E}_h$ can be obtained via Eq. 8-22. Last, C is obtained via Eq. 8-20. Thus $\mathscr{E}_p$, $\mathscr{E}_h$, and C are known for many of these binary compounds. From these values of C and some physical reasoning, an analytic formula is obtained for the $A^N B^{8-N}$ compounds

$$C = b[(Z_A/r_A) - (Z_B/r_B)] \exp[-k_s(r_A + r_B)/2] \quad (8\text{-}23)$$

where Z_A and Z_B are the valence numbers N and 8-N, respectively. The atomic radii r_A and r_B are defined as half the bond length of the group IV element belonging to the same row of the periodic table as atoms A and B, respectively. k_s is the Thomas-Fermi screening factor. The constant b is a dimensionless number that gives best fit to the known values of C; b = 1.5 is obtained. The entire exponential term varies from 0.14 to 0.07 from row 1 to 4 of the periodic table.

We would like to relate the C, $\mathscr{E}_h$, and $\mathscr{E}_p$ to the ionicity of a bond. Figure 8-15 shows how $\mathscr{E}_p$ and ϕ are the polar coordinates corresponding to the covalent and ionic energies, $\mathscr{E}_h$ and C, represented as cartesian coordinates. The ionic phase angle, ϕ, has the value $\tan \phi = C/\mathscr{E}_h$. We want the ionicity, f_i, to be between 0 and 1 and to be zero when C = 0. This suggests

$$f_i = \sin^2\phi = C^2/\mathscr{E}_p^2 \quad (8\text{-}24a)$$

Then the fraction of covalency f_c, with $f_i + f_c = 1$, is given by

$$f_c = \cos^2\phi = \mathscr{E}_h^2/\mathscr{E}_p^2 \quad (8\text{-}24b)$$

Naturally, we must demonstrate that these choices are useful, which will be done presently.

Figure 8-16 summarizes some of the ideas that have been discussed. For Ge there are bonding and antibonding levels separated by $\mathscr{E}_p = \mathscr{E}_h = 4.31$ eV. As some ionic bonding is added, the separation becomes larger for the same reason that Δ_{AB}, in Eq. 8-16, is positive. Thus, even though $\mathscr{E}_h$ is the same for ZnSe as for Ge, $\mathscr{E}_p$ is larger because C is no longer zero. Again we see how this physical picture is similar to that presented at the beginning of the chapter for the very simple molecules.

To test the usefulness of the prediction for f_i in Eq. 8-24, Fig. 8-17 shows a plot of C versus $\mathscr{E}_h$ for a large number of $A^N B^{8-N}$ compounds that have four- and sixfold coordination. As can be seen a critical value of ionicity $f_i^c = \sin^2\phi_c = 0.785$ separates the more covalent fourth-coordinated compounds from the more ionic sixth-coordinated compounds. Several borderline compounds, such as MgS and MgSe, with $f_i = 0.786$ and 0.790, respectively, are metastable in both coordinations, and AgI, with $f_i = 0.770$, transforms from zinc blende to the rock salt structure with a pressure of only several thousand atmospheres. The only competing systematic method of predict-

182   CHAPTER 8   THE COVALENT BOND

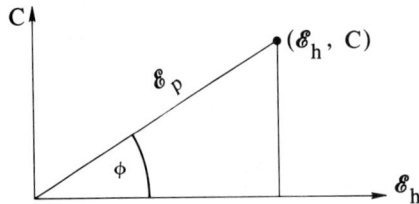

Fig. 8-15 $\mathscr{E}_p$ and ϕ shown in the $\mathscr{E}_h$, C plane.

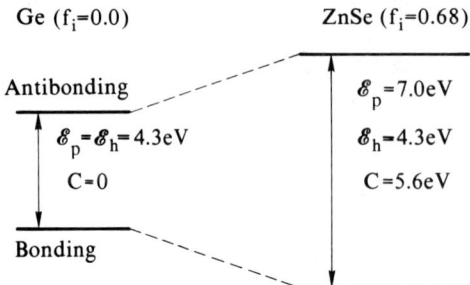

Fig. 8-16 The bonding and antibonding levels for Ge and ZnSe, $\epsilon(0) = 16.0$ and 5.90, respectively.

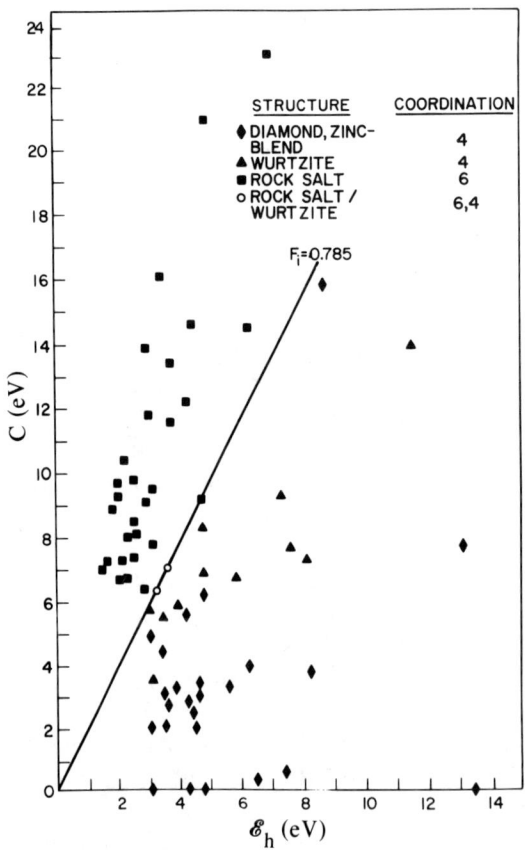

Fig. 8-17 Values of $\mathscr{E}_h$ and C for crystals of the $A^N B^{8-N}$ type. The coordination numbers (four and six) are indicated. This plot is similar to Fig. 8-15. The value of $f_i = 0.785$ separates the four- and six-fold coordinated materials. (From Phillips; see the Notes.)

CHAPTER 8   THE COVALENT BOND   183

ing structures is via calculating the cohesive energy using the Born ionic model discussed in Chapter 7. However, structural predictions using the Born-Mayer model are not very good. Thus, the results shown in Fig. 8-17 are impressive in that they easily separate the covalent four-coordinated structures from the more ionic six-coordinated structures.

As emphasized, covalent bonding means that there are noncentral forces that are important and these give the special angles that are characteristic of this type of bond. Using elastic constant data, we may determine the ratio of the bond bending to bond stretching forces. This ratio decreases linearly as f_i decreases toward 0.785 (see the Notes for references) and shows again that this value of f_i^c has important effects in the covalent to ionic "transition."

One of the nice aspects of the ideas summarized in this subsection is that explicit formulas, Eqs. 8-22 and 8-23, are proposed, relating the parameters to internuclear distances. Thus, other predictions can be made, such as the pressure dependence of various properties, force constant ratios, nonlinear optic properties, and so forth. The book by Phillips (see the Notes) has references to some of this work.

Appendix – All insulators have a band gap; we list $\mathscr{E}_g$ of the highly ionic alkali halides. Also listed are the melting temperatures and the static, $\varepsilon_1(0)$, and optic, ε_∞, dielectric constants. The static dielectric constant enters into Eqs. 8-15, and is called ε. The reason for the more full symbol $\varepsilon_1(0)$ and the meaning of ε_∞ is discussed in Chapter 13. More complete tables of properties of alkali halides, and other materials, can be found in many places, including R. K. Watts, "Point Defects in Crystals" (Wiley, 1977). Also see Table 10-1 in this book. (CsCl, CsBr, and CsI have the CsCl structure; all of the rest have the NaCl structure.)

	$\mathscr{E}_g$ (eV)	T_m (°C)	$\varepsilon_1(0)$	ε_∞		$\mathscr{E}_g$ (eV)	T_m (°C)	$\varepsilon_1(0)$	ε_∞
LiF	13.7	845	9.04	1.92	RbF	10.3	795	5.91	1.93
LiCl	9.4	605	11.86	2.79	RbCl	8.5	718	4.92	2.91
LiBr	7.6	550	13.33	3.22	RbBr	7.2	693	5.0	2.33
LiI	6.1	449	11.0	3.80	RbI	6.3	647	4.94	2.61
NaF	11.5	993	5.07	1.74	CsF	9.8	682	8.08	2.2
NaCl	8.75	801	5.89	2.35	CsCl	8.3	645	6.95	2.67
NaBr	7.1	747	6.40	2.64	CsBr	7.3	636	6.66	2.83
NaI	5.9	661	7.28	3.08	CsI	6.1	626	6.59	3.09
KF	10.8	858	6.05	1.85					
KCl	8.7	770	4.81	2.20					
KBr	7.4	734	4.90	2.39					
KI	6.34	681	5.09	2.68					

Notes

H. Eyring, J. Walter, and G. E. Kimball, Chapter 11 "Quantum Chemistry," (Wiley, 1944) has the integrals required to calculate the energy for H_2^+ and also discusses the Heitler-London calculation for H_2. A discussion of the exact solution is also given. N. F. Mott and I. N. Sneddon, Chapter 7 "Wave Mechanics and its Applications" (Oxford Press, 1948) and S. Sugano, Y. Tanabe, and H. Kamimura, Chapter 10 "Multiplets of Transition Metal Ions in Crystals," (Academic Press, 1970) also are good references.

C. A. Coulson "Valence" (Oxford, 1952), covers many of the areas discussed in this chapter. He also has a table showing the results for many different types of approximate wave functions for H_2.

The Wolfsberg-Helmholtz method [J. Chem. Phys. **20**, 837 (1952)] is used in many molecular orbital calculations. See C. J. Ballhausen, Chapter 7, Appendix I, "Ligand Field Theory," (McGraw-Hill, 1952) and R.S. Mulliken, C.A. Rieke, D. Orloff, [J. Chem. Phys. **17**, 1248 (1949)] for the evaluation of group overlaps. As an example, see J.D. Axe and G. Burns, Phys. Rev. **152**, 331 (1966). There are many books on molecular orbital calculations.

Topics such as the **metal insulator transition** (sometimes called the **Mott transition**) and the related effect in disordered materials, **Anderson transition,** are of high current interest. In the latter electrons below a certain energy are localized in local potential minima while those above this energy are delocalized and can conduct. A good background can be obtained from N. F. Mott, "Metal Insulator Transitions" (Taylor and Francis Ltd., 1974); "The Metal-Non-Metal Transition in Disordered Systems," Ed. L. R. Friedman and D. P. Tunstall (University of Edinburgh: SUSSP Publications, 1978); Mott and Davis in the bibliography; "Search and Discovery" in Physics Today, **34**, No. 5, 19 (1981). For a more recent paper with useful references see T. F. Rosenbaum, et al., Phys. Rev. **B27**, 7509 (1983).

A clear discussion of Pauling's ideas on electronegativity can be found in his book, Sections 9, 11, and 12. For the Phillips approach see his book "Bonds and Bands in Semiconductors" (Academic Press, 1973) or Rev. Mod. Physics **42**, 317 (1970). The details and a summary are in J.A. Van Vechten, Phys. Rev. **182**, 891 (1969) and **187**, 1007 (1969). Values of $\varepsilon_1(0)$, $\mathscr{E}_h$, C, and related quantities for many of the compounds shown in Fig. 8-17 can be found in J. A. Van Vechten in "The Handbook on Semiconductors" Vol. 3, Ed. S. P. Keller (North Holland, 1980). Also see Adams, Chapter 5 as well as C. R. A. Catlow and A. M. Stoneham, J. Phys. C **16**, 4321 (1983).

Calculations of the electronic charge densities in semiconductors (Section 8-7b) are nicely described in M.L. Cohen, Science **179**, 1189

CHAPTER 8   THE COVALENT BOND   185

(1973); J.P. Walter and M.L. Cohen, Phys. Rev. **B4**, 1877 (1971); M.L. Cohen, Physics Today, July 1979.

Problems

1. Evaluate Eq. 8-5 in terms of the internuclear distance. (Hint: see Eyring, Walter, and Kimball.)

2. Draw an energy level diagram for a diatomic AB molecule as you would obtain using the results in Section 8-2b. Show only the 1s-, 2s-, and 2p-levels. (Hint: see Burns Section 10-3.)

3. What is the distance between the closest chlorine atoms that are covalently bonded to different Pt atoms in K_2PtCl_4? Compare this distance to the van der Waals radii for chlorine.

4. Do the same for K_2PtCl_6 as for Problem 3. The details of the structure can be found in Wyckoff, Vol. 3, p. 339.

5. Briefly discuss how **covalent radii** are determined; see Pauling.

6. (a) Assume one cubic centimeter of silicon is doped with 10^{17} arsenic atoms, and sketch a graph for the number of electrons in the conduction band as a function of temperature. (b) The conductivity σ equals $(N/V)e\mu$ where N/V is the number of electrons in the conduction band, V is the volume of the sample, and μ is the mobility of these electrons. Assume that the temperature dependence of $\mu \sim 1/T$, then comment on the temperature dependence of σ.

7. **Tetrahedral hybridization** — Let s be a normalized s-wave function and p_x, p_y, and p_z be the three normalized p-wave functions. Consider the four hybrid functions

$$\phi_1 = \tfrac{1}{2}s + \tfrac{1}{2}p_x + \tfrac{1}{2}p_y + \tfrac{1}{2}p_z$$
$$\phi_2 = \tfrac{1}{2}s - \tfrac{1}{2}p_x - \tfrac{1}{2}p_y + \tfrac{1}{2}p_z$$
$$\phi_3 = \tfrac{1}{2}s + \tfrac{1}{2}p_x - \tfrac{1}{2}p_y - \tfrac{1}{2}p_z$$
$$\phi_4 = \tfrac{1}{2}s - \tfrac{1}{2}p_x + \tfrac{1}{2}p_y - \tfrac{1}{2}p_z$$

(a) Show that under a rotation by 120° and 240° about the [111] direction that ϕ_2, ϕ_3, and ϕ_4 transform into each other. (b) Show that these four functions are the appropriate tetrahedral hybrid functions, that is, their maximum electron density is directed in the four tetrahedral directions. (c) Show that the functions are normalized to one; show that all of the s and p electron densities are used up by these four functions. (d) Show that the angle between the tetragonal directions is $\approx 109°\ 30'$.

8. sp^2 planar hybrid functions – Show that the results given in Eq. 8-13 are correct.

9. Intrinsic conductivity – Calculate the ratios of the intrinsic conductivity with respect to 77°K, at 27°C and 200°C for PbTe, Ge, Si, and GaP.

10. The definition of ionicity, Eq. 8-18, is applicable only to single bonds. For multiple bonds, as in a crystal, the definition must be altered. For an $A^N B^{8-N}$ crystal, such as Ge^{IV} or Zn^{II}Se^{VI}, with coordination number M the fractions of covalent and ionic bonds, f_c' and f_i', respectively, are given by

$$f_c' = 1 - f_i' = (N/M)f_c = (N/M)(1 - f_i)$$
$$= (N/M)\ \exp\ [-(X_A - X_B)^2/4]$$

where f_c and f_i are the single bond fractions. (a) Calculate f_i' and f_i for ZnSe in a crystal and in a gas. (b) What is the physical interpretation for multiple bonds? (See Pauling's first edition (1939), and J.C. Phillips, Rev. Mod. Physics **42**, 317 (1970).)

11. Calculate f_i for GaAs using the Phillips method.

9

Metals

PART A DRUDE'S MODEL

9-1 Drude's Free Electron Theory
9-2 Drude's Assumptions
9-3 DC Conductivity
9-4 Wiedemann-Franz Law
9-5 Frequency-Dependent Conductivity (S,A)
    a Frequency dependence
    b Transparency of metals in the ultraviolet
9-6 Problems of Drude's Model

PART B QUANTUM MECHANICS APPLIED

9-7 Eigenfunctions of Free Electrons in a Metal
    a Wave equation
    b One-dimensional example (two kinds of boundary conditions)
    c Three dimensions
    d Degeneracy
9-8 Fermi Energy, Density of States, and Fermi Surface
9-9 Soft X-rays, Heat Capacities
9-10 Fermi-Dirac Statistics
9-11 Low Temperature Expansion Using F-D Statistics
9-12 Thermal Properties of the Electron Gas
    a $\mathscr{E}_F(T)$
    b Electron heat capacity
    c Electron spin paramagnetism (S)
9-13 DC Conductivity (with F-D Statistics)
9-14 Electron–Electron Collisions (S)
    a Phase space considerations
    b Fermi liquid theory
9-15 Hall Effect (and Other Magnetic Field Effects) (S)
9-16 Landau Levels (S, A)

Notes
Problems

Properties of some elemental metals – Values of $\mathscr{E}_F$ and T_F ($\mathscr{E}_F = k_B T_F$) using Eq. 9-29 and the electron concentrations from Table 9-1. The Fermi velocity v_F is obtained from $\mathscr{E}_F = m v_F^2/2$ and the Fermi wave vector k_F is obtained from $\mathscr{E}_F = \hbar^2 k_F^2/2m$.

El.	Z	N/V (10^{22}/cm^3)	r_s (Å)	$\mathscr{E}_F$(eV)	T_F(°K)	v_F(cm/sec)	k_F(cm^{-1})
Li	1	4.70	1.72	4.74	5.51×10^4	1.29×10^8	1.12×10^8
Na	1	2.65	2.08	3.24	3.77	1.07	0.92
K	1	1.40	2.57	2.12	2.46	0.86	0.75
Rb	1	1.15	2.75	1.85	2.15	0.81	0.70
Cs	1	0.91	2.98	1.59	1.84	0.75	0.65
Cu	1	8.47	1.41	7.00	8.16	1.57	1.36
Ag	1	5.86	1.60	5.49	6.38	1.39	1.20
Au	1	5.90	1.59	5.53	6.42	1.40	1.21
Be	2	24.7	0.99	14.3	16.6	2.25	1.94
Mg	2	8.61	1.41	7.08	8.23	1.58	1.36
Ca	2	4.61	1.73	4.69	5.44	1.28	1.11
Sr	2	3.55	1.80	3.93	4.57	1.18	1.02
Ba	2	3.15	1.96	3.64	4.23	1.13	0.98
Fe	2	17.0	1.12	11.1	13.0	1.98	1.71
Zn	2	13.2	1.22	9.47	11.0	1.83	1.58
Cd	2	9.27	1.37	7.47	8.68	1.62	1.40
Hg	2	8.65	1.40	7.13	8.29	1.58	1.37
Al	3	18.1	1.10	11.7	13.6	2.03	1.75
Ga	3	15.4	1.16	10.4	12.1	1.92	1.66
In	3	11.5	1.27	8.63	10.0	1.74	1.51
Tl	3	10.5	1.31	8.15	9.46	1.69	1.46
Sn	4	14.8	1.17	10.2	11.8	1.90	1.64
Pb	4	13.2	1.22	9.47	11.0	1.83	1.58
Bi	5	14.1	1.19	9.90	11.5	1.87	1.61
Sb	5	16.5	1.13	10.9	12.7	1.96	1.70

METALS

I do not know what effect my accusers have had upon you, gentlemen, but for my own part I was almost carried away by them; their arguments were so convincing. On the other hand, scarcely a word of what they said was true.

Plato, "Apology"

Starting the book here — As discussed in the Preface, use of this book can begin with Chapter 1, Chapter 6, or this chapter. If you begin here, before going on to Chapter 10 you should read, in order, Section 6-1, Section 4-3, and Section 8-6. Section 6-1 summarizes the translational symmetry aspects of crystal structures, Section 4-3 introduces k-space and the reciprocal lattice, and Section 8-6 introduces semiconductors.

Metals background — The most striking characteristics of metals are their excellent electrical and thermal conductivities (Table 6-1). The electrical conductivity immediately leads us to think in terms of a model in which the electrons are relatively free and can easily move under the influence of small electric fields. Many metals are also found to be ductile; they can be forged into useful shapes such as plows, knives, and similar tools. (In fact, the early ages of mankind are known as the Stone, Bronze and Iron Ages, reflecting the fact that people progressed from using stones to using metals.)

More than two thirds of the elements crystallize as metals having a crystal structure that is either face-centered cubic (fcc), hexagonal close packed (hcp), or body-centered cubic (bcc). See Table 3-1. The former two have 12-nearest neighbors. In the bcc structure there are eight nearest-neighbors a distance d away, but the six next-nearest neighbors are only a distance $2d/\sqrt{3} = 1.15d$ so these 14 atoms effectively can be considered nearest neighbors. Thus, all of the elemental metals have a large number of neighbors compared to the six nearest neighbors and four nearest neighbors typical for many ionic and covalent materials respectively.

With these large coordination numbers we cannot expect localized hybrid bonds; many metals have only one valence electron which clearly cannot form a covalent bond with 12-nearest neighbors atoms of the type discussed in Chapter 8. Rather, the valence electrons in metals occupy the space between the ion cores in a fairly uniform manner. For covalent bonds, the space between nearest neighbors is filled with electron density but only in certain directions. Thus, compared to the covalent bonds discussed in Chapter 8, the metallic bonds are much less angle-dependent. This gives rise to weaker bond bending forces and leads to the ductility. Nevertheless, the binding energy per atom (see Table 6-1) in typical metals is several electron volts, which is a fairly strong bond. For example, the binding energy per atom in the covalently bonded Li_2 molecule is 0.6 eV while per atom in metallic Li it is 1.8 eV. However, in the latter case this 1.8 eV can be said to be shared among many more bonds (i.e., nearest-neighbor atoms) so the "energy per bond" in metals is small because there are many "bonds," that is, neighbors.

Another important characteristic of metals is that they are rather "empty" structures from the point of view of the ionic cores. For example, the internuclear distance of metallic Li is 3.04 Å, which is very much larger than two times the ionic radius of Li^+ (2 × 0.60 Å). As a result, there is a great deal of volume available for the valence, or conduction electrons. Thus, the valence electrons can be rather uniformly distributed over most of the crystal in a manner quite different from covalent or ionic crystals. This will lead us to a model where we will assume that the electrons are free to move about in the crystal, that is, the free electron model. To calculate the actual binding energy of a metal requires us to treat the electrons as if they are "smeared out" and this situation will be discussed in Section 10-8.

Nevertheless, we may think of the formation of a "metallic bond" in the following very simple way. For an isolated atom the valence electron(s) is in the potential well due to the nucleus and core electrons. As the various atoms approach each other, the overlap of the atomic potentials causes the valence electrons to be in an effective potential that is lower than in the isolated atoms. A greatly simplifying and useful assumption is then made for metals. This complicated, "hilly" potential (very deep at the position of the ion cores, as in Fig. 9-1a) is assumed to be constant as long as the electron is in the metal. If the electron tries to leave the metal it will be pulled back by the net positive charge left on the metal. Thus, the electron is in a potential energy well, of depth W, as shown in Fig. 9-1b. The valence electrons now are "free," and such free electrons can be treated in several ways, as will be seen in this chapter. The energy of these free electrons can be determined. This energy, which is all kinetic energy in the models used in this chapter, is shown as $\mathcal{E}_m$ in Fig. 9-1b, but a more correct value will be found in Part B in Eq. 9-31. For most of Part A we, following Drude, assume that all of the electrons have the same energy $\mathcal{E}_m$. The use of a Maxwell-Boltzmann distribution has little effect on the results and complicates the mathematics. In either case, W minus this average kinetic energy of the electrons minus the energy of the repulsion of the cores is the binding energy of the metal.

CHAPTER 9   METALS   191

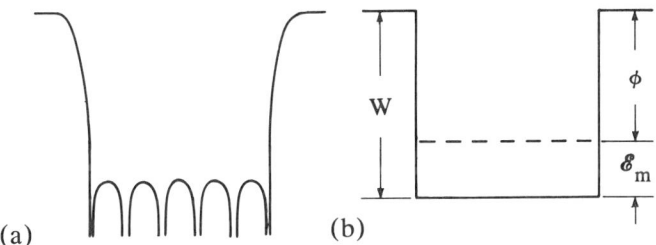

Fig. 9-1 (a) The potential in a crystal. (b) A model of this potential in which the electrons are in a "box" of depth W.

This chapter is divided into two parts, both concerned with the elementary theory of metals. In Part A the Drude theory is discussed. The ideas were published in 1900 and are an application of the kinetic theory of an "electron gas" in a solid. This theory had some striking successes as well as some failures. In Part B, what is often called the Sommerfeld theory is presented. This is the Drude theory with elementary quantum mechanics applied to the free electron gas. Rather than using one average energy and velocity or a Maxwell–Boltzmann distribution for the electron velocities, Sommerfeld used quantum mechanics to find the eigenfunctions and eigenvalues of a free electron gas and then used the Pauli exclusion principle when filling up the quantized states. This has some extremely important consequences and enables us to understand many of the fundamental properties of metals. However, these theories shed no light on a fundamental question: why are some materials metals and some insulators? To answer this question, as well as to understand many other properties of solids, band theory must be used. Band theory takes the translational symmetry of the crystal structure into account with some startling consequences; it is the basis of the modern theory of solids and is discussed in the next chapter.

Part A – Drude's Model

9-1 Drude's Free Electron Theory

The basic assumption in this theory is that within a metal the valence electrons of the atoms are free. That is, each atom gives up its valence electrons, which are then free to move about like a gas of electrons. For example, atomic sodium (Na) has electrons in the $1s^2 2s^2 2p^6 3s$ states. In metallic sodium, the 3s-electron (the valence electron) from each atom becomes free. (The remaining, bonding, $1s^2 2s^2 2p^6$ electrons are called the core electrons, which along with the nuclei, are called the cores.) The deep potential within the cores and relatively flat potential between the cores is assumed to be flat; because the metal is neutral, an electron attempting to leave it will be pulled back into the metal. Figure 9-1b shows the result, that is the electrons are in a potential of depth W in the metal and this well contains all the electrons. In fact, we will at first assume that all the electrons have the same energy given by $\mathscr{E}_m$. It is further assumed that the cores provide a mechanism for collisions with the electrons to bring the electrons to thermal equilibrium. The kinetic theory of gases is then applied to this free electron gas. Drude published this theory [Annalen der Physik **1**, 566 and **3**, 369 (1900)] just three years after the discovery of the electron by J. J. Thomson.

Although the ideal gas equations will be applied, remember that the electron gas is very dense. In copper, for example, taking Avogadro's number as 6.023×10^{23} atoms/mole, 8.92 gm/cm^3 for the density, and 63.5 for the atomic weight of Cu, the electron density is

$$\frac{N_A}{V} = 6.023 \times 10^{23} \frac{\text{atoms}}{\text{mole}} \cdot \frac{8.92 \frac{\text{gm}}{\text{cm}^3}}{63.5 \frac{\text{gm}}{\text{mole}}} = 8.47 \times 10^{22} \frac{\text{electrons}}{\text{cm}^3}$$

assuming that each atom contributes one electron. The high density of this gas can also be appreciated by calculating the pressure, P. Using the ideal gas law $PV = RT$ where V is the volume of one mole of gas, R the gas constant (82.057 cm^3 atmosphere mole^{-1} °K^{-1}, = 1.987 cal mole^{-1} °K^{-1}, which also equals the Boltzmann constant times Avogadro's number) and T the absolute temperature,

$$P = \frac{82.06(\text{cm}^3 \text{atm/mole}°K)(293°K)}{7.11(\text{cm}^3/\text{mole})} = 3381 \text{ atm}$$

Here we have used the fact that the atomic weight/density = (63.5 gm/mole)/(8.92 gm/cm^3) = 7.11 cm^3/mole in order to obtain the molar volume. Such a large pressure should be expected from a gas with this density. In fact, the density and pressure are so large that we would not expect the ideal gas law to hold. This problem is not of fundamental difficulty since corrections to the ideal gas equations could always be invoked. However, a related question is: what holds the electrons in the metal, that is, provides the large value of W (Fig. 9-1b)? The answer is the uniform background of positive charge, which makes the metal on the average neutral. These ideas also lead to the fact that on the surface we expect a thin double layer of positive ion cores and negative electrons, (Chapter 17) but Drude's model ignores surface effects. In fact, the simplicity of the model is one of its attractive features; the electrons are free and in a potential well of

CHAPTER 9   METALS

depth W with some collision process to establish equilibrium. Nothing else is assumed but the kinetic theory of gases.

In order to determine the electron densities for metals in general, we note that the number of electrons per cm³ is given by

$$N_A/V = 6.023 \times 10^{23} (\text{atoms/mole}) \, Z\rho/(\text{At. Wt.}) \qquad (9\text{-}1)$$

as in the preceding calculation for copper except that here we have added a factor Z that is equal to the number of free electrons that each atom contributes. (Although it is obvious for Cu and the alkali metals, for some other elements we must guess at values of Z, for example, for Fe and Pb.) Table 9-1 shows N/V for a number of elemental metals. The results are all in the $10^{22} - 10^{23}$ electrons/cm³ range. These values are sometimes presented in a different way. Let r_s be the radius of the sphere whose volume is equal to the volume per free or conduction electron. Then

$$\frac{V}{N} = \frac{4\pi r_s^3}{3} \qquad r_s = \left(\frac{3V}{4\pi N}\right)^{1/3} \qquad (9\text{-}2)$$

Thus, given the number of electrons/cm³, r_s can be found. Values for various metals are listed in Table 9-1. It can be seen that each electron has considerably more volume in the alkali metals than in most of the other metals. Compare r_s for Na (2.08 Å), for example, to the ionic radius of Na⁺ (0.95 Å). Since $(r_s/0.95)^3 > 10$ we see that an effective volume of a Na atom in the metal is very much larger than the volume of the Na ion. Thus, for the alkali metals at least, most of the volume of the metal crystal is taken by the free electrons; the core takes up very little volume.

The last introductory point we should make about Drude's ideas is concerned with the average energy, average thermal velocity, and specific heat of the free electrons. From the kinetic theory we expect an average energy of $k_BT/2$ for each degree of freedom. The free electrons have kinetic energy and no potential energy so that the average energy, $\mathscr{E}_m$, is $(3/2)k_BT$. (These relations are formally calculated in Chapter 11 from elementary statistical mechanics.) This can be related to a root mean square (rms) velocity v_m by

$$\mathscr{E}_m = (3/2)k_BT = mv_m^2/2 \qquad (9\text{-}3)$$

where m is the free electron mass. At room temperature $v_m \approx 10^7$ cm/sec. Remember that this is an rms thermal velocity for the electrons, while the vector sum of all the velocities, summed over all the electrons, equals zero. Figure 9-1b is a picture of this situation. The electrons are in a potential well of depth W. There is a distribution of energies with an average energy given by $\mathscr{E}_m$, Eq. 9-3. ϕ, the **work function** ($\phi \equiv W - \mathscr{E}_m$), has a value large enough to keep the electrons

Table 9-1 Properties of some of the elemental metals. (Values of N/V for Li and Hg are at 77°K, those for Na, K, Rb, and Cs are at 5°K, and the rest are at room temperature.) The electrical resistivities are in 10^{-6} Ω cm. [The room temperature r_s values are from Wyckoff and the resistivity values from G. W. C. Kaye and T. H. Loby "Table of Physical and Chemical Constants" (Longmass Green, 1966).]

El.	Z	N/V (10^{22}/cm^3)	r_s (Å)	Electrical Resistivity 77°K	Electrical Resistivity 273°K	τ (10^{-14}sec) 77°K	τ (10^{-14}sec) 273°K
Li	1	4.70	1.72	1.04	8.55	7.3	0.88
Na	1	2.65	2.08	0.8	4.2	17.	3.2
K	1	1.40	2.57	1.38	6.1	18.	6.1
Rb	1	1.15	2.75	2.2	11.0	14.	2.8
Cs	1	0.91	2.98	4.5	18.8	8.6	2.1
Cu	1	8.47	1.41	0.2	1.56	21.	2.7
Ag	1	5.86	1.60	0.3	1.51	20.	4.0
Au	1	5.90	1.59	0.5	2.04	12.	3.0
Be	2	24.7	0.99		2.8		0.51
Mg	2	8.61	1.41	0.62	3.9	6.7	1.1
Ca	2	4.61	1.73		3.43		2.2
Sr	2	3.55	1.80	7.	23.	1.4	0.44
Ba	2	3.15	1.96	17.	60.	0.66	0.19
Fe	2	17.0	1.12	0.66	8.9	3.2	0.24
Zn	2	13.2	1.22	1.1	5.5	2.4	0.49
Cd	2	9.27	1.37	1.6	6.8	2.4	0.56
Hg	2	8.65	1.40	5.8	Melted	0.71	
Al	3	18.1	1.10	0.3	2.45	6.5	0.80
Ga	3	15.4	1.16	2.75	13.6	0.84	0.17
In	3	11.5	1.27	1.8	8.0	1.7	0.38
Tl	3	10.5	1.31	3.7	15.	0.91	0.22
Sn	4	14.8	1.17	2.1	10.6	1.1	0.23
Pb	4	13.2	1.22	4.7	19.0	0.57	0.14
Bi	5	14.1	1.19	35.	107.	0.072	0.02
Sb	5	16.5	1.13	8.	39.	0.27	0.05

in the solid. The details of the spatial variation of the potential due to the ion cores are ignored.

The kinetic theory also predicts a specific heat $C_v = (3/2)k_B$ for each free electron or $(3/2)R$ for a mole of electrons. (Note that the free electrons have three degrees of freedom associated with the kinetic energy but none associated with the potential energy since the latter is constant, independent of position.) For each atom in a crystal, on the other hand, we expect $C_v = 3k_B$, or 3R for a mole since each atom has three quadratic terms in its potential energy as well as three in its kinetic energy. (If the atom moves from its equilibrium position

CHAPTER 9   METALS   195

it gains potential energy and is forced to move back towards its equilibrium position.) Thus, for a metal, we expect that the specific heat is the sum of the atom and the free electron contribution. This should be $3R + 3ZR/2$, for Z free electrons per atom and assuming one atom per primitive unit cell.

9-2 Drude's Assumptions

We have assumed that the valence electrons in a metal are free and that these electrons can be treated by the kinetic theory of gases. If these assumptions are valid then the general results discussed in the previous section are appropriate. However, several more specific assumptions must be made concerning the details of the collisions before calculating the dc conductivity.

First, we assume that the electrons experience **collisions** (by an unspecified interaction), or that the electrons are scattered. These collisions are treated as instantaneous scattering events which means the time for the scattering to take place is much shorter than any other times in the problem. It is through these collisions that electrons achieve thermal equilibrium corresponding to the metal temperature T. Thus, it is assumed that the electrons emerge from collisions with no memory of their velocities before and are then randomly directed with velocities appropriate to T. Second, between collisions the electrons travel in straight lines obeying Newton's laws. For example, if an electric field is applied in the x-direction then we have $m\ddot{x} = -eE$ thus the electron will have an additional velocity given by $-(eE/m)t$ for as long as the electric field is applied and as long as the electron is not scattered. However, on the average, the electrons are scattered after a time τ, and since after each scattering event they are in thermal equilibrium, a constant electric field will cause the electrons to have an extra average velocity given by $v_d = -(eE/m)\tau$. This is the **drift velocity**, v_d, that is due to the applied electric field. The root mean square velocity, v_m, which Drude assumed to be due to the thermal distribution at temperature T, in most metals turns out to be very much larger than v_d (i.e., $v_m \gg v_d$). Third, all details of the electron scattering are summarized in a **relaxation time** τ. This is the average time between scattering events; on the average, the probability that an electron will scatter in a time dt is dt/τ. Once an average thermal speed, v_m, and a relaxation time are defined, a **mean free path** ℓ is implied. It is defined as the average distance travelled by an electron between collisions and is given by $\ell = v_m \tau$.

With these assumptions we can calculate several properties of an electron gas. The dc conductivity is discussed first.

9-3 DC Conductivity

Ohm's law is (voltage) = (current) × (resistance). To make the terms independent of shape and length we relate the current density, J, to the electric field, E. Consider a piece of metal with a uniform cross sectional area A and length L. Then, J ≡ current/A and E ≡ voltage/L. The units of J are (charge/time-area) and those of E are (voltage/length). Then we can state Ohm's law in terms of the shape independent resistivity, ρ, or conductivity, σ

$$\mathbf{E} = \rho \mathbf{J} \quad \text{or} \quad \mathbf{J} = \sigma \mathbf{E} \tag{9-4}$$

where, for example, ρ = resistance × A/L. Actually, ρ and σ are second-rank tensors, as discussed in Chapter 5, but Drude ignored this complication and thought in terms of an isotropic metal.

Consider the microscopic picture of the free electrons. They are moving with large velocities of the order of those found from Eq. 9-3, but the directions are random so that the vector sum of all the velocities is zero. However, suppose that owing to an external field there is an additional average drift velocity, v_d, across an area A in a time dt. The charge that flows is $-(N/V)ev_d A dt$ where N is the number of electrons of charge $-e$. Thus, the current density is

$$\mathbf{J} = -(N/V)\, e\, \mathbf{v}_d \tag{9-5}$$

Using the expression for v_d (= $-eE\tau/m$) obtained in the last section, we can substitute it in Eq. 9-5. This results in the linear relation between **J** and **E** and yields an expression for the electrical conductivity σ containing only known quantities except for τ.

$$\sigma = (N/V)e^2\tau/m \tag{9-6}$$

From an experimental measurement of σ, we can use Eq. 9-6 to calculate τ. The resultant value is understood more clearly by noting that the experimental values of conductivity at dc and microwave frequencies ($\sim 10^{10}$ Hz) are the same. Thus, the electrons can respond to time-varying electric fields at microwave frequencies as easily as they can respond to dc fields. This means that τ, obtained via Eq. 9-6, must correspond to a time considerably shorter than 10^{-10} sec if the theory is to have any validity. The 300°K resistivity of Cu is approximately 1.5×10^{-6} ohm-cm or $\sigma = 6.5 \times 10^5$ (ohm-cm)$^{-1}$. Equation 9-6 is true in either cgs-Gaussian units or SI units. As long as we use a consistent set of units for values of σ, N, V, e, and m we will obtain τ in units of seconds. Values of σ are usually quoted in the SI unit of (ohm cm)$^{-1}$. To convert to the cgs-Gaussian unit we can use the conversion table in the appendix at the end of the book and have $\sigma =$

CHAPTER 9   METALS   197

$(6.5 \times 10^5)(9 \times 10^{11}) = 5.8 \times 10^{17}$ sec^{-1}. Using N/V from Table 9-1 to evaluate τ from Eq. 9-6,

$$\tau = (5.8 \times 10^{17})(9.1 \times 10^{-28} \text{gm})/(8.5 \times 10^{22})(4.8 \times 10^{-10})^2$$
$$= 2 \times 10^{-14} \text{ sec} \qquad (9\text{-}7a)$$

Table 9-1 shows experimental values of the resistivity at two temperatures and the corresponding values of τ calculated using Eq. 9-6. As can be seen, the times are all of the order of 10^{-14} to 10^{-16} sec, corresponding to frequencies considerably higher than microwave frequencies. Assuming E = 100 volts/cm and τ from Eq. 9-7a, we can estimate the drift velocity to be

$$v_d = \frac{-eE\tau}{m} = \frac{-(4.8 \times 10^{-10} \text{esu})(10^2 \text{ volts/cm})(2 \times 10^{-14} \text{ sec})}{(9.11 \times 10^{-28} \text{gm})(300 \text{ volts/statvolt})}$$
$$= -3.5 \times 10^3 \text{ cm/sec} \qquad (9\text{-}7b)$$

Hence, $v_d \ll v_m$, which is consistent with the idea that the current flow is a very small perturbation on the electron's normal behavior. (For a metal, a field of 100 V/cm is very large. The current required to maintain this field would ecessively heat the sample.)

This simple theory gives the experimentally found linearity between J and E but the conductivity is given in terms of τ, a quantity that is difficult to measure directly. This aspect of the theory is not very satisfying. However, in the next section we show that when σ is combined with the thermal conductivity, the value of τ drops out.

9-4 Wiedemann-Franz Law

The thermal conductivity κ is the constant of proportionality between flux of thermal energy, Q, (energy/area-time) and the temperature gradient (temperature/distance) where the energy flows in the opposite direction of the temperature gradient. So $Q = -\kappa \nabla T$. From the kinetic theory we know that for gases

$$\kappa = \ell v_m (N/V) C_v /3 \qquad (9\text{-}8)$$

where $\ell = v_m \tau$ is the **mean free path** between collisions and C_v is the electron heat capacity. Thus, the thermal energy flux, Q, is linearly dependent on the following quantities: distance that the carriers can travel before suffering a collision; the velocity of the carriers; the density of the carriers; their ability to carry heat; and of course the temperature gradient. Using σ from Eq. 9-6, $C_v = 3k_B/2$, and $mv_m^2/2 = 3k_B T/2$, we obtain

$$\frac{\kappa}{\sigma} = \frac{3}{2}\left(\frac{k_B}{e}\right)^2 T \tag{9-9}$$

This ratio, known as the Wiedemann-Franz ratio, is independent of any material constants. From Eq. 9-9 the **Lorenz number** $\bar{L} \equiv (\kappa/\sigma T)$ = 1.24 × 10^{-13} esu/deg^2 = 1.11 × 10^{-8} (volt/deg)^2. The experimental values are about twice as large as can be seen in the following table for several metals where σ is listed in units of (ohm-cm)$^{-1}$ and $\bar{L}$ in (volt/deg)^2

	273°K σ	273°K $\bar{L}$	100°K σ	100°K $\bar{L}$
Cu	6.5 × 10^5	2.3 × 10^{-8}	2.9 × 10^6	1.9 × 10^{-8}
Al	4.0 × 10^5	2.2 × 10^{-8}	2.1 × 10^6	1.5 × 10^{-8}
Pb	5.2 × 10^4	2.5 × 10^{-8}	1.5 × 10^5	2.0 × 10^{-8}

The prediction of a parameter-independent ratio and a value fairly close to what is experimentally observed is support for the theory. However, as shown in the table, $\bar{L}$ does depend on temperature even between 273°K and 100°K, with larger variations at lower temperatures. This theory cannot account for such behavior. (See the Notes for references to the σ and κ data.)

9-5 Frequency Dependent Conductivity

9-5a Frequency dependence In a straightforward manner, the relaxation time τ enters into the expression for the conductivity. Thus, it is natural to inquire into the frequency dependence of σ.

To investigate this, consider the time-dependence of the drift velocity v_d under the action of an external force, such as $\mathbf{F} = -e\mathbf{E}$. From the assumptions discussed in Section 9-2 we know that for any electron considered at time t, the probability that it will have a collision before t + dt is dt/τ. Therefore, the probability that it will not have a collision is 1 − dt/τ. Between collisions the electron will be acted on only by the external force **F**. From this straightforward model, with the details left as a problem, the equation of motion for v_d is

$$m\left[\frac{dv_d}{dt} + \frac{v_d}{\tau}\right] = F \tag{9-10}$$

In two limits this equation gives just what is expected. First, in a constant electric field the steady state solution is $v_d = -e\tau E/m$. Second, if F is suddenly turned off, one can solve Eq. 9-10 with F = 0 and obtain $v_d(t) = v_d(0)\exp(-t/\tau)$ so that the drift velocity decays

CHAPTER 9   METALS

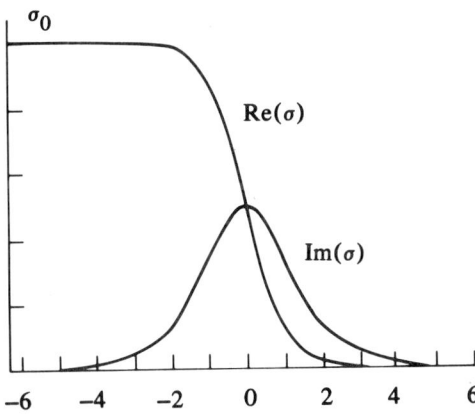

Fig. 9-2 Classical relaxation behavior shown by the real and imaginary parts of the conductivity.

from its steady state value to zero exponentially, with a characteristic time which is just the relaxation time τ.

Now solve Eq. 9-10 with $\mathbf{F} = -e\mathbf{E}$ where $\mathbf{E}$ is a function of time so that the frequency dependence of v_d and hence J and σ can be determined. Consider an oscillating electric field so that the time dependence can be taken as $\mathbf{E}(t) = \text{Re}[\mathbf{E}(\omega)\exp(-i\omega t)]$ where $\mathbf{E}(\omega)$ is complex. We also take $v_d(\omega)$ with this form: $v_d(t) = \text{Re}[v_d(\omega)\exp(-i\omega t)]$. Substituting in Eq. 9-10 we obtain Eq. 9-11 and then 9-12 by using $\mathbf{J}(\omega) = -(N/V)ev_d(\omega)$

$$-im\omega\, v_d(\omega) + m\frac{v_d(\omega)}{\tau} = -e\mathbf{E}(\omega) \qquad (9\text{-}11)$$

$$\mathbf{J}(\omega) = \frac{(Ne^2/Vm)\mathbf{E}(\omega)}{(1/\tau) - i\omega} = \frac{(Ne^2\tau/Vm)\mathbf{E}(\omega)}{1 - i\omega\tau} \qquad (9\text{-}12)$$

Since $\mathbf{J}(\omega) = \sigma(\omega)\mathbf{E}(\omega)$ defines the conductivity, we obtain

$$\sigma(\omega) = \frac{\sigma_0}{1 - i\omega\tau} \qquad \sigma_0 \equiv (N/V)e^2\tau/m \quad (9\text{-}13a)$$

For $\omega\tau \ll 1$, this agrees with Eq. 9-6. By taking the real and imaginary parts of σ we find the part of the current density that is in phase with the applied electric field and leads to the resistive, joule heating part of the current and the part of σ that is $\pi/2$ out of phase with the applied electric field and has inductive character. Thus,

$$\sigma(\omega) = \frac{\sigma_0}{1 + (\omega\tau)^2} + i\frac{\sigma_0\omega\tau}{1 + (\omega\tau)^2} \qquad (9\text{-}13b)$$

Figure 9-2 shows the frequency dependence of the real and imaginary parts of σ; it is the classical relaxation behavior found in this and many related problems. First, at low frequency ($\omega\tau \ll 1$) the conductivity is just the dc value and the current is in phase with the driving field, resulting in purely resistive behavior. Second, when the frequency of the electric field is much larger than $1/\tau$ ($\omega\tau \gg 1$) the electrons cannot follow because they have too much inertia so no energy is absorbed from the field. In fact, the inductive part of $\sigma(\omega)$ is much larger than the resistive part. Third, for $\omega\tau = 1$, the out-of-phase component peaks. This is the frequency at which the electrons move in resonance with $E(\omega)$ to give the maximum displacement.

9-5b Transparency of metals in the ultraviolet The discussion in Section 9-5a is appropriate for low frequencies, but at really high frequencies the spatial variation of the electromagnetic radiation must be taken into account. The preceding discussion assumes that $\mathbf{E}(\omega)$ is instantaneously the same over the entire sample. Also, the magnetic part of the electromagnetic wave was ignored, but that is reasonable since the $\mathbf{v} \times \mathbf{H}$ part of the force on the electrons is very, very much smaller than the $\mathbf{E}$ part. To take the spatial effects into account we must use Maxwell's equations.

Thus, we want to calculate a wave-like response for the electric field, and relate it to $\sigma(\omega)$, which is the only material property for our system. This will result in a definition of the dielectric constant. Assume that the wavelength of light, λ, is long compared to the mean free path of the electrons, $v_m\tau$. Thus, the macroscopic expression for J can be used, $\mathbf{J}(\mathbf{r}, t) = \sigma(\omega)\mathbf{E}(\mathbf{r}, t)$. λ for visible and ultraviolet light, 10^4 to 10^3 Å, is much larger than $v_m\tau$ for most metals at room temperatures, so this local field point of view is valid. Thus, the current density at $\mathbf{r}$ is determined by $\mathbf{E}(\mathbf{r})$. Then Maxwell's equations for free space are

$$\nabla \cdot \mathbf{E} = 0 \qquad \nabla \times \mathbf{E} = -\frac{1}{c}\frac{\partial \mathbf{H}}{\partial t}$$

$$\nabla \cdot \mathbf{H} = 0 \qquad \nabla \times \mathbf{H} = \frac{1}{c}\frac{\partial \mathbf{E}}{\partial t} + \frac{4\pi}{c}\mathbf{J} \qquad (9\text{-}14)$$

(See Section 13-3 for these equations in SI units; here they are in cgs-Gaussian units.) Using an $\exp(-i\omega t)$ time dependence, the preceding relation $\mathbf{J} = \sigma\mathbf{E}$, and the vector identity $\nabla \times (\nabla \times \mathbf{E}) = \nabla(\nabla \cdot \mathbf{E}) - \nabla^2 \mathbf{E}$ where $\nabla \cdot \mathbf{E} = 0$ from Maxwell's equations, we obtain

$$\nabla \times (\nabla \times \mathbf{E}) = -\nabla^2 \mathbf{E} = \frac{i\omega}{c}\nabla \times \mathbf{H} = \frac{i\omega}{c}\left[\frac{4\pi\sigma}{c}\mathbf{E} - \frac{i\omega}{c}\mathbf{E}\right] \qquad (9\text{-}15)$$

or

$$-\nabla^2 \mathbf{E} = \frac{\omega^2}{c^2}\left[1 + \frac{4\pi i\sigma}{\omega}\right]\mathbf{E} \qquad (9\text{-}16)$$

As a consequence of the real part, propagating solutions of the electromagnetic fields can be found; the imaginary part leads to attenuation.

CHAPTER 9 METALS 201

Table 9-2 Experimentally observed threshold wavelength for transparency of alkali metals (from Born and Wolf). Calculated values of λ_p, Eq. 9-19, are also shown.

	Li	Na	K	Rb	Cs
Exp. λ	2000 Å	2100	3150	3600	4400
λ_p	1500	2000	2900	3200	3600

These results are discussed fully in Chapter 13. Equation 9-16 is just the usual wave equation and allows us to define a complex dielectric constant

$$\nabla^2 E = \frac{\omega^2}{c^2}\varepsilon(\omega)E \qquad \varepsilon(\omega) = 1 + \frac{4\pi i \sigma}{\omega} \qquad (9\text{-}17)$$

For frequencies such that $\omega\tau \gg 1$, Eqs. 9-17 and 9-13 can be combined to obtain

$$\varepsilon(\omega) = 1 - \frac{\omega_p^2}{\omega^2} \qquad \omega_p^2 \equiv \frac{4\pi(N/V)e^2}{m} \qquad (9\text{-}18)$$

where ω_p is called the **plasma frequency**. For $\omega < \omega_p$, $\varepsilon(\omega)$ is negative so that solutions of E from Eq. 9-17 decay exponentially and there is no propagation through the metal, only a decay of the radiation in the metal. However, for $\omega > \omega_p$, ε is positive and the radiation can propagate through the metal, that is, for $\omega \geq \omega_p$ the metal is transparent. The critical value occurs at $\varepsilon = 0$ or $\omega = \omega_p$, which is usually quoted in terms of an equivalent wave length, the **plasma wavelength** λ_p. Using $\omega = 2\pi\nu$ and $\lambda\nu = c$

$$\lambda_p \equiv 2\pi c/\omega_p = 2\pi[mc^2/4\pi(N/V)e^2]^{1/2} \qquad (9\text{-}19)$$

Alkali metals have indeed been observed to become transparent in the ultraviolet region of the spectrum. Table 9-2 shows the experimental values for the threshold wavelength for transparency and the values calculated from Eq. 9-19. The agreement is reasonable. This quantity is difficult to determine accurately by experiment due to the width of the cutoff region. The calculated values in Table 9-2 are obtained using the bare mass of the electron (i.e., m = 0.911×10^{-27}gm); however an effective mass, slightly different from the bare mass, should be used. The effective mass concept will be discussed in the next chapter, and the interaction of radiation with solids will be discussed in more detail in Chapter 13.

9-6 Problems of Drude's Model

Because the free electron model is so simple, it has a great deal of appeal. However, many problems arise. There are problems even with its best triumph, the prediction of the Wiedemann-Franz ratio. The value of the Lorenz number $\kappa/\sigma T$ is half what is found experimentally for most metals at room temperature. (See the table in Section 9-4 and Ashcroft and Mermin Table 1.6.) This disagreement is not discouraging by itself. However, even at room temperature there is some temperature dependence to this ratio that is not accounted for by the theory. Further, the Lorenz numbers for some metals at low temperatures are a factor ten smaller than their room temperature values, which is not explained by the Drude model.

Moreover, there is another fundamental problem with the calculation of the Lorenz number. In obtaining it we used $C_v = 3R/2$ for the specific heat of electrons coming from a mole of atoms where each atom contributes one electron to the free electron gas. Thus, a mole of a monovalent metal should have $C_v = (3R/2) + 3R = 9R/2$, where the first contribution comes from the electrons and the second from the ions, as discussed in Section 9-1. We find an experimental value of C_v much closer to $3R$, which is what would be expected if the electrons contributed little or nothing. Thus, we are forced to accept the fact that at most only several percent of the electrons are contributing to C_v and hence to κ. It is particularly difficult to see how so few electrons can contribute to heat conduction while all the electrons apparently contribute to the electrical conductivity. (As we will see in Part B of this chapter, there is a compensating error in σ so that the Lorenz number comes out almost correct.)

Also, the theory does not explain the temperature dependence of σ. Experimentally, at room temperature we find $\sigma \propto T^{-1}$. The only way to obtain this temperature dependence is to assume that τ has this temperature dependence. $\tau = \ell/v_m$, where ℓ is the mean free path between collisions. We know from $\mathscr{E}_m = 3k_B T/2 = mv_m^2/2$ that $v_m \propto T^{1/2}$. Thus, we expect $\sigma \propto T^{-1/2}$ if ℓ were independent of temperature. We are left to conclude that ℓ must depend on temperature as $\ell \approx T^{-1/2}$ in order to reproduce the experimental temperature dependence of σ. However, this strange temperature dependence must be added to the theory as an ad hoc assumption.

Two aspects of the actual crystal structure are ignored. First, in some noncubic metals very large anisotropies in σ are found. (In Chapter 5 the formal symmetry aspects of σ are discussed.) Second, the periodic (translational) character of the crystal is ignored. (This is treated in the next chapter.)

Last, these free electrons should contribute to a paramagnetic susceptibility. An electron has a magnetic moment of one Bohr magneton, $\mu_B \equiv e\hbar/2mc = 0.927 \times 10^{-20}$ erg/gauss. The fraction of electrons with magnetic moment parallel to the magnetic field exceeds the antiparallel fraction by $\approx \mu H/k_B T$. For N free electrons we expect a magnetic moment $\approx N\mu_H^2 H/k_B T$ and a paramagnetic

CHAPTER 9   METALS   203

contribution to the susceptibility of $N\mu_B^2/k_BT$ or about 10^{-5} at room temperature and 10^{-2} at 0.3°K. Instead, in an ordinary metal, we find a temperature-independent value and only 1/100 of the expected room temperature value.

By applying quantum mechanics and the Pauli principle we take care of a number of these problems in a natural way.

Part B — Quantum Mechanics Applied

In this part we apply quantum mechanics to free electrons in a metal and demand that the Pauli principle be satisfied. This is called the Sommerfeld theory of metals. It will lead to the concept of the Fermi energy and to related concepts.

9-7 Eigenfunctions of Free Electrons in a Metal

9-7a Wave equation We shall neglect spin and spin-dependent forces. The electron spin will be added later, in the usual manner, by doubling the spatial degeneracy. In this case, the space part of the nonrelativistic wave equation, for a particle of mass m in a potential energy V(r), is

$$H\psi = \left[-\frac{\hbar^2}{2m}\nabla^2 + V(r)\right]\psi = \mathscr{E}\psi \qquad (9\text{-}20)$$

where $\nabla^2\psi = (\partial^2\psi/\partial x^2) + (\partial^2\psi/\partial y^2) + (\partial^2\psi/\partial z^2)$, ψ is the wave function, and $\mathscr{E}$ the energy. We would like to solve this equation to determine all the possible values of ψ and $\mathscr{E}$.

First, look more carefully at the fundamental approximation that we have been using in this chapter, namely our **independent electron approximation,** in which the electrons interact with one another only through some averaged potential energy. By considering just two electrons we can better appreciate the situation for N electrons. The Hamiltonian for two electrons is H(1, 2), which would contain a term e^2/r_{12} representing the electron-electron interaction. With a term that depends on the inter-electron distances, the solution of Eq. 9-20 becomes very difficult. So we make an approximation that H(1, 2) = H(1) + H(2), that is, H(1, 2) can be approximated by a Hamiltonian containing a term that just depends on electron 1, and a term that just depends on electron 2. (For a many-electron system this approximation is valid because the many electrons reduce the interactions between any two electrons and the strong electron-electron interaction can be approximated by an average interaction energy.) Then the problem is vastly simplified because instead of the difficult-to-solve equation $H(1, 2)\Psi(1, 2) = \mathscr{E}\Psi(1, 2)$, we have

$$[H(1) + H(2)] \,\psi(1)\psi(2) = \underbrace{\psi(2)H(1)\psi(1)}_{\mathscr{E}_1\psi_1} + \underbrace{\psi(1)H(2)\psi(2)}_{\mathscr{E}_2\psi_2}$$
$$= (\mathscr{E}_1 + \mathscr{E}_2) \,\psi(1)\,\psi(2) \quad (9\text{-}21)$$

Now H(1) operates only on the coordinates of electron 1, while H(2) operates only on the coordinates of electron 2. Thus, we see that when the total Hamiltonian is the sum of one electron Hamiltonians, the total energy is the sum of the **one electron energies**, and the total wave function is the product of **one electron wavefunctions** and this results in a vast simplification. The individual one-particle Hamiltonians in Eq. 9-21 typically contain terms that are the averaged electron-electron potential where the average is taken in some self-consistent manner. (Hartree first applied these ideas to atoms and the procedure is sometimes called the Hartree self-consistent approximation.)

We solve Eq. 9-20 for a single particle, an electron, taking $V(\mathbf{r})$ to be a constant, as shown in Fig. 9-1. Thus, as discussed previously, we are assuming that the electron-electron interactions and the interactions with all the cores average out to give a constant potential energy. Further, we take $V(\mathbf{r}) = 0$ without loss of generality since it has no coordinate dependence. Then, for a solution of Eq. 9-20 try the classic plane wave $\psi = A \exp(i\mathbf{k}\cdot\mathbf{r})$. The result is

$$(\hbar^2/2m)k^2\psi = \mathscr{E}\psi$$
or
$$\mathscr{E} = (\hbar^2/2m)k^2 \quad (9\text{-}22a)$$

k is called the **wave vector** since the phase of the plane wave is constant over any plane perpendicular to **k** ($\mathbf{k}\cdot\mathbf{r}$ = constant defines these planes). The wave vector has further significance in that for $\psi = A \exp(i\mathbf{k}\cdot\mathbf{r})$, $\hbar\mathbf{k}$ is an eigenfunction of the quantum mechanical momentum operator

$$\mathbf{p}\psi \equiv \frac{\hbar}{i}\frac{\partial}{\partial \mathbf{r}}\psi \equiv \frac{\hbar}{i}\nabla\psi = \hbar\mathbf{k}\psi \quad (9\text{-}22b)$$

The last equality is obtained by using $\psi = A \exp(i\mathbf{k}\cdot\mathbf{r})$. From the **de Broglie relationship**, $p = h/\lambda$, we know there is a wavelength associated with the particle, and using the preceding

$$k = |\mathbf{k}| = 2\pi/\lambda \quad (9\text{-}22c)$$

Further, since $\mathbf{p} = m\mathbf{v}$, we have $\mathscr{E} = p^2/2m = mv^2/2$. This is the correct classical energy of a free particle of mass m. The eigenfunctions are normalized by taking the integral over the volume of the metal, V, equal to unity. Using the notation of $d\tau$ as a volume element in any convenient coordinate system, we see that

CHAPTER 9   METALS   205

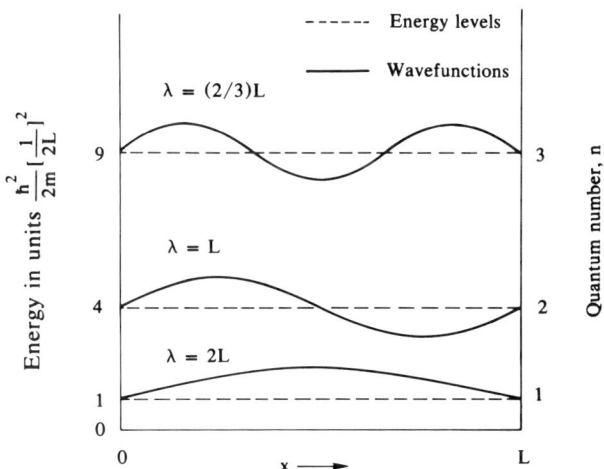

Fig. 9-3 The wave functions and energy levels of the first three eigenstates of a one-dimensional free electron in a box of length L.

$$1 = \int \psi^* \psi \, d\tau = A^2 \int \exp(-i\mathbf{k}\cdot\mathbf{r}) \exp(i\mathbf{k}\cdot\mathbf{r}) \, d\tau = A^2 V$$

so
$$\psi = V^{-1/2} \exp(i\mathbf{k}\cdot\mathbf{r}) \tag{9-23}$$

is the normalized eigenfunction. If we are thinking of a cube of metal of side L, since $L^3 = V$, then $\psi = L^{-3/2} \exp(i\mathbf{k}\cdot\mathbf{r})$ is often used; in the one dimensional case $\psi = L^{-1/2} \exp(ikx)$.

9-7b One-dimensional example (two kinds of boundary conditions) In a real metal suitable boundary conditions must be used; two kinds will be discussed. One leads to standing waves and one to propagating waves. The latter is of more general use because in three dimensions it leads to functions that are considerably easier to use when band theory (Chapter 10) is discussed. At first, for simplicity, the discussion is confined to one-dimensional space.

One reasonable boundary condition for the free electron wave functions is to require the wave function to go to zero at the boundaries, i.e. $\psi(0) = 0 = \psi(L)$. (In fact, this would be required if the well were infinitely deep, $W = \infty$ in Fig. 9-1.) Then the wave functions and corresponding energies are

$$\psi_n = (2/L)^{1/2} \sin(n\pi x/L)$$
$$\mathscr{E}_n = (\hbar^2/2m)(\pi/L)^2 n^2 \qquad n = 1, 2, 3, \ldots \tag{9-24}$$

The three lowest energy eigenfunctions are shown in Fig. 9-3. The functions are the ordinary **standing waves** where various integral multiples of $\lambda/2$ fit into a box of length L. These functions appear whenever standing waves are encountered. That they obey the boundary condition is clear. The probability density, $|\psi^2|$, of these standing waves is $(2/L)\sin^2(n\pi x/L)$.

The other situation we consider is the **periodic boundary condition** (sometimes called the Born-von Kármán boundary condition). This requires that the solutions should be periodic over the distance L so $\psi(x) = \psi(x+L)$. We may think of the one-dimensional box being twisted around back onto itself. This boundary condition allows the use of exponential wave functions, which are the natural ones to use when the periodic potential is considered in the next chapter. Upon substitution we immediately see that the periodic boundary condition demands that $\exp(ikL) = 1$ so $kL = 2\pi\bar{m}$ where $\bar{m}$ is any integer. Then the solutions, for this boundry condition, are

$$\psi_{\bar{m}} = L^{-1/2} \exp(i2\pi\bar{m}x/L)$$
$$\mathcal{E}_{\bar{m}} = (\hbar^2/2m)(2\pi/L)^2 \bar{m}^2 \qquad \bar{m} = 0, \pm 1, \pm 2, \ldots \quad (9\text{-}25)$$

The boundary condition is clearly obeyed since $\psi(x+L) = L^{-1/2}\exp[i2\pi\bar{m}(x+L)/L] = L^{-1/2}\exp(i2\pi\bar{m}x/L)\exp(i2\pi\bar{m}L/L) = \psi(x)$. For these propagating waves the probability density is $\psi^*\psi = 1/L$, that is, it is uniform throughout the sample.

9-7c Three dimensions For three dimensions the periodic boundary condition is $\psi(x+L, y, z) = \psi(x, y, z)$ with similar relations for the y- and z-coordinates. For a free electron the x, y, and z parts of the Hamiltonians are independent of each other (Eq. 9-20 can be factored into the sum of three terms) one for each coordinate; the wave function is the product of the wave functions, in the three directions; and the energy is the sum as discussed previously. Then

$$\psi_{\bar{m}} = L^{-3/2} \exp[(i2\pi/L)(\bar{m}_x x + \bar{m}_y y + \bar{m}_z z)]$$
$$\mathcal{E}_{\bar{m}} = (\hbar^2/2m)(2\pi/L)^2(\bar{m}_x^2 + \bar{m}_y^2 + \bar{m}_z^2) \qquad (9\text{-}26)$$

where $\bar{m}_x$, $\bar{m}_y$, and $\bar{m}_z = 0, \pm 1, \pm 2, \ldots$. These **spatial** (or **orbital**) **quantum numbers** are written as $\bar{m} \equiv (\bar{m}_x, \bar{m}_y, \bar{m}_z)$ a vector in "$\bar{m}$-space" and $\bar{m}^2 = \bar{m}_x^2 + \bar{m}_y^2 + \bar{m}_z^2$, hence the subscripts on the wave function and the energy. Thus, the allowed values of the wave vector **k** are

$$\mathbf{k} = (2\pi/L)\bar{\mathbf{m}} \qquad (9\text{-}27a)$$

Rewriting the wave function and energy in terms of **k** we have

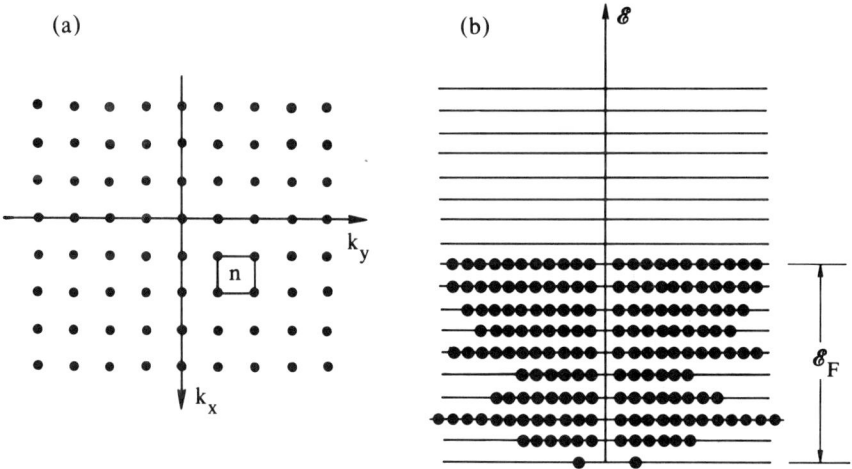

Fig. 9-4 (a) Part of k-space near the origin showing the allowed states for a two-dimensional free electron gas. Each dot represents an allowed orbital (or spatial) state. (b) A schematic representation of energy levels for an electron gas. Each dot represents one electron, separated into spin up and spin down states. $\mathscr{E}_F$ is shown.

$$\psi_k = V^{-1/2} \exp(i\mathbf{k} \cdot \mathbf{r})$$
$$\mathscr{E}_k = (\hbar^2/2m)(k_x^2 + k_y^2 + k_z^2) \tag{9-26}$$

where $V = L^3$, the volume of the sample. A unit cell in $\bar{m}$-space is $1 \times 1 \times 1$ so it has a volume equal to 1. However, a unit cell in **k-space** has a volume

$$\Delta V_k = \frac{(2\pi)^3}{L^3} = \frac{(2\pi)^3}{V} \tag{9-27b}$$

This is a fundamental result and just depends on the periodic boundary conditions. This volume in k-space per allowed **k**-value will be used in many problems. Figure 9-4a shows, in two dimensions, the uniform spacing of the allowed states in k-space. The square (n) that is outlined has an area of $(2\pi)^2/L^2$, obviously resulting from Eq. 9-27a.

9-7d Degeneracy As is well known, more than one state can have the same energy. The number of such states with the same eigenvalue is called the degeneracy.

The wave equation for free electrons in a box has been solved and the energy levels determined. Now we would like to fill the levels with electrons. The **Pauli exclusion principle** requires that no two

Table 9-3 Degeneracies of free electron levels

Typical possibilities			$\overline{m}^2$	orbital degeneracy	total degeneracy
$\overline{m}_x$	$\overline{m}_y$	$\overline{m}_z$			
0	0	0	0	1	2
±1	0	0	1	6	12
±1	±1	0	2	4×3	24
±1	±1	±1	3	8	16
±2	0	0	4	6	12

electrons have the same set of quantum numbers. Thus, for every one orbital quantum number $\overline{m}$ only two electrons, of opposite spin, can be accommodated. Thus for one $\overline{m}$-value the degeneracy is two. We will fill the levels starting from the lowest energy. For $\overline{m} = (0, 0, 0)$ only two electrons can be accommodated. For $\overline{m} = (\pm 1, 0, 0)$ or $(0, \pm 1, 0)$ or $(0, 0, \pm 1)$ all the energies are degenerate; there is a sixfold orbital degeneracy or a twelvefold degeneracy counting spins. So far a total of 14 electrons can be accommodated. The energy of the two electrons with $\overline{m} = 0$ is zero, while that of the next twelve is $(\hbar^2/2m)(2\pi/L)^2$. The next electron must go into a higher energy state, one given by $\overline{m}^2 = 2$. For $\overline{m}^2 = 2$ we can have $\overline{m} = (\pm 1, \pm 1, 0)$, a fourfold orbital degeneracy, but this can be done in three ways, that is, $(\pm 1, 0, \pm 1)$, and so forth. A few of the degeneracies for small values of **m** are given in Table 9-3. While it is not apparent by considering low values of $\overline{m}^2$, for large $\overline{m}$-values the degeneracies go up as $(\overline{m}^2)^{1/2}$. This will be seen in the next section. Figure 9-4 schematically represents the filled energy levels of the free electrons in a metal. Equation 9-27a shows how, given $\overline{m}$-space, one can also define k-space, which is called **wave vector space** or **k-space.**

9-8 Fermi Energy, Density of States, and Fermi Surface

The previous section introduced the useful concept of $\overline{m}$-space. It is the space with orthogonal coordinates, $\overline{m}_x$, $\overline{m}_y$, and $\overline{m}_z$. Any one value of $\overline{m}$ corresponds to one particular point in $\overline{m}$-space. Since the allowed values of $\overline{m}_x$, $\overline{m}_y$, and $\overline{m}_z$ are all integers, the allowed values fill $\overline{m}$-space with points that outline cubes of dimension 1 × 1 × 1. (See Fig. 9-4a, which could be $\overline{m}$-space just as well as k-space.)

Using $\overline{m}$-space we define several very useful terms, but first consider some of the properties of this space. The discussion in this section applies to absolute zero temperature so that we need not be concerned with thermal excitations of the electrons. After Section 9-12, where the proper statistics are discussed, the temperature dependence of these terms will be discussed. When filling states with

CHAPTER 9   METALS   209

electrons, the lowest energy states are occupied first. From Eq. 9-26, we see that this corresponds to the smallest values of $\bar{m}$; thus, $\bar{m}$-space gets filled up spherically, starting from the origin, and neglecting the granularity of the sphere. (The states in Table 9-3 are listed in order of increasing energy.) For a given number of electrons in a metal, a large sphere in $\bar{m}$-space surrounds the values of $\bar{m}$ that correspond to the occupied states. Last, each point in $\bar{m}$-space corresponds to two electrons of opposite spin since $\bar{m}$ is the spatial or orbital quantum number and for each orbital state there are two spin states.

For a given number of electrons we define $\bar{m}_F$ as the largest value of $|\bar{m}|$. Then we ask: what is the energy, temperature, and velocity of the electrons with $\bar{m}_F$? The number of electron states that have $\bar{m}$ less than $\bar{m}_F$ is contained in a sphere in $\bar{m}$-space, and the number of such states is $2(4\pi/3)\bar{m}_F^3$. The factor 2 comes from the two possible spin values per orbital state. If N is the number of electrons in the sample, then $\bar{m}_F$ is given by

$$N = 2\left(\frac{4\pi}{3}\right)\bar{m}_F^3 = 2\left(\frac{4\pi}{3}\right)\frac{V}{(2\pi)^3}k_F^3 \qquad (9\text{-}28)$$

Substituting this value of $\bar{m}_F$ into Eq. 9-26, or using k from Eq. 9-27 or applying $\mathscr{E} = (\hbar^2/2m)k^2$, which holds for one, two, and three dimensions, we calculate the corresponding maximum energy, $\mathscr{E}_F$, which is called the **Fermi energy**. Since this is done at absolute zero temperature it is written as $\mathscr{E}_F(0)$.

$$\mathscr{E}_F(0) = \frac{\hbar^2}{2m}k_F^2 = \frac{\hbar^2}{2m}\left(\frac{2\pi}{L}\right)^2 \bar{m}_F^2 = \frac{\hbar^2}{2m}\left(\frac{3\pi^2 N}{V}\right)^{2/3} \quad (9\text{-}29a)$$

Given an energy, we can always define an equivalent temperature. The **Fermi temperature**, T_F, is a temperature that is equivalent to the Fermi energy, and is given by $k_B T_F \equiv \mathscr{E}_F$. Similarly, given an energy, a velocity can be defined. The **Fermi velocity**, v_F, is defined in terms of $\mathscr{E}_F$ as $mv_F^2/2 = \mathscr{E}_F(0)$, since all the energy is kinetic energy. (Remember the model for this entire chapter assumes a constant potential energy, taken as zero, as in Fig. 9-1.) Last, a **Fermi wave vector** can be defined in terms of $\mathscr{E}_F = (\hbar^2/2m)k_F^2$, or from the identity between $\bar{m}$ and k (Eq. 9-27), and already has been indicated in Eq. 9-28. From Eq. 9-28

$$k_F = (3\pi^2 N/V)^{1/3} \qquad (9\text{-}29b)$$

Values of $\mathscr{E}_F$, T_F, v_F, and k_F are given in Table 9-4 for some of the same metals as in Table 9-1.

Note that the values of $\mathscr{E}_F$ are about 5 eV, which is very much larger than the thermal energies at room temperature (0.025 eV) considered in the Drude theory. Another way to say this is

Table 9-4 Values of $\mathscr{E}_F$ using Eq. 9-29a and the electron concentrations of the metallic elements from Table 9-1. Values of the equivalent Fermi temperatures, Fermi velocities and Fermi wave vectors are also given.

Element	$\mathscr{E}_F$(eV)	$T_F(°K)$	v_F(cm/sec)	k_F(cm^{-1})
Li	4.74	5.51×10^4	1.29×10^8	1.12×10^8
Na	3.24	3.77	1.07	0.92
K	2.12	2.46	0.86	0.75
Rb	1.85	2.15	0.81	0.70
Cs	1.59	1.84	0.75	0.65
Cu	7.00	8.16	1.57	1.36
Ag	5.49	6.38	1.39	1.20
Au	5.53	6.42	1.40	1.21
Be	14.3	16.6	2.25	1.94
Mg	7.08	8.23	1.58	1.36
Ca	4.69	5.44	1.28	1.11
Sr	3.93	4.57	1.18	1.02
Ba	3.64	4.23	1.13	0.98
Fe	11.1	13.0	1.98	1.71
Zn	9.47	11.0	1.83	1.58
Cd	7.47	8.68	1.62	1.40
Hg	7.13	8.29	1.58	1.37
Al	11.7	13.6	2.03	1.75
Ga	10.4	12.1	1.92	1.66
In	8.63	10.0	1.74	1.51
Tl	8.15	9.46	1.69	1.46
Sn	10.2	11.8	1.90	1.64
Pb	9.47	11.0	1.83	1.58
Bi	9.90	11.5	1.87	1.61
Sb	10.9	12.7	1.96	1.70

$T_F \gg 300°K$. These large values of $\mathscr{E}_F$ (and T_F) lead to Fermi velocities of the order of 10^8 cm/sec, which is much faster than thermal velocities. These large numbers come about because of the quantization of the free electron energy levels in the metal and because the Pauli principle restricts the number of electrons in each quantum state. Thus, the electrons fill much higher energy states than are required for a classical gas.

The wave vector — Via Eq. 9-27, $\overline{m}$ and the wave vector **k** are linearly related. Thus, just as in the case of $\overline{m}$-space, we may define **k-space** or **wave vector space**. k-space rather than $\overline{m}$-space will be used in Chapter 10 and Section 9-13. It is within k-space that the Fermi surface is defined. Remember $\overline{m}$ or **k** are just *orbital quantum numbers*. We have already written the wave functions and energies in terms of **k** in Eq. 9-26, and Fig. 9-4 showed the allowed states in k-space. How-

CHAPTER 9   METALS   211

ever, a change in labels from k_x and k_y to $\bar{m}_x$ and $\bar{m}_y$ would make Fig. 9-4 appropriate to $\bar{m}$ space as well. There are two important things to realize. First, k-space is uniformly filled with dots, each of which corresponds to an allowed (quantum) state. Second, within the Fermi sphere there are a huge number of allowed states, $\sim 10^{22}$, and the density is very large. In a linear dimension the spacing between allowed states is ~ 1 cm^{-1} (Eq. 9-27a). Thus, for normal size crystals, on a scale that shows k_F, the allowed states correspond to a quasi-continuum of k-values.

Fermi surface – The **Fermi surface** is the energy boundary between the occupied and unoccupied states in k-space at $T = 0°K$. Clearly it is a surface of constant energy; in fact the energy is $\mathscr{E}_F$. For free electrons the Fermi surface is a sphere (called the **Fermi sphere**) but, as we shall see in Chapter 10, for real metals it can have a more complicated shape. Although we define the Fermi surface at $0°K$, we shall see that $\mathscr{E}_F$ hardly is affected by temperature. Thus, to very high temperatures the surface remains sharp and the same as at $0°K$. This surface is very significant because it is only those electrons that energetically are close to it that can participate in transport processes and thermal excitations. These points are covered in the rest of this chapter. For the more complicated Fermi surfaces encountered in Chapter 10 these same general statements apply; the shapes of the Fermi surfaces may vary but the physics is the same.

Density of states, $g(\mathscr{E})$ – This is defined as the number of states between $\mathscr{E}$ and $\mathscr{E} + d\mathscr{E}$. To find $g(\mathscr{E})$, note that the number of states with energy up to arbitrary value of $\mathscr{E}$ is

$$2\left(\frac{4\pi}{3}\right)\bar{m}^3 = 2\left(\frac{4\pi}{3}\right)\frac{V}{(2\pi)^3}k^3 = \left(\frac{V}{3\pi^2}\right)\left(\frac{2m\mathscr{E}}{\hbar^2}\right)^{3/2}$$

where we have used Eqs. 9-26 and 9-27 to relate $\mathscr{E}$ to $\bar{m}$ or k. This quantity is just

$$\int_0^{\mathscr{E}} g(\mathscr{E}) \, d\mathscr{E} = 2\left(\frac{4\pi}{3}\right)\bar{m}^3 = \left(\frac{V}{3\pi^2}\right)\left(\frac{2m\mathscr{E}}{\hbar^2}\right)^{3/2} \quad (9\text{-}30a)$$

so

$$g(\mathscr{E}) = (2^{1/2}V/\pi^2)(m/\hbar^2)^{3/2}\mathscr{E}^{1/2} \equiv C\mathscr{E}^{1/2} \quad (9\text{-}30b)$$

where the constant C, defined by this expression, is introduced for later convenience. Figure 9-5 shows a plot of $g(\mathscr{E})$ vs. $\mathscr{E}$ for $T = 0°K$ (and for $T > 0°K$, which will be discussed shortly). The parabolic form is clear. At $T = 0°K$ the states are filled up to $\mathscr{E}_F$, as shown. The parabolic form for $g(\mathscr{E})$ is not apparent in Fig. 9-4 (which is closely related to Fig. 9-5) because only a negligibly few of the low energy occupied states are shown.

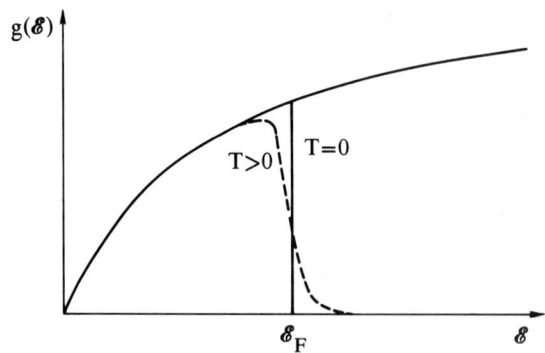

Fig. 9-5 The $\mathscr{E}^{1/2}$ dependence for the density of states g($\mathscr{E}$) of a free electron gas. At T = 0°K the allowed states are occupied up to the Fermi energy. This also is shown, as well as the occupation at higher temperatures.

To calculate the total kinetic energy, U_0, of all the electrons we make use of Eq. 9-30b. Then it follows that

$$U_0 = \int_0^{\mathscr{E}_F} \mathscr{E}\, g(\mathscr{E})\, d\mathscr{E} = (2/5)C\, \mathscr{E}_F^{5/2} = (3/5)\, N\mathscr{E}_F \qquad (9\text{-}31)$$

where $\mathscr{E}_F^{3/2}$ from Eq. 9-29a is used. Note that all of the energy is kinetic energy since there is no potential energy. Thus, in this model U_0 is the total energy of the free electron system. From U_0 other thermodynamic quantities can be calculated. For example, the pressure exerted by the electron gas is given by $P = -(\partial U_0/\partial V)_N$. Since $\mathscr{E}_F \propto V^{-2/3}$, from Eq. 9-29a, we quickly obtain

$$P = (2/3)(U_0/V) \qquad (9\text{-}32)$$

The compressibility, K, or bulk modulus B (B = K^{-1}), can be calculated in the same manner as for the ionic and molecular solids. Thus,

$$B = -V(\partial P/\partial V) = (10/9)(U_0/V) = (2/3)(N/V)\mathscr{E}_F \qquad (9\text{-}33)$$

is found, using $P \propto V^{-5/3}$ from Eqs. 9-29, 9-31, and 9-32. Table 9-5 shows values of the calculated bulk moduli (Eq. 9-33) and experimental results for several metals. While the agreement is not very good, it is satisfying to see that the values are of the correct order of magnitude even though the ion cores are totally neglected in this model and the electrons are treated in a rather simple manner.

Dimensionality effects — For three-dimensional space the density of states is proportional to $\mathscr{E}^{1/2}$ as in Eq. 9-30. For two dimensions it is constant, independent of energy; for one dimension it is proportional to $\mathscr{E}^{-1/2}$. See the Problems. Thus, merely from dimensional considerations we obtain very different energy dependent behavior for the density of states. There are related differences for the Fermi energy, Fermi wave vector, and so on. These dimensionality effects play

CHAPTER 9   METALS   213

Table 9-5 Measured and calculated (via Eq. 9-33) values for the bulk moduli for some metals. The values of B are expressed in units of 10^{10} dynes/cm^2.

Element	Measured B	Calculated B
Li	11.5	23.9
Na	6.42	9.23
K	2.81	3.19
Rb	1.92	2.28
Cs	1.43	1.54
Cu	134.3	63.8
Ag	99.9	34.5
Al	76.0	228

important roles in various areas of science. "Two-dimensional" physical effects are important in surface science (Chapter 17), inversion layers and superlattices (Chapter 18). "One-dimensional" effects are observed experimentally in very narrow "wires", in special inversion layer configurations, and certain organic conductors.

9-9 Soft X-Rays, Heat Capacities

Before proceeding with quantitative calculations for finite temperatures we discuss qualitatively two observations that show that the quantum model of the free electron gas is in reasonable agreement with experiment and an improvement over the Drude model.

$g(\mathscr{E})$ – If aluminum metal is bombarded by electrons with enough energy to knock a 2p-electron from the Al core then x-rays are emitted. Figure 9-6 shows a plot of the measured intensity of x-ray emission vs. the energy of the x-rays. The experiment is simple to visualize. The Al atom has a $1s^2 2s^2 2p^6 3s^2 3p^1$ electron configuration. In the metallic state the $3s^2 3p^1$ electrons form the "free electron gas;" the other electrons remain in the core. The metal is bombarded by electrons that are accelerated from a hot filament. These electrons have just enough energy to impart ionized electrons from the 2p-core level, leaving a hole there. Then any electron in the free electron gas (Fermi sea) can fall into the vacant 2p-level and emit radiation. For Al this radiation falls in the soft x-ray region of the spectrum ~70 eV. The intensity of this radiation is proportional to $g(\mathscr{E})$, the number of conduction electrons in each small energy interval in the band. Thus, the measured intensity should be something like $g(\mathscr{E})$ as is observed (Fig. 9-6). It is difficult to be quantitative since the transition probability matrix element for the free electrons is found to depend on energy (from more detailed calculations than discussed here) and the experimental curve tails off at low energy rather than abruptly termi-

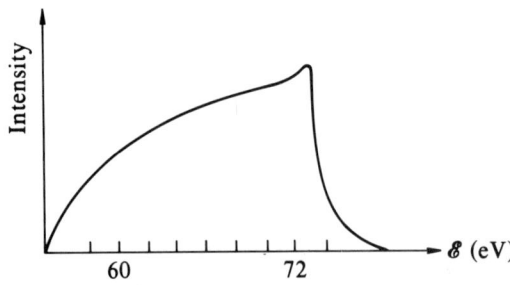

Fig. 9-6 The intensity of x-ray emission for aluminum.

nating, as drawn. From the experiment we obtain $\mathscr{E}_F \approx 16$ eV while ≈ 12 eV is calculated from Eq. 9-29. This agreement is fair, but more importantly the result shows that the calculated shape of the density of states is in reasonable agreement with experiment.

Heat capacity – The total heat capacity of a metal consists of a contribution from the ions in the crystal plus the contribution from the free electrons. The former is discussed in Chapter 11. Here we consider the free electron contribution. For a classical gas, the kinetic energy of N particles is $\mathscr{E} = (3/2)k_B NT$ and the heat capacity at constant volume is $C_v = \partial \mathscr{E}/\partial T = (3/2)k_B N$. This says that if the temperature is raised an amount ΔT, the kinetic energy of the gas is raised an amount $\Delta \mathscr{E}' = (3/2)k_B N \Delta T$; that is, all the particles can increase their kinetic energy. However, this is not possible for an electron gas in a metal. Consider why this is so. Ordinary temperatures correspond to $\mathscr{E}_T = k_B T \approx 25$ meV while $\mathscr{E}_F \approx 5$ eV. Thus, at these temperatures, we see that for most of the electrons the states within $\mathscr{E}_T$ of their energy are occupied and they cannot be excited. Thus, only the states within $\mathscr{E}_T$ of $\mathscr{E}_F$ are available for excitations by a change in temperature. Instead of N states available for excitation, roughly $(\mathscr{E}_T/\mathscr{E}_F)N$ will be available, so that if the temperature is raised an amount ΔT, then the kinetic energy will be raised an amount $\Delta \mathscr{E}' = (3/2)k_B N(\mathscr{E}_T/\mathscr{E}_F)\Delta T$ and $C_v \approx (3/2)k_B N(\mathscr{E}_T/\mathscr{E}_F)$. At room temperature for a Fermi energy of 5 eV this is just $\approx 1/200$ smaller than what would be calculated from the simple Drude model. This result is in good agreement with experimental observation.

After considering Fermi-Dirac statistics, which are the proper statistics for the electron gas, we will quantitatively calculate quantities like C_v. However, the underlying physics is as described previously.

Summary – By quantitizing the space part of the free electron gas, we obtain discrete energy levels. Taking the Pauli principle into account when filling these states, we find that they are occupied to a very high energy, $\mathscr{E}_F$. Besides this Fermi energy ($\mathscr{E}_F$), we also have an equivalent Fermi temperature (T_F), a Fermi velocity (v_F), and a Fermi wave vector (k_F). In the Drude theory discussed in Part A, the mean energy of the electrons is $\mathscr{E}_m \approx 3k_B T/2$. Thus, $\mathscr{E}_F \gg \mathscr{E}_m$, so the Fermi velocity is much larger than expected from the Drude theory. This has

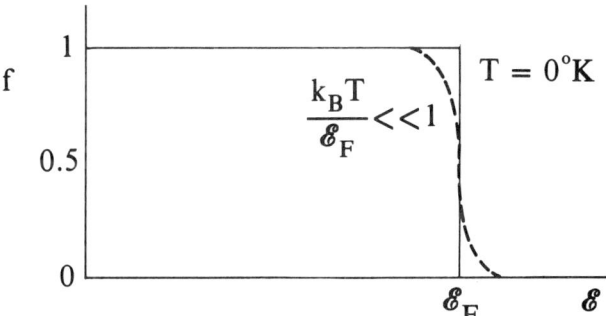

Fig. 9-7 The Fermi-Dirac distribution function for T = 0°K and low temperatures.

consequences as to the relaxation times and mean free paths of the electrons at the Fermi energy.

9-10 Fermi-Dirac Statistics

In the original Drude theory a single average energy, average velocity, and so on, were used rather than any statistical distribution. Lorenz later applied the Maxwell-Boltzmann distribution law to the gas of electrons with no significant changes in the results. As we have seen, however, the use of quantum mechanics and the Pauli exclusion principle gives quite different results at T = 0°K. In order to calculate the properties for T > 0°K, we must have the proper description of the thermal occupation of the allowed quantum states. Particles with a spin of 1/2 obeys Fermi-Dirac statistics, and particles that obey these statistics are called **fermions**. These statistics are derived in any statistical mechanics textbook. For an ideal electron gas in thermal equilibrium with a heat bath at a temperature T, the probability that an allowed state, with energy $\mathscr{E}$, will be occupied is

$$f(\mathscr{E}) = \frac{1}{e^{(\mathscr{E}-\mathscr{E}_F)/k_BT} + 1} \tag{9-34}$$

Here we note that the energy for which the probability of occupancy is one-half (f = 1/2) is $\mathscr{E}_F$, the Fermi energy. In fact, we can take this as the definition of **Fermi energy** for T > 0°K. It will turn out that $\mathscr{E}_F$ is a very weak function of temperature, thus, for the purpose of discussion, the 0°K result can be used.

Figure 9-7 shows $f(\mathscr{E})$ at T = 0 and at T ≪ T_F (or $k_BT \ll \mathscr{E}_F$). The distribution for T = 0°K exactly describes the situation discussed so far in Part B. From Table 9-4 we see that $T_F > 10^{4}$°K for most metals, so for the usual range of temperatures T ≪ T_F, and deviations from $f(\mathscr{E}) = 1$ occur only very, very near $\mathscr{E} = \mathscr{E}_F$. Moreover, it is only for states very near $\mathscr{E}_F(T)$ that electrons can be thermally excited into unoccupied states, resulting in important contributions to physical phenomena. Thus, this region of the distribution must be treated carefully as is done in Section 9-11. The high energy tail of this distribution can be measured experimentally by thermionic emission. See Problem 4.

To obtain the number of electrons, dN, with energy between $\mathscr{E}$ and $\mathscr{E} + d\mathscr{E}$ we multiply the density of states, Eq. 9-30b, by the probability that a state may be occupied.

$$dN = f(\mathscr{E}) \, g(\mathscr{E}) \, d\mathscr{E} = \frac{C\mathscr{E}^{1/2}d\mathscr{E}}{[\exp{(\mathscr{E} - \mathscr{E}_F)/k_BT}] + 1} \tag{9-35}$$

$\mathscr{E}_F(T)$ is determined by the condition that the integral of dN is N, the total number of electrons.

<u>At absolute zero</u> we can determine the Fermi energy $\mathscr{E}_F(0)$ by integrating Eq. 9-35 and obtain agreement with Eq. 9-29.

$$N = C \int_0^{\mathscr{E}_F(0)} \mathscr{E}^{1/2} \, d\mathscr{E} = (2/3) \, C \, [\mathscr{E}_F(0)]^{3/2}$$

$$\mathscr{E}_F(0) = \frac{\hbar^2}{2m} \left(\frac{3\pi^2 N}{V} \right)^{2/3} \tag{9-36}$$

9-11 Low Temperature Expansion using F-D Statistics

The application of the Fermi-Dirac (F-D) distribution to various types of calculations can be handled in a mathematically systematic manner. Almost always, we are in a situation where $T \ll T_F$ so $f(\mathscr{E})$ differs from 1 only very near $\mathscr{E}_F$ (within a width a few k_BT). It is only this region of the electron distribution that will contribute to most of the measured properties. Thus, for any property that depends on energy we expand it about $\mathscr{E}_F$ and the evaluation will be easy. This is called the **Sommerfeld expansion**. Here we discuss this type of approach in general and then apply it in the next section.

Let $\Gamma(\mathscr{E})$ be any function of energy that is zero at $\mathscr{E} = 0$, and consider the integral

$$I = \int_0^\infty f(\mathscr{E}) \left[\frac{d\Gamma(\mathscr{E})}{d\mathscr{E}} \right] d\mathscr{E} \tag{9-37}$$

$f(\mathscr{E})$ is the F-D distribution function, Eq. 9-34, and we use the standard notation for derivatives, that is, $\Gamma' = d\Gamma/d\mathscr{E}$, $f' = df/d\mathscr{E}$, and so on, where the derivatives are taken with respect to energy. Integrating

CHAPTER 9 METALS 217

Eq. 9-37 by parts gives

$$I = [f\ \Gamma]_0^\infty - \int_0^\infty \frac{df}{d\mathscr{E}}\ \Gamma\ d\mathscr{E} \tag{9-38}$$

where the first term on the right is zero because $f(\infty) = 0 = \Gamma(0)$. This result, Eq. 9-38, shows the reason for choosing the form for the integral I in Eq. 9-37. Namely, I will have contributions coming only from the region $\mathscr{E} = \mathscr{E}_F$ because it is only in this region that f' differs from zero, at least for $T \ll T_F$, as also is apparent in Fig. 9-7. Next, expand $\Gamma(\mathscr{E})$ in a Taylor series about $\mathscr{E}_F$

$$\Gamma(\mathscr{E}) = \Gamma(\mathscr{E}_F) + (\mathscr{E} - \mathscr{E}_F)\ \Gamma'(\mathscr{E}_F)$$
$$+ (1/2)(\mathscr{E} - \mathscr{E}_F)^2\ \Gamma''(\mathscr{E}_F) + \ldots \tag{9-39}$$

Substituting this result into Eq. 9-38 we have

$$I = \Gamma(\mathscr{E}_F)\ I_0 + \Gamma'(\mathscr{E}_F)\ I_1 + \Gamma''(\mathscr{E}_F)\ I_2 + \ldots \tag{9-40}$$

$$I_0 = -\int_0^\infty f'\ d\mathscr{E} = 1 \tag{9-40a}$$

$$I_1 = -\int_0^\infty (\mathscr{E}-\mathscr{E}_F)\ f'\ d\mathscr{E} = 0 \tag{9-40b}$$

$$I_2 = -(1/2)\int_0^\infty (\mathscr{E} - \mathscr{E}_F)^2\ f'\ d\mathscr{E} = (\pi^2/6)\ (kT)^2 \tag{9-40c}$$

The numerical result for I_0 is clear. I_1 is zero because $(\mathscr{E} - \mathscr{E}_F)$, which is an odd function of $(\mathscr{E} - \mathscr{E}_F)$, is multiplied by f', which is an even function of $(\mathscr{E} - \mathscr{E}_F)$. Last, for I_2 we may replace the lower limit on the integral by $-\infty$, substitute $x = (\mathscr{E} - \mathscr{E}_F)/k_BT$, and obtain an integral that is found in a standard table of integrals

$$I_2 = \frac{1}{2}(k_BT)^2 \int_{-\infty}^\infty \frac{x^2 \exp x}{(1 + \exp x)^2}\ dx = (\pi^2/6)(k_BT)^2 \tag{9-40d}$$

This gives our desired result, which involves only $\Gamma(\mathscr{E}_F)$ and its second derivative at $\mathscr{E}_F$

$$I = \int_0^\infty f\ \Gamma'\ d\mathscr{E} = \Gamma(\mathscr{E}_F) + \left(\frac{\pi^2}{6}\right)(k_BT)^2\ \Gamma''(\mathscr{E}_F) \tag{9-41}$$

9-12 Thermal Properties of the Electron Gas

The Pauli exclusion principle requires that no two electrons can occupy the same quantum state. As we have seen (at 0°K), this causes the electrons to fill up all of the energy states to the Fermi level. However, only electrons with energies greater than $\approx \mathscr{E}_F - k_B T$ can be thermally excited and hence contribute to properties such as specific heat, spin paramagnetism, scattering, and so on. This is because electrons with energies much below $\mathscr{E}_F$ find that all the states that are within approximately $k_B T$ in energy are occupied and hence the electrons cannot be excited to these states. Table 9-4 shows that T_F varies from $18,400°K$ for Cs to $166,000°K$ for Be. These temperatures are hundreds of times higher than room temperature and even at least ten times higher than the hottest ceramic kiln ($\approx 2000°K$). Therefore only a very small fraction of the free electrons in a metal can be thermally excited. This qualitative picture can be expressed mathematically via Eq. 9-41 for $T \ll T_F$. We use this expression to calculate some thermal properties of metals within the quantitized free electron model.

9-12a $\mathscr{E}_F(T)$ The first property we evaluate is the temperature dependence of the Fermi energy $\mathscr{E}_F(T)$. As mentioned in Section 9-10, $\mathscr{E}_F$ can be determined by the condition that the integral of all the occupied states is equal to N

$$N = \int_0^\infty f(\mathscr{E}) \, g(\mathscr{E}) \, d\mathscr{E} \tag{9-42}$$

In order to use the general result Eq. 9-41, take

$$\Gamma(\mathscr{E}) \equiv \int_0^\mathscr{E} g \, d\mathscr{E} \tag{9-43}$$

so $\Gamma''(\mathscr{E}_F) = g'(\mathscr{E}_F)$. This immediately leads to

$$N = \int_0^\infty f g \, d\mathscr{E} = \int_0^{\mathscr{E}_F} g \, d\mathscr{E} + (\pi^2/6)(k_B T)^2 g'(\mathscr{E}_F) \tag{9-44}$$

To eliminate N we subtract it, in the form of Eq. 9-45, from both sides of Eq. 9-44.

$$N = \int_0^{\mathscr{E}_F(0)} g \, d\mathscr{E} \tag{9-45}$$

Thus,
$$\int_{\mathscr{E}_F(0)}^{\mathscr{E}_F} g(\mathscr{E}) d\mathscr{E} + (\pi^2/6)(k_B T)^2 g'(\mathscr{E}_F) = 0 \tag{9-46a}$$

Since we expect $\mathscr{E}_F(T)$ and $\mathscr{E}_F(0)$ to differ by only a small amount and $g(\mathscr{E})$ is a slowly varying function of $\mathscr{E}$, Eq. 9-46a can be written approximately as

$$[\mathscr{E}_F(T) - \mathscr{E}_F(0)] g(\mathscr{E}_F) + (\pi^2/6)(k_B T)^2 g'(\mathscr{E}_F) = 0 \tag{9-46b}$$

CHAPTER 9   METALS   219

Using Eq. 9-30b for $g(\mathscr{E}) = C\mathscr{E}^{1/2}$, from which $g'(\mathscr{E})$ may also be determined, we obtain the desired result.

$$\mathscr{E}_F(T) = \mathscr{E}_F(0) \left[1 - \frac{\pi^2}{12} \left(\frac{k_B T}{\mathscr{E}_F(0)} \right)^2 \right] \quad (9\text{-}47)$$

Thus, the first correction term for $\mathscr{E}_F(T) - \mathscr{E}_F(0)$ is of order $(T/T_F)^2$, which for most temperatures is indeed rather small. For sodium, where $T_F = 3.75 \times 10^{4}\,°K$, from Table 9-4, the correction term at room temperature is 6.4×10^{-5}. This is very small and is consistent with the approximation made in obtaining Eq. 9-46b.

9-12b Electron heat capacity   Now we calculate the heat capacity due to the electron gas. The total energy per unit volume is

$$U = \int_0^\infty f(\mathscr{E})\, \mathscr{E}\, g(\mathscr{E})\, d\mathscr{E} \quad (9\text{-}48)$$

So
$$\Gamma(\mathscr{E}) \equiv \int_0^{\mathscr{E}} \mathscr{E}\, g\, d\mathscr{E} \quad (9\text{-}49)$$

Using the general formula, Eq. 9-41, for this integral, we have

$$U = \int_0^{\mathscr{E}_F} \mathscr{E}\, g\, d\mathscr{E} + (\pi^2/6)(k_B T)^2 \{d[\mathscr{E}\, g(\mathscr{E})]/d\mathscr{E}\}_{\mathscr{E}_F} \quad (9\text{-}50)$$

Using Eq. 9-30b, the derivative in the last term in curly brackets of Eq. 9-50 immediately yields $(3/2)\, g(\mathscr{E}_F)$. Since the temperature correction to $\mathscr{E}_F$ is so small (Eq. 9-47) this is approximately equal to $(3/2)g[\mathscr{E}_F(0)]$. The integral is also straightforward to evaluate

$$C \int_0^{\mathscr{E}_F} \mathscr{E}^{3/2} d\mathscr{E} = (2/5)\, C\, \mathscr{E}_F^{5/2}$$

$$= (2/5)\, C\, \mathscr{E}_F(0)^{5/2} [1 - (\pi^2/12)\{k_B T/\mathscr{E}_F(0)\}^2]^{5/2}$$

$$= \frac{2}{5} C\, \mathscr{E}_F(0)^{5/2} - \frac{2}{5} C\, \mathscr{E}_F(0)^{5/2}\, \frac{5}{2}\frac{\pi^2}{12}\left\{ \frac{k_B T}{\mathscr{E}_F(0)} \right\}^2$$

$$= \frac{2}{5} C\, \mathscr{E}_F(0)^{5/2} - \frac{\pi^2}{12}(k_B T)^2\, g[\mathscr{E}_F(0)] \quad (9\text{-}51)$$

In the first line Eq. 9-30b is used for $g(\mathscr{E})$. In the second line use Eq. 9-47 to express $\mathscr{E}_F(T)$ in terms of $\mathscr{E}_F(0)$ and in the third expand $(1-x)^{5/2} = 1 - (5/2)x + \dots$ for small x. Collecting terms we find

$$U = (2/5)\, C\, [\mathscr{E}_F(0)]^{5/2} + (\pi^2/6)(k_B T)^2\, g[\mathscr{E}_F(0)]$$

$$= (2/5)\, C\, [\mathscr{E}_F(0)]^{5/2} + (\pi^2/4)N k_B T^2 / T_F \quad (9\text{-}52)$$

Table 9-6 Measured and calculated (via Eq. 9-53) values of γ, the coefficient of the free electron term in the molar specific heat. The values are given in 10^{-4} calories moles^{-1}°K^{-2}. (See Gopal for a more extensive list.)

Element	Meas.	Calc.	Element	Meas.	Calc.
Li	4.2	1.8	Fe	12	1.5
Na	3.5	2.6	Mn	40	1.5
K	4.7	4.0	Zn	1.4	1.8
Rb	5.8	4.6	Cd	1.7	2.3
Cs	7.7	5.3	Hg	5.0	2.4
Cu	1.6	1.2	Al	3.0	2.2
Ag	1.6	1.5	Ga	1.5	2.4
Au	1.6	1.5	In	4.3	2.9
Be	0.5	1.2	Tl	3.5	3.1
Mg	3.2	2.4	Sn	4.4	3.3
Ca	6.5	3.6	Pb	7.0	3.6
Sr	8.7	4.3	Bi	0.2	4.3
Ba	6.5	4.7	Sb	1.5	3.9

where we have used $g[\mathscr{E}_F(0)] = (3N/2)/\mathscr{E}_F(0) = (3N/2k_B)/T_F$ obtained from Eqs. 9-29 and 9-30b. The heat capacity per unit volume, $C_v = (\partial U/\partial T)_v$, is obtained directly from Eq. 9-52. However, the heat capacity per mole usually is measured. If we consider one mole of atoms and if each atom contributes Z electrons to the electron gas, then instead of $k_B N$ in Eq. 9-52 where N is the number of electrons, we write $Zk_B N_A = ZR$ where N_A is the number of atoms in a mole, which is Avogadro's number and R is the gas constant. Then the electronic heat capacity per mole is

$$C_v = (\pi^2/2)ZR\,[T/T_F] \equiv \gamma T \tag{9-53}$$

This is just the form expected as discussed qualitatively in Section 9-9. Instead of a value R, this is reduced to $R(T/T_F)$ because only the fraction (T/T_F) of the electrons, with energies near the Fermi energy, have empty states within a range $k_B T$ to which they can be excited.

Before we can compare the calculated electronic heat capacity to experimental values, the lattice heat capacity (Chapter 11) must be mentioned. For N harmonicly coupled atoms, the classical energy is $3Nk_B T$, since for each atom there is $k_B T/2$ for each degree of freedom and there are three degrees of freedom for the kinetic energy and also three for potential energy. Then $C_v = 3Nk_B$, or 3R, for a mole. However, since the atoms in a crystal are not classic but quantum mechanical, at low temperatures their energy and heat capacity are sugnificantly different from the classical values. Low temperatures for a crystal lattice means temperatures low compared to the so called Debye temperature, which is typically several hundred degrees Kelvin.

CHAPTER 9   METALS   221

At these low temperatures the lattice heat capacity is proportional to the third power of the absolute temperature, AT^3. The total heat capacity, electronic plus lattice, then is $\gamma T + AT^3$. At temperatures within several degrees of absolute zero the T^3 term falls off faster than the T term and the linear term can be determined. Since

$$C_v/T = \gamma + AT^2 \tag{9-54}$$

a plot of C_v/T vs. T^2 should be a straight line with an intercept equal to γ. This is found for most metals.

Table 9-6 lists values of γ obtained from experiment and those calculated from Eq. 9-53. The agreement is fairly good for the alkalies and many other simple metals. Energy band effects, discussed in Chapter 10, must be taken into account in detail for certain metals.

9-12c Electron spin paramagnetism   As mentioned at the end of Section 9-6, N free electrons are expected to exhibit a Langevin paramagnetic susceptibility of $\chi \sim N\mu_B^2/k_BT$ where μ_B is the Bohr magneton. However, the room temperature experimental measurements yield a temperature independent value that is $\approx 10^{-2}$ of this predicted value. The proper theory of the paramagnetic behavior of electrons in a metal is more complicated than we can discuss here. Nevertheless, one of the major contributions to χ can be calculated using the formalism of Section 9-11, and this is done here.

Figure 9-8a shows a plot of total energy vs. density of states. The figure separates the electrons with m_s parallel and antiparallel to the direction of the applied field. Other than this separation, this plot is the same as that shown in Fig. 9-5. Due to the Pauli exclusion principle only those electrons within k_BT of $\mathscr{E}_F$ can reorient, by changing their m_s quantum number, in an applied field. The electrons at lower energy cannot orient their spins parallel to the field since those states are already occupied. Only the fraction T/T_F of the total number of electrons can reorient and thus, qualitatively, a paramagnetic susceptibility of the order of T/T_F times the classical result is expected. This is $N\mu_B^2/kT_F$. This is temperature independent and about $1/100$ the free electron value since $T/T_F \sim 1/100$ at room temperature. This result is in qualitative agreement with experiment.

Figure 9-8b shows the same curves as Fig. 9-8a but shifted by $\pm \mu H$ due to an applied magnetic field and before any spins can reorient. The total number of occupied states is the same with and without an applied field. However, this is a nonequilibrium situation since some of the highest energy electrons with $m_s = \pm 1/2$ have a higher energy than those with $m_s = -1/2$. These electrons have empty states to fall into which they occupy by reversal of their spins. Figure 9-8c shows the result due to the reversal of some of the electron spins with a Fermi energy that corresponds to the equilibrium situation. As

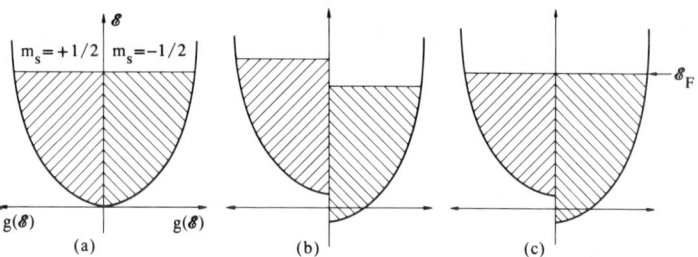

Fig. 9-8 (a) The energy vs. the density of states for electrons with $m_s = \pm 1/2$ shown separately; (b) in a magnetic field before the spins reorient; (c) after equilibrium.

usual, due to the Pauli principle, only those electrons with kinetic energy of the order of $\mathscr{E}_F - \mu H$ can reverse their spin, because only for these electrons are there unoccupied states available. The others at lower energies cannot reverse their spins.

Quantitatively, we can calculate χ using F-D statistics in the following manner. From Fig. 9-8 the net magnetism from the electrons in the metal is given by μ_B times the difference between the numbers of electrons with moments parallel and antiparallel to the field, $M = \mu_B(N_1 - N_2)$, so

$$M = \mu_B \int \{(1/2) g(\mathscr{E} + \mu_B H) - (1/2) g(\mathscr{E} - \mu_B H)\} f(\mathscr{E}) d\mathscr{E} \quad (9\text{-}55)$$

For small magnetic field, H, the intergral can be expanded to give

$$M = \mu_B^2 H \int g'(\mathscr{E}) f(\mathscr{E}) d\mathscr{E} \quad (9\text{-}56)$$

This expression is in the form of Eq. 9-41 if

$$\Gamma(\mathscr{E}) \equiv \int_0^{\mathscr{E}} g' d\mathscr{E} \quad (9\text{-}57)$$

Thus, $\quad M = \mu_B^2 H g(\mathscr{E}_F) \quad (9\text{-}58)$

The result is usually expressed as a susceptibility, $\chi \equiv M/H$ which at low temperatures is given by

$$\chi = \mu_B^2 g[\mathscr{E}_F(0)] = 3N\mu_B^2/2k_B T_F \quad (9\text{-}59)$$

This is just the form expected from the qualitative discussion. For Na, we calculate $\chi = 0.66 \times 10^{-6}$ per gram.

There are several other important terms that also contribute to the electronic susceptibility in a metal. In the preceding discussion it was assumed that the spatial motion of the electron is unaffected by the applied field. This is not the case. A diamagnetic susceptibility of 1/3 of that given in Eq. 9-59 results from this effect. For Na this is -0.22×10^{-6}/gm. The closed-shell, that is, core, electrons also have

CHAPTER 9   METALS

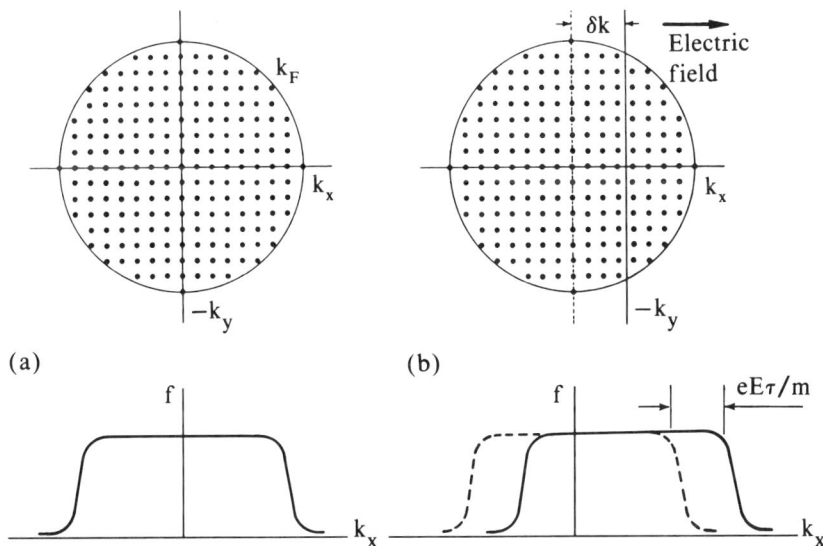

Fig. 9-9 (a) Allowed states in two dimensional k-space filled up to the Fermi surface with wave vector k_F. (b) When an electric field is applied for a time t the wave vectors of all the states change by an amount $\delta k = -eEt/\hbar$. (Assuming no collisions.) The situation is shown assuming collisions and a relaxation time tau. The lower part of (a) and (b) shows the corresponding Fermi distributions.

a diamagnetic susceptibility, which is estimated to be -0.18×10^{-6}/gm for Na. Last, but hardly least, a correction of 0.22×10^{-6}/gm for Na has been calculated from exchange, correlation, and effective mass effects [D Pines, Phys. Rev. **95**, 1095 (1954)]. These four calculated contributions total 0.48×10^{-6}/gm, which should be compared to the measured 0.70×10^{-6}/gm. The agreement for the other alkali metals is similar.

9-13 DC Conductivity (With F-D Statistics)

An important point to consider in this chapter is the effect of Fermi-Dirac distribution on the conductivity. Equation 9-6, obtained by Drude, gives the dc conductivity as $(N/V)e^2\tau/m$. The interesting point is that the F-D distribution does not alter this formula.

The reason that the F-D distribution does not alter the dc conductivity, as calculated in Section 9-3, is that the actual electron distribution never enters into the considerations. Consider this point in more detail (a mathematical proof is left as a problem). Figure 9-9 shows a two dimensional slice of k-space at a time 0 and t later after the application of an electric field **E** in the x-direction. For a free-electron gas the wave vector and momentum are related by $m\mathbf{v} = \hbar\mathbf{k}$. From Newton's law $-e\mathbf{E} = m\, d\mathbf{v}/dt = \hbar\, d\mathbf{k}/dt$. Thus, in the absence of collisions, all of the k-vectors in k-space are displaced as shown in Fig. 9-9b. This is just what happens in the calculation in Section 9-3, except that we need not refer to k-space since velocity space is just as useful. The displacement in k-space is

$$\delta\mathbf{k} = -e\mathbf{E}\,\delta t/\hbar \qquad (9\text{-}60)$$

We could ask, is such a change in k-values allowed since most of the k-values are occupied for $T \ll T_F$? The answer is yes because when the time-dependent wave equation is solved each and every k-value changes by the infinitesimal amount given by Eq. 9-60 so that they all move in unison as shown in the figure. Another way to say this is that they do not move from one quantitized k-value to a different one. Rather all of the quantitized k-values change in accordance with Eq. 9-60 and as shown in Fig. 9-9b.

Due to collisions in an average time τ, the displacement of the Fermi sphere will be $\delta\mathbf{k} = -e\mathbf{E}\tau/\hbar$ in equilibrium. This means that drift velocity is given by $\mathbf{v}_d = -e\mathbf{E}\tau/m$, which is the same result as in Section 9-3 and leads to the same expression for the dc conductivity.

Which electrons can have collisions and change their k-values so that equilibrium can be maintained? For equilibrium to be maintained those electrons with k-values to the left in Fig. 9-9b must have collisions and change their k-values to values on the right of the Fermi sphere. It is just such collisions, and hence changes in **k**, that will restore equilibrium so that in equilibrium the Fermi sphere is displaced $\delta\mathbf{k} = -e\mathbf{E}\tau/\hbar$. And it is just such collisions that are allowed because the k-states on the right are unoccupied. Most other collisions are not allowed, because they would scatter electrons into occupied states.

Although the formula obtained by Drude is the same as that obtained with the Fermi-Dirac distribution, the values of τ expected from the two considerations are different. For the F-D distribution, allowed collisions occur only for electrons at the Fermi surface. The Fermi velocities are much larger than those used by Drude. In the Drude model we estimate an average velocity from the classical equipartition of energy $mv_m^2/2 = 3k_B T/2$. At room temperature this yields $v_m \approx 10^7$ cm/sec. However, values of v_F (Table 9-4) are $\approx 10^8$ cm/sec, an order of magnitude larger than expected from the Drude model. While the mean free paths obtained using v_m are of the order of the interatomic spacing, those obtained from the Sommerfeld model are much larger. Using the value of v_F and τ (Table 9-1) for copper at room temperature we obtain a mean free path ($= v_F\tau$) of 4×10^{-6} cm or 400 Å. That is rather long in the sense that it is many times the Cu-Cu atomic distance. What is even more surprising is that in very pure samples at low temperatures we measure relaxation times $\approx 10^5$

CHAPTER 9   METALS   225

times longer. This leads to mean free paths of 0.4 cm! These distances are of the order of the size of the samples. Indeed, this is in agreement with experiment. Namely, in very pure metals at low temperatures, the conductivity is found to be sample-size dependent.

It is the spatial quantitization of the free electron gas and the application of the Pauli principle to these fermions that has led to the filling of states to these very high energies (to $\mathscr{E}_F$) with large velocities (v_F). These mean free path results show the importance of this quantitization in understanding experimental results.

9-14 Electron-Electron Collisions

An amazing thing about the free electron theory of metals is that it works, at least fairly well. On average free electrons are separated by a distance of only the order of the interatomic distance or smaller, see r_s in Table 9-1. Thus, we might guess that, via the Coulomb interaction, the strong electron-electron interactions would completely break down the one-electron model. However, this is not the case and two reasons are discussed here. First, there is an extremely large decrease in the possibility of electron-electron collisions at ordinary temperatures because of the Pauli principle. Second, there is electron screening so that a given electron does not see the strong, long-range Coulomb force from all of the other electrons. The electrons are accompanied instead by a distorted electron cloud which shields the Coulomb field of distant electrons.

The effects discussed in this section usually go under the name of the **Fermi liquid theory** discussed by Landau (1957) and others. It might be said that up to now we have considered a system of noninteracting (bare) electrons or fermions, which is called a **Fermi gas**. When interactions are allowed we have a **Fermi liquid**.

9-14a Phase space considerations   First consider 0°K and the interaction of electrons **1** and **2** which have wave vectors $\mathbf{k}_1$ and $\mathbf{k}_2$ and energies $\mathscr{E}_1$ and $\mathscr{E}_2$. When they interact it is said that they **scatter**, and the scattered electrons have wave vectors $\mathbf{k}_3$ and $\mathbf{k}_4$, and energies $\mathscr{E}_3$ and $\mathscr{E}_4$ according to

$$\mathbf{k}_1 + \mathbf{k}_2 = \mathbf{k}_3 + \mathbf{k}_4 \tag{9-61a}$$

$$\mathscr{E}_1 + \mathscr{E}_2 = \mathscr{E}_3 + \mathscr{E}_4 \tag{9-61b}$$

(We ignore the possibility of this interaction modulo a reciprocal lattice vector. These types of collisions are discussed in Section 12-6 under phonon-phonon interactions but are of little consequence for electrons because of the occupation probabilities that are considered in this section.)

At 0°K both k_1 and k_2 must be smaller than k_F, but from Eqs. 9-61a at least one of the scattered electrons must be from a state that has $k < k_F$, and all of these states are already occupied. Since the Pauli principle does not allow two fermions to occupy the same quantum state, this scattering process is not allowed. Thus, we see that all of these types of electron-electron scattering are eliminated at 0°K. As we shall see, even at $T > 0°K$ this same phase space argument drastically reduces this type of scattering.

$T > 0°K$ – Referring to Fig. 9-10a, consider ordinary temperatures, so most of the states below k_F are occupied. Consider electron 1 which has a small excitation energy above $\mathscr{E}_F$. This interaction is represented by $1 + 2 \rightarrow 3 + 4$. Since we are assuming ordinary temperatures, essentially all of the electron states below k_F are occupied (Fig. 9-7), then by the Pauli exclusion principle electrons 3 and 4 must lie outside of the Fermi sphere. For convenience, take the zero of energy at $\mathscr{E}_F$, so that energies outside the Fermi sphere are positive and those inside are negative. Thus, $\mathscr{E}_1$, $\mathscr{E}_3$, and $\mathscr{E}_4$ are positive and $\mathscr{E}_2$ is negative but $|\mathscr{E}_2| < \mathscr{E}_1$ or else the left-hand side of Eq. 9-61b would not be positive. Figure 9-10a shows the shell, of thickness $\mathscr{E}_1$, within which electron 2 must lie in order to scatter with electrons 1 and within which electrons 3 and 4 must lie in order to fulfill conservation of energy, Eq. 9-61b. As can be seen only the fraction $\mathscr{E}_1/\mathscr{E}_F$ of electrons fulfill these conditions. This is a very small number; it shows that the number of electrons that can enter into a scattering process with electron 1 is severely reduced by energy considerations after the application of the Pauli principle.

However, a second factor $\mathscr{E}_1/\mathscr{E}_F$ is obtained from the k-conservation, Eq. 9-61a. This also is straightforward, see Fig. 9-10b. A "resultant sphere" of final states is drawn with its center at the "center of k" of electrons 1 and 2 and whose radius extends to the k value of electrons 1 or 2. In order for the pair of electrons 3 and 4 to conserve momentum and energy (Eqs. 9-61), their k-values must be at the opposite ends of a diameter of this resultant sphere, as shown in the figure. However, only a small number of electrons 3 and 4 that satisfy this condition on the resultant sphere are allowed, because, as we have seen from the exclusion principle they must be outside the Fermi sphere by less than $\mathscr{E}_1$. This argument reduces the fraction of electrons that can scatter by another factor $\mathscr{E}_1/\mathscr{E}_F$.

Thus, because of the Pauli exclusion principle, the electron-electron scattering rate is reduced by $(\mathscr{E}_1/\mathscr{E}_F)^2$ from what it would be without the application of the exclusion principle.

If, instead of an electron at energy $\mathscr{E}_1$, we have a thermal distribution of electrons at a temperature $k_B T \ll \mathscr{E}_F$, then the arguments are not changed. We find that the electron-electron scattering reduction is $(k_B T/\mathscr{E}_F)^2$. Since a typical Fermi temperature is $10^5 °K$, at 1°K these

CHAPTER 9   METALS   227

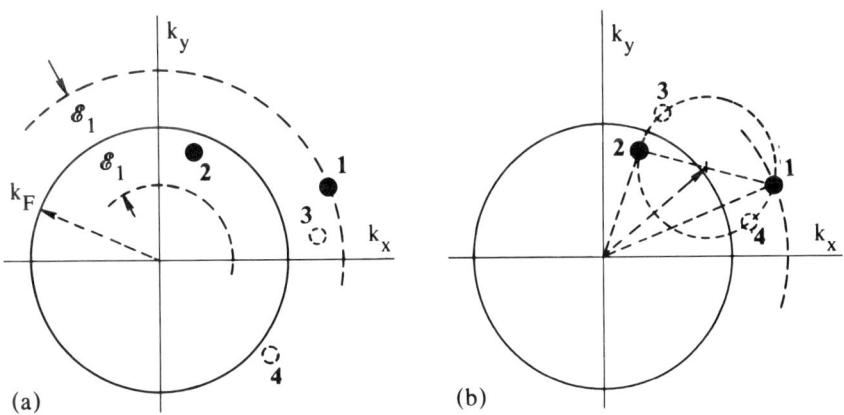

Fig. 9-10 A projection of the Fermi sphere. (a) emphasizing energy conservation. (b) emphasizing **k** conservation.

arguments reduce the scattering rate by $\approx 10^{-10}$, and even at room temperature by $\approx 10^{-5}$.

Summary — With the use of the Pauli exclusion principle on our quantitized free electron system, we have shown that the electron-electron scattering rate is reduced by $\approx (k_B T/\mathscr{E}_F)^2$. Thus, in spite of the strong electron-electron forces at the short distances encountered in ordinary metals, the electron-electron scattering rate is small, and is zero at 0°K. This is one of the principle reasons why the free electron gas (i.e. a gas of noninteracting particles or an ideal gas) approximation has been so successful.

9-14b Fermi liquid theory   In several fundamental papers (1957) Landau considered the free electron gas problem. He considered the electron-electron interaction and showed that the Coulomb interaction between one electron and all of the others is strongly reduced due to screening by the free electrons. For typical metals, the scattering cross section is 10 Å^2 (which is far smaller than one would calculate from Rutherford scattering using an unscreened Coulomb potential). Furthermore, this screened cross section must be multiplied by $(k_B T/\mathscr{E}_F)^2$, determined in Section 9-14a, which comes from phase space considerations. The electron-electron scattering cross sections in a metal are small indeed.

Landau further showed that, as long as electrons within $k_B T$ of $\mathscr{E}_F$ are considered, the free electron gas ideas should work fine with a few subtle changes. Instead of bare electrons we really have quasi-particles (that comes about from the screening) whose mass can be different from that of a free electron. This can effect the $\mathscr{E}$ vs. k results. However, in general the free electron gas ideas are appropriate, and require only small modification if numerical results of the excitation spectrum or the transport properties are to be calculated. For example, both the free electron contribution to the heat capacity (Eq. 9-53) and paramagnetism (Eq. 9-59) are proportional to the mass (i.e., $T_F \propto m^{-1}$ from Eq. 9-36). So, from Fermi liquid theory, both of these terms are proportionately larger.

Since the electron-electron interaction is taken into account in these considerations, we are no longer dealing with a free electron gas but are dealing with a liquid. The ideas discussed in this section (9-13a and 9-13b) are called the Fermi liquid theory and apply to any gas of fermions. See the Notes for references to this deeply theoretical subject.

9-15 Hall Effect (And Other Magnetic Field Effects)

In this section, we discuss a relatively simple effect called the Hall effect (E. H. Hall, 1879). As will be seen, Hall measurements show that there are some dramatic shortcomings in free electron theory for some metals. This should serve as motivation for the more complicated theory in the next chapter. The Hall effect is also very useful for measuring the electronic properties of semiconductors, which is discussed in Chapter 10.

Equation 9-10, which is repeated here, is the equation of motion for the free electron *drift velocity* v_d, under the action of external forces, **F**.

$$m\left[\frac{dv_d}{dt} + \frac{v_d}{\tau}\right] = F \tag{9-10}$$

where τ is the average time between collisions and can be thought of as playing a role similar to a friction term. If the free electron is in a magnetic field H as well as on electric field **E** then the force on any free electron is the Lorenz force,

$$F = -e\left[E + \frac{1}{c}v_d \times H\right] \tag{9-62}$$

Consider a magnetic field in the z-direction. Then combining Eqs. 9-10 and 9-62 for each component of v_d we have

CHAPTER 9 METALS

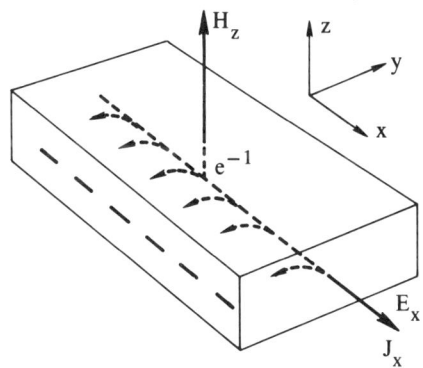

Fig. 9-11 The standard geometry used in measuring the Hall effect. The electrons bunch up on the −y face as shown.

$$\left[\frac{d}{dt} + \frac{1}{\tau}\right]v_{dx} = \frac{-e}{m}\left[E_x + \frac{H}{c}v_{dy}\right] = \frac{-e}{m}E_x - \omega_c v_{dy}$$

$$\left[\frac{d}{dt} + \frac{1}{\tau}\right]v_{dy} = \frac{-e}{m}\left[E_y - \frac{H}{c}v_{dx}\right] = \frac{-e}{m}E_y + \omega_c v_{dx} \quad (9\text{-}63)$$

$$\left[\frac{d}{dt} + \frac{1}{\tau}\right]v_{dz} = \frac{-e}{m}E_z$$

where $\omega_c \equiv eH/mc$, the **cyclotron frequency**, is the frequency of the spatial rotation of the electron orbit around the magnetic field.

The **Hall effect** is usually observed using the geometry shown in Fig. 9-11. An electric field in the x-direction results in a current density in that direction, J_x. When the magnetic field, H, in the z-direction is turned on, the electrons with v_{dx} experience a force given by $-ev_d \times H$; this force is in the negative y-direction (that is, the current J_x is flowing in the direction of $+x$ so the electrons are moving in the direction of $-x$). Thus, they take a circular path and bunch up on the negative y-face of the material as indicated in the figure. This continues until the electric field in the y-direction (the **Hall field**), due to bunching electrons, just cancels the $-ev_d \times H$ force; this is the steady state result. Thus, the Hall field is perpendicular to the current and magnetic field (i.e., it is perpendicular to the $J_x - H_z$ plane). Then in the steady state the average drift velocity in the y-direction is zero and from the first two expressions in Eqs. 9-63 we can determine that this electric field is given by

$$E_y = -\omega_c \tau E_x = -\frac{eH}{mc}\tau E_x \quad (9\text{-}64)$$

(From the middle expression in Eq. 9-64, $E_y = Hv_{dx}/c = -H\mu E_x/c$, where μ is the mobility which will be used in our discussions of semiconductors. The mobility is defined by $v_{dx} = -\mu E_x$. The **Hall angle**, θ, is defined as the angle between the current and the resulting electric

Table 9-7 From Drude theory we would expect a Hall coefficient $R_H(\text{theo}) = -[(N/V)ec]^{-1}$ where N/V values are listed in Table 9-1. We label the experimental measured values $R_H(\text{meas})$. For a few of the simpler metals we list $R_H(\text{theo})/R_H(\text{meas})$. If Drude's theory is correct the result should be one, that is, +1.0. [The values used here as well as many other values can be found in Landolt-Börnstein, Vol. 2, Part 6 pp. 161 (1959). More recent results, using helicon wave techniques, are in J. M. Goodman, Phys. Rev. **171**, 641 (1968).]

Li	0.78	Cu	1.4	Be	−0.1
Na	1.1	Ag	1.2	Mg	0.88
K	1.1	Au	1.5	Ca	0.76
Rb	0.92			Zn	−0.75
				Cd	−1.2

field and is given by $\tan\theta = E_y/E_x$, which is equal to $-H\mu/c$, so $\mu = e\tau/m$. The concept of the Hall angle proves useful for materials with two different current carriers, a situation often found in semiconductors.)

The **Hall coefficient** R_H is defined as

$$R_H \equiv E_y/J_x H \tag{9-65}$$

To evaluate R_H from the free electron model for the preceding geometry we use for J_x Eqs. 9-4 and 9-6, that is, $J_x = \sigma E_x$ where $\sigma = (N/V)e^2\tau/m$, and for E_y we use Eq. 9-64. Thus,

$$R_H = \frac{-\dfrac{eH}{mc}\tau E_x}{\dfrac{N}{V}\dfrac{e^2\tau}{m}E_x H} = -\frac{1}{(N/V)ec} \tag{9-66}$$

For electrons, this is a negative number; since the values of e and c are known accurately, the Hall coefficient should give the carrier concentration. We should expect to find good agreement with the N/V values, which are listed for many elemental metals in Table 9-1.

In Table 9-7 we list values of the **dimensionless Hall coefficient.** As can be seen, reasonable agreement is obtained for the alkali metals and for Cu, Ag, and Au. However, for some other metals values with the opposite signs are obtained. It is as if the electrical current is carried by positively charged carriers. This is rather embarrassing for a free *electron* theory. Furthermore, the experimental results for R_H often depend on the magnetic field (and for some materials change sign with increasing H) and on temperature. This is more surprising since the only material parameter that enters R_H is the carrier concentration; parameters such as τ are not contained. Further, for some metals, notably As, Sb, and Bi, values of the dimensionless Hall coefficient that are very much larger than one are obtained, and of positive and negative signs.

The free electron approximation is good for many metals, for example those in the first two columns of Table 9-7, all of which have one valence electron per atom (Table 9-1). In the next chapter, the effects of translational symmetry on the electron states are discussed. As will be seen, the effects are profound, yet are much less important for cubic crystals with one valence electron per primitive unit cell. Materials with more electrons per primitive unit cell can have: complex Fermi surfaces, Fermi energies very close to discontinuities in the $\mathscr{E}$ vs. k curves, almost full bands where the carriers effectively behave as positively charged current carriers (holes). So for most real metals band theory must be considered.

Transverse magnetoresistance refers to the change in the resistance in the direction of the applied field, E_x, due to a perpendicular (i.e., transverse) magnetic field (Fig. 9-11). For the steady state $v_{dy} = 0$, so from the first expression in Eqs. 9-63 we expect that v_{dx} depends only on E_x. Thus, the transverse magnetoresistance is predicted to be zero. Here again experimentally we find that for many metals there is a nonzero effect and often it can be quite large. There are many reasons for this. The complex Fermi surfaces found in real metals lead to effects that are not expected from spherical Fermi surfaces. Also, note that a distribution of velocities of the carriers is not important for electric fields but it is for magnetic fields, because of the **v** × **H** term. For example, the Hall voltage can stop the average drift velocity in the y-direction but not the entire distribution of carriers in that direction so there still are electrons that interact with the magnetic field.

Longitudinal magnetoresistance refers to the change in resistance with magnetic field when **J** and **H** are colinear. In this case the Lorenz force $-e\mathbf{v} \times \mathbf{H}$ is zero. However, effects are measured. These effects result from the nonspherical Fermi surfaces and related properties discussed in the previous two paragraphs.

Cyclotron resonance is associated with the $\omega_c \equiv eH/mc$ term, in Eqs. 9-63. Unlike the preceding dc topics it is an ac phenomenon. For magnetic field H in the z-direction, due to the **v** × **H** force an electron will have a circular orbit in the xy-plane. Using the ordinary mass of the electron, the frequency around this circular orbit ν_c ($\equiv \omega_c/2\pi$) is 2.80×10^9 Hz/kilogauss. For ≈3 kilogauss (0.3 tesla) we obtain ≈10^{10} Hz, which is a microwave frequency (x-band). Now, if $\omega_c \tau \gg 1$, the electron will make many circular orbits before being scattered; if $\omega_c \tau \ll 1$ it will be scattered while making only a small fraction of an orbit, and then after being scattered it will start a new orbit only to be scattered rapidly again. For $\omega_c \tau \approx 1$ an electron will complete approximately one orbit before being scattered.

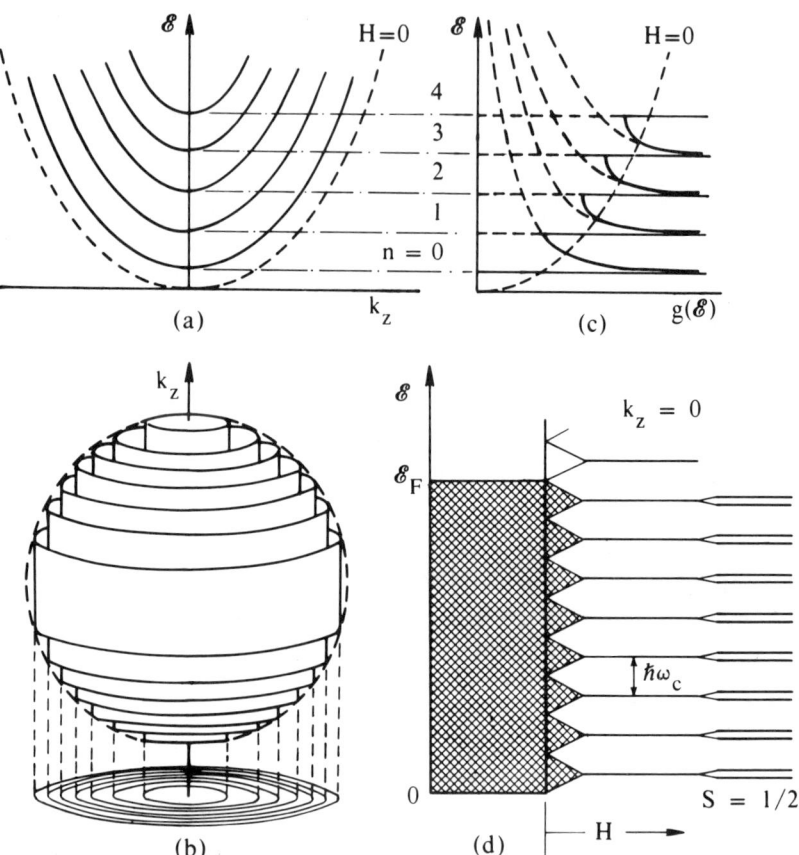

Fig. 9-12 Various aspects of Landau levels. (a) $\mathscr{E}$ vs. k_z for the low energy Landau levels. The H = 0 parabola refers to the ordinary free electron case with zero magnetic field. (b) k-space showing Landau levels. The allowed k-values lie on the concentric cylinders, and the spherical Fermi surface cuts these cylinders. (c) The solid line is the density of states for the Landau levels. H = 0 refers to the results in zero field, and shows the expected $\mathscr{E}^{1/2}$ behavior. (d) A schematic diagram showing how the states, filled up to the Fermi energy, split into Landau levels when H is applied. The diagram also shows the electron-spin splitting.

Using standard microwave spectroscopy techniques, we can apply radiation at a frequency ω. By varying the magnetic field the resonance condition occurs at $\omega = \omega_c$, at which frequency energy is absorbed from the microwaves. The absorption is sharp only if $\omega_c \tau > 1$; for $\omega_c \tau \gg 1$ it is very sharp (that is, at a very well defined H). This technique, called cyclotron resonance, is used as a method of measuring the effective mass of the charged carriers, but this is a topic of the next chapter.

CHAPTER 9   METALS

SI units — Throughout this section on the Hall effect the equations are correct in SI units simply by setting c = 1.

9-16 Landau Levels

In the previous section, using classical physics, we discussed some free electron magnetic field effects and introduced the cyclotron frequency, $\omega_c \equiv eH/mc$. For a magnetic field in the z-direction (Fig. 9-11) the electron motion is circular in the xy-plane at the cyclotron frequency. When the mean free path of the electrons in the xy-plane is much larger than the circular path length, and $\hbar\omega_c > k_B T$ so thermal motion is not dominant the effect no longer should be discussed from the simple classical point of view (as in Section 9-15). (Or when $\omega_c \tau \gg 1$, where τ is the lifetime of the carriers, which is related to the circular frequency by $\omega = 2\pi/\tau$.) In this case quantization of the motion must be considered.

Quantization — The quantum problem is not difficult, but it is beyond the level of our presentation at this point. (Consult the Notes for references.) However, the physics is straight forward and many of the results can be understood easily. The magnetic field has no effect on the electron motion in the k_z-direction, so the $\mathscr{E}$ vs. k_z behavior should be unaltered and still should be $\mathscr{E} = \hbar^2 k_z^2/2m$. However, in the xy-plane, the electron motion is no longer free electron-like, rather it is circular about H_z. Thus, in the xy-plane the electron behaves like a two-dimensional simple harmonic oscillator, and we can expect quantum mechanical simple harmonic oscillator energy levels. These are just the results that are found. The energy levels are given by

$$\mathscr{E} = (n + 1/2)\hbar\omega_c + (\hbar^2/2m)k_z^2 \qquad (9\text{-}67a)$$

where $n = 0, 1, 2, ...$

ω_c is the cyclotron frequency defined previously, n is a quantum number describing the degree of excitation, and the zero point energy $\hbar\omega_c/2$, typical of quantitized harmonic oscillators, appears. This energy expression shows that nothing has changed for motion along the magnetic field direction (the z-direction); the plane wave solution with k_z is still appropriate. However, the electron motion in the xy-plane is simple harmonic-like giving quantized simple harmonic energy levels, which are called **Landau levels**. Various aspects of these results are shown in Fig. 9-12. In fact, the Landau levels are not discrete levels but one-dimensional bands called **subbands**. The bottom of each subband is given by Eq. 9-67a with $k_z = 0$. Higher states in each subband correspond to k_z greater than zero.

In Fig. 9-12a, $\mathscr{E}$ vs. k_z is plotted showing the parabolic behavior for H = 0, as well as for the different subbands corresponding to the n

= 0, 1, 2, ... Landau levels. Studying Fig. 9-12b may help to appreciate this diagram better. Figure 9-12b shows how the allowed k-values in k-space are altered; in the z-direction there still is a quasi-continuous set of allowed k_z-values, but in the xy-plane the allowed k-values are bunched so that they all lie on circles. Basically, the magnetic field has changed the three-dimensional free electron gas; parallel to the magnetic field direction there is one-dimensional free electron gas behavior, and in the other two directions there are discrete (i.e., quantitized) states.

Density of state – A little thought about the density of states for the energy levels in Eq. 9-67a will make us realize that something very peculiar must happen. For the free electron case in three-dimensions, we know that $g(\mathscr{E}) \propto \mathscr{E}^{1/2}$ (Eq. 9-30b and Fig. 9-5) where the zero of energy is always taken at the lowest energy for the allowed states (i.e., at $\mathbf{k} = 0$). Since each subband corresponding to a different Landau level can be considered as a one-dimensional free electron gas (in the z-direction), from Problem 10 we see that $g(\mathscr{E}) \propto \mathscr{E}^{-1/2}$ (i.e. a divergence as the bottom of the subband is approached). Figure 9-12c shows this result and compares it to the free electron case. As can be seen, an $\mathscr{E}^{-1/2}$ curve has been drawn for each subband. Naturally, the total number of states, with and without the magnetic field, is conserved, but the distributions are very different.

Spin effect – However, Eq. 9-67a is not complete. Only the orbital quantization due to the magnetic field was accounted for. An electron also has a spin ($S = 1/2$ so $m_s = \pm 1/2$). Including the magnetic field spin quantization, we write the complete quantized energies as

$$\mathscr{E} = (n + 1/2)\hbar\omega_c + g\mu_B H m_s + (\hbar^2/2m)k_z^2 \qquad (9\text{-}67b)$$

where g is the electron g-value and μ_B is the Bohr magneton. A surprising result is found for free electrons. For free electrons $g = 2$, and we know that $\mu_B \equiv e\hbar/2mc$, then the total spin splitting is the same as the total orbital splitting. In semiconductors different results are found. For semiconductors it turns out (Chapter 10) that the effective mass of electrons near the bottom of the conduction band is much smaller than that of free electrons. This enters into the expression for ω_c (which results from orbital motion of the electron) causing it to be much larger. Hence the separation between the subbands is much larger than would be expected for free electrons. However, being a fundamental constant, μ_B contains the ordinary free electron mass. Thus, typically for semiconductors, the $g\mu_B H m_s$ term in Eq. 9-67b is much smaller than the $\hbar\omega_c$ term. (The g-values of electrons in the conduction bands of many semiconductors are close to 2.) It is this (semiconductor) situation that is shown in Fig. 9-12d.

CHAPTER 9   METALS   235

Now a proper explanation of cyclotron resonance can be given. Absorption occurs with quantum number selection rules of $\Delta n = \pm 1$ and $\Delta m_s = 0$ (Eq. 9-67b or Fig. 9-12a). Thus, absorption occurs between neighboring Landau levels at an energy of $\hbar\omega_c$.

Oscillatory behavior – Let us consider the position of the Landau levels with respect to the Fermi energy; looking at Fig. 9-12d will help. As H is varied the separation between levels changes and hence their position with respect to $\mathscr{E}_F$. Let $\mathscr{E}_T$ be the total energy of the electron system; with H = 0, this quantity was $(3/5)\mathscr{E}_F$ but now it varies slightly with H. Suppose H is such that $\mathscr{E}_F$ is exactly midway between two Landau levels; then in the dependence of $\mathscr{E}_T$ vs. H there is a local minimum, since half of the population in the highest filled Landau level has been shifted down from higher energy values (compared to the H = 0 situation). The average shift is $\hbar\omega_c/4$. As H is increased, the highest filled Landau level moves up in energy, and $\mathscr{E}_T$ increases reaching a maximum when the Landau levels is just at $\mathscr{E}_F$. Once this Landau level is above the Fermi energy, it will start to be emptied into lower energy Landau levels again lowering $\mathscr{E}_T$.

This oscillatory phenomenon repeats itself whenever $\mathscr{E}_F/\hbar\omega_c = mc\mathscr{E}_F/\hbar eH$ is integral. Thus, any experimental measurement that can sense this effect should have constant period in H^{-1} given by

$$\Delta(H^{-1}) = \frac{e\hbar}{mc\mathscr{E}_F} \tag{9-68}$$

This oscillatory, in 1/H, can be seen in a variety of magnetic and galvanomagnetic properties of metals and semiconductors. This effect is discussed in the next chapter when the determination of Fermi surfaces is considered.

Notes

Most solid state books have a discussion on the Drude and Sommerfeld theories. Also see the article by A. A. Abrikosov, "Introduction to the Theory of Normal Metals," Solid State Physics, Supplement **12** (1972); this discussion is advanced, as is N. Wiser, Contemp. Phys. **25**, 211 (1984). The book by Gopal, referred to in Chapter 11, has a good discussion on the heat capacity of an electron gas.

Values for the conductivity of metals can be found in G. T. Meaden, "Electrical Resistance of Metals" (Plenum Press, 1965) and thermal conductivity data can be found in the "American Institute of Physics Handbook" (McGraw-Hill, 1963). A complete reference to Born and Wolf's book can be found in the Notes in Chapter 13. See their Optics of Metals chapter for a discussion of the ultraviolet transparency problem. Also see Section 13-16 of this book.

An assembly of particles can be such that under the interchange of any two particles the wave function is antisymmetric (the result of the exchange multiplies the wavefunction by -1). This is called the **antisymmetry principle** and particles that obey this principle are called **fermions** and the average thermal equilibrium probability of any state being occupied is given by Fermi-Dirac statistics, Eq. 9-34. We can show the particles with half integral spin (1/2, 3/2, ...) are fermions. The **Pauli exclusion principle** (1925), which was originally introduced empirically to explain atomic spectra, says that no two electrons can have the same quantum numbers. This is a weaker principle than the antisymmetry principle, but often the two terms are used interchangeably. See the Notes in Chapter 12 (and Problem 8 there) for more information on bosons and fermions.

Landau levels are covered in many solid state and semiconductor books. The quantitization of the motion, as well as related magnetic effects, is discussed in Ziman (Chapter 9), Pererls (Chapter 7), and D. Shoenberg, "Magnetic Oscillation in Metals" (Cambridge University Press, 1984).

g-values (and related electronic spin resonance properties) of electrons and holes in semiconductors is the subject of a review article by G. W. Ludwig and H. H. Woodbury in Solid State Physics **13**, 223 (1962). The g-values of the shallow donors approximately apply to free electrons in the conduction band. This is because the shallow donors wave function extends over very many unit cells (Eq. 8-15b), thus their properties are very similar to free electrons.

"The **Hall Effect** in Metals and Alloys" by C. M. Hurd (Plenum Press, 1972) is the subject of an entire book. Landau levels, the two carrier Hall effect, as well as many other topics are discussed.

For a discussion of **Fermi liquid theory** see: L. Landau, Soviet Physics JEPT **3**, 920 (1957), ibid. **5**, 10 (1957), P. Nozières, "Theory of Interacting Fermi Systems" (Benjamin, 1964), D. Pines and P. Nozières, "Theory of Quantum Liquids" (Benjamin, 1966) Vol. **1**.

Problems

1. From the kinetic theory of gases derive the expression for the thermal conductivity (Eq. 9-8).

2. Derive an expression for the time dependence of the drift velocity, v_d, in the presence of an electric field and collisions, Eq. 9-10. (Hint: see Ashcroft and Mermin.)

CHAPTER 9   METALS   237

3. For the alkali metals (Li to Cs) the unit cell lengths of these bcc crystals are 3.50, 4.28, 5.33, 5.62, and 6.05 Å, respectively. Using the ionic radii given in Chapter 7, show that the fraction of space outside the core region decreases in going from Li to Cs. What might be a reason for this decrease? What are the plasma frequencies?

4. **Thermionic Emission** – (a) Consider Fig. 9-1. At a given temperature some free electrons have an energy larger than W and can partake in thermionic emission. Show that the current density is $j = AT^2 \exp(-\phi/k_B t)$ where $A = 4\pi m e k_B^2/h^3$. This is known as the **Richardson-Dushman equation.** (b) ϕ can also be obtained from the photoelectric effect. Compare the value of ϕ obtained from these two methods. (c) We can measure W by the following technique. Electrons in a vacuum that are excited across a potential V have a wavelength, given by the de Broglie relationship $\lambda = h/mv = h/(2meV)^{1/2}$. When these electrons enter a metal crystal they have a potential V + W, so we can measure W. Discuss this experiment. (Hint: See Dekker p. 220, Kittel early editions, and Zhdanov Chapter 8.)

5. (a) Calculate the degeneracies of the free electron levels, using periodic boundary conditions, for $\overline{m}^2 = 10, 11,$ and 12, as well as 100, 101, and 102. (b) Consider the two kinds of boundary conditions discussed in Section 9-7b. Show that in real (that is, large) systems both conditions lead to the same physics, that is, the same density of states. (c) It is only in systems that are limited to the first few quantum numbers that we can distinguish between the two boundary conditions. Can you think of such a system? (See Chapter 18.)

6. (a) Calculate the energy of the most energetic free electrons in equilibrium at 0°K for Na and Al. Compare the results to those given in Table 9-4. (b) For Al, what is the relativistic correction for the mass of these electrons? (c) What is the average kinetic energy of the free electrons in a 1 cm³ piece of aluminum? (d) Using this block as a projectile, what would its velocity be to have the same kinetic energy as the free electrons?

7. Assume that Na has a linear thermal expansion between 0°K and 300°K of $1.5 \times 10^{-6}/°K$. Between these temperatures what is the percentage change in $\mathscr{E}_F$, T_F, v_F, and k_F? At 300°K, approximately how many and what fraction of the electrons are above $\mathscr{E}_F$?

8. Using the conditions that $\psi = 0$ on the boundary of a cube of side L calculate the wave functions and energies of free electrons. What is the degeneracy of the first five levels? What is the density of states at 0°K? Compare this to the result in Eq. 9-30b.

9. (a) For $\mathscr{E} \gg \mathscr{E}_F$ show that the Maxwell-Boltzmann and the Fermi-Dirac distribution have the same functional form. (b) Show mathematically that using the Maxwell-Boltzmann or the Fermi-Dirac distribution we obtain the same result for the dc conductivity. (Hint: see the early editions of Kittel.)

10. **One-, two- and three-dimensional free electron gases** – Consider the Sommerfeld model for the free electron gas in one, two and three dimensions. Take the crystal to have a length L, area A, or volume V, respectively. Take the orbital degeneracy as one (as in the three-dimensional case in the text) and the spin degeneracy as g_s. Then, in the different dimensional spaces show that k_F, $\mathscr{E}_F$, and $g(\mathscr{E})$ are as given. (Notice that the density of states is independent of energy for the two-dimensional case.)

1-D	2-D	3-D
$k_F = \left[\dfrac{\pi}{g_s}\dfrac{N}{L}\right]$	$\left[\dfrac{4\pi}{g_s}\dfrac{N}{A}\right]^{1/2}$	$\left[\dfrac{6\pi^2}{g_s}\dfrac{N}{V}\right]^{1/3}$
$\mathscr{E}_F = \dfrac{\pi^2 \hbar^2}{2g_s^2 m}\left(\dfrac{N}{L}\right)^2$	$\dfrac{2\pi \hbar^2}{g_s m}\dfrac{N}{A}$	$\dfrac{\hbar^2}{2m}\left[\dfrac{6\pi^2}{g_s}\dfrac{N}{V}\right]^{2/3}$
$g(\mathscr{E}) = \dfrac{g_s L}{\pi \hbar}\left(\dfrac{m}{2}\right)^{1/2}\mathscr{E}^{-1/2}$	$\dfrac{g_s m A}{2\pi \hbar^2}$	$\dfrac{g_s V}{2^{1/2}\pi^2}\left(\dfrac{m}{\hbar^2}\right)^{3/2}\mathscr{E}^{1/2}$

11. Assume that a monolayer of Cu atoms are epitaxially deposited on 1 cm² of an insulator. If the Cu atoms have the same structure and spacings as a normal (100) plane of a Cu crystal, calculate $\mathscr{E}_F$, T_F, v_F, and k_F of the free electrons in the Sommerfeld model. What is the radius of the Fermi circle in cm^{-1}? The average kinetic energy $\langle KE \rangle$ of the electrons at the Fermi surface is $\mathscr{E}_F$ while the average potential energy is

$$\langle PE \rangle \approx e^2/r_0 = e^2 \pi^{1/2}(N/A)^{1/2}$$

since $N/A \approx (\pi r_0^2)^{-1}$. Then calculate the ratio $\langle PE \rangle / \langle KE \rangle$ and compare this result to that obtained in Problem 13.

12. **Surface states of electrons on liquid helium** – The surface of liquid helium is inherently clean and planar. It turns out that "free" electrons can be trapped "on" this surface. Due to the Pauli exclusion

CHAPTER 9   METALS   239

principle the free electrons cannot penetrate the He atoms, but they are attracted to the liquid He surface by their classical image potential energy given by

$$\text{PE} = -\frac{1}{4}\left(\frac{\varepsilon-1}{\varepsilon+1}\right)\frac{e^2}{z} \equiv -Q\frac{e^2}{z} \qquad z>0$$

where z is the distance from the liquid surface, and ε is the dielectric constant of liquid He. Since $\varepsilon = 1.057$, $Qe \approx 7 \times 10^{-3}e$. Assume an infinite potential for $z \leq 0$ and show that the resulting one-dimensional problem has the same form as the s-states of a hydrogen atom multiplied by z/z_0 where z_0 is the Bohr length of this one-dimensional hydrogen atom (proton charge Qe) given by $z_0 = a_0/Q$ where a_0 is the Bohr radius of hydrogen. Thus, $z_0 \approx 76$ Å.

The energy eigenvalues are given by

$$\mathcal{E}_n = -(Q^2/n^2)\,\mathcal{E}_H \cong -0.65/n^2 \text{ meV}.$$

(Compare to Eq. 13-57b.) Calculate the splittings between the ground state and the first two excited states. Compare these splittings with experiment [C. C. Grimes, T. R. Brown, M. L. Burns, and C. L. Zipfel, Phys. Rev. B **13**, 140 (1976)]. How would you account for the differences? What determines the line widths of these transitions? [Several review-type articles on the subject of electrons on liquid He surface are all in Surface Science: R. S. Crandall, **58**, 266 (1976); C. C. Grimes, **73**, 379 (1978); C. C. Grimes and G. Adams, **98**, 1 (1980); volume **113**, (1982). Also see the subsequent proceedings of the conference on the "Electronic Properties of Two-Dimensional Systems" in this journal every two years.]

13. **Wigner crystallization** – In 1938 Wigner considered the problem of a collection of electrons moving in a uniform background of positive charge. He found that a crystallization of the electrons will occur when their kinetic energy is small compared to their potential energy. For crystallization to occur the average distance between electrons must be $\approx 10a_0$ (a_0 is the Bohr radius $=0.529$ Å). For electrons in a solid this is difficult to achieve along with the constraint of a uniform background of positive charge. However, it has been proposed [R. S. Crandall and R. Williams, Physics Letters **34A**, 404 (1971)] and observed [see the article by Grimes and Adams referenced in the previous problem] that electrons on the surface of liquid He could crystallize. The relevant parameter is $\Gamma \equiv \langle PE\rangle/\langle KE\rangle$, the ratio of the average potential to kinetic energy. Using the relationship for $\langle PE\rangle$ in Problem 11, and using the high temperature approximation for $\langle KE\rangle$, $\Gamma = e^2\pi^{1/2}n_s^{1/2}/k_BT$. [Wigner crystallization has been reported in a three-dimensional system, but these results are controversial. T. F.

Rosenbaum, et al. Phys. Rev. Letters **54**, 241 (1985) and G. Nimtz, et al. Sol. State Commun. **32**, 669 (1979).]

Crystallization is observed for $\Gamma = 137$. (a) At $0.5°K$ what is the average spacing between electrons? (b) Normally the KE of a free electron system is given by $\mathscr{E}_F$. What is $\mathscr{E}_F$ for this electron system and why can the high temperature expansion be used? (c) Why would you expect the crystal structure to be one with an electron at each lattice point of a hexagonal lattice? (d) Discuss the method of observation of this crystallization.

14. In order to observe cyclotron resonance of the free electrons in Cu what magnetic field must be used if your microwave apparatus operates at 30 GHz (that is, the Q-band)? What is the value of $\omega_c \tau$? At resonance, the electrons absorb energy from the microwave bridge. What happens to this energy?

10

Band Theory

PART A   QUALITATIVE DISCUSSION

10-1   Nearly Free Electrons
    a   Free electrons
    b   Addition of the periodic potential
    c   Wave functions
    d   Magnitude of the gap
    e   Summary
10-2   Classifications of Solids
10-3   Effective Mass

PART B   WAVE FUNCTIONS AND ENERGY LEVELS

10-4   Bloch Functions
    a   Bloch's theorem
    b   Reduced zone scheme
    c   Band Index
    d   Wave equation for Bloch functions
10-5   Nearly Free Electrons
10-6   Brillouin Zones
    a   Review of the reciprocal lattice
    b   Reduced zone
    c   Brillouin zones
10-7   Examples of Brillouin Zones
    a   Two-dimensional examples
    b   Construction of Fermi surfaces
    c   Three-dimensional examples
    d $\partial \mathcal{E}/\partial$ at zone surfaces
    e   A real Fermi surface (S)
10-8   Wigner-Seitz Approximation— The Binding Energy (S)
10-9   The Tight Binding Approximation (S)
    a   Introduction
    b   Energy calculation
    c   Examples
10-10   Crystal Momentum

PART C   SEMICONDUCTORS, REAL BANDS, AND RELATED CONCEPTS

10-11   Holes
    a   Introduction
    b   Effective masses
    c   Carrier density for an intrinsic material
    d   Conductivity
10-12   Band Preliminaries (A)
10-13 $\mathcal{E}(k)$ for a Two-Dimensional Square Lattice
    a   Introduction
    b   Behavior at special points and lines (S, A)
    c   Compatibility relations (S, A)
    d   Summary
10-14   Body-Centered Cubic Lattice — Sodium (S, A)
10-15   Si, Ge, GaAs, and GaP
    a   Bonds (and band gaps)
    b   Optical absorption
    c   Effective masses
10-16   Carrier Concentration at Thermal Equilibrium
    a   Law of mass action
    b   Intrinsic semiconductors
    c   Extrinsic semiconductors
    d   Impurity band conductions
10-17   p-n Junctions
    a   Introduction
    b   Equilibrium conditions
    c   Quantitative considerations at equilibrium
    d   Biased junctions
    e   Capacitance of a p-n junction
    f   Reverse bias breakdown
    g   Light emission
10-18   Metal-Semiconductor Junctions
    a   Schottky barriers
    b   Ohmic contacts
10-19   The Gunn Effect (S)
10-20   Other Topics (S)
10-21   Summary
Notes
Problems

Table 10-1 Some properties of semiconductors. Values are for room temperature unless otherwise indicated and are uncertain by one or two units in the last significant figure. The structures are: Si and Ge – Diamond; GaP, GaAs and InSb – zinc blende; PbTe – rock salt; CdS – wurtzite. Elastic constants are in Table 12-2; other optical phonon energies are in Table 13-1. The references are: (a) ε_∞ and $\varepsilon(0)$ refer to the optic and clamped static dielectric constants, respectively (see Section 13-9c); (b) the units are meV, also see Table 13-1; (c) the units are $cm^2 V^{-1} sec^{-1}$; (d) c = 6.716 and a = 4.137 Å (based on a compilation by F. Stern).

		Si	Ge	GaP	GaAs	InSb	PbTe	CdS
Lattice Const. (Å)		5.431	5.646	5.451	5.653	6.479	6.488	(d)
Atoms/Vol. ($\times 10^{22}/cm^3$)		5.00	4.44	4.94	4.42	2.94	2.94	4.00
Density (gm/cm^3)		2.33	5.33	4.13	5.31	5.78	8.25	4.82
Melting Temp. (°C)		1412	937	1467	1238	536	905	1550
Debye Temp. (°K)		650	370	450	340	200	180	220
Dielectric Const.[a]	ε_∞	11.9	16.0	9.0	11.0	15.7	32	5.2
	$\varepsilon(0)$	11.9	16.0	11.1	13.0	17.7	800	8.7
Optical phonon energy at k=0[b]	TO	63	37	45	33	22.3	2.8	29
	LO	63	37	50	36	23.7	14	38
Electron mobility[c]	300°K	1500	3800	200	9000	90000	1700	400
	77°K	25000	35000	3000	2×10^5	2×10^6	30000	5000
Hole mobility[c]	300°K	500	1800	140	400	750	800	15
	77°K	6000	35000	2000	7000	10000	20000	–
$\mathscr{E}_g$ (eV)	300°K	1.124	0.663	2.267	1.424	0.18	0.31	2.42
	77°K	1.167	0.738	2.338	1.510	0.228	0.21	2.57
	0°K	1.170	0.744	2.350	1.519	0.235	0.190	2.582
Conduction Band								
Valley degeneracy		6	4	6	1	1	4	1
Density of states mass (m*)	per valley	0.32	0.22	0.4	0.07	0.014	0.05	0.2
	total	1.06	0.56	0.8	0.07	0.014	0.13	0.2
Valence band								
Den. of St. mass (m*)		0.6	0.4	0.8	0.55	0.45	0.14	1.3

BAND THEORY

*We shall not cease from exploration
And the end of all our exploring
Will be to arrive where we started
And know the place for the first time.*

T. S. Eliot, "Four Quartets"

The theory discussed in this chapter forms the basis for the modern theory of electrons in solids. It arises from the consideration of the periodicity of the crystal structure. This periodicity leads to the formation of bands. An underlying assumption is that the entire electron-electron interaction can be taken into account via the **independent electron approximation**. Namely, the interaction of one electron with all of the others can be approximated via some average effective periodic potential. It turns out that this assumption is valid; the periodic potential energy is more important than the energies associated with the breakdown of this approximation. At the end of Section 10-9 some aspects of this assumption are discussed.

The chapter is broken up into three major parts. Part A discusses bands from a qualitative point of view, showing that the ideas are easy to understand and arise directly from the periodic symmetry of the crystal potential. Part B discusses electronic wave functions and their energy levels, including the mathematical formulation. Semiconductors and real bands are discussed in Part C.

Part A – Qualitative Discussion

10-1 Nearly Free Electrons

10-1a Free electrons As shown in the last chapter (Eqs. 9-26), for a running (travelling) free electron wave, the wave function is given by $\exp(i\mathbf{k}\cdot\mathbf{r})$, where $\mathbf{k}$ is the wave vector. The energies and k-values of the free electrons in a cube of side L are

$$\mathcal{E} = \left(\frac{\hbar^2}{2m}\right)(k_x^2 + k_y^2 + k_z^2) = \frac{\hbar^2}{2m}k^2$$

$$\mathbf{k} = \frac{2\pi}{L}\overline{\mathbf{m}} = \frac{2\pi}{L}(\overline{m}_x, \overline{m}_y, \overline{m}_z)$$

(10-1)

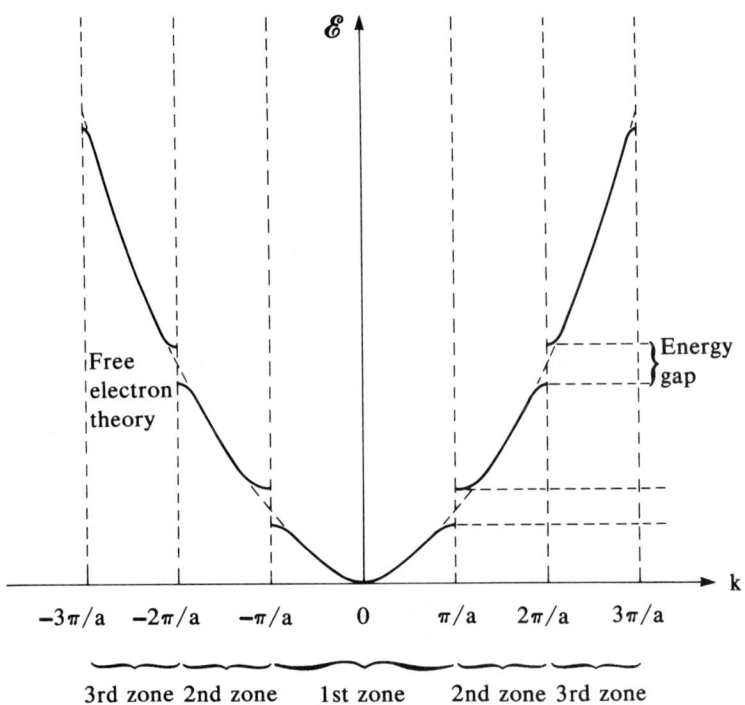

Fig. 10-1 Energy vs. wave vector for free electrons (dashed parabola) and for almost free electrons (solid lines). Gaps develop at k = pπ/a where p = ±1, ±2, and so forth.

where $\bar{m}_i = 0, \pm 1, \pm 2, ...$, resulting from the periodic boundary conditions that allow these discrete values of **k**. (Remember the wave vector $k = 2\pi/\lambda$.) As discussed at length in the last chapter, the Pauli principle and Fermi-Dirac distribution must be applied to the free electrons in a metal. This leads to the concepts of the Fermi surface, the Fermi energy, $\mathscr{E}_F$, as well as to the Fermi temperature, velocity, and wave number (Table 9-4). This distribution, which describes the possible occupation of electron states, affects many properties of the electron gas in important ways. However, all the calculations and discussions in Chapter 9 were for free electrons; that is, the periodic arrangement of atoms was totally ignored.

10-1b Addition of the periodic potential In this section we qualitatively consider the effects due to a periodic crystal structure. The changes are profound. Working in one dimension will be emphasized; the results for three dimensions usually follow straightforwardly.

Electrons, besides acting as discrete particles, have wave-like behavior. This property was used in considering the number of al-

CHAPTER 10 BAND THEORY

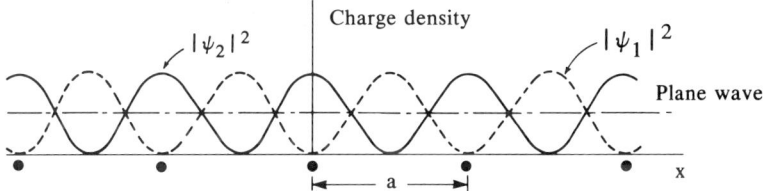

Fig. 10-2 The spatial dependence of the charge density for a plane wave and for the two standing waves from Eq. 10-3.

lowed states in Section 9-7. Consider a one-dimensional lattice in which the energy of an electron is being slowly increased so that k increases, Eq. 10-1. When k becomes large enough (λ small enough) the electron wave will suffer a Bragg reflection. Recall from Section 4-2 that Bragg's law is $2d \sin\theta = p\lambda$. For the one-dimensional case d = a (the spacing between the atoms) so Bragg reflections occur at

$$k = p\pi/a \qquad p = \pm 1, \pm 2, \ldots \qquad (10\text{-}2)$$

The same Bragg reflection will occur as k approaches $p\pi/a$ from above. Figure 10-1 shows the $\mathscr{E}$ versus k plot for the free electron case and for the case with a periodic crystal. As we shall see, at $k = p\pi/a$ energy gaps develop due to these reflections.

10-1c Wave functions To understand the gaps better we must appreciate that the constructive interference (the reflections) occurs at $k = p\pi/a$ for waves traveling to the left as well as to the right. That is, the wave with $k = +\pi/a$ will be reflected and thus have $k = -\pi/a$, which will in turn be reflected, so there is a phase change of π between waves reflected by the atoms. Thus, the solutions for these values of k are made up of equal components of waves traveling to the left, $k = -\pi/a$, and to the right, $k = +\pi/a$. We will see later that waves with $k = \pm\pi/a$ are connected (by perturbation theory) by the periodicity of the crystal. From the traveling waves $\exp(\pm ikx)$ linear combinations can be formed to obtain two standing waves

$$\begin{aligned}(1/L^{1/2})(e^{i\pi x/a} - e^{-i\pi x/a}) &= (2i/L^{1/2}) \sin \pi x/a \equiv \psi_1 \\ (1/L^{1/2})(e^{i\pi x/a} + e^{-i\pi x/a}) &= (2/L^{1/2}) \cos \pi x/a \equiv \psi_2\end{aligned} \qquad (10\text{-}3)$$

The fact that one of those two wave functions is imaginary is no problem because the charge density, $\rho(x) = -e|\psi|^2$, is real. Figure 10-2 shows that for a plane wave the charge density is independent of x. Figure 10-2 also shows that $\rho(x)$ for the two standing waves in Eq. 10-3 are the same except shifted along the x-axis by $a/2$ so that their sum is exactly the same as the value shown for the plane wave. The charge density for ψ_1 peaks in between the atomic sites while that of

ψ_2 peaks at the atomic sites. In the free electron approximation both of these wave functions have the same energy. However, when the periodicity of the crystal is considered, they differ in energy. As to which is lower depends on the details of the potential energy and is calculated in the next subsection. Covalently bonded crystals have the valence charge density between the atoms while in ionic crystals it is on the atomic sites.

10-1d Magnitude of the gap Now we show that an energy gap actually exists. The potential energy due to the crystal can be approximated as $V(x) = V_1 \cos 2\pi x/a$. This potential has the periodicity of the lattice, $V(x) = V(x + a)$, and is the first Fourier component of an arbitrary crystal potential. (The zero component, which has no x-dependence, is the averaged potential energy that was used in Chapter 9 for the completely free electron case.) The two different normalized wave functions are taken from Eq. 10-3. These functions are the "extreme" case for the linear combinations since they center the charge either at the nuclear site or in between the sites. Then, by first-order perturbation theory, the energy difference is

$$\mathcal{E}_g = \frac{2V_1}{L} \int_0^L \cos(2\pi x/a) [\cos^2 \pi x/a - \sin^2 \pi x/a] \, dx = V_1 \quad (10\text{-}4)$$

By taking the potential energy as a cosine term we have, of course, immediately decided which state has lower energy. Note that for either a sine or cosine term for the crystal potential, the resultant energy gap is equal to the magnitude of the Fourier component of $V(x)$. For our choice, the eigenstate ψ_2 is lower in energy than ψ_1 by $\mathcal{E}_g = V_1$ at $k = \pm \pi/a$. The same type of energy gap will appear at $k = p\pi/a$ (Eq. 10-2) as long as the periodicity of the crystal structure is taken into account. (Higher order terms in the Fourier analysis of the potential energy are required for the higher p-values.) Thus, at $k = p\pi/a$, *electrons with certain values of energy cannot propagate in the crystal as indicated in Fig. 10-1*. This is consistent with the Bragg law for reflection of waves from crystal planes.

10-1e Summary The periodic potential of the crystal perturbs free electron states and results in energy gaps at $k = \pm \pi/a$ (and $\pm 2\pi/a$, and so on). The $\mathcal{E}$ vs. k plot in Fig. 10-1 shows these gaps. In the free electron case these states at $k = \pm \pi/a$ have the same energy. After the periodic potential energy is taken into account the wave functions at $k = +\pi/a$, for example, are made up of equal parts from the free electron functions at $k = +\pi/a$ and from $k = -\pi/a$. (These values of k differ by a reciprocal lattice vector as will be discussed later.) The relationship between Bragg scattering and the periodic potential energy should be noted.

10-2 Classification of Solids

From the simple ideas discussed in the previous section a rather general classification of solids into metals and nonmetals can be understood. Consider the number of the allowed k-values between the Bragg reflections at $k = \pm \pi/a$. For the one-dimensional case, k-values with integral multiples of $2\pi/L$ are allowed, Eq. 10-1. Thus, there are a huge number of allowed k-values because $L \gg a$. For example, in Fig. 10-1 the $\mathscr{E}$ vs. k curve is quasicontinuous, being made up of a huge number of points, one for each of the allowed k-values. Since there is a k-value, or allowed state, at every integer multiple of $2\pi/L$ the total number of allowed states between the Bragg reflections at $k = -\pi/a$ and $+\pi/a$ is

$$\frac{(2\pi/a)}{(2\pi/L)} = L/a = \text{number of primitive cells!} \qquad (10\text{-}5)$$

(Or we may solve for the largest value of p in Eq. 10-2, that is, $2\pi p/L = \pi/a$ or $p = L/2a$. Thus, $2p = L/a$.) For the three-dimensional case, using Eq. 10-1, L^3/a^3 will be obtained, still the number of primitive cells in the crystal. Thus, *if there are N primitive cells in the crystal, there are N allowed values of k* for each band. Due to the electron spin degeneracy each band can hold 2N electrons.

This value for the allowed states in a band allows us to understand the classification of solids from a fundamental point of view. Consider a crystal with an odd number of valence electrons per primitive unit cell. Then the uppermost band can be only half full because the band can hold 2N electrons and there are only N electrons. Such a crystal *must* be a metal since there are empty states extremely close to the filled states to which electrons can be excited by an electric field, as discussed in Chapter 9. This explains why the alkali elements are metals. These elements have the body-centered cubic crystal structure, Table 3-1, thus there is one atom and one valence electron per primitive unit cell. This also explains why many other elements in Table 3-1 are metals.

Now consider what may happen if the crystal has an even number of valence electrons per primitive unit cell. For the one-dimensional case at 0°K all the states below a band gap are exactly filled and the material is an insulator. An electric field applied to the material will not be able to excite electrons because there are no available unoccupied states close in energy to the occupied states. NaCl or CsCl crystals are examples. For both of these materials there are six valence electrons per primitive unit cell (one from Na or Cs and five from Cl). These crystals indeed form insulators. However, for a three-dimensional crystal there are certain other possibilities. The top of the filled states in one k-direction may be lower in energy than the "empty" states in another direction. Materials that have this property are called semimetals; bismuth is one example.

Semiconductors are a subclass of insulating materials that have an energy gap smaller than ≈ 1 eV. There is nothing exact about this definition. The idea is that if the energy gap is small enough, charge carriers can be thermally excited across the gap. Materials that have this property are called semiconductors. (Most technologically useful semiconductors are materials that can be doped, in a controllable manner, with donors and acceptors, as discussed in Section 8-5.) For example, silicon has two atoms per primitive unit cell, therefore eight valence electrons. The criterion for an insulator is met and the band gap is 1.12 eV (Table 8-3).

10-3 Effective Mass

Mass is defined by Newton's law as the constant of proportionality between force and acceleration, $F = ma$. The interaction of the electron with the metal causes the effective mass of the "no longer free electron" to be different from that of a free electron.

We should expect a value for the effective mass to be different from the bare mass of the electron. For example, consider a ball with charge Z and mass M being accelerated by an external electric field, E, in a frictionless fluid. The force on the ball is ZE, yet the ratio of ZE to the acceleration of the ball will not be M because the ball will also push a certain amount of fluid ahead of it. Thus, if we focus only on the ball, Newton's law will give an effective mass different from M because the acceleration of the fluid is ignored. An electron in a crystal behaves in a similar manner. However, as will be seen, the mass not only can be larger than m, it can be smaller and even negative. This is because the electron-lattice interaction is more complicated than that of a ball and a frictionless fluid.

Consider the group velocity, v_g, of an electron in an energy band in a crystal under an applied electric field, E. The electron can be described by a wave packet whose wave vector is centered at k. From physical optics $v_g = d\omega/dk$ where ω is the angular frequency of the wave. Since $\mathscr{E} = \hbar\omega$, we have for v_g and its time derivative,

$$v_g = \hbar^{-1} d\mathscr{E}/dk \tag{10-6a}$$

$$\frac{dv_g}{dt} = \frac{1}{\hbar}\frac{d^2\mathscr{E}}{dt\,dk} = \frac{1}{\hbar}\frac{d^2\mathscr{E}}{dk^2}\frac{dk}{dt} \tag{10-6b}$$

CHAPTER 10   BAND THEORY   249

The electron acceleration, dv_g/dt, is related to $d^2\mathscr{E}/dk^2$, the curvature of the band and the variation of k with time under the influence of an external force. An external electric field, E, applied for a time interval δt will do an amount of work on an electron of charge $-e$, given by $\delta\mathscr{E}$ (all the work goes into increasing the energy of the electron)

$$\delta\mathscr{E} = -eE\, v_g\, \delta t \qquad (10\text{-}7a)$$

However, the work is also $\delta\mathscr{E} = (d\mathscr{E}/dk)\delta k$; using Eq. 10-6a we have

$$\delta\mathscr{E} = (d\mathscr{E}/dk)\,\delta k = \hbar v_g\, \delta k \qquad (10\text{-}7b)$$

Equating Eqs. 10-7a and 10-7b, and taking $\delta k/\delta t = dk/dt$, we obtain

$$d(\hbar k)/dt = -eE \qquad (10\text{-}8)$$

This says that $d(\hbar k)$ is equal to the external force on the electron while for a free particle we obtain $d(mv)/dt$, and this equation is the same as Eq. 9-60. We can see more clearly the result of the particle not being free when considering the effective mass; substituting dk/dt into Eq. 10-6b. The result is just like Newton's law ($a = F/m$).

$$\frac{dv_g}{dt} = \frac{1}{\hbar^2}\frac{d^2\mathscr{E}}{dk^2}(-eE) \qquad (10\text{-}9)$$

From this we define an **effective mass**, m*, which is simply related to the curvature of the energy bands

$$\frac{1}{m^*} \equiv \frac{1}{\hbar^2}\frac{d^2\mathscr{E}}{dk^2} \quad \text{or} \quad \left(\frac{1}{m^*}\right)_{ij} \equiv \frac{1}{\hbar^2}\frac{d^2\mathscr{E}}{dk_i\, dk_j} \qquad (10\text{-}10)$$

The second expression is the effective mass tensor generalized to three-dimensional anisotropic energy surfaces, to be discussed later.

By substituting values for the free electron from Eq. 10-1 into Eq. 10-10 we see immediately that m* = m as expected. However, consider the values of m* for electrons in bands. Figure 10-3 shows the results for the various derivatives of $\mathscr{E}$ vs. k for two cases. Let EB be the width of the band itself; then Fig. 10-3a schematically represents the general case, while in Fig. 10-3b $\mathscr{E}_g/EB \gg 1$. First, consider the nearly free electron case in Fig. 10-3a. Remembering the inverse relationship between m* and $d^2\mathscr{E}/dk^2$, we see that as k increases from zero, m* = m throughout most of this lowest band. However, just below $k = \pi/a$, m* gets very small and negative. Second, in the other extreme (the wide gap, narrow band case where $\mathscr{E}_g/EB \gg 1$), is shown in Fig. 10-3b, where m* tends to be greater than m since there is relatively little curvature in the $\mathscr{E}$ vs. k curve. For this case m* very different from m should be expected since the $\mathscr{E}$ vs. k bears little resemblance to the free electron case. This is the **tight binding** situa-

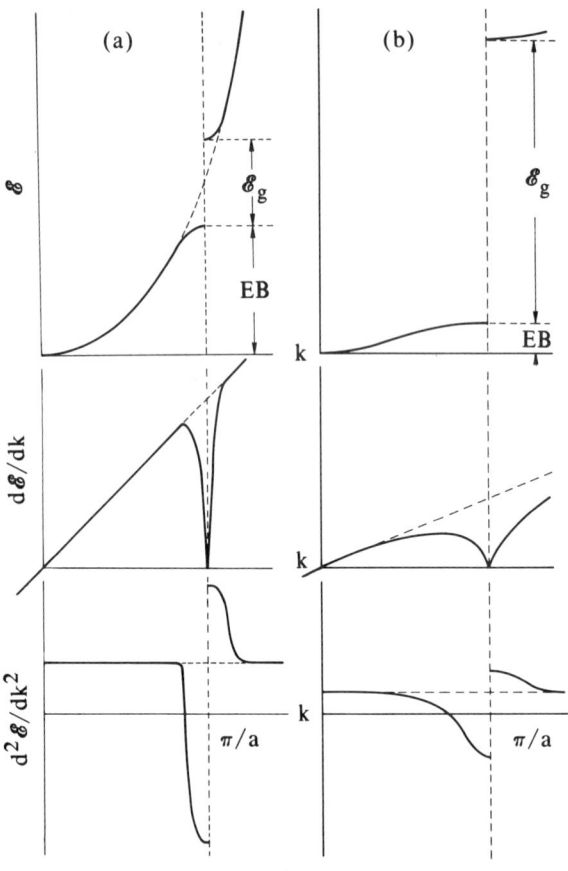

Fig. 10-3 Plots of $\mathscr{E}$ and various derivatives with respect to k showing what happens near k = π/a. (a) is for narrow band gaps as well as for wide gaps. (b) is the same but emphasizes what the situation for wide band gaps, that is, for $\mathscr{E}_g >> EB$.

tion to be discussed in Section 10-9. It arises when the valence electrons are strongly localized (or tightly bound) at the atomic core sites and hence the electrons are not very "free." Under the influence of an external force, they will not be accelerated rapidly across the crystal, which means that their effective mass is large.

A simple *gedanken* experiment helps to understand physically the meaning of a negative mass. Figure 10-4a shows an electron beam incident on and passing through a thin crystal. The wave vector of the electrons in the crystal is just slightly too low to satisfy the Bragg reflection condition, that is, $k \lesssim \pi/a$. Then, in Fig. 10-4b a small voltage is applied to the grid so that the electron beam is slightly accelerated, increasing its k-value in the crystal to that required for a Bragg reflection. Thus, when the electrons are in the right region of the $\mathscr{E}$ vs. k curve, a small increase in energy can completely reverse their momenta. This corresponds to a negative mass and to Bragg scattering since the wave length of the electron is equal to the period of the periodic arrangement in the crystal. Thus, the intimate connec-

CHAPTER 10   BAND THEORY   251

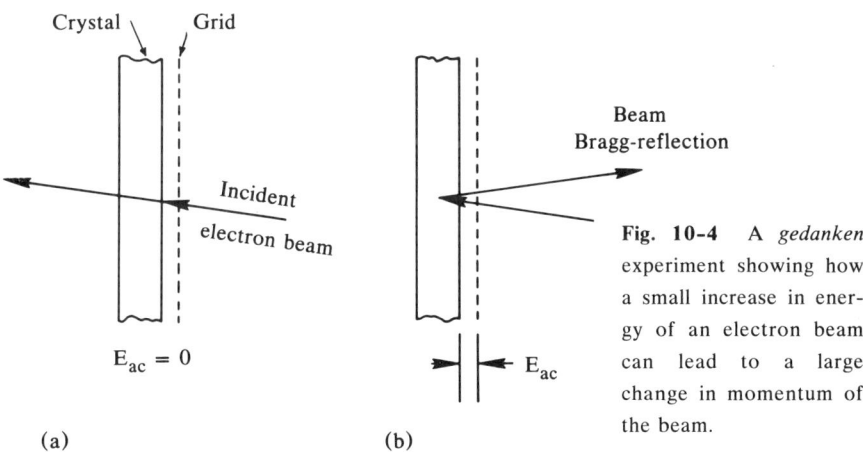

Fig. 10-4 A *gedanken* experiment showing how a small increase in energy of an electron beam can lead to a large change in momentum of the beam.

tion between Bragg scattering and m* can be appreciated. A negative effective mass means that as the electron goes from a state of k to one of k + Δk the momentum transfer from the electron to the lattice is larger than the transfer from the applied electric field to the electron.

Only near the top of the bands, where the deviations from the free electron behavior in the $\mathscr{E}$ vs. k can be large, does the possibility of a negative m* become important. For most simple metals the bands are half filled and a negative m* is not possible. However, this concept is of great importance in materials with nearly filled bands. In fact, the effect is emphasized since most of the observable electron transport properties are determined by those electrons at the top of the band, that is, at the Fermi energy. Semiconductors and semimetals usually have $\mathscr{E}_F$ just near the band edges.

The last point concerns metals with narrow bands and large band gaps, that is, $\mathscr{E}_g/EB \gg 1$, as sketched in Fig. 10-3b. Materials such as the transition metals and the rare earth metals tend to have this property. The small values of EB occur because the overlap of the valence electrons (3d or 4f electrons, respectively) are smaller than the overlap of the valence electrons of the alkali metals. Since the curvature of the $\mathscr{E}$ vs. k band is relatively small, $m^*/m \gg 1$. This is in agreement with experiments for these metals. For example, the electronic heat capacity and Pauli spin susceptibility are proportional to the electron mass, Eqs. 9-53 and 9-59, since the Fermi temperature is proportional to the inverse of the mass. Most of the contribution to these two quantities comes from electrons at $\mathscr{E}_F$ as discussed in Chapter 9. Thus, to a good approximation, m in these formulas can be replaced by m*. Experiments show m*/m as high as 10, in agreement with these ideas.

Part B – Wave Functions and Energy Levels

In this part the qualitative results of the previous sections are made more quantitative. However, the physics is the same. First Bloch functions are discussed. These functions have the correct translational symmetry and are the eigenfunctions of the electrons. Then the translational symmetry of the lattice is examined in more detail leading to a deeper understanding of the k wave vector. This leads to Brillouin zones, an important concept that enables us to understand the dynamics of the electrons.

10-4 Bloch Functions

10-4a Bloch's theorem A fundamental theorem concerning electrons in a crystal was proved by Bloch in 1928. It states that the wave functions of the electrons in a crystal have the **Bloch form**

$$\psi_k(r) = e^{i k \cdot r} u_k(r) \tag{10-11}$$

where $u_k(r)$ is a function that has the periodicity of the lattice, that is, $u_k(r) = u_k(r + t_m)$ where t_m is any primitive lattice translation (Eqs. 1-10 or 2-1). This function, in general, depends on the wave vector **k**. The Bloch function is just a free electron wave function, $\exp(i k \cdot r)$, modulated by a function that has the periodicity of the lattice. Thus, it is a *modulated plane wave*.

We present a proof of Bloch's theorem in the one-dimensional case. Consider a crystal of length $L = Na$ where there are N primitive unit cells each of length a. The wave equation is

$$-\frac{\hbar^2}{2m} \frac{d^2 \psi}{dx^2} + V(x)\psi = \mathscr{E}\psi \tag{10-12}$$

with $V(x) = V(x + a)$. The periodic boundary condition demands that $\psi(x) = \psi(x + L)$ and the translational symmetry demands that the charge density is the same at x as at $x + a$ and at $x + ma$ where m (not to be confused with the mass in Eq. 10-12) is any integer, that is, $|\psi(x + ma)|^2 = |\psi(x)|^2$. Therefore

$$\psi(x + a) = A \, \psi(x) \tag{10-13a}$$

and $\quad \psi(x + ma) = A^m \, \psi(x) \tag{10-13b}$

where A is a complex number such that $A^*A = 1$ (the charge density

CHAPTER 10 BAND THEORY 253

is proportional to $\psi^*\psi$). By applying the translation N times we have

$$\psi(x + Na) = A^N \psi(x) = \psi(x) \qquad (10\text{-}14a)$$
$$A^N = 1 \qquad (10\text{-}14b)$$
$$A = \exp(2\pi i m/N) \qquad m = 0, \pm 1, \pm 2, \ldots \qquad (10\text{-}14c)$$

Equation 10-14b shows that A is just one of the roots of unity, which is written explicitly in Eq. 10-14c. Thus, the wave functions must be of the form

$$\psi(x + a) = \exp(2\pi i m/N) \psi(x) \qquad (10\text{-}15)$$

which is the **Bloch condition** on the wave function. A function that satisfies this condition has the requirements demanded by translational symmetry. This result suggests that in seeking an electron wave function (i.e., a function that satisfies Eq. 10-12) we should try a modulated free electron function, $\psi(x) = \exp(ikx) u(x)$, and determine the conditions that allow this function to satisfy the Bloch condition, Eq. 10-15. Substituting this modulated plane wave in both sides of Eq. 10-15 we see that the equality is kept if

$$k = 2\pi m/Na \qquad (10\text{-}16a)$$
and $\quad u(x + a) = u(x) \qquad (10\text{-}16b)$

If $u(x)$ is a constant, then the free electron wave function with the same condition on k, is recovered, Eq. 10-1 or Eq. 10-16a. Thus, a solution of the wave equation with the lattice periodicity, Eq. 10-12, can be written in one dimension in the form

$$\psi(x) = \exp(ikx) u_k(x) \qquad (10\text{-}17)$$

The value of k is given in Eq. 10-16a where m is any integer and $u(x)$ has the periodicity of the lattice, Eq. 10-16b. This proves Bloch's theorem in one dimension. The extension to three dimensions, Eq. 10-11, follows straightforwardly. (It is clear from Eq. 10-17 that $\psi(x)$ depends on k, and there is nothing in the preceding proof that restricts $u(x)$ from also depending on k.)

Note that we have not as yet put $\psi(x)$, Eq. 10-17, into the wave equation, Eq. 10-12, to find eigenfunctions and eigenvalues. This is done in Section 10-4d. First, we investigate the geometry of k-space.

10-4b Reduced zone scheme We now pursue the idea that in one band the wave vector k is limited to N values where N is the number of primitive cells in the crystal, as discussed in Section 10-2. (This applies to three dimensions as well; namely the number of allowed k-values is equal to the number of primitive unit cells.) Again consider a one-dimensional crystal, but the extension to three dimensions is straightforward.

Replace k in the Bloch function by $k' + 2\pi m/a$, where m is an integer, and see what happens. First, in Eq. 10-11 or 10-17 replace x with x + a to yield

$$\psi(x + a) = \exp(ika)\psi(x) \qquad (10\text{-}18a)$$

This equation follows because u(x) is periodic in x + a, Eq. 10-16b. Second, in this equation replace k by $k' + 2\pi m/a$.

$$\exp(ika)\psi(x) = \exp(ik'a)\exp(i2\pi m)\psi(x) \qquad (10\text{-}18b)$$

$$= \exp(ik'a)\psi(x) \qquad (10\text{-}18c)$$

where $\qquad k = k' + 2\pi m/a \qquad (10\text{-}18d)$

Equation 10-18c is obtained because $\exp(i2\pi m) = 1$ since m is an integer. So the same Bloch condition holds for k or k'. Thus, the k-value is periodic in k-space modulo $2\pi/a$ and from Eq. 10-16a we know that there are just N values of k in this interval. *Thus, we can restrict the k-values to an interval of length $2\pi/a$ and need not consider k beyond this interval since no new information is gained.* For convenience, the interval is usually chosen to be symmetric about k = 0, so that

$$-\pi/a \leq k \leq \pi/a \qquad (10\text{-}19)$$

The reasons for this choice will become clear shortly, particularly when Brillouin zones are discussed but the fundamental point is that any interval of $2\pi/a$ will be sufficient.

We would like to relate the idea expressed in Eq. 10-19 to the almost free electron behavior shown in Fig. 10-1. Figure 10-5a shows an $\mathscr{E}$ vs. k curve for a nearly free electron. Band gaps have been inserted at $k = p\pi/a$ for p equal to any integer except zero. However, as in Eq. 10-18, translational symmetry allows the addition to or subtraction from the free electron k, a quantity that is any integral multiple of $2\pi/a$. Figure 10-5b shows the results of such the procedure restricting the k-values as in Eq. 10-19. Since k can always be replaced by $k + 2\pi m/a$ (where m is any integer), in Fig. 10-5c we sketch the results in Fig. 10-5b but repeated to show this periodicity. These three different presentations, Figs. 10-5a, b, and c, represent the same physical phenomenon and are respectively called the **extended zone scheme,** the **reduced zone scheme,** and the **repeated zone scheme.** *All three approaches represent the identical physical behavior and any one of these schemes can be used for convenience.* For example, when discussing electronic states energy may be viewed as a continuous function of **k**; the repeated zone scheme would then be the natural one to use. However, the reduced zone scheme is usually used because the proper interval for the k-values is most obvious.

CHAPTER 10   BAND THEORY   255

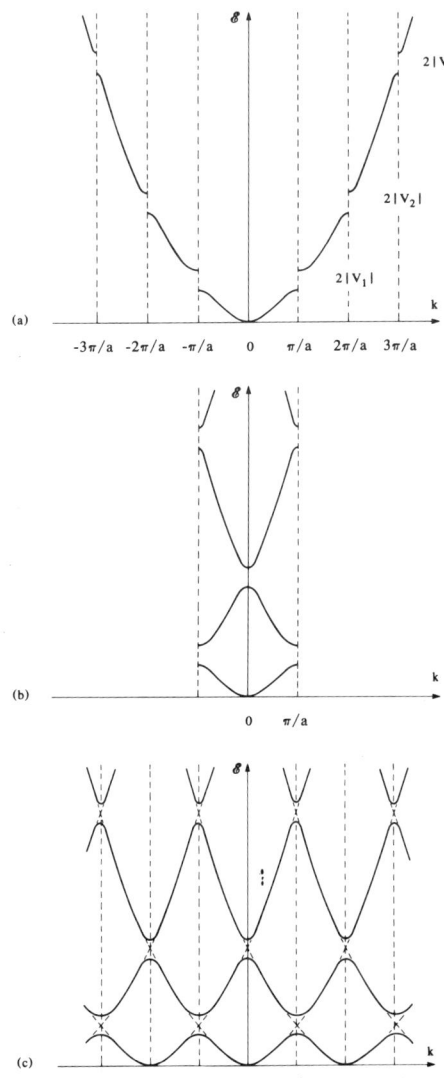

Fig. 10-5 Energy as a function of wave vector for a one-dimensional crystal, shown in the (a) extended, (b) reduced, and (c) repeated zone scheme.

10-4c Band index The interval in Eq. 10-19 is the one-dimensional analogue of what is usually called the first Brillouin zone, a concept that will be discussed at length. With a very simple example, we show that the crystal periodicity gives the Bloch functions a new index. This band index is analogous to the principal quantum number in the hydrogen atom solutions.

We shall now require the electronic wave functions to be expressed with a k-value in the reduced zone scheme. As will be seen a new band index will come about in a natural way. Another way to think of the process of transferring the wave functions from outside to within the first Brillouin zone is: the transfer incorporates the excess periodicity of the wave function into the $u_k(r)$ part of the Bloch function. Working through an example will make this clear.

Consider a free electron wave function in one dimension, $\exp(ikx)$ for all k-values. To get all of the wave functions between $-\pi/a$ and π/a as in Fig. 10-5b, add and subtract integral multiples of $K = 2\pi/a$. To obtain the lowest, or first, band nothing need be added. Thus,

$$\psi_{k1}(x) = e^{ikx} \tag{10-20a}$$

To obtain the second band, $2\pi/a$ must be added to and subtracted from the various pieces. To obtain the third band $4\pi/a$ must be used. The results for the second and third band $\psi_{k2}(x)$ and $\psi_{k3}(x)$ are in Eqs. 10-20b and 10-20c, respectively

$$\left. \begin{array}{ll} -\pi/a \leq k \leq 0 & \psi_{k2}(x) = e^{ix(k+2\pi/a)} = e^{ikx} e^{i2\pi x/a} \\ 0 \leq k \leq \pi/a & \psi_{k2}(x) = e^{ix(k-2\pi/a)} = e^{ikx} e^{-i2\pi x/a} \end{array} \right\} \tag{10-20b}$$

$$\left. \begin{array}{ll} -\pi/a \leq k \leq 0 & \psi_{k3}(x) = e^{ix(k+4\pi/a)} = e^{ikx} e^{i4\pi x/a} \\ 0 \leq k \leq \pi/a & \psi_{k3}(x) = e^{ix(k-4\pi/a)} = e^{ikx} e^{-i4\pi x/a} \end{array} \right\} \tag{10-20c}$$

These results are similar to those in Eq. 10-18, except the expressions are written explicitly for several integral values of $2\pi/a$. Had we started with complete Bloch functions, that is, Eqs. 10-11 or 10-17, which also contain the $u_k(x)$ part, the results in Eq. 10-20 merely would be multiplied by $u_k(x)$, so these equations actually are general. Notice that all the results in Eq. 10-20 are of the Bloch form $\psi_k(x) = \exp(ikx) u_k(x)$ where $u_k(x)$ has the periodicity of the lattice, that is, in the u part of any of the functions in Eq. 10-20 replacement of x by $x + ma$ (where m is any integer) does not change the function since $\exp(im2\pi) = 1$. The transformation by $K = m2\pi/a$ merely changes $u_k(r)$ from one function with the periodicity of the lattice to another one also with the lattice periodicity. Also notice that for any k-value there are an infinite number of wave functions corresponding to the infinite number of bands that appear in Fig. 10-5b. At any one k and at any one x-value the $u_k(x)$ functions are in general different for the different bands. Thus, we add a **band index** n to the Bloch function to describe to which band the function belongs. In three dimensions

$$\psi_{kn}(r) = e^{ik \cdot r} u_{kn}(r) \tag{10-21a}$$

and $\quad \psi_{kn}(r + t_m) = e^{ik \cdot t_m} \psi_{kn}(r) \tag{10-21b}$

Eq. 10-21b follows from Eq. 10-21a by trivial substitution and the fact that $u(r + t_m) = u(r)$, since this function has the periodicity of the

CHAPTER 10   BAND THEORY

lattice for any lattice translation vector. The converse also follows, namely, if Eq. 10-21b is true then $\psi(\mathbf{r})$ can be written in the form of Eq. 10-21a. *Thus, a wave function that has either form is an appropriate Bloch function.* The band index is similar to the principle quantum number in the hydrogen atom problem. Both of the indices take on an infinite number of values from 1 to ∞.

10-4d Wave equation for Bloch functions As a last point about Bloch functions we determine the equation for $u_{kn}(x)$. To do this, substitute the Bloch function, Eq. 10-21, into the wave equation, Eq. 10-12. Again the one-dimensional example is used for simplicity

$$\frac{-\hbar^2}{2m}\left[\frac{d^2}{dx^2} + 2ik\frac{d}{dx} + k^2\right]u_{kn}(x) + V(x)\,u_{kn}(x) = \mathscr{E}_n(k)\,u_{kn}(x)$$

(10-22)

In order for the Bloch function to be an eigenfunction, u_{kn} must satisfy this differential equation over a primitive cell and the solutions must join smoothly to the same functions in the next cell. This same result for the three-dimensional case can be easily determined. Using the fact that the momentum $\mathbf{p} = -i\hbar\nabla$, the solution is

$$\left[\frac{-\hbar^2}{2m}\nabla^2 + \frac{\hbar}{m}\mathbf{k}\cdot\mathbf{p} + V(\mathbf{r})\right]u_{kn}(\mathbf{r}) = \left[\mathscr{E}_n(k) - \frac{\hbar^2}{2m}k^2\right]u_k(\mathbf{r})$$

(10-22)

The $\mathbf{k}\cdot\mathbf{p}$ term plays a role similar to that of the orbital angular momentum term $-2\ell(\ell + 1)/r^2$ for the hydrogen atom. For small $\mathbf{k}$, the $\mathbf{k}\cdot\mathbf{p}$ term can be treated as a perturbation. This term leads to the concept of the "group of k" when point symmetry operations as well as translational symmetry are considered. Applying the two transformations $u_{kn} \to u_{kn}^*$ and $k \to -k$ in Eq. 10-22, it can be seen that

$$u_{-kn} = u_{kn}^*, \qquad \psi_{-kn} = \psi_{kn}^*, \qquad \mathscr{E}_n(-k) = \mathscr{E}_n(k) \quad (10\text{-}23)$$

These symmetry properties are of importance in actual calculations when time reversal is considered. The result in Eq. 10-23 for the energy is another reason to choose the fundamental interval for k to be symmetric about the origin as shown in Fig. 10-5b and Eq. 10-19.

10-5 Nearly Free Electrons

We will now determine analytically the energy gaps, starting from the free electron case and considering a weak crystalline periodic potential. Since free electrons are the starting point, the extended zone, Fig. 10-5a, is being used. However, we can always shift the various pieces of the $\mathscr{E}$ vs. k curve back so that they fit into the reduced zone scheme. As before, for simplicity the calculations are performed in one dimension.

The free electron unperturbed energy and wave functions are

$$\mathscr{E} = \frac{\hbar^2}{2m} k^2 + V_0, \qquad \psi_k(x) = L^{-1/2} e^{ikx} \qquad (10\text{-}24)$$

where the constant potential V_0 is independent of x and is set equal to zero. L is the length of the crystal, which has N primitive unit cells of length a, that is, L = Na, and k goes from $-\infty$ to $+\infty$.

Consider a small periodic potential added to the free electron Hamiltonian, as in Eq. 10-12, and treat this extra potential as a perturbing term $H' = V(x)$. From standard perturbation theory,

$$\mathscr{E} = \mathscr{E}_k + H'_{kk} + \sum_{k'}' \frac{|H'_{kk'}|^2}{\mathscr{E}_k - \mathscr{E}_{k'}}$$

$$\mathscr{E}_k = \hbar^2 k^2 / 2m \qquad H'_{ij} \equiv \int_0^L \psi_i^* \, H' \, \psi_j \, dx \qquad (10\text{-}25)$$

The periodic part of the potential, H', can be Fourier analyzed

$$V(x) = \sum_{p=-\infty}^{\infty}{}' V_p \, \exp(2\pi i p x / a) \qquad (10\text{-}26)$$

where the prime means that the p = 0 term is excluded. This is a general expression for V(x) and it is clear that Eq. 10-26 has the periodicity of the lattice, $V(x) = V(x + a)$, as required. The Fourier coefficients V_p are in general complex and have the property that $V_{-p} = V_p^*$, which is required since V(x) is real. Evaluating the matrix elements of Eq. 10-25 we have

$$H'_{kk'} = L^{-1} \int_0^L V_p \, \exp\left[i(k' + \frac{2\pi p}{a} - k)x\right] dx \qquad (10\text{-}27)$$

All the matrix elements are zero unless $k' = k - 2\pi p/a$, in which case the matrix element clearly is V_p. These results are summarized here.

$$\left. \begin{array}{l} H'_{kk'} = V_p \quad \text{if } k' = k - 2\pi p/a \\ H'_{kk'} = 0 \\ H'_{kk} = 0 \end{array} \right\} \text{ for other k values} \qquad (10\text{-}27)$$

These results show that the first-order correction term in Eq. 10-25 is zero and the second-order term is proportional to the absolute square of the Fourier coefficients. However, this latter statement applies only if the energy denominator is large compared to $|V_p|$. Whenever k =

CHAPTER 10 BAND THEORY   259

$\pi p/a$ and $k' = -\pi p/a$ then $\mathscr{E}_k = \mathscr{E}_{k'}$. It is for these values of k and k' that $k' = k - 2\pi p/a$, so ordinary perturbation theory breaks down and degenerate perturbation theory is required.

In degenerate perturbation theory two different linear combinations of the two unperturbed functions are sought such that the matrix elements of H' between these new wave functions will be zero. Thus, the wave functions are of the form

$$\psi = c_0 e^{ikx} + c_p e^{ik'x} \tag{10-28}$$

where there are two distinct ratios for c_0/c_p. (This problem was treated in Section 10-1c where for a real potential $c_0/c_p = \pm 1$). For degenerate perturbation theory the secular equations are

$$c_0 (H_{kk} - \mathscr{E}) + c_p H_{kk'} = 0$$
$$c_0 H_{k'k} + c_p (H_{k'k'} - \mathscr{E}) = 0 \tag{10-29}$$

where H_{ij} are matrix elements of the total Hamiltonian. $H_{kk} \equiv \mathscr{E}_0 = \hbar^2 k^2/2m$ and $H_{k'k'} \equiv \mathscr{E}_p = \hbar^2(k')^2/2m$. The off-diagonal matrix elements are related by Eq. 10-27, giving $H_{kk'} = H'_{kk'} = (H'_{k'k})^* = V_p$. The secular equations have solutions when the secular determinant equals zero.

$$(\mathscr{E}_0 - \mathscr{E})(\mathscr{E}_p - \mathscr{E}) = |V_p|^2$$

or $\quad \mathscr{E} = \left(\dfrac{\mathscr{E}_0 + \mathscr{E}_p}{2}\right) \pm \dfrac{1}{2}[(\mathscr{E}_0 - \mathscr{E}_p)^2 + 4|V_p|^2]^{1/2} \tag{10-30}$

Thus, the degeneracy $\mathscr{E}_0 = \mathscr{E}_p$ at $k = p\pi/a$ is removed and the unperturbed levels split, giving two energies of

$$\mathscr{E} = \mathscr{E}_0 \pm |V_p| \tag{10-31}$$

These splittings are shown in Fig. 10-5a. The splitting at $k = \pi p/a$ is given by two times the appropriate Fourier coefficient of that k-value, as defined in Eq. 10-26. As mentioned previously the various pieces of the $\mathscr{E}$ vs. k plot can be brought back into the reduced zone scheme by translations by amounts $k = 2\pi p/a$, Fig. 10-5b.

For the simple case where the V_p values are real we have already discussed how the two wave functions are made up of waves traveling in opposite directions to give two standing waves shifted in phase from one another by half a lattice spacing. Further, in Section 10-1d, it was shown that the two waves have different energies as calculated here.

10-6 Brillouin Zones

In this section we discuss the three-dimensional geometry of the reduced zone in k-space. This polyhedron in k-space is called the first Brillouin zone. Then in the next section examples of these zones are given. (Examples are also given in Part C of this chapter.) As will be seen, the shape of the Brillouin zone can be complicated and can affect the electron paths in unusual ways.

Since the construction of Brillouin zones is intimately related to the reciprocal lattice, it will be reviewed first. Also see Section 4-3 and the problems in Chapter 4.

10-6a Review of the reciprocal lattice The fundamentals of the reciprocal lattice were discussed in Section 4-3b in connection with x-ray scattering. These points are summarized here since they are required for the discussion of Brillouin zones.

In ordinary, or direct, space we defined the **lattice** or **direct lattice** as the end points of the vector $\mathbf{t}_m = m_1 \mathbf{a} + m_2 \mathbf{b} + m_3 \mathbf{c}$ where m_1, m_2, and m_3 separately take on all integer values. In k-space there is a set of vectors, labeled $\mathbf{K}_{m'}$, that satisfies

$$e^{i\mathbf{K}_{m'} \cdot \mathbf{t}_m} = 1 \tag{10-32}$$

for all $\mathbf{t}_m$ where m'_i is any integer. The points determined by the end points of this set of $\mathbf{K}_{m'}$ vectors are called the **reciprocal lattice**. Thus, each direct space Bravais lattice defines one reciprocal lattice.

In k-space, the reciprocal lattice vectors $\mathbf{a}^*$, $\mathbf{b}^*$, and $\mathbf{c}^*$ may be generated by three primitive lattice vectors as follows, with $V_p \equiv \mathbf{a} \cdot (\mathbf{b} \times \mathbf{c})$ the volume of the primitive cell in direct space,

$$\mathbf{a}^* \equiv \frac{2\pi \, \mathbf{b} \times \mathbf{c}}{V_p}, \qquad \mathbf{b}^* \equiv \frac{2\pi \, \mathbf{c} \times \mathbf{a}}{V_p}, \qquad \mathbf{c}^* \equiv \frac{2\pi \, \mathbf{a} \times \mathbf{b}}{V_p} \tag{10-33a}$$

Clearly $\mathbf{a}^*$ is perpendicular to $\mathbf{b}$ and $\mathbf{c}$, also $\mathbf{b}^*$ to $\mathbf{a}$ and $\mathbf{c}$, and $\mathbf{c}^*$ to $\mathbf{a}$ and $\mathbf{b}$; note that the dimensions of reciprocal lattice vectors are the inverse to those of the direct lattice. Any reciprocal lattice vector and any vector in k-space can be written in terms of these vectors:

$$\mathbf{K}_{m'} = m'_1 \mathbf{a}^* + m'_2 \mathbf{b}^* + m'_3 \mathbf{c}^* \tag{10-33b}$$

$$\mathbf{k} = k_1 \mathbf{a}^* + k_2 \mathbf{b}^* + k_3 \mathbf{c}^* \tag{10-33c}$$

where m'_i is any integer while k_i is any number (including a fraction of an integer). Thus, $\exp(i \mathbf{k} \cdot \mathbf{t}_m)$ has some complex value but when $\mathbf{k}$ is a reciprocal lattice vector this expression equals unity (Eq. 10-32).

In Section 4-3c we showed that Bragg scattering occurs when the component of the incident wave vector $\mathbf{k}$ along the reciprocal lattice vector $\mathbf{K}_m$ is half the length of $\mathbf{K}_m$, that is,

CHAPTER 10   BAND THEORY   261

$$\mathbf{k} \cdot \mathbf{K}_m = (1/2) \, K_m^2 \qquad (10\text{-}34)$$

(Both **m** and **m'** can be any integer values so either symbol may be used as convenience dictates.) Thus, Bragg reflections occur whenever a k-vector lies on the perpendicular bisector plane of any reciprocal lattice vector. Figure 4-5 shows this construction.

10-6b Reduced zone   In Section 10-4b we discussed the reduced zone scheme. In the one-dimensional example, all of the eigenstates can be written with k-values in the interval $-\pi/a \le k \le \pi/a$. If any k-value is outside, it can be brought into this interval by replacing the k with $k + p2\pi/a$ where p is any integer. This is just a reciprocal lattice vector translation in one dimension since $2\pi/a$ is the basis vector of the reciprocal lattice. We would like to determine the corresponding algorithm in three dimensions.

First, note that $\exp(i\,\mathbf{K}_m \cdot \mathbf{r})$ has the periodicity of the lattice,

$$e^{i\mathbf{K}_m \cdot (\mathbf{r}+\mathbf{t}_{m'})} = e^{i\mathbf{K}_m \cdot \mathbf{r}} \, e^{i\mathbf{K}_m \cdot \mathbf{t}_{m'}} = e^{i\mathbf{K}_m \cdot \mathbf{r}} \qquad (10\text{-}35)$$

Second, if a Bloch function, $\psi_{k'}(\mathbf{r}) = \exp(i\mathbf{k}' \cdot \mathbf{r}) u_{k'}(\mathbf{r})$, is encountered that is outside the proper reduced zone then it can be moved into the proper zone by a suitable reciprocal lattice vector translation. Consider $\mathbf{k} = \mathbf{k}' + \mathbf{K}$ in a Bloch function

$$\psi_{k'}(\mathbf{r}) = e^{i\mathbf{k}' \cdot \mathbf{r}} u_{k'}(\mathbf{r}) = e^{i\mathbf{k} \cdot \mathbf{r}} \left[e^{-i\mathbf{K} \cdot \mathbf{r}} u_{k'}(\mathbf{r}) \right]$$
$$\equiv e^{i\mathbf{k} \cdot \mathbf{r}} u_k(\mathbf{r}) = \psi_k(\mathbf{r}) \qquad (10\text{-}36)$$

If $u_{k'}(\mathbf{r})$ has the periodicity of the lattice then $[\exp(-i\mathbf{K}\cdot\mathbf{r})]u_{k'}(\mathbf{r})$ also has this periodicity. Thus, by the addition of a reciprocal lattice vector to any k-value in a Bloch function, another Bloch function is obtained that has a k-value in the appropriate region of k-space.

10-6c Brillouin zones   We would like to obtain the conditions for the k-vector in three dimensions so that the reduced zone can be found and its geometry studied.

First review the one-dimensional example of a crystal of length, $L = Na$. The periodic boundary conditions require that

$$\psi(x) = \psi(x + Na) \qquad (10\text{-}37)$$

This is the same as Eq. 10-14a. Similarly the Bloch function requires that (Eq. 10-13a or Eq. 10-21)

$$\psi(x + Na) = e^{ikNa} \psi(x) \tag{10-38a}$$

therefore
$$e^{ikNa} = 1 \tag{10-38b}$$

with
$$k = \frac{m}{N}\frac{2\pi}{a} \tag{10-38c}$$

Along with these conditions on k notice that the allowed k-values are equally spaced (as in Eq. 9-27, which is the same as Eq. 10-38c), and the reduced zone in one dimension has $N/2 \leq m \leq N/2$, which is the same as $-\pi/a \leq k \leq \pi/a$. All the information needed for any problem can be determined for k-values in this reduced zone. Further, there are just N unique values of k given by Eq. 10-38c, which is identical to the reduced zone idea. The fact that there are $N + 1$ values of k in the reduced zone is no problem. The outer boundaries of this zone have k differing by $\mathbf{K} = 2\pi/a$, and we recall that k-values that differ by a reciprocal lattice vector give the same information.

The extension of these ideas to three dimensions is straightforward. Again assume that the bulk properties of the crystal do not depend on the choice of boundary conditions. The Born-von Karman periodic boundary conditions are used and it is assumed that the crystal can be taken to have the shape of the primitive cell of the Bravais lattice. These boundary conditions give

$$\psi(\mathbf{r} + N_1\mathbf{a}) = \psi(\mathbf{r}) \qquad \psi(\mathbf{r} + N_3\mathbf{c}) = \psi(\mathbf{r})$$
$$\psi(\mathbf{r} + N_2\mathbf{b}) = \psi(\mathbf{r}) \tag{10-39}$$

Again the application of Bloch's theorem, Eq. 10-21b, gives

$$\psi_{\mathbf{k}n}(\mathbf{r} + N_1\mathbf{a}) = e^{i\mathbf{k}\cdot N_1\mathbf{a}} \psi_{\mathbf{k}n} \text{ etc.} \tag{10-40a}$$

therefore
$$e^{i\mathbf{k}\cdot N_1\mathbf{a}} = 1 = e^{i\mathbf{k}\cdot N_2\mathbf{b}} = e^{i\mathbf{k}\cdot N_3\mathbf{c}} \tag{10-40b}$$

Writing **k** in terms of reciprocal lattice vectors (Eq. 10-33b) in order to satisfy Eq. 10-40b, **k** must be of the form

$$\mathbf{k} = \frac{m_1}{N_1}\mathbf{a^*} + \frac{m_2}{N_2}\mathbf{b^*} + \frac{m_3}{N_3}\mathbf{c^*} \tag{10-41}$$

for any integer m_i. Within a primitive cell in reciprocal space, notice that there is a very high, uniform density of allowed k-values. In fact, the volume ΔV_k associated with each k-value is just that of the parallelepiped with edges $\mathbf{a^*}/N_1$, $\mathbf{b^*}/N_2$, and $\mathbf{c^*}/N_3$ is

$$\Delta V_k = \mathbf{a^*}\cdot(\mathbf{b^*}\times\mathbf{c^*})/N \tag{10-42a}$$

where $N = N_1N_2N_3$, the total number of primitive cells in the crystal. The volume of the unit cell in reciprocal space, $\mathbf{a^*}\cdot(\mathbf{b^*}\times\mathbf{c^*})$, can be evaluated from Eq. 10-33a and is $(2\pi)^3/(V_p)$ where V_p is the volume of the primitive cell in direct space. Thus, ΔV_k is

$$\Delta V_k = (2\pi)^3/V \tag{10-42b}$$

This is the same result as for a free electron gas (Eq. 9-27b).

Within any primitive cell in reciprocal space there are N allowed values of k. Further, within the first primitive cell near the origin, all the information about the Bloch functions can be expressed using the reduced zone scheme. Last, any value of **k** that is outside this cell can be brought into it by a translation of some reciprocal lattice vector. This is all fine and true except there is a more natural cell in reciprocal space in which the reduced zone scheme is always displayed. This cell is one that is symmetric about **k** = 0. This cell is the Wigner-Seitz cell of the reciprocal lattice and it is called the **first Brillouin zone**.

Recall how the Wigner-Seitz cell was formed in the direct lattice (Section 2-4b). Start at any lattice point (the origin) and draw vectors to all neighboring lattice points constructing planes perpendicular to and passing through the midpoints of these vectors. The Wigner-Seitz cell is the cell with the smallest volume about the origin bounded by these planes. This cell contains just one lattice point and has the same volume as the original primitive cell. When this is done for the reciprocal lattice we obtain a cell that is called the first Brillouin zone. Clearly it is centered at k = 0, displays the symmetry of k-space (which is the same as direct space), and contains N values of k just as does the original reciprocal lattice primitive cell. Thus, in this cell the reduced zone scheme can be displayed. Moreover, an important point has been glossed over. The construction to determine the planes that bound the first Brillouin zone is the same as the construction of the planes from which we obtain Bragg reflections. That is, the k-values on these planes satisfy Eq. 10-34.

In summary – The first Brillouin zone is the cell in reciprocal space in which the reduced zone scheme (the energy bands) is displayed. This cell has the correct number of k-values, displays the proper symmetry, and is centered at **k** = 0. Further, opposite planes (sides) of this cell differ in k-value by a reciprocal lattice vector, so discontinuities in the energy versus k behavior will be observed across these planes as calculated in Section 10-5. The examples in the next section will help you to understand these results better.

10-7 Examples of Brillouin Zones

In this section, examples of first and higher Brillouin zones are shown. Two-dimensional structures are discussed first because they are easily represented diagramatically. Then the Brillouin zones for the simple three-dimensional structures are shown. In Section 4-3 and the problems in Chapter 4, the general derivation of a reciprocal lattice starting from a direct space lattice was discussed. Last, we discuss the filling of the states with electrons and the Fermi surface.

10-7a Two-dimensional examples Consider the reciprocal lattice in Fig. 10-6a. The lines are the perpendicular bisectors of the reciprocal lattice vectors $\pm \mathbf{a}^*$, $\pm \mathbf{b}^*$, $\pm(\mathbf{a}^* - \mathbf{b}^*)$. These lines determine the first Brillouin zone. The area of this cell is the same as that of the primitive cell in the upper right portion of the figure whose area is $\mathbf{a}^* \cdot \mathbf{b}^*$. Notice that the first Brillouin zone has the same twofold symmetry as the reciprocal lattice. Each of the values of $\mathbf{k}$ in the first Brillouin zone corresponds to an allowed state; two electrons per $\mathbf{k}$ are allowed corresponding to opposite spins. We will discuss the filling of states with electrons but now continue with the geometry of the zones.

Figure 10-6b shows how various "bits and pieces" of the primitive cell in the upper right can be translated by various reciprocal lattice vectors to the first Brillouin zone. This is the scheme for obtaining all the k's in the symmetric cell around $\mathbf{k} = 0$, which is called the first Brillouin zone. The translations are performed by $\mathbf{k} = \mathbf{k}' + \mathbf{K}$ as discussed in Section 10-6b. The translation of the parts labeled $1'$, $2'$, and $3'$ requires $\mathbf{K} = -\mathbf{b}^*$, $-\mathbf{a}^* - \mathbf{b}^*$, and $-\mathbf{a}^*$, respectively.

In Fig. 10-6c many other solid lines are shown. Each line is the perpendicular bisector of a reciprocal lattice vector. The first Brillouin zone is shown in white; all the N allowed k-values are in this zone. To obtain the second Brillouin zone find the area enclosed between the first zone and these various perpendicular bisectors. Each of these pieces of the second zone can be translated into the first zone by some $\mathbf{K}_m$ (Eq. 10-33b). The effects of these translations are indicated. The area (number of allowed k-values) is the same for the two zones. Notice that there are no perpendicular bisectors between the pieces of the second zone, since just as for the one-dimensional case, there are energy discontinuities across these boundaries (Bragg scattering occurs and an energy gap is obtained as in Eq. 10-31). The third Brillouin zone is also shown.

As a second example consider a two-dimensional rectangular lattice with $a = 2b$; the reciprocal lattice, Fig. 10-7a, has $2\mathbf{a}^* = \mathbf{b}^*$. The first Brillouin zone is trivial to obtain; the second, third, and fourth are easily determined and are shown.

The last two-dimensional example discussed is that of the square lattice, which has a square reciprocal lattice, Fig. 10-7b. The first through fourth complete zones are shown as well as pieces of the fifth and sixth. In the book by Brillouin the complete zones up to the tenth are shown as well as other examples of Brillouin zones of two- and three-dimensional lattices.

CHAPTER 10   BAND THEORY   265

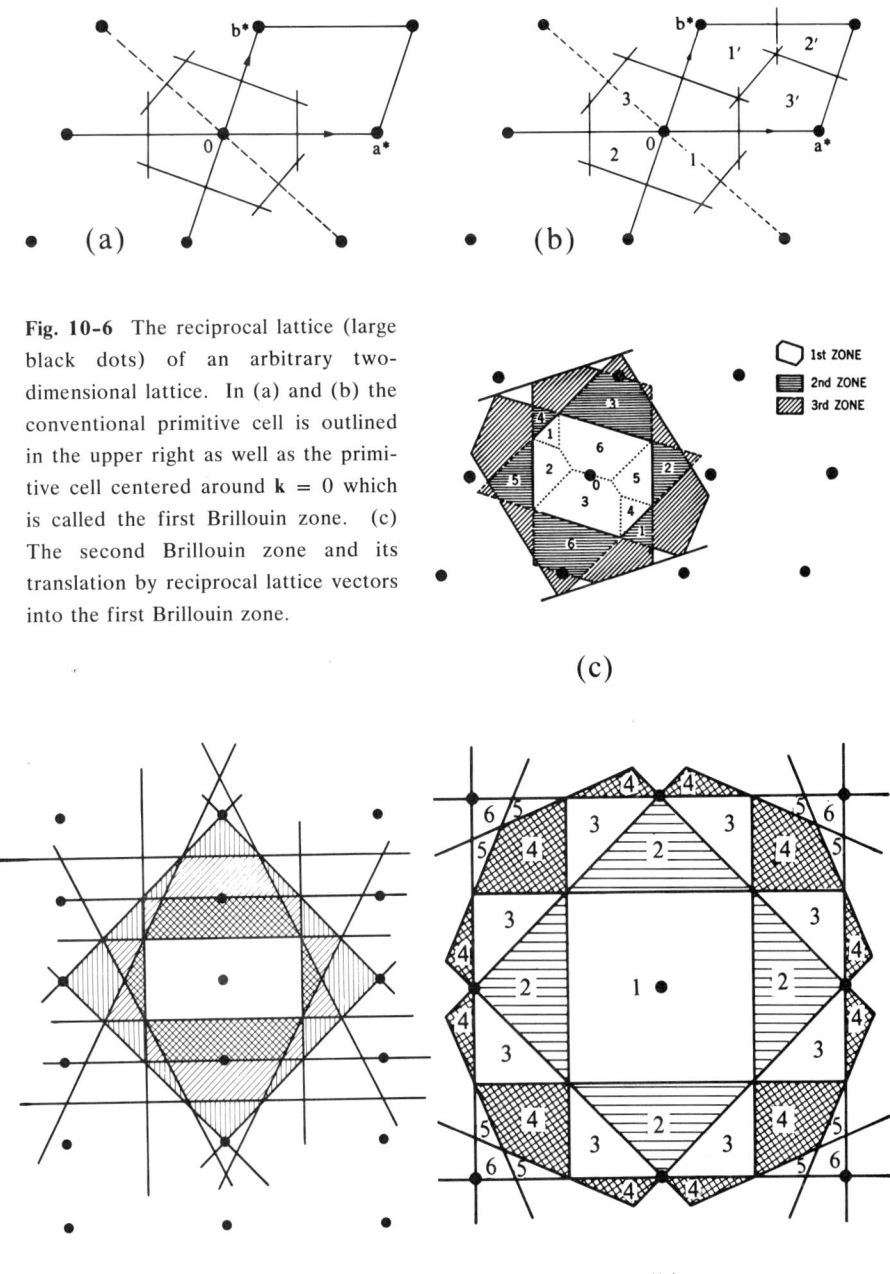

Fig. 10-6 The reciprocal lattice (large black dots) of an arbitrary two-dimensional lattice. In (a) and (b) the conventional primitive cell is outlined in the upper right as well as the primitive cell centered around **k** = 0 which is called the first Brillouin zone. (c) The second Brillouin zone and its translation by reciprocal lattice vectors into the first Brillouin zone.

Fig. 10-7 (a) The Brillouin zones of a two-dimensional rectangular lattice. (b) The complete first through fourth Brillouin zones of a two-dimensional square lattice. Parts of the fifth and sixth zones are also shown.

10-7b Construction of Fermi surfaces We now add electrons to the solid using the nearly free electron approximation.

Remember that a Brillouin zone is a geometric construction determined by the structure of the translation lattice. (It is the locus of points in k-space that describes Bragg scattering.) The **Fermi surface** is defined as the surface in k-space that bounds the occupied electronic states (at $T = 0°K$). For the free electron case the Fermi surface is a sphere centered at $\mathbf{k} = 0$ (Chapter 9 or see Fig. 10-8a). The almost free electron case is shown in Fig. 10-8b. These results are similar to the one-dimensional case shown in Figs. 10-1 and 10-5. Figure 10-8c shows the results for a reciprocal lattice of a square lattice as more electrons are added; these electrons successively occupy states of increasing energy. The curve labeled 1 represents a low concentration of electrons where the radius of the Fermi circle is much smaller than the distance to any Brillouin zone surface. For curve 2 there is a bigger perturbation of some of the electron states that are nearest the Brillouin zone boundary. Curve 3 represents electron states for which there is a large perturbation for some of the electron states yet all the filled electron states remain within the first Brillouin zone. More electrons are added to obtain curve 4, which shows that some of the electron states have high enough energy to overcome the energy separation between the first and second Brillouin zone. Remember, the lines are constant energy curves so the electrons in the $\mathbf{a}^*$ direction in the second zone have the same energy as those in the $\mathbf{a}^* + \mathbf{b}^*$ direction.

Figure 10-9 is the first Brillouin zone of a primitive orthorhombic lattice. Rather clearly, the three examples show what can happen as there are more electrons per primitive unit cell. First, in the left, the Fermi sphere is undistorted because the k-values are far from the zone surfaces. Second, k-values are larger so there is an energy discontinuity in the $\mathbf{c}^*$-direction. Last, the electron concentration is large enough so that energy discontinuities exist in two directions but not in the third direction. From this figure anisotropic electrical conductivity can be visualized.

As a last example reconsider the reciprocal lattice of a two-dimensional square direct space lattice (Fig. 10-7b) and move the various "bits and pieces" of the extended zone into the reduced zone. Figure 10-10a shows the reciprocal lattice in question with electrons filled to a k-value just larger than $|\mathbf{a}^*/2|$. The electrons fill parts of the first and second zones. Figure 10-10b shows the electrons almost filling the first zone. The few filled electron states in the second zone are shown in Fig. 10-10c. Remember, there are always energy gaps between the electrons in the different zones as in Fig. 10-5 and Fig. 10-8 so that all of the sharp corners of the Fermi surfaces become rounded. We might think in terms of the Figs. 10-10b and 10-10c as

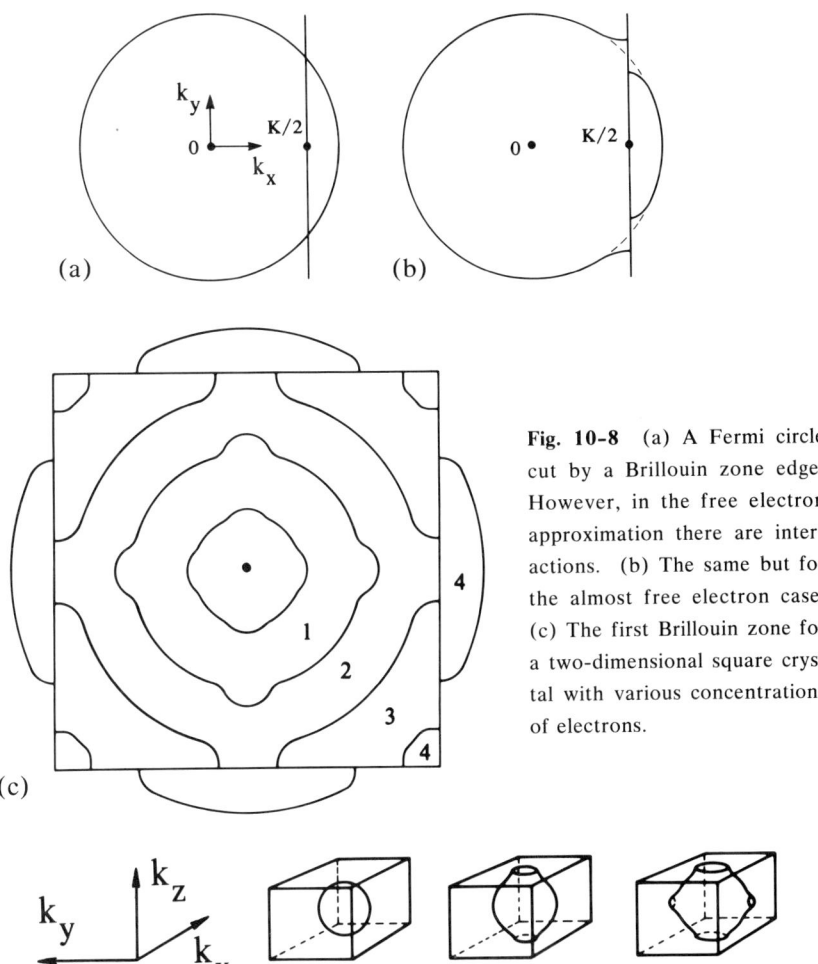

Fig. 10-8 (a) A Fermi circle cut by a Brillouin zone edge. However, in the free electron approximation there are interactions. (b) The same but for the almost free electron case. (c) The first Brillouin zone for a two-dimensional square crystal with various concentrations of electrons.

Fig. 10-9 The Brillouin zone for a simple orthorhombic crystal structure showing constant energy contours of successively higher energies.

being placed on top of one another with a separation in between them as in Fig. 10-5b. However, while in one dimension the electrons in higher zones have larger energies than those in lower zones, it is not necessarily so in two and three dimensions. (Problem 5 is instructive in this matter.)

Consider the changes for the Fermi surfaces of free electrons (Fig. 10-10) when they are drawn for the nearly free electron case. The nearly free electron Fermi surfaces can be drawn freehand using the following facts:

268 CHAPTER 10 BAND THEORY

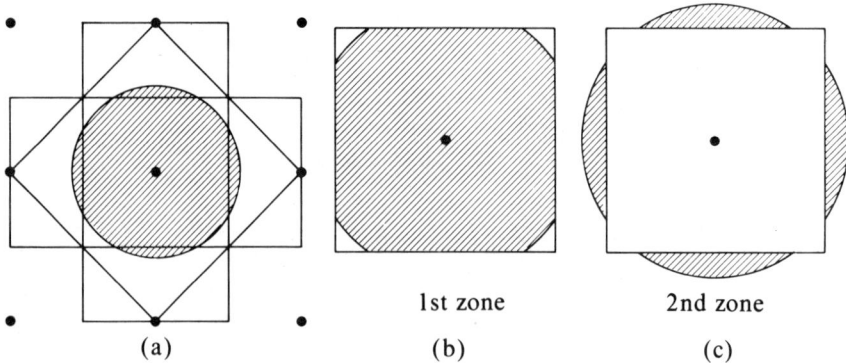

Fig. 10-10 (a) Same as Fig. 10-8 but showing a Fermi circle of occupied electrons for the free electron case. In (b) and (c) the reduced zone scheme for the free electron Fermi surface of the first and second zone taken from part a.

1. Energy gaps at the zone boundaries will occur.
2. The Fermi surface will almost always intersect the zone boundaries perpendicularly. (Discussed in Section 10-7d.)
3. Sharp corners of the Fermi surface will be smoothed by the crystalline potential.
4. The total number of states enclosed by the Fermi surface depends only on the electron concentration and not on the details of the interaction.

Figure 10-11 shows the application of these rules to the first and second zones of the square reciprocal lattice problem discussed previously. In particular the effects of rules 2 and 3 can be seen. Figures 10-8 and 10-9 also show the effect of these rules.

10-7c Three-dimensional examples A **simple cubic lattice** has a simple cubic reciprocal lattice. So $\mathbf{K}_m$ is of the form $(2\pi/a)$ (m_1, m_2, m_3). The volume of the direct lattice unit cell is a^3 so that of the reciprocal lattice unit cell is $8\pi^3/a^3$. The first Brillouin zone is a cube with the six faces given by Eq. 10-33c with $k_1 = \pm\pi/a = k_2 = k_3$. These planes are the perpendicular bisectors of $\mathbf{K} = (2\pi/a)$ $(1, 0, 0)$. To find the second Brillouin zone we realize that the next shortest $\mathbf{K}$ vectors are those of the form $(2\pi/a)$ $(1, 1, 0)$. Figure 10-12 shows the first and second Brillouin zones. The six pyramids of the second zone can be translated by a reciprocal lattice vector to fill the same volume as the first zone.

The **body-centered cubic (bcc)** lattice, Fig. 10-13a, has primitive lattice vectors and reciprocal lattice vectors (Eq. 10-33) given by

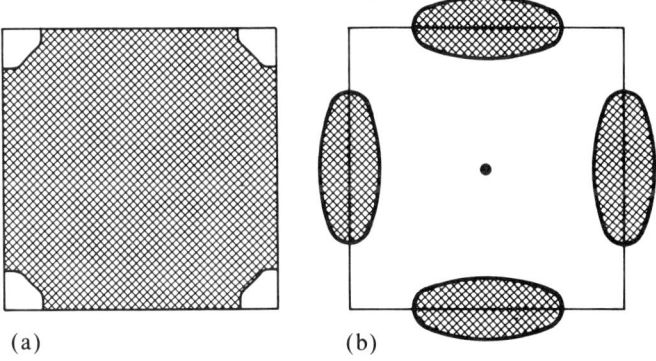

Fig. 10-11 The first and second Brillouin zones from the previous figure, showing the rounding effect due to a weak periodic potential. The second Brillouin zone is shown in the periodic zone scheme while the first is in the reduced zone scheme.

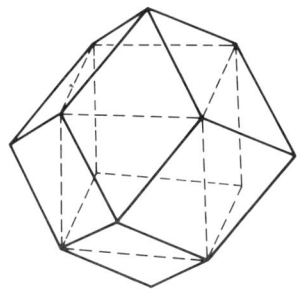

Fig. 10-12 The first and second Brillouin zones for a simple cubic direct space lattice. (It is obvious how the bits and pieces of the second Brillouin zone can be shifted into the first zone by different values of **K**.)

$$\begin{aligned}
\mathbf{a} &= (a/2)(1, 1, -1) & \mathbf{a}^* &= (2\pi/a)(1, 1, 0,) \\
\mathbf{b} &= (a/2)(-1, 1, 1) & \mathbf{b}^* &= (2\pi/a)(0, 1, 1) \\
\mathbf{c} &= (a/2)(1, -1, 1) & \mathbf{c}^* &= (2\pi/a)(1, 0, 1)
\end{aligned} \quad (10\text{-}43)$$

Thus, the reciprocal lattice is face-centered cubic. From consideration of the Wigner-Seitz cell in direct space for the fcc lattice, we immediately know that Fig. 10-13b shows the first Brillouin zone for a direct space bcc lattice. This dodecahedron is bounded by the twelve planes at $(\pi/a) \{(\pm 1, \pm 1, 0), (\pm 1, 0, \pm 1), (0, \pm 1, \pm 1)\}$ and has a volume $16\pi^3/a^3$. (The direct space primitive cell volume $a^3/2$.)

A sphere of radius k_i can be inscribed in this first Brillouin zone. The distance from the origin to the (1, 1, 0) planes is $k_i = (\pi/a) \sqrt{2}$. If there is one electron per atom per lattice point, and a spherical energy surface is assumed, then the radius of the sphere, k_e, is

$$4\pi k_e^3/3 = 8\pi^3/a^3$$

Since $k_i/k_e = 1.14$ it is reasonable to assume an approximately spherical Fermi surface for one valence electron per atom in a bcc structure. (The inscribed sphere can accommodate up to 1.48 electrons.) Many metals including the alkalis have this structure and one valence electron per atom.

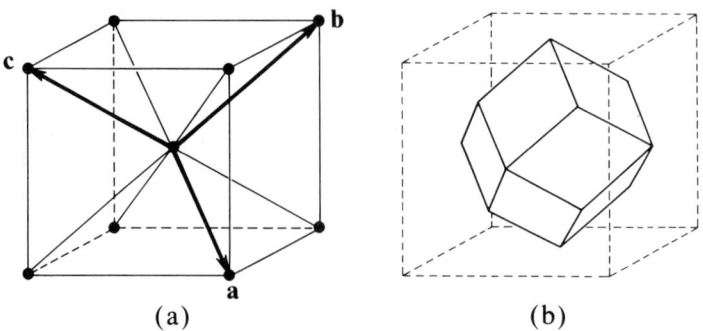

Fig. 10-13 (a) The direct lattice, showing the primitive lattice vectors, and (b) first Brillouin zone of a body centered cubic lattice. (The primitive cell also is in Fig. 2-8b.)

The **face-centered cubic (fcc) lattice**, Fig. 10-14a, has primitive lattice vectors and reciprocal lattice vectors given by

$$\mathbf{a} = (a/2)(1, 0, 1) \qquad \mathbf{a}^* = (2\pi/a)(1, -1, 1)$$
$$\mathbf{b} = (a/2)(1, 1, 0) \qquad \mathbf{b}^* = (2\pi/a)(1, 1, -1) \qquad (10\text{-}44)$$
$$\mathbf{c} = (a/2)(0, 1, 1) \qquad \mathbf{c}^* = (2\pi/a)(-1, 1, 1)$$

Thus, the reciprocal lattice is body centered and the first Brillouin zone is as shown in Fig. 10-14b. The volume of the direct lattice primitive cell is $a^3/4$ while that of the first Brillouin zone is $32\pi^3/a^3$.

A sphere inscribed in the first Brillouin zone will touch the boundary first on one of the hexagonal faces, so $k_i = \pi\sqrt{3}/a$. If we assume spherical energy surfaces, then the Fermi radius for one electron per atom, k_e, is

$$4\pi k_e^3/3 = 16\pi^3/a^3$$

Thus, $k_i/k_e = 1.11$ is still larger than one, but smaller than the result for the bcc metals. (This inscribed sphere could accommodate 1.36 electrons.) Thus, larger perturbations due to the Brillouin zone boundaries might be anticipated in fcc monovalent metals with a fcc structure than in monovalent metals with a bcc structure.

In general, the first Brillouin zones in three dimensions can be handled readily and for the simple structured monovalent metals the effects of filling the allowed k states can be understood. The problem of high zones is also straightforward, but just harder to visualize. When there are enough electrons to have k-values at the zone boundaries, detailed calculations must be performed to determine the energy gaps. There are many sophisticated numerical methods of performing such calculations. The approaches go under the names the cellular method; the APW method (augmented plane wave); KKR method; OPW method (orthogonalized plane wave); and the pseudopotentional method. See the Notes for references.

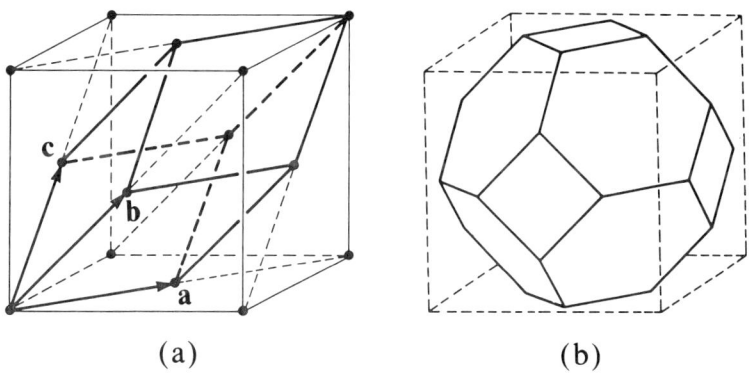

Fig. 10-14 (a) The direct lattice and (b) first Brillouin zone of a face-centered cubic lattice. (The primitive cell is the same as in Fig. 2-8a.) Notice that the direct space fcc lattice has a first Brillouin zone and reciprocal lattice that is a bcc lattice (and a direct space bcc lattice has a first Brillouin zone and reciprocal lattice that is fcc, Fig. 10-13).

10-7d $\partial \mathscr{E}/\partial \mathbf{k}$ **at zone boundaries** Very briefly, for a crystal with a mirror plane perpendicular to $\mathbf{k}$, we show that $\partial \mathscr{E}/\partial \mathbf{k} = 0$ at the zone boundary. This was the second point in Section 10-7b.

Since $\mathbf{k}$ is assumed to be perpendicular to a mirror plane, $\mathscr{E}(\mathbf{k})$ must be symmetric along this direction. Also $\mathscr{E}(\mathbf{k})$ is a periodic function of the reciprocal lattice vectors. These two conditions lead to

$$\mathscr{E}(\mathbf{k}) = \mathscr{E}(-\mathbf{k}), \qquad (\partial \mathscr{E}/\partial \mathbf{k})_{\mathbf{k}} = -(\partial \mathscr{E}/\partial \mathbf{k})_{-\mathbf{k}} \qquad (10\text{-}45a)$$

$$\mathscr{E}(\mathbf{k}) = \mathscr{E}(\mathbf{k} + \mathbf{K}), \qquad (\partial \mathscr{E}/\partial \mathbf{k})_{\mathbf{k}} = (\partial \mathscr{E}/\partial \mathbf{k})_{\mathbf{k}+\mathbf{K}} \qquad (10\text{-}45b)$$

However, at the zone boundary $\mathbf{k} = \pm \mathbf{K}/2$, and Eqs. 10-45a and 10-45b lead to different results

$$(\partial \mathscr{E}/\partial \mathbf{k})_{\mathbf{K}/2} = -(\partial \mathscr{E}/\partial \mathbf{k})_{-\mathbf{K}/2} \qquad (10\text{-}45c)$$

$$(\partial \mathscr{E}/\partial \mathbf{k})_{\mathbf{K}/2} = +(\partial \mathscr{E}/\partial \mathbf{k})_{-\mathbf{K}/2} \qquad (10\text{-}45d)$$

The only way that both of these equations can be true is if $(\partial \mathscr{E}/\partial \mathbf{k}) = 0$ at the zone boundary.

10-7e A real Fermi surface We now indicate how we deal with the experimental determination of a Fermi surface of a real metal, taking fcc copper as an example. Powerful methods have been developed for their determination, and this science, and art, is sometimes referred to as **fermiology**. We give a flavor of the field without going into detail.

Figure 10-15a shows the first Brillouin zone of the fcc lattice. It is the same as shown in Fig. 10-14b but now has some labels for orientation purposes. In Fig. 10-15b a (110) section of the Brillouin zone is shown; the use of the letters will help you become oriented.

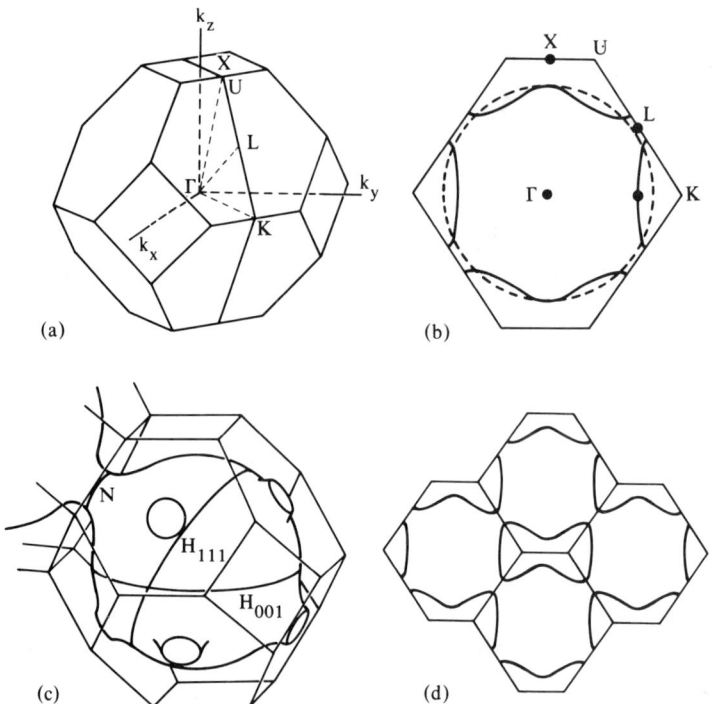

Fig. 10-15 Various aspects of the Fermi surface of Cu. (a) The Brillouin zone of an fcc lattice with some special points labeled. (b) A (110) section of the Brillouin zone. See the text for the meaning of the internal curves. (c) The proposed Fermi surface of Cu. (d) The extended zone picture of a (110) section of the Fermi surface showing the dog bone orbits.

The dashed circle represents a sphere whose volume is one-half that of the Brillouin zone. It is close to the L-point, but farther from the X-point. Cu metal has one free electron per atom, so the first Brillouin zone should be half filled.

However, the half-filled Fermi sphere is strongly perturbed near the L-point due to its proximity to the face of the Brillouin zone. (This result is similar to what is seen in Figs. 10-8, 10-9 and 10-11.) For Cu, Fig. 10-15c shows the expected results. "Necks" appear near the L-point in the same way as shown in Figs. 10-8 and 10-9.

Many experimental techniques have been applied to determine Fermi surfaces. One of the simpler and more general is the **de Haas-van Alphen (dHvA) technique,** in which we measure the oscillations of the (dia-) magnetic moment of the metal sample as a function of magnetic field, H. These oscillations occur as the Landau levels pass through the Fermi surface causing slight changes in the total energy (and related properties) of the conduction electrons. In Section 9-16 we showed that these changes in the total energy will have a constant period in H^{-1}, given by $\Delta(H^{-1}) = e\hbar/mc\mathscr{E}_F$ (Eq. 9-68). Using $\mathscr{E}_F = (\hbar^2/2m)k_F^2$, we obtain

$$\Delta(H^{-1}) = \frac{2\pi e}{c\hbar A_k} \tag{10-46}$$

where A_k ($= \pi k_F^2$) is the area in k-space of a cut in the Fermi surface perpendicular to the applied magnetic field. Thus, by measuring the periods of oscillation for **H** applied in different directions, A_k can be determined. This enables one to piece together what the Fermi surface looks like.

For example, if a magnetic field is applied in the [001] direction, the Landau levels cut the Fermi surface in the circle called H_{001} in Fig. 10-15c. From the oscillation frequency, the area of this orbit in k-space can be determined. (For subtle phase relationship reasons, orbits that are parallel but have smaller areas tend to cancel. The dominant response comes from orbits of the type drawn and labeled H_{001}; such orbits are called **extremal orbits**.) By applying the magnetic field in the [111] direction, two extremal orbits are obtained and are labeled N and H_{111} in Fig. 10-15c. In the dHvA measurements oscillations corresponding to the areas of these two extremal orbits are found, and we must separate the two. With such results we can appreciate how the areas of the H_{111} and H_{001} orbits can be separated and then compared to the neck orbit. With this sort of information the Fermi surface can be pieced together.

A check on the proposed Fermi surface for Cu is shown in Fig. 10-15d. The section of the Fermi surface from Fig. 10-15b is shown using the repeated zone scheme. For the magnetic field in the [1$\bar{1}$0] direction, the "dog bone" extremal orbits shown can be measured.

Thus, the measurements show that the Fermi surface of fcc copper is as indicated in Fig. 10-15. Many other metals have been studied with the dHvA and related techniques and their Fermi surfaces have been determined. See the Notes.

10-8 Wigner-Seitz Approximation – the Binding Energy

The Wigner-Seitz method of calculating the cohesive energy of metals was the first serious attempt (1933) to determine this energy. The approach is easy to understand and the calculation is performed in direct space.

Consider a monovalent body-centered cubic metal, such as sodium (Na). The Wigner-Seitz cell for the bcc lattice is the polyhedron that looks like the first Brillouin zone of a face-centered cubic direct lattice. Thus, the Wigner-Seitz cell looks like the polyhedron in Fig. 10-15b. In the center of this polyhedron is the positively charged Na$^+$

ion core. We would like to calculate the wave function for the 3s valence electron. Near the center of the polyhedron the potential seen by the valence electron is spherically symmetric, and is small near the boundaries. Wigner and Seitz made the following approximations: (1) within the polyhedron the valence electron sees a potential energy, V(r), of a Na$^+$ free ion; (2) the effects on V(r) due to other ions and free electrons in neighboring polyhedrons are ignored (because each polyhedron is on the average electrically neutral and the free electron tends to screen out the effect of the positive ion cores); (3) it is still difficult to impose the translation symmetry for the actual shape of the polyhedrons, so the polyhedron is replaced by a sphere of radius r_0 such that $4\pi r_0^3/3$ equals the volume of the polyhedron (which is the volume of a primitive unit cell); the boundary condition on the lowest energy wave function is $(\partial \psi/\partial r)_{r_0} = 0$. (This last assumption replaces the usual atomic boundary condition $\psi \to 0$ as $r \to \infty$.)

In Bloch language, these assumptions mean we solve the wave equation for the usual Bloch function $\psi_k = \exp(i\mathbf{k}\cdot\mathbf{r})\, u_k(r)$, but takes $\mathbf{k} = 0$. Thus, there is no modulation of the wave functions from cell to cell. The wave equation to be solved is the radial form of Eq. 10-22 with $\mathbf{k} = 0$,

$$\frac{1}{r}\frac{d^2}{dr^2}(r\psi) + \frac{2m}{\hbar^2}[\mathscr{E} - V(r)] = 0 \qquad (10\text{-}47)$$

where V(r) is the potential energy of a singly charged ion and the boundary condition at r_0 is as stated previously.

Figure 10-16 shows the resultant wave function. First, notice that the wave function is flat for r/a_0 between two and four. Thus, the $\mathbf{k} = 0$ wave function is flat over about 90% of the atomic volume, that is, it behaves like a free electron. Again this is a good indication that for Na, and for the other alkali metals as well, the free electron approximation is fairly reasonable. Second, the binding energy of this solution is $\mathscr{E}_0 = -8.2$ eV.

The binding energy of the 3s electron in atomic sodium is -5.15 eV. However, the difference, $-8.2 - (-5.15)$, may not be taken as the cohesive energy of metallic sodium because we must account for electron states other than $\mathbf{k} = 0$. To do this the Bloch function of the form $\psi_k = \exp(i\mathbf{k}\cdot\mathbf{r})\, u_0(r)$ is used where $u_0(r)$ is the wave function calculated previously. Thus, $\mathscr{E}_0$ is the energy of the bottom of the conduction band. The total energy of the metal also contains a contribution due to the interaction of one polyhedron with another, but since the polyhedron are neutral this term is ignored. Last, the total energy contains the kinetic energy of the electrons. For sodium the Fermi energy, $\mathscr{E}_F$, is 3.24 eV (Table 9-4). The average kinetic energy is $(3/5)\,\mathscr{E}_F$ (Eq. 9-28), or 1.9 eV. Thus, the average cohesive energy

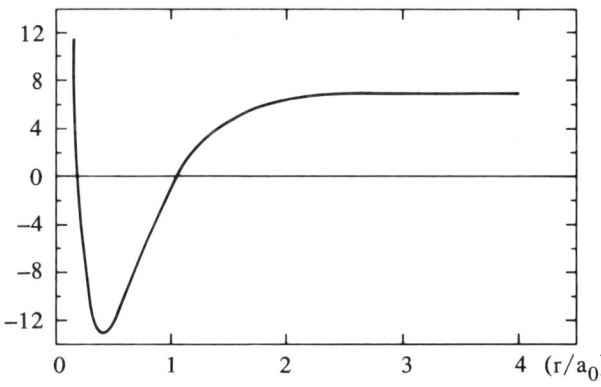

Fig. 10-16 The 3s electron wave function for metallic sodium calculated by the Wigner-Seitz method. The radial distance is given in atomic units ($a_0 = 0.529 \times 10^{-8}$ cm).

of sodium metal is -6.3 eV, which is 1.1 eV lower than a free Na atom, and the stability of Na metal compared to atomic Na is explained.

This calculated result is fortuitously close to the experimental result of 1.13 eV. We say fortuitous because there are several significant terms that have been neglected. These include the repulsive Coulomb interaction between electrons in different polyhedra; the Pauli exclusion principle, which requires that electrons with parallel spins will avoid each other; the van der Waals interaction between neutral polyhedra. Yet compressibility calculations for sodium using the Wigner-Seitz approach also give a surprisingly good results, namely 12.0 and 12.3×10^{-12} cm^2/dyne are obtained for the calculated and measured values, respectively. The cohesive energy results for some of the other alkali metals are given here, in units of eV/atom.

	Calc.	Exp.
Li	1.6	1.69
Na	1.1	1.13
K	0.72	1.00

To <u>summarize</u>, the binding energy of a metal is the sum of a kinetic energy (repulsive force) and a potential energy (attractive force). First, there is an increase in the energy (decrease in the binding energy) associated with the electron gas. In our approximation, this term increases as the volume decreases as the two-thirds power of the density, that is, the kinetic energy $\propto \mathscr{E}_F \propto (N/V)^{2/3}$, Eq. 9-29a. Second, a decrease in the potential energy arises when the atoms approach each other and the electrons move into neighboring atoms. In the calculation here, as the radius of the Wigner-Seitz sphere decreases, $\mathscr{E}_0$ decreases. The sum of these two energies has a minimum near the interatomic spacing in Na metal and gives a reasonable estimate of the binding energy. (See Section 10-18 for some comments on other contributions to the binding energies of metals.)

10-9 The Tight Binding Approximation

10-9a Introduction A great deal has been said about the free electron and nearly free electron approximations. These ideas are basic to our understanding of electrons in metals. The eigenfunctions of these are Bloch functions whose form is dictated by the translation symmetry. Then we might ask how the tightly bound electrons should be treated. The answer is: they should be treated in a manner quite similar to the nearly free electrons. This leads to the tight binding approximation, where to a good approximation single electron states in a crystal keep their atomic-like behavior, implying that the overlap of an electron state of one atom with that of its neighbors is relatively small. The reason is that these valence electrons are in relatively deep potential wells and are thus less "free electron-like" than the valence electrons of sodium. Some aspects of this approximation have already been discussed in Section 10-3. The tight binding approximation is useful for the 3d electrons of the transition metal atoms, for the inner electrons of metals, and for electrons in insulators. Thus, the wave functions of nearly free electrons in metals and the wave functions of electrons in insulators are both Bloch functions.

Take $\phi(\mathbf{r})$ as the eigenstate of an isolated atom with eigenvalue $\mathscr{E}_0$ and assume that it is normalized and nondegenerate, for example, an s-state. (Degenerate wave functions cause complications that only cloud the important issues.) Our basic assumption is that the overlap of this atomic state $\phi(\mathbf{r})$ with its neighbors is small (the **tight binding approximation**) and that the extra potential energy seen by the electrons in the crystal is small compared to the atomic potential energy. The Hamiltonian for the electron is $H = H_{atom} + H_{cry}$ where $\phi(\mathbf{r})$ is an eigenfunction of H_{atom} with eigenvalue $\mathscr{E}_0$. We treat the effect of the crystal, H_{cry}, as a perturbation.

Assume that the atoms lie on a lattice of points given by $\mathbf{t}_m$. When an electron is near the atom at $\mathbf{t}_m = 0$ its eigenfunction is given approximately by $\phi(\mathbf{r})$. Similarly when the electron is near the atom at the lattice point $\mathbf{t}_m$ its wave function is approximately $\phi(\mathbf{r} - \mathbf{t}_m)$. Thus, the wave function for one electron in the whole crystal is

$$\psi_{\mathbf{k}}(\mathbf{r}) = \sum_m C_{\mathbf{k}m} \phi(\mathbf{r} - \mathbf{t}_m) \tag{10-48a}$$

where the sum is over all the lattice points. This wave function is a **linear combination of atomic orbitals (LCAO)**. Since this wave function is to be in the Bloch form, take $C_{\mathbf{k}m} = N^{-1/2} \exp(i\mathbf{k} \cdot \mathbf{t}_m)$ where N is the number of atoms in the crystal.

$$\psi_{\mathbf{k}}(\mathbf{r}) = N^{-1/2} \sum_m e^{i\mathbf{k} \cdot \mathbf{t}_{m'}} \phi(\mathbf{r} - \mathbf{t}_m) \tag{10-48b}$$

CHAPTER 10 BAND THEORY   277

First, we must show that this function satisfies the Bloch condition, and second that the N in Eq. 10-48b is the correct normalizing factor.

To see if this wave function satisfies the Bloch condition (Eq. 10-21b) in place of **r**, substitute **r** + **T** where **T** is one of the primitive lattice vectors, $\mathbf{t}_m$.

$$\psi_k(\mathbf{r} + \mathbf{T}) = N^{-1/2} e^{i\mathbf{k}\cdot\mathbf{T}} \sum_m e^{i\mathbf{k}\cdot\mathbf{t}_m - \mathbf{T}} \phi[\mathbf{r} - (\mathbf{t}_m - \mathbf{T})] \quad (10\text{-}49a)$$

$$= e^{i\mathbf{k}\cdot\mathbf{T}} N^{-1/2} \sum_{m'} e^{i\mathbf{k}\cdot\mathbf{t}_{m'}} \phi(\mathbf{r} - \mathbf{t}_{m'}) \quad (10\text{-}49b)$$

$$= e^{i\mathbf{k}\cdot\mathbf{T}} \psi_k(\mathbf{r}) \quad (10\text{-}49c)$$

Since the sum in Eq. 10-49a is over all the primitive lattice vectors, the sum in Eq. 10-49b is equivalent. Thus, the last expression follows, showing that Eq. 10-48b is indeed a Bloch function (Eq. 10-21b).

Now consider the normalization. Evaluate

$$\langle \psi_k(\mathbf{r}) | \psi_k(\mathbf{r}) \rangle$$

$$= N^{-1} \sum_m \sum_j e^{i\mathbf{k}\cdot(\mathbf{t}_m - \mathbf{t}_j)} \int \phi^*(\mathbf{r} - \mathbf{t}_j) \phi(\mathbf{r} - \mathbf{t}_m) d^3r \quad (10\text{-}50a)$$

$$= N^{-1} \sum_m \int \phi^*(\mathbf{r} - \mathbf{t}_m) \phi(\mathbf{r} - \mathbf{t}_m) d^3r \quad (10\text{-}50b)$$

$$= N^{-1} N = 1 \quad (10\text{-}50c)$$

Equation 10-50b is obtained from Eq. 10-50a since $\phi(\mathbf{r} - \mathbf{t}_j)$ only has an appreciable value very close to $\mathbf{t}_j$. Similarly $\phi(\mathbf{r} - \mathbf{t}_m)$ only has an appreciable value close to $\mathbf{t}_m$. Thus, to the extent that we can neglect the overlap of neighboring charge distributions, there is a contribution to the integral only when j = m in the summation. The integral in Eq. 10-50b is unity since normalized atomic orbitals are used. The sum over m in this expression gives N because we assume there is one electron at every lattice point. Thus, the LCAO wave function in Eq. 10-48b is correctly normalized.

It is instructive to sketch the tight binding wave function of Eq. 10-48b, and this is done in Fig. 10-17 for a one-dimensional case for several values of k. For k = 0 a 3s atomic wave function centered at each lattice site can be seen. The amplitudes of all the functions are identical. For k = π/a the other extreme is shown. The phase of each neighboring wave function is opposite because of the modulation term in this Bloch function, that is, exp $[i(\pi/a)(a)]$ = exp $[i\,\pi]$ = -1. Also shown is the result for an intermediate value of k. The results in Fig. 10-17 are, in principle, the same as the results for the nearly free electron case. In the latter instead of a 3s wave function to start with, a straight line of constant amplitude represents the free electron. The results look similar to those in Fig. 10-17 except for all the oscillations due to the 3s function.

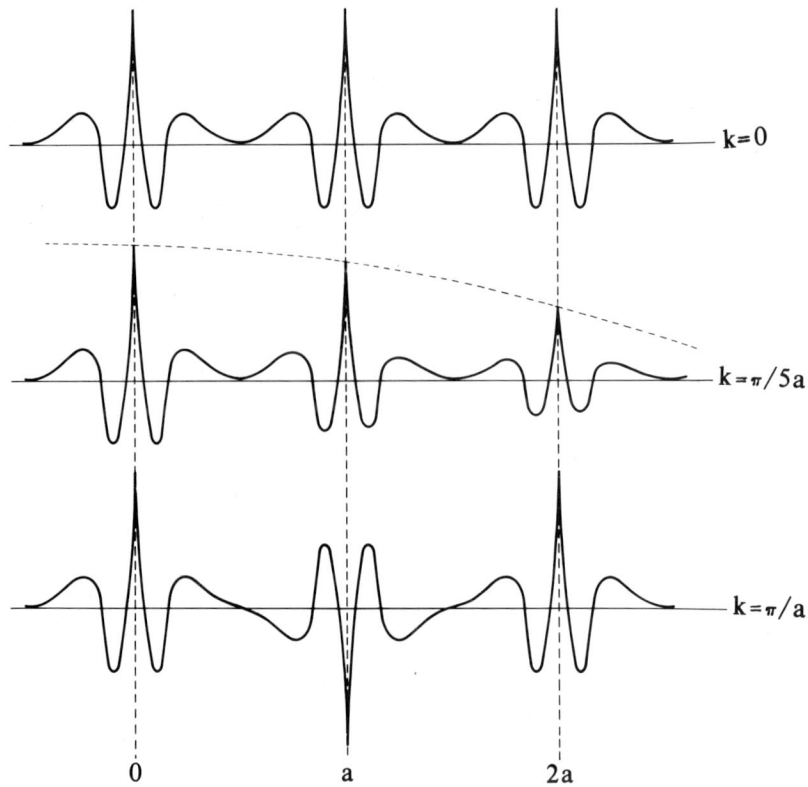

Fig. 10-17 A one-dimensional tight binding wave function for several k-values, including zero and the Brillouin zone boundary. A 3s atomic wave function is used.

10-9b Energy calculation We calculate the energy shifts for the LCAO function in Eq. 10-48b using first-order perturbation theory.

$$\langle \psi_k(\mathbf{r}) | H_{cry} | \psi_k(\mathbf{r}) \rangle$$
$$= N^{-1} \sum_m \sum_j e^{i\mathbf{k} \cdot (\mathbf{t}_m - \mathbf{t}_j)} \int \phi_j^* H_{cry} \phi_m \, d^3r \quad (10\text{-}51)$$

where $\phi_m \equiv \phi(\mathbf{r} - \mathbf{t}_m)$. Again the tight binding approximation leads us to keep only terms in the double sum when $j = m$ and terms where j and m refer to nearest neighbors. Thus,

$$\mathcal{E}_k = \mathcal{E}_0 - \alpha - \gamma \sum_m e^{i\mathbf{k} \cdot (\mathbf{t}_m - \mathbf{t}_j)} \quad (10\text{-}52a)$$

$$\alpha \equiv -\langle \phi_m | H_{cry} | \phi_m \rangle \quad (10\text{-}52b)$$

$$\gamma \equiv -\langle \phi_k | H_{cry} | \phi_m \rangle \quad (10\text{-}52c)$$

where the sum in Eq. 10-52a is over nearest neighbors, which is also implied for the integral in Eq. 10-52c. In going from Eq. 10-51 to

CHAPTER 10 BAND THEORY 279

10-52 the sum over the N lattice points just cancels the normalization. The α and γ terms involve detailed calculations of a crystalline Hamiltonian. These two terms are usually taken as parameters, and the sum in Eq. 10-52a is performed to yield the wave vector dependence of the energy.

10-9c Examples As an example of a tight binding calculation, consider a simple cubic lattice. The sum in Eq. 10-52a involves the six-nearest neighbor terms

$$t_m - t_j = (\pm a, 0, 0); (0, \pm a, 0); (0, 0, \pm a) \quad (10\text{-}53)$$

from which we immediately determine that for our tightly bound s-state electrons the $\mathscr{E}$ vs. k is

$$\mathscr{E}(k) = \mathscr{E}_0 - \alpha - 2\gamma(\cos k_x a + \cos k_y a + \cos k_z a) \quad (10\text{-}54a)$$

Focusing on the k-dependence, notice that the minimum energy, -6γ, occurs at $k = 0$ and the maximum, $+6\gamma$, at the Brillouin zone edge $\pm \pi/a$ for all three directions. The width of the band, 12γ, is dependent on the overlap of the neighboring wave functions, Eq. 10-52c, which varies rapidly with the internuclear distance (Section 6-4).

For small k-values Eq. 10-54a can be expanded in terms of ka

$$\mathscr{E}(k) = \mathscr{E}_0 - \alpha - 6\gamma + \gamma k^2 a^2 \quad (10\text{-}54b)$$

where $k^2 = k_x^2 + k_y^2 + k_z^2$, to yield a quadratic k-dependence independent of the k-direction. Thus, at the bottom of the band, at $k = 0$, we have spherical energy surfaces. The effective mass for these electrons can be evaluated via Eq. 10-10, from which $m^* = \hbar^2/2\gamma a^2$ is obtained. Notice that as the overlap decreases, m^* increases, so for isolated atoms m^* is infinite and an external force cannot accelerate an electron. This is expected hence in this limit an electron cannot move from one atom to another.

Consider the general solution of $\mathscr{E}(k)$ in Eq. 10-54a. Figure 10-3b shows the shape of the k-dependence (the zero is shifted). As found previously, the effective mass at $k \approx 0$ was found to be positive. From Eq. 10-54a, at $k \approx \pi/a$ we obtain $m^* = -\hbar^2/2\gamma a^2$, a negative value as expected. Thus, the result sketched in Fig. 10-3b is obtained.

In summary we see that even tightly bound atomic states form bands in crystals. The width of the band depends sensitively on the amount of overlap of the individual atomic states. The wave function for one electron, Eq. 10-48b, shows that this electron has equal probability of being found in any cell throughout the crystal. It is interesting to relate this to the group velocity, Eq. 10-5a. For completely isolated atoms $v_g = 0$ because $d\mathscr{E}/dk = 0$ (the energies of all the atoms is the same). However, once a band is formed, $d\mathscr{E}/dk \neq 0$, and although the group velocity is small and m* large the electron is able to move throughout the crystal. We may think in terms of a small overlap resulting in a small tunneling probability, which gives a small v_g, that is, from Eq. 10-5a, $v_g = (2\gamma a^2/\hbar)k$. The $\mathscr{E}$ vs. k results for the bcc and fcc structures are given in the problems.

Failure of the independent electron approximation – From the preceding discussion you might imagine that as the overlap of the valence electrons decreases the conductivity would decrease continuously to zero. However, it is unlikely that this is true. For some systems, as the nearest-neighbor separation increases, the conductivity drops to zero abruptly. (See the discussion of the Mott transition, Section 8-6c.) The problem is not that the tight bonding approximation fails; rather it is the independent electron approximation that fails. When the overlap is very small, this latter approximation is unable to treat properly the repulsion of a second electron at a given atomic site when one electron already is at that site.

Anderson localization is another effect that becomes important when the bands become narrow. There is a certain amount of disorder in all crystals (Section 11-8) that gives rise to a nonperiodic variation of the potential energy, ΔV. When the bandwidth is of the order of ΔV the electrons can become localized and thus confined to a small region in space (the wave function falls off exponentially with distance from the localized position). This can decrease abruptly the conductivity. Thus, we can have a continuous but localized density of states resulting in no conductivity. (In disordered systems this effect is important but it may be important in normal systems as well.)

10-10 Crystal Momentum

A short, elementary discussion of the meaning of the **k** in the Bloch function is in order. Consider the connection between **k** and the momentum. Take the momentum operator $\mathbf{p} = (-i\hbar)\nabla$ acting on a Bloch function ψ_{kn}

$$-i\hbar \nabla \psi_{kn} = -i\hbar \nabla [e^{i\mathbf{k}\cdot\mathbf{r}} u_{kn}(\mathbf{r})]$$
$$= \hbar \mathbf{k} \psi_{kn} + e^{i\mathbf{k}\cdot\mathbf{r}} (-i\hbar) \nabla u_{kn}(\mathbf{r}) \qquad (10\text{-}55)$$

If u_{kn} is constant, the second term is zero and ψ_{kn} is an eigenstate of the momentum operator with eigenvalue $\hbar\mathbf{k}$. Thus, for free electrons the momentum is $\mathbf{p} = \hbar\mathbf{k}$. However, if the electrons are not free, $\hbar\mathbf{k}$ is no longer the true momentum.

Nevertheless, **k** is the natural extension of $\mathbf{p}/\hbar$ and is called the **crystal momentum**. (This is covered more explicitly in Section 12-6.) For example, **k** enters into the conservation laws. Suppose an electron in state ψ_{kn} collides with a phonon (lattice vibration) of wave vector **q** and the collision scatters the electron to a new state state $\psi_{k'n}$ and the phonon is absorbed. Then we write

$$\mathbf{k} + \mathbf{q} = \mathbf{k}' + \mathbf{K}$$

where **K** can be any reciprocal lattice vector. This expression shows the conservation of crystal momentum.

Part C – Semiconductors, Real Bands, and Related Concepts

In the last part of this chapter properties of semiconductors, real bands and some related concepts are covered. First we discuss holes. Then aspects of the band gap, intrinsic carrier concentration, and conductivity are discussed for electrons and holes. Other aspects of semiconductors and real bands are discussed in the rest of the chapter.

As has already been seen, there are huge qualitative differences between the spherical Fermi surface in the free electron model and the situation in real metals. Basically this occurs because the Brillouin zones are not spherical and the k-values of some points on the boundaries are appreciably smaller than other points. Since $\mathscr{E} \propto k^2$ the energy differences can be large. Thus, the energies of electron states at the surface of a given Brillouin zone vary considerably. For semiconductors this is very important since we are dealing with filled or almost full bands. In Sections 10-12 and 10-13 we deal with the simplest band problem that has the complexities encountered in the bands of the ordinary semiconductor. We recommend for a first reading the beginning of Section 10-12, the summary in Section 10-13d and a look at Figs. 10-22 and 10-23a. A qualitative appreciation that the energies are rather different at the X- and M-points is enough to carry away from these sections. In Section 10-15 the bands and other aspects of the common semiconductors are discussed.

Properties of semiconductors also are discussed in Chapters 8, 13 and 18. Selected information on some typical semiconductors is in Table 10-1 (page 242). The meaning of most of the entries will become clear as the chapter progresses. Note from the high melting temperatures that these materials have strong bonding forces.

10-11 Holes

10-11a Introduction The concept of holes is very useful, particularly in dealing with semiconductors when there are a few unoccupied electron states at the top of the valence band. It is much easier to consider the few unoccupied states than the entire electron band minus

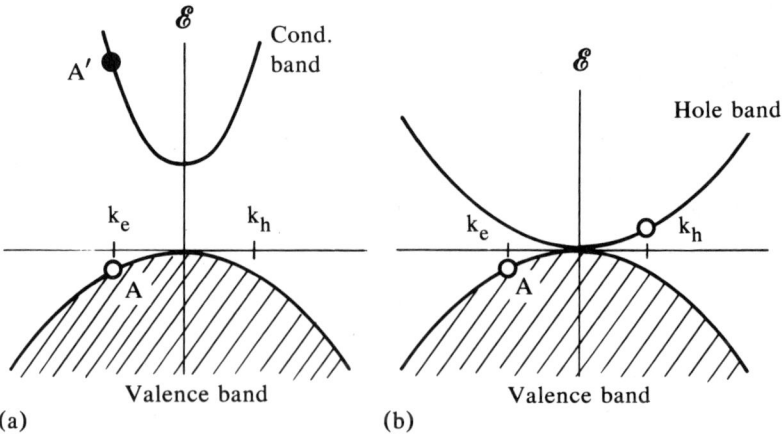

Fig. 10-18 (a) The conduction and valence band states of electrons in a semiconductor. $\mathscr{E}$ is the energy for the electrons. We show a valence band with one electron missing from position A. (The electron is shown in the conduction band at position A′.) (b) The same valence band with one electron missing from position A and also the corresponding hole band with the one hole shown.

these few states. This introduction focuses on the basic concepts, followed by a discussion of the effective masses, conductivity, and position of the Fermi level.

Suppose that a given band is completely full; then the total momentum of the band is zero, that is $\mathbf{k}_T \equiv \Sigma \mathbf{k}_i = 0$. Now focus your attention on one electron in the valence band (at position A in Fig. 10-18a) which in a moment will be ejected from this band but right now is still in the band and has a wave vector $\mathbf{k}_e$, energy $\mathscr{E}_e(\mathbf{k}_e)$, velocity $\mathbf{v}_e$, effective mass $m^* = m_e$, and a charge $-e$. Now, excite this particular electron out of the valence band (perhaps by light), and consider how to describe the valence band with the missing electron. Since the electron is no longer in this valence band, the total momentum of the valence band is $\mathbf{k}_T = \Sigma \mathbf{k}_i - (\mathbf{k}_e) = -\mathbf{k}_e$. This situation is shown in Fig. 10-18a where the electron initially located at position A (it happens now to be at position A′ but that is irrelevant for this discussion as it could just as well be outside the sample). If we want to describe the entire valence band (with the electron ejected) with one particle, which we shall call a hole, then

$$\mathbf{k}_h = -\mathbf{k}_e \qquad (10\text{-}56a)$$

Then the total momentum of the valence band is $\mathbf{k}_h$. Also, by definition this hole has energy $\mathscr{E}_h(\mathbf{k}_h)$, velocity $\mathbf{v}_h$, and effective mass m_h. We would like to relate these quantities to those of the electron before it was ejected from the valence band.

CHAPTER 10   BAND THEORY   283

Now focus on the energy of this valence band. The zero of energy is taken at the top of the band as indicated by the horizontal line in Fig. 10-18a. An electron at a position B (not shown in the figure), which is lower in energy than position A, requires more energy to be ejected from the valence band than the electron at position A. Thus, the valence band with an electron missing at B is in a more highly excited state than one with an electron missing at A, that is, the lower the energy of the electron being ejected the more highly excited is the valence band with the missing electron. When we describe the valence band with a missing electron in terms of a hole, the lower the energy of the state from which the electron is ejected, the higher the energy of the hole. Thus, in describing this situation by a hole we want to have $\mathscr{E}_h$ as far above zero as $\mathscr{E}_e$ is below zero, so

$$\mathscr{E}_h = -\mathscr{E}_e \qquad (10\text{-}56\text{b})$$

The upper part of Fig. 10-18b shows the situation that we have created; the hole band is the inverse of the valence band with the energies and wave vectors related by inversion through the origin (Eqs. 10-56a and 10-56b). Thus, the dynamics of the valence band with some missing electrons can be treated directly or can be treated totally by focusing on the hole band. Clearly, we cannot treat both when explicitly calculating results since they both represent the same phenomenon.

(Note that it is not necessary to define a hole band in the preceding manner, as shown in the upper part of Fig. 10-18b. Some people prefer the following. Take positive $\mathscr{E}_h$ to point down and the positive k-values for the holes going to the left. This inverts both coordinates for the holes compared to the coordinates for the electrons and automatically takes care of Eqs. 10-56a and 10-56b. Thus, the hole band will exactly overlap the electron band and the hole will appear in the same place in the figure as the missing electron. For equilibrium properties this picture has certain advantages.)

The other fundamental properties of the hole band will be determined but note that these *other properties of holes are obtained by algebra from Eq. 10-56a and 10-56b.*

The group velocity of a particle is given by $\hbar^{-1}\nabla_k\mathscr{E}(k)$, Eq. 10-6a. If v_e is the group velocity of the electron at position A before ejection and v_h is the group velocity of the hole (in the hole band) that describes the valence band with an electron missing at position A, then since the curvatures of $\mathscr{E}$ vs. k are the same

$$v_h = v_e \qquad (10\text{-}56\text{c})$$

This is obtained directly by using Eqs. 10-56a and 10-56b. (To calculate this take $\mathscr{E} = \pm\hbar^2k^2/2m$, using + for the hole band and − for the electron band.) Equation 10-56c also is clear from Fig. 10-18b; the slope of $\mathscr{E}$ vs. k is the same in the valence band at $\mathbf{k}_e$ as in the hole band at $\mathbf{k}_h$.

The effective mass of an electron in a band is proportional to the inverse of the second derivative of $\mathscr{E}$ vs. k, Eq. 10-10, so

$$m_h = -m_e \quad (10\text{-}56d)$$

This is clear from Fig. 10-18b since the hole band and electron band have upward and downward curvature, respectively.

The last concern is the charge assigned to the hole. For a band the total current is $\mathbf{j}_T \equiv q \Sigma \mathbf{v} = -e \Sigma \mathbf{v}$ where $-e$ is the charge of the electrons, $\mathbf{v}$ their velocity, and the sum is over all the electrons in the band. For a full band the sum is zero, just as $\mathbf{k}_T = 0$. However, if one electron is missing from the band at $\mathbf{k}_e$ with $\mathbf{v}_e$ then it does not enter the sum and does not compensate the electron with $-\mathbf{v}_e$, in which case $\mathbf{j}_T = (-e)(-\mathbf{v}_e) = e\mathbf{v}_e$. (This is the same situation that leads to Eq. 10-56a.) Now to describe this in terms of the hole band, $\mathbf{j}_T = q_h \mathbf{v}_h$ where q_h is the charge of the hole. Therefore $q_h \mathbf{v}_h = e\mathbf{v}_e$, but since $\mathbf{v}_h = \mathbf{v}_e$, Eq. 10-56c, $q_h = +e$ or

$$\text{the charge of the hole} = +e \quad (10\text{-}56e)$$

In further considerations we should use either the valence band with a few missing electrons or the hole band with the holes in it. Both cannot be used at the same time as they represent the same physical processes.

10-11b Effective masses In the next few subsections the preceding properties of holes will be applied. Figure 10-19a shows a schematic band structure for GaAs and related semiconductors. The details and degeneracies of these bands will be discussed in Section 10-15. The zero of energy is conventionally taken at the top of valence band. With respect to this zero the dispersion relations for electrons in these three bands, respectively, are

$$\mathscr{E} = \mathscr{E}_g + (\hbar^2/2m_e)k^2$$
$$\mathscr{E} = -(\hbar^2/2m_{v1})k^2, \qquad \mathscr{E} = -(\hbar^2/2m_{v2})k^2 \quad (10\text{-}57)$$

Parabolic bands are assumed, that is, $\mathscr{E} \propto k^2$. (We use the most popular **convention**; m_e and m_h indicate the effective mass of the electron and hole. In some books these are referred to, respectively, as m_n and m_p, which comes from negative and positive charges. Often in the semiconductor literature the symbols n and p are used to refer to the density of electrons and holes. These symbols are used, for example, in Eqs. 10-60.)

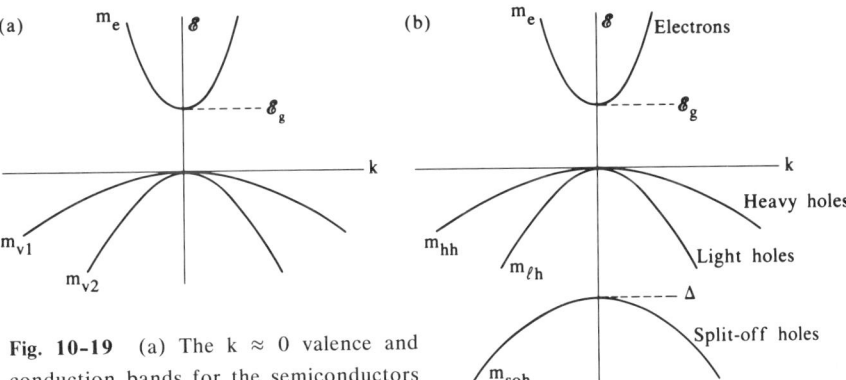

Fig. 10-19 (a) The $k \approx 0$ valence and conduction bands for the semiconductors shown in Fig. 10-27. (b) The same but with spin-orbit coupling allowed.

To find the effective masses of electrons in these bands one should use Eq. 10-10, $(m^*)^{-1} = (\hbar^{-2})(d^2\mathscr{E}/dk^2)$. Thus

$$m^* = m_e, \quad -m_{v1}, \quad -m_{v2} \qquad (10\text{-}58a)$$

respectively, for the various bands. The signs should be no surprise; electron bands with upward curvature have positive effective masses and those with downward curvature have negive effective masses.

What are the hole masses near the top of the valence band? Using Eq. 10-56d and the results in Eq. 10-58a we obtain the effective masses of the holes in what is called the heavy-hole (hh) and light-hole (ℓh) bands, respectively:

$$m_{hh} = m_{v1} \qquad \text{and} \quad m_{\ell h} = m_{v2} \qquad (10\text{-}58b)$$

Note that the effective masses of the holes are positive quantities (corresponding to the negative values for electrons in the valence band, in agreement with Eq. 10-56d, and consistent with the results in Fig. 10-3 just below the band gap). Also note that $m_{hh} > m_{\ell h}$ because the light-hole band has larger curvature than the heavy-hole band; hence the names! For GaAs, $m_{hh} = 0.45m$ and $m_{\ell h} = 0.082m$ where m is a free electron mass (and $m_e = 0.067m$). These results and Fig. 10-19b will be discussed in more detail in Section 10-15.

10-11c Carrier density for an intrinsic material We calculate the equilibrium number of electrons and holes in a pure or intrinsic semiconductor. An **intrinsic semiconductor** is defined as a semiconductor in which the electronic properties (such as dc conductivity) are not determined by impurities. An **extrinsic** or **doped** semiconductor is one in which impurities determine the electronic behavior.

In an intrinsic semiconductor, for every electron thermally excited into the conduction band a hole is left in the valence band. Thus, the number of conduction electrons is equal to the number of holes. **Conventionally** the density of electrons in the conduction band and holes in the valence band are given by n and p, respectively; m_e and m_h, respectively, are their effective masses. For an intrinsic semiconductor with a wide band gap, n = p.

To calculate the equilibrium number of electrons and holes, recall that the Fermi-Dirac function, Eq. 9-34, gives the probability of a state at energy $\mathscr{E}$ being occupied. For a wide band gap material such that $(\mathscr{E}_c - \mathscr{E}_F) \gg k_B T$ the probability of a state in the conduction band being occupied is small (f ≪ 1) and the semiconductor is called **nondegenerate**. A **degenerate** semiconductor is one where $\mathscr{E}_F$ is in the conduction or valence band; for this case f ≈ 1 for some of the states.

The full Fermi-Dirac function and the approximation in the nondegenerate case, respectively, are given by

$$f(\mathscr{E}) = \{ \exp[(\mathscr{E} - \mathscr{E}_F)/k_B T] + 1 \}^{-1} \quad (10\text{-}59a)$$

$$f(\mathscr{E}) \approx \exp[-(\mathscr{E} - \mathscr{E}_F)/k_B T] \quad (10\text{-}59b)$$

Using Eq. 10-59b and the quadratic energy dependence of the density of states bands, one may readily calculate (Section 10-16)

$$n = 2(m_e k_B T/2\pi\hbar^2)^{3/2} \exp[(\mathscr{E}_F - \mathscr{E}_g)/k_B T] \quad (10\text{-}60a)$$

$$p = 2(m_h k_B T/2\pi\hbar^2)^{3/2} \exp[-\mathscr{E}_F/k_B T] \quad (10\text{-}60b)$$

Most of the temperature dependence for n or p comes from the exponential part of the function; the dependence in the prefactor is relatively weak. For Si, where $\mathscr{E}_g$ = 1.12 eV and m_e = 0.26m, the value of n varies from 10^{10} to 10^{13} electrons/cm³ from approximately 295°K to 403°K.

Multiplying Eqs. 10-60a and 10-60b, we find that the np product is independent of the Fermi level,

$$np = 4(k_B T/2\pi\hbar^2)^3 (m_e m_h)^{3/2} \exp(-\mathscr{E}_g/k_B T) \quad (10\text{-}60c)$$

For Si, m_h = 0.49m, close to that of the electrons, so $(np)^{1/2}$ is close to the values quoted previously for n.

Last, since n = p for an intrinsic semiconductor we can equate Eqs. 10-60a and 10-60b and obtain a relation for the Fermi energy

$$\mathscr{E}_F = (\mathscr{E}_g/2) + (3k_B T/4) \ln(m_h/m_e) \quad (10\text{-}61)$$

At T = 0°K the Fermi energy is midway in the gap. If the effective masses of the holes and electrons are the same, then the Fermi energy is always midway in the gap. However, when these masses are not the same, $\mathscr{E}_F$ is displaced from this position for $T > 0°K$. In fact, at room temperature intrinsic InSb is almost a degenerate semiconductor because the electron and hole effective masses are so different (0.015m and 0.39m, respectively) and $\mathscr{E}_g$ is so small (0.24 eV).

In the technologically useful semiconductors the materials are doped with controlled kinds and numbers of impurities that determine the electrical properties. The simple hydrogen-like theory of shallow **donors** and **acceptors** discussed in Section 8-6b accounts semiquantitatively for the energy levels. Since the binding energies of some of these impurities are smaller than k_BT at room temperature (Table 8-4), the density and binding energies greatly influence the number of electrons and holes in the conduction and valence bands. The temperature dependence of $\mathscr{E}_F$ for the case where the semiconductor is doped is discussed in Section 10-16b.

10-11d Conductivity The minimum energy state in the conduction band is called the **conduction band edge**. Suppose that an electron is excited from the valence band across the band gap into a state above the conduction band edge. Then it will emit phonons and rapidly (on a picosecond scale) thermalize, that is, will "fall" to the conduction band edge. Similarly a hole in the valence band will rapidly thermalize toward the valence band maximum, called the **valence band edge**. If $T > 0°K$ then the electron or hole will not be exactly at the band edges but will be within $\approx k_BT$ of the band edges.

We next consider the current under the influence of an applied electric field E in a semiconductor that has electrons in the conduction band and holes in the valence band at thermal equilibrium. The easiest way to think about the problem is to consider the electrons and holes at the respective conduction and valence band edges. However, this is not necessary as the electron (and hole) can be anywhere in its band and the same results will apply; the drift velocity will obey the following equations, but if $\mathbf{k} \neq 0$ then there is another velocity, given by $\nabla_k \mathscr{E}(\mathbf{k})$, that is not a drift velocity. This is the velocity discussed in Section 10-11a; since it does not depend on an applied electric field, we ignore it and consider only thermalized carriers at $\mathbf{k} = 0$.

Consider the current density **J** in a semiconductor under the influence of an external electric field E with electrons and holes at the respective band edges. If there are ρ charges per unit volume, each with charge q and drift velocity **v** (which is due to an applied field), then the current density is $\mathbf{J} = \rho q \mathbf{v}$. For electrons there are n electrons per unit volume, charge $-e$, and a drift velocity $\mathbf{v}_{nc}$; then $\mathbf{J}_n = -ne\mathbf{v}_{nc}$. We use the subscript nc on the electron velocity as a reminder that (1) the electrons are in the conduction band, and (2) this drift velocity has nothing to do with the velocity given in Eq. 10-56c. Similarly there are p holes per unit volume, with charge $(+e)$, drift

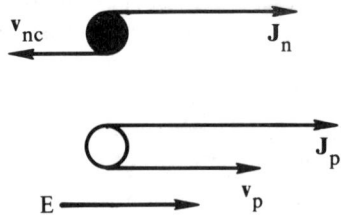

Fig. 10-20 For the application of an electric field, E, the direction of the drift velocities and currents for electrons at the conduction band edge and holes at the valence band edge are shown.

velocity v_p ($v_p \equiv v_h$ as both symbols are often used), and a current J_p. Under the influence of the external applied electric field, E, there is a force, qE, on the electrons and the holes that is equal and opposite since the charges are equal and opposite. Thus, the drift velocities v_{nc} and v_p are in *opposite directions*! The current density and related quantities are given by

$$J_n = -nev_{nc} = \sigma_n E = ne\mu_n E \qquad (10\text{-}62a)$$
$$J_p = pev_p = \sigma_p E = pe\mu_p E \qquad (10\text{-}62b)$$

Focus on the first equality in these equations. Figure 10-20 shows the directions. The drift velocities of these thermalized electrons and holes are opposite so the current densities are in the same directions (opposite velocity and opposite charge). (Note that in most of this book the number of particles per unit volume is expressed as N/V but for the thermal equilibrium densities of electrons and holes we use n or p, the standard notation in semiconductor physics.)

The second equality in these equations is a defining relation for **conductivity**, $J \equiv \sigma E$. The third equality comes from the definition of a **mobility** $v \equiv \mu E$. For electrons this relationship has a minus sign since the drift velocity and the applied electric field are in opposite directions. Notice that the values of σ and μ for the electrons and holes are all positive quantities. Some values of electron and hole mobilities are given in Table 10-1. From the definition of mobility the units are $(cm/sec)/(V/cm) = cm^2\ V^{-1}\ sec^{-1}$. The values depend strongly on temperature and will be discussed. To a lesser extent they can also depend on doping (see Sze, Chapter 1).

The total current is the sum of the contribution from the electrons and holes $J_t(=J_n + J_p)$. This current is related to the total conductivity $J_t = \sigma E$ which is

$$J_t = (ne\mu_n + pe\mu_p)E = (\sigma_n + \sigma_p)E = \sigma E \qquad (10\text{-}62c)$$

so σ is just the sum of the individual electron and hole conductivities.

In Eq. 9-6 we found that the conductivity and collision time, τ, are related by

$$\sigma = (N/V)e^2\tau/m$$

CHAPTER 10   BAND THEORY   289

Using $\mu_n = \sigma_n/ne$ from Eq. 10-62a, and the corresponding relationship for holes, the expressions for mobility can be written in terms of a relaxation time as

$$\mu_n = e\tau_n/m_e \qquad \mu_p = e\tau_p/m_h \qquad (10\text{-}63)$$

We have used the effective mass for the electrons and holes in their respective bands. Relaxation times depend in complicated ways on the details of the band structures. Thus, it is difficult to make generalizations except to say that if the valence and conduction bands are similar then the relaxation times will be similar. This is the reason for the similar mobility results in PbTe (Table 10-1). For the III-V materials the simpler conduction bands yield smaller effective masses and larger mobilities than the more complicated valence bands. At room temperature phonons (lattice vibrations) are usually the dominant scattering mechanism that determines the values of τ. This is the reason that the mobilities increase appreciably (Table 10-1) upon lowering the temperature from room temperature to 77°K.

Usually, semiconductors with small gaps have small effective masses (due to a large curvature of $\mathscr{E}$ vs. k caused by repulsion of the eigenstates) and hence high mobility from Eq. 10-63. This accounts for some of the variation found for μ_n in Table 10-1. The situation for holes is less clear because of the complex valence bands (to be discussed later but already indicated in Fig. 10-19).

10-12   Band Preliminaries

Some details of the subjects covered in Sections 10-12 and 10-13 are beyond the general scope of this text. However, the topics are sufficiently important for an understanding of the physics of solids that even a cursory reading will benefit the reader.

The information presented so far in this chapter is fundamental to an understanding of the subject. The use of Bloch functions is the backbone of solid state physics as it presently stands. The concept and application of Brillouin zones enable us to treat insulators and metals from the same fundamental point of view. However, to go further into Brillouin zones and to understand the electrical conduction processes in metals and semiconductors, the full use of symmetry and the mathematics of symmetry (group theory) must be used. A study of elementary group theory, although simple, will not be covered here. Instead, we will discuss briefly the reason that there are more symmetry operations in the problem than have been considered and what might result from these symmetry operations. In Section 10-13 the energy bands and wave functions for a simple two-dimensional problem are worked out. This latter example will familiarize you with the notation, which is complex but logical. A familiarity of the notation for bands and wave functions in Brillouin zones enables us to understand much new information.

Essentially everything that has been discussed so far in this chapter applies to lattices that have only translational symmetry (i.e., a triclinic lattice). At no point did we interject effects due to the other symmetries of the crystal structure. Yet most of the crystals that we discuss have one of the cubic structures. Thus, we want to get some understanding of the effects due to the extensive symmetry possessed by those crystals most often studied.

The first thing to realize is that the set of symmetry operations (the group) in real space is also the set of symmetry operations in reciprocal space. This comes about from the construction of the lattice in reciprocal space (Eq. 10-33); an example will help to show this.

Consider a two-dimensional square space lattice. The reciprocal lattice and Brillouin zones are shown in Fig. 10-7b. For simplicity assume that this lattice has point symmetry C_{4v}(4mm) with the eight symmetry operations $\{E, C_4, C_2, C_4^3, 2\sigma, 2\sigma_d\}$. The 2σ refers to mirror planes perpendicular to the x- and y-axes; $2\sigma_d$ are the two diagonal mirror planes; and the first four symbols refer to rotations about the axis out of the paper by 0, $\pi/2$, π, and $3\pi/2$. These symmetry operations are applied to the crystal in real space. However, since the reciprocal lattice is mathematically "tied" to the real space lattice, when a symmetry operation R operates on the real space lattice its effect is felt on the reciprocal lattice. Mathematically we can show that $(R\mathbf{k})\cdot\mathbf{r} = \mathbf{k}\cdot(R^{-1}\mathbf{r})$, which means that if a symmetry operation acts on real space, its inverse operation acts on reciprocal space. (In a group, the inverse of every symmetry operation is also a member of the group.) In summary, the eight symmetry operations of this point group in real space are also symmetry operations of the first Brillouin zone in reciprocal space. This can be seen by applying these symmetry operations to the (first) Brillouin zone in Fig. 10-8.

Moreover, there are other related symmetry considerations in this problem. Consider $\mathbf{k}$ along $\mathbf{k}_x$ as shown in Fig. 10-21a with the end point of the k-vector labeled Δ. By applying all eight of the symmetry operations to this k-vector, we obtain the four k-values shown in this figure by the vectors with solid lines. Notice that under two of the symmetry operations, E and σ^y, the original k-vector, labeled Δ, transforms into itself, that is, $R\mathbf{k} = \mathbf{k}$ when $R = E, \sigma^y$. These two symmetry operations form the **group of** Δ. We will discuss its use but first let us look for other special points in the Brillouin zone.

Consider $\mathbf{k}$ along the diagonal $\mathbf{k}_x + \mathbf{k}_y$, with the end point of the vector labeled Σ in Fig. 10-21a. Again apply all eight symmetry operations to the Σ-vector and the four k-values shown in the figure

CHAPTER 10 BAND THEORY

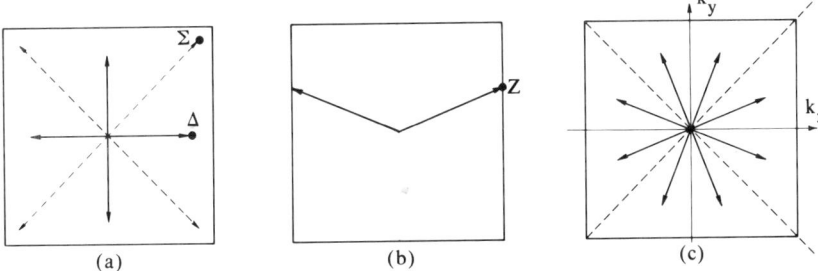

Fig. 10-21 The first Brillouin zone of a two-dimensional square lattice showing special k-values: (a) the Δ- and Σ-lines; (b) the Z line; (c) a general point.

with dashed vectors are obtained. Again we notice that this k-value, labeled Σ, transforms into itself under the symmetry operations E and σ^d, that is, R**k** = **k** when R = E, σ^d. These two symmetry operations are called the **group of Σ**.

Finally, consider **k** to be along the line perpendicular to $\mathbf{k}_x$ at $k_x = \pi/a$, labeled Z in Fig. 10-21b, that is, the right boundary of the Brillouin zone. Under the symmetry operation $\sigma^x \mathbf{k}$ the right boundary becomes the left boundary. Modulo a reciprocal lattice vector, **K** = **a***, both of these boundaries are equivalent in that any wave function at these two k-values are the same because the wave vector differs only by a reciprocal lattice vector. This equivalence is expressed mathematically as

$$R\mathbf{k} = \mathbf{k} + \mathbf{K}_m \tag{10-64}$$

where R is a symmetry operation of the crystal. R**k** and **k** are said to be **equivalent** (or invariant) if they differ by a reciprocal lattice vector. Or we say R**k** = **k** modulo a reciprocal lattice vector. Thus, with this definition the k-value labeled Z transforms into itself or its equivalent under the symmetry operations E and σ^x. These two symmetry operations are called the **group of Z**.

All three of these lines, Δ, Σ, Z, are called **special lines** because more than one symmetry operation of the group transforms them into an equivalent k-vector ($\mathbf{K}_m \neq 0$ in Eq. 10-64) or themselves ($\mathbf{K}_m = 0$). Figure 10-22 shows the first Brillouin zone with these three lines labeled Δ, Σ, and Z.

Consider a **k** close to the Δ-line but with a small value of k_y so that this **k** is not on any special line. Then under all eight symmetry operations of this group other nonequivalent k-values are generated as shown in Fig. 10-21c. Such a k-value is called a **general point**.

There are also **special points**, shown in Fig. 10-22, labeled Γ, X, and M. For example, the X-point under σ^y transforms into itself ($\sigma^y X = X$) and under C_2 and σ^x transforms into itself modulo a reciprocal lattice vector, that is, $C_2 X = X + (2\pi/a)$. These four symmetry operations (E, C_2, σ^y, σ^x) are called the **group of X**. The Γ, $k = 0$, and M, $k = (a^*/2) + (b^*/2)$, points transform into themselves (modulo a K_m for the M-point) under *all* symmetry operations of the group. Thus, the **group of** Γ and **group of M** consist of all eight symmetry operations.

What is the importance of all this? Along these special lines and at these special points there are some degeneracies that do not split no matter how strong the crystalline potential energy may be (an **essential degeneracy**). For the general points there is no degeneracy. The general shape of the energy contours are to a large extent determined by the splittings, degeneracies, and shapes of the bands at the special points and lines. Also note (Fig. 10-22) that the bands need be calculated only in 1/8 of k-space; the rest can be determined by symmetry.

The group of k and their irreducible representations — There is one more important point that must be mentioned, although it may be skimmed over on a first reading. There is a group, in the mathematical sense, of symmetry operations that transforms each of these special points or special lines into themselves. This group is called the **group of k**, that is, the group of Δ or the group of Σ, and so on. Associated with every group is a set of irreducible representations that precisely describes how all functions (defined in ordinary space) transform under all of the symmetry operations of the group. (For example, there are eight symmetry operations in C_{4v} and five irreducible representations.) The different irreducible representations are labeled (i.e., given a symbol to distinguish one from the other). Since the electron wave functions are functions defined in ordinary space, the irreducible representations are used to serve as labels for the wave functions. The labels describe precisely how the wave functions transform under all of the symmetry operations of the group. Thus, the labels contain all of the exact information, the angular dependence, about the wave functions. This information is of critical importance for calculating matrix elements. (For example, for hydrogen-like atoms all of the exact information is contained in the spherical harmonic part of the wave functions and none in the radial part. The spherical harmonics transform as different irreducible representations and are so labeled.) Further, wave functions that transform as different irreducible representations (i.e. have different labels) have different energies except that their energy levels (vs. k, for example) may cross. Also, wave functions that transform as the same irreducible representation (have the same label) repel each other and may not cross. Moreover, some of these irreducible representations are of dimension 2 (or 3), in which case under *all* circumstances there is a two- or threefold degeneracy. The **conventional notation** for the irreducible representations in the group of Δ, or group of Σ, and so on, are subscripts, that is, Δ_1 and Δ_2

CHAPTER 10   BAND THEORY   293

or Σ_1 and Σ_2 and so on, as in Table 10-2. These labels are "powerful magic." The appropriate character tables are included in Table 10-2. Perhaps on a second reading this will come together and make sense. For the present, just plow ahead.

10-13 $\mathscr{E}(\mathbf{k})$ for a Two-Dimensional Square Lattice

10-13a Introduction   We will now calculate the **k** dependence of the energy in the **empty lattice approximation**. Only free electron Bloch functions are used, but the correct symmetry behavior at special points and lines are calculated. Thus, the proper splittings and remaining degeneracies will be known when the crystal potential is turned on, no matter how small the potential may be. For a general k-value the solution is trivial; the interesting properties are found along special lines and at special points. Thus, after the general results are shown, we shall calculate $\mathscr{E}(\mathbf{k})$ and the wave functions along the special line Δ from the special points Γ to X.

For the empty lattice approximation, the general solution for the Bloch function is

$$\psi_{\mathbf{k}n}(\mathbf{r}) = e^{i\mathbf{k}\cdot\mathbf{r}} u_{\mathbf{k}n}(\mathbf{r})$$

where $u_{\mathbf{k}n} = e^{i\mathbf{K}\cdot\mathbf{r}}$ (10-65a)

and $\mathbf{r} = x\mathbf{a} + y\mathbf{b}$. As has been seen before, the reciprocal lattice vector **K** plays the role of the band index. When k-values are brought back into the first Brillouin zone by the appropriate **K**-vectors, higher bands are formed. The examples will make this clearer. The energy of the above wave function is

$$\mathscr{E}_n(\mathbf{k}) = \frac{\hbar^2}{2m} (\mathbf{k} + \mathbf{K})^2 \qquad (10\text{-}65b)$$

and we are writing $\mathbf{k} = k_x\mathbf{a}^* + k_y\mathbf{b}^*$ with the reciprocal lattice vectors $|\mathbf{a}^*| = |\mathbf{b}^*| = 2\pi/a$.

10-13b Behavior at special points and lines   We will now calculate the Bloch functions and band diagrams along the special lines and at the special points.

Δ-line – Start from $\mathbf{k} = 0 = \mathbf{K}$, which is the lowest energy possible (the Γ-point). Along the Δ-line the Bloch function is $\exp(i2\pi k_x x/a)$ with energy $(\hbar^2/2m)(2\pi/a)^2 k_x^2$ for $0 < k_x < 1/2$. (Under both of the symmetry operations of the group of Δ, Table 10-2, this Bloch function transforms into itself so it transforms as the Δ_1 irreducible representation.) For convenience define $\mathscr{E}_0 \equiv (\hbar^2/2m)(2\pi/a)^2$ and plot reduced energy, that is, divided by this quantity. When $k_x = 1/2$

Table 10-2 Character tables and compatibility tables for the various special points and lines of the square lattice as defined in Fig. 10-18.

Γ, M	E	C_2	$2C_4$	$2\sigma_v$	$2\sigma_d$
Γ_1, M_1	1	1	1	1	1
Γ_2, M_2	1	1	1	-1	-1
Γ_3, M_3	1	1	-1	1	-1
Γ_4, M_4	1	1	-1	-1	1
Γ_5, M_5	2	-2	0	0	0

X	E	C_2	σ^y	σ^x
X_1	1	1	1	1
X_2	1	1	-1	-1
X_3	1	-1	1	-1
X_4	1	-1	-1	1

Δ	E	σ^y
Σ	E	σ_d
Z	E	σ^x
Δ_1, Σ_1, Z_1	1	1
Δ_2, Σ_2, Z_2	1	-1

Compatibility Relations

Representation	Compatible with
Δ_1	Γ_1, Γ_3, Γ_5; X_1, X_3
Δ_2	Γ_2, Γ_4, Γ_5; X_2, X_4
Σ_1	Γ_1, Γ_4, Γ_5; M_1, M_4, M_5
Σ_2	Γ_2, Γ_3, Γ_5; M_2, M_3, M_5
Z_1	X_1, X_4; M_1, M_3, M_5
Z_2	X_2, X_3; M_2, M_4, M_5

Γ_5 Reduces into $\Delta_1 + \Delta_2$ or $\Sigma_1 + \Sigma_2$
M_5 Reduces into $\Sigma_1 + \Sigma_2$ or $Z_1 + Z_2$

along this line, the energy is $\mathcal{E}_0/4$. This is the X-point, which we treat as a special case of the Z-line.

Z-line – For the special Z-line there are two equivalent values of **k** with the same energy. One is at the point labeled Z in Fig. 10-22 and the other is related by the mirror image point across the reflection plane σ^x. This latter point is equivalent, that is, related by a reciprocal lattice vector as in Eq. 10-64. These two points were discussed previously and can be seen in Fig. 10-23b where the extended zone scheme is shown. The first Brillouin zone is the lightly outlined square in the center, where the special points are labeled. The k-values and Bloch

CHAPTER 10   BAND THEORY   295

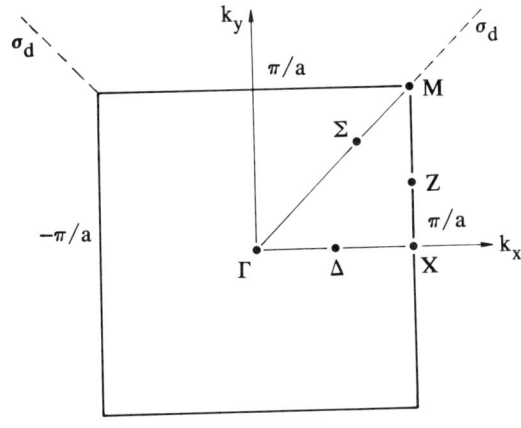

Fig. 10-22 The first Brillouin zone of a two-dimensional square lattice showing the special points and lines. Writing the positions with a z-coordinate always equal to zero for this two-dimensional problem, $\mathbf{a}^* = [1, 0, 0]\, 2\pi/a$ and $\mathbf{b}^* = [0, 1, 0]\, 2\pi/a$. Thus, k_x and k_y are in the $[1, 0, 0]$ and $[0, 1, 0]$ directions, respectively. The X and M points are at $[1, 0, 0]\, \pi/a$ and $[1, 1, 0]\, \pi/a$, respectively.

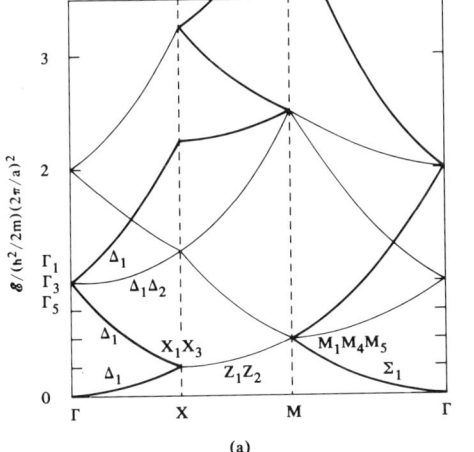
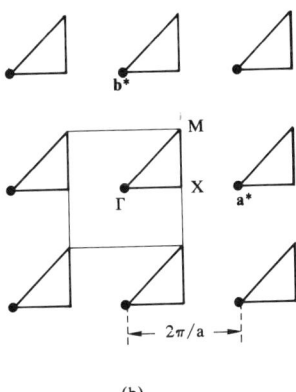

Fig. 10-23 (a) $\mathscr{E}$ vs. k for the two-dimensional lattice. The energy is divided by $\mathscr{E}_0 = (\hbar^2/2m)(2\pi/a)^2$. (b) The region in reciprocal space surrounding the first Brillouin zone. The dots are the reciprocal lattice points. The extended zone scheme is helpful in calculating $\mathscr{E}$ vs. k.

functions are

$$\mathbf{k}_a = \mathbf{a}^*/2 + \alpha \mathbf{b}^*/2 \qquad \psi_a = e^{i\pi x/a}\, e^{i\alpha 2\pi y/a}$$
$$\mathbf{k}_b = -\mathbf{a}^*/2 + \alpha \mathbf{b}^*/2 \qquad \psi_b = e^{-i\pi x/a}\, e^{i\alpha 2\pi y/a} \qquad (10\text{-}66)$$

where α is a positive number between 0 and 1. The energy evaluated via Eq. 10-65b is $(\hbar^2/2m)(\mathbf{a}^*/2 + \alpha \mathbf{b}^*/2)^2 = \mathscr{E}_0(1/4 + \alpha^2/4)$. Thus, it varies between $\mathscr{E}_0/4$ and $\mathscr{E}_0/2$ and is shown in Fig. 10-23a. (Under the symmetry operations of the group of Z, Table 10-2, these two Bloch functions transform between themselves since σ^x transforms ψ_a

into ψ_b and vice versa, forming a reducible representation with characters 2 and 0. This reduces to the irreducible representations $Z_1 + Z_2$.) The appropriate linear combination of the two Bloch functions in Eq. 10-66 that transform as these representations can be found by inspection (or through the use of projection operator techniques). The result is given here, and it is made up of standing waves, as in the one-dimensional case.

$$\psi(Z_1) = \psi_a + \psi_b = 2e^{i\alpha 2\pi y/a} \cos \pi x/a$$

$$\psi(Z_2) = \psi_a - \psi_b = 2i\, e^{i\alpha 2\pi y/a} \sin \pi x/a \qquad (10\text{-}67)$$

X-point – When $\alpha = 0$ in Eq. 10-67 we have the X-point. In this case the two functions transform as $\psi(X_1)$ and $\psi(X_3)$, respectively. In Fig. 10-23a, these two irreducible representations are listed at the X-point at the energy equal to $\mathscr{E}_0/4$. This leads to a very important conclusion. Once the empty lattice approximation is dropped and a small periodic potential energy is allowed, these two functions each will have different energies. This point was covered in a much less sophisticated approach in Section 10-1c. *This energy difference is a band gap.*

Δ-line – We continue along the Δ-line but for k-values outside the first Brillouin zone in the extended zone scheme. The k-values to the right of the X-point in Fig. 10-22 have the same energy as those just to the left of the first Brillouin zone. These latter k-values are $-\beta a^*$ with $1/2 \leq \beta \leq 1$, which are clearly outside the first Brillouin zone. In order to write them with a k-value in the first Brillouin zone (as required for Eq. 10-65), we have $-\beta a^* = \alpha a^* - a^*$ where $0 \leq \alpha \leq 1/2$ so $\mathbf{k} = \alpha a^*$ and $\mathbf{K} = -a^*$. Then the Bloch function and energy, from Eqs. 10-65, are

$$\psi(\Delta_1) = e^{i\alpha 2\pi x/a}\, e^{-i2\pi x/a} \qquad \mathscr{E} = \mathscr{E}_0(\alpha - 1)^2 \qquad (10\text{-}68)$$

(We have already written $\psi(\Delta_1)$ because under both symmetry operations of the group of Δ this function transforms into itself, hence as the Δ_1 representation.) Notice that the energy varies from $\mathscr{E}_0/4$ to $\mathscr{E}_0$ and is shown in Fig. 10-23a. At $k = 0$, the Γ-point where the energy is $\mathscr{E}_0$, there are several other Bloch functions that get mixed together and will be considered later. First, let us continue along the Δ-line to higher energies.

At higher energies along the Δ-line something new comes into the problem. We can have $\mathbf{k} = \alpha a^*$ and $\mathbf{K} = \pm b^*$ where $0 \leq \alpha \leq 1/2$. Notice that this traces out a Δ-line but displaced by $\mathbf{K}$ in reciprocal space above and below the first Brillouin zone (Fig. 10-23b). The Bloch functions and energies, from Eq. 10-65, are

CHAPTER 10   BAND THEORY   297

$$\psi_a = e^{i\alpha 2\pi x/a} e^{i2\pi y/a} \qquad \mathscr{E} = (\hbar^2/2m)(\alpha a^* + b^*)^2$$

$$\psi_b = e^{i\alpha 2\pi x/a} e^{-i2\pi y/a} \qquad \mathscr{E} = (\hbar^2/2m)(\alpha a^* - b^*)^2 \quad (10\text{-}69\text{a})$$

The energy for both functions is $\mathscr{E}_0(\alpha^2 + 1)$, which varies from $\mathscr{E}_0$ to $5\mathscr{E}_0/4$. (Under the symmetry operation E of the group of Δ the two functions transform into themselves but under σ^y they transform into each other; thus the characters of the reducible representation are 2 and 0. This immediately reduces to the irreducible representations $\Delta_1 + \Delta_2$.) The wave functions are

$$\psi(\Delta_1) = \psi_a + \psi_b = 2\, e^{i\alpha 2\pi x/a} \cos 2\pi y/a$$

$$\psi(\Delta_2) = \psi_a - \psi_b = 2\, i\, e^{i\alpha 2\pi x/a} \sin 2\pi y/a \quad (10\text{-}69\text{b})$$

Standing waves again are obtained (as in Eq. 10-67), but this time on the y-coordinate. In Fig. 10-23a this $\mathscr{E}$ vs. k-line is labeled $\Delta_1\, \Delta_2$. Remember that when the periodic potential energy is nonzero the energies of these two lines will be different and a splitting along this entire special line will be observed.

There is another Bloch function along the Δ-line with higher energy that also starts at $\mathscr{E} = \mathscr{E}_0$ at the Γ-point. This is the function with wave vector βa^* with $1 \leq \beta \leq 3/2$, which can be written as $\alpha a^* + a^*$ with $0 \leq \alpha \leq 1/2$. This wave vector can be seen in Fig. 10-23b. Thus $\mathbf{k} = \alpha a^*$ and $\mathbf{K} = a^*$ and the Bloch function and energy are

$$\psi(\Delta_1) = e^{i\alpha 2\pi x/a}\, e^{i2\pi x/a} \qquad \mathscr{E} = \mathscr{E}_0\, (1 + \alpha)^2 \quad (10\text{-}70)$$

(Clearly this function transforms into itself under the symmetry operations of the group of Δ, hence Δ_1 as written.) The energy varies from 1 at the Γ-point to 9/4 at the X-point as shown in Fig. 10-23a.

Γ-point − Now go back to the Γ-point at $\mathscr{E} = \mathscr{E}_0$. It is clear from the preceding discussion that for this point there are four values of $\mathbf{K}$ = $\pm a^*$, and $\pm b^*$ with $\mathbf{k} = 0$. This results in the following four functions all with $\mathscr{E} = \mathscr{E}_0$.

$$\psi_a = e^{i2\pi x/a} \qquad\qquad \psi_c = e^{i2\pi y/a}$$

$$\psi_b = e^{-i2\pi x/a} \qquad\qquad \psi_d = e^{-i2\pi y/a} \quad (10\text{-}71\text{a})$$

Under the symmetry operations of the group of Γ these functions transform among themselves and the character of the reducible representation for each symmetry operation is

E	C_2	$2C_4$	$2\sigma_v$	$2\sigma_d$		
4	0	0	2	0	→	$\Gamma_1 + \Gamma_3 + \Gamma_5$

where the reduction into the irreducible representations is shown. By fairly obvious guessing (or by projection operator techniques) the linear combinations of the Bloch functions that transform as the appropriate Γ_i can be written. The results are

$$\psi(\Gamma_1) = \psi_a + \psi_b + \psi_c + \psi_d = 2\cos 2\pi x/a + 2\cos 2\pi y/a$$

$$\psi(\Gamma_3) = \psi_a + \psi_b - (\psi_c + \psi_d) = 2\cos 2\pi x/a - 2\cos 2\pi y/a$$

$$\psi(\Gamma_5)_i = \psi_a - \psi_b = 2i\sin 2\pi x/a$$
$$\psi(\Gamma_5)_j = \psi_c - \psi_d = 2i\sin 2\pi y/a \qquad (10\text{-}71\text{b})$$

Again we have something that is new in our understanding. At the Γ-point at energy $\mathscr{E}_0$ there will be a splitting of the energy levels when the empty lattice approximation is dropped. However, the two functions that transform as partners of the Γ_5 irreducible representation will remain degenerate. This is called an **essential degeneracy**. Thus, the splitting will be into two singlets and one doublet. The essential degeneracy is due to the point symmetry, C_{4v}(4mm) in this case, and will split only when the symmetry is lowered. For example, the symmetry can be broken by an external strain, electric field, or magnetic field.

Some interesting physics can be learned by expanding the wave functions in Eq. 10-71b to lowest order in x and y. Use the expansion $\cos x \sim 1 + x^2/2 + ...$, and $\sin x \sim x + ...$ Then

$$\psi(\Gamma_1) \sim 1 \qquad\qquad \psi(\Gamma_5)_i \sim x$$
$$\psi(\Gamma_3) \sim x^2 - y^2 \qquad \psi(\Gamma_5)_j \sim y$$

This result shows that $\psi(\Gamma_1)$ behaves like an s-wave function, $\psi(\Gamma_3)$ like one of the d-wave functions, and $\psi(\Gamma_5)$ like a planar p-wave function made up of x and y parts which will be degenerate. This is in agreement with sp²d hybrid functions for square planar bonding as discussed in Chapter 8.

Σ-line – We now consider what happens along the Σ-line. This leads to the M-point, which like the Γ-point has interesting behavior. Along the Σ-line $\mathbf{k} = \alpha(\mathbf{a^*} + \mathbf{b^*})$ where $0 \leq \alpha \leq 1/2$. The Bloch function and energy are

$$\psi(\Sigma_1) = e^{i\alpha 2\pi(x+y)/a} \qquad \mathscr{E} = 2\alpha^2\mathscr{E}_0 \qquad (10\text{-}72)$$

The function transforms as the Σ_1 irreducible representation and the energy goes from 0 to $\mathscr{E}_0/2$ (Fig. 10-23a).

M-point – For this point the energy is the same for four different k-values as shown in Fig. 10-23b. This leads to the following four Bloch functions, all of which have energy $\mathscr{E}_0/2$.

CHAPTER 10   BAND THEORY   299

$$k_1 = (a^* + b^*)/2 \quad \psi_1 = e^{i\pi(x+y)/a}$$
$$k_2 = (-a^* + b^*)/2 \quad \psi_2 = e^{i\pi(-x+y)/a}$$
$$k_3 = (-a^* - b^*)/2 \quad \psi_3 = e^{-i\pi(x+y)/a}$$
$$k_4 = (a^* - b^*)/2 \quad \psi_4 = e^{i\pi(x-y)/a} \quad (10\text{-}73a)$$

Under the symmetry operations of the group of M these four functions transform among themselves. The characters of the reducible representation are given here as well as the reduction to the irreducible representations

$$\begin{array}{ccccc} E & C_2 & 2C_4 & 2\sigma_5 & 2\sigma_d \\ 4 & 0 & 0 & 0 & 2 \end{array} \rightarrow M_1 + M_4 + M_5$$

The functions that transform as these various irreducible representations can be determined by inspection (or projection operator techniques). They are

$$\begin{aligned}
\psi(M_1) &= \psi_1 + \psi_2 + \psi_3 + \psi_4 \\
&= 2\cos\pi(x+y)/a + 2\cos\pi(x-y)/a \\
\psi(M_4) &= \psi_1 - \psi_2 + \psi_3 - \psi_4 \\
&= 2\cos\pi(x+y)/a - 2\cos\pi(x-y)/a \\
\psi(M_5)_i &= \psi_1 - \psi_2 - \psi_3 + \psi_4 \\
&= 2i\sin\pi(x+y)/a + 2i\sin\pi(x-y)/a \\
\psi(M_5)_j &= \psi_1 + \psi_2 - \psi_3 - \psi_4 \\
&= 2i\sin\pi(x+y)/a - 2i\sin\pi(x-y)/a
\end{aligned} \quad (10\text{-}73b)$$

At the M-point there is a splitting of the energy bands into three different energy values when the periodic potential is applied. This is similar to what happens at the Γ-point as previously discussed. However, the essential degeneracy of the two levels that transform as the doubly degenerate M_5 irreducible representation remains under all values of the periodic potential.

In a manner similar to the preceding we can calculate the energy for all the **k**-values shown in Fig. 10-23b. The results of these calculations for higher lying levels are shown in Fig. 10-23a. There are larger **k**-values outside the small piece of the extended Brillouin zone shown in Fig. 10-23b. This results in more bands than shown in Fig. 10-23a.

10-13c Compatibility relations   The preceding calculations can be extended to higher energy. The considerations get complicated in that more values of **K** are used, but nothing new is involved. Here we discuss simple relationships that help with this problem and also give a deeper understanding of the solutions.

The compatibility relations used in band theory are the same as the correlation tables used in crystal field theory. The relations are

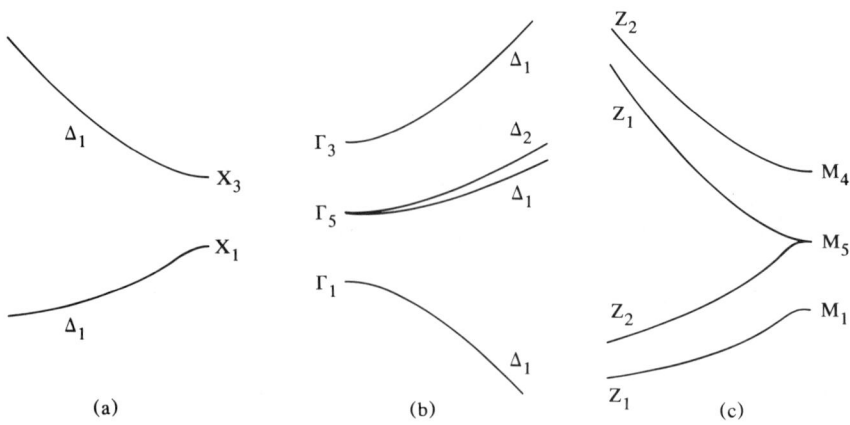

Fig. 10-24 A schematic diagram of $\mathscr{E}$ vs. **k** near some of the special points showing compatibilities.

obtained by simple subgroup considerations.

As an example, consider the irreducible representation Δ_2 in the group of Δ. Under the symmetry operation σ^y it transforms into -1 times itself, as shown in Table 10-2. This means that for any function that transforms as Δ_2 then $\sigma^y \psi(\Delta_2) = -\psi(\Delta_2)$. When this function is evaluated at the end of the Δ-line at the X-point, it still must have this property. In the group of X, there are only two irreducible representations that have this property. They are X_2 and X_4 because only for these representations does $\sigma^y \psi(X_2) = -\psi(X_2)$ and the same for $\psi(X_4)$. Thus we say that Δ_2 in the group of Δ is compatible with X_2 and X_4 in the group of X. From this simple procedure the compatibility relations shown in Table 10-2 are determined for all the one-dimensional irreducible representations. For the two-dimensional irreducible representations the determinations are just as simple but take a small amount of group theory jargon, so we leave them as a problem.

Let us use the compatibility relations. Consider the X-point at energy $1/4$ in Fig. 10-23a where the functions transform as X_1 and X_3. Proceeding along the Δ-line both these irreducible representations are seen to correlate with Δ_1. First, this agrees with what was previously calculated independently. Second, when the lattice potential energy is nonzero, the energies of X_1 and X_3 will be different, and each of these levels will "take along" one Δ_1 level with it. If X_1 is lower in energy than X_3, then the lower Δ_1 will be joined with it and the higher Δ_1 will be joined with X_3 as shown in Fig. 10-24a. This result must occur because the Δ_1 levels cannot cross. Consider the Γ-point at energy 1 where the functions transform as the Γ_1, Γ_3, and Γ_5 irreducible representations. Γ_5 is compatible with $\Delta_1 + \Delta_2$, while Γ_1

CHAPTER 10   BAND THEORY   301

and Γ_3 are compatible only with Δ_1. Assume that the levels have different energies. Most likely the s-like level will be lowest followed by the p-like then the d-like level. Remembering the no-crossing rule for wave functions that transform as the same irreducible representations, we see in Fig. 10-24b the only possible arrangement (except the order of Δ_1 and Δ_2 that come from Γ_5 is not determined). Consider the M point at energy 1/2 and the Z-line. M_1 and M_4 are compatible with Z_1 and Z_2, respectively, while M_5 is compatible with $Z_1 + Z_2$. If the M-levels occur as in Fig. 10-24c, the levels must look as shown. Note that the correlation tables show that the Z-line going from energy 1/2 at the M-point to 5/4 at the X-point must contain functions that transform as Z_1 and Z_2. Thus, these tables are used not only to check but also to predict results.

10-13d Summary We summarize some of the things that have been learned by working out the details for this two-dimensional crystal in the empty lattice approximation lattice. First, it should be remarked that this is about the simplest problem that displays all the complexities that may arise. Three-dimensional lattices involve nothing new, just more computation.

Perhaps the most important concept that has been covered is that at special points and lines the wave functions are in general linear combinations of the Bloch functions, and the different energy levels and remaining degeneracies are governed by the group of **k**. All the calculations were performed in the empty lattice approximation but with the symmetry properly taken into account. Thus, the actual energy separations depend on the details of the periodic potential energy, but the separations are determined by symmetry. (Some separations are shown in Fig. 10-24.) One of the unexpected things that was shown is that there will be energy degeneracies at some of these special points and lines.

The complexity of the band is another result (Fig. 10-23a). Examples of even more complex bands will be seen in the next two sections. Point symmetry causes these complexities but it also has the following simplifications. From Fig. 10-22 the energy needs to be calculated only in 1/8 of the first Brillouin zone. This piece of the zone is surrounded by the special lines. The point symmetry indicates that the result in the other 7/8 of the zone is then known. (See Fig. 10-21c, for example.) The fraction of the unit cell that is unique is $1/h$ where h is the order of the point group of the space group (i.e., h is the number of symmetry operations in the point group). For C_{4v}(4mm) h = 8. For the cubic groups discussed in the next sections the point groups of the space groups are both O_h(m3m) with h = 48. This results in a considerable computational saving.

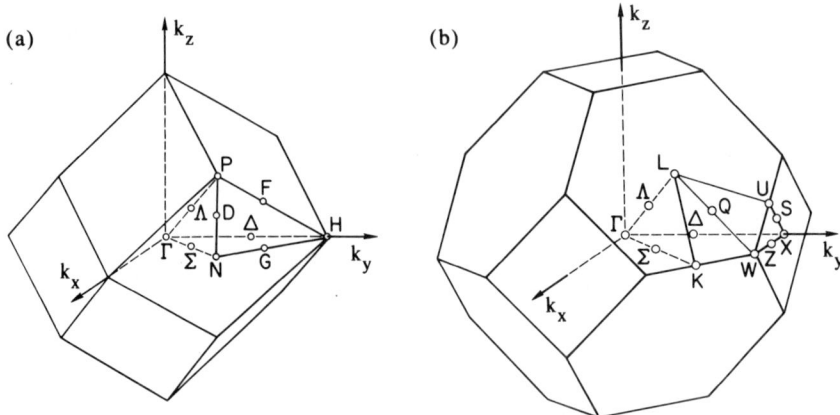

Fig. 10-25 The first Brillouin zone of (a) a bcc lattice and (b) a fcc lattice with the special points and lines labeled. k_x, k_y, and k_z are in the [1, 0, 0], [0, 1, 0], and [0, 0, 1] directions, respectively.

A last important point is that in two or three dimensions some of the points on the Brillouin zone boundaries have considerably larger k-values than others. This leads to the energies being rather different; in Fig. 10-23 we can see this for the X- and M-points. For a spherical Fermi surface this does not arise, but in real crystals these effects are very important.

10-14 Body-Centered Cubic Lattice—Sodium

Figure 10-25a shows the first Brillouin zone of a body-centered cubic (bcc) lattice with the special points and lines labeled. The special point at **k** = **0** point is always called Γ; the labeling is arbitrary but is consistent in its arbitrariness. The **k**-values of the special points, in terms of the unit cube of side a, are

$$H = [0, 1, 0]2\pi/a \qquad P = [½, ½, ½]2\pi/a$$
$$N = [½, ½, 0]2\pi/a$$

The Bloch functions and the $\mathscr{E}$ vs. **k** for the lowest energies along the Δ, Σ, and Λ special lines can be determined in the same manner as in the two-dimensional example. The lowest, nontrivial Γ-point, just as in the two-dimensional example, is clearly composed of Bloch functions, from twelve values of **K** of the type **a*** + **b***, **a*** + **c***, **b*** + **c***, and so on. This gives a twelve-dimensional representation, which, under the group of Γ, O_h(m3m), reduces to a number of irreducible representations. Thus, the basic problem is the same as discussed previously. Figure 10-26a shows the results in the empty lattice approximation. For example, at the Γ-point there are 1, 1, and 3 one-,

CHAPTER 10   BAND THEORY   303

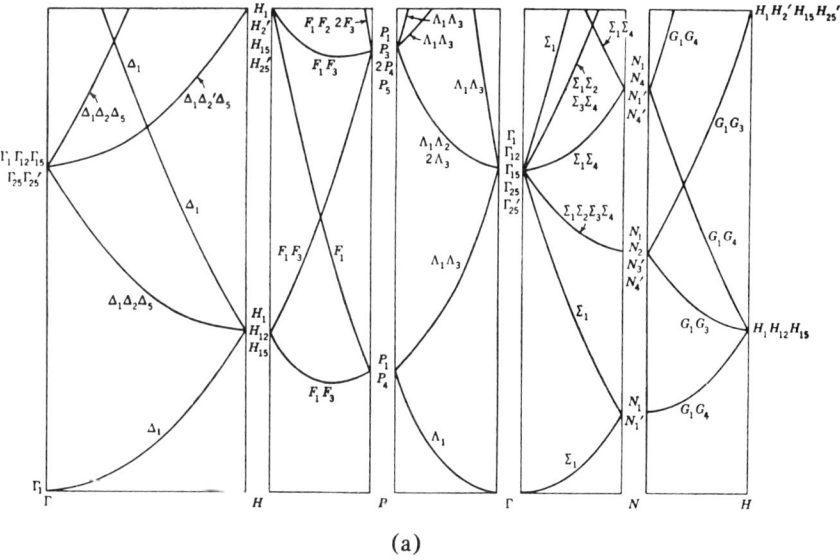

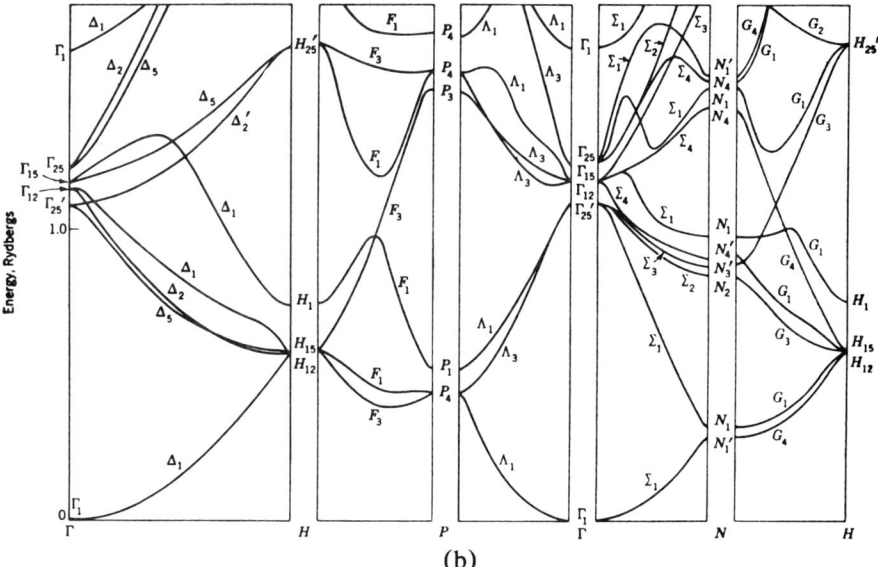

Fig. 10-26 (a) $\mathscr{E}$ vs. **k** for a bcc lattice in the empty lattice approximation. (b) The same but for sodium. By studying these figures along with the Brillouin zone in Fig. 10-25a it is clear how the zone is traversed in different directions starting at the Γ-point. (From "Quantum Theory of Molecules and Solids" by J. C. Stater. Copyright 1963. Used with permission of McGraw-Hill Book Company.)

two- and three-dimensional irreducible representations, respectively. We leave as a problem the determination of some energy levels in a simple cubic lattice, which is easier to handle than a bcc or fcc lattice.

Figure 10-26b shows the results for an energy band calculation for sodium metal, which has a bcc crystal structure. A great deal can be learned by comparing Figs. 10-26a and 10-26b. Consider the Γ-point. Note that all the Bloch functions that transform as different irreducible representations have different energies, as will always happen barring accidental degeneracy. However, notice that one of the energies associated with Γ_1 has a much larger shift than the rest. These shifts depend on the details of the potential energy. Similar effects occur at the other special points. Consider the Δ special line; the lowest energy line is the Δ_1 Δ_2 Δ_5 line in Fig. 10-26a. Notice that for sodium the energies of the different Bloch functions along this line are different and that Δ_2 and Δ_5 cross. This is permissible from a group theory point of view: the energies of eigenstates that transform as different irreducible representations can cross. However, the energies of wave functions that transform as the same irreducible representation cannot cross each other (the **no-crossing rule**) because there would be an allowed off-diagonal matrix element between them that will cause repulsion. A good example of no-crossing can be seen along the F special line. In Fig. 10-26a an energy level labeled F_1 crosses one labeled F_1 F_3 in the empty lattice approximation. However, once some lattice potential energy is allowed, Fig. 10-26b shows that the F_3 level is not affected appreciably but a strong repulsion of the F_1 levels can be seen. Other examples of the no-crossing rule appear in many places in Fig. 10-26b, but the one along the F-line is the clearest. *From this example we can see that extrema can occur in band energies at positions in the middle of special lines and not only at their end points.* This occurs in silicon as will be seen in the next section.

10-15 Si, Ge, GaAs, and GaP

10-15a Bands (and band gaps) The crystal structures of these four materials are shown in Fig. 3-7 and on page 148. The first two have the diamond structure and the last two have the zinc blende structure. However, all four materials have a face-centered cubic (Bravais) lattice and thus, the same Brillouin zone, which is shown in Fig. 10-25b where the special points and lines are labeled. Naturally the Γ-point is at $\mathbf{k} = (0,0,0)$; in terms of the side a of the unit cube in direct space, the k-values of the other important points are located at (only Γ, X, L, and W are special points)

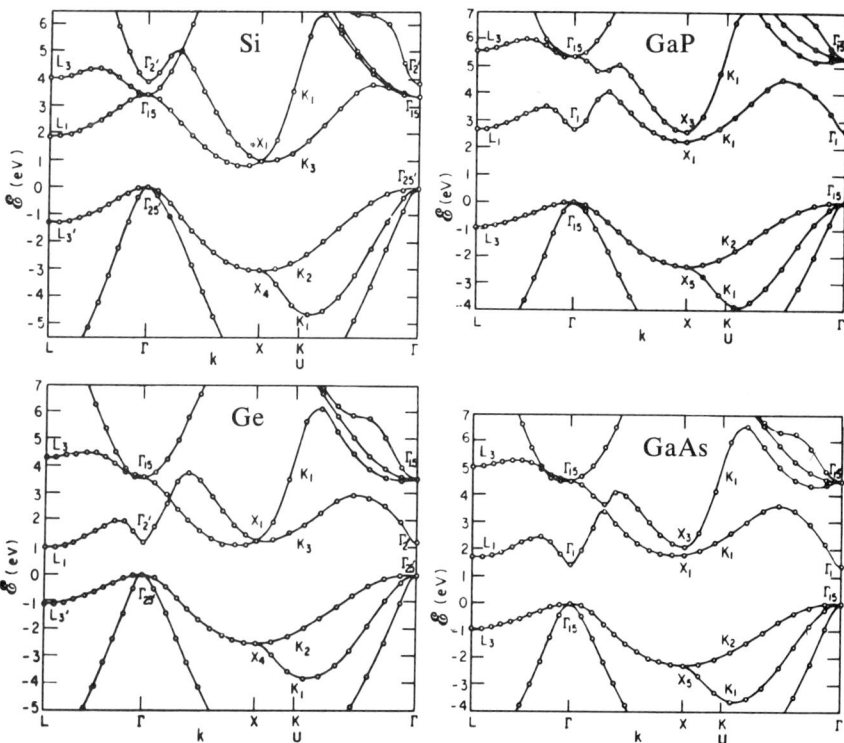

Fig. 10-27 The band structure for several semiconductors. From M. L. Cohen and T. K. Bergstresser, Phys. Rev. **141**, 789 (1966). (The finer details of the valence bands at the Γ-point can be seen in Fig. 10-19b.)

$$X = [0, 1, 0]2\pi/a \quad L = [½, ½, ½]2\pi/a$$
$$K = [¾, ¾, 0]2\pi/a \quad W = [½, 1, 0]2\pi/a$$
$$U = [¼, 1, ¼]2\pi/a$$

To conform to Fig. 10-25b the X-point is written as $[0, 1, 0]2\pi/a$, but it could just as well be written as $[1, 0, 0]2\pi/a$ or $[0, 0, 1]2\pi/a$. All three are *equivalent* X-points (Eq. 10-64). Similar statements are appropriate for the other special points. For example, using Eq. 10-64 it is easy to show that the K and U points are equivalent.

Figure 10-27 shows the calculated bands along some of the special lines for these crystals. For all of these materials there are two atoms per primitive unit cell and hence eight valence electrons per primitive unit cell. Since two electrons per primitive unit cell can be accommodated in a band (one spin up and the other spin down), the eight electrons can fill four bands.

GaAs — Consider GaAs first. The most important thing to notice is that between the occupied bands and the empty bands there is a gap, $\mathscr{E}_g$, of about 1.5 eV. This is the energy difference between the full and empty electron states. These filled states are called the **valence band**, and the energy at the top of the valence band is conventionally the **zero of energy** and is called the **valence band edge**. The empty states above the gap are called the **conduction band**. The lowest point in the conduction band is called the **conduction band edge**. For GaAs the conduction band edge is at k = 0, the Γ-point, which is also the k-value of the valence band edge. Since for GaAs the valence band and the conduction band edges occur at the same k-values, the material is called a **direct band gap** semiconductor. (The labeling is discussed at the end of this section.)

Discussion of Fig. 10-19 — Now that we have some useful vocabulary let us consider, in general, the k = 0 conduction bands and the k = 0 valence band edges for all four materials. Figure 10-19a shows these bands more clearly and Fig. 10-19b shows the effect of spin-orbit coupling. (In Fig. 10-27 the spin-orbit coupling is taken as zero, but in real materials it is finite and the valence bands are as shown in Fig. 10-19b.) In Fig. 10-19 the conduction at k = 0 is composed of s-like wave functions so the orbital degeneracy is one. However, the valence band edge is p-like so has threefold orbital degeneracy. Just as in the atomic case, when spin-orbit coupling is considered, p-states split into $J = 3/2$ and $J = 1/2$; the former is doubly orbitally degenerate and the latter is singly orbitally degenerate. The split-off hole band (Fig. 10-19b) corresponds to the $J = 1/2$ state and cannot split further unless a magnetic field is applied to lift the spin degeneracy. The higher valence band remains doubly degenerate at **k** = 0. However, at **k** ≠ 0 the symmetry is lower (it is governed by the group of k) and the orbital degeneracy is lifted as can be seen in Fig. 10-19b. The band with less curvature has a larger effective mass and is thus called the **heavy-hole band**. The smaller mass band is called the **light-hole band**.

GaP — The band structure of GaP is shown in Fig. 10-27. The structure of the valence band is similar to that of GaAs with the valence band edge at the Γ-point. The conduction bands have one important difference: the conduction band edge is at the X-point. A material where the top of the valence band and the bottom of the conduction band are at different k-values is called an **indirect band gap** material. $\mathscr{E}_g \approx 2.3$ eV for GaP. Thus, even though the band structure of these two materials is quite similar, the fact that GaP is an indirect band gap material causes it to behave quite differently with respect to electrical conduction and optical properties. This will be discussed shortly and in Chapter 13. Note that there apparently are six equivalent conduction band minima at each of the six equivalent X-points in the first Brillouin zone, whereas in GaAs there is a single conduction band minima at the **k** = 0, or Γ-point. Really there are only three equivalent X-points since the ones on opposite faces (Fig. 10-25b)

CHAPTER 10   BAND THEORY   307

differ only by a reciprocal lattice vector. Actually, the conduction band minimum is not exactly at the X-point but along the Δ-line about 92% of the way to the X-point, that is, $k_c \approx 0.92[1,0,0]2\pi/a$. As a result the conduction band valley degeneracy is six, as listed in Table 10-1 (page 242).

Since the difference between a direct and indirect band gap semiconductor is very important, we reemphasize the distinction. To take an electron from the valence band edge to the conduction band edge in GaAs, a direct band gap semiconductor, no change in **k** is involved. While in GaP, an indirect band gap semiconductor, this process occurs with a change in **k**. When no change of **k** occurs the transition (from the valence to conduction bands) is called a **vertical transition**. Note than in GaP vertical transitions can occur but not from the valence band edge to the conduction band edge.

Ge – Ge (Fig. 10-27) is similar to GaAs. (All the elements come from the same row of the periodic table.) Ge is a group IV material while GaAs is a III-V compound. Thus, the similarity of the band gaps is expected but there is one important difference. In Ge the top of the valence band is at the Γ-point in k-space but the bottom of the conduction band is (exactly) at the L-point, causing Ge to be an indirect band gap semiconductor with $\mathscr{E}_g \approx 0.7$ eV.

Si – Figure 10-27 shows the band structure of Si. The valence band edge is at the Γ-point (k=0) but the bottom of the conduction band is along the Δ-line 85% of the way to the X-point. Thus, in Si there are six equivalent k-values for the conduction band minima as listed in Table 10-1. (The band structure of diamond is similar, with the conduction band minimum along the Δ-line about 75% of the way to the X-point.)

For the diamond structure there is a center of inversion at [⅛, ⅛, ⅛] halfway between the Ge atoms at [0, 0, 0] and [¼, ¼, ¼]. For GaAs this symmetry operation does not exist since there is a Ga atom at one position but an As atom at the other. We can ask where the effect of the lower symmetry can be seen in the bands of GaAs compared to those in Ge. The most important effect of this symmetry difference occurs at the X-point where for GaAs two bands are at different energies (Fig. 10-27) while for Ge they are degenerate. This degeneracy is required by symmetry. The different symmetries also affect the labeling. For example, the labeling of the Γ-point irreducible representations appears to be rather different for GaAs than for Ge. The difference stems from the fact that the point group of the space group of the GaAs is $T_d(\bar{4}3m)$ while that of Ge is $O_h(m3m)$. The labels of the irreducible representations are different (although they are related since T_d is a subgroup of O_h).

Table 10-3 Values of the energy gap $\mathscr{E}_g$, in eV, at 0°K. The i and d refer to indirect and direct gaps. Table 8-3 also has values of $\mathscr{E}_g$ at 300°K.

Diamond	i	5.48	AlSb	i	1.70	CdTe	d	1.61
Si	i	1.17	SiC	i	3.03	ZnO	d	3.44
Ge	i	0.74	Te	d	0.33	ZnS	d	3.85
αSn	d	0.0	CaO	i	1.09	SnTe	d	0.3
InP	d	1.42	PbS	d	0.286	AgCl	i	3.25
InAs	d	0.42	PbSe	d	0.145	AgI		3.02
InSb	d	0.24	PbTe	d	0.187	Cu_2O	d	2.172
GaP	i	2.35	CdS	d	2.58	TiO_2	d	3.03
GaAs	d	1.52	CdSe	d	1.85			
GaSb	d	0.81						

Table 10-3 lists the band gaps for some semiconductors. Also noted is an indication of whether the material is a direct or indirect gap material if this fact is known.

10-15b Optical absorption In a crystal, the crystal momentum of an electron with wave vector **k** is $\hbar\mathbf{k}$, as discussed in Section 10-10. For Brillouin zones $\mathscr{E}$ vs. k goes from 0 to $k \approx \pi/a \approx 1$ Å^{-1} (= 10^8 cm^{-1}). The momentum of light is $p = \mathscr{E}/c = h\nu/c = h/\lambda$. The wave vector of visible light, $\lambda^{-1} \approx 10^{-4}$ Å^{-1}, is very small compared to the range of wave vectors in the Brillouin zone.

Now consider visible light incident on a semiconductor such that the energy of the light is larger than the separation between the valence and conduction band edges. This beam of light can excite electrons from the valence band to the conduction band. However, in order to conserve crystal momentum the k-value of the electron in the conduction band must be essentially the same ($\pm 10^{-4}$ Å^{-1}) as in the valence band. Such an electronic transition is called a **vertical transition**, that is, for a vertical transition the change of momentum of the electron between the valence and conduction bands is $\approx \hbar k_i$ where k_i is the wave vector of the incident light, which ≈ 0 on the scale of the Brillouin zone. For a direct band gap material such transitions can excite electrons from states at the top of the valence band to the bottom of the conduction band, that is, across $\mathscr{E}_g$ as shown in Fig. 10-28a.

However, in indirect band gap semiconductors there is a large crystal momentum difference between the valence and conduction band edges. In order for light to be absorbed at an energy $\mathscr{E}_g$, the crystal momentum must be taken up by lattice vibrations. Lattice vibrations are discussed in Chapters 11 and 12; here it is sufficient to say that they too have crystal momentum. Thus, light absorption at $\mathscr{E}_g$ of an indirect band gap semiconductor is a two-step process shown in

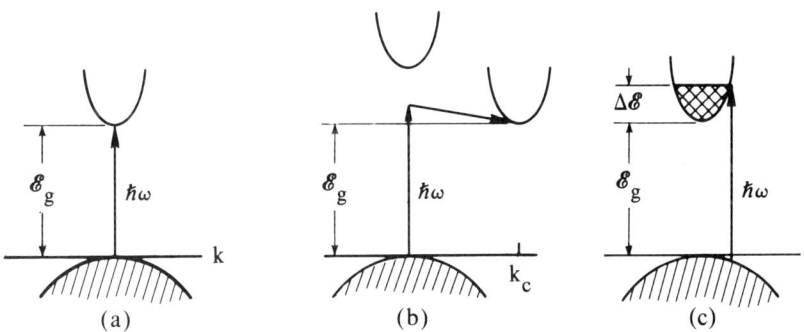

Fig. 10-28 Optical absorption processes for (a) a direct band gap (b) an indirect band gap (c) a direct band gap with the conduction band filled to the level shown.

Fig. 10-28b. The photon excites the electron vertically (in k-space) to a virtual state; then a quantum of lattice vibration (a phonon) is emitted, enabling the electron to make the transition to the conduction band edge. The energy of the light, $\hbar\omega$, which induces the indirect transition equals $\mathscr{E}_g + \hbar\omega_{vib}$, where ω_{vib} is the frequency of the lattice vibration that has $\mathbf{k}_{vib} \approx \mathbf{k}_c$, where k_c is the wave vector of the lowest conduction band.

In summary, in an indirect band gap semiconductor light absorption for energies $\approx \mathscr{E}_g$ is a two-step process. The result is that an electron near the valence band edge at $\mathbf{k} \approx 0$ ends up at $\mathbf{k}_c$ ($\neq 0$). The lattice vibration typically is dissipated as heat. Actually, what has been described should be modified slightly at higher temperatures where lattice vibrations with energies $\hbar\omega_{vib}$ exist in the crystal. In such a situation the lattice vibration can be absorbed and the transition occurs at $\mathscr{E}_g - \hbar\omega_{vib}$. These effects are covered in Section 13-19c. For the direct band gap semiconductors absorption is a one-step process and hence has a larger probability of occurring, that is, a larger matrix element for the transition, which means a larger absorption constant. Of course, in either type of semiconductor vertical transitions can occur from any filled state to any empty state throughout the Brillouin zone. These are strong transitions but occur at energies above $\approx \mathscr{E}_g$. The optical techniques used to measure $\mathscr{E}_g$ and the optical absorption at higher energies are discussed in Chapter 13.

The Burstein–Moss effect (1954) is an interesting example of the consequences of direct (vertical) transitions. The experimental observation is that in some materials (first seen in InSb) the optical band gap increases in energy with increasing impurity concentrations. This is illustrated in Fig. 10-28c where for high donor doping the electrons fill up states in the conduction band. Since the density of states $\propto (m^*)^{3/2}$, Eq. 9-30b, the band filling is more pronounced for materials with small effective masses. (For InSb $m_e = 0.014$.) Figure 10-28c shows that in optical absorption the transitions are vertical so there also is a small contribution to the optical band gap from the valence band as well as the filled conduction band.

The number of conduction electrons required to fill the band to a level $\Delta\mathscr{E}$, shown in Fig. 10-28c, is

$$N = \int_0^{\Delta\mathscr{E}} g(\mathscr{E})d\mathscr{E} = \left(\frac{V}{2\pi^2}\right)\left(\frac{2m^*}{\hbar^2}\right)^{3/2} \int_0^{\Delta\mathscr{E}} \mathscr{E}^{1/2} d\mathscr{E} \quad (10\text{-}74)$$

$$N/V = (3\pi^2)^{-1}(2m^*/\hbar^2)^{3/2}(\Delta\mathscr{E})^{3/2}$$

where the density of states from Eq. 9-30b is used. As can be seen the number of conduction electrons required goes as $(m^*)^{3/2}$.

This effect is most easily seen for donors in InSb. There are two closely related reasons for this. First, the donor binding energy in InSb is only 0.7 meV (Table 8-4c). This is an extremely small binding energy and means that at very low donor density the hydrogen-like electron orbits overlap. This leads to a metal-nonmetal transition where the donor electrons are delocalized in the conduction band. (See Eqs. 8-15.) Second, once the electrons are in the conduction band, they rapidly fill it up to relatively high energies since the density of states is so small. (The density of states is small because of the very small effective mass, Table 10-1.) For acceptors in InSb the binding energy is ten times larger than that of the donors and the weighted effective mass that enters the equations is much larger than that of the electrons. Effective masses are discussed in the next section.

10-15c Effective masses In many semiconductors the effective mass of the carriers can be measured by **cyclotron resonance** techniques. This was discussed briefly in Section 9-15. In a static applied magnetic field, H, the electrons move in helical orbits. The angular frequency, ω_c, is called the **cyclotron frequency** and is given by

$$\omega_c = eH/m^*c \quad (10\text{-}75)$$

where m^* is the effective mass of the charge carriers. Normally, as high a value of ω_c as possible is preferred, because this effect is observable only when $\omega_c \tau \geq 1$, where τ is the collision relaxation time. $\omega_c \gg 1/\tau$ means that the carriers make many orbits before losing their phase coherence by collision, leading to a narrow experimental line, while in the opposite limit the line is not observable. Another way to think about this is that in order to see sharp lines the mean free path (mfp) of the carriers should be large compared to the circumference of an orbit. Since we observe cyclotron resonance from the carriers at the Fermi surface then relevant mfp = $v_F\tau$. We can estimate these values from the free electron values in Tables 9-1 and 9-4. For semiconductors, the experiments are normally performed on very pure samples at low temperature to increase τ. Taking $\nu_c = \omega_c/2\pi = 24$ GHz (24×10^9 cycles/sec), a common microwave frequency, and m*/m = 0.1, we obtain H = 860 gauss, which is a readily accessible magnetic field.

The lowest conduction band for GaAs, and many other III-V semiconductors, is s-like, singly degenerate with spin of 1/2. Thus, there are no complications. However, the valence bands of the four semiconductors in Fig. 10-27 are actually more complicated than shown. The reason is the coupling between the spins and the orbitals. When the spin-orbit interaction is considered there are some noticeable modifications to the band structure. Specifically, the degeneracy at the valence band edges will split. The general effect is similar to obtained from the atomic spin-orbit term $\lambda\mathbf{l}\cdot\mathbf{s}$ resulting in the total angular momentum **j** being a good quantum number.

To understand the effects of spin-orbit coupling, consider GaAs as an example. The top of the valence band of the direct gap semiconductor GaAs in Fig. 10-27 behaves like a threefold degenerate p-like function. Thus, when spin-orbit interactions are considered, two different energy levels are expected, corresponding to j = 3/2 and 1/2 (Fig. 10-19b). The $p_{3/2}$ level is higher in energy than the $p_{1/2}$ levels and the separation is called Δ_{so}, the **spin-orbit splitting energy**. Furthermore at **k** = 0, the $p_{3/2}$ is still fourfold degenerate ($m_j = \pm 3/2$ and $\pm 1/2$), but for **k** $\neq$ 0 it splits further and the various bands are labeled. m* in the light hole band is smaller than in the heavy-hole band because the curvature is larger and m* $\propto (d^2\mathscr{E}/dk^2)^{-1}$, Eq. 10-10. Table 10-4 lists values of m*/m for the electron, light and heavy hole, and split-off hole, as well as Δ_{so}. The quantities are all defined with respect to zero energy at the top of the valence band as follows:

$$\mathscr{E}_c = \mathscr{E}_g + (\hbar^2 h^2/2m_e) \qquad \mathscr{E}_v(\text{lh}) = -\hbar^2 k^2/2m_{\ell h} \quad (10\text{-}76a)$$

$$\mathscr{E}_v(\text{hh}) = -\hbar^2 k^2/2m_{hh} \qquad \mathscr{E}_v(\text{soh}) = -\Delta - (\hbar^2 k^2/2m_{soh})$$

As can be seen, for the direct gap materials m_e is smallest for InSb, the narrowest band gap material; however, the effective masses of the electrons in the conduction bands of all of these direct band gap semiconductors are small. The light-hole masses are also fairly small while the other holes are heavier, but all have m*/m < 1. For density of states considerations the heavy-hole mass is more important than the light-hole mass. (See Table 10-1.) Thus, for the direct gap materials for a given concentration of impurities it is easier to fill up the conduction band than the valence band. (Note that there are other

Table 10-4 Values of the effective masses of electron and holes in direct gap semiconductors, and the spin-orbit splitting. The values for electrons in the conduction band for indirect band gap materials are given below (see the Landolt-Börnstein tables referenced in the Notes).

Crystal	Electron m_e/m	Heavy hole m_{hh}/m	Light hole $m_{\ell h}/m$	Split-off Hole m_{soh}/m	Spin-orbit Δ_{so}(eV)
Diamond	$k \neq 0$	2.18	0.7	1.06	--
Si	$k \neq 0$	0.54	0.15	0.24	0.04
Ge	$k \neq 0$	0.28	0.04	0.09	0.29
GaP	$k \neq 0$	0.67	0.17	--	0.08
GaAs	0.067	0.45	0.082	0.15	0.34
GaSb	0.041	0.30	0.052	0.12	0.80
InP	0.079	0.65	0.12	0.12	0.11
InAs	0.023	0.41	0.025	0.14	0.38
InSb	0.014	0.40	0.016	0.43	0.81

tetrahedrally bonded materials, for example, CuCl and ZnO, that have the opposite sign for the spin-orbit constant — and the reversed order for the valence band levels. This reversal of sign is due to effects of the d-electrons.)

For the indirect gap materials discussed previously the conduction band edges all occur away from $\mathbf{k} = 0$ and have other interesting properties. All of these conduction band edge states are single non-degenerate bands and hence cannot be further split by spin-orbit interaction. However, the band edges are not spherical but are spheroidal energy surfaces oriented along the $\langle 111 \rangle$ for germanium and the $\langle 100 \rangle$ directions in silicon, GaP, and diamond. The constant energy surfaces are shown in Fig. 10-29. Clearly the effective mass tensor, Eq. 10-10, must be used for these anisotropic bands. Consider Si, for example, and focus on one of the six conduction band edges, which is an ellipsoid with two axes of equal length. These two axes correspond to the lighter transverse mass (m_T) while the other axis corresponds to the larger longitudinal mass (m_L). Here we list the k-value of the conduction band edge and the longitudinal and transverse electron masses (m*/m)

		(m_L/m)	(m_T/m)
Diamond	$\approx 0.75X$	–	–
Si	$0.85X$	0.92	0.19
Ge	L point	1.58	0.081
GaP	$0.92X$	7.25	0.21

For example, let us apply this to one of six equivalent conduction band minima in Si. The conduction band edge in the k_z-direction is at $(0, 0, k_0)$, where k_0 has a value that is approximately 85% of the

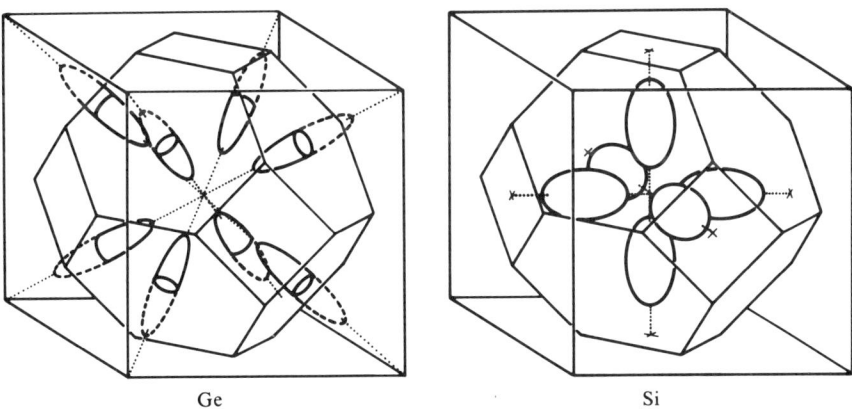

Fig. 10-29 Constant energy surfaces in the conduction bands of germanium and silicon. For Ge the dashed parts of the ellipses of revolution are outside the first Brillouin zone. Note how they differ by a reciprocal lattice from the surface that is inside the zone. Thus, in the first Brillouin zone there are eight, half-complete elliptical conduction band minima or four complete ones. The latter terminology is usually used. For Si the six equivalent entire conduction band minima are inside the first Brillouin zone.

value at the X-point. Near this particular minimum, the ellipsoidal constant energy surfaces are given by

$$\mathscr{E}(\mathbf{k}) = \mathscr{E}(0, 0, k_0) + \frac{\hbar^2}{2m_T}(k_x^2 + k_y^2) + \frac{\hbar^2}{2m_L}(k_z - k_0)^2 \quad (10\text{-}76\text{b})$$

The transverse and longitudinal masses can be measured separately using cyclotron resonance techniques combined with the application of stress to the sample in appropriate directions. The stress lowers some of the constant energy surfaces with respect to the others allowing the effective mass tensor to be measured directly.

When the conduction band effective mass depends on direction, as in these cases, an average that gives the correct density of states is required for m*. An example is given in Section 18-6c, and values are given in Table 10-1.

10-16 Carrier Concentration at Thermal Equilibrium

In this section we treat the carrier concentrations in the conduction and valence bands. First the intrinsic case (no-doping) is discussed and then the important case is discussed where the material contains donor or acceptor impurities. The former case was sketched in Section 10-11c but with only single effective masses in the conduction and valence bands. Now we use the appropriate expressions for the density of states masses whose values are given in Table 10-1, and the equations for these masses in terms of the band masses can be found in Smith, Sze, or Kireev.

Convention — Usually in semiconductor books and papers instead of writing N/V for a density, as has been used throughout this book, n and p are used for the densities of electrons in the conduction band and holes in the valence bands. N_D and N_A are the *densities* of the donors and acceptors. Thus, the volume V does not explicitly appear in any of the equations in this section. Also, we use $\mathscr{E}_g = \mathscr{E}_c - \mathscr{E}_v$ without taking the zero of energy at the top of the valence band, making the relations more symmetric.

10-16a Law of mass action The number of occupied conduction band states (the electron density) is the integral of Eq. 9-35

$$n = V^{-1} \int_{\mathscr{E}_c}^{\mathscr{E}_{top}} f(\mathscr{E}) \, g(\mathscr{E}) \, d\mathscr{E} = \int \frac{C(\mathscr{E} - \mathscr{E}_c)^{1/2} \, d\mathscr{E}}{1 + \exp[(\mathscr{E} - \mathscr{E}_F)/k_B T]}$$

where $C \equiv (2)^{1/2} (m_{de})^{3/2} / \pi^2 \hbar^3$

$\mathscr{E}_{top}$ is the top of the band which, for ordinary temperature, can be taken as infinity, and m_{de} is the density of states mass for electrons

$$m_{de}^3 = M_c^2 (m_1^* m_2^* m_3^*) \tag{10-77b}$$

where m_1^*, m_2^*, and m_3^* are the effective masses along the principal axes of the ellipsoidal energy surface and M_c is the number of equivalent minima in the conduction band. $M_c = 4$ for Ge and 6 for Si (Fig. 10-29). Then, for example, for silicon $m_{de} = 6^{2/3}(m_l^* m_t^{*2})^{1/3}$. Explicit values for these masses are given in Table 10-1. We can evaluate Eq. 10-77a to be

$$n = N_c \frac{2}{\pi^{1/2}} Fn\left[\frac{\mathscr{E}_F - \mathscr{E}_c}{k_B T}\right] \tag{10-78a}$$

where $$N_c \equiv 2(2\pi m_{de} k_B T / h^2)^{3/2} \tag{10-78b}$$

and $$Fn(x_f) \equiv \int_0^\infty \frac{x^{1/2} dx}{1 + \exp(x - x_f)} \tag{10-78c}$$

This latter integral, a function of $x_f \equiv (\mathscr{E}_F - \mathscr{E}_c)/k_B T$, is called the Fermi-Dirac integral and can be evaluated numerically for all x_f. The usual case is that the Fermi level is within the forbidden gap by more

CHAPTER 10   BAND THEORY   315

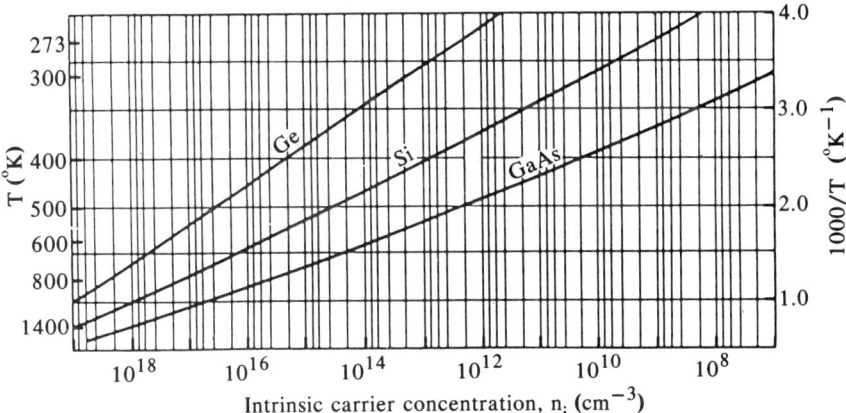

Fig. 10-30 The intrinsic carrier density n_i for the three semiconductors as indicated. Note that the lines are not straight in this plot but would be in a plot of log $(n_i/T^{3/2})$ vs. T^{-1}.

than $3k_BT$, that is, $\mathscr{E}_v + 3k_BT \leq \mathscr{E}_F \leq \mathscr{E}_c - 3k_BT$. This is the so-called **nondegenerate** case and is appropriate in most practical situations. In this case Eq. 10-78 simplifies to

$$n = N_c \exp[-(\mathscr{E}_c - \mathscr{E}_F)/k_BT] \tag{10-78d}$$

In a similar manner we obtain the number of unoccupied states near the top of the valence band, the hole density, except in Eq. 10-77a, f is replaced by $1 - f$ for the unoccupied states.

$$p = N_v \frac{2}{\pi^{1/2}} Fn\left[\frac{\mathscr{E}_v - \mathscr{E}_F}{k_BT}\right] \tag{10-79a}$$

where
$$N_v \equiv 2(2\pi m_{dh} k_B T/h^2)^{3/2} \tag{10-79b}$$

The density of states mass in the valence band is

$$m_{dh} = (m_{\ell h}^{3/2} + m_{hh}^{3/2})^{2/3} \tag{10-79c}$$

Values of the light- and heavy-hole masses are in Table 10-4 and numerical values of m_{dh} are given in Table 10-1. Again for the nondegenerate case a simple relation is obtained

$$p = N_v \exp[-(\mathscr{E}_F - \mathscr{E}_v)/k_BT] \tag{10-79d}$$

N_c and N_V are called the **effective density of states of the valence and conduction bands,** respectively. This is because in the nondegenerate

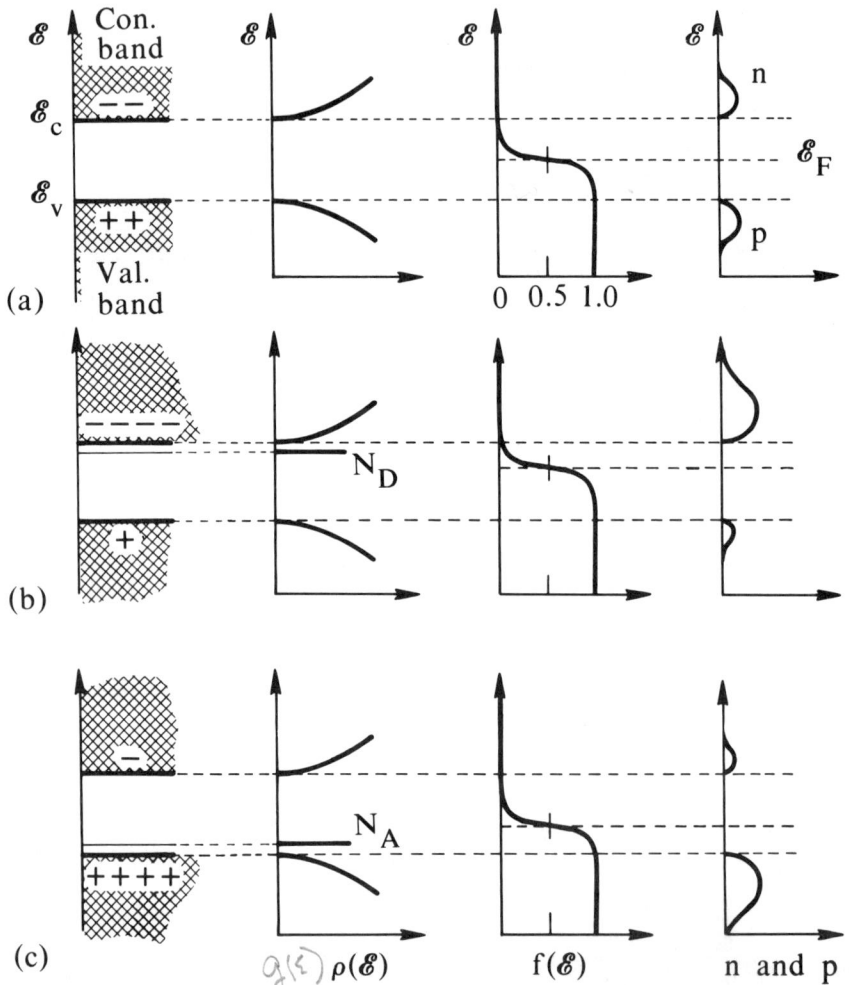

Fig. 10-31 Schematic diagrams showing the bands, density of states, Fermi distribution for T>0°K, the carrier concentrations for T>0°K. (a) An intrinsic semiconductor. (b) An n-type (donor-doped) material. (c) A p-type (acceptor-doped) material.

case the density electrons (holes) is N_c (N_v) multiplied by the exponential, Eqs. 10-78d or 10-79d. Numerically, we have

$$N_v = 4.82\left(\frac{m_{dh}}{m}\right)^{3/2} T^{3/2} \times 10^{15}/\text{cm}^3$$
$$= 2.50\left(\frac{m_{dh}}{m}\right)^{3/2} \left(\frac{T}{300}\right)^{3/2} \times 10^{19}/\text{cm}^3 \quad (10\text{-}79\text{e})$$

with the same expression for N_c except with m_{de} replacing m_{dh}.

The product of n and p is obtained from Eqs. 10-78d and 10-79d

CHAPTER 10 BAND THEORY   317

$$np = N_c N_v \exp(-\mathscr{E}_g/k_B T) \tag{10-80a}$$

This is an interesting and important result. It shows that for a given semiconductor, at a given temperature, the product of the carriers in the conduction band (the electrons with density n) and carriers in the valence band (holes with density p) is a constant. This result (np = constant), valid for the nondegenerate case, is known as the **law of mass action**. Thus, an intrinsic carrier density can be defined as

$$n_i^2 = np \tag{10-80b}$$

$$n_i = (N_c N_v)^{1/2} \exp(-\mathscr{E}_g/2k_B T)$$

$$= 2.50 \left[\frac{m_{de}}{m} \frac{m_{dh}}{m}\right]^{3/4} \left[\frac{T}{300}\right]^{3/2} e^{-\mathscr{E}_g/2k_B T} \times 10^{19}/cm^3$$

n_i depends only on the temperature and band gap and is shown in Fig. 10-30 for the three most studied and used semiconductors.

It is important to realize that in deriving the law of mass action, Eqs. 10-80, no where did we have to assume how the carriers are obtained. Only Fermi-Dirac statistics and a Fermi level were used. In subsequent sections where n might be increased by doping, we can determine p from the fact that the np product is constant.

10-16b Intrinsic semiconductors An intrinsic semiconductor is one in which there are no donors or acceptors in the sample. Thus, for every electron that is thermally excited from the valence band to the conduction band, there is a hole left in the valence band so n = p. For this case it is easy to find the position of the Fermi energy. Using Eqs. 10-77a and 10-78a (or Eqs. 10-78d and 10-79d) an expression similar to Eq. 10-61 can be obtained.

$$\mathscr{E}_F = \frac{\mathscr{E}_c + \mathscr{E}_v}{2} + \frac{3k_B T}{4} \ell n\left(\frac{m_{dh}}{m_{de}}\right) \tag{10-81}$$

We commented in Section 10-11c that the Fermi level for intrinsic semiconductors is in the middle of the band gap at T = 0°K and will stay close to that position unless there is a large difference between the density of states mass for holes and electrons. Figure 10-31a shows various aspects of the intrinsic case.

10-16c Extrinsic semiconductors When impurities contribute a significant fraction of the electrons or holes to the respective bands then we have an extrinsic semiconductor. The Fermi level must adjust so that overall charge neutrality is maintained.

Consider a uniformly doped semiconductor and let N_D and N_A be the density of donor and acceptor sites per cm^3 (whose binding energies are $\mathscr{E}_D$ and $\mathscr{E}_A$). The ionized densities are N_D^+ and N_A^-; an ionized donor gives up an electron so is positively charged as indicated and an ionized acceptor gives up a hole so is negatively charged. When a donor becomes ionized its electron is in the conduction band. Thus, the **electric neutrality equation** is

$$p - n + N_D^+ - N_A^- = 0 \qquad (10\text{-}82)$$

For some semiconductors the binding energies of the donors and acceptors are sufficiently small (shallow donors and acceptors) so that at room temperature they are all ionized, that is, $N_D^+ = N_D$ and $N_A^+ = N_A$.

The use of these ideas can be illustrated in the following example. Assume only donor doping and room temperature so that $N_D = N_D^+$ and $N_A = 0$. Then from charge neutrality (Eq. 10-82) and the law of mass action (Eq. 10-80b)

$$n = (N_D/2) + [n_i^2 + (N_D/2)^2]^{1/2} \qquad (10\text{-}83\text{a})$$

Clearly $n > n_i$, as we would expect for a donor-doped material. Correspondingly, the density of holes is decreased ($p = n_i^2/n$). Also, if the number of donors is very much greater than the intrinsic carriers ($N_D \gg n_i$) then

$$n = N_D$$
$$p = n_i^2/N_D \qquad (10\text{-}83\text{b})$$

showing how small the hole density can become.

Under equilibrium conditions, given any one quantity n, p, or $\mathscr{E}_F$ the other two can be determined. However, more generally none of these three are known, rather $N_D - N_A$, $\mathscr{E}_D$, $\mathscr{E}_A$, and $\mathscr{E}_g$ are known so the determination of, for example, $\mathscr{E}_F$ as a function of temperature is more complicated. We treat this latter case here and the simple case is given as a problem.

First, consider a donor-doped material. Assume that there is a donor impurity concentration of N_D (in units of cm^{-3}). We will discuss the details, but the qualitative results are clear. At 0°K none of the donor states are ionized, that is, the electrons are all attached to the donors, so $\mathscr{E}_F$ is between $\mathscr{E}_D$ and $\mathscr{E}_c$. This is because the Fermi function must be unity at $\mathscr{E}_D$ since at this energy the probability of the electron states being filled is one. As the temperature increases $\mathscr{E}_F$ will change in a way that depends on the comparison between N_D and the density of states in the conduction band. At temperatures high enough so that there are more electrons thermally excited across the

CHAPTER 10 BAND THEORY 319

band gap than N_D, then the discussion and equations appropriate to the intrinsic semiconductor case follow and $\mathscr{E}_F$ approaches the middle of the gap, Eq. 10-81.

Now let us make these considerations more quantitative. To preserve electrical neutrality, the total number density of negative charges (electrons in the conduction band, n, and ionized acceptors, N_A^-, which is zero in our case) must equal the total number density of positive charges (holes in the valence band, p, and ionized donors, N_D^+). For our problem

$$n = N_D^+ + p \tag{10-84a}$$

while the density of ionized donors is given by the density of donors minus the density of occupied donor levels, so

$$N_D^+ = N_D \left[1 - \frac{1}{1 + \frac{1}{g}\exp\left(\frac{\mathscr{E}_D - \mathscr{E}_F}{k_B T}\right)} \right] \tag{10-84b}$$

where g is the degeneracy of the donor impurity ground state. Since the donor state can have an electron with spin up or down, g = 2. Presumably $\mathscr{E}_g$, $\mathscr{E}_D$, and N_D are known and we would like to determine the temperature dependence of $\mathscr{E}_F$, n, and p. Into Eq. 10-84a substitute N_D^+ from Eq. 10-84b and the expressions for n and p from Eqs. 10-78d and 10-79d (the nondegenerate case). We can use these expressions for n and p because they are general, coming from Eq. 10-77a, which is just a statement of Fermi statistics. $\mathscr{E}_F$ naturally adjusts, as will be seen. Making these substitutions into Eq. 10-84a we have

$$N_c e^{-(\mathscr{E}_c - \mathscr{E}_F)/k_B T} = \frac{N_D}{1 + 2e^{(\mathscr{E}_F - \mathscr{E}_D)/k_B T}} + N_v e^{(\mathscr{E}_v - \mathscr{E}_F)/k_B T} \tag{10-84c}$$

In this equation everything is known except the Fermi energy. From it $\mathscr{E}_F$ can be found by graphical or computer methods. At intermediate and low temperatures the hole term on the right is negligible, n being controlled by N_D. This is called the **saturation range**. The **intrinsic range** occurs at high temperatures when the N_D term is small compared to the hole term and $\mathscr{E}_F$ goes back to the center of the band. The temperature (and concentration) dependence of $\mathscr{E}_F$, numerically obtained via Eq. 10-84c, is shown in Fig. 10-32a. As N_D is increased a higher temperature is required to bring the Fermi level back to the center of the gap. Note that at a given temperature $\mathscr{E}_F$ approaches $\mathscr{E}_c$ as N_D increases.

We can now understand Fig. 10-31b. As mentioned previously, at very low temperatures $\mathscr{E}_F$ is between $\mathscr{E}_D$ and $\mathscr{E}_c$. As the tempera-

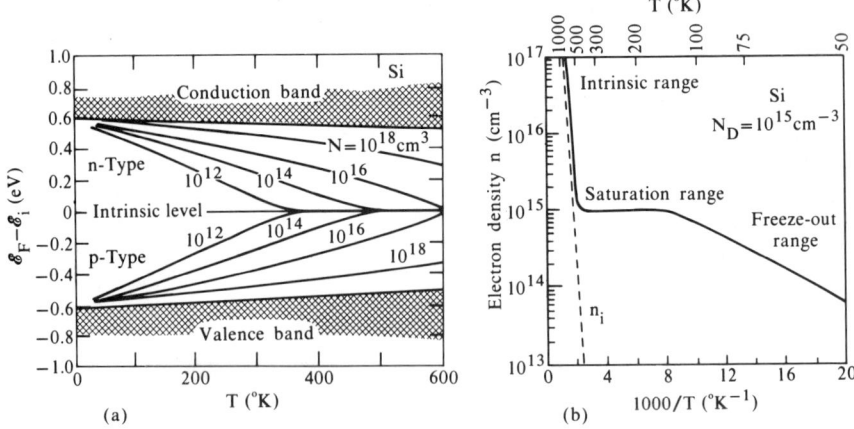

Fig. 10-32 (a) The temperature dependence of the Fermi level for Si for different impurity concentrations (as marked). Note that the intrinsic temperature dependence of the gap is incorporated. [A. S. Grove, "Physics and Technology of Semiconductor Devices" (Wiley, 1967).] (b) The temperature dependence of the electron density for Si with 10^{15} cm^{-3} donors.

ture is increased the donor levels get ionized so the Fermi level moves down. At still higher temperatures, electrons in valence band also begin to get excited into the conduction band. This is the situation shown in Fig. 10-31b. There are holes in the valence band (p ≠ 0) and $\mathcal{E}_F$ is below $\mathcal{E}_D$ and beginning to approach the center of the gap. The temperature is just above the saturation range.

Expressions for holes can be obtained in the same manner, but we shall not go into this case here. The density of acceptor states is N_A so that for the ionized ones, N_A^-, we have the relationship $N_A^-/N_A = \{1 + g \exp[(\mathcal{E}_A - \mathcal{E}_F)/k_BT]\}^{-1}$ where g = 4 for Si, and so on; 2 from the spins and 2 from the heavy and light holes. Now Fig. 10-31c is clear. It is the complement to Fig. 10-31b but with acceptor doping. The temperature is high enough so that there are some electrons in the conduction band and the Fermi energy is toward the center of the band gap.

Returning to the donor-doped case, once $\mathcal{E}_F(T)$ is obtained from Eq. 10-84c we can determine n(T). The results are shown in Fig. 10-32b. For the very low temperature case we obtain

$$n = (N_D N_c/2)^{1/2} e^{-\mathcal{E}_d/2k_BT} \qquad (10\text{-}85a)$$

where $\mathcal{E}_d \equiv \mathcal{E}_c - \mathcal{E}_D$; this is a small energy and accounts for the small slope on the right of the figure where n is determined by the thermal ionization of the donors. This is called the **freeze-out range**. The left of the figure is the intrinsic range where the thermal activation of carriers across the gap produces more carriers than N_D, so the slope is $-\mathcal{E}_g/2$ (Eq. 10-80b). In the intermediate temperature range below the intrinsic range, all of the donors are ionized ($\mathcal{E}_d \ll k_BT$) so $n \approx N_D$.

In real semiconductors even when they are only intentionally donor doped there are always some residual acceptor states in the sample. In this case the expression for the electron density at very low temperatures is

$$n = \left(\frac{N_D - N_A}{2N_A}\right) N_c\, e^{-\mathcal{E}_d/k_BT} \qquad (10\text{-}85b)$$

Note that the factor two is not present in the exponents here as compared to Eq. 10-85a. Thus, at very low temperatures when most of the donor states are "frozen out," that is, $N_D^+ \ll N_D$, the acceptor states become noticeable and the slope of n in Fig. 10-32b becomes $-\mathcal{E}_d$; this is not indicated in the figure. These few acceptor states also affect the position of the Fermi energy at very low temperature, because some of the electrons from donors get captured at acceptor sites, resulting in some ionized donors. To describe this, the Fermi level must be very close to the donor binding energy. As the temperature increases, more donors get ionized with their electrons filling acceptor sites, keeping $\mathcal{E}_F$ at $\mathcal{E}_D$. This is called **Fermi level pinning.** Eventually, at high enough temperatures the intrinsic conditions set in and $\mathcal{E}_F$ approaches the middle of the gap.

Note that several complications arise when detailed calculations or measurements are performed for some semiconductors. For silicon the binding energies of typical donors and acceptors are $\approx$ 45 meV; for example, phosphorus and boron are typical (Table 8-4). These binding energies are the order of k_BT at room temperature (25 meV). Thus, the simple results in Eq. 10-83 are not totally appropriate. Also the spin-orbit split-off band is only 40 meV below the valence band edge (Table 10-4). Thus, away from $\mathbf{k} = 0$, the bands perturb each other, causing deviations from the parabolic band approximation. This causes the effective mass of the holes to depend on k_BT and/or the acceptor concentration for heavily doped p-type Si. However, these considerations are important only for detailed device considerations.

10-16d Impurity band conduction At very low temperatures the fraction of ionized carriers approaches zero, so n approaches zero. One would expect the conductivity to approach zero. However, it does not since the electron wave function of the bound donor impurity has considerable spatial extent. This is discussed in Section 8-6b but now we can appreciate that the effective mass in Eqs. 8-15 should be $m^* = 3[m_1^{*-1} + m_2^{*-1} + m_3^{*-1}]^{-1}$. There is considerable overlap for these extended wave functions even at fairly low impurity concen-

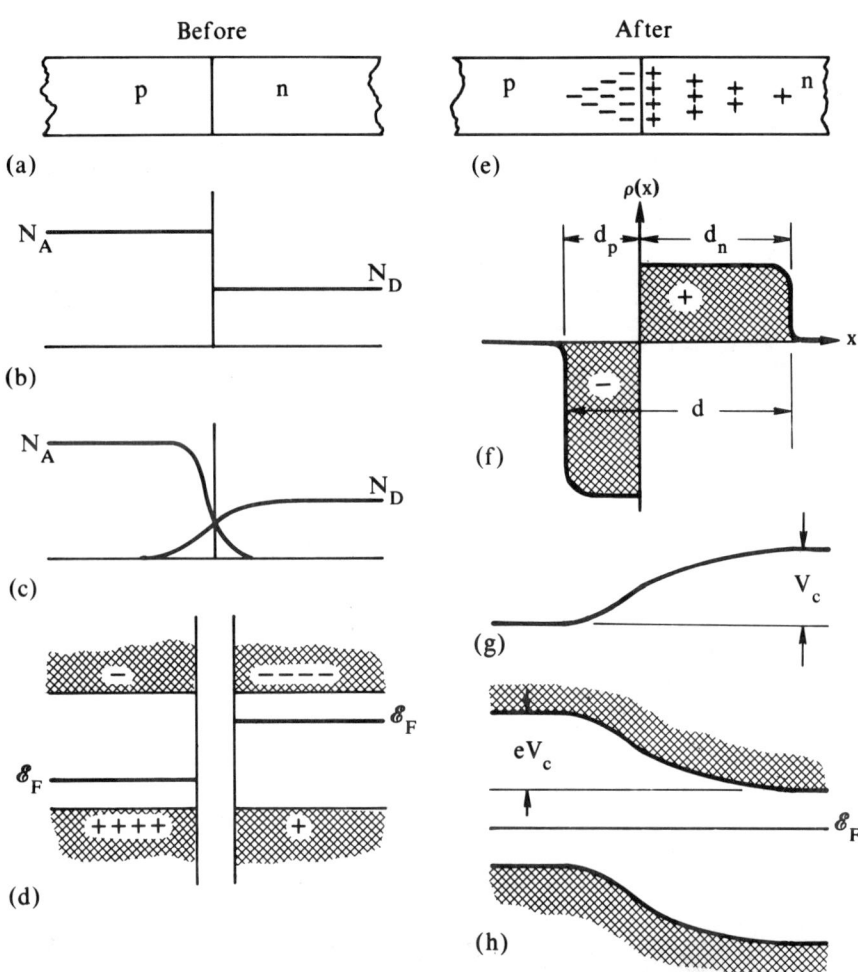

Fig. 10-33 Various aspects of a p-n junction. (a) through (d) Before any charge flows. (e) through (h) After the flow of charge. (a) The junction of p-type and n-type materials. (b) The spatial dependence of the acceptor and donor density in a step junction. (c) The same but in a graded junction. (d) The electron energy levels before contact. (e) After contact, the ionized acceptors and donors. (f) The spatial dependence of the ionized acceptors and donors. (g) The spatial dependence of the electrostatic potential. (h) The same but for the electron energy levels.

trations, which allows the electrons to tunnel between sites, leading to a finite conductivity. This transport is usually called **impurity band conduction.** Since the donors are randomly distributed and not periodically arranged, impurity band conduction is not really the same as that in tight binding bands. See the comments at the end of Section 10-9c.

For larger donor doping there is enough overlap of the wave functions to form an impurity band that is wide enough to merge with the conduction band. For Si this occurs for $n > 3 \times 10^{18}$ cm^{-3}, and in this situation freeze-out at low temperatures (Fig. 10-32b) is not observed.

10-17 p-n Junctions

In this section we discuss p-n junctions and, in the next section, metal-semiconductor junctions. These two junctions are used widely in semiconductor devices so discussing their basic physics will help us to understand other, more specialized devices. For example, a bipolar transistor is two p-n junctions back to back (a p-n-p transistor).

10-17a Introduction In principle we could make a p-n junction by mechanically connecting a p-type semiconductor to an n-type semiconductor. However, this is not practical because of the huge numbers of defects and impurities on the surface and the severe flatness requirements. Moreover, it would be impossible to maintain the lattice periodicity at the boundary. Useful p-n junctions are obtained only when they are formed as an *internal boundary in a single crystal semiconductor,* as indicated in Fig. 10-33a. A **step junction** is shown in Fig. 10-33b where the donor and acceptor dopings are uniform on each side of the junction. In a **graded junction** (Fig. 10-33c) $N_D - N_A$ varies over a significant distance on either side of the junction. A step junction will be assumed in the discussion here; the corrections to cover graded junctions are important only for detailed device design. First we discuss methods of obtaining p-n junctions.

Grown junctions is a technique in which the dopant in the melt is abruptly changed. For example, assume an n-type Si crystal with $N_D = 10^{14}$ phosphorus atoms/cm^3 is being pulled from a melt. To obtain a p-n junction sufficient aluminum atoms are added to the melt so that 3×10^{14} aluminum atoms/cm^3 occur in the crystal. This Al concentration is sufficient to compensate the donors and provide a net acceptor concentration of $N_A - N_D = 2 \times 10^{14}$/cm^3. Such a **counter doping** procedure could be repeated, with even more dopant being required in the melt. This procedure has been replaced by more flexible techniques such as **vapor** (or **liquid**) **phase epitaxial** growth, often called **VPE** (or **LPE**). With these techniques a layer of semiconducting material of either type can be grown on a substrate and then a layer of the other type can be grown on top of the first. All this is done **epitaxially** (i.e., maintaining the substrate single crystal nature). VPE is used in making some **integrated circuits** (**IC**) in the semiconductor industry.

Alloyed junctions are formed, for example, by placing a small piece of indium on an n-type germanium substrate and heating it to only several hundred degrees (perhaps in an argon or hydrogen atmosphere to prevent oxidation). Some of the Ge dissolves in the In and as it cools the Ge saturated with In crystallizes as a continuation of the Ge-substrate. The regrown Ge is p-type since it is In-doped, and the In dot alloyed to the Ge is a convenient ohmic contact.

Diffused junctions are obtained by allowing a high concentration of dopant atoms to diffuse into a substrate. The boundary between the diffused dopant and the other type of doped substrate is the p-n junction. For example, an n-type Si substrate in a furnace at 1000°C is in an atmosphere with a high concentration of boron atoms. The B atoms that stick to the Si surface diffuse into it. The depth of the junction below the surface is controlled by the time and temperature of the diffusion. The pattern of the junction on the surface of the substrate is controlled by photoresist masks that cover the substrate where diffusion is not desired. This technique is widely used to make ICs.

Ion implantation junctions are obtained by the direct implantation of ions into a semiconductor substrate. The ions, obtained from an accelerator, are in the keV to MeV range, and the distance that they travel into the substrate is determined by their mass and kinetic energy as well as the atomic mass of the substrate. Again the position of the junctions can be controlled by photoresist layers, and the low temperature of the process means that previously diffused regions will not be disturbed very much. The semiconductor substrate crystal structure is damaged because of collisions between the energetic ions and the crystal atoms. This damage is removed by annealing the substrate at an appropriate temperature and for appropriate time.

10-17b Equilibrium conditions Imagine an n-type and a p-type semiconductor before the two pieces are in contact with each other. In the uniformly doped n-material the donor concentration is N_D, the Fermi level is near the conduction band, most of the donors are ionized, and the density of electrons in the conduction band is n_n (Figs. 10-33b and 10-33d). Naturally, the material has zero net charge density. In the uniformly doped p-type material the acceptor concentration is N_A, the Fermi level is near the valence band, and most of the acceptors have captured an electron leaving a hole density in the valence band of p_p. The electrons in the conduction band in the n-type material and the holes in the valence band in the p-type material are called the **majority carriers**. We have added a subscript to the density of majority carrier, n_n and p_p, in the n-type and p-type material, respectively. The reason for this is that in the n-type material there also can be a smaller density of holes, given by p_n, called the **minority**

CHAPTER 10   BAND THEORY

carriers. Similarly, in the p-type material there can be minority electrons whose density is n_p.

Now consider what happens when these two homogeneous materials are in intimate contact with each other (Figs. 10-32e to 10-32h). There are large concentration gradients of the majority carriers across the junction; thus, diffusion occurs. The electrons diffuse from the n-region to the p-region, and holes diffuse from the p-region to the n-region. (Diffusion driven by concentration gradients is discussed in Section 11-9. However, physically it is clear that the carriers will move in a way to decrease the concentration gradient. The same concentration gradients exist for the donors and acceptor atoms. However, the impurities cannot diffuse because the temperature is too low for atomic diffusion.) When electrons diffuse from the n-type material into the p-type, they are **minority carriers**. Naturally the minority electrons recombine at a certain rate with the majority holes in the p-region; similarly, the minority holes and majority electrons recombine in the n-region. The average time for the minority carriers to recombine is called the **minority carrier lifetime**. As the electrons flow out of the n-type material, a static positive space charge layer of positively charged ionized donors is left behind near the junction. This gives rise to the **space charge region** or **depletion width** (sometimes called **depletion length**) d_n shown in Figs. 10-33e and 10-33f. Also driven by the concentration gradient, holes flow from the p-material into the n-material leaving behind negatively charged acceptors extending over an acceptor depletion width d_p. The sum of these lengths (d = $d_n + d_p$) gives the **depletion region,** a region on both sides of the junction that is depleted of majority carriers.

Figure 10-33f is a plot of density of unbalanced charge vs. distance. There is a positive charge in the n-type material and a negative charge in the p-type material, but the total charge on both sides of the junction is the same (larger density, smaller depletion width). Notice that outside the space charge region the material is electrically neutral so the potential is constant.

A double layer of charge in the space charge region raises the potential of the n-side relative to the p-side by an amount V_c; this is called the **contact potential, built-in potential,** or **diffusion potential,** and is shown in Fig. 10-33g. Remember that by electromagnetic theory **convention** the potential is measured with a positive test charge. However, *in energy band diagrams we always plot the potential energy of electrons*. Since the charge of an electron is −e the potential energy is −eV(x). Thus, the potential, V(x), and the electron band diagram, −eV(x), are the horizontal mirror images of each other (Figs. 10-33g and 10-33h). To get to its lowest energy state, on a band diagram an electron "falls down" while a hole "floats up."

Due to the dipole layer, and hence the potential difference, the flow of charge continues until the Fermi levels on both sides of the junction are equal (Fig. 10-33h). When this occurs, thermal equilibrium is established and there is no further net flux of electrons into the p-region or holes into the n-region. (However, carriers do continue to flow to make up for the recombination of the minority carriers with the majority carriers.)

The fact that at thermal equilibrium the Fermi level is the same on both sides of the junction is really a thermodynamic property. $\mathscr{E}_F$ is the energy such that above it the probability of the occupation of the states is 50%. Thus, carriers must flow until the Fermi levels are the same spatially.

Notice that a potential energy barrier has been created. The number of electrons that can flow from the n- to the p-side is reduced by approximately $\exp(-eV_c/k_BT)$. In typical p-n junctions the acceptor and donor doping is large enough for the Fermi level to be close to the valence and conduction bands, respectively (Fig. 10-32a). Thus, eV_c can be of the order of the band gap; this exponential term is very small at ordinary temperatures.

10-17c Quantitative considerations at equilibrium It is straightforward to calculate the contact potential, its distance dependence, hence d_n and d_p, and related quantities. Most of the calculations involve integrating Poisson's equation and the relation between electric field and the potential. For the one-dimensional geometry of our problem these equations are

$$\frac{d^2V}{dx^2} = -\frac{4\pi\rho(x)}{\varepsilon}, \qquad E = -\frac{dV}{dx} \qquad (10\text{-}86)$$

where $\rho(x)$ is the charge density and ε is the dielectric constant of the semiconductor. (In **SI units** the 4π is replaced by ε_0^{-1} in Poisson's equation.) The usual approximation is to take the charge density as constant in the depletion regions but with different signs at the appropriate values of x, so from Eqs. 10-86 we expect $E \propto x$ and $V \propto x^2$. The details are discussed later; however, first the value of the total contact potential (V_c) is obtained, not its x-dependence. This is done by understanding the physics of the equilibrium current flow.

Contact potential — Formation of the p-n junction involves diffusion of charge that is proportional to the gradient of the charge density (dn/dx). This is called the diffusion current $J_{\text{diffusion}}$. This diffusion of charge sets up an electric field that acts to drive the carriers in the opposite direction and creates a drift current, J_{drift}, Eq. 10-62. At equilibrium the sum of these two currents is zero, hence for the electron current

CHAPTER 10   BAND THEORY   327

$$J_{drift} + J_{diffusion} = ne\mu_n E + eD_n(dn/dx) = 0 \quad (10\text{-}87a)$$

where μ_n is the electron mobility and D_n is the electron diffusivity (Section 11-9b). Solving for the electric field and using the Einstein-Nernst relation $D/\mu = k_B T/e$, Eq. 11-52e, we obtain

$$E = -\frac{k_B T}{e} \frac{1}{n} \frac{dn}{dx} \quad (10\text{-}87b)$$

Using this expression for the electric field in Eq. 10-86 and integrating x from minus to plus infinity the contact potential is given by

$$V_c = -\int_{x=-\infty}^{x=+\infty} E\, dx = \frac{k_B T}{e} \int_{n(-\infty)}^{n(+\infty)} \frac{dn}{n} \quad (10\text{-}87c)$$

For n at $+\infty$ we have the majority carrier density and assume high enough temperature so that $n_n = N_D$. For n at $-\infty$ we have the minority carrier density and use the law of mass action (Eq. 10-80b) so $n_p = n_i^2/N_A$. Thus,

$$V_c = \frac{k_B T}{e} \ln\left[\frac{N_A N_D}{n_i^2}\right] \quad (10\text{-}87d)$$

For Si at room temperature ($k_B T/e \approx 0.025$ eV/e ≈ 0.025 V), from Fig. 10-32, $n_i \approx 10^{10}$ cm^{-3}. If the material is doped so that $N_A = 10^{16}$ cm^{-3} = N_D then $V_c = 0.72$ V, which is a reasonable fraction of $\mathscr{E}_g$ (1.12 eV).

Distance dependences – To determine the x-dependence of the electric field and voltage (and electron energy = $-eV$) we solve Eqs. 10-86. Only $\rho(x)$ need be known. Usually the **depletion approximation** is made. It is assumed that within their respective depletion regions all of the donors and acceptors are fully ionized, and outside these regions the material is neutral, as indicated in Fig. 10-33f. Hence, the charge density is constant, although different, on each side of the junction, that is,

$$\rho(x) = -eN_A \quad \text{for} \quad -d_p \le x \le 0 \quad (10\text{-}88a)$$
$$\rho(x) = +eN_D \quad \text{for} \quad 0 \le x \le d_n \quad (10\text{-}88b)$$

With these two constant values of the charge density, the electric field can be obtained immediately. From Eqs. 10-86, $dE/dx = 4\pi\rho/\varepsilon$ and integrating gives

$$E(x) = -(4\pi e N_A/\varepsilon)(d_p + x) \quad (10\text{-}88c)$$
$$E(x) = -(4\pi e N_D/\varepsilon)(d_n - x) \quad (10\text{-}88d)$$

corresponding to the two sides of the junction as just shown, and we assume that $E = 0$ outside the depletion region, that is, in the bulk.

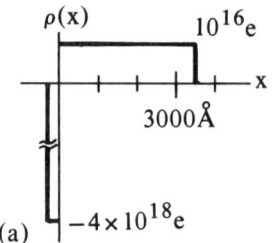 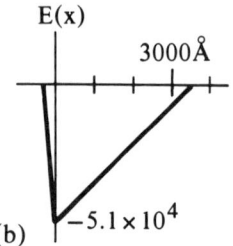

Fig. 10-34 $\rho(x)$ and $E(x)$ (in volts/cm) for a typical abrupt, room temperature p-n junction. The parameters are $N_A = 4 \times 10^{18}$ cm^{-3}, $N_D = 10^{16}$ cm^{-3}, $A = 0.508$ mm. As indicated schematically $d_p = 8.3$ Å and $d_n = 3333$ Å.

Since there is no layer of charge at the junction, the field must be continuous so Eqs. 10-88c and 10-88d can be equated at $x = 0$ and an important charge neutrality relation is obtained

$$d_p N_A = d_n N_D \qquad (10\text{-}88\text{e})$$

If we multiply both sides by eA, where A is the area of the junction, this equation shows that the total negative charge equals the total positive charge. Results from the preceding equations are illustrated in Fig. 10-34 for a typical junction. Note that for $N_A \gg N_D$ the n-side depletion region is large (Eq. 10-88e).

The potential can be obtained by integrating $E(x)$ from Eqs. 10-88c and 10-88d. Conventionally, the bulk p-region is taken at zero potential, $V(-d_p) = 0$, so $V(d_n) = V_c$. Then the integrations yield

$$V(x) = (2\pi e N_A/\varepsilon)(d_p + x)^2 \qquad (10\text{-}88\text{f})$$

$$V(x) = V_c - [(2\pi e N_D/\varepsilon)(d_n - x)^2] \qquad (10\text{-}88\text{g})$$

On the p-side, starting at $x = -d_p$, the potential increases parabolically with distance as the junction is approached. On the n-side, starting at $x = d_n$, the potential decreases parabolically from a value of V_c as the junction is approached. These effects can be seen in Fig. 10-33g.

The electron potential energy diagram, the band diagram, is given by $-eV$ and can be seen in Fig. 10-33h to be the horizontal mirror image of the $V(x)$ diagram.

Depletion widths – Since there is no dipole layer at $x = 0$ the potential must be continuous. Setting Eqs. 10-88f and 10-88g equal at $x = 0$ gives one relation between d_n and d_p. Another relation comes from the charge neutrality relation (Eq. 10-88e). Solving these two simultaneous equations for the **depletion** or **transition width**, where d ($= d_n + d_p$) is the total depletion width, we obtain

$$d = \left[\frac{\varepsilon V_c}{2\pi e} \frac{N_A + N_D}{N_A N_D} \right]^{1/2}$$

$$d_n = \frac{N_A}{N_A + N_D} d, \qquad d_p = \frac{N_D}{N_A + N_D} d \qquad (10\text{-}89)$$

Note that the transition width varies as the square root of the contact potential, V_c. So far only the equilibrium situation has been considered, that is, no applied fields. However, in the next section we shall see that an applied voltage can add to or subtract from V_c increasing or decreasing the potential across the transition region, which, via Eq. 10-89, changes the transition widths.

From Eq. 10-89 we can appreciate that the transition width extends farther into the side of the junction with the lighter doping. For example, for the p^+-n case (i.e., very heavily acceptor-doped p-side) $N_A \gg N_D$, then almost all of the transition width is in the n-type material, that is, $d_n \gg d_p$. This makes sense physically because the total space charge on both sides of junction must be equal. So in the lighter doped material, doping impurities farther from the junction must be ionized to provide the same space charge as for the more heavily doped side. The example in Fig. 10-34 shows these effects.

10-17d Biased junctions Now consider the current voltage (I-V) characteristics of a p-n junction. As will be seen, the junction acts as a **rectifier** (conducts high current for one sign of applied voltage but not for the opposite sign). By **convention** a junction is **forward-biased** when a battery is connected with its positive to the p-side and negative to the n-side. When the battery is connected the opposite way, it is **reversed-biased.**

Figure 10-35a shows an unbiased p-n junction. On the right is the potential V(x), and the contact potential, V_c, and on the left the electron energy level diagram is shown. These are the same as Figs. 10-33g and 10-33h.

A forward-biased junction is shown in Fig. 10-35b. The barrier voltage is lowered to $V_c - V$ and now many more electrons on the n-side have sufficient energy to diffuse from the n-side to become minority carriers in the p-side. Thus, n_p increases and the same situation occurs for holes. With no bias applied the number of electrons that can overcome the potential energy barrier $-eV_c$ is very small. Thus, the probability of occupation of these states is determined by the very high energy tail of the Fermi-Dirac distribution function and is given by a Maxwell-Boltzman exponential dependence (Eq. 10-59b). Even when the barrier is lowered to $-e(V_c - V)$ the electrons that can overcome the barrier are in the high energy tail of the F-D distribution

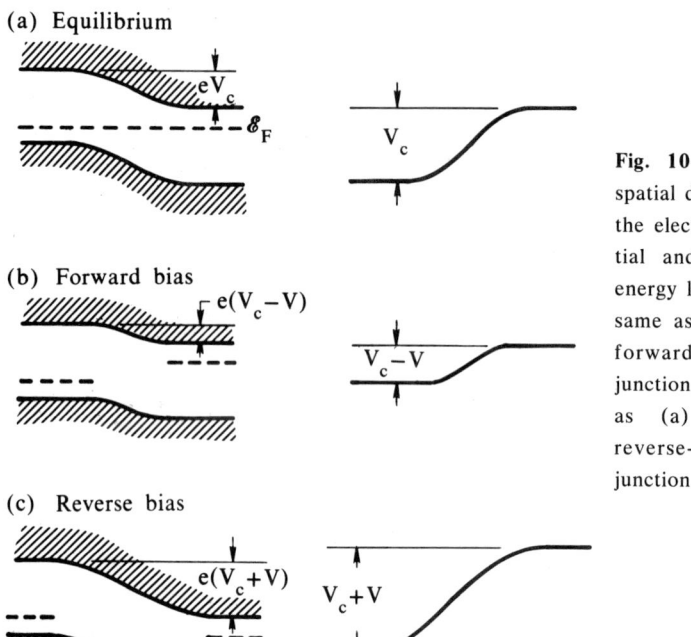

Fig. 10-35 (a) The spatial dependencies of the electrostatic potential and the electron energy levels. (b) The same as (a) but for a forward-biased p-n junction. (c) The same as (a) but for a reverse-biased p-n junction.

(Eq. 10-59b). Thus, we expect that the charge of the minority carriers, and hence the total current, to depend exponentially on applied voltages, V, that is, as $\exp(eV/k_BT)$ from Eq. 10-59b. The holes behave similarly.

A reversed-biased junction is shown in Fig. 10-35c. The situation is analogous to the previous case except now the barrier is raised and the minority diffusion decreases as $\exp(-eV/k_BT)$.

The current voltage character for either bias can be summarized with the same equation as

$$I = I_0 (e^{eV/k_BT} - 1) \qquad (10\text{-}90)$$

where V is positive for forward bias and negative for reverse bias. For $V = 0$ then $I = 0$; I_0 depends on minority carrier diffusion length, lifetime, and equilibrium density. Figure 10-36 is a plot of Eq. 10-90 showing rectifying properties with typical values encountered for the I-V curve of Si p-n junctions.

The expression for the transition width when the junction is biased is the same as Eq. 10-89 except that V_c is replaced by $V_c - V$. The transition width decreases under forward bias and increases under reverse bias as shown in Fig. 10-35.

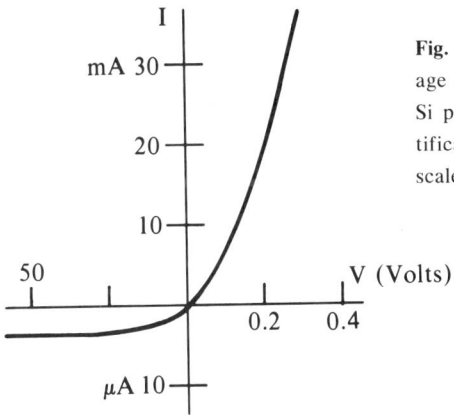

Fig. 10-36 The current voltage (I-V) characteristic of a Si p-n junction showing rectification. Note the different scales.

10-17e Capacitance of a p-n junction In applications of p-n junctions the capacitance is an important design factor. It is an easy quantity to measure and can give information about the structure of the junction.

For a normal, parallel plate capacitor, $C = (Q/V)$ where Q is the charge on either plate. For a p-n junction the more general definition $C = |dQ/dV|$ is used. When there is a small increase in forward-bias voltage, the depletion width decreases a small amount. This is equivalent to a positive charge layer at $x = -d_p$ (which cancels the negative charge at this position, Fig. 10-33f) and a negative charge layer at $x = +d_n$. In reverse-bias the depletion width increases in a similar way (but with a negative charge layer at $x = -d_p$, and so on). Thus, the capacitance C_j of a p-n junction is obtained from the parallel plate capacitance result $\varepsilon A/t$ where t, the distance between the plates, is d which is where dQ/dV changes. Thus, for the p-n junction

$$C_j = \frac{\varepsilon A}{d} = \varepsilon A \left[\frac{2\pi e}{\varepsilon(V_c - V)} \frac{N_A N_D}{N_A + N_D} \right]^{1/2} \quad (10\text{-}91)$$

where A is the area of the junction and d is the depletion width, Eq. 10-89. This result is completely analogous to the parallel plate capacitor with the width of the transition region replacing the plate separation.

One example of the use of this result is in a p^+-n junction $(N_A \gg N_D)$; the N_A term drops out, and from a measurement of C_j we can determine N_D.

For time-dependent applied voltages there are other contributions to C_j from charge storage effects, but these are not discussed here.

10-17f Reverse bias breakdown   As shown in Fig. 10-36, when a p-n junction is biased in the reversed direction there is a small, approximately voltage-independent current. However, reverse breakdown occurs at some critical voltage, for which large currents can flow with little increase in voltage. (This is not shown in the figure.) Breakdown is not inherently destructive to the junction as long as heating is kept under control (typically by a series resistor which limits the current); useful devices, called **breakdown diodes,** operate in this regime. There are two mechanisms for reverse breakdown and we discuss them briefly.

Zener breakdown – This occurs for heavily doped junctions and is due to quantum mechanical tunneling of electrons from the valence band on the p-side to empty states in the conduction band on the n-side. This is the opposite direction of the electron flow when the diode is forward-biased. By using Fig. 10-35c, we can see that for a larger reverse bias than shown, filled states in the valence band energetically line up with empty states in the conduction band, and electron tunneling can occur. For heavily doped junctions the depletion width is relatively narrow (Eq. 10-89), so for a reversed bias voltage of the order of $\mathscr{E}_g/e$, the electric field in the junction is $\approx 10^6$ V/cm. This is large enough to field ionize electrons from their covalently bonded states in the valence band.

The tunneling observed in Zener breakdown occurs in reverse biased p-n junctions. Tunneling in forward biased, degenerately doped p-n junctions also can be observed, is used in devices (**tunnel** or **Esaki** diodes), and is discussed in Problem 10.

Avalanche breakdown – This occurs for lightly doped p-n junctions at much larger voltages (a few volts to a few thousand volts) than observed in Zener breakdown (up to a few volts). Unlike Zener breakdown, it is not due to tunneling.

For lightly doped junctions, electrons also are ionized from the covalent bonds, but the tunneling current is negligible. Instead, ionized electrons acquire large kinetic energies by being accelerated in the large electric field. These high energy electrons collide with covalently bonded electrons in the valence band, causing the latter to be ionized to the conduction band and leaving a hole in the valence band. This newly created electron-hole pair, along with the initial electron, are further accelerated by the electric field (in the opposite direction, due to the opposite charges). This **impact ionization** process repeats many times causing a multiplication of the carriers, an avalanche of carriers, which leads to large currents.

10-17g Light emission   If a p-n junction is forward-biased, electrons are injected into the p-side and holes into the n-side. For convenience, consider the electrons in the conduction band on the p-side (similar arguments apply to the holes on the n-side). After an average time, given by the minority carrier lifetime, the electrons in the conduction band recombine with holes in the valence band. However, when this recombination takes place, what happens to the energy of the electron-hole system, which is $\approx \mathcal{E}_g$? This energy can be emitted from the electron-hole system in the form of photons or phonons (lattice vibrations which are discussed in Chapter 12). It turns out that for most direction band gap semiconductor **radiative recombination** occurs (i.e., photons are emitted).

The energy gap of many semiconductors is 1 eV to 2.5 eV, corresponding to wavelengths 12,000 Å to 5000 Å – near infrared to the visible part of the spectrum. For direct band gap semiconductors radiative recombination can be a very efficient process; that is, for almost every one minority carrier electron crossing the junction there is a recombination with one majority carrier hole resulting in the emission of one photon of energy $\mathcal{E}_g$ (i.e., $\approx 100\%$ quantum efficiency). Thus, we have an efficient means of transforming dc electrical energy into optical energy (light). Furthermore, the emitted light can be modulated at quite high frequencies if desired. These junctions are called **light emitting diodes** (**LED**s), and usually are made from III-V semiconductors such as GaAs or GaP.

Even though the internal quantum efficiency for photons with energies near $\mathcal{E}_g$ is $\approx 100\%$ the very large indices of refraction of these semiconductors (>3) cause most of the light to be totally reflected internally, and absorbed in the heavily doped parts of the junction. Therefore, the external efficiency of LEDs typically is only $\approx 1\%$ (external efficiency ≡ light energy out of the structure divided by the electrical energy in).

Lasers (light amplification by stimulated emission of radiation) can be made from direct band gap semiconductors, utilizing this high quantum efficient recombination radiation. In a laser mode of operation, the external efficiency can be much larger (>20%) and the modulation more rapid ($\approx 10^{10}$ Hz). Lasers are fabricated from these p-n junctions by making two opposite ends of the diode, perpendicular to the junction, flat and parallel to each other; this can be accomplished by cleaving. Because of the large index of refraction, the reflectivity from the flat ends is large; this creates an efficient optical cavity, between the parallel ends, in which optical gain can build up and lead to laser action. LEDs and semiconductor lasers, in conjunction with optical fibers, have found many applications in high frequency communications.

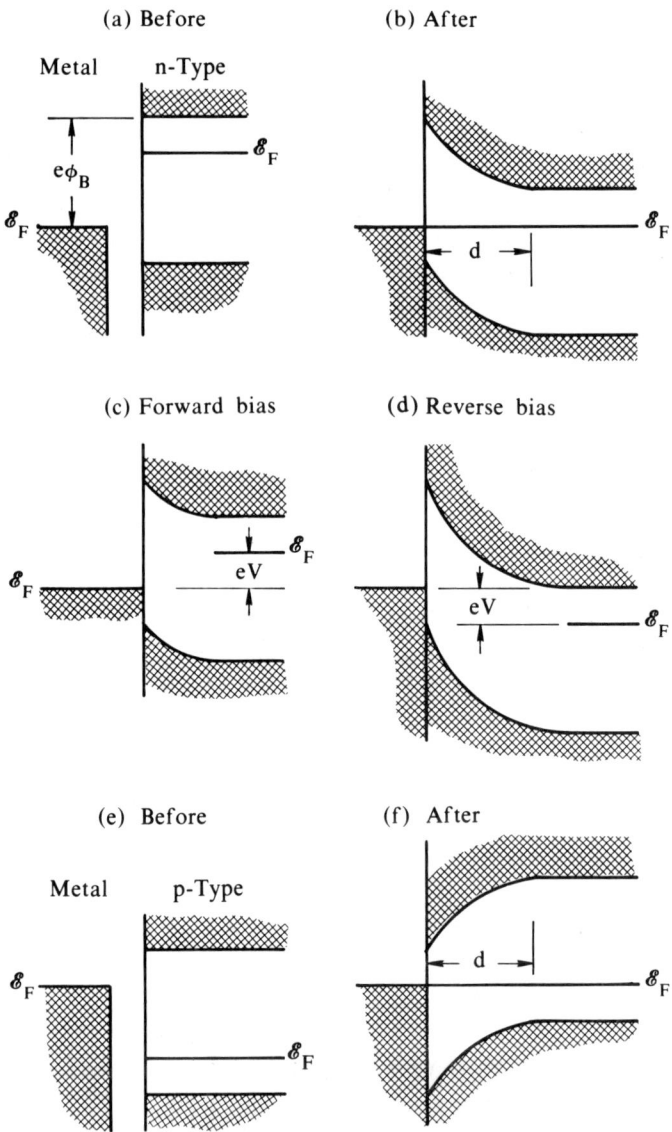

Fig. 10-37 Various aspects of a metal-semiconductor junction. (a) A metal and n-type semiconductor before contact. (b) The same as (a) but after contact. (c) and (d) The same as (b) but forward and reversed biased. (e) and (f) The same as (a) and (b) but the semiconductor is p-type.

10-18 Metal-Semiconductor Junctions

Rectification can be obtained from a metal-semiconductor contact as well as with a p-n junction. A metal-semiconductor contact with rectifying properties is called a **Schottky barrier**, and with a linear I-V

CHAPTER 10   BAND THEORY   335

curve it is called an **ohmic contact**. Both are very important and are discussed here briefly. The third type of important contact to a semiconductor is a **metal–oxide–semiconductor (MOS)** contact, that is, there is a thin oxide layer between the metal and the semiconductor. An MOS structure is useful since it can support a large voltage drop in either direction. It is discussed in Chapter 18.

10-18a Schottky barriers   Figure 10-37a shows the energy level diagram of a metal and n-type semiconductor before they are brought into contact. After contact (Fig. 10-37b) the Fermi levels must line up for the same reasons discussed in the p-n junction case. Electrons flow from the n-type semiconductor to the metal leaving behind positively charged donors over a transition width d. Thus, the electrostatic potential in the semiconductor is raised relative to the metal, and the electron energies are lowered as shown. The electron density in the metal is very, very much larger than that in the semiconductor (n_n) so one can calculate d, C_j (Eqs. 10-89 and 10-87), and related quantities assuming a p^+-n junction (i.e. $N_A \gg N_D$). This effect is similar to that shown in Fig. 10-34 and is why essentially the entire transition width is in the semiconductor.

Thus, the relations derived previously for the p-n junction for $E(x)$, $V(x)$, and d can be applied to the metal-semiconductor junction by treating the metal as an extremely heavily doped semiconductor. Also the parabolic dependence found in Eq. 10-88g for the electrostatic potential, V, and hence the electron energy, $-eV$, found for the p-n junction case, applies to the metal-semiconductor case and can be seen in Fig. 10-37b.

Over the transition region in the n-type semiconductor (Fig. 10-37b) the majority carriers are depleted, leaving behind positively charged donors. Hence this region is called the **depletion region**.

In Fig. 10-37c the bands are shown for a forward-biased metal-semiconductor junction (semiconductor biased positively with respect to the metal). The depletion width decreases in accordance with Eq. 10-89e ($V_c - V$ replaces V_c and $N_A \gg N_D$). The effect of reverse bias is shown in Fig. 10-37d with a resulting increase in the depletion width. As can be seen, the forward and reverse bias effects in a metal-semiconductor junction are just like those in a p-n junction, and the result is rectifying behavior as in Fig. 10-36 and Eq. 10-90. For the metal-semiconductor case I_0 can be expected to be proportional to $\exp(-e\phi_B/k_B T)$ where ϕ_B is the potential barrier that an electron must surmount in order to flow from the metal to the semiconductor (in the conduction band); it is shown in Fig. 10-37a.

Note that the forward current in the metal-semiconductor junction is due to the injection of *majority carriers* from the semiconductor into the metal. This results in much shorter time scales, higher frequency

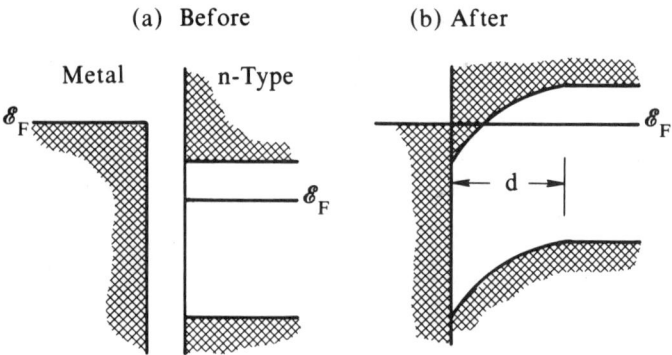

Fig. 10-38 A metal-semiconductor ohmic contact with an n-type semiconductor before and after contact.

properties, and switching speeds than for a p-n junction where minority carriers and their lifetimes control the behavior.

p-type – A metal-semiconductor junction with rectifying properties using p-type material is shown in Figs. 10-37e and 10-37f. As in the n-type case, there is a depletion width but now the region is depleted of holes so there is a negatively charged acceptor region in the semiconductor next to the junction. Thus, forward biasing requires the opposite voltage as in the n-type case. However, except for this change in sign, the I-V curves are the same as shown in Fig. 10-36.

10-18b Ohmic contacts Besides rectifying contacts, we would like to have a metal-semiconductor contact with linear I-V behavior and low resistance, that is, an ohmic contact.

Ohmic contacts can be obtained; Fig. 10-38 shows the result for an n-type semiconductor. In order to line up the Fermi levels, electrons from the metal flow into the semiconductor and accumulate near the junction. This gives an **accumulation region** in the semiconductor of the *majority carrier* (electrons in the n-case). This transfer of electrons from the metal to the semiconductor lowers the electrostatic potential and raises the electron energies in the semiconductor relative to the metal (Fig. 10-38b). With an increase or decrease of an externally applied voltage, electrons can flow in either direction without having to overcome any barriers (i.e., an ohmic contact). Clearly the difference between the junctions in Figs. 10-37 and 10-38 are the relative work function of the metals and semiconductors.

A typical ohmic contact is formed by alloying indium to p-type Si as discussed in Section 10-17a. As a practical point we note that since metal-semiconductor junctions are formed between two different materials, they are sensitive to surface states and oxide layers at the interface. The surface states tend to pin the Fermi surface and the

CHAPTER 10   BAND THEORY   337

oxide layer presents a thin barrier through which the current carriers must tunnel. Ohmic contacts also can be formed by making a very heavily doped Schottky barrier. The very heavy doping decreases the depletion width so that carriers can tunnel through the barrier rather than going over it. The resistance of such contacts can be quite low.

10-19 The Gunn Effect

The Gunn effect (1963) is of practical importance and can be understood simply in terms of the band structure of semiconductors. Gunn observed microwave oscillations in a sample of GaAs with ohmic contacts in an applied pulsed electric field of several thousand volts per cm. Gunn oscillators are commercially available as microwave oscillators. Their low power (milliwatts) is more than offset by their low cost, small size, and ruggedness. Actually, these oscillators are one of a larger class of devices that are of interest because they have bulk negative differential conductivity. Such a device is inherently unstable to oscillation because a random fluctuation of the carrier density produces a space charge that grows exponentially in time.

In metals, due to the very large conductivities, a field of the order of 10 V cm^{-1} produces very large currents that can result in power $\approx 10^8$ W cm^{-3} due to I^{2}R heating. Such large power densities would cause vaporization. However, semiconductors have much lower conductivities, so that large electric fields can be applied, particularly for short times, that is, in a pulsed mode.

Gunn oscillations can be observed in GaAs because of the special features of the bands that lead to negative differential conductivity. Although this material is a direct band gap semiconductor, the L-point conduction band minima are just 0.31 eV above the value at the Γ-point, Fig. 10-27. The effective mass of electrons in an L-point conduction minimum is m*(L) $\approx$ 0.55m while at the Γ-point is m*(Γ) = 0.07m. Thus, the density of states is much larger at the L-point, Eq. 9-30b. Moreover there are four equivalent minima at L as opposed to one at the Γ-point.

Gunn oscillations are observed when large electric fields are applied to n-type GaAs causing the electrons in the conduction band to gain enough energy to be transferred (i.e., scattered) into the four L-point conduction minima. As long as the electric field is maintained, the probability of the electron being scattered back into the Γ-point is small because the total density of states at the L-point is considerably larger. Most importantly, the mobility of electrons in the L-point conduction minimum is much smaller than that in the Γ-point, Eq. 10-63. Thus, once the electrons have enough energy to start to populate the L-point conduction minimum the average electron mobility

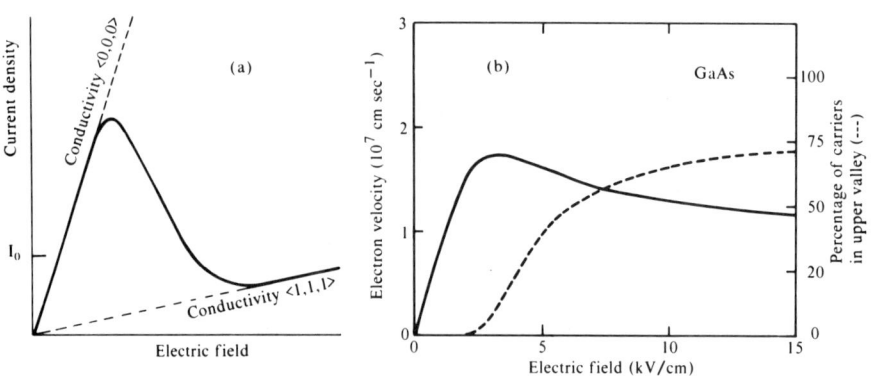

Fig. 10-39 (a) A schematic diagram of current density vs. applied electric field in a Gunn oscillator showing a negative differential conductance. (b) Calculated v_d and fraction of the electrons in the L-point valleys (dashed line).

decreases; hence the conductivity and drift velocity also decrease. Figure 10-39a is a schematic diagram showing the electric field-dependent current. At ≈ 3200 V cm^{-1} there is a decrease in conductivity due to this intervalley electron transfer from Γ to L. (The Gunn effect is sometimes referred to as the **transferred electron effect** since there is a transfer of electrons from a high mobility conduction band valley to lower mobility valleys.) The negative sloping region implies negative differential conductivity. In Fig. 10-39b room temperature calculations of the electron drift velocity in n-type GaAs are shown. Also shown is the percentage of the carriers in the L-point conduction band valley divided by the total number of carriers. As can be seen the velocity decreases above 3300 V cm^{-1}, where it is $\approx 2.1 \times 10^7$ cm/sec, and the L-point conduction valleys begin to be populated.

How does the negative differential conductivity, shown in Fig. 10-39a, lead to oscillations? This is more difficult to explain. Consider the situation in which the electric field is large enough to enter the negative differential conductivity region. The largest field will be at the cathode due to the discontinuity of the voltage between the electrode and the sample. Thus, at this position in space the conductivity drops due to the decrease in the mobility ($\sigma = ne\mu$), the mobility being decreased because the carriers are transferred to the low mobility valleys. Then most of the electric field drops across this small spatial region (a domain). In this domain we have low μ and high E. Once the domain is formed, the electric field in the rest of the sample is low since most of the field drops across the domain. This domain propagates across the crystal with a velocity equal to the drift velocity of the electrons. However, when the domain is annihilated at the anode, because the carriers flow into the contact, the fields in the sample rise

CHAPTER 10   BAND THEORY   339

to their previous large values and the process is repeated. The repetitiveness of the domain motion across the sample determines the frequency of the oscillations ($\nu = v_d/d$); for $v_d = 10^7$ cm/sec and sample thickness of $d = 10^{-3}$ cm, we have a convenient high frequency of ν equals 10^{10} Hz. These domains can be observed directly by several techniques. (See the Notes for more details on these devices.)

Since Gunn oscillation depends sensitively on the positions of the bands, it has been observed in only a few materials besides GaAs, namely in InP, CdTe, and Ge (at low temperatures). The following criteria must be met: (a) The band gap should be larger than 1 eV so that at room temperature there will be very few intrinsic carriers that can interfere with the operation; (b) The separation in energy between the lower and upper conduction band minima, $\Delta\mathscr{E}$, should be greater than $\approx 4k_BT$ so that the upper minimum will not be occupied at low electric fields; (c) $\Delta\mathscr{E}$ should be smaller than $\mathscr{E}_{gap}/2$ so that at the critical electric field for Gunn oscillation, impact ionization across the gap will be small. (Impact ionization would increase the conductivity); (d) Last, we must also be able to make ohmic contacts to the materials so that carriers can be injected.

Negative differential conductivity due to other mechanisms has been observed. For example, in CdS the capture rate of electrons in certain negatively charged deep traps is found to be increased in large fields. (Apparently there is a potential barrier that the electrons must surmount before they can be trapped.) When the capture rate increases sufficiently fast with increasing E, the number of available carriers decreases, resulting in negative conductivity. This results in domain formation and oscillations as discussed previously.

10-20 Other Topics

We discuss, in less detail, several other topics related to band theory. Many others can be found in the references in the Notes.

Semimetals – A semimetal is a metal that has a small overlap in energy values between the top of the highest energy valence band (which is in one k-direction) and the bottom of the lowest conduction band (which is in another k-direction). Thus, the electrons that would have filled the valence band if there were no overlap of energies, spill over to the conduction band. This results in an equal number of electrons in the conduction band and holes in the valence band.

For normal metals the number of conduction electrons per unit volume is greater than 10^{22} cm^{-3} (Table 9-1) while for semimetals it is in the 10^{20} to 10^{17} cm^{-3} range. These values lead to electronic conductivities of 10^5 to 10^6(ohm cm)$^{-1}$ for normal metals and $\approx 10^4$(ohm cm)$^{-1}$ for semimetals [compared to $\approx 10^{-4}$ to 10^{-1}(ohm cm)$^{-1}$ for intrinsic Si and Ge at room temperature]. However, for both normal metals and semimetals the conductivity increases as the temperature is lowered (while in semiconductors the conductivity decreases as the temperature is lowered).

The group V elements arsenic, antimony, and bismuth are the classic examples of semimetals. They have five valence electrons per atom, for example, As has ([Ar]$3d^{10}4s^{2}4p^{3}$), so if there were one atom per primitive cell the material would be a metal. However, there are two atoms or ten valence electrons per primitive cell. With these ten valence electrons the crystal could be an insulator but there is small overlap in energy values between the fifth and sixth bands, resulting in semimetal behavior. Besides low carrier concentrations, these semimetals have small effective masses, high diamagnetic susceptibilities, high lattice dielectric constants, and huge electron g-values.

Semimetals, like semiconductors, can be doped with suitable impurities to vary the number of electrons and holes. The electronic properties of these materials are also very sensitive to pressure. This is because pressure changes the internuclear distances, which sensitively changes the amount of overlap of the bands. This can cause large changes of the carrier concentrations.

Graphite is another well-known semimetal with an electron (and hole) carrier density $\approx 3 \times 10^{18}/cm^{3}$. The structure is a layer type with a plane of covalently bonded carbon atoms in a hexagonal arrangement (as in Fig. 2-1b). The bonding of atoms between planes is very weak. Between the planes the C-C distance is about 2.4 times larger than that in the planes (which allows one plane to "slip" easily with respect to its neighbors, accounting for the lubricating properties of graphite). There are four C-atoms per primitive unit cell so the crystal could be an insulator, but it is a semimetal.

If a semiconductor is doped very heavily with donors so that the Fermi energy is in the conduction band at all temperatures, then it behaves as a semimetal. (Or it could be doped with acceptors and the Fermi energy would be in the valence band.) It has a carrier ioncentration in the correct range and the conductivity, which is due to extended states as opposed to hopping, increases as the temperature is decreased. However, for this case the number of electrons does not equal the number of holes. For this reason, sometimes this type of material is called an **extrinsic semimetal**; then those described previously, such as As, Sb, Bi, and graphite, are referred to as **intrinsic semimetals**.

Transition metals – So far we have not mentioned metals in which d-electrons play a significant role. They are discussed briefly here. The behavior of these metals varies widely, and many interesting properties have been studied.

CHAPTER 10   BAND THEORY   341

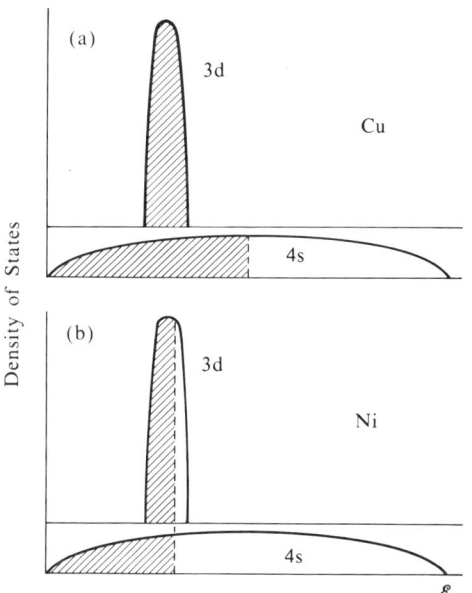

Fig. 10-40 Schematic diagrams of the bands in transition metals.

Consider the metal copper. An isolated Cu atom has an outer electron configuration $3d^{10}4s$, so that there is one s-electron outside the closed shells. However, the 3d electrons cannot be regarded simply as inner shell non-bonding electrons (as would be reasonable for the 2p electrons in Na, i.e., $1s^2 2s^2 p^6 3s$) because the 3d electrons have considerable overlap with neighboring charge clouds at the atomic separation of the metal. Thus, the 3d electrons must be taken into account from a band point of view. However, one of the most interesting properties of the d-electrons is their magnetic behavior, which is discussed in Chapter 15.

Mott and Jones developed a simple Bloch-type picture for these metals. They proposed that the 3d and 4s electrons could be thought to form separate bands. The 4s-band would be wide and the 3d-band narrower because of the larger overlap of the atomic wave functions for the former than for the latter. This situation is shown in Fig. 10-40a where the 3d-band is completely full (from the $3d^{10}$ electrons) and the 4s-band is half full from the one 4s electron. This situation for Cu should be compared with that in metallic nickel, which has one fewer electron and a $3d^9 4s$ atomic configuration. The result is shown in Fig. 10-40b; calculations show that (per atom) there are about 9.4 electrons in the d-band (0.6 holes) and 0.6 electrons in the s-band.

The distinction between Cu and Ni emphasized in Fig. 10-40 accounts for their magnetic properties. The basic idea is that electrons in complete shells (i.e., $3d^{10}$) or electrons that are engaged in bonds have their spins paired and show no resultant spin paramagnetism. On the other hand, the unpaired electrons behave atomic-like and tend to obey Hund's rules. Thus, a complete $3d^{10}$ shell shows no resultant magnetic moment but the incomplete $3d^{9.4}$ will have one. Taking a simple approach, if 60% of the Ni atoms have one missing d electron, then the saturation moment of this metal should be 0.6 Bohr magnetons per atom, which is in agreement with experiment. On the other hand, Cu metal should be nonferromagnetic, as is found.

A fuller theory of the transition metals requires realistic consideration of the mixing of the s- and d-orbitals. However, the ideas associated with Fig. 10-40 form a basis for a qualitative understanding of the transition metals and is discussed in Chapter 15.

Electrical conductivity of a Bloch function – Ignoring the electron-electron interactions but taking translation symmetry into account, Bloch functions are solutions of the one-electron Hamiltonian. The conductivity of a band electron describable by a Bloch function is infinite. In other words, a Bloch function, which is a solution of the wave equation for a perfectly periodic potential, can sustain an electrical current even in the absence of an applied electric field; another way to say this is that the Bloch function electron has an infinite mean free path. This is an interesting point, namely that with this vast array of deep potential wells due to the positive ion cores, this electron wave function has no attenuation from one end of a macroscopic size sample to the other end. The reason for this situation is that the ion cores are assumed to be arranged *perfectly* periodically. Thus, the finite electron conductivity of metals is due to deviations from periodicity. The resistivity (equal to the inverse of the conductivity) of many materials can be due to extrinsic effects such as impurities, defects, and surfaces. Or, more fundamentally, the resistance can be due to thermal vibration of the ion cores about their equilibrium positions. This motion destroys the perfect periodicity of the crystal. Vibration of ions is treated in the next two chapters but we mention a few of the resistivity results here.

From the ion vibration mechanism we can calculate that at low temperatures the resistivity $\rho \sim T^5$, while at high temperatures $\rho \sim T^1$. The temperatures "low and high" are with respect to the Debye temperature, a concept that is discussed in the next chapter. It is of the order of several hundred °K for most elemental metals.

At very, very low temperatures there are fewer and fewer vibrational deviations from perfect periodicity; even in pure metals other contributions to ρ becomes noticeable. Certainly in very pure materials electron scattering from surfaces become noticeable (i.e., ρ becomes dependent on sample size). However, the effect of impurities also becomes noticeable even in very pure samples. For example, the low temperature resistivity of Cu samples with an impurity level of 20ppm (parts per million) is $\rho \approx 1.7 \times 10^{-9}$ ohm cm. The **resistivity ratio**, the ratio of ρ at room temperature to that at low temperature, for such a sample is $\approx 10^3$. However, Cu samples with resistivity ratios as high as 10^6 have been reported.

Amorphous semiconductors – The term amorphous materials is often used synonymously with the term glassy (or vitreous) materials. However, there is a tendency to use the latter for materials obtained by supercooling a melt. The former term is used for materials that are in the form of thin films and are obtained by sputtering, rapid vacuum evaporation, ion bombarding a single crystal surface, or related catastrophic approaches. For example, **splat cooling** can be used to make amorphous materials. A liquid drop of the material is allowed to fall between two flat pieces of metal, which are rapidly forced together, splattering the liquid into a thin specimen. The rapid cooling rate (as high as $10^{6} °K/\text{sec}$) prevents crystallization. In another approach, amorphous Ge and Si can be produced by vacuum depositing these materials onto cold substrates (room temperature is quite cold for this purpose).

Most of the understanding of the properties of amorphous materials comes directly, with some modifications, from the theory of crystalline solids. However, the wave vector selection rules are relaxed since the lack of lattice periodicity prevents the k-vector from being an exact concept. In amorphous solids it is believed that the band gap is modified so that there are long tails of allowed states in the valence and conduction bands that overlap each other. However, in these states the carriers may be described as localized in real space, rather than as modulated plane waves in a crystalline solid. Then conduction takes place by thermally activated hopping from one localized state to another. The mobility of these states in the gap is much lower than we observe in the crystalline material, allowing us to define a **mobility edge** that is related to the conduction and valence band edges of crystalline materials. See the Notes for further reading.

It is interesting to discuss the structure of amorphous materials. In crystals, the interatomic spacings and angles to nearest neighbors, (nn), next-nearest neighbors (nnn), and so on, are fixed due to the long-range order. However, in amorphous materials these values are not fixed, but there are strong similarities between the two states of matter. For amorphous materials the structure is usually described by a radial distribution function $F(r) = 4\pi r^2 \rho(r)$ where $\rho(r)$ is the electron density at a distance r from an arbitrary atom, and it is assumed that the density can be taken as spherically symmetric on the average. For example, in crystalline silicon there are 4 nn at a distance of 2.35 Å (at the tetrahedral angle 109 °28' from each other) and 12 nnn at 3.86 Å. Figure 10-41a shows an experimental plot of $F(r)$ for crystalline (and amorphous) Si. If the peaks are well separated the area under the $F(r)$ curve could be used to determine the number of neighbors that contribute to each peak. Also shown is $F(r)$ for amorphous

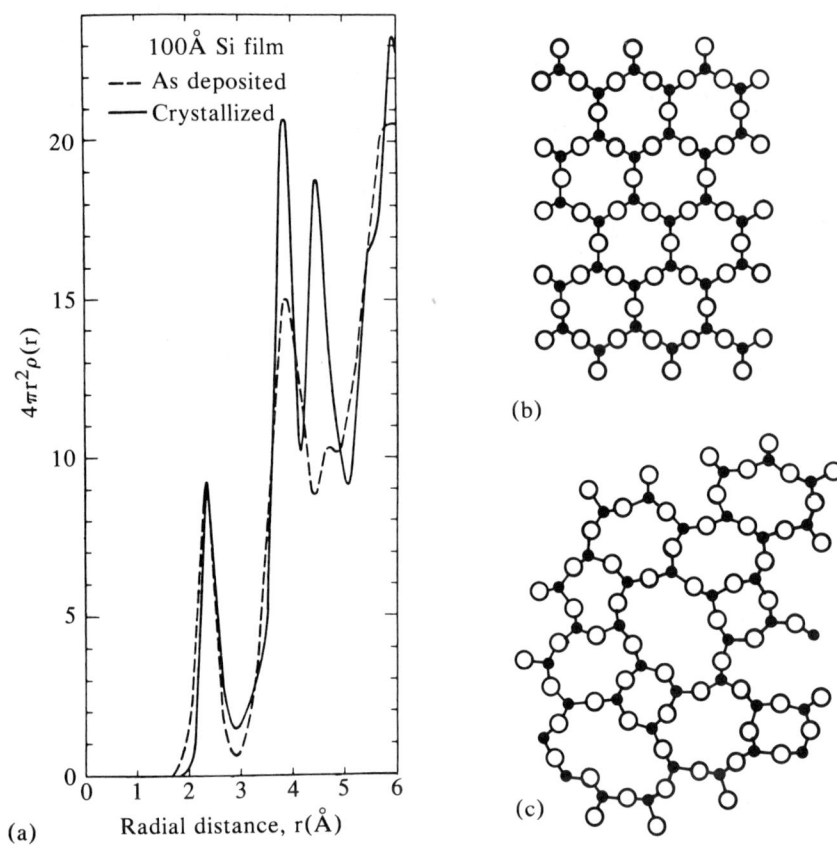

Fig. 10-41 (a) The radial distribution function. (b) Crystalline As_2S_3. Each As is surrounded by 3S atoms and each S by 2As atoms. (c) A model for disordered As_2S_3 where only the angles are varied.

films. The results are very similar for the nn and nnn. For the nn the area under F(r) gives four for both materials and for the next peak 12 nnn are found for both materials. The broadening in the amorphous results is interpreted as due to a distortion of the tetrahedral bond angles by ±20°. The large difference between the two materials occurs in the region of the third neighbor; this is a consequence of the lack of long range order in the amorphous materials.

Amorphous Si is usually thought of in terms of a random network of Si atoms each bonded to 4 nn's at approximately the tetrahedral angle. However, neighboring tetrahedral units can be rotated with respect to each other about a common bond so that five- and six-membered rings can be obtained. This amorphous structure is difficult to show in two dimensions, so in Fig. 10-41b and 10-41c we show a two-dimensional model of crystalline and amorphous As_2S_3. The nn bonds are kept close to $120°$ but the distortions allow rings with various numbers. For silicon a related model of slightly distorted tetrahedral bond angles is consistent with the results in Fig. 10-41a with respect to the similarity of the nn, nnn, and lack of similarity of the third neighbors. See the Notes for further references.

10-21 Summary

A great deal of material has been covered in this chapter so we summarize some of the important basic topics.

In the Sommerfeld free electron theory (Chapter 9) the wave vector **k** was extensively used. The momentum of a free electron is $\hbar\mathbf{k}$, the energy $\mathscr{E}(\mathbf{k}) = \hbar^2 k^2/2m$ where m is the mass of a free electron and **k** can take on an infinite number of values consistent with the Born-von Karman periodic boundary conditions. In electron band theory discussed in this chapter, the periodic potential of the lattice is considered. The wave vector is still of fundamental importance. $\hbar\mathbf{k}$ is the crystal momentum, an important and useful concept (although it is not the real momentum). The wave vector can take on any of the N allowed values in the first Brillouin zone of the reciprocal lattice; N is the number of primitive cells in the crystal. The band index, n, can take on an infinite number of integer values. The energy $\mathscr{E}_n(\mathbf{k})$ usually has a complicated form (for example in Fig. 10-26) but it does have periodicity in reciprocal space with the periodicity of the reciprocal lattice, that is, $\mathscr{E}_n(\mathbf{k}) = \mathscr{E}_n(\mathbf{k} + \mathbf{K})$.

For free electrons, in a crystal of volume V, the normalized wave function is $\psi_\mathbf{k}(\mathbf{r}) = V^{-1/2} \exp(i\,\mathbf{k}\cdot\mathbf{r})$ where V is the volume of the crystal. In band theory the Bloch function is $\psi_{\mathbf{k}n} = \exp(i\,\mathbf{k}\cdot\mathbf{r})\, u_{\mathbf{k}n}(\mathbf{r})$ where the $u_{\mathbf{k}n}(\mathbf{r})$ has the periodicity of the direct lattice, that is, $u_{\mathbf{k}n}(\mathbf{r}) = u_{\mathbf{k}n}(\mathbf{r} + \mathbf{t}_m)$.

The point symmetry operations cause many splittings and result in essential degeneracies along special lines and at special points. These can make the band structure complicated. Also spin-orbit interactions can cause splittings as seen at the valence band edges in the materials in Fig. 10-19b.

The filling of the allowed states by electrons takes place by filling the lowest $\mathscr{E}_n(\mathbf{k})$ levels first. The first Brillouin zone has N allowed spatial states, which can accommodate 2N electrons since each state can accommodate an electron with a spin up and down. Thus, crystals with an odd number of electrons per primitive unit cell will always be metals since the highest energy band will be only half full. Figure 10-42 shows the possibilities for filling. (a) is a metal because there is no gap between the filled states and empty ones. (b) and (d) are

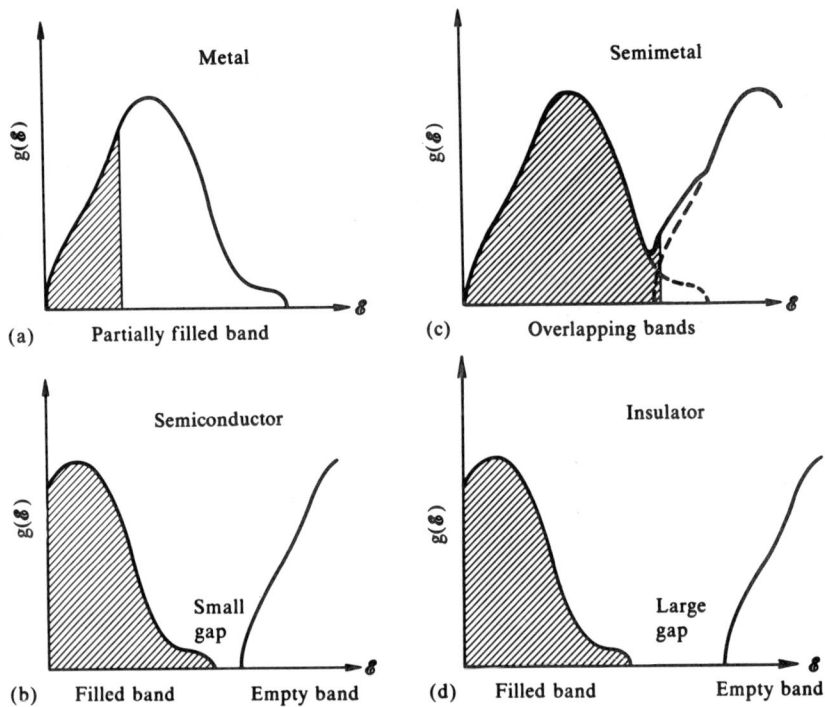

Fig. 10-42 A schematic $\mathscr{E}$ vs. k showing different band possibilities.

insulators because there is an energy gap between the valence and conduction band edges, but the small gap insulator is called a semiconductor. (c) is a semimetal because there is no gap but the bands overlap in different **k** directions. This usually causes the conduction electron concentration at the Fermi energy to be relatively small, $\sim 10^{18}$ to 10^{20} electrons cm^{-3}, compared to $\sim 10^{22}$ to 10^{23} electrons cm^{-3} for ordinary metals. This compares to $\sim 10^{13}$ electrons cm^{-3} for pure Ge at 300°K due to thermal excitation across the gap.

A new point will be discussed here that was used a number of times but never stated in its generality, yet it is important. If V(**r**) has the period of the direct lattice then it can be expressed as a Fourier series in terms of the reciprocal lattice vectors

$$V(\mathbf{r}) = \Sigma_{\mathbf{K}} A(\mathbf{K}) e^{i\mathbf{K}\cdot\mathbf{r}} \qquad (10\text{-}92a)$$

The sum is over all reciprocal lattice vectors. The fact that the right side has the periodicity of the lattice follows immediately by replacing **r** by **r** + $\mathbf{t}_m$ and realizing that exp $(i\mathbf{K}\cdot\mathbf{t}_m) = 1$. Thus, the expression has the correct periodicity and A(**K**) is given by

CHAPTER 10   BAND THEORY   347

$$A(K) = V_p^{-1} \int_{cell} V(r)\, e^{-iK \cdot r}\, d\tau \qquad (10\text{-}92b)$$

where V_p is the volume of a primitive cell and the integral is over one primitive unit cell. The theorem was used in Eq. 10-26 and is the backbone of x-ray crystal structural determinations. The charge density $\rho(r)$ certainly has the periodicity of the direct lattice. Thus

$$\rho(r) = \Sigma_K\, \rho(K)\, e^{iK \cdot r} \qquad (10\text{-}93)$$

If $\rho(K)$ can be measured at every reciprocal lattice vector in k-space, or at least at a large number of values of K, then the charge density, or structure, can be determined immediately. In fact, this is just what the crystallographer does. However, there is one catch. $\rho(K)$ in Eq. 10-93 is complex, and the intensity of the diffraction pattern at the reciprocal lattice vector spots is proportional to $|\rho(K)|^2$. Thus, only the magnitude, and not the phase, of $\rho(K)$ is determined. [If $c = a + ib$ then $c = (a^2 + b^2)^{1/2} \exp(i\Theta)$ where $\tan \Theta = b/a$ and Θ is the phase and $(a^2 + b^2)^{1/2}$ is the magnitude of c.] This is the famous **phase problem** in diffraction.

Another function that does not in general have the periodicity of the lattice but is closely related is the Bloch function. A useful form that is used in many calculations is

$$\psi_{km} = e^{ik \cdot r}\, u_{km}(r) = \Sigma_K\, A_m(k + K)\, e^{i(k+K) \cdot r} \qquad (10\text{-}94)$$

Clearly for $k = 0$ the function has the lattice periodicity but for $k \neq 0$ the Bloch function does not (see Fig. 10-17). For example, in a one-dimensional lattice for $k \approx \pi/a$, the wave function can be approximated by two plane waves $A(k) \exp(ikx) + A(k - K) \exp i(k - K)x$ where $K = 2\pi/a$. Thus, the expansion in Eq. 10-94 is often severely truncated.

The one-electron approximation – As mentioned at the beginning of this chapter, there is a fundamental assumption underlying band theory. It is that all electron-electron repulsions can be taken into account in a periodic potential energy $V(r)$ that appears in the one-electron Schrödinger equation

$$\left[-\frac{\hbar^2}{2m}\, \nabla^2 + V(r) \right] \psi_k(r) = \mathcal{E}(k)\, \psi_k(r)$$

Even without the electron-electron repulsion problem there are impurities, missing atoms, interstitials, and lattice vibrations that destroy the perfect periodicity. Most of these are small effects. However, the potential energy of two electrons at a distance equal to that of neighboring atoms in a crystal is a large effect. **Landau's Fermi liquid theory** provides at least a qualitative way to determine how the electron-

electron interactions affect the electron properties of metals. Furthermore, this theory offers an explanation of why the independent electron approximation is so successful. (See Section 9-14.)

A discussion of the methods that are used to produce a self-consistent Hartree V(r) are outside the scope of this book. Even then the more complicated Hartree-Fock approximation, which includes "exchange" of the identical electrons, is more appropriate. The difference between the true energy of the electron system and the Hartree-Fock energy is defined as the **correlation energy** and various many-body corrections are involved in this energy (see, for example, Chapters 5 and 6 of Kittel's "Quantum Theory of Solids").

Notes

All solid state physics books have at least one chapter on the band theory of solids. These books have other approaches and some other topics. You should not get the impression that all valence bands have a maximum at k = 0. Just like the conduction bands, they can be at other k-values. For example, see the work on AgCl by F. Bassani, R. S. Knox and B. Fowler, Phys. Rev. **137**, A 1217 (1965) and P. S. Scop, Phys. Rev. **139**, A 934 (1965).

The original papers of Wigner and Seitz on cohesive energy calculations are worth reading. They are in Phys. Rev. **43**, 804 (1933) and **46**, 509 (1934). Also see J. C. Slater, Phys. Rev. **45**, 794 (1934). The original paper on the group of **k** was written by Bouckaert, Smoluckowski, and Wigner [Phys. Rev. **50**, 58 (1936)], referred to as **BSW**. It is clear and well written and is worth reading. The article by Koster in Solid State Physics **5**, 173 (1957) is more complete.

Tables 10-1, 10-3, 8-3 and 8-4 compile some properties of the standard semiconductors. An extensive compilation can be found in Landolt-Börnstein; New Series, Ed. in Chief. K. H. Hellewege; Vol. 17, Semiconductors, Eds. O. Maclelang, M. Schulz, H. Weiss; Subvolume a, Physics of Group IV Elements and III-V Compounds (Springer-Verlag, 1982).

The application of Brillouin zones to understanding binary metallic alloys is discussed in Altman, Chapter 5, and the first two editions of Kittel. References to the work of Jones and Hume-Rothery on this subject can be found in these books.

There is a new class of metals called **heavy fermion metals** where the effective mass of electrons (m*) at the Fermi energy is two to three orders of magnitude greater than that of a free electron. At $\mathscr{E}_F$ these materials have a very large density of electron states. This leads to an enormous γT term in the specific heat (Eq. 9-52), a very flat $\mathscr{E}$ vs. k (Eq. 10-10), and an enormous electron spin susceptibility (Eq.

9-59). Several of these metals also are superconducting (with huge dH_c/dT values, Eq. 16-1) and are the last three entries in Table 16-1. For further information see G. R. Stewart, Rev. Mod. Phys. **56**, 755 (1984) and the Notes in Chapter 16.

There are many good books on semiconductor physics. Some of them are listed in the bibliography. Dalvan, Kireev, Smith and Sze cover the subjects discussed here. Tunneling in solids of all kinds from p-n junctions to the Josephson effect is covered in C. B. Duke, Solid State Physics Supplement 10, (1969).

J. B. Gunn, Sol. State Commun. **1**, 88 (1963) is the original publication of the Gunn effect. Other discussions can be found in semiconductor texts. Also see Elliott and Gibson.

An excellent, readable overview of Group V **semimetals** is J-P. Issi, Aust. J. Phys. **32**, 585 (1979). Also see V. S. Edelman, Adv. in Phys. **25**, 555 (1976), and M. S. Dresselhaus J. Phys. Chem. Solids **32**, Suppl. 1, page 3 (1971).

The book by N. F. Mott and E. A. Davis is an excellent place to start reading about **amorphous semiconductors**. A more general reference on glasses is G. O. Jones, "Glass" (Chapman and Hall, 1971, 2nd ed.) and R. Zallen, "The Physics of Amorphous Solids" (Wiley, 1983). The original work on glasses was done by W. H. Zachariasen, J. Am. Chem. Soc. **54**, 3841 (1932).

Problems

1. Bloch functions – (a) Using the normalized free electron wave function $L^{-1/2}$ exp(ikx), with $L \neq 1$, show that the value for $\mathscr{E}_g$ given in Eq. 10-4 is obtained. What order of magnitude would you expect for V_1 in an ionic crystal and in a molecular crystal? (b) Show that Eq. 10-22 is correct for a Bloch function in three dimensions. By making the appropriate transformations show that if u_{kn} is a solution then so is $u_{-kn}*$, as pointed out in Eqs. 10-23.

2. Consider the results in Eq. 10-31 for the nearly free electron in a periodic lattice. Consider $k = \pi/a$ where the unperturbed energy is $\mathscr{E}_0$ and the energy of the upper gap is $\mathscr{E}_u = \mathscr{E}_0 + \Delta\mathscr{E}/2$. Expand the energy about $\mathscr{E}_u$, letting $k_{eff} = k - (\pi/a)$, and show that

$$\mathscr{E} - \mathscr{E}_u = \frac{\hbar^2}{2m^*}(k_{eff})^2$$

where $m/m^* = 1 + (4\mathscr{E}_0/\Delta\mathscr{E})$. Thus, for $\mathscr{E}_0/\Delta\mathscr{E} \gg 1$ a very small effective mass is expected at this point in the $\mathscr{E}$ vs. k diagram. (Hint: drawing the picture with a repeated zone scheme will help you to visualize the problem and results.)

350   CHAPTER 10   BAND THEORY

3. Reciprocal lattice and first Brillouin zones – Do problems 4, 5, 6, and 7a in Chapter 4.

4. Tight binding – Using the tight binding results in Eq. 10-52, show that for a bcc structure (eight neighbors) and a fcc structure (12 neighbors) the k-dependence of the energies are, respectively,

$$-8\gamma(\cos k_x a/2)(\cos k_y a/2)(\cos k_z a/2)$$
$$-4\gamma[(\cos k_y a/2)(\cos k_z a/2) + (\cos k_z a/2)(\cos k_x a/2) + (\cos k_x a/2)(\cos k_y a/2)]$$

For small k determine $\mathscr{E}(k)$ and the effective mass for both structures. At which points in k-space are the minimum and maximum energy values?

5. Band filling – Starting from Fig. 10-7b, we let $|k|$ be larger than $|a^*/2|$ in order to obtain the results shown in Fig. 10-10. Using this same approach, do the following. (a) Using $|k|$ a little smaller than $|a^*|$, show as in Fig. 10-10 the filled bands either in the reduced or periodic zone scheme, whichever is more appropriate. Notice how electrons are in the fourth Brillouin zone. For small crystal potentials, which electrons in the fourth zone have energies lower than those in the second zone? (b) Then take $|k|$ a little larger than $|a^*|$ so that the other "bits and pieces" of the fourth zone are occupied. Moving the fourth zone into the reduced zone scheme, notice the double valued band. Find some bands in Fig. 10-27 that have this property.

6. Energy bands for a square lattice – Consider the empty lattice approximation for the square lattice discussed in Section 10-13. (a) What is the energy of the next higher Γ band? Determine the appropriate linear combination of Bloch functions that transform as the correct irreducible representations. Expand these functions for small x and y. (b) Determine the Bloch function appropriate along the Z-line between energies 1/2 and 5/4 in the units used in Fig. 10-23a. From compatibility tables, what is the result at the X-point? Check this result with the functions just calculated.

7. Energy bands for a simple cubic lattice – Using the empty lattice approximation for a crystal with a simple cubic crystal structure determine the $\mathscr{E}$ vs. k curves along the Δ-line between Γ- and the X-point only up to and including energy $(\hbar^2/2m)(2\pi/a)^2$. Expand the Γ-functions for small x, y, and z.

8. Energy bands for a square crystal structure – Assume that the lattice potential energy of a two-dimensional square crystal structure is $-V_0$

cos(2πx/a) cos(2πy/a). Calculate the energy shifts from the empty lattice approximation at the lowest nontrivial Γ- and M-points.

9. Fermi level with doping — Assume that the binding energy of donors in Si is low enough so that at room temperature $n = N_D$. Show that the Fermi energy has a linear temperature dependence given by

$$\mathscr{E}_F = \mathscr{E}_c - k_B T \ln(N_c/N_D)$$

What is the maximum value of N_D in order to stay in the nondegenerate limit? What is the corresponding value of N_D if the material is at 77°K? What is this value if GaAs instead of Si is considered?

10. p-n junctions — The Si p-n junction illustrated in Fig. 10-34 has $N_A = 4 \times 10^{18}$ cm^{-3}, $N_D = 10^{16}$ cm^{-3} and an area of 2×10^{-3} cm^2. Calculate the contact potential and the depletion widths. Determine $E(x)$ and compare to the figure. What is the capacitance of the junction? What is the minority carrier density on the n-side? Why is Eq. 10-80b inappropriate to use to determine the minority carrier density on the p-side? What happens if this junction is at 77°K?

11. Tunnel (or Esaki) diodes — Consider the I-V characteristics of a degenerately doped p-n junction with the Fermi level in the conduction band on the n-side and in the valence band on the p-side. Note that for a small forward bias there is a current but for a slightly larger bias the current decreases. Sketch the band diagram and the corresponding point on the I-V curve for: zero voltage, a small forward voltage, a larger forward voltage in the negative differential conductivity region, a larger forward voltage in the ordinary forward biased conductivity region, a reverse bias showing Zener breakdown. This junction is called a **tunnel diode** or **Esaki diode** and its negative conductivity characteristic has been used for device applications.

12. The **Hall effect** is a powerful tool for studying properties of semiconductors, but it is more complicated than Eq. 9-65 where both electrons and holes are present. (a) Using the directions in Fig. 9-11, show that the transverse current density is

$$J_y = -ne\mu_e\theta_e E_x + pe\mu_p\theta_p E_x$$

where it assumed that the **Hall angle** is small. Notice that the electrons and holes are deflected to the same side. (b) Since the field required to reduce this field to zero is given by $E_y = -J_y/\sigma$ show that the **two carrier Hall effect** is given by

$$R = \frac{1}{ec} \frac{p\mu_p^2 - n\mu_e^2}{(p\mu_p + n\mu_e)^2}$$

(c) For intrinsic GaAs calculate the Hall coefficient at 300°K and 77°K. Notice how the higher mobility carrier dominates the value of R. (In **SI units** set c = 1.) (Note $\mu_e = \mu_n$.)

13. X-ray diffraction — In no more than two pages, outline the theory of x-ray diffraction from a crystal. Start from Eq. 10-94 and be sure to define the structure factor and atomic form factor. Calculate the former for a body-centered cubic structure and show that the missing reflections are in agreement with those shown in Section 4-2. Discuss how, and why, the form factor varies with scattering angle. (For example, see Ashcroft and Mermin or Kittel.)

14. The random close-packed (rcp) structure — In Section 3-7 we discussed that in the close-packed (fcc and hcp) structures hard spheres fill 0.74 of the space. There is a well-defined and reproducible structure that is a satisfactory model for an **amorphous metal,** called the rcp structure. The packing density is found to be 0.637 or 86% of the crystalline close-packed density. This result is found in amorphous metals or when ball bearings are shaken into bumpy-walled containers. In one paragraph for each topic discuss the following aspects of the rcp structure: (a) experiments that determine the packing fraction; (b) the radial distribution function, especially the value of the first peak to the internuclear distance and the splitting of the second peak; (c) describe a physical model for this splitting. [Hint: see R. Zallen, "The Physics of Amorphous Solids," (John Wiley and Sons, 1983), G. S. Cargill III, Solid State Phys. **30**, 227 (1975), H. J. Güntherodt, Festkörperprobleme **17**, 25 (1977).]

11

Some Thermal Effects in Solids

PART A HEAT CAPACITY

- 11-1 Specific Heat at Constant Volume and Pressure
  - a Theory
  - b Classical results
- 11-2 Energy and C_V from Statistical Mechanics
  - a Elementary statistical mechanics
  - b A two-level system
- 11-3 Classical Results for C_V
  - a Classical statistical mechanics
  - b C_V of free particles
  - c Potential energy of an atom in a crystal
  - d C_V of a harmonic oscillator
- 11-4 Einstein's Model
  - a General results for an oscillator
  - b The Einstein distribution
- 11-5 Debye's Calculation of C_V
  - a ω vs. **k**
  - b The density of states
  - c Results
  - d Shortcomings

PART B EFFECTS ASSOCIATED WITH DISORDER

- 11-6 Orientational Disorder in Molecular and Ionic Crystals
  - a Qualitative discussion
  - b Quantitative discussion
  - c Related systems (S)
- 11-7 Polarization by Orientation (S)
  - a Polarization by orientation in gases and liquids
  - b Polarization by orientation in solids
  - c Anomalous dispersion in solids
- 11-8 Point Imperfections in Crystals
  - a Introduction
  - b Point defects in crystals
  - c Density of defects
  - d Other point imperfections
- 11-9 Diffusion (S)
  - a Fick's first law
  - b The Einstein-Nernst equation
  - c Fick's second law and experiments
  - d A simple model
  - e Mechanisms of diffusion
- 11-10 Color Centers in Ionic Crystals (S)
- 11-11 Localized Vibrational Modes (S)

Notes
Problems

Table 11-1 The Debye temperature, Θ_D, in °K for a number of elements and compounds. The values are determined from measurements for $T \sim \Theta_D/2$.

A	90	Dy	155	Li(^{7}Li)	337	Rn	400
Ac	100	Er	165	Mg	330	Sb	140
Ag	220	Fe	460	Mn	420	Se	150
Al	385	Ga(rhom)	240	Mo	375	Si	635
As	275	Ga(tetra)	125	N	70	Sn(grey)	260
Au	180	Gd	160	Na	150	Sn(wh.)	170
B	1220	Ge	360	Nb	265	Ta	230
Be	940	H(para)	115	Nd	150	Tb	175
Bi	120	H(ortho)	105	Ne	60	Te	130
C(diam.)	2200	H($\eta-D_2$)	95	Ni	440	Th	140
C(graphite)	760	He	30	O	90	Ti	355
Ca	230	Hf	195	Os	250	Tl	90
Cd(hcp)	280	Hg	100	Pa	150	U	207
Cd(bcc)	170	I	105	Pb	85	V	280
Ce	110	In	140	Pd	275	W	315
Cl	115	Ir	290	Pr	120	Y	230
Co	440	K	100	Pt	225	Zn	250
Cr	430	Kr	60	Rb	60	Zr	240
Cs	45	La	130	Re	300		
Cu	310	Li(^{6}Li)	367	Rh	350		

LiF	670	RbCl	194	BN	600
LiCl	420	RbBr	149	CaF_2	470
LiBr	340	RbI	122	$CrCl_2$	80
LiI	280	CsF	245	$CrCl_3$	100
NaF	445	CsCl	175	Cr_2O_3	360
NaCl	297	CsBr	125	FeS_2	630
NaBr	238	CsI	102	InSb	200
NaI	197	AgCl	180	MgO	800
KF	335	AgBr	140	MoS_2	290
KCl	240	As_2O_3	140	SiO_2(quartz)	255
KBr	192	As_2O_5	240	TiO_2(rutile)	450
KI	173	$AuCu_3$(ord)	200	ZnS	260
RbF	267	$AuCu_3$(disord)	180		

SOME THERMAL EFFECTS IN SOLIDS

Nothing at first can appear more difficult to believe than that the complex organs and instincts have been perfected, not by means superior to, though analogous with, human reason, but by the accumulation of innumerable slight variations, each good for the individual possessor.

C. Darwin, "The Origin of Species"

So far our concern has been with static properties of crystals. In Chapter 9 electron motion was discussed but the atoms were considered to be fixed and essentially ignored. In the next chapter we shall look into details of atomic motion where the displacement from the equilibrium position is small compared to the internuclear distance. Such motions are called lattice vibrations (phonons). In this chapter we are concerned with effects of atomic motion where the details of the motion are not important. Part A is devoted to the specific heat of a solid where the model assumes that the atoms behave like harmonic oscillators with small displacements. In Part B we consider other thermal effects such as diffusion and orientational disorder; the common thread is that the displacement from the equilibrium positions is comparable to the internuclear distance.

Part A - Heat Capacity

11-1 Specific Heat at Constant Volume and Pressure

11-1a Theory The first law of thermodynamics (essentially the law of conservation of energy) states that the increase in energy, dU, of a system is equal to the amount of heat absorbed by the system, δQ, minus the amount of work done by the system, δW.

$$dU = \delta Q - \delta W \tag{11-1}$$

dU is an exact differential. This means that there is a quantity called the **internal energy**, U, which is a function of the state of the system (i.e. the temperature and volume), and that the difference between two values of U is independent of the "path" or the manner of getting from one state to another. Hence, U is the integral of an exact differential dU. However, the amount of heat absorbed by the system and the amount of work done by the system depends on the path taken to arrive at a new state of the system. If the energy of the system can be described by the thermodynamic variables pressure, temperature, and volume, then for an infinitesimal quasistatic reversible process, the work done by the system is PdV and the heat absorbed by the system is TdS (where dS is the change in entropy).

The heat capacity of such a system is defined as the amount of heat absorbed by a system per unit change in temperature. For a reversible infinitesimal process at constant volume (which defines the path) the heat capacity $C_V \equiv (dQ/dT)_V$ which is also given by $C_V \equiv (\partial U/\partial T)_V$. At constant pressure $C_P \equiv (dQ/dT)_P$ and from Eq. 11-1, $(\partial U/\partial T)_P = C_P - P(\partial V/\partial T)_P$. From these relations and standard thermodynamic manipulations we can show that

$$C_P - C_V = -T(\partial V/\partial T)_P^2 \, (\partial P/\partial V)_T \qquad (11\text{-}2a)$$

This expression can be written in terms of experimentally measured quantities. Recall that the volume compressibility of a solid is $K \equiv -(1/V)(\partial V/\partial P)_T$ and the volume thermal expansion coefficient $\beta \equiv (1/V)(\partial V/\partial T)_P$. Then Eq. 11-2a can be written as

$$C_P - C_V = \beta^2 \, T \, V/K \qquad (11\text{-}2b)$$

Figure 11-1 shows, for copper, experimental results for C_P and those calculated for C_V via Eq. 11-2b. Below 200°K there is very little difference between the two quantities, but above this temperature there is a difference (which is proportional to temperature over most of the range). The compressibility is always positive, as required for mechanical stability, so $C_P \geq C_V$. This inequality can be understood physically, since if the temperature of a system is increased at constant pressure the amount of heat absorbed by the system must be enough to increase the internal energy and also do external work in expanding the substance against the pressure of the surroundings. However, if the temperature is increased at constant volume, only enough heat need be absorbed to increase the internal energy of the substance.

Equation 11-2b is useful since the experimental measurements of specific heat for solids are normally done at constant (atmospheric) pressure, yielding C_P. However, theory calculates C_V since constant internuclear distances are used. From Eq. 11-2b, we can relate C_V to C_P by easily measurable quantities.

11-1b Classical results For solids at high temperatures, independent of the substance, we find experimentally that the vibrational contribution to $C_V \approx 3R = 6$ cal mole^{-1}°K^{-1} = 25.1 joule mole^{-1}°K^{-1} where R is the gas constant; see the Notes in Chapter 6. This is known as the **law of Dulong and Petit** (1819). (R = 1.987 cal°K^{-1} mole^{-1} is called the gas constant and is the product of the Boltzman constant k_B and Avogadro's number, 6.022×10^{23} mol^{-1}.) As will be shown, this temperature independent value is predicted by

CHAPTER 11   SOME THERMAL EFFECTS IN SOLIDS   357

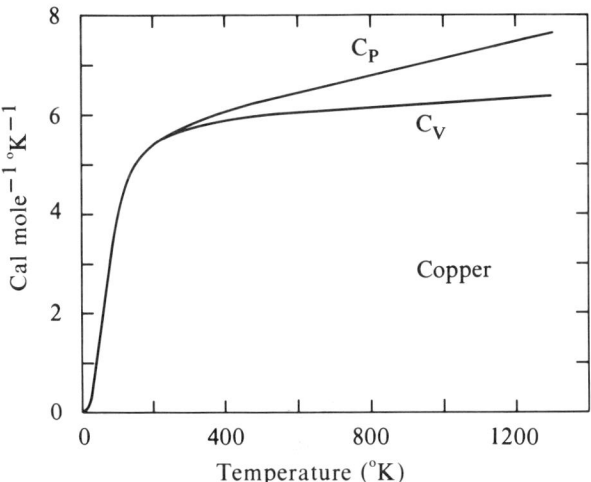

Fig. 11-1   The variation of C_P and C_V for copper with temperature. (From M. W. Zemansky, Heat and Thermodynamics, 3rd ed. McGraw-Hill, New York).

classical statistical mechanics and in the 1890's was considered to be well understood.

11-2 Energy and C_V from Statistical Mechanics

In order to calculate the vibrational contribution to C_V for solids we first review elementary quantum statistical mechanics and calculate C_V for a simple two-level system. This example provides a general understanding of the temperature dependence of C_V. In the next section a similar program is carried out using classical statistical mechanics, for a free particle and for an atom in a crystal. The latter result gives the law of Dulong and Petit.

11-2a Elementary statistical mechanics   For either one atom or a large system of interacting particles there are certain energy levels $\mathscr{E}_i$. It is easier to think of the various $\mathscr{E}_i$'s as associated with discrete energy levels; however, the results apply equally well to systems with continuous values of $\mathscr{E}$ but with integrals replacing the sums over $\mathscr{E}_i$. The probability of finding the ith level occupied (i.e., the probability of finding the system in the ith state) is

$$p_i = \frac{g_i\, e^{-\mathscr{E}_i/k_B T}}{\sum_j g_j\, e^{-\mathscr{E}_j/k_B T}} \equiv \frac{g_i\, e^{-\mathscr{E}_i/k_B T}}{Z} \qquad (11\text{-}3)$$

$$N_1 \rule{4cm}{0.4pt} \mathcal{E}_1 = \Delta\mathcal{E}$$

Fig. 11-2 The energy levels for a two-level system.

$$N_0 \rule{4cm}{0.4pt} \mathcal{E}_0 = 0$$

where g_i is the degeneracy of the ith level and the sum is over all the levels of the entire system. The sum in the denominator is called the **partition function**, Z. The mean internal energy of the system is

$$U = \sum_i p_i \mathcal{E}_i = \frac{\sum_i \mathcal{E}_i g_i \exp(-\mathcal{E}_i/k_B T)}{Z} \quad (11\text{-}4a)$$

Writing both U from Eq. 11-4a and $C_V = (\partial U/\partial T)_V$ in terms of the partition function, Z, we have

$$U = -\partial(\ln Z)/\partial(1/k_B T) \quad (11\text{-}4b)$$

$$C_V = T \, \partial^2 (k_B T \ln Z)/\partial T^2 \quad (11\text{-}4c)$$

Thus, evaluation of quantities like U and C_V can be done in a straightforward manner in terms of derivatives of the partition function. The problem is that to evaluate Z all of the values of $\mathcal{E}_i$ must be known.

11-2b A two-level system A system with two energy levels is an important system often met in practice. Also the simplicity makes the calculation and understanding transparent.

Figure 11-2 shows the two-level system, with a ground state at energy $\mathcal{E}_0 = 0$ and an upper state $\mathcal{E}_1 = \Delta\mathcal{E}$. N_0 and N_1 are the number of ground and upper occupied states. (We have not said what is occupying these two states and it is not necessary to specify anything. One might think of a large number of atoms each with two electronic states, or of impurity molecules in a solid that have only two possible orientations.) Setting $N = N_0 + N_1$, and taking all the degeneracies equal to one, from Eq. 11-3 we have

$$\frac{N_1}{N} = \frac{\exp(-\Delta\mathcal{E}/k_B T)}{1 + \exp(-\Delta\mathcal{E}/k_B T)} = \frac{1}{1 + \exp(\Delta\mathcal{E}/k_B T)} \quad (11\text{-}5a)$$

The energy of the system is $N_1 \Delta\mathcal{E}$ and the heat capacity is the derivative of this energy with respect to temperature. Thus,

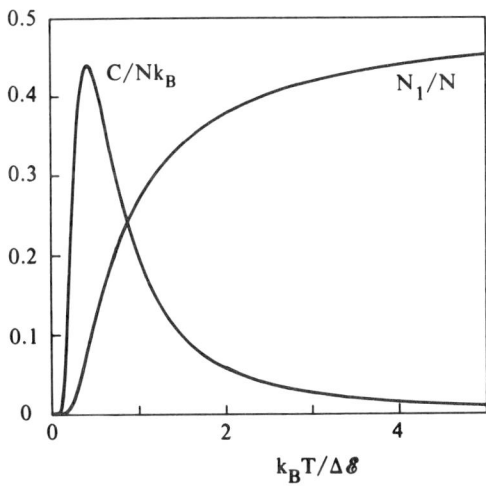

Fig. 11-3 The heat capacity and fraction of the population in the upper state for a two-level system as a function of reduced temperature.

$$U = N_1 \Delta\mathscr{E} = N \Delta\mathscr{E} / [1 + \exp(\Delta\mathscr{E}/k_B T)] \quad (11\text{-}5b)$$

$$C_V = \frac{\partial U}{\partial T} = Nk_B \frac{(\Delta\mathscr{E}/k_B T)^2 \exp(\Delta\mathscr{E}/k_B T)}{[1 + \exp(\Delta\mathscr{E}/k_B T)]^2} \quad (11\text{-}5c)$$

Figure 11-3 shows the temperature dependence of the heat capacity and that of N_1. Notice that at a temperature of less than one-tenth of the energy separation there is hardly any contribution to C_V. However, at slightly higher temperatures, but still at temperatures well below those corresponding to the energy separation ($\Delta\mathscr{E}/k_B$), the heat capacity starts to rise rapidly (approximately exponentially). This is because energy (heat) is required as the upper level begins to be filled. The heat capacity peaks close to, but below, that corresponding to the energy separation.

Peaks in the heat capacity of the type shown in Figure 11-3 are called **Schottky anomalies** and are observed at low temperatures in some solids. For some materials, $\Delta\mathscr{E}$ is so small that the experimental measurements yield only the high temperature tail. To analyze this situation note that the high temperature expansion of the heat capacity in Eq. 11-5c is $(Nk_B/4)(\Delta\mathscr{E}/k_B T)^2$, so the experiments yield $N\Delta\mathscr{E}^2$. Schottky anomalies are observed in a variety of solids; for example, for paramagnetic ions where there is a crystal field splitting of the ground state and in ionic crystals that have OH^- impurities with rotational states.

11-3 Classical Results for C_V

11-3a Classical statistical mechanics We now write the average energy for a classical system. Recall from elementary statistical mechanics that for a collection of N particles there are 3N position and 3N momentum coordinates, which are

$$\mathbf{q} = (x_1, y_1, z_1, x_2, ..., x_N, y_N, z_N)$$
$$\mathbf{p} = (p_{x1}, p_{y1}, p_{z1}, p_{x2}, ..., p_{xN}, p_{yN}, p_{zN}) \tag{11-6a}$$

Any measurable quantity, A, depends on the position and momentum of all the particles. This is expressed as $A(\mathbf{p}, \mathbf{q})$, and the energy of the system is given by $\mathscr{E}(\mathbf{p}, \mathbf{q})$. Then the thermal average is

$$\langle A \rangle = \frac{\int A \exp(-\mathscr{E}/k_B T) \, d\mathbf{p} \, d\mathbf{q}}{\int \exp(-\mathscr{E}/k_B T) \, d\mathbf{p} \, d\mathbf{q}} \tag{11-6b}$$

where the $\langle \; \rangle$ brackets are used for thermal averages. The similarity between this expression, when A equals $\mathscr{E}$, and that in Eq. 11-4a is obvious. The sum in Eq. 11-4a for discrete levels goes over to an integral when the measurables depend continuously on the coordinates.

11-3b C_V of free particles Consider a single free particle in one dimension, the x-direction. Then $\mathscr{E} = p_x^2/2m$. Using Eq. 11-6b to calculate the average thermal one-particle energy,

$$\langle \mathscr{E} \rangle = \frac{\frac{1}{2m} \int_{-\infty}^{\infty} p_x^2 \exp(-p_x^2/2mk_B T) \, dp_x}{\int_{-\infty}^{\infty} \exp(-p_x^2/2mk_B T) \, dp_x} \tag{11-7a}$$

The integrals over $\mathbf{q}$ in the numerator and denominator cancel. Setting $t^2 = p_x^2/2mk_B T$, the energy is evaluated to be

$$\langle \mathscr{E} \rangle = \frac{k_B T \int_{-\infty}^{\infty} t^2 \exp(-t^2) \, dt}{\int_{-\infty}^{\infty} \exp(-t^2) \, dt} = k_B T \frac{\pi^{1/2}/2}{\pi^{1/2}} = \frac{k_B T}{2}$$

$$\tag{11-7b}$$

This is just the classical result that each particle has an energy of $k_B T/2$ for each degree of freedom.

If we have N free particles in three dimensions, the expression looks more complicated. However, since the particles are not coupled, the results reduce to N expressions of the same type as already stated. The energy of the particles is then

CHAPTER 11   SOME THERMAL EFFECTS IN SOLIDS   361

$$\mathscr{E} = \frac{1}{2m} \sum_{i=1}^{N} (p_{xi}^2 + p_{yi}^2 + p_{zi}^2) \tag{11-8a}$$

At thermal equilibrium the thermal average energy can be written out just as the preceding

$$\langle \mathscr{E} \rangle = \frac{\frac{1}{2m} \sum_{i=1}^{3N} \int_{-\infty}^{\infty} p_i^2 \prod_{j=1}^{3N} \exp(-p_j^2/2mk_BT) \, d\mathbf{p}}{\int_{-\infty}^{\infty} \prod_{j=1}^{3N} \exp(-p_j^2/2mk_BT) \, d\mathbf{p}} = \frac{3}{2} Nk_BT$$

(11-8b)

The integrals over d**q** cancel and the 3N integrals are just the same as those in the preceding equations. As shown, the result is $3Nk_BT/2$ or just 3N larger than the result in Eq. 11-7b. The 3 occurs because the particles have three degrees of freedom (three directions) rather than one, and the N occurs because there are this many particles.

Thus, from the kinetic energy for N free particles the specific heat is $3Nk_B/2$. This is half of the Dulong and Petit value. We shall see that the other half comes from the potential energy which holds the atoms at their equilibrium positions in the solid, that is, the energy depends on the 3N positions as well as the 3N momenta.

11-3c   Potential energy of an atom in a crystal   In Chapters 6 and 7, for molecularly and ionicly bonded materials, we determined the static internuclear distance dependence of the energy. For example, see Figure 7-3. Actually, the general shape near the minimum of energy is always the same for any type of bonding. At the equilibrium distance R_0, $dU/dR = 0$ and U increases for smaller as well as larger R. Thus, about R_0 the energy can always be expanded and as a power series in $R - R_0$; besides a constant energy, the first nonzero term depends on $(R - R_0)^2$, which is a **harmonic oscillator potential.** In such a potential the atoms in a crystal will oscillate with amplitudes small compared to the internuclear distance. The energy of one harmonic oscillator in one dimension, with oscillator frequency ω, is

$$\mathscr{E} = p^2/2m + m\omega^2 q^2/2 \tag{11-9a}$$

11-3d C_V of a harmonic oscillator   Using this expression for the energy of a one-dimensional harmonic oscillator, the average thermal energy can be evaluated using Eq. 11-6b.

$$\langle \mathscr{E} \rangle = \frac{\frac{1}{2m}\int_{-\infty}^{\infty} p^2 \exp(-p^2/2mk_BT)dp}{\int_{-\infty}^{\infty} \exp(-p^2/2mk_BT)dp} \tag{11-9b}$$

$$+ \frac{\frac{m\omega^2}{2}\int_{-\infty}^{\infty} q^2 \exp(-m\omega^2 q^2/2k_BT)dq}{\int_{-\infty}^{\infty} \exp(-m\omega^2 q^2/2k_BT)dq}$$

where the other factors in the numerator and denominator cancel. The first term is identical with Eq. 11-7a and yields $k_BT/2$. The second term also evaluated, via Eq. 11-7b also yields $k_BT/2$. This proves the result that the contribution from each degree of freedom, momentum or position, to the average thermal energy is $k_BT/2$. The equality of these two terms is called the **equipartition theorem**.

For N harmonic oscillators in three dimensions the average internal thermal energy is $3Nk_BT$ so $C_V = 3Nk_B$, which for a mole of oscillators is just $3R$, the Dulong and Petit value! This value is independent of temperature, and classical physics offers no possibility of a temperature dependence. However, as will be presently seen, once the oscillators are quantized we obtain very different results at low temperatures while retaining these high temperature values.

11-4 Einstein's Model

In 1907 Einstein, in one of the first papers on quantum effects in solids, formulated a microscopic theory of C_V. This paper, following his 1905 paper on the photoelectric effect, is another beautiful example of the application of Planck's quantization rule. The results derived below by quantizing the lattice harmonic oscillators are completely general and serve as the basis of all calculations of the average thermal energy and C_V. In Section 11-4b a rather simple specific model of the frequency distribution of the oscillators is used to evaluate C_V, and in Section 11-5 the Debye model is used.

11-4a General results for an oscillator As a model of a solid, Einstein assumed that the atoms are vibrating as harmonic oscillators, but instead of taking the classical expression for the energy of an oscillator, Eq. 11-9a, he assumed Planck's quantization rule for each oscillator $\mathscr{E}_i = n_i h\nu = n_i \hbar\omega$. n takes on all integer values from 0 to ∞ and the subscript is dropped. Using Eq. 11-4a for the average thermal energy of an oscillator ($g_i = 1$ for all i) then

CHAPTER 11   SOME THERMAL EFFECTS IN SOLIDS   363

$$\langle \mathscr{E} \rangle = \frac{\sum_{n=0}^{\infty} n\hbar\omega \exp(-n\hbar\omega/k_BT)}{\sum_{n=0}^{\infty} \exp(-n\hbar\omega/k_BT)} \quad (11\text{-}10)$$

The denominator is just the partition function, Z, Eq. 11-3, which for $x \equiv \exp(-\hbar\omega/k_BT)$ is

$$Z = \sum_{n=0}^{\infty} x^n = \frac{1}{1-x} = \frac{1}{1 - \exp(-\hbar\omega/k_BT)} \quad (11\text{-}11)$$

Using Eq. 11-4b to evaluate $\mathscr{E}$ from the partition function, we have

$$\langle \mathscr{E} \rangle = \frac{\hbar\omega \exp(-\hbar\omega/k_BT)}{1 - \exp(-\hbar\omega/k_BT)} = \frac{\hbar\omega}{\exp(\hbar\omega/k_BT) - 1} \quad (11\text{-}12)$$

Note that in the high temperature limit, $k_BT \gg \hbar\omega$, this expression reduces to k_BT, which is just the classical result.

For an assembly of N oscillators, each of which can oscillate in three dimensions, there are 3N vibrational frequencies given by ω_i. We must sum over all discrete frequencies, or the sum can be replaced by an integral if the frequency distribution is continuous. Thus,

$$U = \sum_{i=0}^{3N} \frac{\hbar\omega_i}{\exp(\hbar\omega_i/k_BT) - 1} = \int_0^{\omega_m} \frac{\hbar\omega}{\exp(\hbar\omega/k_BT) - 1} g(\omega) \, d\omega$$

with

$$\int_0^{\omega_m} g(\omega) \, d\omega = 3N \quad (11\text{-}13)$$

where $g(\omega)$ is the density of oscillators between ω and $\omega + d\omega$, and ω_m is the maximum frequency of the assembly of oscillators. As indicated, the integral over $g(\omega)$ must be the 3N. The derivative of these expressions with respect to T gives C_V.

$$C_V = \sum_0^{3N} k_B \left(\frac{\hbar\omega_i}{k_BT}\right)^2 \frac{\exp(\hbar\omega_i/k_BT)}{[\exp(\hbar\omega_i/k_BT) - 1]^2} \quad (11\text{-}14)$$

$$= \int_0^{\omega_m} k_B \left(\frac{\hbar\omega}{k_BT}\right)^2 \frac{\exp(\hbar\omega/k_BT)}{[\exp(\hbar\omega/k_BT) - 1]^2} g(\omega) d\omega$$

As T approaches zero, C_V approaches zero; as discussed previously, C_V approaches $3k_BN$ at very high temperatures. These limits agree with experiment. What remains is to pick a reasonable frequency distribution for the oscillators. The results of the Einstein assumption are presented next, followed by Debye's approximation. (It is now realized that for a simple harmonic oscillator $\mathscr{E} = (n + 1/2)\hbar\omega$ rather than $n\hbar\omega$ as was used by Einstein. The term $\hbar\omega/2$ is the **zero-point energy**. The addition of this term only adds a constant $\hbar\omega/2$ to the energy expressions and makes no contribution to the expressions to C_V. See Problem 3.)

11-4b The Einstein distribution   For simplicity Einstein assumed that all the atoms in the solid vibrate independently and at the same frequency ω_E. Then the terms in the sums in Eqs. 11-13 and 11-14 are all the same, can be taken outside of the summation, and the sum gives just 3N. Then these equations give

$$U = 3N \frac{\hbar\omega_E}{\exp(\hbar\omega_E/k_B T) - 1} \quad (11\text{-}15a)$$

$$C_V = 3k_B N \left(\frac{\hbar\omega_E}{k_B T}\right)^2 \frac{\exp(\hbar\omega_E/k_B T)}{[\exp(\hbar\omega_E/k_B T) - 1]^2}$$

$$= 3zR \left(\frac{\Theta_E}{T}\right)^2 \frac{\exp(\Theta_E/T)}{[\exp(\Theta_E/T) - 1]^2} \quad (11\text{-}15b)$$

We have written the molar heat capacity in the second equation for C_V. This is done by setting N to be the number of formula units in a mole (Avogadro's number) and z is the number of atoms in the formula unit. For example, in the molar heat capacity of $BaTiO_3$, $z = 5$. In the second expression for C_V we also have defined an **Einstein temperature** from the frequency, as can always be done, $k_B \Theta_E \equiv \hbar\omega_E$.

As mentioned before, for $T \gg \Theta_E$, $C_V = 3zR$, the correct high temperature value. For $T \ll \Theta_E$, $C_V = 3R(\Theta_E/T)^2 \exp(-\Theta_E/T)$, essentially an exponential falloff. This is more rapid than the $\sim T^3$ dependence that is observed experimentally. Figure 11-4 shows the experimental C_V data of silver fitted to an Einstein curve (Eq. 11-15b). Although the agreement is reasonable, it is better for the Debye model.

The similarity between the Einstein quantized oscillator model and the two-level system, Eq. 11-5b, should be noted. For $T \ll \Theta_E$ and $\Delta\mathscr{E} = \hbar\omega$ the same results are obtained and the reason is clear. In this temperature range most of the quantized oscillators are in the n = 0 state, a few in the n = 1 state, and the upper states are not occupied. Compare the equations for C_V and Figs. 11-4 and 11-3. However, at higher temperatures the upper n-states are occupied and the behavior of the quantized oscillator is different from that of a two-level system.

CHAPTER 11   SOME THERMAL EFFECTS IN SOLIDS   365

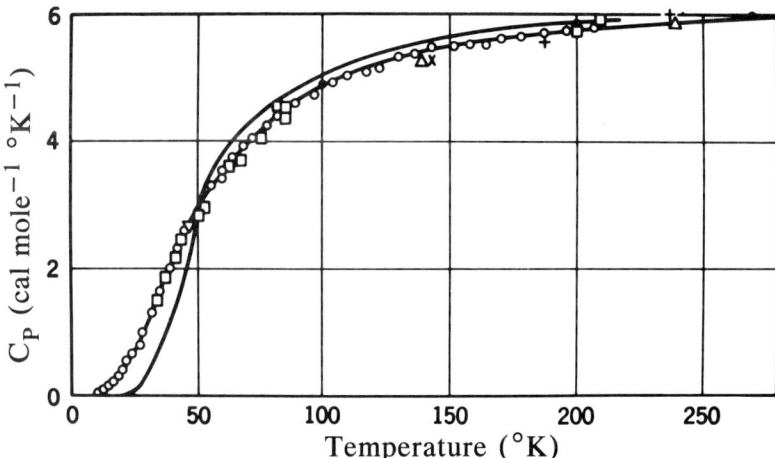

Fig. 11-4 Specific heat data (points) for silver. The lines are the fits from the Einstein and Debye results. The Debye curve goes through the data points. (From Blatt and Kittel, 2nd ed.)

11-5 Debye's Calculation of C_V

Debye realized that there are not only high-energy modes, as in the Einstein distribution model, but also very low-energy modes as well. For example, sound waves have very low energy and very long wavelength. The two-level problem shows that there will be a contribution to C_V at temperatures corresponding to that of the energies of the modes. Thus, the low-energy modes cannot be ignored.

11-5a ω vs. k Debye suggested that a more accurate way to characterize the vibrational modes of a crystal would be to treat the crystal as a continuous medium that is isotropic. For a one-dimensional continuum of length L (e.g., a string) the wave equation is

$$\frac{\partial^2 u}{\partial x^2} = \frac{1}{v^2} \frac{\partial^2 u}{\partial t^2} \tag{11-16a}$$

where u(x, t) is the deflection of the string at a time t and position x, and v is the velocity of propagation (equal to the square root of the tension divided by mass per unit length). For fixed ends at x = 0 and L, the standing wave solutions (i.e. $n\lambda/2 = L$) are

$$u(x, t) = A \sin(n\pi x/L) \cos(\omega_m t) \tag{11-16b}$$

where n is an integer ≥ 1 and the angular frequency $\omega = 2\pi v/\lambda = vk = 2\pi v n/2L$. It is important to notice that for the continuous medium there is a linear relation between ω and k, the wave number, and the frequency spectrum is discrete. The spatial part of Eq. 11-16b is, of course, just the same as in Eq. 9-24 for electron waves where the same type of problem was encountered, namely fitting an integral number of half wavelengths into a sample of length L.

For elastic waves in a three-dimensional isotropic solid, one longitudinal and two transverse vibrational modes can propagate. (The displacements are parallel to the propagation direction for longitudinal vibrations and perpendicular for transverse vibration.) The wave equations and running wave solutions are

$$\nabla^2 u_\ell = v_\ell^{-2}(\partial^2 u_\ell/\partial t^2), \qquad \nabla^2 u_t = v_t^{-2}(\partial^2 u_t/\partial t^2) \quad (11\text{-}17a)$$

$$u = A \exp i(\mathbf{k}\cdot\mathbf{r} - \omega t) \quad (11\text{-}17b)$$

where the subscripts ℓ and t are for longitudinal and transverse waves, respectively, and the displacement equation is appropriate for either with the correct subscripts on u, k, and ω, but $k_i = \omega_i/v_i$ always.

The linear dependence of ω on k, that is, $\omega = vk$, in the Debye model for both transverse and longitudinal waves should be emphasized. In the Einstein distribution ω ($=\omega_E$) is a constant high frequency value that represents the independent oscillation of all the atoms; ω_E has no k dependence. In the Debye model the atoms play no direct role; the solid is assumed to behave as a continuous medium. (The behavior of a real solid is in between these extremes.)

Since the Debye approximation considers very long wavelength, that is, small k modes, the velocities in Eq. 11-17a can be related to the elastic constants of the solid. Assuming, as Debye did, that the material is isotropic, then v can be related to the elastic constants

$$v_\ell = (c_{11}/\rho)^{1/2} \qquad v_t = [(c_{11} - c_{12})/2\rho]^{1/2} \quad (11\text{-}18)$$

where ρ is the density and c_{ij} are the two independent elastic constants written using the reduced notation (Chapter 5). v_ℓ is always greater than v_t for real crystals.

11-5b The density of states Now that ω vs. k is determined for the Debye approximation, the density of states must be found so that Eq. 11-14 can be used to determine C_V.

The calculation of the density of states is the same as in Section 9-8 for electrons waves in a solid. The only difference is that for electrons every allowed k-value could accommodate two electrons (spin up and spin down), while for the vibrational problem three vibrational modes can be accommodated (two transverse and one longitudinal) for each k. By assuming periodic boundary conditions

CHAPTER 11   SOME THERMAL EFFECTS IN SOLIDS   367

for the displacement u(**r**, t) in Eq. 11-17b for a cube of side L then

$$e^{i(k_x x + k_x L)} = e^{ik_x x} \quad \text{and so on for y and z}$$

$$k_x L = 2\pi m_x \quad \text{where } m_x = 0, \pm 1, \pm 2 \quad (11\text{-}19a)$$

$$\mathbf{k} = (2\pi/L)\mathbf{m} \quad \text{where } \mathbf{m} = (m_x, m_y, m_z) \quad (11\text{-}19b)$$

This last relation is a simple way to write the allowed **k** and is the same as Eq. 9-27. Writing k for $|\mathbf{k}|$, the density of modes with a k-value between k and k + dk is g(k) and is given by the integral in the three-dimensional k-space

$$\int_0^k g(k)\,dk = \frac{4\pi}{3} m^3 = \frac{4\pi}{3} k^3 \left(\frac{L}{2\pi}\right)^3$$

$$g(k) = 4\pi k^2 \frac{V}{(2\pi)^3} = \frac{V}{2\pi^2} k^2 \quad (11\text{-}20)$$

where $L^3 = V$, the volume of the sample. To change this quantity into $g(\omega)$, as in Eq. 11-14, use $g(\omega)\,d\omega = g(k)\,dk$ and $\omega = v_i k$ so

$$g(\omega) = g(k)\frac{dk}{d\omega} = \frac{V}{2\pi^2 v_i} k^2 = \frac{V}{2\pi^2 v_i^3} \omega^2 \quad (11\text{-}21a)$$

This expression is applicable to any one of the three modes. The results can be written out explicitly or combined with an average velocity v_0

$$g(\omega) = \frac{\omega^2 V}{2\pi^2}\left(\frac{1}{v_l^3} + \frac{2}{v_t^3}\right) = \frac{\omega^2}{2\pi^2}\frac{3V}{v_0^3} \quad (11\text{-}21b)$$

The former expression is used if v_l and v_t are to be evaluated from the elastic constants via Eq. 11-18, but more usually the latter expression is used to obtain a one-parameter fit to experimental data. To normalize $g(\omega)$, as in Eq. 11-13, Debye assumed that the linear dependence of ω vs. k continues up to some maximum cutoff value, which is called the **Debye frequency** ω_D, so that the total number of modes is 3N.

$$3N = \int_0^{\omega_D} g(\omega)\,d\omega = \int_0^{\omega_D} \frac{\omega^2}{2\pi^2}\frac{3V}{v_0^3}\,d\omega = \frac{\omega_D^3}{2\pi^2}\frac{V}{v_0^3} \quad (11\text{-}22a)$$

$$\omega_D = (6\pi^2 N/V)^{1/3} v_0 \quad (11\text{-}22b)$$

$$\Theta_D = \frac{\hbar \omega_D}{k_B} = \frac{\hbar v_0}{k_B}\left(\frac{6\pi^2 N}{V}\right)^{1/3} \quad (11\text{-}22c)$$

where the last equation defines the **Debye temperature**, Θ_D. (A temperature can always be defined from a frequency via $\hbar\omega = k_B T$.) Now the density of states result, Eq. 11-21b, can be directly substituted into the expressions Eqs. 11-13 and 11-14, for U and C_V.

$$U = \frac{3V(k_B T)^4}{2\pi^2 v_0^3 \hbar^3} \int_0^{\omega_D} \frac{\left(\frac{\hbar\omega}{k_B T}\right)^3}{\exp(\hbar\omega/k_B T) - 1} \frac{\hbar\, d\omega}{k_B T}$$

$$= 9Nk_B T \left(\frac{T}{\Theta_D}\right)^3 \int_0^{x_D} \frac{x^3 dx}{\exp(x) - 1} \quad (11\text{-}23a)$$

$$C_V = 9Nk_B \left(\frac{T}{\Theta_D}\right)^3 \int_0^{x_D} \frac{x^4 \exp(x)\, dx}{[\exp(x) - 1]^2} \quad (11\text{-}23b)$$

where $\quad x \equiv \hbar\omega/k_B T, \quad x_D \equiv \hbar\omega_D/k_B T = \Theta_D/T \quad (11\text{-}23c)$

The terms in the first expression for U are grouped so that the variable $\hbar\omega/k_B T$ can be changed to x, making integration simpler. (See Problem 8 in Chapter 12 for a general way to treat U and C_V.)

11-5c Results Figure 11-4 gives the results derived from the Debye model for $\Theta_D = 210°K$ as well as the experimental results. As can be seen the fit is very good, and it is useful to compare these results to the fit using the Einstein distribution. At low temperatures the latter falls more rapidly than the Debye calculation.

The T^3 dependence of C_V at low temperature for the Debye model comes naturally. In this region x_D in Eq. 11-23b can be taken as ∞ and then the definite integral can be explicitly evaluated.

$$\int_0^\infty \frac{x^3 dx}{\exp(x) - 1} = \int_0^\infty x^3 dx \sum_{p=1}^\infty e^{-px} = \sum_{p=1}^\infty \int_0^\infty e^{-px} x^3 dx$$

$$= 6 \sum_{p=1}^\infty \frac{1}{p^4} = \frac{\pi^4}{15} \quad (11\text{-}24)$$

where the sum can be found in standard tables. Thus, for $T \ll \Theta_D$ we obtain the famous Debye T^3 law.

$$C_V = \left(\frac{12\pi^4}{5}\right) Nk_B \left(\frac{T}{\Theta_D}\right)^3 \quad (11\text{-}25)$$

and in this same temperature region a T^4 dependence for the internal energy, that is, $U = 3\pi^4 Nk_B T^4/5\Theta_D^3$. These results are reasonably accurate for temperatures below $\Theta_D/10$. In the preceding, N is always the number of atoms in the sample. To write this expression in moles, we take Avogadro's number of molecules or formula units $N_A =$

CHAPTER 11   SOME THERMAL EFFECTS IN SOLIDS   369

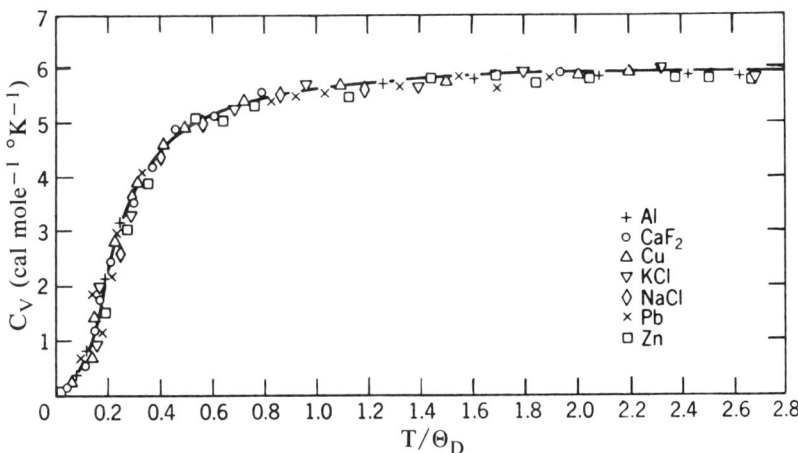

Fig. 11-5   Heat capacity vs. reduced temperature for a number of materials.

6.022×10^{23} mol^{-1} for N and obtain $C_V = (12\pi^4/5)zR(T/\Theta_D)^3$ where z is the number of atoms in one molecule or formula unit and $R = k_B N_A$, the gas constant. For $z = 1$ and for the units that are usually used $C_V = 464.4(T/\Theta_D)^3$ cal/mol°K = 1944 $(T/\Theta_D)^3$ J/mol°K.

In the other limit $T \gg \Theta_D$, it can be shown that $C_V = 3Nk_B[1 - (1/20)(\Theta_D/T)^2 + ...]$. At intermediate temperatures tables must be used to evaluate the integrals. (See the Notes.)

The Debye model has been very useful for calculating the specific heats of many solids. Figure 11-5 shows C_V vs. T/Θ_D plotted for a number of materials. As can be seen this one universal formula, Eq. 11-23b, correlates a great deal of data with just one parameter. Table 11-1 (page 354) lists Θ_D for a selected number of elements and compounds. (The Debye model also works for non-crystalline solids.)

11-5d Shortcomings   Although the Debye theory of specific heats is extremely useful in fitting a broad sweep of crystalline solids, it does have defects. Even in the 1920's it was appreciated that Θ_D obtained from fitting C_V was not always in agreement with values derived from the elastic constants (Eqs. 11-18 and 11-22c). See Gopal's Table 2.IV, referenced in the Notes, for a comparison of Θ_D obtained by these two approaches. As more accurate values of C_V were determined experimentally, it was seen that the calculated C_V did not fit the experimental results exactly. This is usually demonstrated by calculating, at each temperature, the effective values of Θ_D that are necessary to fit the experimental data. For most materials, we find that as the temperature is lowered, Θ_D is constant to about $\Theta_D/2$; below this temperature it starts to decrease, usually having a minimum $\approx \Theta_D/10$, then increasing at lower temperatures and again becoming

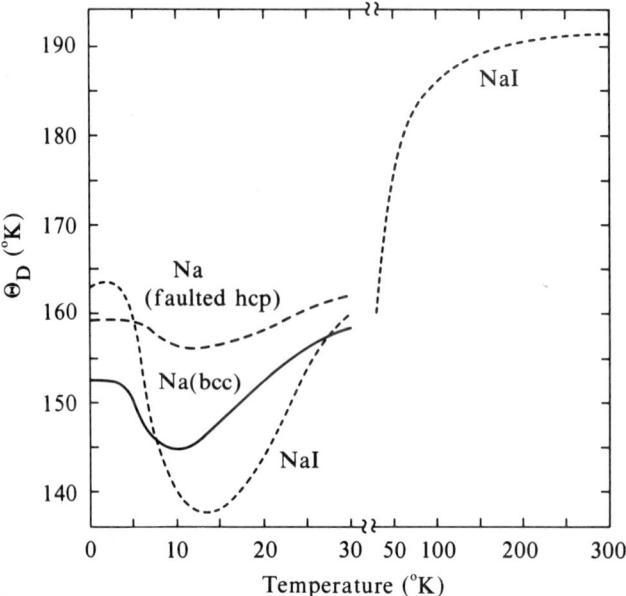

Fig. 11-6 The temperature dependence of Θ_D for two phases of metalic sodium [J. D. Filby and D. L. Martin, Proc. Roy. Soc. (London) **276A**, 187 (1963).] and for NaI [W. T. Berg and J. A. Morrison, Proc. Roy. Soc. (London) **242A**, 467 (1957)].

constant at very low temperatures. The values obtained at high and low temperatures are usually not the same; Fig. 11-6 shows some typical results. Naturally the Debye theory implies a temperature independent Θ_D. Note that Θ_D is also structure dependent. This comes via the dependence of Θ_D on the parameters in Eq. 11-22b.

At one time such variations, Figure 11-6, were attributed to experimental errors. However, in 1937 Blackman reopened the discussion by showing that a real crystal can have a $g(\omega)$ that deviates considerably from an ω^2 dependence. The fundamental problem is that the idea of treating a solid as an elastic continuum is inadequate and the atomic nature must be taken into account. When this is done, a complicated ω vs. k and $g(\omega)$ results. Using this $g(\omega)$ in Eq. 11-14 the heat capacity can be fit very well; alternatively the temperature dependence of Θ_D can be predicted using more realistic $g(\omega)$ results.

It is amusing to note that Born and von Karman developed a fairly rigorous lattice dynamic method to compute correctly the modes of vibration of a solid in the same year, 1912, as Debye's publication (see the Notes for references). However, the Debye model fit the data very well, so the more complicated approach was not investigated. Now ω vs. k dispersion curves of the lattice vibrations are of great

CHAPTER 11   SOME THERMAL EFFECTS IN SOLIDS   371

interest for many reasons besides accurate calculation of C_V. Methods to obtain these results are discussed in the next chapter.

Part B-Effects Associated with Disorder

So far we have been discussing a number of aspects of crystalline solids where perfect long-range order has been assumed. Another way to say this is that the translation vector, Eq. 1-10 or 2-1, perfectly describes the lattice, and that the basis associated with every lattice point is the same throughout the crystal structure. However, real crystalline solids are never perfect. To begin with, they are finite in size, so they have boundaries. However, the surface effects can be rather small; if the crystal has N_1, N_2, and N_3 primitive translations for t_m in the three directions ($N_1 N_2 N_3 = N$) then the ratio of the surface area to volume is approximately $N^{-1/3}$, a very small number since normally $N \approx 10^{23}$. Thus, when considering bulk properties, we can ignore the surface. In this chapter we will cover a few topics concerning disorder in the bulk of the crystal. These topics can be divided, in general, into orientational disorder of the basis and disorder of the structure due to some missing atoms. Either of these two effects creates a disordered crystal structure, but the division is pedagogically and qualitatively useful although quantitatively both effects can be treated similarly.

11-6 Orientational Disorder in Molecular and Ionic Crystals

11-6a Qualitative discussion Figure 11-7 shows two unit cells of the high temperature disordered NH_4Cl crystal structure. The averaged crystal structure is a simple cubic lattice with the basis of Cl at (0, 0, 0) and NH_4 at (1/2, 1/2, 1/2). The averaged space group is then $O_h^1(Pm3m)$. However, notice that the NH_4 group in cell B is rotated by 90° along one of the $\langle 100 \rangle$ directions with respect to that in cell A. In the high temperature phase the probability of finding the NH_4^+ ion in position A, in any randomly chosen cell, is equal to the probability of finding it in position B. Thus, with respect to these two rotational states, the crystal structure is disordered. This disorder leads to the term "average" crystal structure and "average" space group. If the orientation of the NH_4 ion in A and B were superimposed, or averaged, then H atoms would be at the eight corners of a cube leading to point group of $O_h(m3m)$, and the averaged space group $O_h^1(Pm3m)$.

In NH_4Cl and related crystals there is a transition temperature, T_c, below which the crystal orders. Below T_c all the NH_4^+ ions as-

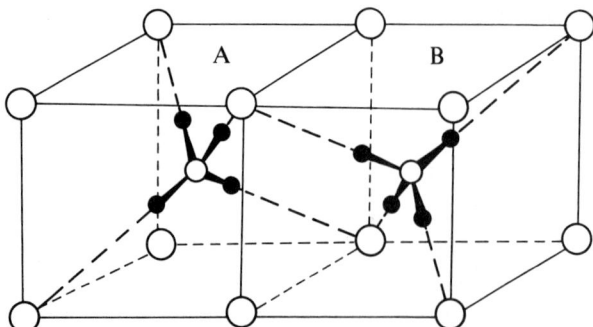

Fig. 11-7 Two unit cells of NH_4Cl in the high temperature phase showing the two possible NH_4^+ ion orientations.

sume the orientational position of the one in cell A (or it could be that of cell B), which leads to an ordered crystal structure.

This type of orientational order-disorder transition is fairly common in ionic crystal that contain highly symmetric polyatomic groups such as NH_4, CH_3, etc., because the hydrogen atom is very small and at high temperature the "bumps" due to its charge density have only a small effect in keeping the ion from rotating from one orientation to another. Other examples of this phenomenon will be given.

We should note that the preceding discussion has been missing one important aspect. We do not have perfect order immediately below T_c in all phase transitions. Rather the situation can be described by a **long-range order parameter**, Δ, where Δ goes from zero above T_c to a value >0 below T_c. This parameter is usually chosen so that it is equal to one for complete order. In the present situation Δ is usually defined in the following way: there are a total number N of NH_4 ions, some of which have the orientation A of Fig. 11-7 and some with the orientation B. Then the number of ions with orientation of the A type is given by $(1 + \Delta)N/2$; the number with orientation B must therefore be $(1 - \Delta)N/2$. With this definition, $\Delta = \pm 1$ gives perfect long-range order, where $\Delta = +1$ means all the NH_4 ions will have the orientation of the A type, and $\Delta = -1$ they all will be of the B type. If $\Delta = 0$ there is equal probability of finding any NH_4 ion with either orientation. There is an equivalent definition of Δ. Fix your attention on one lattice site; suppose orientations of the A type in Fig. 11-7 are of the "correct type" (to give $\Delta = 1$ at $0°K$). If the probability of having the correct type of orientation is c and the wrong type is w, then $\Delta \equiv (c - w)/(c + w)$ gives the same result as stated previously.

11-6b Quantitative discussion With the help of a model we can find the temperature dependence of the long-range order parameter, Δ. The calculation is simple with very little clutter of mathematical detail, yet it has the essence of all the more complicated theories.

CHAPTER 11   SOME THERMAL EFFECTS IN SOLIDS   373

The procedure is straightforward. We calculate the free energy, F. Since $F = U - TS$, where U is the internal energy of the system and S the entropy, we determine U and S as functions of Δ. For stability the internal energy must depend quadratically on Δ, while S depends on Δ in a more complicated manner. Then we minimize F with respect to the long-range order parameter (i.e., $\partial F/\partial \Delta = 0$). The minimization condition will result in an equation relating Δ to T. This is the essence of the calculation. (We really should minimize the **Gibbs free energy**, $G = U + PV - TS = H - TS$ where H is the enthalpy. However, since the pressure is very low, we ignore the PV term and just minimize the **Helmholtz free energy**, $F = U - TS$.)

Note that at high enough temperatures the TS term in the free energy will *always* be dominant and will tend to produce a disordered system because S is larger for a more disordered system.

We now proceed with the help of a simple model. Assume that the internal energy depends only on the *relative* orientation of the nearest neighbors NH_4 ions. This is written as

$$U = N_{AA}\mathscr{E}_{AA} + N_{BB}\mathscr{E}_{BB} + N_{AB}\mathscr{E}_{AB} \tag{11-26}$$

N_{AA} is the average number of nearest-neighbor pairs where both have A orientation and $\mathscr{E}_{AA}$ is the energy of one such pair, and so on.. Since each NH_4 ion has 6 nn, for a completely ordered sample $N_{AA} = 3N$ and $N_{AB} = 0$, where N is the number of molecules in the sample. To determine N_{AA} when $\Delta \neq 1$, take the number of NH_4 ions with A orientation as $(1 + \Delta)N/2$ multiplied by one-half of the probability of finding one of the 6 nn with an A orientation, which is $(1/2)6(1 + \Delta)/2$. The $(1/2)$ enters so that we do not count the pairs twice. (Note the correct values in the limit of $\Delta = \pm 1$, and 0.) Similarly the probability of finding one of the 6 nn with a B orientation is $(1/2)6(1 - \Delta)$. Then the values of N_{ij} are

$$N_{AA} = [(1 + \Delta)N/2][3(1 + \Delta)/2] = (3N/4)(1 + \Delta)^2$$

$$N_{BB} = [(1 - \Delta)N/2][3(1 - \Delta)/2] = (3N/4)(1 - \Delta)^2 \tag{11-27}$$

$$N_{AB} = [(1 + \Delta)N/2][3(1 - \Delta)] = (3N/2)(1 - \Delta^2)$$

with these values of N_{ij} we substitute into Eq. 11-26 to obtain

$$U/(3N/4) = (\mathscr{E}_{AA} + \mathscr{E}_{BB} + 2\mathscr{E}_{AB}) + \Delta^2(\mathscr{E}_{AA} + \mathscr{E}_{BB} - 2\mathscr{E}_{AB}) \tag{11-28}$$

$$\equiv \mathscr{E}_0 + \Delta^2 \mathscr{E}_{Ave}$$

Note that besides the energy $\mathscr{E}_0$, there is a quadratic dependence of U on the long-range energy parameter, which is a characteristic of these simple equilibrium models.

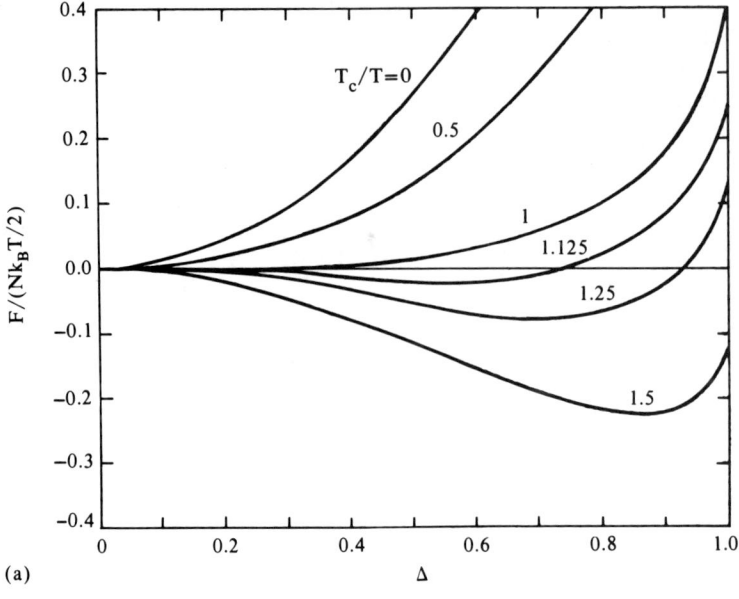

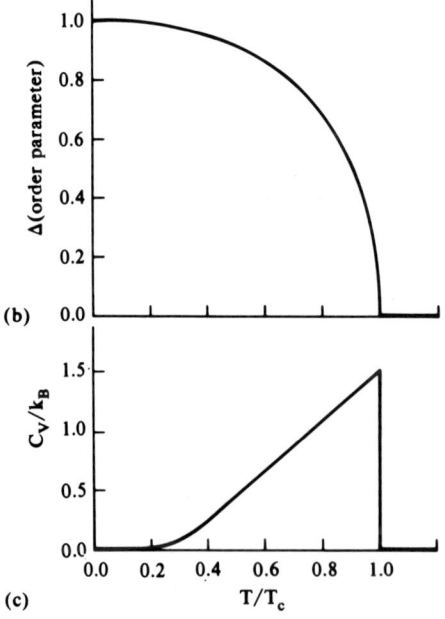

Fig. 11-8 These are schematic diagrams associated with the NH$_4$Cl rotation phase transition. (a) The free energy vs. order parameter is plotted for a number of values of the ratio T_c/T. (b) The order parameter vs. reduced temperature. (c) The configurational heat capacity vs. reduced temperature.

Next, we calculate the entropy, given by $S = k_B \log w$, where w is the total number of arrangements of the NH$_4$ ions. For a given site there is a probability $(1 + \Delta)/2$ for an NH$_4$ ion with A orientation and a probability $(1 - \Delta)/2$ with a B orientation. For the N sites there are

CHAPTER 11   SOME THERMAL EFFECTS IN SOLIDS   375

N! arrangements but this must be reduced by the number of ions with A type of orientation and the number of ions with B type of orientation. Thus, the number of arrangements of the $(1 + \Delta)N/2$ ions with A orientation and the $(1 - \Delta)N/2$ ions with B orientation on N sites is

$$w = \frac{N!}{[(1 + \Delta)N/2]! \, [(1 - \Delta)N/2]!} \quad (11\text{-}29)$$

With the use of Stirling's formula, $\log x! \sim x \log x - x$ (which is good for large x) and straightforward algebra, the entropy is given by

$$S = k_B N \log 2 - (k_B N/2)[(1 + \Delta) \log (1 + \Delta) \\ + (1 - \Delta) \log (1 - \Delta)] \quad (11\text{-}30)$$

Note that for the completely ordered case, when $\Delta = \pm 1$, $S = 0$. On the other hand, for the random high temperature case, when $\Delta = 0$, then $S = k_B N \log 2$ as would be expected. This entropy term just takes into account the possible orientational configuration and is referred to as the **configurational entropy**. It neglects contributions from vibrational effects, strain effects, nearest neighbors or short-range order, and other subtleties.

Our program can now be finished because all the ingredients of the free energy

$$F = U - TS \quad (11\text{-}31)$$

are at hand. U is given by Eq. 11-28 and S by Eq. 11-30. To determine the values of Δ as a function of temperature we minimize the free energy with respect to the order parameter, that is, $\partial F/\partial \Delta = 0$. Straightforwardly, we obtain

$$0 = \frac{3N\mathscr{E}_{Ave}}{2} \Delta + \frac{N}{2} k_B T \log \left[\frac{1 + \Delta}{1 - \Delta}\right] \quad (11\text{-}32)$$

This transcendental equation can be solved graphically to determine Δ as a function of T. We may also use this equation to determine T_c by expanding the log term for small Δ. For this expansion

$$0 = (3N/2)\mathscr{E}_{Ave} \Delta + (N/2)k_B T_c[2\Delta]$$

or $T_c = -3 \mathscr{E}_{Ave}/2k_B \quad (11\text{-}33)$

Note that to have a phase transition with long-range order, $\mathscr{E}_{Ave}$ (Eq. 11-28) must be negative. In other words, if there is a phase transition to an ordered state, then the similar neighbor energy $\mathscr{E}_{AA} = \mathscr{E}_{BB}$ must be less than the dissimilar neighbor energy $\mathscr{E}_{AB}$.

The free energy in Eq. 11-31 can be found explicitly in terms of Δ for any value of temperature. The result, Fig. 11-8a, shows rather beautifully how the minimum of F shifts from $\Delta = 0$ for $T > T_c$ toward

$\Delta = \pm 1$ at low temperatures. Only half of the figure is presented because F is an even function of Δ so the part from -1 to 0 is the mirror image. Solving Eq. 11-32 is equivalent to determining the minimum of F as a function of temperature. Figure 11-8b shows the results. (The book by Dekker, Chapter 4, or the review article by Muto and Takagi referred to in the Notes may be consulted for details of the solution of the transcendental equation, Eq. 11-32 or see Fig. 15-7a where a similar type of transcendental equation is solved.) As is seen in either Fig. 11-8a or 11-8b, Δ goes smoothly to zero at T_c with no discontinuity. This is a **second-order phase transition.** The characteristic of a **first-order phase transition** is that at T_c there is a discontinuity in Δ from zero to a finite value. (See Chapter 14.)

The specific heat due to the disordering of the NH_4 ions can be calculated using our simple model.

$$C_V = \left(\frac{\partial U}{\partial T}\right)_V = \frac{dU}{d\Delta}\frac{d\Delta}{dT} = \frac{3N}{2}\mathscr{E}_{Ave}\Delta\frac{d\Delta}{dT} \quad (11\text{-}34)$$

where Eq. 11-28 has been used. The subscript V is a reminder that the volume has been assumed to be constant through the calculation. Figure 11-8c shows the result for the model. C_V increases to a maximum at T_c then drops abruptly to zero. This is a characteristic of order-disorder transition, and experimental results show this kind of behavior. The integral under this configurational specific heat curve is

$$\int_0^\infty C_V dT = N k_B T_c/2 \quad (11\text{-}35)$$

which is obtained simply via Eqs. 11-34 and 11-33. Thus, each NH_4 ion contributes $k_B T_c/2$ to this integral. Again there is fair agreement with experiment. The article by Muto and Takagi, mentioned in the Notes, should be consulted for details of this and related systems.

11-6c Related systems There are many systems that have order-disorder transitions of the type just described. We mention a few that are closely related to NH_4Cl in the sense of orientational disorder. In Problem 11 phase transitions in metallic alloys are discussed. Although not immediately apparent, the ideas and mathematics are very similar to those in the preceding simple model.

However, before going further we should define the term **order-disorder phase transition** as a phase transition, in which below T_c the displacements of the atoms from their high temperature position are of the order of magnitude of the internuclear distances. A **displacive phase transition** is one in which the displacements are very much smaller than the internuclear distances. In fact, for a second-order displacive phase transition they are infinitely small at T_c.

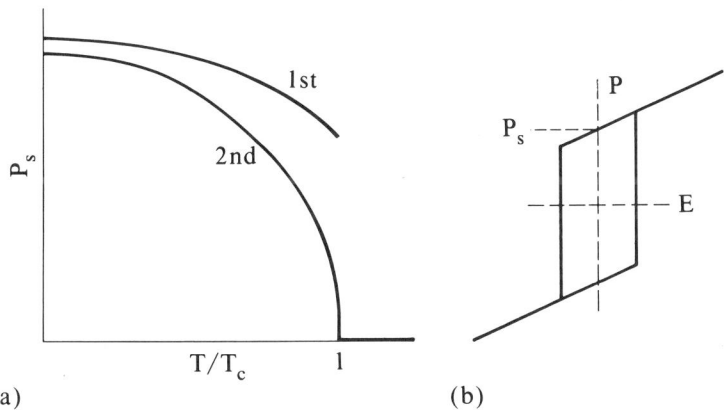

Fig. 11-9 (a) A schematic diagram for the temperature dependence of polarization for a first- and second-order phase transition. (b) P_s vs. applied electric field showing hysteretic behavior below T_c.

An example of a (non-orientational) displacive phase transition is the ferroelectric transition in $BaTiO_3$, a unit cell of which is shown in Fig. 3-1c. At T_c the central Ti ion moves a very small distance along the c-axis. Actually, $BaTiO_3$ has a first-order, displacive phase transition, and the movement of the Ti ion is much smaller than the internuclear distances ($\approx 0.1\text{Å}/2\text{Å}$). The resultant displacement causes the center of gravity of the positive charge distribution to be located at a different position from that of the negative charge distribution. This results in a net dipole moment, μ, within the unit cell. The net dipole moments from every unit cell add, resulting in a macroscopic polarization $P \equiv$ dipole moment per unit volume. The crystal has a spontaneous polarization, P_s, below T_c. Further, a rather small external applied electric field can reverse the direction of P_s, leading to the statement that the crystal has a **reversible spontaneous polarization.** Materials with this property are called **ferroelectrics** since they are the electrical analogues of ferromagnetic materials. In the electric case there is a hysteresis in the dependence of P on the applied electric field, while in the magnetic case it occurs in the magnetic moment vs. the applied magnetic field. Figure 11-9a shows the behavior for P_s vs. temperature for a first- and a second-order phase transition. Both types are observed among ferroelectric materials. Figure 11-9b shows P vs. applied field at a temperature below T_c.

Let us come back to our rotationally disordered systems (which is an order-disorder phase transition). An interesting effect is observed in NH_4Br. Instead of all the NH_4 ions ordering in type A states, the material orders with alternating A and B types in neighboring cells (Fig. 11-10). Apparently the interaction energy for this type of pairs is lower than for the AA type of pairs. The crystal structure is no

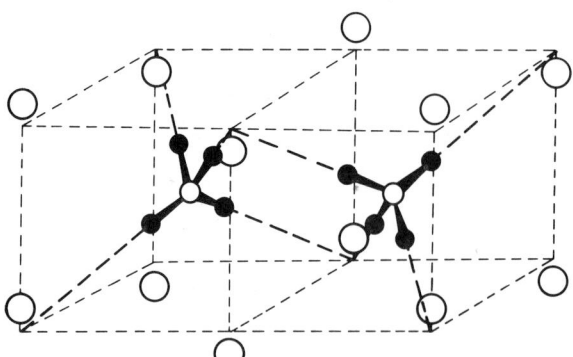

Fig. 11-10 A unit cell of ordered NH_4Br showing the AB pairs and the Br displacements. In this low temperature phase the Br ions are displaced up and down alternately in lines from their high temperature positions.

longer cubic since the Br ions are displaced from their original positions and the low temperature space group is D_{4h}^7(P4/nmm).

$NaNO_2$ is another interesting example of an orientational order-disorder transition. Figure 11-11a shows this material in the high temperature disordered form. Along the b-axis of this orthorhombic crystal structure the NO_2 ions are disordered with their dipole moments in the ±**b** direction. Below a transition temperature, T_c, the NO_2 ions all point in the same direction, Fig. 11-11b, and the crystal is ferroelectric. A treatment such as that in Section 11-6b is not appropriate for this material because unlike the NH_4 ion the NO_2 ion has a large dipole moment and thus strong long-range orientational forces; the preceding calculation is good only for weak nearest-neighbor orientational interactions. The phase transition in $NaNO_2$ is a cooperative phenomenon produced by long-range forces.

As a last example of a material that exhibits an orientational order-disorder transition we consider adamantane, $C_{10}H_{16}$. This is a molecular crystal; one molecule is shown in Figure 8-7. This material was mentioned briefly in Section 6-4d since it is a "plastic crystal." The point symmetry of the molecule is $T_d(\bar{4}3m)$ which is the same as NH_4 so the first nonzero electric moment is an octupole moment which leads to very short-range, nearest neighbor, electrical interactions. This molecule, like NH_4, has two possible orientational states within the high temperature crystal structure, which is an averaged cubic structure similar to NH_4Cl, Figure 11-7. Below $T_c = 208°K$ an ordered crystal structure occurs in which the adamantane molecules all assume an A type of orientation as in NH_4Cl. However, there is one important difference here. The molecules have no charge so the fundamental bonding of the solid is rather different from that of NH_4Cl. The crystal structures of molecular crystals are often determined by **conformational effects** (fitting together due to the geometry of the molecule) as discussed in Section 6-5c. Above T_c the averaged crystal structure consists of one $C_{10}H_{16}$ molecule on each lattice point of a fcc lattice. Below T_c, when all the molecules assume the same

CHAPTER 11   SOME THERMAL EFFECTS IN SOLIDS   379

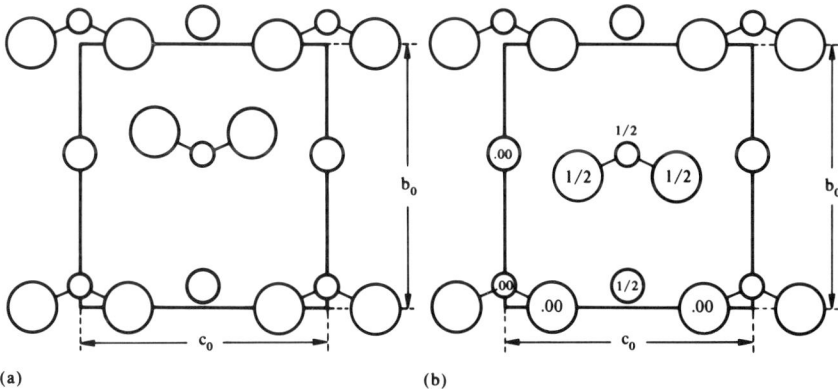

Fig. 11-11 (a) The disordered, high temperature structure of sodium nitrite. (b) The low temperature ordered form. The numbers on the atoms represent the coordinates of the atoms in the a_0 direction, which is out of the plane of the paper.

orientational state, conformational effects cause the molecules to rotate slightly (9°) with respect to each other (similar to a large number of gears that are meshing); this lowers the symmetry. See the Notes in Chapter 6 for references to plastic crystals.

Some discussion of phase transitions due to the "freezing in" of rotational motion can be found in Section 6-5b. There are many other examples, but we shall go on to related matters. The discussion in this section is meant only to show that movements of atoms of the order of magnitude of the internuclear distance and ordering can lead to a variety of interesting solid state effects, including occurrence of an ordered phase, lower symmetry due to the order, and interesting macroscopic effects such as ferroelectricity.

11-7 Polarization by Orientation

11-7a Polarization by orientation in gases and liquids This subsection is devoted not to solids but to gases and liquids. Besides its intrinsic interest, it serves as an introduction to the next subsection, which is on the same topic but describes effects in solids. The effect is also a nice example of the statistical averaging of Section 11-3a.

Consider a gas or liquid composed of molecules that have a permanent electric dipole moment. (The molecule CH_4, with $T_d(\bar{4}3m)$ point symmetry, cannot have a dipole moment, but CH_3Cl, with $C_{3v}(3m)$ point symmetry, has one.) In the presence of an external applied electric field the dipole moment of each molecule will tend to line up parallel to the field. The forces that hinder this alignment are

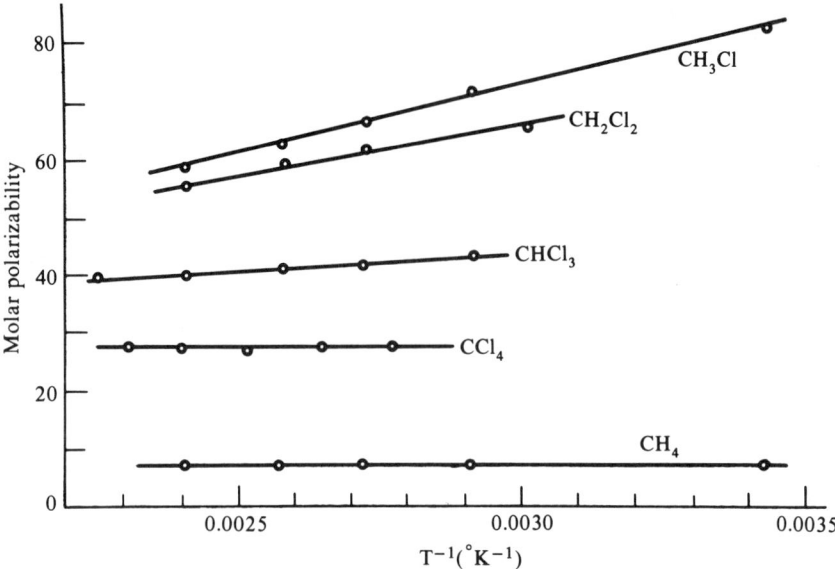

Fig. 11-12 The molar polarizability vs. T^{-1} for several polar and nonpolar substituted methane compounds. (After Debye's book referenced in the Notes.)

thermal agitation, steric effects, and long-range electric forces. Figure 11-12 shows the molar polarizability (the induced polarization divided by the applied electric field) vs. $1/T$. The two molecules that have no permanent dipole moment show results that are temperature independent, while the other materials show a linear variation with $1/T$. This $1/T$ temperature dependence suggests that thermal agitation keeps the electric moments from completely aligning in the electric field. With this in mind we carry out the following calculation.

The potential energy, V, of a permanent dipole moment, μ, in an external electric field, E, and the resulting polarization, P, is

$$V = -\boldsymbol{\mu}\cdot\mathbf{E} = -\mu E \cos\theta \tag{11-36a}$$
$$P = \mu \langle \cos\theta \rangle \tag{11-36b}$$

where θ is the angle between μ and E, $\langle \cos\theta \rangle$ is the Boltzmann thermal average as in Eq. 11-6b. We neglect other effects that might hinder the alignment. Thus, $\langle \cos\theta \rangle$ is

$$\langle \cos\theta \rangle = \int e^{-V/k_BT} \cos\theta\, d\Omega \div \int e^{-V/k_BT}\, d\Omega$$

$$= \int_0^\pi \cos\theta \sin\theta\, e^{-\mu\varepsilon\cos\theta/k_BT} d\theta \div \int_0^\pi \sin\theta\, e^{-\mu\varepsilon\cos\theta/k_BT} d\theta$$

$$= x^{-1} \int_x^{-x} ze^z dz \div \int_x^{-x} e^z dz$$

$$= \text{ctnh}(x) - 1/x \equiv L(x) \tag{11-37}$$

where $x \equiv \mu E/k_B T$, $z \equiv x \sin\theta$, $d\Omega$ is the usual solid angle $\sin\theta \, d\theta \, d\phi$. The $d\phi$ integration cancels in the numerator and denominator. Ctnh x is the usual hyperbolic cotangent of x and is equal to $[\exp(x) + \exp(-x)]/[\exp(x) - \exp(-x)]$ and L(x) is known as the **Langevin function** since this expression was first obtained by him, in 1905, in connection with a very closely related problem – that of the mean magnetic moment of a gas of molecules carrying a permanent magnetic moment. For ordinary materials at ordinary temperatures $x \ll 1$ (see Problem 6), so in this limit $L(x) \approx x/3 = \mu E/3k_B T$. Figure 11-13 shows L(x) vs. x where the behavior at small x can be seen. Also note the saturation behavior at large x.

From Eq. 11-36b, for small x we obtain the polarization of the molecule $P = (\mu^2/3k_B T)E$. An orientational polarizability can be defined from the linear P vs. E behavior as $P = \alpha E$ so $\alpha \equiv \mu^2/3k_B T$ for one molecule. This response to an external electric field considers only the orientational behavior of the permanent dipole. To it we add the deformation of the electron cloud by the applied electric field. This gives the electronic polarizability α_e, which Debye shows is $(\alpha_{11} + \alpha_{22} + \alpha_{33})/3$ in terms of the polarizability tensor described in Section 5-4b. Thus, the complete expression for one molecule is

$$P = (\alpha_e + \mu^2/3k_B T)E \tag{11-38}$$

Figure 11-12 is a plot of the molar polarizability for different kinds of molecules. The linear dependence on $1/T$ for polar molecules is clear. α_e can be obtained from the very high temperature intercept ($1/T = 0$) and μ^2 is obtained from the slope. See Debye or the other books mentioned in the Notes for details.

11-7b Polarization by orientation in solids We have already discussed, in Section 11-6, various aspects of molecules and ions rotating in solids, including the rotation of ions with a dipole moment, namely, $NaNO_2$ in Section 11-6c. In this subsection we perform a calculation similar to that in Section 11-7a for gases with a dipole moment, but here we consider a solid.

Assume that an isolated dipole, such as H_2O or NO_2^- in a solid, has two allowed states, one parallel (A) and one antiparallel (B) to the external applied field, E. The probability of finding the ion in state A is given by the Boltzmann distribution function, $p_A = \exp(\mu E/k_B T)/Z$, as in Eq. 11-3, and the probability for state B is $p_B = \exp(-\mu E/k_B T)/Z$ with the interaction energy given by Eq. 11-36a.

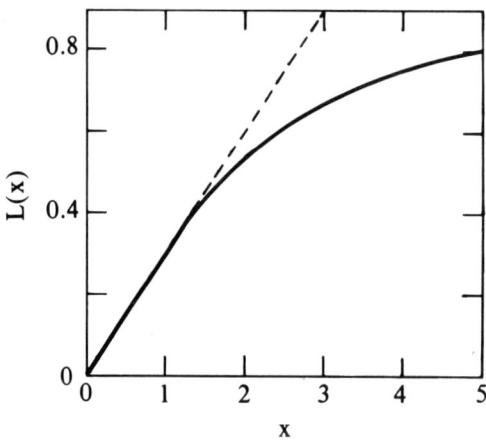

Fig. 11-13 The Langevin function L(x) vs. x where $x = \mu E/k_B T$. Note the initial linear behavior at small x.

Then the excess fraction (if there are N dipoles we must multiply the probabilities by N) of dipoles orientated parallel to the field is

$$\frac{p_A - p_B}{p_A + p_B} = \frac{e^x - e^{-x}}{e^x + e^{-x}} = \tanh x \approx x \qquad (11\text{-}39a)$$

where $x = \mu E/k_B T$. Expanding for small x, the polarization is

$$P = (\mu^2/k_B T)E \qquad (11\text{-}39b)$$

This result and that given in Eq. 11-38 are identical except for a factor three. This factor arises because of the restriction to only two orientations here while in the gas all orientations are allowed.

Thus, the polarization by rotation of permanent electric dipoles is very similar in a gas and in a solid. The next question is just how rapidly can these dipoles follow an alternating field? After all, the NO_2^- ion, for example, has a large moment of inertia. Furthermore, in a solid there are strong steric forces that can hinder rotation. We consider the relaxation of the dipoles in a solid in the next subsection.

11-7c Anomalous dispersion in solids Figure 11-14a shows the measured temperature dependence of the dielectric constant of ice at several frequencies. Consider the frequency dependence at $-10°C$. At 300Hz and lower frequencies, a value close to the full dielectric constant is measured. Apparently the frequency is low enough so that the dipoles follow the oscillating electric field and contribute completely to the measured dielectric constant. Another way to say this is that the molecules can react fast compared to the rate of change of the applied electric field. However, as the frequency of the applied field increases, the measured dielectric constant decreases. At 6×10^4Hz the dielectric constant is just the electronic part with no orientational

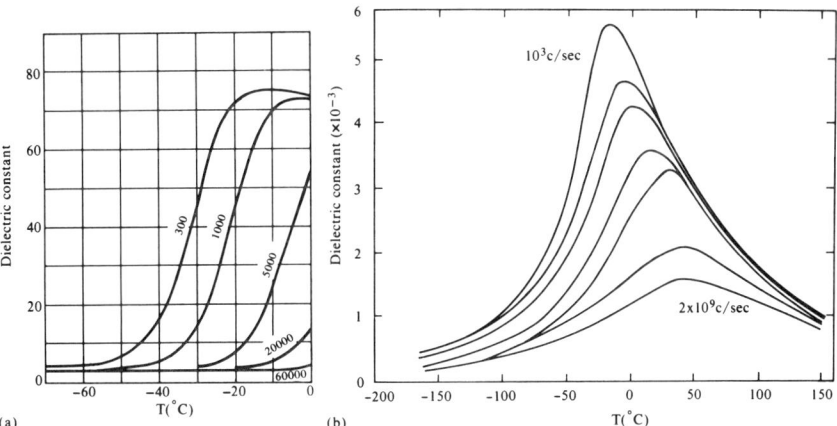

Fig. 11-14 (a) The temperature dependence of the dielectric constant of ice at several different frequencies (in Hz) showing anomalous dispersion (frequency dependence). [Smyth and Hitchcock, J. Am. Chem. Soc. **54**, 4631 (1932).] (b) Similar results for $PbMg_{1/3}Nb_{2/3}O_3$. The frequencies of measurement between the largest response which is taken at lowest frequency and the smallest response are: 10^3; 4.6×10^4; 4.5×10^5; 1.3×10^7; 5×10^7; 1.5×10^9 all in Hz [G. A. Smolensky, J. Phys. Soc. Jap. **28** (Supplement), 26 (1970).]

contribution. At this high frequency the permanent dipoles cannot follow the oscillating field, since their inertial effects are too large.

To help understand these effects Debye has given a simple mathematical model, which we discuss now. Consider a number of molecules, N, each with dipole moment μ, that can point in only two directions A and B, along the applied field E or opposite to it. Let n_A and n_B be the number of molecules with direction A and B, respectively. Molecules can interchange between A and B partly due to thermal effects and partly due to the applied field; a molecule pointing along the applied field can rotate to be in the opposite direction (and vice versa). The rate of change of n_A during a time δt is $(dn_A/dt)\delta t$; it has a negative contribution from those leaving direction A and a positive contribution from those assuming direction A (from direction B). Consider one molecule. The probability that it makes a transition from A to B in one second is taken as p_{AB}. Thus, the negative contribution to $(dn_A/dt)\delta t$ is $p_{AB}n_A\delta t$. Determining the other terms in a similar manner leads to

$$dn_A/dt = -p_{AB}n_A + p_{BA}n_B$$
$$dn_B/dt = +p_{AB}n_A - p_{BA}n_B$$
(11-40)

(These types of equations occur often in physics and a few minutes should be spent to make sure that they are understood.) For thermal equilibrium $dn_A/dt = 0 = dn_B/dt$ which immediately gives

$$\frac{p_{AB}}{p_{BA}} = \frac{n_B}{n_A} \tag{11-41}$$

Since the Boltzmann distribution law applies to n_A and n_B, Eq. 11-3,

$$n_A = e^{\mu E/k_B T}/Z \qquad n_B = e^{-\mu E/k_B T}/Z \tag{11-42}$$

which, from Eq. 11-41, leads us to try expressions for the transition probabilities of the following form

$$p_{AB} = (1/2\tau)e^{-\mu E/k_B T} \qquad p_{BA} = (1/2\tau)e^{\mu E/k_B T} \tag{11-43}$$

where τ has the dimensions of time and, as we shall see, plays the role of a relaxation time. While τ might be any function of T and E, it turns out experimentally to be independent of these two quantities to a first approximation. Considering $\mu E \ll k_B T$ and combining Eqs. 11-40 and 11-43 we obtain

$$(2\tau)(dn_A/dt) = -(n_A - n_B) + (\mu E/k_B T)(n_A + n_B)$$
$$(2\tau)(dn_B/dt) = (n_A - n_B) - (\mu E/k_B T)(n_A + n_B) \tag{11-44}$$

Assuming that the applied electric field is $\propto \exp(i\omega t)$, subtracting the second from the first equation, we can see that a solution is $(n_A - n_B) \propto \exp(i\omega t)$, which yields

$$n_A - n_B = \frac{n_A + n_B}{1 + i\omega t} \frac{\mu E}{k_B T} \tag{11-45}$$

For N molecules ($N = n_A + n_B$) a polarizability can be defined in the usual manner via $P = N\alpha E$, and taking $P = \mu(n_A - n_B)E$, we have

$$\frac{P}{E} = N\alpha = \frac{N\mu^2}{k_B T} \frac{1}{1 + i\omega\tau} = N\frac{\alpha_0}{1 + i\omega\tau} \tag{11-46}$$

This is the classical relaxation behavior already encountered in Section 9-5a where the real and imaginary parts of α can be seen in Fig. 9-2. Re(α) corresponds to the polarizability that can follow (be in phase with) the applied field. As can be seen in Fig. 9-2, at very low ω it has the value α_0 in agreement with the static calculation of Section 11-7b. However, as ω increases, Re(α) decreases and is equal to $\alpha_0/2$ at $\omega\tau = 1$, decreasing further at larger ω. Im(α) is the out-of-phase component, which is proportional to the energy loss due to the inertial "friction" of the dipole. It peaks at $\omega\tau = 1$ as shown in the figure.

Figure 11-14b show results similar to those in 11-14a but for an entirely different system. These results are near a ferroelectric phase transition and show strong anomalous dispersion. The results, of the type shown in Eq. 11-46, appear in many branches of solid state science besides the relaxation of permanent dipoles discussed here. *There is always some frequency above which the response can no longer follow the applied field.* Any simple linear behavior of the type that can be given in terms of rate equations, Eq. 11-40, will lead to Eq. 11-46.

11-8 Point Imperfections in Crystals

11-8a Introduction to the rest of the chapter   The term imperfection in a crystal refers to any deviation of the crystal from a perfect periodic structure. Thus, any excitation, such as a lattice vibration, causes the atoms to deviate from their proper positions and can be considered as an imperfection. However, we will concern ourselves with more distinct imperfections, the point imperfections (in this section) and, in succeeding sections, several effects due to these point imperfections. As the name implies a **point imperfection** is an imperfection (irregularity) in the crystal structure localized at one point. This could be a missing atom, an additional atom at a site in the structure that should not be occupied, or a foreign atom in the structure. In the technologically interesting semiconductor materials, such as silicon, impurity atoms control the electronic properties.

There are other types of imperfections such as lines and planes of missing atoms called dislocations. These often play the major role in crystal growth and the mechanical strength of the material. However, we will not address these interesting topics. In this section we shall concern ourselves with the rather simple imperfections corresponding to point defects, namely, lattice vacancies and interstitial atoms. Then in Section 11-9 we discuss diffusion, describing the actual movement of atoms through the structure. The kinetics of diffusion is very dependent on the imperfections. Section 11-10 deals with the color and luminescence of crystals which is almost always associated with point imperfections. The last section shows how we may detect the vibrations of certain impurities.

11-8b Point defects in crystals   Figure 11-15 shows three of the simplest types of point defects. The first shows missing atoms called **Schottky defects**. The creation of such a defect can be visualized by imagining removing an atom from the interior of the crystal and placing it on the surface. The atom has a higher binding energy in the interior than on the surface because in the bulk it has a larger number of bonds. The energy of formation of a Schottky defect is difficult to

386 CHAPTER 11 SOME THERMAL EFFECTS IN SOLIDS

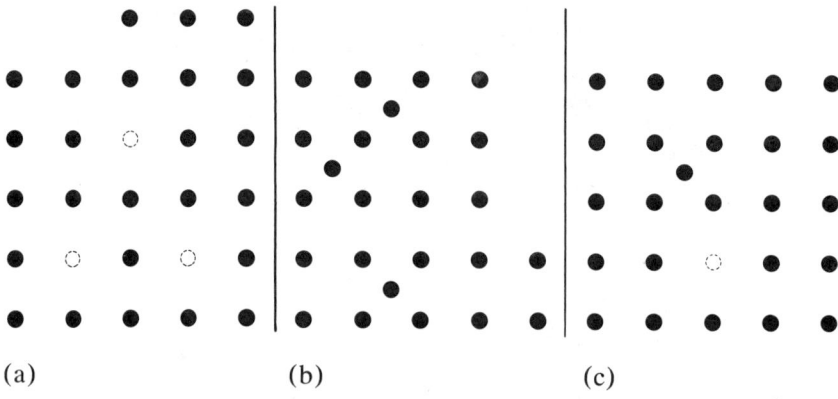

Fig. 11-15 (a) Schottky defects. (b) Interstitial atoms. (c) Frenkel defects.

calculate because of the relaxation of the atoms around the vacancy, but values of the order of 1eV per vacancy are estimated. As we shall see, in solids there is an equilibrium number of these and other similar vacancies. Thus, crystals can never be perfect!

Interstitial atoms are shown in Fig. 11-15b. There is a certain amount of distortion around each interstitial atom, which again makes quantitative calculations of the formation energy difficult.

Figure 11-15c shows an interstitial atom in conjunction with a vacant atom site. This is called a **Frenkel defect**. A certain number of these defects are expected in thermal equilibrium and this number increases with increasing temperature. However, the number of defects can be increased in several nonequilibrium ways. By very rapidly lowering the temperature of the sample (i.e., by quenching) we can obtain an excess number of defects. By the bombardment of the sample with electrons or nuclear particles (i.e., by radiation damage) atoms will be displaced from both direct collisions and local heating. Severely cold working the samples also produces excess defects.

These defects are quite general. However, in strongly ionic crystals, that is, NaCl type, the defects tend to occur in pairs of plus and minus ions to preserve the electrical neutrality of the sample and of the local region of the crystal. Figure 11-16 shows examples of some important defects in ionic crystals. Besides the Schottky and Frenkel defects in Figs. 11-16a and 11-16b, a vacancy pair is shown in Fig. 11-16c. If a small fraction of CaF_2, for example, is added to an NaCl melt and a crystal is grown, then some Ca^{2+} will replace Na^+ at some of the normal Na^+ ion sites. However, to compensate charge there will be some missing Na^+ ions. Figure 11-16d shows a complex made up of a divalent impurity (enclosed by dashes) and a positive ion vacancy. Figure 11-16e is a reminder that site symmetry of an interstitial and substitutional site can be quite different. The site symmetry

CHAPTER 11   SOME THERMAL EFFECTS IN SOLIDS   387

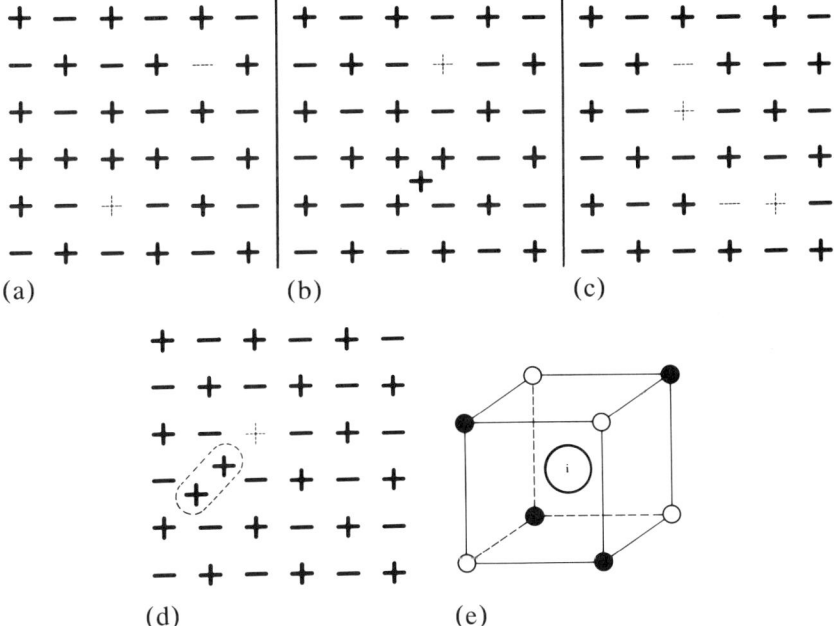

Fig. 11-16 Defects in ionic crystals. (a) Schottky defects. (b) Frenkel defect. (c) Vacancy pairs. (d) Divalent impurity-vacancy complex. (e) A possible interstitial site, labeled i, in a NaCl crystal structure.

of an Na or Cl site in the NaCl structure is O_h(m3m), Figs. 3-5 and 3-6. However the interstitial position with the highest symmetry has $T_d(\bar{4}3m)$ site symmetry (the 8c position in Fig. 3-6). Of course, depending on the bonding, the interstitial could be slightly displaced from this high symmetry position. For example it could be randomly at the 32f position (Fig. 3-6) with $x \approx 1/4$. Then the site symmetry would be C_{3v}(3m).

Also note that the vacancy pairs in Fig. 11-16c have a dipole moment, which should exhibit a polarization by orientation for low frequency electric fields and which should have relaxation effects as discussed in Section 11-7.

11-8c Density of defects   Now we calculate the number of Schottky defects as a function of temperature. As in Section 11-6b, we minimize the Helmholtz free energy $F = U - TS$, rather than the Gibbs free energy, that is, we ignore the PV term because the pressure is low. Let $\mathscr{E}_s$ be the energy required to take one atom from the N atoms and move it to the surface, that is, to form a Schottky defect. The energy to produce n isolated, noninteracting Schottky defects is $n\mathscr{E}_s$. We use $S = k_B \log w$ to calculate the entropy where w is the total number of ways that we may pick these n atoms from a crystal that contains N atoms. Then

$$w = \frac{N(N-1)\ldots(N-n+1)}{n!} = \frac{N!}{(N-n)!\,n!} \tag{11-47}$$

where the n! in the denominator occurs because the order in which the vacancies are formed is immaterial. Thus, the free energy is

$$F = n\mathscr{E}_s - k_B T \log \frac{N!}{(N-n)!\,n!} \tag{11-48}$$

Using Stirling's formula $\log x! = x \log x - x$, which is good for large x, and minimizing the free energy with respect to n we have,

$$w = N \log N - (N-n) \log (N-n) - n \log n$$

$$(\partial F/\partial n)_T = 0 = \mathscr{E}_s - k_B T \log [(N-n)/n] \tag{11-49a}$$

$$n = N \exp(-\mathscr{E}_s/k_B T) \tag{11-49b}$$

The last expression is the desired result and is obtained by assuming that $n \ll N$, which is true because for $\mathscr{E}_s = 1\text{eV}$ and $T = 1000°\text{K}$ $n/N \sim \exp(-12) \sim 10^{-5}$. This shows that a number of Schottky defects can be expected on thermodynamic grounds and that this number has an activated form (Eq. 11-49b) as expected.

The calculation for the number of Frenkel defects proceeds along very similar lines. Let $\mathscr{E}_f$ be the energy necessary to remove an atom from a normal position to an interstitial position. Take N to be the number of normal atomic sites, N' the number of interstitial sites, and n be the number of Frenkel defects. Then

$$S/k_B = \log \frac{N!}{(N-n)!\,n!} + \log \frac{N'!}{(N'-n)!\,n!} \tag{11-50a}$$

By minimizing the free energy we obtain

$$n = (NN')^{1/2} \exp(-\mathscr{E}_f/2k_B T) \tag{11-50b}$$

again assuming that $n < N, N'$. Thus, as expected, the form of the result is very similar to that obtained for the Schottky defects.

The existence of such defects is important in that they provide a mechanism for diffusion. Diffusion is discussed in the next section.

11-8d Other point imperfections There are many other important point imperfections. We have already mentioned in Section 3-5 two simple structures that by their very nature have a large number of vacancies. In fact, the number of vacancies is about equal to the number of atoms. The materials discussed there, the high temperature forms of AgI and CuI, have the property that the Ag and Cu ions can

CHAPTER 11   SOME THERMAL EFFECTS IN SOLIDS   389

very rapidly diffuse through the crystals (about as rapidly as in liquid at the same temperature). These materials are called **fast ion conductors** or **super ionic conductors**. However, there are other crystals that have structures with many vacancies but are poor ionic conductors. The difference arises because of different energy barriers between the vacancies. See the Notes for further reading.

11-9 Diffusion

11-9a Fick's first law   Diffusion is a process in which atoms actually move from one site in a crystal to another site. It is easiest to visualize the process by considering some foreign atoms on one surface of a crystal. Then by actually jumping from one site to another, some of these atoms will get to the other side of the crystal. We will discuss the phenomenological laws that describe diffusion driven just by a concentration gradient and then discuss some of the microscopic mechanisms. Let n be the concentration of impurity atoms in a solid (number of atoms/cm^3). Assume that the +z direction is to the right and that n increases in this direction so that $\partial n/\partial z$ is a positive number. If this solid is held at a fixed temperature we would expect that there would be a net flux of atoms to the left due to diffusion. The flux of atoms, j, is defined as the number of atoms crossing a unit area in a unit time. $\partial n/\partial z$ is positive to the right but j flows to the left, and if $\partial n/\partial z$ is reversed in sign then j will reverse in sign. Thus, j is proportional to the odd power of $\partial n/\partial z$ rather than the even power. Putting these words into the form of an equation we have a phenomenological relation known as **Fick's first law**.

$$j = -D\,(\partial n/\partial z) \qquad (11\text{-}51a)$$

where D is the **diffusion constant** or **diffusivity**, with units of cm^2/sec, and the minus sign expresses the fact that $\partial n/\partial z$ and j have opposite signs. Writing this law for a concentration gradient in a general direction produces $\mathbf{j} = -D\nabla n$. We immediately recognize, from Neumann's principle in Chapter 5 that since D relates two vectors it is not a scalar but a second-rank tensor. For example, in tetragonal, hexagonal, or trigonal crystal systems D will have one value along the unique c-axis and a different value perpendicular to this axis. However, we shall ignore the tensor aspects of D and treat it as a scalar, or equivalently, consider cubic crystals where D is isotropic.

Actually, the driving force for the flux is not simply the concentration gradient, as implied previously, but the chemical potential. However, for the simple systems these two quantities are equal.

For most materials, the diffusion constant is found to vary with temperature in an activated manner. Thus,

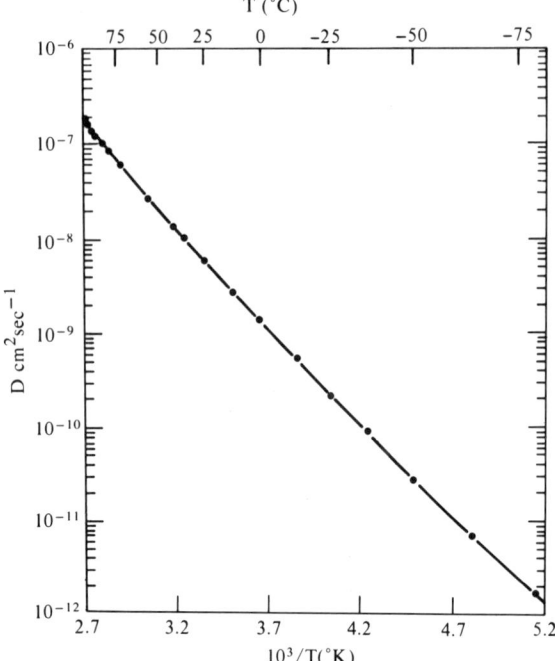

Fig. 11-17 The self diffusion D vs. T^{-1} for ^{22}Na from J. N. Mundy, Phys. Rev. B3, 2431 (1971). This paper also reports data on the pressure dependence of the diffusivity.

$$D = D_0 \exp(-\mathscr{E}_A/k_B T) \qquad (11\text{-}51b)$$

where $\mathscr{E}_A$ is the **activation energy** and D_0 is called the prefactor. A simple model, presented in Section 11-9b, shows that this form for D is expected. Figure 11-17 shows experimental data for the temperature dependence of the diffusion constant. Equation 11-51b predicts that log D vs. 1/T should be a straight line; as can be seen the data is fairly close to a straight line over about five decades of D and $\mathscr{E}_A$ and D_0 can be obtained. A better straight line is found for most materials. However, we picked this particular experimental result to show that deviations from Eq. 11-51b can be found near the melting temperature. Of course, such deviations are highly interesting. This particular deviation is thought to be caused by a correlated event of two atoms jumping in unison as opposed to the normal one atom random event. Table 11-2 lists values of, the preexponental, D_0 and $\mathscr{E}_A$ for some representative metals. Note that $\mathscr{E}_A$ is of the order of 1eV and D_0 is of the order of 1 cm²/sec but that the diffusion constant, D in Eq. 11-51, varies over many orders of magnitude (Fig. 11-17).

11-9b The Einstein–Nernst equation Using Fick's first law and a simple model we can derive the Einstein–Nernst equation, which relates ionic conductivity to diffusion. It is not surprising that the two are related since ionic conduction occurs by ions diffusing.

CHAPTER 11   SOME THERMAL EFFECTS IN SOLIDS

Table 11-2 Some representative diffusion constants and activation energies. Instead of $\mathscr{E}_A$ the more correct enthalpy term is used. The table is divided into values for self diffusion and values for impurity diffusion in pure Ag and pure Cu. These tables are adapted from Shewmon (see the Notes). (23.05 kcal/mole = 1 eV/molecule.)

Metal	Cryst. Struct.	ΔH $\left(\frac{kcal}{mole}\right)$	D_0 $\left(\frac{cm^2}{sec}\right)$	Impurity	Solvent	ΔH $\left(\frac{kcal}{mole}\right)$	D_0 $\left(\frac{cm^2}{sec}\right)$
Cu	fcc	47.1	0.20	Silver	Pb	38.1	0.22
Ag	fcc	44.1	0.40		Sn	39.3	0.25
Ni	fcc	66.8	1.30		Zn	41.7	0.54
Au	fcc	41.7	0.091		Hg	38.1	0.08
Pb	fcc	24.2	0.28		Cu	46.1	1.2
Mg	hcp	32.5	1.5		Au	45.5	0.26
α-Fe	bcc	67.2	118	Copper	As	42.0	0.12
γ-U	bcc	27.5	0.0018		Hg	44.0	0.35
β-Zr	bcc	38	0.0024		Ag	46.5	0.63
Nb	bcc	105	12		Zn	45.6	0.34
Na	bcc	10.5	0.24		Au	49.7	0.69
Ge	dia.	68.5	7.8		Fe	51.8	1.4
					Ni	56.5	2.7

Suppose an electric field, E, in the z-direction is applied to an ionic crystal via blocking electrodes so that ionic carriers cannot be injected into or taken from the crystal. A drift current will flow in the z-direction due to E. At equilibrium the net drift current is equal and opposite to the current due to diffusion, which is caused by the concentration gradient of the ions. The net electric current density $j = \sigma E = (ne\mu)E$ where the density of carriers is n and each carrier has a charge e and a mobility μ. Thus, the conductivity is given by $\sigma = ne\mu$. At equilibrium, this flow of ions is opposed by the diffusion current, which is ej, where j, which is given by Eq. 11-51a, is multiplied by a charge to change it into an electric current. At equilibrium the two current densities are equal.

$$(ne\mu)E = -eD(\partial n/\partial z) \tag{11-52a}$$

On integrating, we obtain

$$n = (const)\, exp\, (-\mu Ez/D) \tag{11-52b}$$

Also, for an electrical potential of Ez, from Boltzmann statistics

$$n = (const)\, exp\, (-eEz/k_B T) \tag{11-52c}$$

Equating the exponents in Eqs. 11-52b and 11-52c yields the Einstein-Nernst equation, which in terms of the mobility or conductivity is

$$\frac{\mu}{D} = \frac{e}{k_B T}, \qquad \frac{\sigma}{D} = \frac{ne^2}{k_B T} \qquad (11\text{-}52d)$$

These equations find extensive use in ionic systems. For example, it is often much easier to measure the ionic conductivity, which is a nondestructive experiment, than to measure D. From these equations we can determine D from a measurement of σ.

11-9c Fick's second law and experiments We have not indicated how D is measured. We can see how this is done from Fick's second law. The equation of continuity, which is just an expression of the conservation of the number of particles is

$$\nabla \cdot \mathbf{j} = -(\partial n/\partial t) \qquad (11\text{-}53)$$

By applying the continuity equation to the first law we obtain

$$\partial n/\partial t = D \nabla^2 n \qquad (11\text{-}54a)$$

which is Fick's second law. In Eq. 11-54a D is assumed to not depend on position, that is, on concentration, which is usually true.

Equation 11-54a can be solved for a variety of experimental boundary conditions. For example, suppose a quantity c_0 of solute is plated as a thin film onto the end of a long rod of solute-free material. Then a similar rod is welded to the plated end of the first rod with minimal diffusion taking place. The composite rod is then annealed for a time t, at a temperature T. The concentration along the rod as a function of distance z, where z = 0 is the position of the plated film, given by Fick's second equation, is

$$n = \frac{c_0}{2(\pi Dt)^{1/2}} \exp(-z^2/4Dt) \qquad (11\text{-}54b)$$

Remember n is a function explicitly of t and z as well as implicitly of T via D. This result shows the solute will spread in the $\pm z$ directions and essentially form a gaussian distribution about z = 0. Solutions for the more usual experimental case where a second rod is not considered are straightforward but more complicated.

In experiments to determine D as a function of temperature, a radioactive solute material is usually plated on the end of a rod of a crystal. The crystal is heated at a certain temperature for a certain time, then rapidly cooled to room temperature, and n is measured as a function of z. (This might be done by actually sectioning the crystal and measuring the radioactivity in each section along the z-axis.) This gives D at that one temperature. To obtain D at other temperatures, the experiment must be repeated. Of course the ingenuity of scientists has led to other ways of making these measurements. For ionic crystals ionic conductivity and the Einstein-Nernst equation can be used.

For certain materials, particularly some metals, nuclear magnetic resonance (NMR) techniques can be used to measure the activation energy for **self diffusion**. (Self diffusion is the diffusion of the atoms of the material itself as opposed to any impurities.) At a given NMR frequency, ω_0, the line width of the NMR signal will narrow drastically when the jump frequency of the atoms between the sites is equal to or greater than the measured line width. (This occurs because the local magnetic fields, which cause the line width, are averaged out for rapidly jumping atoms. This is called **motional narrowing**.) Thus, the sample is heated slowly and the width of the resonance line observed; when it narrows we know the jump frequency at that temperature. Values of self diffusion in good agreement with those determined by direct measurements are obtained. One advantage of the NMR technique is that it is nondestructive.

11-9d A simple model We sketch a very simple model that gives some appreciation of the physics involved in diffusion. In order for an atom to change its position from one site to another, it must pass over some energy barrier of height $\mathscr{E}_B$. It is such a barrier that keeps an atom in its normal position. The diffusing atom will have enough energy to surmount the barrier for a fraction $\exp(-\mathscr{E}_B/k_BT)$ of the time that it impinges on the barrier. If the atom is oscillating in its potential well with a frequency ν_0, called the **attempt frequency**, then the probability per unit time that it will get over the barrier is $r \sim \nu_0 \exp(-\mathscr{E}_B/k_BT)$.

Suppose that a constant concentration gradient of diffusing particles exists in the z-direction; two planes of atoms perpendicular to the z-direction are separated by a small distance a_B. If there are S impurity atoms in one plane per unit area, then due to the concentration gradient there are $(S + a_B dS/dz)$ in the next. The net numbers of atoms diffusing between the planes is just $-ra_B\, dS/dz$. Since $S = a_B n$, where n is the concentration of diffusing atoms, then

$$j \approx -ra_B^2 (dn/dz)$$

or $$D \approx \nu_0 a_B^2 \exp(-\mathscr{E}_B/k_BT) \qquad (11\text{-}55)$$

From this expression for D for this simple model, we can appreciate what parameters are important for diffusion.

11-9e Mechanisms for diffusion The equations and discussion in this section have been phenomenological so far. Now we discuss how the atoms actually get from one site to another. Ideally one would like to calculate the activation energy $\mathscr{E}_A$ and the prefactor D_0 for the diffusion constant in Eq. 11-51b. Even for the simplest realistic diffusion mechanism this is a difficult problem because the relaxation of the neighboring atoms is difficult to take into account properly. Thus, we will discuss only qualitatively a few of the more important diffusion mechanisms.

Interstitial mechanism – Figure 11-18a shows a (100) plane of a face-centered cubic crystal structure. Many elemental metals have this crystal structure (Table 3-1). There is also an interstitial atom labeled 1 which we will think of as trying to diffuse to the interstitial position labeled 2. An atom is said to diffuse via an interstitial mechanism if it goes from one interstitial to another interstitial site without permanently displacing any of the host atoms of the crystal structure. In order for the diffusing atom to jump from interstitial site (1) to (2) the atoms labeled A and A' must move apart. It is this local distortion or dilation that constitutes the barrier for diffusion leading to the activation energy. (Actually, in the fcc structure, there is a geometrically larger channel from position 1 to 2 that is slightly above the plane shown.) An actual calculation of $\mathscr{E}_A$ and D_0 must take into account carefully the strain and coulomb energies involved with the relaxation of atoms A and A' as well as their neighbors. The result depends sensitively on the size of the interstitial channel.

Vacancy mechanism – Figure 11-18b shows the close-packed plane, (111), of the same crystal as in Fig. 11-18a. However, a normally occupied atomic position is vacant. If Δ is the diameter of one of the atoms, then the diameter of the channel between atoms A and A' is 0.73Δ. Thus, if atom 1 is to move into the unoccupied site the distortions of the neighboring atoms can be considerably less than the distortions in the interstitial mechanism for the same size diffusing atom. For example, γ – Fe has the fcc structure and the distortion energy required to move an Fe-atom into a neighboring vacancy is about the same as that required to move a much smaller C-atom from one interstitial site to a neighboring one. This gives us an idea of the effect of size on the energies. However, interstitial carbon atoms diffuse through iron much more rapidly than the self diffusion of iron in iron. This is because each C-atom always has several nearest neighbor empty interstitial sites to diffuse into, while very few Fe-atoms have a neighboring Fe-atom vacancy.

Ring mechanism – The distortions required for the simple exchange of two nearest-neighbor atoms is very large. However, if three or four atoms rotate as a group, Fig. 11-18c, then the required distortion is considerably reduced. It is thought that some of the apparent anomalies observed for D in the bcc metals (the structures are relatively open) can be explained by these ring mechanisms.

Other diffusion mechanisms – There are a number of other mechanisms that are more specialized than those mentioned previously.

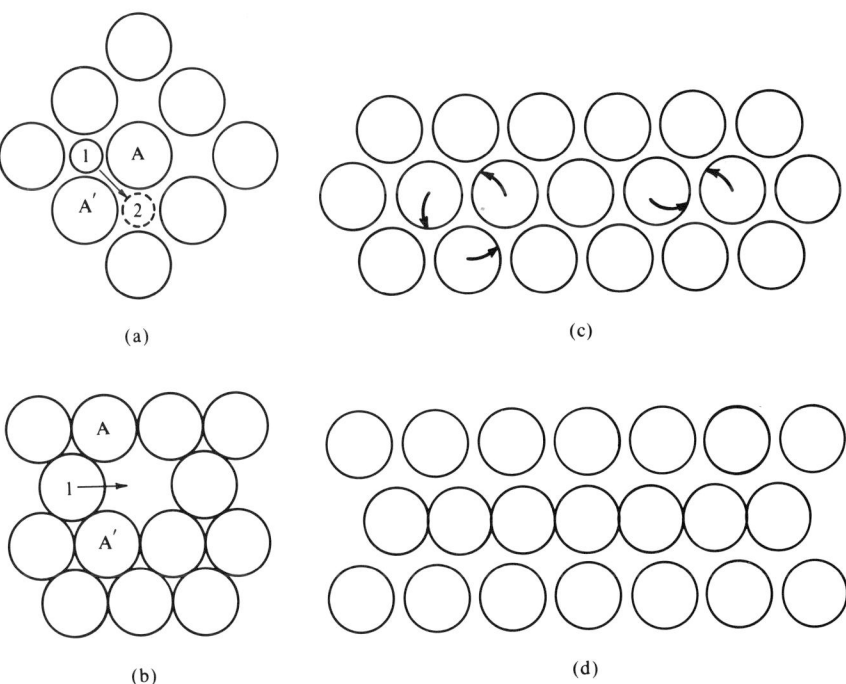

Fig. 11-18 (a) A (100) plane of an fcc structure showing an interstitial atom diffusing. (b) A(111) plane of the same structure showing vacancy diffusion. (c) The ring mechanism of diffusion showing rings of two and three. (d) A (111) plane of an fcc structure with an extra atom in the middle row (i.e., a crowdion).

The interstitialcy mechanism may be envisioned as an interstitial atom diffusing by replacing one of its nearest neighbor normal atoms, which is then pushed into the next interstitial position. The distortion involved can be quite small. Note that if this diffusion mechanism is important in ionic materials (and it is thought to be so in AgCl) then D measured directly can be different from D obtained via conductivity and the Einstein-Nernst equation. This is because the interstitial atom moves only one step while the interstitial hole moves a distance equivalent to two steps (in a colinear process). This is taken care of via a correlation factor f that is added to Eq. 11-52e.

Figure 11-18d shows a **crowdion configuration** where an extra atom has been placed in a close-packed plane and the distortion is taken up by displacing several atoms rather than just nearest neighbors. The energy to move such a configuration, once formed, is small. Often by spreading out the displacements over many atoms the energy required to move a configuration is smaller than if all the displacement is localized in a single atom. The configuration shown here resembles an edge dislocation that can have a low energy for motion. In super ionic conductors such as Naβ aluminate the Na diffusion is thought to occur by this mechanism. Other types of diffusion mechanisms are discussed in the references in the Notes.

11-10 Color Centers in Ionic Crystals

Color centers in ionic crystals (alkali halides are the most studied) represent interesting examples of a property associated with disorder. For most color centers the imperfection responsible for the colorations is a lattice defect such as a vacancy rather than an impurity atom, and an electron may be trapped at the defect.

Pure alkali halide crystals have band gaps larger than 5eV making them transparent to visible light. However, by a variety of methods, these crystals can be colored, that is, absorption bands can be produced in the visible region of the spectrum. They can be colored by exposure to radiation such as x-rays, gamma rays, or strong ultraviolet light; heating the crystal in an alkali metal vapor also colors the crystals; they may be colored electrolytically by passing a current through them at high temperatures. These centers were discovered in 1894 by Goldstein, who used an electron beam to color alkali halides, and the best known is the F-center (from the German Farbzentren).

Figure 11-19a shows the F-center absorption spectrum for KBr, whose band gap is greater than 6.5eV. Even at low temperatures the line is broad. In this crystal the absorption is in the red causing the crystal to appear blue in transmitted light. The center consists of an electron trapped at a negative ion vacancy. Figure 11-19b shows the position of the peak of the absorption band plotted against the lattice constant, a, of the host crystal. From the slope of the curve we obtain a dependence proportional to a^{-2}.

A simple model for the F-center that fits this and other data has been developed. The vacancy is replaced by a box with sides at $\pm a$. The box, or cage, is caused by the cubic terms of the potential due to the six positive ions surrounding the vacancy at distances of $\pm 2a$. The energy levels for a "particle in a box" are well known.

$$\mathscr{E} = (\pi^2 \hbar^2 / 2ma^2)(n_x^2 + n_y^2 + n_z^2), \qquad n_i = 1, 2, \ldots \qquad (11\text{-}56)$$

The F-center absorption is due to excitation of the electron in the ground state to the first excited state; the predicted dependence of a^{-2} is apparent. It is believed that excitations to higher quantum states appear as a shoulder on the F-band, the so-called K-band, which is not visible in Fig. 11-19a. The large width of the F-band is due to zero-point vibrations; thus it is intrinsic. While this is certainly a crude model, it is a starting point for more complicated ones.

(a)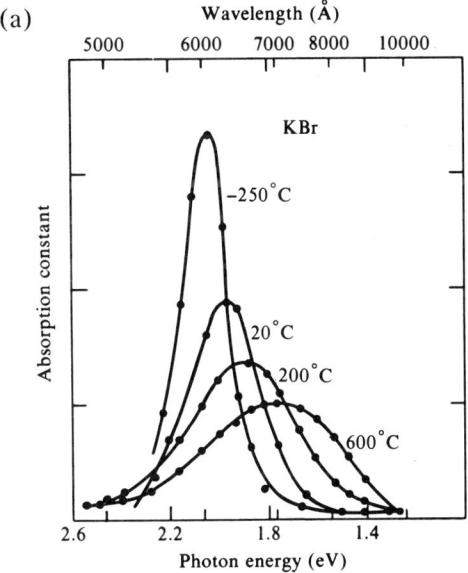

(b)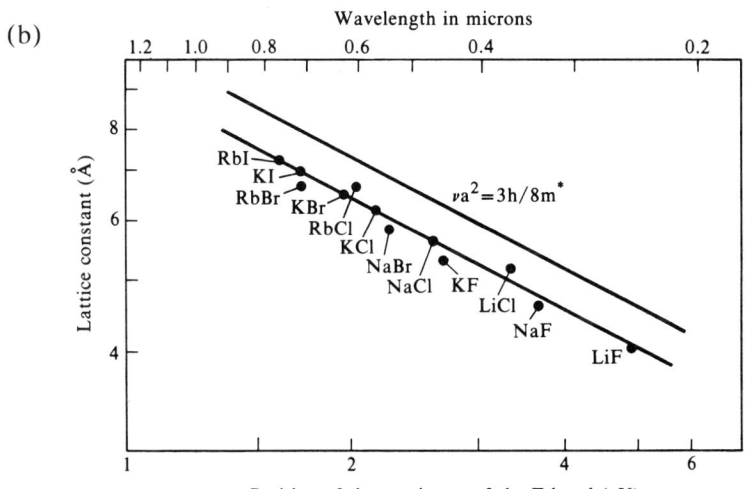

Fig. 11-19 (a) The F-center absorption band in KBr. [R. W. Pohl, Proc. Phys. Soc. (London), **49** (supplement), 3 (1937)]. (b) The F-band absorption peak vs. the lattice constant of the host alkali halide crystal. [H. Pick, Nuovo Cimento, **7** (Supplement), 498 (1958)].

Many other color centers in alkali halides have been studied by optical and electron spin resonance techniques. For example, crystals can be colored by adding negative hydrogen ions. This results in a U-band absorption at 5.5eV in KBr. The interactions between the U-center and the F-center, as well as many complex centers involving more than one atomic site such as aggregates of F-centers, have been studied. In the Notes references can be found to other books and articles on this interesting subject.

11-11 Localized Vibrational Modes

Impurity atoms constitute another type of point defect and they can be studied via their vibration spectra. Details of vibrations are covered in the next chapter but here we mention some effects appropriate to this chapter.

Assume that an impurity atom of mass M' is randomly substituted for normal atoms of mass M in a crystal and the impurities are "connected" to the rest of the lattice by a force constant α. For a single independent oscillator (which a crystal is not) the frequency of oscillation is $\omega_0 = (\alpha/M')^{1/2}$. Often the relative change in force constant for the impurity atoms is smaller than the relative change in mass, so the former can be ignored for discussion purposes. If $M' < M$ the impurity atom will have a characteristic frequency higher than that of the host crystal. This will give rise to a **localized mode**. The mode is localized in that the spatial extent of the displacements due to this mode will decrease as the distance from the impurity atom increases. If $M' \ll M$, as for hydrogen impurities, the localization will be extreme and the local mode will behave similar to an independent oscillator.

Figure 11-20 shows some optical data for localized modes. In Fig. 11-20a the mode is that for Li, which substitutes for the heavy Ag atom in AgBr. By measuring the isotope effect, we can very clearly identify atoms causing the vibration as well as obtain information about the effective mass of the entire localized mode. The infrared absorption spectra of H^- ions replacing Cl^- ions in KCl is shown at several temperatures in Fig. 11-20b. This is a U-center, and the measurements are another way that color centers can be studied. Since the mass difference is so large in this case, the localization is extreme and the H^- ions vibrate in a nearly stationary crystalline environment. Great detail is obtained from such systems, and transitions to higher lying oscillator states as well as the effects due to the anharmonic potential for this system are observed. See the Problems.

When $M' > M$ we have the other case where the characteristic frequency of the impurity atom lies within the vibrational bands of the host crystal. The amplitudes of the band modes will be enhanced by a resonance near the characteristic frequency of the impurity while the amplitudes will be attenuated at other frequencies. Such modes are more difficult to observe because of the tendency for the energy to "leak away" into the host crystal vibration bands. However, they can be observed in a number of cases; they are called **quasi-localized** or **resonance states**. See the Notes for references.

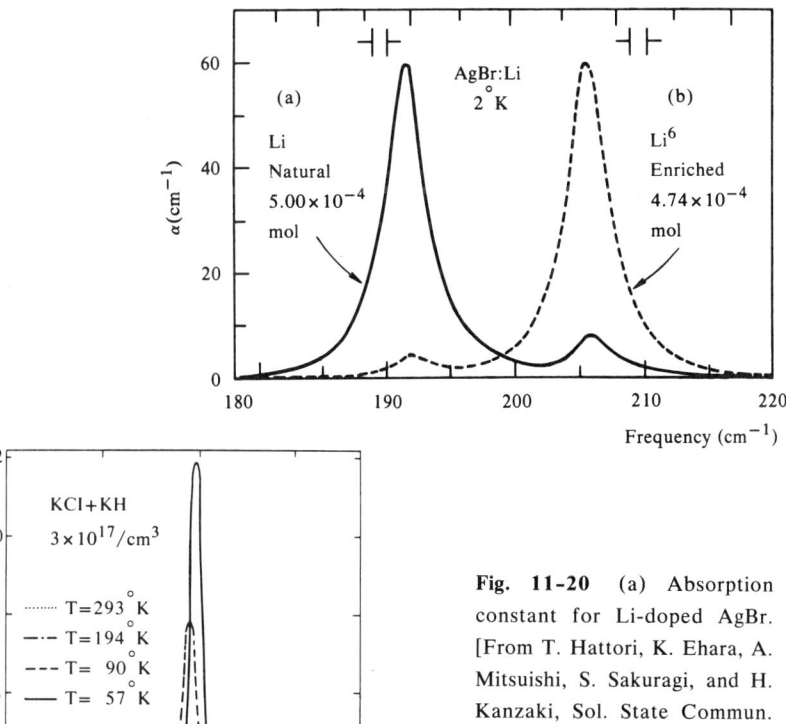

Fig. 11-20 (a) Absorption constant for Li-doped AgBr. [From T. Hattori, K. Ehara, A. Mitsuishi, S. Sakuragi, and H. Kanzaki, Sol. State Commun. **12**, 545 (1973).] (b) The infrared absorption due to H^- ions in KCl. [G. Shaefer, J. Phys. Chem. Solids **12**, 233 (1960).]

Notes

For the first law of thermodynamics and the definitions and relations between C_P and C_V see any textbook on thermodynamics.

J. DeLaunay [Solid State Physics **2**, 219 (1956)] has a nice historical review of the early heat capacity work including references to Dulong and Petit's work; Einstein [Ann. Physik **22**, 180 (1907); **31**, 679 (1911)]. Also see: Debye [Ann Physik **39**, 789 (1912); an English translation of this paper is in "The Collected Papers of Peter J. W. Debye" Interscience, 1954]; Born and von Kármán [Physik Z. **13**, 297 (1912)]; Blackman [Reports of Progress in Physics **8**, 11 (1941)].

The book by E. S. R. Gopal, "Specific Heats at Low Temperatures" (Plenum Press, 1966) is a nice reference. Many other solid state texts have a chapter on specific heat.

T. Muto and Y. Takagi's article in Solid State Physics **1**, 193 (1955) is a fine article on the theory of order-disorder transition of the type discussed in this chapter. Dekker and Kittel (especially the earlier editions) also have discussions of this topic.

Debye's little book "Polar Molecules" (Dover Publications, 1945) is a delight to thumb through. See his Figs. 4 and 5 for a picture of a quadrupole and octupole moment. The material in Section 11-7 is adapted from Debye's book and dates back to his original work in this field in 1912. See also the earlier editions of Kittel. Two related books also worth looking at are C. J. F. Böttcher, "Theory of Electric Polarization" (Elsevier, 1952), H. Fröhlich, "Theory of Dielectrics" (Oxford, 1949), and C. P. Smyth, "Dielectric Constant and Molecular Structure" (Chem. Catalog, 1931).

Brown has a large chapter devoted to color centers in alkali halide crystals and is a good reference for point defects. Two earlier review articles by F. Seitz [Rev. Mod. Phys. **18**, 384 (1946); **26**, 7 (1954)] should also be looked at as well as B. S. Gourary and F. R. Adrian [Solid State Physics **10**, 127 (1960)]. Also see A. M. Stoneham, "Theory of Defects in Solids" (Oxford University Press, 1975).

For diffusion the book by P. G. Shewmon, "Diffusion in Solids" [McGraw Hill, 1963], is useful, as is D. Lazarus, Solid State Physics **10**, 71 (1960), and Dekker, Chapter 3.

The field of superionic conductivity is a fast-moving one (no pun intended) and thus it is difficult to give one reference. There are several edited books and conference proceedings summarizing the field. For reference to Na diffusion in Naβ aluminate by a crowdion mechanism see C. Wang, M. Gaffari, and S. Choi, J. Chem. Phys. **63**, 772 (1975). The essence of the idea can be understood simply.

The article by A. J. Barker Jr. and A. J. Sievers [Rev. Mod. Phys. **47** (supplement 2) 1975] on local and resonance vibrational modes is useful. It also covers vibrational modes in disordered systems, a subject we have not discussed.

A fine overview and review of "Long range order in solids" has been written by R. M. White and T. H. Geballe and appears as Supplement **15** of Solid State Physics (1979). The book has extensive coverage of order in the fields of superconductivity and magnetism but covers other areas as well.

Problems

1. Derive Eq. 11-2a. (Hint: See a thermodynamics textbook.)

2. (a) For a **two-level system** write an expression for the specific heat with arbitrary degeneracies for both levels. (b) Find the expression that gives the temperature, T_m, at which the specific heat attains a maximum value. (c) Show that at T_m the specific heat is given by $C = (RT_m/4\Delta\mathscr{E}) [(\Delta\mathscr{E}/T_m)^2 - 4]$. (d) Roughly sketch the specific heat for the ratio of the degeneracies equal to 1/2, 1, and 2.

3. "Modern" quantum mechanics (after 1927) shows that the energy levels of a harmonic oscillator are $\mathscr{E} = (n + 1/2) \hbar\omega$ rather than $n\hbar\omega$. The $\hbar\omega/2$ term is called the **zero-point energy**. (a) Show that the heat capacity is still given by Eq. 11-14, but that the expression for the energy is different but in a very simple way. Why is C_V unaffected by the zero-point energy? (b) For a free HCl molecule the vibrational force constant $\approx 10^5$ dyne cm^{-1}. At room temperature, what is the probability of the $n = 2$ vibrational state being occupied and what is the value of C_V? For the I_2 molecule, $\omega = 214$ cm^{-1}. What is the probability of the $n = 1$ and $n = 2$ states being occupied?

4. Quantum mechanics shows that the energy levels of the **rotational states of a diatomic molecule** are $\mathscr{E} = J(J + 1) \hbar^2/2I$ where J is the rotational quantum number (J = 0, 1, 2, ...) and I the moment of inertia. (a) Determine the rotational partition function. Don't forget that the degeneracy is $2J + 1$. (b) Assuming that k_BT is much larger than the spacing between the rotational levels, replace the sum with an integral and do the integration. [Hint: Let $x = J(J + 1)$.] Also determine the rotational energy and heat capacity in this limit. Why would you expect to obtain $C_V = k_B$ all along? (c) For the HCl molecule the internuclear distance is 1.28 Å. What is the spacing between the J = 0 and 1 state? How does this compare to room temperature? (d) Water has a molar heat capacity of 18 cal mole^{-1}°K^{-1}. What does this say about H_2O molecules freely rotating?

5. How does the **Debye temperature** depend on the density? Compare this result with experimental measurements for Li isotopes, the alkali metals, and several alkali halides with the NaCl and CsCl structure. When these comparisons are made within a series of similar materials, what is the basic assumption?

6. The replacement of the sum over k by an integral, for example in Eqs. 11-13 and 11-14, is justified if there are many modes with energy less than k_BT. If measurements are being made at 10^{-6}°K what size sample is required to justify this replacement?

7. Estimate the magnitude of the maximum displacements due to thermal energy of Na and Cl in NaCl. Compare the result at 300°K to that at 30°K.

8. For an electric field of 10^4V/cm and a dipole moment of 10^{-18}esu cm (this is called a **Debye unit**) calculate $\mu E/k_B T$ at room temperature and show that in the Langevin function, $L(x)$ in Eq. 11-37, $x \sim 10^{-3} \ll 1$. Estimate μ of an HCl and an H_2O molecule.

9. In no more than three pages discuss the experiments and theory of localized H^- ion vibrations in ionic crystals. See the article by Barker and Sievers in the Notes and R. J. Elliott, W. Hayes, G. D. Jones, H. F. MacDonald and C. P. Sennett, Proc. Royal Soc. **A289**, 1 (1965).

10. Consider a defect such as Li^+ in AgBr where the defect has O_h(m3m) point symmetry. For O_h(m3m) or $T_d(\bar{4}3m)$ the first term of the potential energy seen by the defect has spherical symmetry $V = M^*\omega^2 r^2/2$, where M^* is the defect mass. If the motion of the defect were truly localized, what would you expect the ratios of the local mode frequencies of Li^6 and Li^7 to be? Compare this to the observed values in AgBr:Li. In AgCl:Li the observed ratio is 1.043. Comment about the meaning of the differences from the expected value of 1.080. Using Barker and Sievers as a reference compare these results to those found for hydrogen and its isotropes in alkali halides.

11. AB alloy – The problem of an order-disorder phase transition in a simple AB alloy, such as CuZn, can be treated similarly to the orientational disorder problem in Section 11-6. Consider an AB alloy with the high temperature structure where either A or B atoms are distributed randomly on a bcc lattice. At low temperatures this structure orders so that the A atoms are at the corners of the cubes and the B atoms are at the centers. This is often called two interpenetrating sc lattice. One lattice we call a or the other b and there are N atoms of type A and N of type B. The **long-range order parameter** Δ is defined so that the number of A atoms on the a lattice is $(1 + \Delta)N/2$ and the number of A atoms on the b lattice is $(1 - \Delta)N/2$. As in Eq. 11-26 the internal energy

$$U = N_{AA}\mathscr{E}_{AA} + N_{BB}\mathscr{E}_{BB} + N_{AB}\mathscr{E}_{AB}$$

(a) Then in a manner similar to Eq. 11-27 show that
$$N_{AA} = 8[(1+\Delta)N/2][(1-\Delta)/2] = 2(1-\Delta^2)N = N_{BB}$$
$$N_{AB} = 8N[(1+\Delta)/2]^2 + 8N[(1-\Delta)/2]^2 = 4(1+\Delta^2)N$$
$$U/2N = (\mathscr{E}_{AA} + \mathscr{E}_{BB} + 2\mathscr{E}_{AB}) + \Delta^2(2\mathscr{E}_{AB} - \mathscr{E}_{AA} - \mathscr{E}_{BB})$$
$$\equiv \mathscr{E}_0 + \Delta^2 \mathscr{E}_{Ave}$$

CHAPTER 11   SOME THERMAL EFFECTS IN SOLIDS   403

(b) Since the entropy $S = k_B \log w$ where w is the total number of arrangements of the atoms, show that

$$w = \frac{N!}{[(1+\Delta)N/2]![(1-\Delta)N/2]!}$$

Check the values of S for the completely ordered case $\Delta = \pm 1$, and for the completely random case of $\Delta = 0$.
(c) For equilibrium show that

$$0 = (4N\mathscr{E}_{Ave})\Delta + (Nk_BT/2) \ln\left[\frac{1+\Delta}{1-\Delta}\right]$$

$$k_BT_c = -4\mathscr{E}_{Ave}$$

so that $\mathscr{E}_{Ave}$ must be negative in order to have a phase transition. From these equations, we can show that the $\Delta(T)$ is just like that shown in Fig. 11-8b. The classic paper by F. C. Nix and W. Shockley, Revs. of Modern Phys. **10**, 1 (1938) is worth looking at. Also see L. Guttman, Solid State Physics **3**, 145 (1956).

12. Dislocations – (a) Describe an **edge dislocation.** Drawing a picture of a simple edge dislocation, show the **Burgers vector** and explain why half of the region around this dislocation is under compressive pressure and half is under expansive pressure. What is the relation between edge dislocations and plastic flow? Between edge dislocations and strain layered superlattices (Chapter 18). Find the edge dislocation in Fig. 16-6b. (b) Describe a **screw dislocation** showing that the Burgers vector is parallel to the screw axis. What is the connection between this dislocation and crystal growth? [See A. H. Cottrell, "Dislocations and Plastic Flow in Crystals" (Claredon Press, 1953) and W. T. Read, "Dislocations in Crystals" (McGraw-Hill, 1953).

13. Heat capacity in low dimensions – The heat capacity in a three-dimensional (3-D) solid was determined in the text. Now consider a 2-D and a 1-D solid. Calculate $g(\omega)$ and θ_D and show that the low temperature heat capacity is as follows. Note that if in the 2-D problem one allows out-of-plane vibrations, as well as the two in-plane, then C_V is 3/2 larger; in the 1-D problem C_V is three times larger if out-of-line vibrations are allowed. The general **Riemann zeta function** is:

$$\int_0^\infty \frac{x^{s-1}dx}{e^x - 1} = (s-1)! \sum_{p=1}^\infty \frac{1}{p^s}$$

	3-D	2-D	1-D
$g(\omega)$	$\dfrac{3V}{2\pi^2}\dfrac{\omega^2}{v_0^3}$	$\dfrac{2A}{2\pi}\dfrac{\omega}{v_0^2}$	$\dfrac{L}{2\pi}\dfrac{1}{v_0}$
Θ_D	$\dfrac{\hbar v_0}{k_B}\left[\dfrac{6\pi^2 N}{V}\right]^{1/3}$	$\dfrac{\hbar v_0}{k_B}\left[\dfrac{4\pi N}{A}\right]^{1/2}$	$\dfrac{\hbar v_0}{k_B}\dfrac{2\pi N}{L}$
C_V	$\dfrac{12\pi^4}{5}\dfrac{Nk_B T^3}{\Theta_D^3}$	$28.85\dfrac{Nk_B T^2}{\Theta_D^2}$	$3.290\dfrac{Nk_B T}{\Theta_D}$

14. C_V of glasses – (a) If the energy levels of a two-level system (Section 11-2b) are at $\pm\varepsilon$, then show that the energy and heat capacity are

$$U = -\varepsilon \tanh(\varepsilon/k_B T),$$

$$C_V = k_B(\varepsilon/k_B T)^2 \operatorname{sech}^2(\varepsilon/k_B T)$$

(b) If all values of ε are equally probable up to ε_0, then show, for $k_B T \ll \varepsilon_0$, that $C_V \propto T$. This is a model for the low temperature anomalous heat capacity observed in glasses and certain disordered crystals. See "Amorphous Solids," Ed. W. A. Phillips, Springer-Verlag's series, "Topics in Current Physics" (1981).

12

Lattice Vibrations

12-1 Introduction
12-2 Vibrations of a One-Dimensional Monatomic Chain
    a A monatomic linear chain
    b Group velocity
    c Non-nearest neighbor forces
    d The Brillouin zone
    e The number of allowed k-values
12-3 Vibrations of a One-Dimensional Diatomic Chain
    a $m \ne M$ case
    b Brillouin zone changes at a phase transition
    c The molecular approximation (S)
12-4 Real Crystal Systems
    a Transverse modes
    b The number of branches
    c Simple crystal systems
    d Modes throughout the Brillouin zone
    e Density of states
12-5 Phonons (A)
12-6 Crystal Momentum (A)
    a Conservation of crystal momentum
    b Selection rules
12-7 Neutron Diffraction from Phonons
12-8 Thermal Conductivity (S)
    a Phonon mean free path
    b Anharmonic effects
    c Normal and umklapp processes
    d Thermal conductivity due to defects
Notes
Problems

LATTICE VIBRATIONS

Man needs not six feet of earth, not a farm, but the whole globe, all of Nature, where unhindered he can display all the capacities and peculiarities of his free spirit.

A. Chekhov, *"Gooseberries"*

12-1 Introduction

In discussing heat capacity in Chapter 11, two dynamic models of a solid were considered. One model assumed that all the atoms vibrate independently of each other and at a single frequency, ω_E. This model, due to Einstein in 1907, removed the major difficulty in understanding the experimental C_V data. Namely, it explained why C_V decreased at low temperatures below the classical Dulong and Petit value. In 1912 Debye proposed a more realistic dynamic model of a solid. He treated it as an elastic continuum. Debye's model is particularly appropriate for a solid in the low energy, small wave vector region because in this long wavelength approximation $\lambda = 2\pi/k$ the details of the forces between the individual atoms are not important; only the average forces are important. The **dispersion curve** of the crystal, that is, ω and k, is determined by this average force. Then C_V vs. T can be expressed in terms of one parameter, the Debye temperature, Θ_D.

The Debye model successfully predicts that $C_V \propto T^3$ at low temperatures in agreement with experiment. However, as C_V measurements became more accurate, it was found that Θ_D is not really a constant but varies with temperature, Fig. 11-6. This is just another way of saying that the Debye model is not accurate enough. In fact, also in 1912 Born and von Karman published their basic paper that describes the lattice vibrations in a crystal. However, the immediate success of the Debye theory caused this paper, which is relatively complex, to be largely ignored. It was not until 1935 that Blackman reopened this area of work. He showed that even in simple systems, where the forces between individual atoms are considered, there can be large deviations from the Debye results and that the more general force constant model of Born and von Karman is appropriate.

We proceed in a very simple way, using the latter model. The basic assumptions are that (1) each atom is located at an equilibrium position in the crystal structure and (2) it can oscillate about this equilibrium position with an amplitude that is small compared to the internuclear distances. These assumptions have been used in Section 11-4 to derive the basic equation for the energy of a lattice of oscillators. From these assumptions it follows (see the Notes) that the potential energy of the system depends quadratically on the displacements of the atoms from their equilibrium positions for small displacements (Fig. 8-3b). Hence, the force in Newton's equation depends linearly on the displacements. Thus, the system behaves as if all the particles are joined by harmonic springs, and the motion is reduced to a classical problem. The solution, for small oscillations, may be described in terms of **normal modes**. (A normal mode is a correlated motion of the atoms that has a characteristic wave vector and frequency. If this motion is started, it will continue forever, ignoring friction, i.e., anharmonic terms.) We first consider monatomic and diatomic chains of atoms, that is, the one-dimensional case.

12-2 Vibrations of a One-Dimensional Monatomic Chain

Although vibrations can be treated classically, the problem is difficult even in simple crystals with high symmetry. This is partly because there are significant force constants (springs) between each atom and many more neighbors than just its nearest neighbors. In three dimensions the many subscripts that are required to keep track of these forces tend to obscure the physics. For these reasons, we first treat a one-dimensional chain of similar atoms so there is one atom per primitive unit cell. In a mathematically simple way the results show the essentials of the lowest energy dispersion curve. This **dispersion curve** is just the function ω vs. k (i.e., energy vs. wave number).

In this section and the next section we assume further that the forces act only between nearest neighbors and that the motion is only along the direction of the chain (longitudinal waves). The first assumption is made only to reduce the subscripts. If this assumption is not made the results are very similar. If motion is allowed perpendicular to the chain (transverse waves) the same results are obtained except with different force constants. Thus, both of these assumptions are not restrictive from the point of view of the physics.

12-2a A monatomic linear chain Figure 12-1a shows a linear chain of identical atoms, each with mass m, an equilibrium distance a apart, and connected to each nearest neighbor by a spring with spring constant α. When these atoms are displaced longitudinally along the chain from their equilibrium positions by an amount u_n as in Fig. 12-1b, the potential energy of the system is

CHAPTER 12   LATTICE VIBRATIONS

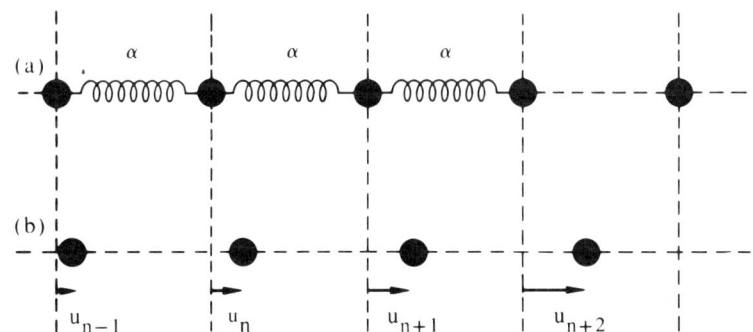

Fig. 12-1 A linear chain of like atoms in (a) equilibrium; (b) displaced from equilibrium by amounts u_n.

$$PE = \Sigma_n (\alpha/2)(u_{n+1} - u_n)^2 \tag{12-1}$$

where the sum is over the N atoms. Forces between only nearest neighbors are assumed. This equation is very simple, being just the sum of $\alpha x^2/2$ type terms for each spring. Differentiation with respect to u_n gives minus the force on the nth atom, so $F = -d(PE)/dx$.

$$m\, d^2 u_n/dt^2 = \alpha[(u_{n+1} - u_n) - (u_n - u_{n-1})] \tag{12-2}$$

There is a similar equation for each of the N atoms. The two terms come from the two springs that are attached to the nth atom. We expect wave-like behavior and try traveling wave solutions of the form

$$u_n = u\, \exp[i(nka - \omega t)] \tag{12-3}$$

This has wave-like behavior along the chain, $\exp[i(nka)]$, with amplitude u, and a time dependence $\exp(i\omega t)$. Substituting this form into Eq. 12-2 gives

$$-m\omega^2 u = \alpha[\exp(ika) + \exp(-ika) - 2]\,u \tag{12-4}$$

Clearly this holds for any value of u provided that

$$\omega^2 = (2\alpha/m)(1 - \cos ka) = (4\alpha/m)\sin^2(ka/2) \tag{12-5a}$$

or $$\omega = 2(\alpha/m)^{1/2}\,|\sin(ka/2)| \tag{12-5b}$$

This is called the **dispersion relation** because it relates ω to k and Fig. 12-2 shows the result. The positive root of Eq. 12-5a is taken because physically we want real frequencies.

There are many important points that should be noticed about the ω vs. k result in Eq. 12-5b (Fig. 12-2). First, there is symmetry between k and $-k$. This corresponds to the fact that waves traveling to the right and left are identical, $\omega(k) = \omega(-k)$. Second, the results repeat with a period of $2\pi/a$. (Later, we shall return to this point.)

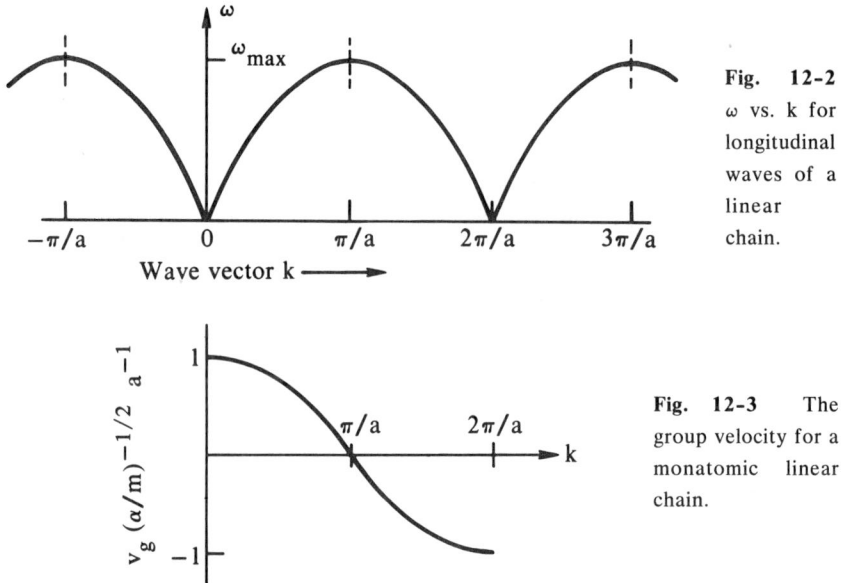

Fig. 12-2 ω vs. k for longitudinal waves of a linear chain.

Fig. 12-3 The group velocity for a monatomic linear chain.

Third, there is a maximum or cutoff frequency, ω_{max}, above which the chain of atoms cannot sustain a traveling wave. This frequency occurs at $k = (\pi/a) + (m2\pi/a)$ where m is any integer and its value is $\omega_{max} = 2(\alpha/m)^{1/2}$. Fourth, the index n does not appear, so a single relation gives ω vs. k. Last, the linear relation between ω and k as in the Debye model does not hold here. However, expanding Eq. 12-5b for small ka, we have

$$\omega \approx (\alpha/m)^{1/2} ak$$
$$\omega/k \approx \text{velocity} = (\alpha/m)^{1/2} a \qquad (12\text{-}6)$$

Thus, in the long wavelength limit, ω is linear with k. This can be seen in Fig. 12-2, and the velocity is given by the spring constant, the mass, and the lattice constant. This makes physical sense since for small k we expect "not to notice" the individual atoms, which corresponds to the Debye approximation (i.e., treating the system as a continuous medium). However, as k becomes larger (λ shorter), the discreteness of the chain affects the dispersion relation.

12-2b Group velocity The group velocity (i.e., the velocity of a wave packet) is $v_g = d\omega/dk$. For this monatomic linear chain, between $k = 0$ and π/a, it is given by

$$v_g = (\alpha/m)^{1/2} a \cos(ka/2) \qquad (12\text{-}7)$$

CHAPTER 12   LATTICE VIBRATIONS   411

This result is plotted in Fig. 12-3. For small k, v_g starts with the value given by Eq. 12-6 and for $k = \pi/a$ it is zero. More generally v_g is zero whenever $k = \pi/a + m2\pi/a$ where m is any integer. This k-periodic dispersion curve is analogous to the electronic case in earlier chapters. Both are results of the translational symmetry of the crystal.

The vibrational modes discussed here are called **acoustic modes** and the ω vs. k is called the **acoustic branch** because at small k there is a constant nonzero group velocity as in a continuous medium.

12-2c   Non-nearest neighbor forces   If forces between other than nearest neighbors are included in deriving Eq. 12-2, then

$$md^2u_n/dt^2 = \Sigma_p \, \alpha_p(u_{n+p} - u_n) \qquad (12\text{-}8a)$$

where α_p is the force constant between the nth atom and its pth neighbor. Trying a solution $u_{n+p} = u \exp i[(n+p)ka - \omega t]$, we obtain effectively the same expression as Eq. 12-5, that is,

$$\omega^2 = (2/m) \Sigma_{p>0} \, \alpha_p(1 - \cos pka) \qquad (12\text{-}8b)$$

which gives the same type of dispersion as in Fig. 12-2 and a group velocity that is constant at small k and zero at $k = \pi/a$. (The details are left as a problem.) Of course, depending on the actual values of α_1, α_2, and so on, it is possible to obtain an ω vs. k curve with more structure, but the periodicity of $2\pi/a$ in k is unaltered.

12-2d   The Brillouin zone   Like electrons in a periodic potential (Chapter 10) any interval in k of length $2\pi/a$ contains all the information. Consider k' such that

$$k' = k + (m2\pi/a) \qquad m = 0, \pm 1, \pm 2, \ldots \qquad (12\text{-}9a)$$

Since $\exp(i\,2\pi m) = 1$,

$$u_n(k) = u_n(k') \quad \text{and} \quad \omega(k) = \omega(k') \qquad (12\text{-}9b)$$

Since *the motion is defined only at the atomic sites*, u_n, an interval in k of $2\pi/a$ gives all the information that there is in the dynamic solution. Figure 12-4 stresses the point. Two different values of k are shown, and both fit the wave motion for the same displacements (a transverse wave is shown for clarity). Thus, clearly the range of k can be restricted. The range of k is usually chosen to be

$$-\pi/a \leq k \leq \pi/a \qquad (12\text{-}9c)$$

so that the interval is $2\pi/a$ and waves that propagate to the right and left are treated symmetrically. This interval of k-space is just the first Brillouin zone for a one-dimensional lattice. (See Section 10-4.) There are N atoms and one degree of freedom for the atomic motion

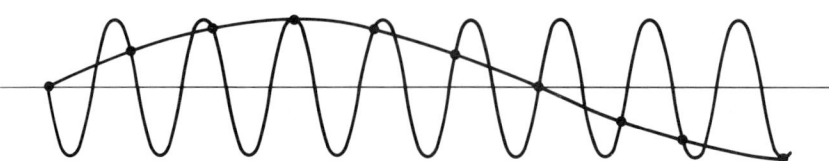

Fig. 12-4 Waves with different wave numbers that represent the same atomic displacements. Note how the wave numbers obey Eq. 12-9a.

(longitudinal motion along the chain), so there are N k-values in this zone. If we allow transverse, as well as longitudinal, vibrations then there are 3N degrees of freedom. Since we assume that the chain is rotational symmetric the two transversely polarized vibrations are degenerate; their ω vs. k curves are identical to each other and they have a shape that is similar to the longitudinal case. However, the transverse force constants are generally smaller so their ω values are less than corresponding longitudinal values.

This same very important result for the Brillouin zones applies to lattice vibrations in three dimensions. That is, in three dimensions, a crystal with one atom at each of the N lattice points has 3N degrees of freedom, and the 3N k-values all can be confined to the first Brillouin zone with one longitudinal and two transverse branches. This will be developed further when real crystals are discussed.

12-2e The number of allowed k-values There are N values of k that are allowed for a lattice with N primitive unit cells. This results from the periodic boundary conditions and was discussed in Section 9-7 so is not repeated here.

Thus, the ω vs. k curve for vibrational motion, just as for electron states, should be thought of as made up from discrete k-values $k = (2\pi/Na)p = (2\pi/L)p$, where $L = Na$ is the length of the crystal, and $p = 0, \pm 1, \pm 2, ..., \pm N/2$ to keep within the first Brillouin zone, Eq. 12-9c. Of course, when N is very large the density of points (k-values) in reciprocal space is very large, and the ω vs. k result is quasi-continuous. These aspects of the ω vs. k results and the density of points in k, or reciprocal, space are discussed in Section 9-8.

12-3 Vibrations of a One-Dimensional Diatomic Chain

In this section we treat a similar problem, but now let every other atom have a mass M. Thus, we consider a chain of 2N atoms with N unit cells but with two atoms in each cell so there are 2N atoms. Each unit cell is of length $a = 2d$, where d is the distance between the atoms with dissimilar masses, and the masses are m and M. For two per unit

CHAPTER 12   LATTICE VIBRATIONS   413

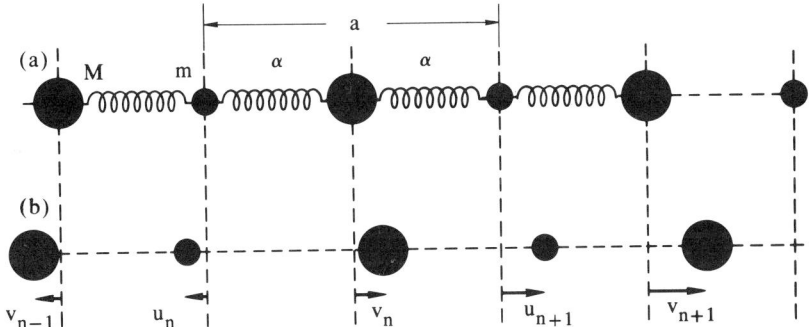

Fig. 12-5 A linear chain of atoms of masses m and M. The primitive unit cell has length (a = 2d where d is the nearest neighbor separation between the atoms). The spring constant between the atoms is the same. (a) At equilibrium. (b) Displaced from equilibrium as noted.

cell we will find two dispersion curves. One is just the acoustic branch discussed previously; the other, called the **optic branch,** has a finite frequency at k = 0, as we will see. This simple problem has the essentials of any vibrational problem. Again the forces are taken to be just between nearest neighbors. See Fig. 12-5. Letting u_n and v_n be the displacements of the atoms of mass m and M, respectively

$$\text{PE} = \sum_{n=1}^{N} (\alpha/2)(u_n - v_n)^2 + (\alpha/2)(v_n - u_{n+1})^2 \qquad (12\text{-}10)$$

This equation is essentially the same as that in Eq. 12-1. Proceeding similarly as in the monatomic lattice the equations of motion for each atomic type are

$$\begin{aligned} m\, d^2u_n/dt^2 &= \alpha(v_n + v_{n-1} - 2u_n) \\ M\, d^2v_n/dt^2 &= \alpha(u_{n+1} + u_n - 2v_n) \end{aligned} \qquad (12\text{-}11)$$

Again determine the traveling wave solutions but now allow the different types of atoms to have different amplitudes. Try

$$\begin{aligned} u_n &= u\, \exp i(2nka - \omega t) \\ v_n &= v\, \exp i[(2n+1)ka - \omega t] \end{aligned} \qquad (12\text{-}12)$$

where u and v are the amplitudes of the particle displacements. When substituted into Eq. 12-11 we find

$$\begin{aligned} -m\omega^2 u &= \alpha[v(e^{ika} + e^{-ika}) - 2u] \\ -M\omega^2 v &= \alpha[u(e^{ika} + e^{-ika}) - 2v] \end{aligned} \qquad (12\text{-}13)$$

Solutions to these homogeneous linear equations exist, if and only if the determinant of the coefficients of u and v vanish. Thus,

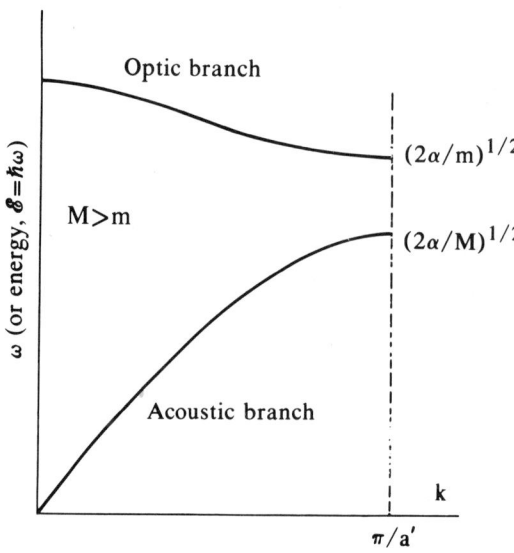

Fig. 12-6 The ω vs. k (optic and acoustic branches) for a linear diatomic chain.

$$\begin{vmatrix} m\omega^2 - 2\alpha & \alpha[1 + \exp(-ika)] \\ \alpha[1 + \exp(ika)] & M\omega^2 - 2\alpha \end{vmatrix} = 0 \quad (12\text{-}14a)$$

or $\quad mM\omega^4 - 2\alpha(m + M)\omega^2 + 2\alpha^2 (1 - \cos ka) = 0 \quad (12\text{-}14b)$

Equation 12-14b can be solved for ω^2 which yields Eq. 12-15, which can be substituted into one of the expressions in Eq. 12-13 to give the particle displacements.

$$\omega^2 = \alpha\left(\frac{m + M}{mM}\right) \pm \alpha\left[\left(\frac{m + M}{mM}\right)^2 - \frac{2(1 - \cos ka)}{mM}\right]^{1/2} (12\text{-}15)$$

The solution of ω vs. k is plotted in Fig. 12-6. The ± signs give two solutions of ω for each k; only half of the first Brillouin zone is shown because ω(k) = ω(−k) as before. Also note the first Brillouin zone goes from $-\pi/a \leq k \leq \pi/a$ where a (= 2d) is the length of the primitive unit cell.

It is instructive to look at the k ≈ 0 limit. For small ka, cos ka ≈ 1 − (ka)²/2, and we obtain

$$\omega \approx \left(\frac{2\alpha}{\mu}\right)^{1/2}, \qquad \frac{u}{v} \approx -\frac{M}{m} \qquad \textbf{optic branch}$$
(12-16)
$$\omega \approx \left(\frac{\alpha/2}{m + M}\right)^{1/2} ka, \qquad \frac{u}{v} \approx 1 \qquad \textbf{acoustic branch}$$

where the reduced mass $\mu \equiv mM/(m + M)$. The bottom equation, with ω ∝ k, reduces to Eq. 12-6 when m = M (a = 2d). The top equation, which comes from the + sign in Eq. 12-15, has ω indepen-

CHAPTER 12   LATTICE VIBRATIONS   415

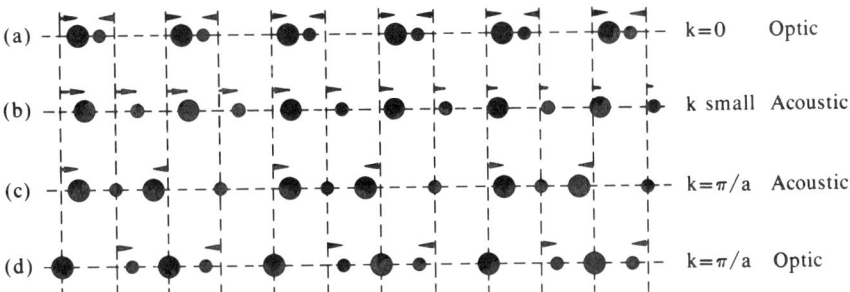

Fig. 12-7 A schematic diagram showing the atomic displacements for (a) k = 0 optic branch; (b) acoustic branch for small but certainly not zero k; (c) and (d) show the zone boundary modes for the diatomic chain at k = π/a, that is, the zone boundary, for diatomic chain.

dent of k to a first approximation, Fig. 12-6. Now the nomenclature of the two branches can be clarified. For the acoustic branch u ≈ v, so the light and heavy atoms in each unit cell are moving in phase as in a sound wave. Of course, if all unit cells moved simultaneously to the right then the crystal would move to the right; this happens at k identically zero. However, in the limit of small (but not zero) k, there are large regions where the light and heavy atoms are moving to the right and then other regions where they are both moving to the left. Figure 12-7b is a poor attempt to show this. For the optic branch at k ≈ 0 the phase of the light and heavy atoms are opposite, mu = −Mv. Any two neighboring atoms will oscillate around their equilibrium positions, toward and away from each other (Fig. 12-7a). Thus, if the m and M atoms have opposite charges, as in an ionic crystal, this motion will set up a long wavelength (small k) oscillating dipole moment, which can interact strongly with light of the same frequency. (For atoms in crystals, this is an infrared frequency.) Hence, the name optic branch and the motion is called the optic mode.

At the zone boundary, where k = π/a, the solutions are

$$\omega = (2\alpha/m)^{1/2}, \qquad u/v = \infty, \text{ i.e., } v = 0$$
$$\omega = (2\alpha/M)^{1/2}, \qquad u/v = 0, \text{ i.e., } u = 0 \qquad (12\text{-}17)$$

In each case, one type of atom is stationary while the other one oscillates. The mass of the oscillating atom determines the frequency of the mode. This is shown in Figs. 12-7c and 12-7d. The mode involving motion of the lighter atom gives the higher frequency oscillation.

For a general k (e.g., away from these two limits) no simple distinctions between the two branches exist. Nevertheless, the entire lower branch is called the acoustic branch and the entire upper branch is called the optic branch as labeled in Fig. 12-6.

12-3a m → M case It is interesting to consider what happens as the value of m approaches M. Some important general conclusions are obtained about the folding of Brillouin zones.

So far we have not made much use of the fact that a = 2d, where d is the smallest distance between atoms. Let us think about the problem by starting with the m = M case. Then Fig. 12-8a shows the solution. It is just the result for the monatomic chain where ω vs. k is shown in the entire first Brillouin zone between $\pm \pi/d$. The dashed lines at k = $\pm \pi/2d$ are for future use. Notice the confusing nomenclature. In the monatomic crystal, discussed in Section 12-2, we used a as the distance between neighboring atoms, while now we call this distance d.

Now let m ≠ M. The primitive cell is twice as large, which halves the size of the Brillouin zone. The new zone boundaries are now the dashed lines in Fig. 12-8a, at $\pm \pi/2d$. But what happened to the other half of the degrees of freedom (lattice vibrations) since the number of atoms is the same? Figures 12-8b and 12-8c show what happens. Figure 12-8c is just the solution of the diatomic chain, remembering that for the diatomic crystal the size of the primitive cell is 2d. However, Fig. 12-8b shows the upper branch of Fig. 12-8c, which is in the second Brillouin zone, translated in reciprocal space by plus and minus $2\pi/a$ (which is a reciprocal lattice vector and equals π/d). Thus, by this translation to inside the first Brillouin zone we obtain the optic branch just as previously found.

This effect is similar to what was found in Chapter 10 when the translational symmetry is added to the free electron wave functions. In that case the various "bits and pieces" of the free electron ω vs. k curve from the various Brillouin zones were translated into the first Brillouin zone by reciprocal lattice vectors. It is the same situation when the masses become unequal; the Brillouin zone becomes half as large as when the masses are equal, and the various pieces of the ω vs. k are translated into this new Brillouin zone by the new reciprocal lattice vectors $2\pi/a$ (= π/d).

12-3b Brillouin zone changes at a phase transition This halving of the Brillouin zone and the resulting "folding back" of the ω vs. k curves (Fig. 12-8) is encountered in some phase transitions. Of course, this is not a result of changing the masses of the atoms. Usually there is an atomic displacement which doubles the unit cell; this halves the size of the Brillouin zone, which results in folding of the branches into the new smaller zone.

CHAPTER 12 LATTICE VIBRATIONS   417

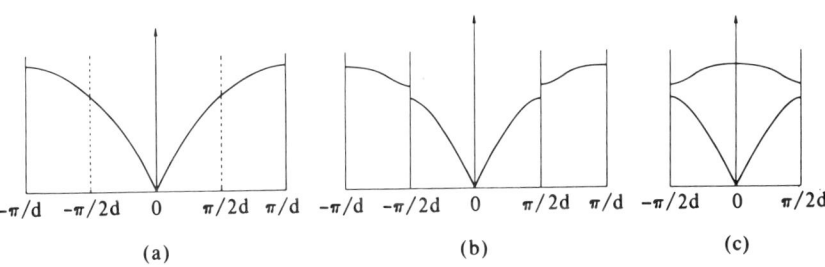

Fig. 12-8 Diagrams showing the folding back of a Brillouin zone. (a) ω vs. k in the first Brillouin zone k between $\pm\pi/d$. (b) The primitive unit cell has become twice as large so the Brillouin zone is twice as small. (c) The same as (b) but the pieces of the ω vs. k curve are translated into the first Brillouin zone.

Consider the monatomic chain at high temperatures. The result of ω vs. k is as in Figs. 12-2 or 12-8a. Now, as the temperature is lowered, assume that a phase transition occurs at T_c to a unit cell twice as large. (The transition could be due to every other atom displacing to the right so that at equilibrium it is closer to the atom on the right than the atom on the left.) Then the Brillouin zone will become half the size and the result will be as shown in Fig. 12-8c. Above T_c there is no optic mode but below there is an optic branch.

If at some lower temperature there is another phase transition where the primitive cell again becomes twice as large, then the Brillouin zone again halves and there are now three optic branches.

The results described here are common in phase transitions where the primitive cell changes in size in at least one direction. These k ≈ 0 modes can be measured by various techniques (Chapter 13). Their sudden appearance or disappearance at T_c, and the temperature dependence of their frequencies, is of interest.

12-3c The molecular approximation As is evident in the discussion about phase transitions, folding over of the Brillouin zone depends only on changing the size of the primitive cell. Letting m ≠ M is only one convenient, simple way to accomplish this result. Another way to achieve this aim is to form a diatomic chain by changing every other spring constant. We do this here and in one limit have a simple model for a molecular crystal.

Consider a diatomic chain as in Fig. 12-5, but now take the masses to be equal and let the alternate spring constants be α and β. The length of the primitive cell is a (where the separations between the atoms are d and a-d). To help visualize the problem, let the force constant β be larger than α and the distance between the atoms with the β spring be less than α, that is, the chain is made up of tightly bound diatomic molecules that are loosely coupled to each other. With or without this helpful picture, the potential energy is

$$PE = \sum_{n=1}^{N} (\beta/2)(u_n - v_n)^2 + (\alpha/2)(v_n - u_{n+1})^2 \qquad (12\text{-}18)$$

which should be compared to Eq. 12-10. The solution is left as an exercise, but for $k \approx 0$ we obtain

$$\omega \approx \left[\frac{2(\alpha+\beta)}{m}\right]^{1/2}, \qquad u \approx -v \qquad \textbf{optic mode}$$

$$\omega \approx \left[\frac{\alpha\beta}{2m(\alpha+\beta)}\right]^{1/2} ka, \qquad u \approx v \qquad \textbf{acoustic mode}$$
(12-19)

Comparing these solutions to those in Eq. 12-16, we see that the same type of optic and acoustic motion is observed.

Now consider the $k = \pi/a$ limit. For this case it can be seen that the solutions are

$$\omega = (2\beta/m)^{1/2}, \qquad u = -v$$
$$\omega = (2\alpha/m)^{1/2}, \qquad u = v$$
(12-20)

These solutions have a different physical picture from those in Eq. 12-17, as can be seen by the values of u and v. For example, for $u = -v$ the β spring is alternately stretched and compressed while the α spring is at equilibrium. The $u = v$ case produces the opposite results. However, the dispersion curves have the same shape as in Fig. 12-6.

Thus, we see that when the primitive unit cell contains two atoms due to alternating spring constants rather than alternating masses the same type of dispersion curves are obtained. The fundamental reason for the appearance of an optic mode is that there are two atoms per primitive unit cell.

Molecular case — Now consider the case when $\beta \gg \alpha$, so that two atoms are very tightly coupled compared to their much looser coupling to their nearest neighbors. Then, ignoring terms that are of order α/β and smaller, the solutions, for small k, are

$$\omega \approx (2\beta/m)^{1/2}, \qquad u \approx -v \quad \textbf{optic} \qquad (12\text{-}21a)$$

$$\omega \approx (\alpha/2m)^{1/2} ka \qquad u \approx v \quad \textbf{acoustic} \qquad (12\text{-}21b)$$

Comparing this result for the acoustic branch to that in Eq. 12-5b, we see that the two are the same, remembering that the primitive cell mass and length are 2m and a (= 2d). compared to m and a for a monatomic chain. Only the weak spring enters the dispersion relation for the acoustic branch. However, the optic branch frequency is independent of k and is just determined by β, the very rigid spring.

We have a model of a molecular crystal as discussed in Chapter 6. The molecule is tightly bound with a spring β, which causes the frequencies of vibration to be very high, and the two atoms in the molecule vibrate against each other (i.e., u = −v). The motion of one molecule against the neighboring ones is negligible (for $\beta \gg \alpha$) in the optic branch. The crystalline forces are relatively weak so there is no dispersion in the optic branch, to the order of (α/β). However, in the acoustic branch the effect of the weak coupling, via the α spring, can be seen. In this branch the frequencies of vibration, even at the zone edge, are very low compared to those of the molecule itself and the molecules vibrate as rigid bodies with a mass equal to that of the entire molecule (2m in this case).

In molecular solids this is approximately what is observed. The molecular vibration frequencies (intermolecular frequencies) in the gas phase are typically quite high and are found to be practically unaltered in the liquid and crystalline states. Also, vibrations due to the relatively weak intermolecular forces are found in the crystalline state but at much lower frequencies than for the intramolecular modes. Further, the dispersion of the optical branch is found to be very small.

12-4 Real Crystal Systems

Everything that has been said about the simple monatomic and diatomic chains is appropriate to real crystals. However, real crystals are more complicated. First, they have transverse, as well as longitudinal, waves. This is no problem, but the restoring forces (spring constants) are different for these two kinds of modes. Thus, the branches are not degenerate. Second, with more than one atom in the primitive unit cell, optic branches are obtained. This we have encountered already. Also, if the atoms are charged with respect to each other (i.e., NaCl as opposed to Si, each of which has two atoms in their primitive cells), then a further splitting can occur due to long-range electrostatic forces. However, this is a relatively small overall problem and presents no difficulty in understanding the ω vs. k spectrum. Last, the Brillouin zone of a three-dimensional crystal is much more complicated than that of a one-dimensional chain. For example, as in the case of electron bands, Fig. 10-22, there are complicated bands throughout the Brillouin zone. This can lead to complicated-looking ω vs. k curves. These complications will be addressed one at a time.

12-4a Transverse modes We have already mentioned that the vibrational problem for transverse modes in the one-dimensional chain can be solved just as easily as for longitudinal ones. In fact, the wave-like motion is easier to display in a figure. The only problem is a conceptual one, since springs between two neighboring atoms do not stretch, to first order, for displacements transverse to the chain axis. In real crystals there is no problem since there are neighboring atoms

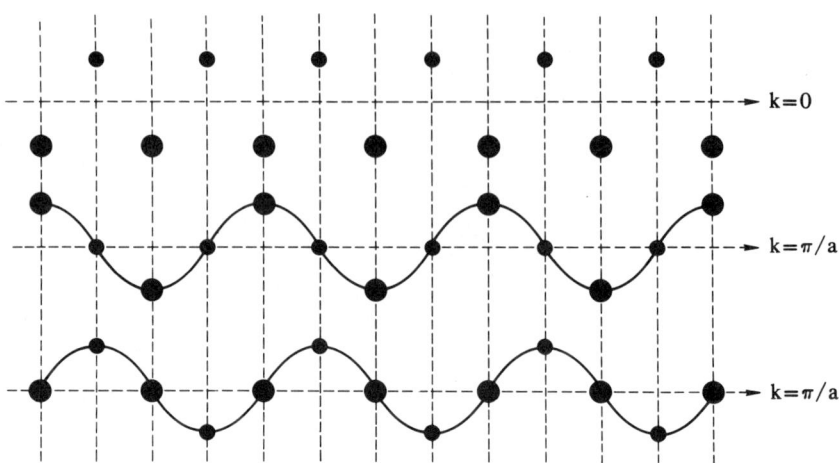

Fig. 12-9 Transverse modes for a one-dimensional diatomic chain of ions. The motion is shown at the center and edge of the Brillouin zone. Note that k points to the right and the displacements are perpendicular to k.

in transverse directions so there are springs that stretch. Also, noncentral forces in crystals can be important. In any case, when transverse solutions are obtained for the diatomic chain, Fig. 12-9 shows the results for $k = 0$ and for the two modes at the zone edge. The similarities to the longitudinal motion, Fig. 12-7, are obvious.

For transverse modes a dispersion curve is obtained just like that shown in Fig. 12-6. However, there is one important difference between the transverse and longitudinal branches. The transverse branches are doubly degenerate for the linear chain. That is, displacements in either of the two directions perpendicular to the chain axis have the same energy. However, in real crystals these two transverse modes are degenerate only in special high-symmetry directions. In general directions of propagation, each of these two transverse modes and the one longitudinal mode all have different frequencies. In fact, in a general k-direction the distinction between longitudinal and transverse waves no longer has meaning since the displacements are no longer exactly parallel or perpendicular to the wave vector.

12-4b The number of branches As discussed here and in Chapters 10 and 11, if there are N primitive unit cells and z atoms in each cell, since each atom has three degrees of freedom the crystal has $3zN$ degrees of freedom, and these must appear in the dispersion curves. The number of allowed k-values in any single branch is N. Thus, for every allowed value of k there are $3z$ different vibrational excitations and the ω vs. k surface has $3z$ curves in any one k-direction.

CHAPTER 12   LATTICE VIBRATIONS   421

We already have seen this in the monatomic, z = 1, chain case. When only one degree of freedom (longitudinal motion) was allowed, we found one branch. Allowing all three degrees of freedom results in three branches; one is the longitudinal branch and a second and third are the doubly degenerate transverse branches. These are called the TA and LA (transverse and longitudinal acoustic) branches.

For z = 2 in Section 12-3 we calculated the LA and LO (longitudinal acoustic and optic) modes by allowing motion in only one direction. If all three degrees of freedom are allowed, six branches are obtained: the doubly degenerate TA; the doubly degenerate TO (transverse optic); and the LA and LO branches. So for one value of **k** there are three acoustic and three optic branches. Figure 12-9 shows the TA and TO modes at the zone center and edge, and Fig. 12-7 shows the corresponding LA and LO modes.

In general, for any k-direction there are 3z branches. Three are acoustic branches and 3z − 3 are called optic branches. Thus, for z = 2, there are three optic and also three acoustic branches.

12-4c Simple crystal systems In real crystals the transverse branches remain degenerate in special symmetry directions. These are directions in which entire planes of atoms move in phase with their displacements perpendicular to **k**. Some examples of this will be seen. The dispersion curves for phonons in various directions in reciprocal space can be measured by inelastic scattering techniques using neutrons from a reactor. The $k \approx 0$ modes usually can be measured very accurately by infrared and Raman techniques (Chapter 13).

Figure 12-10 shows ω vs. k curves for lead in two directions in reciprocal space. Pb has a face-centered cubic (fcc) crystal structure; therefore, there is one atom per primitive unit cell. Since z = 1 three branches are expected in any one k-direction. The particular directions shown in the figure are high-symmetry ones so the transverse modes are degenerate. Notice how in the [111] direction the results are quite similar to those shown in Fig. 12-2, but in the [100] direction the curves tend to turn down at large k-values.

The figure shows similar results for sodium, which has a body-centered cubic (bcc) crystal structure. Since z = 1, again we expect the same number of modes as in the Pb case. However, in the [110] direction the transverse branches are no longer degenerate. Nevertheless, the branches have a shape similar to those in Fig. 12-2.

The results for silicon, where z = 2, are shown in Fig. 12-10e. It has, of course, the diamond structure, so z = 2. As can be seen, optic branches are observed. Again notice that the general shape is similar to that shown in Fig. 12-2. However, at the zone edge there is a very much larger splitting between the transverse modes than between the longitudinal ones. Detailed understanding of behavior throughout the

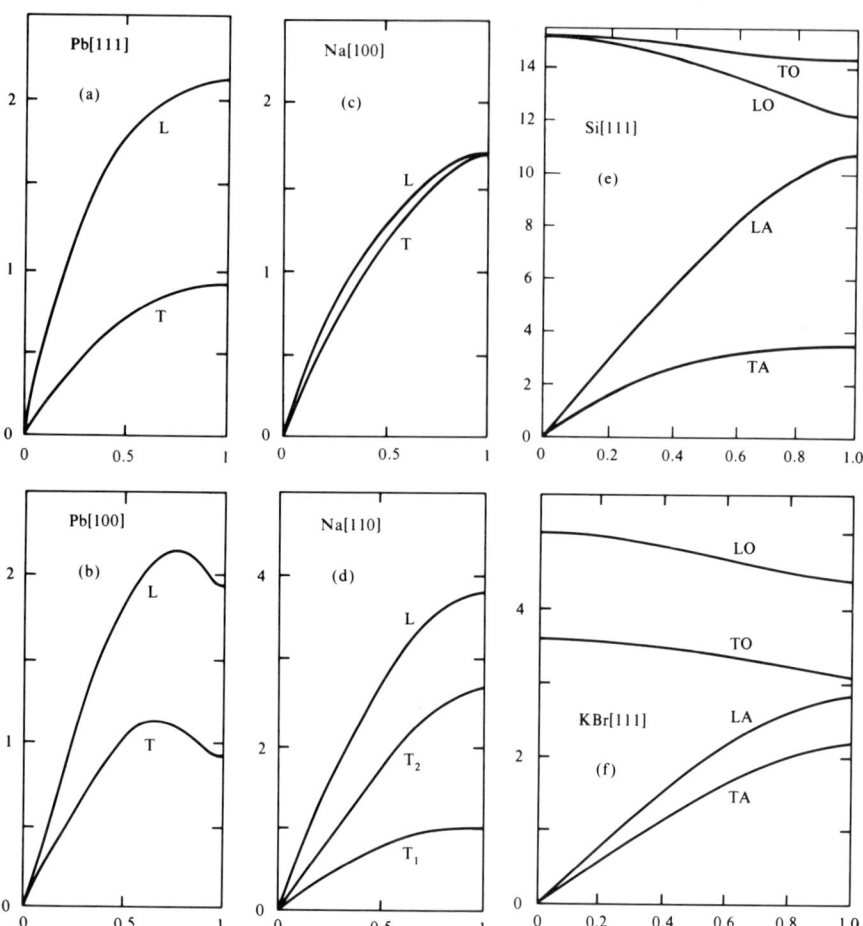

Fig. 12-10 ν vs. k for vibrational modes in different directions in reciprocal space and for different crystals as labeled. The k-values are all given in reduced values, that is, k/k_{max} where k_{max} is the Brillouin zone edge in that particular direction. Thus all the k-values go from 0 to 1. ν is plotted in units of 10^{12} Hz (10^{12} Hz = 4.136 meV = 33.35 cm^{-1}). T and L stand for transverse and longitudinal branches, respectively.

Brillouin zone requires more appropriate models than those considered in this chapter.

As a last simple example, Fig. 12-10f shows the ω vs. k curves for KBr. This has the NaCl structure, Fig. 3-5, also with z = 2 as for Si. For comparison purposes the results for KBr and Si are presented in the same direction in k-space. The general shapes of the curves are similar, at least in the directions shown, but there is one important difference. At k = 0 the optic modes are split, $\omega(LO) > \omega(TO)$, for KBr while there is no splitting for Si. This is a very important, general, result for ionic crystals and it will be discussed in Chapter 13 when the interaction of light with crystals is considered. At present, suffice it to say that oscillating longitudinal electric dipoles set up an additional restoring force that does not occur for transverse modes. This causes the LO mode in KBr to have a higher frequency than the TO at k = 0. The LO and TO splitting is intimately connected with their observability by infrared spectroscopic techniques.

It should not be concluded from the preceding discussion that in ionic crystals all of the optic modes split at k = 0. Most do split, but as in CaF_2 some optic modes do not involve an oscillating dipole moment; such modes will not have an LO-TO splitting and will not be observed by infrared techniques. Also, the optic mode in Si does not involve an oscillating dipole moment and does not split at k = 0. However, this is expected since Si is not an ionic crystal.

12-4d Modes throughout the Brillouin zone Figure 12-10 shows some dispersion curves in a few directions in the Brillouin zone. In general the results are complex and Fig. 12-11a shows values for sodium. The results along special lines and at special points are displayed and the letters are the labels of special points and lines of the Brillouin zone for the bcc crystal, as in Fig. 10-25. The ω vs. k for the [100] and [110] directions shown in Figs. 12-10c and 12-10d were taken from these more complete results.

The measurement and understanding of such curves, and those for materials where $z > 1$, is an important branch of solid state physics. The interpretations of such curves in terms of detailed atomic models (**shell models**) is important and rather sophisticated and is not discussed here. (See the problems.)

12-4e Density of states The density of vibrational states, $g(\omega)$, as a function of energy (or ω) is the number of states with frequency between ω and $\omega + d\omega$. To obtain $g(\omega)$ even data as detailed as that shown in Fig. 12-11a are not sufficient. This is because most of the vibrational states are at general k-values, not along the special lines and special points. Thus, to determine $g(\omega)$ (to calculate C_V, for example), it is necessary either to measure ω vs. k in many more directions throughout the Brillouin zone or to have a "good" lattice dynamically model from which $g(\omega)$ can be calculated. A "good" model is usually taken to be one that agrees with the experimental results along the special lines and at the special points. There has been a great deal of effort in this direction using the shell model. $g(\omega)$ for Na obtained from such a model is shown in Fig. 12-11b. Arrows indicate the critical points of the ω vs. k curve. A **critical point** is defined as a point in the Brillouin zone where $d\omega/dk = 0$ in at least one direction of k. It is interesting to compare the strength of $g(\omega)$ at the critical points to the dispersion results in Fig. 12-11a.

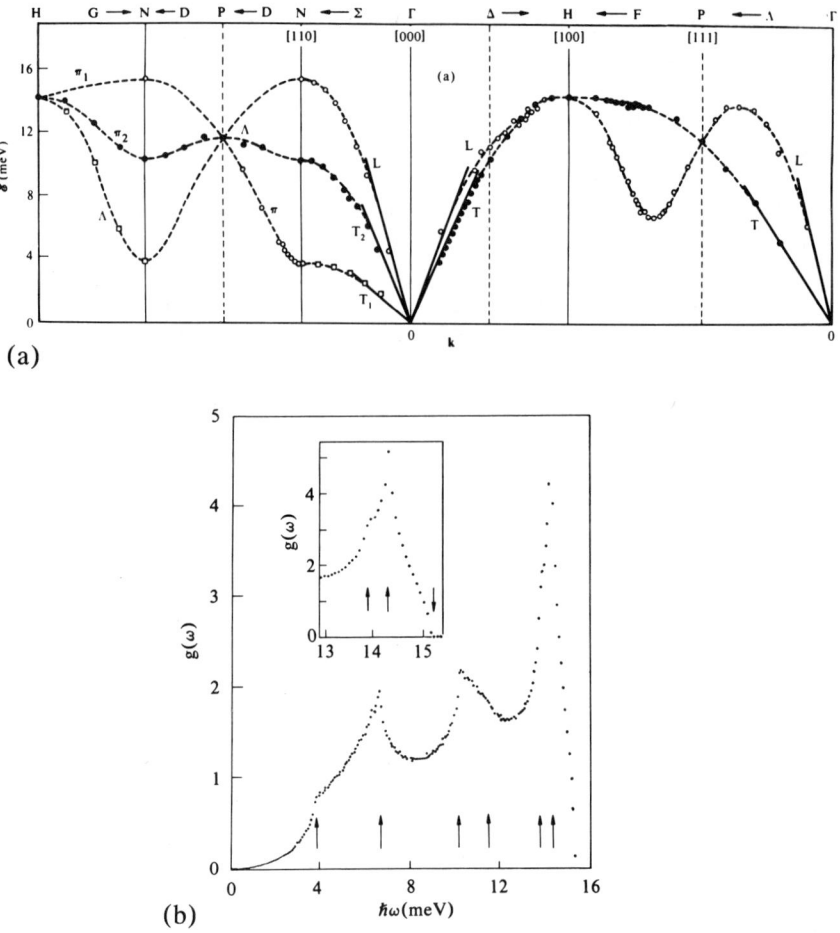

Fig. 12-11 (a) The measured vibrational spectrum for sodium along special lines of the Brillouin zone for this bcc material. [A. D. B. Woods, B. N. Brockhouse, R. H. March, A. T. Stewart, and R. Bowers Phys. Rev. **128**, 1112 (1962).] (b) The density of vibrational states for Na. The arrows indicate the critical points. [A. E. Dixon, A. D. B. Woods, and B. N. Brockhouse, Proc. Phys. Soc. **81**, 973 (1963).]

As an exercise, draw, in Fig. 12-11b, a Debye, $g(\omega)$. According to Debye, $g(\omega) \propto \omega^2$ up to a sharp cutoff so that the total number of modes is correct, that is, the area under the Debye curve and that shown should be the same. Also you should start the Debye curve at small ω with the same ω^2 dependence as that shown, since in the very long wavelength limit the Debye theory is fairly good. Clearly the two curves are not the same.

12-5 Phonons

In Section 12-2 we wrote the displacements from equilibrium of the individual atoms, u_p, in terms of traveling waves. These waves are characterized by a wave vector **k** and a branch index λ, which was stated but not used as a subscript. Reviewing, if there are N primitive cells in the crystal and z atoms per primitive cell, then there are N values of **k** and 3z branches, that is, for each k-value there are 3z different frequencies for the 3z dispersion curves. For z = 1 we discussed the one LA and two TA branches (Figs. 12-10a to 12-10d and 12-11a) and for two atoms per primitive cell we discussed the three optic branches as well as the three acoustic branches (Figs. 12-10e and 12-10f). For any one atom, a general (i.e., arbitrary) displacement involves all 3zN different modes. Thus, a description of the motion in terms of individual atomic displacements is complicated. However, the problem is drastically simplified by using **normal modes;** these are the appropriate linear combinations of the atomic displacements that completely diagonalize the terms in the Hamiltonian up to second order in the displacements. Once these are found, the total harmonic Hamiltonian of the system is the sum of individual (i.e., uncoupled) simple harmonic Hamiltonians, one for each of the 3zN normal modes (all of which commute with one another).

The normal mode concept is very important. Classically, if only one mode is set in motion it continues to oscillate indefinitely while the others remain unexcited. Quantum mechanically the eigenstates of this zN-body system can be written as a product of one-dimensional harmonic oscillator wave functions, one for each normal mode amplitude. This normal mode decomposition is possible only because the 3zN equations of motion for the atomic displacements are linear in all of the displacements. As discussed at the beginning of this chapter, this is only an approximation, arising from the second-order term in a Taylor series expansion of the potential energy about the equilibrium internuclear distance. For most purposes it is an excellent starting approximation. However, higher order terms are required for the understanding of many important processes. The apparent breakdown of the harmonic approximation leads to measurable effects, such as phonon-phonon interactions, and is discussed in the next section.

Since each normal mode yields a single simple harmonic Hamiltonian, the energy for one normal mode as well as the sum of all of them can be written immediately. The energy of the kth normal mode $\mathscr{E}_k$ is quantized and given by

$$\mathscr{E}_k = (n_k + \frac{1}{2})\hbar\omega_k \qquad (12\text{-}22a)$$

where $\quad n_k = 0, 1, 2, \ldots$

while the total energy of the system is just the sum of the individual energies, $\mathscr{E}_{total} = \Sigma_k \mathscr{E}_k$.

One peculiarity of the harmonic oscillator solution is the occurrence of the **zero-point energy**.

$$\mathscr{E}_{k,0} = \frac{1}{2}\hbar\omega_k \qquad (12\text{-}22b)$$

However, this constant amount of energy for each normal mode does not affect most physical results. For example, since $\mathscr{E}_{k,0}$ is independent of temperature, the calculation of C_V is independent of the zero-point energy. This is typical of non-relativistic mechanics, where only energy differences are important. For most discussions we will take the zero of energy as $\mathscr{E}_{k,0}$.

From Eq. 12-22a, we see that the energies of the normal modes are quantized and equally spaced. Thus, the n_kth excited state of the kth normal mode can be viewed as a state of n_k identical excitations in this mode, each having energy $\hbar\omega_k$. These excitations or quanta we call **phonons** in strict analogy with the quanta in an electromagnetic field which we call photons. A quantum state of a crystal with the set of quantum numbers $\{n_k\}$ is said to have "n_k phonons in the kth mode," each phonon contributing an energy $\hbar\omega_k$. Thus, *a phonon is a quantum of excitation in a normal mode.* When exciting the kth normal mode from n_k to $n_k + 1$ phonons, a phonon is "created" (or emitted) and the energy of the vibrational system increases by $\hbar\omega_k$. When deexciting a normal mode from n_k to $n_k - 1$, a phonon is "annihilated" (or absorbed) and the energy of the vibrational system decreases by $\hbar\omega_k$. The notion of phonons as particle-like is strengthened because they possess another particle-like property, crystal momentum $\hbar\mathbf{k}$, in analogy with the field momentum $\hbar\mathbf{k}$ of a photon. Thus, phonons are often called **quasiparticles**. However, for phonons this crystal momentum is a subtle matter, which we discuss in the next section.

Electrons, because they have a spin of $s = 1/2$, have antisymmetric wave functions with respect to the interchange of two of them. Thus, they obey the Pauli exclusion principle, which requires that only one electron can occupy a single quantum state. The statistics of such particles, which are called **fermions,** are described by the Fermi-Dirac distribution function discussed in Section 9-10; in thermal equilibrium the probability of a one-particle state with energy $\mathscr{E}$ being occupied is $\{[\exp(\mathscr{E} - \mathscr{E}_F)/k_B T] + 1\}^{-1}$. This also is the average number of particles in this state and, of course, has a maximum value of one.

For a simple harmonic oscillator in contact with a heat bath at a temperature T, we can calculate the thermodynamic average occupation number $\langle n_k \rangle$ of phonons in the kth mode. From the general expression for a thermal average, Eq. 11-6b,

CHAPTER 12   LATTICE VIBRATIONS   427

$$\langle n_\mathbf{k} \rangle = \frac{\sum_{n=0}^{\infty} n_\mathbf{k} \exp(-n_\mathbf{k} \hbar \omega_\mathbf{k}/k_B T)}{\sum_{n=0}^{\infty} \exp(-n_\mathbf{k} \hbar \omega_\mathbf{k}/k_B T)} \quad (12\text{-}23a)$$

Since this expression is essentially the same as Eq. 11-10, the result can be determined easily (see Eq. 11-12),

$$\langle n_\mathbf{k} \rangle = \frac{1}{\exp(\hbar \omega_\mathbf{k}/k_B T) - 1} \quad (12\text{-}23b)$$

The result, Eq. 12-23b, is the same as the average number of particles of wave vector **k** for a system of particles obeying Bose-Einstein statistics and having a vanishing chemical potential (corresponding to the fact that the number of "particles" is not conserved). Equation 12-23b, of course, is just the same as that found for a system of non-interacting photons and this expression is called the thermal occupation factor obtained for **Bose-Einstein statistics** or just the **Bose-Einstein factor**. Obtaining the same thermal occupation for phonons as for photons is not a surprise. The creation and annihilation operators for photons at two different positions obey the same Bose-Einstein commutation relations as do those of phonons. This, in a fundamental manner, leads to the same statistics.

A fundamental characteristic of bosons is that there is no limit on the number of particles in the same quantum state. For such particles the average thermal energy $\langle \mathscr{E}_\mathbf{k} \rangle$, defined as $\langle n_\mathbf{k} \rangle \hbar \omega_\mathbf{k}$, is given by

$$\langle \mathscr{E}_\mathbf{k} \rangle = \frac{\hbar \omega_\mathbf{k}}{[\exp(\hbar \omega_\mathbf{k}/k_B T)] - 1} \quad (12\text{-}23c)$$

This result can also be obtained directly via Eq. 11-6b, the general expression for a thermal average. In fact, it is the same result as Eq. 11-12. This result for $\langle \mathscr{E}_\mathbf{k} \rangle$ was first obtained by Planck for photons, and it is known as **Planck's law.**

For the Bose-Einstein factor it is useful to appreciate the high and low temperature limits. If $\hbar \omega_\mathbf{k} \gg k_B T$ then $\langle n_\mathbf{k} \rangle \approx \exp(-\hbar \omega_\mathbf{k}/k_B T)$. Thus, there is an exponentially small probability for a phonon to be excited. $\langle n_\mathbf{k} \rangle$ becomes equal to unity at a temperature $k_B T_1 = \hbar \omega_\mathbf{k}/\ln 2 \approx 1.44 \hbar \omega_\mathbf{k}$. At very high temperatures the oscillator is excited to very high quantum states since for $k_B T \gg \hbar \omega_\mathbf{k}$, $\langle n_\mathbf{k} \rangle$ and $\langle \mathscr{E}_\mathbf{k} \rangle$ become

$$\langle n_\mathbf{k} \rangle \approx k_B T / \hbar \omega_\mathbf{k} \quad (12\text{-}23d)$$

$$\langle \mathscr{E}_\mathbf{k} \rangle = \langle n_\mathbf{k} \rangle \hbar \omega_\mathbf{k} \approx k_B T \quad (12\text{-}23e)$$

The latter result is, of course, expected since at high temperatures quantum effects should be unimportant. Since a classical oscillator has two degrees of freedom (kinetic and potential energy) we expect $k_B T/2$ for each degree of freedom.

12-6 Crystal Momentum

Fundamental to our discussion of phonons has been the fact that the potential energy of the system depends quadratically on the atomic displacements. This leads to the harmonic Hamiltonian, and phonons. However, it must be remembered that when the potential energy of the crystal is expanded in a Taylor series of $\mathbf{u}_n$ about the atomic equilibrium positions, higher order cubic, quartic, terms (i.e., **anharmonic terms**) exist, and give rise to important processes in real crystals such as thermal expansion, finite lifetime of phonons (and hence finite mean free paths), sound absorption, and others. We shall not consider these anharmonic processes in any detail, but will consider what a general anharmonic interaction term must look like and in this way be led to the very important concept of **crystal momentum** and **conservation of crystal momentum**. This conservation law, along with the conservation of energy, plays an important role in phonon-phonon interactions, electron-phonon interactions, and even the interaction of phonons with neutrons and photons (both x-rays, and light).

12-6a Conservation of crystal momentum In the harmonic approximation to the Hamiltonian, the potential energy contains terms that involve the products of atomic displacements two at a time (i.e., $u_i u_j$ as in Eqs. 12-1 or 12-10). For cubic anharmonic interactions the potential contains terms like $u_i u_j u_k$. This gives rise to **three phonon processes** and, for example, allows for the possibility of phonons with indices i and j to interact, annihilate one another, while producing a phonon with index k. (In the harmonic approximation such an interaction is not allowed.) If the normal modes are taken as traveling waves (Eqs. 12-3 or 12-12), then the Hamiltonian will have terms of the following form:

$$\Sigma_{k_i k_j k_k} \Sigma_p A_{ijk} \exp[i(\mathbf{k}_i + \mathbf{k}_j + \mathbf{k}_k) \cdot \mathbf{t}_p] \qquad (12\text{-}24a)$$

where the p-sum is over all of the unit cells, and the A-terms are products of three complex normal mode amplitudes. The A-terms, which arise from the Taylor expansion of the potential, are independent of the unit cell index. The sum over the unit cells is zero unless

$$\mathbf{k}_i + \mathbf{k}_j + \mathbf{k}_k + \mathbf{K}_q = 0 \qquad (12\text{-}24b)$$

CHAPTER 12   LATTICE VIBRATIONS   429

where K_q is some reciprocal lattice vector including zero, remembering from Eq. 10-32 that $\Sigma_p \exp(K_q \cdot t_p) = 1$.

By analogy with a free particle, where the momentum is $\hbar k$ (Section 10-10), Eq. 12-24b can be multiplied by $\hbar$ and the result appears like a conservation of momentum relationship. The term $\hbar k$ is called the **crystal momentum** and thus Eq. 12-24b is a conservation law for crystal momentum and, with this interpretation, appears almost like the conservation of ordinary momentum. Further, the general form of Eq. 12-24a is independent of any details of the interactions. The result for the conservation of crystal momentum depends only on the fact that the sum is over all of the unit cells and that the interaction strength, the A-term, is independent of the unit cell index.

Equation 12-24a was written to describe the interaction of three phonons via a cubic anharmonic term. However, a similar term could describe the interaction of the lattice vibrations with neutrons. Thus, a neutron of wave vector k_i can be scattered to k_k with the annihilation of a phonon of wave vector k_j. Then the conservation of crystal momentum still results. And instead of neutrons (giving neutron scattering), there could be electrons (giving an electron-phonon interaction), or visible light (giving Raman scattering), or x-ray radiation (which gives inelastic x-ray scattering), and so on. The particles, waves, or quasiparticles (phonons, magnons, etc.) all have a wave vector associated with them so, when they interact, the conservation of crystal momentum law applies. Also note that Eq. 12-24b only involves the crystal momentum of three interacting particles, waves, or quasiparticles. However, an expression like Eq. 12-24a can be written for any number of interacting entities with a conservation of crystal momentum law resulting.

While $\hbar k$ is called the crystal momentum and it is conserved (modulo a reciprocal lattice vector) when various entities interact, it is not a true momentum. For example, a phonon coordinate involves only relative motion of the atoms, so that for any finite wavelength the center of mass of the entire system does not even move. Only for $k = 0$ on the acoustic branch is there a motion of the center of mass and a true momentum, but then the crystal momentum vanishes.

Nevertheless, the quantity $\hbar k$ is often seen carelessly referred to as the momentum, as opposed to crystal momentum. And the result in Eq. 12-24b often is called carelessly the law of conservation of momentum rather than of crystal momentum.

12-6b Selection rules   It is worthwhile to look into the structure of the A-term in Eq. 12-24a. This term can be shown (Ziman's "Electrons and Phonons") essentially to be

$$A_{k_ik_jk_k} = V_{k_ik_jk_k}(b_{k_i}{}^\dagger + b_{-k_i})(b_{k_j}{}^\dagger + b_{-k_j})(b_{k_k}{}^\dagger + b_{-k_k}) \quad (12\text{-}24c)$$

where the V-term is a constant that only depends on the k-values, and the b† and b are phonon creation and annihilation operators, respectively. These important operators, discussed in problem 9, have the following properties: $b_{k_i}{}^\dagger$, when operating on the phonon system, creates a phonon with k_i; b_{-k_i}, when operating on the phonon system, annihilates (or destroys) a phonon with $-k_i$. Notice that as a result of using either operator in the $(b_{k_i}{}^\dagger + b_{-k_i})$ factor, the total crystal momentum is increased by $+\hbar k_i$; this is why there is a $+k_i$ in Eqs. 12-24a and 12-24b. The other two bracketed factors in Eq. 12-24c have a similar effect and give similar results, as can be seen.

There are eight different three-phonon terms in Eq. 12-24c; let us evaluate a typical one so that the correct signs for the crystal momentum relationship will be obtained. A typical interaction is two phonons, with k-vectors k_m and k_n, interact, annihilating one another, and create a phonon with k-vector k_o. In the language of Eq. 12-4c this is described by the term $b_{k_m} b_{k_n} b_{k_o}{}^\dagger$. But notice the signs of the k-values of the annihilation operators in Eq. 12-24b and the corresponding ones in Eq. 12-24a. Thus, the conservation of crystal momentum for the process under consideration is

$$-\mathbf{k}_m - \mathbf{k}_n + \mathbf{k}_o = \mathbf{K}_q \quad (12\text{-}24d)$$

In the same simple manner any of the eight processes described by Eq. 12-24c can be handled.

So far, the discussion in this section has been concerned with phonon-phonon coupling due to cubic terms in the potential energy, which leads to three phonon processes. This leads to the conservation of crystal momentum law in Eq. 12-24b. In a totally analogous manner, the quartic terms in the potential energy can be considered. This leads to four phonon processes, for which the law of conservation of crystal momentum is

$$\mathbf{k}_i + \mathbf{k}_j + \mathbf{k}_k + \mathbf{k}_\ell + \mathbf{K}_q = 0 \quad (12\text{-}24e)$$

and the expression corresponding to Eq. 12-24c can be written by inspection. Thus, the anharmonic results from the quartic term is similar to those obtained from the cubic term.

12-7 Neutron Diffraction from Phonons

Neutron diffraction is the one technique that can be used to measure the phonon dispersion curves throughout the Brillouin zone. Thermal neutrons (i.e., in the meV range) have both the right energy

CHAPTER 12   LATTICE VIBRATIONS   431

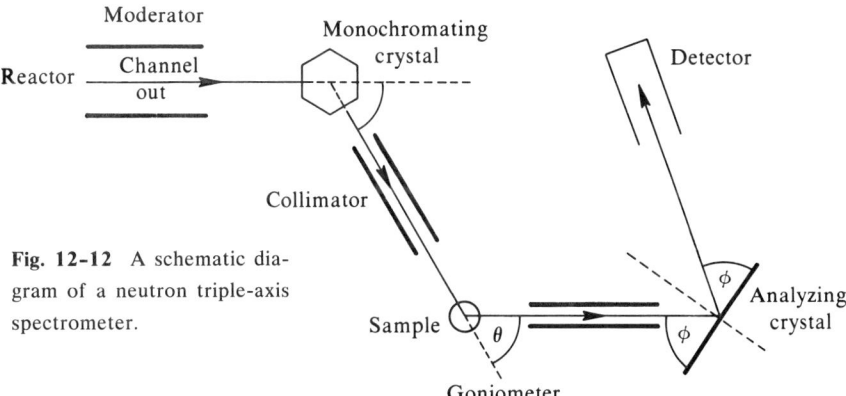

Fig. 12-12 A schematic diagram of a neutron triple-axis spectrometer.

(i.e., they have an energy similar to that of the phonons) and the experiment can be arranged so that the difference of the input and scattered $\hbar k$ of the neutrons matches that of the phonons throughout the entire Brillouin zone. (This is in contrast with infrared radiation where the energy is similar to that of phonons but $\hbar k$ is a very small fraction of the Brillouin zone boundary. As a consequence only phonons with $k \approx 0$ can be measured. X-rays still have $\hbar k$ that is a small fraction of the zone boundary and the energy is several orders of magnitude larger than that of phonons.)

The experimental procedure used for most of these inelastic neutron scattering experiments is simple, Figure 12-12. Monochromatic neutrons, k_i, obtained from a monochromating crystal and collimator, are incident on a sample. The direction of the scattered neutrons, k_s, is fixed by the position of the analyzing crystal. By varying the positions of the analyzing crystal and detector, many values of k_s can be measured and the energies associated with each value can be determined. If a phonon of wave vector k is emitted or absorbed, then conservation of crystal momentum and energy require

$$\hbar k_i - \hbar k_s = \pm \hbar k + \hbar K \qquad (12\text{-}25a)$$

$$\hbar \omega_i - \hbar \omega_s = \pm \hbar \omega = (\hbar^2/2m)(k_i^2 - k_s^2) \qquad (12\text{-}25b)$$

where m is the neutron mass, the positive sign is for the creation of a phonon, and the negative sign is for the annihilation of a phonon. If the sample is at $0°K$ then phonons only can be created, $|k_i| > |k_s|$. However, above $0°K$ there is a probability that phonons already exist, Eq. 12-23a, so a phonon may be annihilated, in which case $|k_i| < |k_s|$. For the case of phonon creation or annihilation the phonon crystal momentum is given by the neutron momentum transfer $\hbar k_i - \hbar k_s$ modulo a reciprocal lattice vector $\hbar K$ (Eq. 12-25a). However, K does not appear in the energy conservation relation (Eq. 12-25b) because the phonon energy is a periodic function in reciprocal space, that is, $\omega(\mathbf{k} \pm \mathbf{K}) = \omega(\mathbf{k})$.

Figure 12-12 shows a neutron triple axis spectrometer of the type that is usually used to measure phonon dispersion curves. The monochromating crystal is set at the proper angle for elastic Bragg scattering so that the wave vector k_i is known. Then, k_s is measured by rotating the analyzing crystal to produce Bragg scattering so that $\mathbf{k}$ and the phonon energy can be determined. Using this technique, curves of phonon energies throughout the Brillouin zone can be determined. Because each point in k-space must be done separately, and moderately long counting times often are required, the entire system is highly automated and runs continuously. Usually the ω vs. $\mathbf{k}$ surfaces are determined along the special lines of the Brillouin zone at about ten k-values for each branch.

Elastic neutron scattering also can be observed. This is exactly analogous to Bragg scattering of x-rays (Chapter 4). For *elastic scattering* no phonons are involved, so the magnitudes of $\mathbf{k}_s$ and $\mathbf{k}_i$ are the same but they differ in direction. Thus, from elastic scattering, $\mathbf{K}$'s (Eq. 12-25a) can be determined. When used to determine ω vs. $\mathbf{k}$ of phonons it is the *inelastically scattered* neutrons that are of interest and are measured. The word inelastic refers to the fact that the energy of the scattered neutron, $\hbar\omega_s$, is different from the energy of the incident neutron, $\hbar\omega_i$. In fact, it is just this energy difference that gives ω of the phonon, Eq. 12-25b.

In a typical inelastic scattering process two peaks are observed. If $\hbar\omega_N$ is the energy of the incident neutron, then the peak in the scattered wave intensity vs. energy at $\hbar(\omega_N - \omega)$ is called the **Stokes scattering** (or Stokes peak) and corresponds to the neutron giving up energy $\hbar\omega$ to the phonon system. Assuming one-phonon scattering, the case observed most often, $\hbar\omega$ is the phonon energy. Similarly, the peak at energy $\hbar(\omega_N + \omega)$ is due to **anti-Stokes scattering** (and is called the anti-Stokes peak). It results from the neutron's absorption of a phonon. Repeating, in a Stokes process, phonons are created and in an anti-Stokes process phonons are annihilated (Fig. 12-13).

Consider the ratio of the experimental intensities due to these two processes. At $T = 0°K$ the phonon states are unoccupied so the anti-Stokes process cannot occur. In fact, classically we expect that, on average, the probability that a phonon will be annihilated is proportional to the amplitude squared of the phonon mode. Quantum mechanically, the average probability turns out to be proportional to $\langle n \rangle$, Eq. 12-23b, for the mode. So the intensity of the signal on the anti-Stokes, or high-energy, side of ω_N is proportional to $\langle n \rangle$. However, the probability of creating a phonon is higher; phonons can be created even at $T = 0°K$ when $\langle n \rangle = 0$. For this case, quantum mechanical calculations yield an intensity proportional to $\langle n \rangle + 1$ (instead of $\langle n \rangle$).

CHAPTER 12   LATTICE VIBRATIONS   433

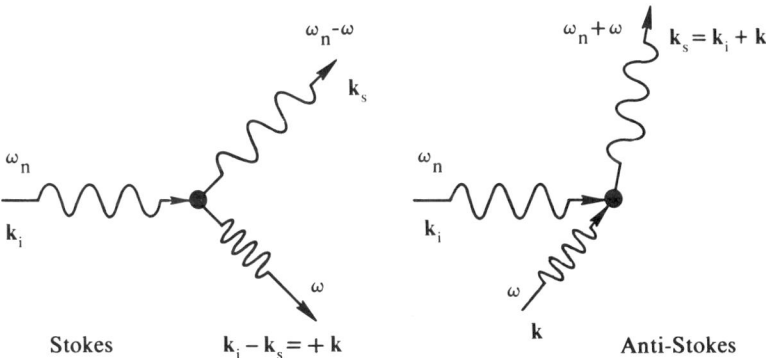

Fig. 12-13 A graphical representation of neutron scattering processes involving the emission and absorption of a phonon.

Thus, the ratio of the neutron scattering signal on the high-energy side of the unscattered beam to that on the low-energy side is

$$\frac{I(\omega_N + \omega)}{I(\omega_N - \omega)} = \frac{\langle n \rangle}{\langle n \rangle + 1} = e^{-\hbar\omega/k_B T} \qquad (12\text{-}26)$$

There is always more "signal" on the Stokes side, but at high temperatures ($k_B T \gg \hbar\omega$) this ratio approaches unity. At very low temperatures, the anti-Stokes intensity goes to zero.

The same considerations occur in Raman spectroscopy where photons are used instead of neutrons. In a similar manner the light at frequency ω_L can be scattered by a phonon of frequency ω. The phonon is annihilated and the scattered light has energy $\hbar(\omega_L + \omega)$. A phonon can also be created, in which case the scattered light has energy $\hbar(\omega_L - \omega)$.

12-8 Thermal Conductivity

We now can discuss several aspects of the microscopic mechanisms of thermal resistivity in crystals due to phonons. In general, heat (i.e., energy) can be transmitted through a crystal by phonons, photons, free electrons or holes, and electron-hole pairs. In metals the free electrons are the best conductors of heat. In insulators phonons are the principal transporters of heat and this will be discussed here.

A phonon with a definite wave vector **k** is an excitation of a normal mode of the entire crystal involving the atoms in the entire crystal. Thus, a single phonon cannot be used to describe a deviation from equilibrium in one region of the crystal. However, such a deviation can be described by using a **wave packet** consisting of phonons

Table 12-1 The lattice thermal conductivity and phonon mean free path for some materials. κ_ℓ values are listed in units of watt/cm-deg and those for Λ in units of cm. The SiO_2 results refer to crystal quartz and ^{7}LiF refers to isotopically pure lithium. [See Blakemore and κ_ℓ values can be found in R. Berman, Cryogenics **5**, 297 (1965).]

	273°K		77°K		20°K	
	κ_ℓ	Λ	κ_ℓ	Λ	κ_ℓ	Λ
Si	1.5	4.3×10^{-6}	15	2.7×10^{-4}	42	4.1×10^{-2}
Ge	0.7	3.3×10^{-6}	3.0	3.3×10^{-5}	13	4.5×10^{-3}
SiO_2	0.14	9.7×10^{-7}	0.66	1.5×10^{-5}	7.6	7.5×10^{-3}
CaF_2	0.11	7.2×10^{-7}	0.39	1.0×10^{-5}	0.85	1.0×10^{-3}
NaCl	0.064	6.7×10^{-7}	0.27	5.0×10^{-6}	0.45	2.3×10^{-4}
^{7}LiF	0.10	3.3×10^{-7}	1.5	4.0×10^{-5}	80	1.2×10^{-1}

within a small range $\Delta \mathbf{k}$ about $\mathbf{k}$. This allows the phonons to be localized in space within a distance $\Delta x \approx 1/\Delta k$. Consider what happens if we heat up one part of a crystal. We can describe a temperature variation localized in space within a region small compared to sample dimensions but large compared to a unit cell dimension (i.e., Δk must be small compared to the Brillouin zone dimensions). In a perfectly harmonic crystal such a wave packet of phonons travels unaltered and thus the thermal conductivity is infinite. However, in real crystals there is **phonon scattering** and this results in a finite thermal resistivity. Phonons can be scattered by other phonons (Section 12-6), static imperfections, surfaces, or by electrons. The latter is important only in metals.

12-8a Phonon mean free path The details of thermal conduction by phonons are best approached via a macroscopically defined mean free path. We have in mind a long bar of sample, with the length much larger than the crossectional dimensions and the two ends at different temperatures. As in Section 9-4, the thermal conductivity κ is defined as the constant of proportionality between a temperature gradient ∇T and the rate of energy flow per unit area Q as in Eq. 12-27.

$$Q = -\kappa_\ell \nabla T \quad (12\text{-}27)$$

$$\kappa_\ell = \Lambda v C_V / 3 \quad (12\text{-}28)$$

The subscript on κ reminds us that we are considering only contributions from phonons in a crystal, the so-called lattice contribution. Equation 12-28 is suggested by considering phonons as a "phonon gas" and using the kinetic theory of gases much as it is used for an "electron gas." Here Λ is the **phonon mean free path** (the average distance traveled by phonons between scattering events), v is the phonon velocity (average speed of sound), and C_V is the lattice specific heat, which is a measure of the phonon density. In interpreting

CHAPTER 12   LATTICE VIBRATIONS   435

these equations, it is important to understand that they imply an *equilibrium situation* in which the energy transfer is a random process, in local thermal equilibrium. The phonons must diffuse through the sample, suffering frequent collisions, as opposed to propagating ballistically from the higher temperature to the lower temperature end (in the latter case Q depends only on the temperature difference, ΔT, between the ends).

In the equilibrium situation Λ decreases rapidly with increasing phonon energy. However, for any distribution of phonons, we may define a nominal mean free path via Eq. 12-28. Table 12-1 lists κ_ℓ and Λ for representative materials. As can be seen, Λ varies over many orders of magnitude from tens of angstroms at room temperature to mm's at lower temperature. Particularly in this latter range, establishing equilibrium is not simple. This will be discussed.

It is useful to compare the order of magnitude of κ_ℓ for an insulator, due to phonons, to that of a metal, due to electrons. At room temperature, typical phonon mean free paths, velocities, and heat capacities are: 3×10^{-6} cm (Table 12-1), 10^5 cm/sec, and 3R (Section 11-1), so $\kappa_\ell \approx 3 \times 10^{-1}$ R. For metals, the same kinetic theory of gas arguments applies to the electron gas as it does to the phonon gas. Thus, the thermal conductivity of the electron gas (κ_e) also should be given by Eq. 12-28, but now the mean free path, velocity, and heat capacity refer to electrons at the Fermi energy. Typical room temperature values are: 10^{-5} cm, 10^8 cm/sec (Table 9-4), and 6×10^{-2} R (Section 9-12b), so $k_e \approx 30$ R. Thus, at room temperature, the electrons in metals conduct heat about 100 times better than the phonons in insulators.

Returning to insulators, and phonons, Fig. 12-14 schematically shows the temperature dependence of Λ and κ_ℓ. We discuss the various temperature regions in terms of higher order phonon processes.

12-8b Anharmonic effects   Consider Eqs. 12-1 and 12-2. To a good approximation the potential energy is quadratic in the relative displacements of the atoms, so the force depends linearly on the relative displacements (i.e., Hooke's law). For a three-dimensional crystal Hooke's law may appear complicated because there are non-negligible "springs" between any one atom and its nearest, and non-nearest neighbors. Nevertheless, the forces are *linear* in the displacements. This leads to the harmonic Hamiltonian and phonons as discussed. In this case the normal mode decomposition of the atomic motion is exact and phonons cannot interact (collide) with another, so in a perfect crystal there would be no phonon scattering and no thermal resistivity (the thermal conductivity would be infinite).

However, the quadratic term in Eq. 12-1 is just the first nonzero term in a Taylor expansion of the potential energy about the equilibri-

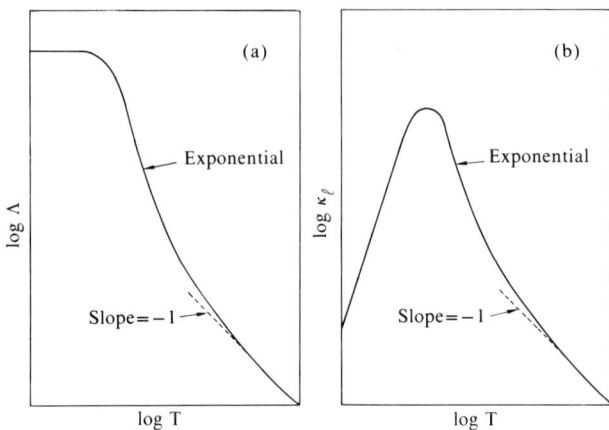

Fig. 12-14 Schematic curves of the variation of (a) the phonon mean free path and (b) the lattice thermal conductivity. These are log-log plots.

um positions (Fig. 8-3b). Third, fourth, and higher order terms of the relative displacements are expected. These *anharmonic terms* couple one phonon to another (phonon-phonon interaction), causing phonon "collisions" and a finite mean free path (Section 12-6) and, by several different processes, establishing equilibrium.

We now can understand qualitatively the high and low temperature portions of the curves in Fig. 12-14. At very low temperatures, the atomic displacements are small so the atoms only vibrate at the bottom of their potential wells where the anharmonic terms are negligible. Thus, Λ becomes large and will be determined by boundary scattering (i.e., limited by the sample dimensions) or scattering from impurities. In either case the distances are fixed so Λ is temperature independent. For most insulators this range is usually below $10°K$. In the high temperature limit, Λ should vary inversely with the total number of phonons with which any given phonon can interact. This number is just $\langle n \rangle$ in Eq. 12-23b, which in the high temperature expansion is $\langle n \rangle \approx k_B T/\hbar\omega$. Thus, at high temperatures, above Θ_D, Λ is expected to decrease as T^{-1} as in Fig. 12-14a.

Now consider the thermal conductivity κ_ℓ via Eq. 12-28. At low temperatures, since Λ is constant, the temperature dependence will be determined by C_V, which goes as T^3 (the Debye result, Eq. 11-25). At high temperatures C_V has very little temperature dependence, Fig. 11-1, so κ_ℓ will vary as T^{-1} due to the temperature variation of Λ. To consider the intermediate temperature range and better understand the attainment of equilibrium, phonon scattering must be considered more carefully.

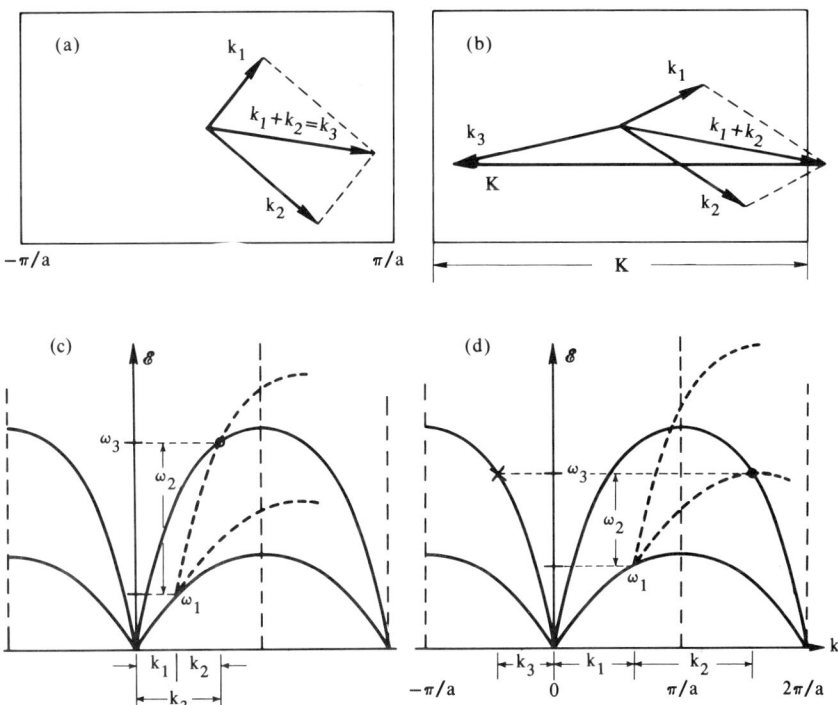

Fig. 12-15 (a) and (b) k-vectors in the first Brillouin zone where (a) $\mathbf{k}_1 + \mathbf{k}_2$ is within this zone and (b) $\mathbf{k}_1 + \mathbf{k}_2$ lies outside this zone and is brought back by a reciprocal lattice vector. (c) Using the ω vs. k from Fig. 12-10a, a construction used to determine k-values that satisfy the conservation of energy and crystal momentum laws. An N-process is shown. (d) The same but a U-process is found. Note, another N-process can be seen but is not discussed. It consists of a TA + LA → LA.

12-8c Normal and umklapp processes Consider the three phonon process, described by Eq. 12-24d. Phonons with wave vectors $\mathbf{k}_1$ and $\mathbf{k}_2$ collide, annihilate one another, and create a phonon with wave vector $\mathbf{k}_3$. Then, from the conservation of crystal momentum, as well as the conservation of energy, we have

$$\mathbf{k}_1 + \mathbf{k}_2 = \mathbf{k}_3 + \mathbf{K} \tag{12-29a}$$

$$\hbar\omega_{\mathbf{k}_1} + \hbar\omega_{\mathbf{k}_2} = \hbar\omega_{\mathbf{k}_3} \tag{12-29b}$$

Suppose $\mathbf{k}_1$ and $\mathbf{k}_2$ are small compared to the Brillouin zone edge and directed to the right, as shown in Fig. 12-15a. These phonons can interact and annihilate one another resulting in a phonon with a wave vector $\mathbf{k}_3$. The result is also shown in Fig. 12-15a where $\mathbf{k}_3$ is within the first Brillouin zone so Eqs. 12-29 apply with $\mathbf{K} = 0$. This is called the **normal process** or **N-process**. Note that before the interaction, energy was flowing to the right, and after the interaction, the same amount of energy is still flowing to the right. Then, we conclude that an N-process does not alter the direction of energy flow, so it cannot contribute to the thermal resistance of a crystal. Thus, if only these normal process collisions occur, there is no phonon-phonon effect on κ_ℓ.

However, it is important to appreciate that normal processes do help to maintain thermal equilibrium. To understand this consider a typical mechanism that give phonons a finite mean free path, Λ. Point defects exist in all crystals (Section 11-8) and, being deviations from perfect translational symmetry, they scatter phonons and contribute to Λ; for this scattering mechanism it is found that $\Lambda \propto \omega^{-4}$, where ω is the energy of the phonon. Thus, for very low energy, small k (long wavelength) phonons, this mechanism is ineffective in establishing an average Λ. However, the N-process enhances this effect as follows. A very low energy phonon (ω_{small}) can scatter from a medium energy phonon (ω_{1m}) and give a second medium energy phonon (ω_{2m}) so that $\omega_{small} + \omega_{1m} = \omega_{2m}$, which conserves energy (crystal momentum also is conserved). Thus, the N-process strongly enhances the ability of defect scattering to maintain thermal equilibrium in the crystal. However, the N-process per se does not provide any thermal resistance.

In 1929 Peierls pointed out that, while still obeying the laws of conservation of energy and crystal momentum, the direction of energy flow may be altered by an **umklapp (flipping over) process** or U-process, and this provides a mechanism for thermal resistance. To appreciate this process, consider a situation very similar to the one described previously. However, now take k_1 and k_2 either larger or more colinear; their vector sum is still to the right, but now it is outside of the first Brillouin zone. By the addition of an appropriate reciprocal lattice vector (in Eq. 12-29a), k_3 lies in the first Brillouin zone but now points to the left, see Fig. 12-15b, with its energy flow (Eq. 12-29b) also going to the left. Thus, while the energy flow for k_1 and k_2 was to the right, the energy flow for k_3 is to the left. *The energy flow has been reversed.* Thus, the umklapp process (having a nonzero **K** in Eq. 12-29a) can provide a thermal resistance to phonon flow (i.e., provides for a finite thermal conductivity).

Let us consider this important concept more deeply by understanding the nature of the solutions admitted by the conservation laws, Eqs. 12-29. As will be seen, these laws sharply restrict the possible solutions. A graphical construction allows us to understand the solutions rather easily. In Fig. 12-15c the solid lines show a typical ω vs. k for acoustic phonons along some crystallographic principal axis (the TA branch is doubly degenerate and the higher energy LA branch is singly degenerate). The entire first Brillouin zone (between $-\pi/a$ and π/a) is shown as well as some of the extended zone to the right. On, let us say, the TA branch an arbitrary point is picked, which then

CHAPTER 12   LATTICE VIBRATIONS   439

corresponds to k_1. Using this point as the origin, again draw the same ω vs. k curves, using dashed lines for ease of viewing (Fig. 12-15c). The intersection (marked with an open circle) of the solid and dashed curves is the required solution. That is, for this particular arbitrary k_1-value, there is just one k_2-value that gets picked out. For this particular value we can have a phonon of wave vector k_3 and energy $\hbar\omega_3$ that satisfies the conservation laws in Eqs. 12-29. Also notice that the intersection is within the first Brillouin zone, so the conservation laws are fulfilled with $\mathbf{K} = 0$; it is an N-process.

By taking the same ω vs. k for the solid and dashed curves in Fig. 12-15c, we are, of course, considering only colinear values for $\mathbf{k}_1$ and $\mathbf{k}_2$. For quantitive purposes ω vs. k curves in different directions must be considered. However, our purpose is qualitative only, so we continue in this vein.

Now consider an arbitrary, but larger, value for $\mathbf{k}_1$ and proceed in the same manner as before. A value of $\mathbf{k}_2$ is obtained (marked with a circle in Fig. 12-15d) that satisfies the conservation laws. However, now the intersection is outside of the first Brillouin zone, and by the addition of a reciprocal lattice vector, $-2\pi/a$, the intersection can be brought back into the first Brillouin zone to the point marked with an × and k_3. For this phonon, with wave vector k_3, the conservation laws are satisfied (using $\mathbf{K} \neq 0$), and the flow of energy is reversed in direction; this is a U-process.

In the same qualitative manner, consider the temperature dependence of the probability of the occurrence of umklapp processes. For very low temperatures only acoustic phonon states with small energies and hence small wave vectors are occupied. Thus, only phonon states close to the center of the Brillouin zone will be populated, making umklapp processes improbable since a U-process cannot occur unless $k_1 + k_2$ extends beyond the first Brillouin zone. To have an umklapp process both $\mathbf{k}_1$ and $\mathbf{k}_2$ should be of the order of $\mathbf{K}/2$. Peierls anticipated that the probability of a U-process would fall off as $\exp(-\Theta_u/T)$ where Θ_u is comparable to $(\Theta_D/2)$. Thus, $\Lambda \propto \exp(\Theta_u/T)$. This is observed experimentally for many crystals at intermediate temperatures and is shown in Fig. 12-14a.

12-8d Thermal conductivity due to defects   The discussion so far has been concerned principally with the fundamental phonon-phonon scattering processes. Actually, defects of one sort or another also are important in determining the thermal conductivity as mentioned. These defects can be (1) point defects, (2) dislocations, (3) rough crystal surfaces, (4) disorder due to alloying, (5) different isotopes of a given chemical species.

Figure 12-16a shows an example of the effect on κ_ℓ of isotopic disorder. Normal germanium is 37% ^{74}Ge, with the rest consisting of

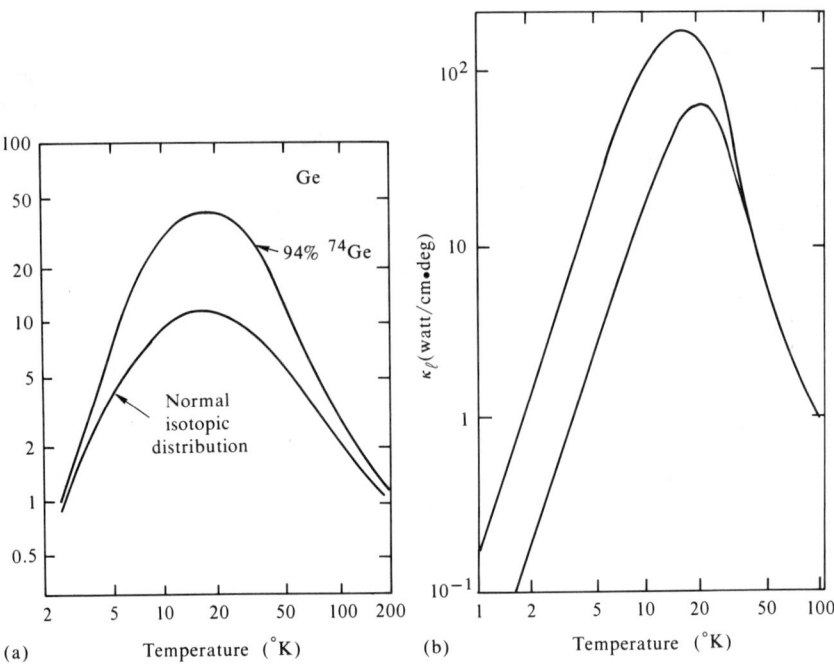

Fig. 12-16 (a) Temperature dependence of the thermal conductivity for normal and isotopically enriched Ge. [T. H. Geballe and G. W. Hull, Phys., Rev. (**110**), 773 (1958)]. (b) The same but for two different size samples of LiF isotopically enriched to 99.9% ^{7}Li. The lower curve has a cross section of 0.123 × 0.091 cm² and the upper 0.755 × 0.697 cm². [R. Berman, Cryogenics **5**, 297 (1965)].

isotopes with mass 70, 72, 73, and 76. With a sample that is isotopically enriched to 94% ^{74}Ge, the thermal conductivity is increased considerably at low temperatures, causing the temperature dependence of the thermal conductivity for an isotopically enriched crystal to be sharper than that of a crystal with naturally abundant isotopes. Thus, crystals that are normally isotopically pure have a steeper rise in κ_ℓ at low temperatures than those with several isotopes.

A rather different effect is shown in Fig. 12-16b for LiF with 99.9% ^{7}Li. In the low temperature T^3 region of κ_ℓ, the size effect is shown; namely, at low temperatures the thermal conductivity is larger in larger samples. This observation is testimony to the fact that the crystals are free of phonon scattering centers, so the mean free path is only limited by the crystal boundaries. Such phonon transport is called **ballistic transport** because the phonon travels in a straight line, without scattering, from one side of the crystal to another.

In pure crystals at low temperatures these ballistic properties can be used to measure the velocity of longitudinal and transverse acoustic phonons. This is done in **heat pulse experiments** by generating phonons on one side of a crystal, perhaps by a short, intense pulse of heat caused by a laser pulse impinging on the sample. The arrival of the phonons on the other side of the sample can be detected by a temperature increase, perhaps with a superconducting bolometer. The time between the generation and arrival yields the velocity. The longitudinal phonons arrive before the transverse ones because their velocities are larger. The velocities of these acoustic waves can be measured in various crystallographic directions, which determines the initial slope of the ω vs. k (Fig. 12-10). See the Notes for references.

Notes

An excellent sourcebook on lattice dynamics is Born and Huang. It should be looked at, if only to appreciate the wealth of information that it includes. A recent book that is easy to read is by P. Brüesch, "Phonons: Theory and Experiments I" (Springer-Verlag, 1982). Also see J. A. Reissland, "The Physics of Phonons" (Wiley and Sons, 1973). Most of the solid state physics books in the bibliography have a chapter on lattice vibrations. Ashcroft and Mermin, Chapter 22, have a good formal discussion on how the basic assumptions of Section 12-1 (pictured in Fig. 8-3b) lead to a harmonic potential energy. The **adiabatic approximation** is also discussed there.

Brillouin's little book is a delight and should be looked at by all. It has a great deal of information on lattice vibrations, electron states, and, of course, on Brillouin zones. For an overview of **neutron scattering** see Physics Today, the January 1985 issue, "Condensed Matter Research Using Neutrons," Ed. S. W. Lovesey and R. Scherm (Plenum Press, 1984), and the references in the Notes in Chapter 14.

As discussed in Section 12-3b and Fig. 12-8, a phase transition can enlarge the unit cell (in one or several directions), which causes the Brillouin zone to decrease in size. This leads to the useful concept of **"folding" of the Brillouin zone** which really refers to **"folding" of the dispersion relations in the Brillouin zone.** For example, this leads to "new" electronic states or vibrational states appearing at the zone center (i.e., the Γ-point). For certain materials, another way to enlarge the unit cell is to make related compounds. For example, two zinc atoms in ZnS can be replaced by one copper and one iron atom to form $CuFeS_2$, which is called **chalcopyrite.** $CuFeS_2$ is an ordered structure and the dispersion curves in it can be treated as folded with respect to the "parent" ZnS Brillouin zone. ZnS is also the "parent" of other ordered, tetrahedrally coordinated compounds such as Cu_2FeSnS_4 (**stannite**) and Cu_3SbS_4 (**famatinite**), as well as defect compounds. The folding of the Brillouin zones in these compounds with respect to their parent compounds with the zinc blende structure is treated by A. Miller, A. MacKinnon, and D. Weaire, Solid State Physics **36**, 119 (1981). Folding of the Brillouin zone also can be seen by observing different **polytypes** of the same material. Polytypism is observed typically in materials that form **layered structures** in which the bonding between the layers is much weaker than within a layer. This allows the layers to stack upon each other in different ways in a manner similar to what was discussed in Section 3-7. Observations of this effect in SiC, PbI_2, and CdI_2 can be found in W. J. Choyke and L. Patrick, Phys. Rev. **172**, 769 (1968); W. M. Sears, M. L. Klein and J. A. Morrison, Phys. Rev. B **19**, 2305 (1979); H. Katahama, S. Nakashima, M. Hanzyo, A. Mitsuishi and B. Palosz, Solid State Commun. **49**, 547 (1984). Also see Section 18-5e.

Huntington's article on the **elastic constants** of crystals, in Solid State Physics **7**, 213 (1958), covers all aspects of this field and tabulates the constants for many crystals. Also tables of the symmetry allowed constants are given. Table 12-2 comes from this reference and shows the values of the c_{ij}'s for a number of elements and compounds that have a cubic crystal structure. The temperature dependence of many c_{ij}'s and data for other materials can be found in H. Bilz, D. Strauch, and R. K. Wehner, Handbuch der Physik, Vol. 25/2d, Licht und Materie, Id. (1984). This book also discusses many aspects of the infrared and Raman spectra of nonmetals.

Launay's article on "Specific Heats and Lattice Vibrations" in Solid State Physics **2**, 219 (1956) summarizes a good deal of information on both of these subjects. He also has a section on the elastic constants and their relationship to the velocity of sound in cubic crystals and in isotopic materials. In a cubic crystal, the velocity of sound is independent of direction when $s = (c_{11} - c_{12})/2c_{44}$ is equal to one. From Table 12-2 it can be seen that this condition is not often met. If the interactions between the atoms of a monatomic crystal can be described by central forces, then certain relations exist between the elastic constants. These are known as the **Cauchy relations**. For a cubic crystal the Cauchy relation is $c_{12} = c_{44}$. Again this condition is not often met. (See Chapter 5 for a discussion of the c_{ij}'s.)

Peierls' original paper on the **umklapp process** is in Ann. Physik **3**, 1055 (1929). His book, "Quantum Theory of Solids" (Oxford University Press, 1955), is an excellent source. The book by R. Berman, "Thermal Conduction in Solids" (Oxford University Press, 1976) is a good general reference on this subject. More detailed considerations of the N-process, showing that it too can affect the thermal conductivity, are discussed in Chapter 8 of J. M. Ziman, "Electrons and Phonons" (Oxford, 1960). For a review of **heat pulse experiments** see R. J. von Gutfeld, Physical Acoustics **5** (Academic Press, 1968) p. 233. Heat pulse experiments were the first phonon experiments to measure time resolved phonon dynamics. For more recent reviews of this

CHAPTER 12   LATTICE VIBRATIONS   443

Table 12-2 Elastic constants of some cubic crystals. c_{ij} are in units of 10^{11} dyne cm^{-2}. Values for CdS are in Section 5-4d. s is defined in the Notes. (From Huntington.)

Material	c_{11}	c_{44}	c_{12}	s
Ag	12.40	4.61	9.34	0.33
Au	18.6	4.20	15.7	0.35
Al	10.56	2.85	6.39	0.73
C	95	43	39	0.65
Cu	16.84	7.54	12.14	0.31
Ge	12.89	6.71	4.83	0.60
K (83°K)	0.457	0.263	0.374	0.16
Li (78°K)	1.48	1.08	1.25	0.11
Mo	46	11.0	17.6	1.29
Na (90°K)	0.945	0.618	0.779	0.13
Ni	24.65	12.47	14.73	0.40
Pb	4.66	1.44	3.92	0.26
Si	16.57	7.96	6.39	0.64
Th	7.53	4.78	4.89	0.28
W	50.1	15.14	19.8	1.00
AgBr	5.63	0.720	3.3	1.12
AgCl	6.01	0.625	3.62	1.91
KI	2.67	0.421	0.43	2.66
LiI	11.12	6.28	4.20	0.55
NaBr	3.87	0.97	0.97	1.49
NaCl	4.87	1.26	1.24	1.44
GaAs	1.192	0.538	0.60	0.55
GaSb	8.85	4.33	4.04	0.56
InSb	6.72	3.02	3.67	0.50

active field see the book "Nonequilibrium Phonon Dynamics" Ed. W. E. Bron (Plenum Press, 1985), and W. E. Bron, Rep. Prog. Phys. **43**, 301 (1980).

The **statistics of bosons** and **fermions** are discussed in P. A. M. Dirac, "The Principles of Quantum Mechanics" (Oxford University Press, 1958) Chapter 9. Also see A. Messiah, "Quantum Mechanics," (North Holland Pub. Co., 1963) Chapter 14.

Problems

1. For the **monatomic linear chain** of Section 12-2, show that the expressions in Eq. 12-8 are correct. Show that $v_g = 0$ at $k = \pi/a + m2\pi/a$ where m is any integer. For first- and second-neighbor forces, as in Section 12-2c, sketch ω vs. k for $\alpha_2 = 2\alpha_1$ with all other force constants equal to zero.

2. For the **diatomic linear chain** of Section 12-3 expand ω and the displacements to the next order in ka for small k (as in Eq. 12-16). For the acoustic branch, in what fraction of the Brillouin zone are the deviations from the linear ω vs. k less than 20%? If this ω vs. k were appropriate to a three-dimensional crystal, what fraction of the volume of the Brillouin zone would be covered in this case?

3. For the **diatomic linear chain** with two different spring constants, show that the expressions in Section 12-3c are correct.

4. **Shell model** – In a page or two discuss the shell model that has been successful in fitting vibrational dispersion curves. Give several examples. (See Elliot and Gibson or Brüesch.)

5. Consider the **monatomic linear chain**, Section 12-2. Define a continuous function u(x) so that u(na) = u_n and u(x) interpolates smoothly for other values of x. This changes u from a function defined only at discrete points to a continuous function. Now by expanding $u_{n\pm1}$ = u[(n±1)a] in powers of a around u(na), show that as a → 0 (which is the long wavelength limit)

$$\frac{1}{v^2}\frac{d^2u}{dt^2} = \frac{d^2u}{dx^2}$$

where v is the velocity of sound = $(\alpha a^2/m)^{1/2}$. If this wave propagates in a [001] direction of a cubic crystal, the longitudinal and transverse sound velocities are $v_\ell = (c_{11}/\rho)^{1/2}$ and $v_t = (c_{44}/\rho)^{1/2}$ where ρ is the density and the c_{ij} are the elastic constants.

6. Using the measured elastic constants for sodium, calculate the longitudinal and transverse sound velocities in the [100] direction. Compare the results to those in Fig. 12-10.

7. The **average thermal internal energy for a collection of bosons** (photons, phonons, magnons, etc.) can be calculated in a general, simple manner. From this energy U, the heat capacity can be calculated as $C_V = (\partial U/\partial T)_V$. The particles in an assembly of identical particles for which the wave function is symmetric under the interchange of any two of them are called **bosons**. (It can be shown that particles with integral spin are bosons. See the Notes.) For such particles the average thermal occupation number for any state is given by Bose-Einstein statistics as $\langle n_k \rangle = [\exp(\hbar\omega_k/k_BT) -]^{-1}$. Therefore, U for bosons is

CHAPTER 12   LATTICE VIBRATIONS   445

$$U = \Sigma_k \langle n_k \rangle \hbar \omega_k = \Sigma_k \frac{\hbar \omega_k}{e^{\hbar \omega_k / k_B T} - 1}$$

These are just Eqs. 12-22b and 12-23c with the zero-point energy left out. If desired, it can always be added. For all systems, the advantage of thinking in terms of **k** as opposed to ω is that in k-space the k-axes are divided uniformly so the conversion from sums to integrals over d^3k is trivial. Since the volume in k-space for each allowed **k** value is $\Delta V_k = (2\pi)^3/V$ (Eq. 9-27b or see Eqs. 11-19 and 11-20)

$$U = \frac{V}{(2\pi)^3} \int d^3k \frac{\hbar \omega_k}{\exp(\hbar \omega_k / k_B T) - 1}$$

where the previous sum or integral here is over the first Brillouin zone. We have been assuming that there is just one branch. However, for many branches in the phonon curves, a branch index α should be added giving $\omega_{k,\alpha}$. Then the sum should be summed over **k** and α, and the integral should be performed on d^3k for each branch (i.e., $\Sigma_\alpha \int d^3k$). However, often the interest is in the results for low temperatures where only the lowest branches of the ω vs. **k** are occupied and even in these branches only the states near **k** = 0 are occupied. In this low temperature limit there may be a very simple direction independent relation between ω and k. In this case the angular integration is trivial and the result is

$$U = \frac{V}{(2\pi)^3} 4\pi \int k^2 dk \frac{\hbar \omega_k}{\exp(\hbar \omega_k / k_B T) - 1}$$

In this low temperature limit take $\omega_k = Ak^p$ and calculate U and C_V. For the **Debye solid** $A = v_0$ and $p = 1$. Check that your C_V result agrees with Eq. 11-25. Also determine the low temperature C_V for **spin waves** where $\omega_k = (2JSa^2/\hbar)k^2$, Eq. 15-28b. For the spin wave case, a $T^{3/2}$ result should be obtained.

8. Stress strain – In no more than three pages discuss, with equations, the relation between the stress and the conventionally chosen, rotationally invariant strains. Also cover the following points: by determining the energy density show that for the elastic constants $c_{ij} = c_{ji}$; for cubic crystals show that **Young's modulus**, Y, and the **bulk modulus**, B, are given by

$$Y = (c_{11} - c_{12})(2c_{12} + c_{11})/(c_{11} + c_{12})$$
$$B = (c_{11} + 2c_{12})/3$$

Last, if the cubic crystal is isotopic, show that $c_{44} = (c_{11} - c_{12})/2$ so that $Y = c_{44}(3c_{12} + 2c_{44})/(c_{12} + c_{44})$ and $B = c_{12} + (2c_{44}/3)$. (See the references in the Notes and see Brown.) ($B \equiv K^{-1}$, where K is the **compressibility**, see Section 7-3c.)

9. Anharmonic terms – Consider a classical harmonic oscillator with small anharmonic terms so that the potential energy is

$$V(x) = ax^2 + bx^3 + cx^4$$

Using the partition function approach, show that the mean thermal energy $\langle \mathscr{E} \rangle$ and mean thermal displacement from equilibrium $\langle x \rangle$ are

$$\langle \mathscr{E} \rangle \approx k_B T + \left[\frac{15b^2}{16a^3} - \frac{3c}{4a^2} \right] (k_B T)^2, \quad \langle x \rangle \approx -\left(\frac{3b}{4a^2} \right) k_B T$$

The former leads to a high temperature contribution to the specific heat that is linear in temperature. The latter is an indication of the origin of thermal expansion (and the proper sign of the b coefficient).

10. Creation and annihilation operators – For the monatomic linear chain (Section 12-2), show how the transformation to the amplitudes (and conically conjugate momenta) of complex waves $\exp(ikx)$ simplifies the Hamiltonian. Then transform to the creation and annihilation operators and obtain for the Hamiltonian $\Sigma_k \, \hbar \omega_k (b_k^\dagger b_k + 1/2)$. For a normal mode containing n_k phonons we write the phonon state as $|n_k\rangle$, which has energy $\mathscr{E}_k = (n_k + 1/2)\hbar\omega_k$. Show that

$$b_k^\dagger | n_k \rangle = (n_k + 1)^{1/2} | n_k + 1 \rangle$$

$$b_k | n_k \rangle = (n_k)^{1/2} | n_k - 1 \rangle$$

hence the name creation and annihilation operators. Show that the **number operator** $b_k^\dagger b_k$ has the property

$$b_k^\dagger b_k | n_k \rangle = n_k | n_k \rangle$$

hence $\quad \hbar \omega_k (b_k^\dagger b_k + 1/2) | n_k \rangle = \mathscr{E}_k | n_k \rangle$

(Many quantum mechanics books and Brown discuss this problem.)

13

Optical Properties of Crystals

PART A MACROSCOPIC THEORY

13-1   Dielectric Polarization
    a   Definition of terms
    b   Effects of crystal symmetry
13-2   Oscillating Fields
13-3   Electromagnetic Waves in Solids
13-4   Reflectivity at an Interface
13-5   Kramers-Kronig Relations (S, A)
13-6   Damped Harmonic Oscillator
13-7   Dielectric Response of a Quantum System

PART B LATTICE VIBRATIONS

13-8   Introduction
13-9   Long Wavelength Optical Vibrations
    a   Transverse waves
    b   Longitudinal waves
    c   Lyddane Sachs Teller (LST) relationship
    d   Generalizations of LST (S)
13-10   Measurements and Results
    a   Infrared measurements
    b   Results
    c   Raman effect (S)
    d   Problems (S)
13-11   Polaritons (S)
    a   Theory
    b   Experiment
13-12   A Microscopic Model (S)
    a   Local field problem
    b   The calculation
    c   Szigeti charges
13-13   Clausius-Mossotti (Lorenz-Lorentz) Equations (S)

PART C FREE CARRIER ABSORPTION

13-14   Introduction
13-15   Oscillator Model
    a   The model
    b   Frequency dependent parameters
13-16   Experimental Results
    a   Metals
    b   Semiconductors
13-17   Transverse and Longitudinal Free Electron Modes (S)
    a   Transverse free electron modes
    b   Longitudinal free electron modes

PART D INTERBAND TRANSITIONS

13-18   Introduction
13-19   Fundamental Absorption Near $\mathscr{E}_g$
    a   Optically induced vertical transitions
    b   Theory (A)
    c   Experiment
    d   Heavily doped materials (S)
    e   Alloys (S)
    f   Donors and acceptors (and photoconductivity)
13-20   Excitons (Mostly Weakly Bound Excitons)
    a   Introduction
    b   Weakly bound excitons
    c   Exciton condensation (S)
    d   Excitonic complexes (S)
    e   Frenkel excitons (S)
    f   Polaritons (S)
    g   Donor-acceptor pairs (S)
    h   Magnetooptical absorption
13-21   Fundamental Absorption Above $\mathscr{E}_g$
13-22   Urbach Edge (S)
    Notes
    Problems

Table 13-1 The transverse and longitudinal optic vibration frequencies listed in units of cm^{-1}. (1 meV = 8.0658 cm^{-1} and 1 cm^{-1} = 1.439°K.) The crystals on the left have the NaCl structure. Those on the right have the CsCl structure (CsCl to TlI), and the zincblende structure (CuCl to SiC), or the diamond structure (C to α-Sn), in which case the TO and LO modes are degenerate in the zone center. A. S. Barker Jr. and A. J. Sievers, Rev. of Mod. Phys. **47**, Sup. 2 1975, have a more extensive list. Also see W. Richter, Springer Tracts Mod. Phys. **78**, 174 (1976). However, values for many of the III-V semiconductors have been corrected according to more recent work.

	ω_{TO}	ω_{LO}		ω_{TO}	ω_{LO}
LiF	307	662	CsCl	99	165
LiCl	191	398	CsBr	73	112
LiBr	159	325	CsI	62	85
LiI	144	–	TlCl	63	158
NaF	239	414	TlBr	43	101
NaCl	164	264	TlI	52	–
NaBr	134	209	CuCl	161	207
NaI	117	181	CuBr	125	161
KF	192	330	ZnS	271	352
KCl	142	214	ZnSe	205	250
KBr	113	165	ZnTe	177	205
KI	101	139	AlSb	319	340
RbF	156	286	CdTe	140	167
RbCl	116	173	GaP	367	403
RbBr	88	127	GaAs	269	292
RbI	75	103	GaSb	224	233
AgCl	106	196	InAs	220	241
AgBr	79	138	InP	303	345
PbS	71	212	InSb	180	191
PbTe	23	113	SiC	794	962
CsF	127	–	C		1331
			Si		520
			Ge		300
			α-Sn		197

OPTICAL PROPERTIES OF CRYSTALS

Physics always develops in two directions. One front pushes forward towards phenomena which do not yet fit into the general picture, and the victories on this front are marked by important changes in our fundamental concepts . . . . But, in addition to this search for new concepts, there is a constant effort directed toward the deepening and broadening of our knowledge of phenomena which, we believe, **can** *be understood on the basis of existing concepts and theories.*

E. P. Wigner, Scientific Monthly, Jan. 1936

This chapter deals with the interaction of light (photons) with solids. After discussing the general macroscopic theory of the interaction of light with matter (Part A), we treat several major areas where this interaction yields invaluable and often unique information. These areas include the properties of phonons, free electron excitations, and last, the electronic excitations in solids. These and related areas of spectroscopy have been continuously emerging important areas of science, and the discovery of the laser in the 1960s certainly has helped to spur the field. We touch only on the elementary aspects. Reference to the books and review articles in the Notes will help you appreciate the extensive effort in this field.

In developing the basic concepts, use is made of the "length" of the probes and excitations. The wavelength of visible light is very much larger than the dimensions of the unit cells (i.e., $\lambda \gg a$). We ignore surface effects and consider only bulk properties (i.e., take $\lambda \ll$ sample dimensions). Thus, the interaction of the electromagnetic wave with the solid is averaged over many unit cells with the result that λ and the wave vector of the light wave, k, are well defined in the solid ($k = 2\pi/\lambda$). Similarly, most of the important excitations in solids are spread out over many unit cells and hence have a well defined wave vector. For example, this results in wave vector conservation applying to the interaction of the photons with phonons (Section 12-6).

Since the wavelength is large compared to the unit cell size, the optical properties of solids can almost always be described in terms of the macroscopic optical constants that we are familiar with, such as the index of refraction and the extinction coefficient. Using Maxwell's equations, we shall see how these frequency-dependent "constants" are related to the other macroscopic parameters such as the optical dielectric constant and optical conductivity.

As a last point of introduction, the similarity of optical spectroscopy in solids and that of atoms and molecules should be noted. In the latter, optical spectroscopy has provided a large body of very detailed knowledge about the energy level scheme. In atoms such knowledge has, of course, led to Bohr's atomic theory. The energy levels in solids are much broader and there are many different types of excitations whose energies can be measured. Nevertheless, optical spectroscopy has provided invaluable information, as will become apparent in this chapter.

Part A — Macroscopic Theory

13-1 Dielectric Polarization

13-1a Definition of terms We shall assume that the solid has no net macroscopic charge. However, it is composed of positively and negatively charged entities. Under the application of an applied electric field E these entities move in opposite directions producing a dipole moment. Differently charged entities, such as bound electrons, conduction electrons, impurities, bound ionic cores, and so on, move by different amounts but the net effect of their motion can be summarized in a polarization that is the electric dipole moment per unit volume, $\mathbf{P} \equiv \mu/V_p$,

$$\mathbf{P} \equiv V_p^{-1} \int \rho \, \mathbf{u} \, dxdydz \equiv V_p^{-1} \Sigma_i \, n_i \, q_i \, \mathbf{u}_i \tag{13-1}$$

where V_p is the unit cell volume, ρ the charge density, $\mathbf{u}$ the displacement of the charge due to $\mathbf{E}$, and the integration is over one primitive unit cell. We have also defined $\mathbf{P}$ in terms of the number n_i of discrete charges q_i within each primitive unit cell that undergo displacement $\mathbf{u}_i$ due to the applied electric field $\mathbf{E}$. This latter definition is sometimes more appropriate, but both definitions give the dipole moment per unit volume.

In general the displacement of charge can be written in terms of a power series in E, that is, $\mu \propto \alpha E + \beta E^2 + \gamma E^3 + ...$, but we shall use only the first term, $\mathbf{P} \propto \mathbf{E}$, since for normally available electric fields (10^4 volts/cm) the higher terms can be ignored. The use of high-power lasers has enabled the effects of these higher order terms to be investigated. This area, called nonlinear optics, is mentioned in Chapter 5. Considering only the linear term, the **dielectric susceptibility** χ which relates $\mathbf{P}$ to $\mathbf{E}$ is defined as

$$\mathbf{P} = \chi \, \mathbf{E} \tag{13-2}$$

When time varying electric fields are applied, there is a velocity associated with each element of charge, $\dot{\mathbf{u}}$ ($\dot{\mathbf{u}} = \partial \mathbf{u}/\partial t$ as usual). A current density $\mathbf{J}$ then is defined as

CHAPTER 13   OPTICAL PROPERTIES OF CRYSTALS   451

$$\mathbf{J} \equiv V_p^{-1} \int \rho \dot{\mathbf{u}} \, dxdydz \equiv V_p^{-1} \Sigma_i \, n_i \, q_i(\partial u_i/\partial t) \quad (13\text{-}3a)$$

so $\mathbf{J} = \partial \mathbf{P}/\partial t$ (13-3b)

From Gauss' theorem the average electric field is related to the polarization; for externally applied charge density, ρ_{ext}

$$\nabla \cdot \mathbf{E} = -4\pi \nabla \cdot \mathbf{P} \quad (13\text{-}4)$$

This result leads to the definition of a new field **D**, called the electric displacement.

$$\mathbf{D} \equiv \mathbf{E} + 4\pi \mathbf{P} \quad \text{and} \quad \nabla \cdot \mathbf{D} = 0 \quad (13\text{-}5a)$$

Since in the low field approximation $\mathbf{P} \propto \mathbf{E}$, and $\mathbf{D} \propto \mathbf{E}$, we define the dielectric constant ε via

$$\mathbf{D} \equiv \varepsilon \mathbf{E} \quad (13\text{-}5b)$$

so $\varepsilon = 1 + 4\pi\chi$ (13-5c)

Thus, ε and χ contain the same information. The existence of two intimately related quantities arises because of the definition of the electric displacement field in Eq. 13-5a and the **constitutive relations** $\mathbf{D} = \varepsilon \mathbf{E}$ and $\mathbf{P} = \chi \mathbf{E}$. The other useful aspect of **D** is that, in general, $\nabla \cdot \mathbf{D} = \nabla \cdot \varepsilon \mathbf{E} = 4\pi \rho_{ext}$ while **E** is related to the total charge density, $\rho = \rho_{ext} + \rho_{ind}$ where ρ_{ind} is the induced charge-density, which is induced by ρ_{ext}. Thus, $\nabla \cdot \mathbf{E} = 4\pi\rho$. For the fields around a circuit with no external charge, **D** is continuous but **E** is not. Thus, $\varepsilon = 1 + \chi$.

The last parameter that should be defined is the conductivity σ which relates J to E, and is called **Ohm's law**.

$$\mathbf{J} = \sigma \mathbf{E} \quad (13\text{-}6)$$

In **SI units** we have $\mathbf{D} = \varepsilon_0 \mathbf{E} + \mathbf{P} = \varepsilon\varepsilon_0 \mathbf{E}$. Also $\nabla \cdot \mathbf{D} = \nabla \cdot (\varepsilon\varepsilon_0 \mathbf{E}) = \rho_{ext}$ and $\nabla \cdot (\varepsilon_0 \mathbf{E}) = \rho$. Last $\mathbf{P} = \chi\varepsilon_0 \mathbf{E}$. Ohm's law is the same in both sets of units.

13-1b Effects of crystal symmetry   Since χ, ε, and σ are "parameters" that relate two vectors (first-rank tensors) to each other, they are second-rank tensors. This is covered in Chapter 5, where it is also shown how to determine which tensor components are zero and which are related to each other. In most crystal systems the eigenvectors of the tensor parameters are parallel to the mutually perpendicular crystal symmetry axes. Thus, the principal axes of χ, ε, and σ coincide, which greatly simplifies the formulation of the problem of propagation of light. For example, if light propagates along one of the principal axes, the propagation problem is really a one-dimensional one. Therefore χ, σ, and ε can be treated as scalars. Further, if cubic crystals are considered, the three principal values of these second-rank tensors are all equal and the crystal is optically isotropic. In this chapter we will usually consider only cubic crystals; thus χ, σ, and ε each will have all diagonal components equal, all off-diagonal components zero, and the tensors can be treated as scalars.

13-2 Oscillating Fields

Now we want to consider explicitly what happens when the driving forces (for example the applied electric fields) vary with time. For zero or very low frequency the polarization completely adjusts to the instantaneous applied field, and thus a real χ describes the situation. However, as the angular frequency ω is increased, some of the various types of charges that contribute to **P** in Eq. 13-1 will not be able to follow the applied electric field. For example, in crystals as ω is increased we expect that the electrons will be able to follow the oscillating electric field to much higher frequencies than the nuclei. Thus, at very low ω both the electrons and nuclei can move in an applied **E** and contribute to **P** and χ. At some higher ω the motion of the nuclei will become much smaller and only the electrons will be able to be displaced by **E** and contribute appreciably to **P** and χ. In fact, the study of the frequency dependence of χ, and similarly ε and σ, is very important in helping us understand exactly what is contributing to the polarization.

Thus, when the fields vary in time there is, in general, a phase shift between them, for example between **P** and **E** as discussed. To account for this phase shift, we do what is done in most branches of science; we take the constant of proportionality between the two fields as a complex quantity. For example, the proportionality between **P** and **E** defines a **susceptibility** χ such that

$$\mathbf{P} \equiv \chi \mathbf{E} = (\chi_1 + i\chi_2)\mathbf{E} \tag{13-7}$$

Writing the susceptibility in terms of a real part with a subscript 1 and an imaginary part with a subscript 2 is standard notation and is followed here. This equation shows that the part of the polarization that is in phase with the electric field is $\chi_1 \mathbf{E}$, and the part that is $\pi/2$ out of phase is $\chi_2 \mathbf{E}$. The ratio of the in-phase to out-of-phase part of the polarization is χ_1/χ_2.

In general, we take the sinusoidial time variation of the fields to be given by $\exp(-i\omega t)$ so

$$\mathbf{E} = \mathbf{E}_0 e^{i(\mathbf{k}\cdot\mathbf{r} - \omega t)} \tag{13-8}$$

This is a particularly simple form to use with Maxwell's equations. However, the actual fields are the real parts of the complex quantities.

CHAPTER 13   OPTICAL PROPERTIES OF CRYSTALS   453

For example, the energy density associated with this field involves the factor $E^2 = E_0^2 \cos^2(\mathbf{k}\cdot\mathbf{r}-\omega t)$, the square of the real part of Eq. 13-8. So the time average of E^2, which is written as $\langle E^2 \rangle$, is just $E_0^2/2$. The time average of the product of the real parts of any two arbitrary vectors, both of which vary as $\exp(-i\omega t)$, is one half of the real part of one vector times the complex conjugate of the other vector. Thus,

$$\langle \text{Re}(\mathbf{A})\cdot\text{Re}(\mathbf{B})\rangle = (1/2)\text{Re}(\mathbf{A}\cdot\mathbf{B}^*) = (1/2)\text{Re}(\mathbf{A}^*\cdot\mathbf{B}) \quad (13\text{-}9)$$

Problem one, in this chapter, makes use of this time average concept.

While **P**, **E**, and **J** are physical observables, χ (and ε and σ) is not, by itself, a physical observable. Rather, it represents a relationship between two physical observables and describes the magnitude and phase of one compared to the other; hence, both the real and imaginary parts are equally meaningful, as can be seen in Eq. 13-7.

As for the complex susceptibility, we define a complex **dielectric constant** in terms of the displacement field, **D**.

$$\mathbf{D} \equiv \mathbf{E} + 4\pi\mathbf{P} = \varepsilon \mathbf{E} = (\varepsilon_1 + i\varepsilon_2)\mathbf{E} \quad (13\text{-}10)$$

Notice that ε and χ are not independent since, using Eq. 13-9a,

$$\varepsilon = 1 + 4\pi\chi$$

so $\quad \varepsilon_1 = 1 + 4\pi\chi_1 \qquad \varepsilon_2 = 4\pi\chi_2 \quad (13\text{-}11)$

Via the current density **J**, a complex **conductivity** σ can be defined

$$\mathbf{J} = \sigma \mathbf{E} = (\sigma_1 + i\sigma_2)\mathbf{E} \quad (13\text{-}12a)$$

From Eqs. 13-3b we obtain

$$\mathbf{J} = -i\omega\mathbf{P} = -i\omega\chi\mathbf{E}$$
$$\mathbf{J} = (\omega\chi_2 - i\omega\chi_1)\mathbf{E} \quad (13\text{-}12c)$$

Thus, the current density has components in and out of phase with the applied electric field in a similar manner to the polarization. From the definition of the conductivity $\sigma = \sigma_1 + i\sigma_2$ in Eq. 13-12a, and using Eqs. 13-11 and 13-12c we find

$$\begin{array}{ll} \sigma_1 = \omega\chi_2 & \sigma_2 = -\omega\chi_1 \\ \sigma_1 = \omega\varepsilon_2/4\pi & \sigma_2 = -\omega(\varepsilon_1-1)/4\pi \end{array} \quad (13\text{-}13)$$

Equations 13-13 and 13-11 show how the various constants are all related to one another.

To determine these relations in **SI units**, we use the expression in the paragraph after Eq. 13-6, which are in **SI units**, and obtain

$$\begin{array}{ll} \sigma_1 = \omega\chi_2\varepsilon_0 & \sigma_2 = -\omega\chi_1\varepsilon_0 \\ \sigma_1 = \omega\varepsilon_2\varepsilon_0 & \sigma_2 = -\omega(\varepsilon_1-1)\varepsilon_0 \end{array}$$

These relations have sometimes been a source of confusion. In many textbooks and papers χ is ignored and a random choice of the

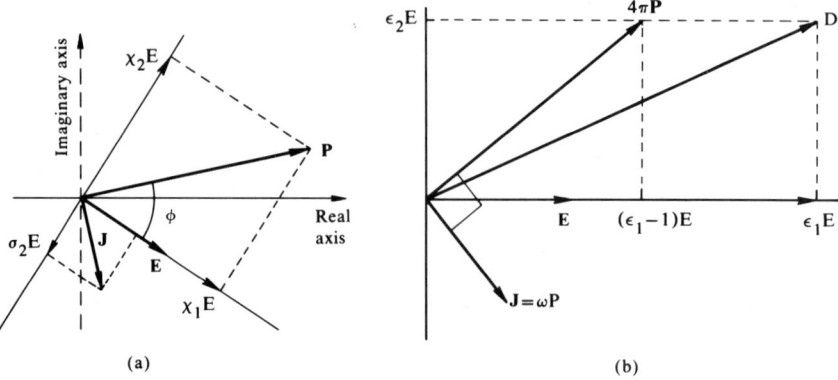

Fig. 13-1 Attempts to summarize the relationship among the fields, currents and so on.

following possible pairs is used (i) a complex dielectric constant; (ii) a complex conductivity; or (iii) real dielectric constant and a real conductivity. From Eq. 13-13 it can be seen that all the constants can be expressed in terms of the quantities of any of these three pairs. In this chapter we shall, in most cases, use the real and imaginary parts of the dielectric constant, and the real part of the conductivity i.e., ε_1, ε_2, and σ_1, respectively. If we keep the subscripts and know the interrelationships, there should be no confusion even though one more parameter than necessary is used. Also, when the frequency of the electromagnetic wave is in the optical region, then the relation between **P** and **E** is usually written in terms of the **optical polarizability**, that is, $\mu = \alpha E$, so $\alpha = V_p \chi$. While this may be confusing, it is standard procedure. A last possible source of confusion is that some authors use a time dependence $\exp(+i\omega t)$ in Eq. 13-8. If so, in the equations here, i should be replaced by $-i$ to compare the equations. You should always check carefully the time variation used.

Figure 13-1 summarizes some of the concepts discussed in this section. In (a) the complex plane E, P, and J are shown. The physically measured quantities are the projections onto the real axis. The phase angle between E and P is arbitrary and depends on the values χ, but the angle between J and P is always $\pi/2$ as can be seen from Eq. 13-3b. Figure 13-1b displays the same ideas but with different emphasis and E is taken along the real axis.

13-3 Electromagnetic Waves in Solids

Maxwell's equations are fundamental and hold even on the atomic scale with the sources and fields varying very strongly between the atomic sites. However, we want to use the equations for the average

CHAPTER 13   OPTICAL PROPERTIES OF CRYSTALS   455

fields in the sense described in the introduction. In terms of the average fields Maxwell's equations are

$$\nabla \cdot \mathbf{D} = 4\pi\rho_{ext} \qquad \nabla \cdot \mathbf{B} = 0$$
$$\nabla \times \mathbf{E} = -\frac{1}{c}\frac{\partial \mathbf{B}}{\partial t} \qquad \nabla \times \mathbf{H} = \frac{1}{c}\frac{\partial \mathbf{D}}{\partial t} + \frac{4\pi}{c}\mathbf{J} \qquad (13\text{-}14)$$

We shall assume that the magnetic permeability μ, (not related to the electric dipole moment) in the constitutive relation $\mathbf{B} = \mu\mathbf{H}$, is equal to one (which is appropriate for nonmagnetic crystals). Thus, $\mathbf{B}$ can be replaced by $\mathbf{H}$. In Part B we shall assume that the crystal is an insulator and there are no external sources of charge or current (i.e., currents are due only to the oscillatory motion of the ions). Thus, we set $\rho_{ext} = 0 = \mathbf{J}$. (In Part C, when metals are considered, we shall take $\mathbf{J} = \sigma\mathbf{E}$.)

To obtain the various properties of waves, such as those given on the next several pages, in **SI units** it is best to start from Maxwell's equation as given here in **SI** units.

$$\nabla \cdot \mathbf{D} = \rho_{ext} \qquad \nabla \cdot \mathbf{B} = 0$$
$$\nabla \times \mathbf{E} = -\frac{\partial \mathbf{B}}{\partial t} \qquad \nabla \times \mathbf{H} = \frac{\partial \mathbf{D}}{\partial t} + \mathbf{J}$$

and the important magnetic quantities are related via $\mathbf{B} = \kappa_m \mu_0 \mathbf{H} = \mu_0(\mathbf{H} + \mathbf{M})$ and $\mathbf{M} = \chi_m \mathbf{H}$, where $\mathbf{M}$ is the magnetization per unit volume. Then $\kappa_m = 1 + \chi_m$. These relations are used in Chapter 15.

To solve Maxwell's equations, plane waves with wave vector $\mathbf{k}$ are assumed,

$$\mathbf{E} = \mathbf{E}_0 \, e^{i(\mathbf{k}\cdot\mathbf{r}-\omega t)}, \qquad \mathbf{H} = \mathbf{H}_0 \, e^{i(\mathbf{k}\cdot\mathbf{r}-\omega t)} \qquad (13\text{-}15)$$

As discussed in Section 13-1b, we consider only cubic crystals and thus the "constants" χ, ε, σ, and k can be treated as complex scalars rather than tensors. Substituting Eq. 13-15 into Maxwell's equations we see that $\partial/\partial t$ introduces a factor $-i\omega$ and ∇ brings in a factor $i\mathbf{k}$. Therefore, the field equations become

$$\mathbf{k}\cdot\mathbf{D} = \varepsilon\mathbf{k}\cdot\mathbf{E} = 0 \qquad \mathbf{k}\cdot\mathbf{H} = 0 \qquad (13\text{-}16a)$$
$$\mathbf{k}\times\mathbf{E} = (\omega/c)\mathbf{H} \qquad \mathbf{k}\times\mathbf{H} = -(\omega/c)\varepsilon\,\mathbf{E} \qquad (13\text{-}16b)$$

These equations are separately correct for both the real and imaginary parts. From Eq. 13-16a we see that Eq. 13-15 can be a solution if $\mathbf{k}$ is perpendicular to $\mathbf{E}$ and $\mathbf{H}$, and from Eq. 13-16b we see that they are all mutually perpendicular.

Using the equality $\mathbf{k}\times\mathbf{k}\times\mathbf{E} = \mathbf{k}(\mathbf{k}\cdot\mathbf{E}) - (\mathbf{k}\cdot\mathbf{k})\mathbf{E}$, we can eliminate $\mathbf{E}$ in Eq. 13-16b, yielding the very useful **dispersion relationship**

$$\mathbf{k}\cdot\mathbf{k} = k^2 = \varepsilon(\omega/c)^2 \qquad (13\text{-}17a)$$

which relates the spatial variation of **k** to the time variation ω. (In **SI units** this equation is $k^2 = \varepsilon\varepsilon_0\mu_0\omega^2$.) In optics it is customary to use, instead of $\mathbf{k} = \mathbf{k}_1 + i\mathbf{k}_2$, a complex **index of refraction**, n, defined in terms of the magnitude of k

$$n \equiv k(c/\omega) = n_1 + in_2 \qquad (13\text{-}18)$$

In frequency regions where ε (and n) is real, the wave vector is real with magnitude $k = \varepsilon^{1/2}(\omega/c)$ and $n_1 = \varepsilon^{1/2}$ and $H = \varepsilon^{1/2}E$; the **phase velocity** is $(\omega/k) = c/n_1$. (This is also the **group velocity**, $d\omega/dk$, since ω and k are linearly related.) Remember, all the parameters are functions of ω. Since n_1 is just the usual index of refraction in regions where the crystal is nonabsorbing, the velocity of propagation in a crystal, c/n_1, is slower than in vacuum, as is well known.

In general ε and therefore k are complex, so the dispersion relationship Eq. 13-17a can be written in the form

$$(c/\omega)^2(k_1^2 - k_2^2 + 2i\mathbf{k}_1\cdot\mathbf{k}_2) = \varepsilon \equiv (n_1 + in_2)^2 \qquad (13\text{-}17b)$$

As we will show, this corresponds to damped waves. For our isotropic, or cubic, case $\mathbf{k}_1$ and $\mathbf{k}_2$ are parallel, which is taken as the z-direction with no loss of generality. Then, in general

$$\varepsilon_1 = n_1^2 - n_2^2 \qquad \varepsilon_2 = 2n_1n_2 = 4\pi\sigma_1/\omega \qquad (13\text{-}19a)$$

The converse relationships are

$$n_1^2 = (1/2)[\varepsilon_1 + (\varepsilon_1^2 + \varepsilon_2^2)^{1/2}]$$
$$n_2^2 = (1/2)[-\varepsilon_1 + (\varepsilon_1^2 + \varepsilon_2^2)^{1/2}] \qquad (13\text{-}19b)$$

Note, the indices of refraction are always positive. (In **SI units** ε_1 in Eq. 13-19a is the same expression but $2n_1n_2 = \sigma_1/\varepsilon_0\omega$ so the converse relationships, Eq. 13-19b, are altered slightly.)

The usefulness of the complex n is clear when we write the spatial dependence of the fields (Eq. 13-15) in terms of n_1 and n_2

$$\exp(ikz) = \exp[i(\omega/c)n_1z]\exp[-(\omega/c)n_2z] \qquad (13\text{-}20)$$

The imaginary component of n determines the spatial decay of the wave; the real part determines the phase velocity $\text{Re}(\omega/k) = c/n_1$.

The **attenuation coefficient** for the intensity loss of a propagating wave is defined as

$$\alpha \equiv -\frac{1}{I}\frac{dI}{dz} \qquad \text{or} \qquad I(z) = I(0)e^{-\alpha z} \qquad (13\text{-}21)$$

Since the intensity depends on the square of the fields,

$$\alpha = 2\omega n_2/c \qquad \text{and} \qquad \sigma_1 = n_1\alpha c/4\pi \qquad (13\text{-}22)$$

The second equation follows from the complex dispersion relation Eq. 13-19a, and is sometimes a useful way to think about the conductivity. In frequency regions where n_1 does not vary very much, the frequency dependence of σ_1 is similar to that of α. Thus, a complex wave vector takes decaying plane waves into account in a simple way. (In **SI units** $\sigma_1 = n_1 \alpha c \varepsilon_0$.)

Note that the wave is attenuated as long as $n_2 > 0$. This usually means that ε_2 and $\sigma_1 > 0$ (Eq. 13-19), and thus there is a loss of power that accompanies the attenuation (Problem 1). However, it is possible, as will be seen later, to have $n_2 > 0$ but $n_1 = 0$ so $\varepsilon_2 = 0 = \sigma_1$ and there is attenuation but no power loss. This happens for reflection from a metal where the wave is attenuated not by power loss but by backward reradiation (i.e., reflection).

The dispersion relation, Eq. 13-17, which relates the various "parameters" of the macroscopic theory, will prove extremely useful in the rest of the chapter. It relates the macroscopic optical parameters (the k part) to quantities that can be calculated by microscopic theory (the ε part). The Lorentz oscillator in Section 13-6 is an excellent example of this use of Eq. 13-17.

Before determining equations for reflectivity in terms of the wave vector k, two subtle points should be made. First, in the plane wave expression, Eq. 13-15, we can show (see the Notes) that planes of constant phase are perpendicular to k_1 while planes of constant amplitude are perpendicular to k_2. Thus, if the propagation direction is not along a principal axis of ε, σ, and k, where k_1 and k_2 are parallel to one another, complications can arise. However, these complications do not arise in cubic crystals, which are optically isotropic. Second, when ε is not only frequency dependent but is also implicitly dependent on **k**, we have **spatial dispersion** of ε. This occurs whenever the relation between **D** and **E** is not local, that is, **D** at a point is not determined solely by **E** at that point. Excitons provide an excellent example of spatial dispersion, and shall be discussed. Optical rotation of polarized light (in quartz, for example) is due to the dependence of D on the spatial gradient of the electric field. The so-called "anomalous skin effect" observed in some metals, is also an example; the electromagnetic wave is reflected so strongly that the mean free path of the conduction electrons is much greater than the penetration depth of the wave. This leads to spatial dispersion.

13-4 Reflectivity at an Interface

One useful experimental method to determine the frequency dependence of the optical constants is to measure the reflectivity of a sample as a function of frequency. Examples of this will be given in

Part B. Here the reader is reminded of the relevant equations.

Fresnel's equations give values for the amplitudes of the electromagnetic fields of the reflected and refracted waves at a boundary between two media. If the boundary is the $z = 0$ plane then the tangential component of the electric and magnetic fields must be continuous across $z = 0$. For normal incidence, the ratio of the electric field of the reflected wave, E_r, to the incident one, E_i, is

$$r = \frac{E_r}{E_i} = \frac{n_c - n_a}{n_c + n_a} \qquad (13\text{-}23a)$$

where n_c and n_a are, respectively, the indices of refraction of the reflecting medium (a crystal with $n_c = n_1 + in_2$ in this case) and of the medium in which the incident and reflected waves are measured (which is usually air, i.e., $n_a = 1$). The measured reflectivity, R, for normal incidence depends on the square of the fields and is

$$R = rr^* = |r|^2 = \frac{(n_1 - 1)^2 + n_2^2}{(n_1 + 1)^2 + n_2^2} \qquad (13\text{-}23b)$$

By measuring the frequency dependence of the attenuation coefficient, n_2 can be obtained (Eq. 13-22). Then, n_1 can be obtained from the frequency dependence of R. From these two quantities all the other constants can be determined. However, often a different procedure is used in conjunction with reflectivity measurements. This will be discussed now.

13-5 Kramers-Kronig Relations

Given some physical model, for example, an oscillator or some absorbing centers in a solid, ε_1 and ε_2 can be calculated. A general relationship between these two quantities, which allows one to be calculated if the other is known over a sufficiently wide frequency spectrum, is described by the following equations. These equations, known as the Kramers-Kronig (or KK) relations, relate the real and imaginary parts of the dielectric constant. They are

$$\varepsilon_1(\omega) - \varepsilon_\infty = \frac{2}{\pi} \mathscr{P} \int_0^\infty \frac{\omega' \varepsilon_2(\omega')}{\omega'^2 - \omega^2} d\omega' \qquad (13\text{-}24a)$$

$$\varepsilon_2(\omega) = \frac{2\omega}{\pi} \mathscr{P} \int_0^\infty \frac{\varepsilon_1(\omega')}{\omega^2 - \omega'^2} d\omega' \qquad (13\text{-}24b)$$

where we have explicitly written the frequency dependence of the dielectric constant, that is, $\varepsilon_1(\omega)$ and $\varepsilon_2(\omega)$. ε_∞ is the real part of the

CHAPTER 13   OPTICAL PROPERTIES OF CRYSTALS   459

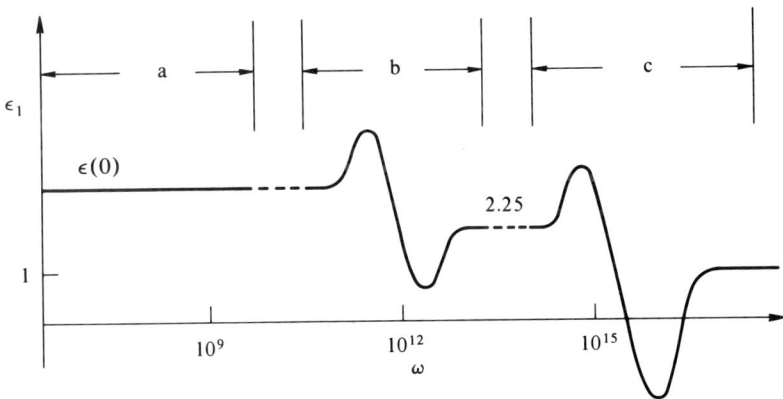

Fig. 13-2 A schematic diagram of the real part of the dielectric constant, ε_1, vs. frequency showing the different important frequency regions. Region a is drawn flat but it could show relaxation of permanent electric dipoles (or perhaps impurities). Region b is due to lattice vibrations. Region c is due to the valence electrons.

dielectric constant at $\omega = \infty$; this is more of an experimental statement since we do not make measurements to $\omega = \infty$. There is usually some upper frequency limit in which there is no longer any loss, $\varepsilon_2 \approx 0$, and the physics of the problem suggests that we have gone "far enough." Then the measurements are stopped and the real part of the dielectric constant is then called ε_∞. (If the integration is to be truly taken to $\omega = \infty$ then ε_∞ must be replaced by 1.) The example of lattice vibrations will make this clearer. Last, $\mathscr{P}$ stands for the Cauchy principle value of the integral—the singularity in the integral at $\omega = \omega'$ is omitted from the integration.

From these relations we see that if $\varepsilon_2(\omega)$ is known over a sufficiently wide frequency range, $\varepsilon_1(\omega)$ can be determined, and vice versa. The term "sufficiently wide" is of course difficult to define, but it must be wide enough to include the regions where ε_1 and ε_2 are strongly frequency dependent.

Figure 13-2 should help to make this and several related points clearer. The real part of the dielectric constant is plotted vs. the log of the frequency. In the 10^8 cycles per sec region the application of **E** results in a **P** which is due to distortions of the electron clouds and displacements of nuclei (or charged ionic cores). $\varepsilon_1(\omega \approx 10^8 \text{ sec}^{-1})$ is usually measured in a capacitance bridge. At some higher frequency the nuclei can no longer follow the applied electric field, but the valence electrons, being much lighter, can still follow the field. This is the high frequency part of region (b). Thus, the polarization is less (it is only due to the electron contribution) and $\varepsilon_1(\omega \approx 10^{14} \text{ sec}^{-1})$ is smaller. In this frequency region the index of refraction, n_1, is measured by several different possible techniques and, where there is no loss, $n_1^2 = \varepsilon_1$ (i.e., Eq. 13-19a). If the frequency is increased further, the valence electrons eventually cannot follow the field and ε_1 will be due to the contributions from the inner core electrons. At still higher frequency, when none of the electrons can follow the time varying field, $\varepsilon_1 \approx 1$.

Thus, we can see the region over which ε_2 must be measured to give $\varepsilon_1(\omega)$ via Eq. 13-24. It might be 10^{11} to 10^{13} sec^{-1} and the $\varepsilon_\infty = 2.25$ would be used. From the Kramers-Kronig relations the static dielectric constant (i.e., $\omega = 0$) is given by

$$\varepsilon_1(0) - \varepsilon_\infty = \frac{2}{\pi} \int_0^\infty \frac{\varepsilon_2(\omega')}{\omega'} d\omega' \tag{13-25a}$$

If there is a large lattice vibrational contribution to $\varepsilon_1(0)$, that is, $\varepsilon_1(0) - \varepsilon_\infty$ is large, then there must be a frequency region in which ε_2 is large; and since $\varepsilon_2 \propto \sigma_1$ there is a high conductivity in this frequency region. This emphasizes how insulators can have very large conductivities at high frequencies. It arises from a polarization due to charges that are bound to atomic positions. These bound charges cannot contribute to the dc current but at the appropriate frequency result in very large σ_1 values. This will become clearer when the model in the next section is presented.

At frequencies lower than 10^9 sec^{-1} there might be other contributions to ε_1 not shown in Fig. 13-2. For example, there could be an orientational contribution of the type discussed in Section 11-7, which is usually of the relaxation type. This could arise from impurities (OH$^-$ impurities, for example) or from the bonding of the crystal (as in H$_2$O). There is a resonance of the type shown in Fig. 13-2 every time the frequency corresponds to the natural resonant frequency of the entity that gives rise to the polarization.

From reflectivity measurements, the real quantity R in Eq. 13-23b is obtained, but we would like r in Eq. 13-23, which is complex. Writing $r = R^{1/2} \exp(i\Theta)$ or $\ln r = \ln R^{1/2} + i\Theta$ we can show, using the Kramers-Kronig relations, that the phase angle is given by

$$\Theta(\omega) = -\frac{\omega}{\pi} \mathscr{P} \int_0^\infty \frac{\ln R(\omega')}{\omega'^2 - \omega^2} d\omega' \tag{13-25b}$$

$$= -\frac{1}{2\pi} \int_0^\infty \ln \left| \frac{\omega' + \omega}{\omega' - \omega} \right| \frac{d \ln R(\omega')}{d\omega'} d\omega' \tag{13-25c}$$

The second expression is obtained from the first by integrating by parts. If $R(\omega')$ is measured then $\Theta(\omega)$ can be calculated, resulting in two equations for the two quantities n_1 and n_2, from which any of the constants may be calculated. An appreciation of the meaning of the limits of integration can be obtained from Eq. 13-25c. Whenever

CHAPTER 13   OPTICAL PROPERTIES OF CRYSTALS   461

$R(\omega')$ is independent of ω', there is no contribution to the integral. Similarly, when ω' is $\gg$ or $\ll \omega$, the contribution is very small.

13-6 Damped Harmonic Oscillator

In this section we consider a simple model, that of a harmonic oscillator with a damping term. When used in the context of this chapter it is often called the **Lorentz oscillator** (after the Lorentz and Drude classical theory of absorption and dispersion of light). The electrons are bound to their cores by harmonic forces, in the model, and, in an externally applied field, their motion is given by

$$m\frac{d^2r}{dt^2} + m\gamma\frac{dr}{dt} + m\omega_0^2 r = -e\mathbf{E} \tag{13-26}$$

where m and e are the mass and charge of the bound electron, ω_0 is the natural oscillator frequency, **E** is the field acting on the electron (i.e., the local field), and the usual velocity dependent damping term, which represents an energy loss mechanism, has been added. For free electrons the loss arises from radiation damping, but in solids the losses usually arise due to anharmonic coupling to other allowed excitations in the solid. It is found that a (γ dr/dt) term can be used to fit a large variety of phenomena.

Letting $\mathbf{E} = \mathbf{E}_0 \exp(-i\omega t)$, we solve for the displacement **r** of the form $\mathbf{r} = \mathbf{r}_0 \exp(-i\omega t)$

$$\mathbf{r} = \frac{-e\mathbf{E}/m}{(\omega_0^2 - \omega^2) - i\gamma\omega} \tag{13-27a}$$

The response can be obtained also in terms of the polarization, **P** = qrN/V from Eq. 13-1, of N/V oscillators per unit volume, where q = $-e$. Also the dielectric response ε can be obtained.

$$\mathbf{P} = \frac{(e^2 N/mV)\mathbf{E}}{(\omega_0^2 - \omega^2) - i\gamma\omega} = \chi \mathbf{E} = (\chi_1 + i\chi_2)\mathbf{E} = \frac{\varepsilon - 1}{4\pi}\mathbf{E} \tag{13-27b}$$

$$\varepsilon = 1 + \left(\frac{4\pi e^2 N}{mV}\right)\frac{1}{(\omega_0^2 - \omega^2) - i\gamma\omega} \tag{13-27c}$$

In Eq. 13-27b the reader is reminded of the definition of the susceptibility (Eq. 13-9) and the relation to the dielectric constant (Eq. 13-11). Writing ε in its real and imaginary parts, and keeping in mind the connection to the optical constants (Eq. 13-19a) we obtain

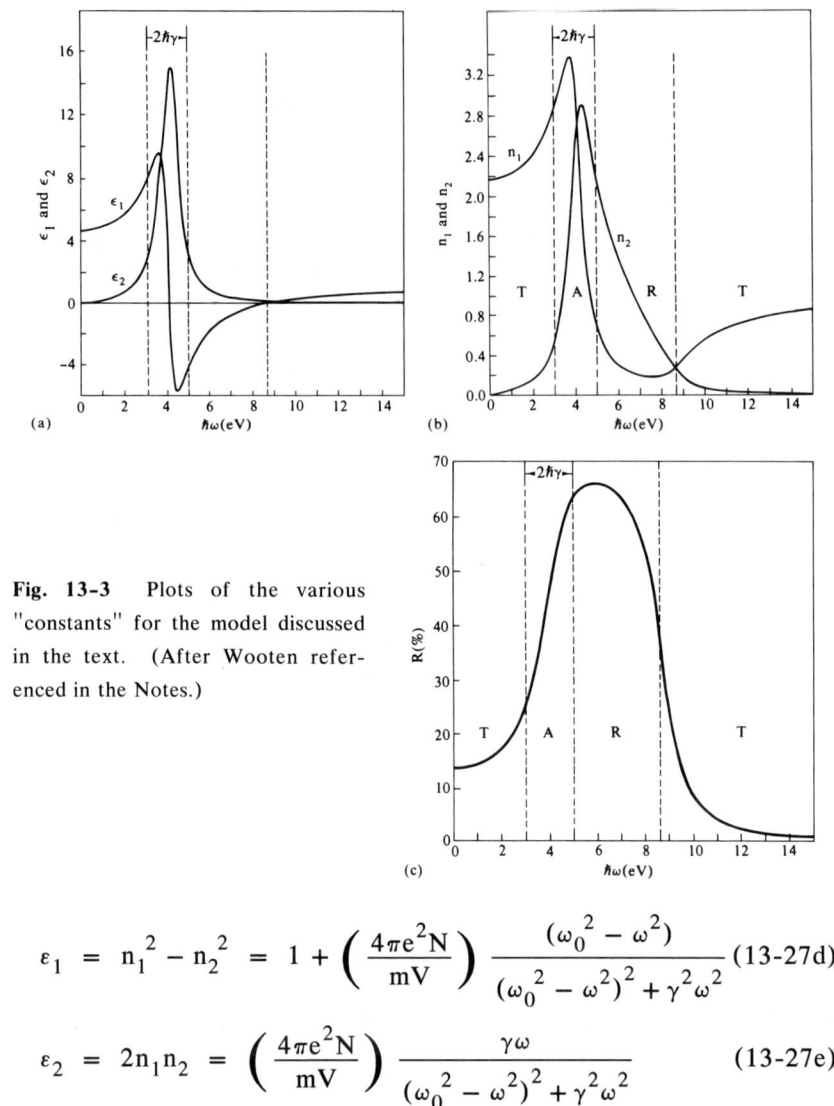

Fig. 13-3 Plots of the various "constants" for the model discussed in the text. (After Wooten referenced in the Notes.)

$$\varepsilon_1 = n_1^2 - n_2^2 = 1 + \left(\frac{4\pi e^2 N}{mV}\right) \frac{(\omega_0^2 - \omega^2)}{(\omega_0^2 - \omega^2)^2 + \gamma^2 \omega^2} \quad \text{(13-27d)}$$

$$\varepsilon_2 = 2n_1 n_2 = \left(\frac{4\pi e^2 N}{mV}\right) \frac{\gamma \omega}{(\omega_0^2 - \omega^2)^2 + \gamma^2 \omega^2} \quad \text{(13-27e)}$$

To these straightforward results we add some physical insight. First, we want to replace the 1 in the real part of the dielectric constant Eq. 13-27d by ε_∞. This comes about by considering other types of oscillators whose resonant frequencies are much higher than the region of ω under study. For oscillators at much higher frequencies there is no contribution to the imaginary part of ε, just to the real part. Second, instead of just one type of oscillator with frequency ω_0 and damping γ, we consider more than one type of oscillator with different binding and different loss mechanisms. Then ε would be

CHAPTER 13   OPTICAL PROPERTIES OF CRYSTALS   463

$$\varepsilon_1 = \varepsilon_\infty + \Sigma_i \left(\frac{4\pi e^2 N_i}{mV}\right) \frac{(\omega_{0i}^2 - \omega^2)}{(\omega_{0i}^2 - \omega^2)^2 + \gamma_i^2 \omega^2} \quad (13\text{-}28\text{a})$$

$$\varepsilon_2 = \Sigma_i \left(\frac{4\pi e^2 N_i}{mV}\right) \frac{\gamma_i \omega}{(\omega_{0i}^2 - \omega^2)^2 + \gamma_i^2 \omega^2} \quad (13\text{-}28\text{b})$$

where the sum is over the i different types of oscillating electrons. The amazing thing about these equations and this model is that they are applicable to many different phenomena; part of the reason for this will be discussed in the next section. For example, when Eqs. 13-28 are applied to lattice vibrations, the values of the charge and mass should have a subscript and are very different from those of electrons. In fact, when applied to lattice vibrations the entire term $(4\pi N_i e_i^2/mV_i)$ is usually written as S_i to cover our ignorance of the individual parts. In **SI units** the 4π is replaced by ε_0^{-1} (Section 13-1a) so this term in Eqs. 13-28 and other similar equations is $(e^2 N/\varepsilon_0 mV)$.

Now we go back to Eqs. 13-27 and plot the results of the frequency dependence of the various "constants" that have been considered in this chapter. For this purpose, we take a system for which $\hbar\omega_0 = 4$ eV, $\gamma/\omega_0 = 1/4$ (modestly damped), and $(4\pi e^2 N/mV) = 60$. Figure 13-3 shows the results for ε, k, and R the reflectivity.

First, notice that ε_1 has the shape seen in Fig. 13-2. If the 1 were replaced by ε_∞, as in Eq. 13-28a, then the entire ε_1 curve would be raised by $\varepsilon_\infty - 1$. At high frequencies $\varepsilon_1 = \varepsilon_\infty$; at zero frequency the result is

$$\varepsilon_1(0) = \varepsilon_\infty + (4\pi e^2 N/mV)/\omega_0^2 \quad (13\text{-}28\text{c})$$

If γ were much smaller, then ε_1 and the corresponding polarization would become much larger just below ω_0 (they become infinite if $\gamma = 0$). This is why the loss term must be included in the model.

Second, both ε_2 and the loss peak at approximately ω_0. Actually $\sigma_1 (=\omega\varepsilon_2/4\pi)$ peaks exactly at ω_0 for any value of γ/ω_0, and the peak of ε_2 is slightly shifted from ω_0, the shift increasing as γ/ω_0 increases. (See the Problems.)

Third, notice that at frequencies below ω_0 the quantity n_1 has the same type of behavior as ε_1 and that $\varepsilon_1(0) = [n_1(0)]^2$. Also $n_1 = n_2$ when $\varepsilon_1 = 0$ as in Eq. 13-19a.

Last, Fig. 13-3c shows the reflectivity for the parameters chosen. For the case where $\gamma = 0$ we obtain 100% reflectivity between ω_0 and the frequency where $\varepsilon_1 = 0$. This latter frequency will be shown to correspond to the longitudinal mode frequency while ω_0 corresponds to transverse mode frequency. Thus, for $\gamma = 0$, R = 100% at frequencies between the transverse and longitudinal mode frequencies. For $\gamma \neq 0$ this statement must be modified, as can be seen.

Qualitatively speaking, the various frequency regions display different types of behavior. The material is primarily absorbing in the region $\omega = \omega_0 \pm \gamma$. Between $\omega_0 + \gamma$ and ω_ℓ, the frequency where $\varepsilon_1 = 0$, the absorption is low and the material is primarily reflecting the incident radiation. For $\omega > \omega_\ell$ and $\omega < \omega_0 - \gamma$ the material is primarily transmitting the incident radiation. These regions are separated by dashed lines in the figure.

ε_∞ **effect** – In the discussion here we have added the ε_∞ term in an ad hoc manner. First, we should point out that a more complete theory (including oscillators at much higher energy) will give this term automatically. Second, the need for this term will be seen, very clearly, when experimental reflectivity spectra are examined. Consider the high energy results in Fig. 13-3. For large value of ω, the figure shows that ε_2 and n_2 approach zero, while ε_1 and n_1 approach one. Naturally, this also can be seen from the equations (Eqs. 13-27d and 13-27e then use Eq. 13-19b). This means that the reflectivity approaches zero at high energies as shown in the figure (or Eq. 13-23b). Later, we will encounter the same result in the discussion of free electron effects (as in Fig. 13-9). However, the experimental reflective curves do not look like this and the reason is ε_∞.

With ε_∞ instead of 1 in Eq. 13-28a, at high energies ε_2 and n_2 still approach zero but $\varepsilon_1 = n_1^2 \approx \varepsilon_\infty$. This causes the reflectivity to go to ≈ 0 or little above where ε_1 goes from negative to positive (Fig. 13-3a) and then increase to a value of $R = (n_1 - 1)^2/(n_1 + 1)^2$ where $n_1 = (\varepsilon_\infty)^{1/2}$ as in Eq. 13-23b. This is seen very clearly in experimental results and demonstrates the need for ε_∞ instead of 1. (For example, see Figs. 13-5 and 13-13.)

13-7 Dielectric Response of a Quantum System

One of the impressive things about the simple Lorentz oscillator discussed in the last section is that it is so generally applicable to real systems. These include the optical response from phonon systems, band systems, and free electron systems in metals. The reason for this wide applicability is that the corresponding quantum mechanical equation for ε has the same form as the classical equation, calculated previously, for the damped harmonic oscillator.

For any quantum mechanical system that is characterized by a ground state ψ_0 with energy $\mathscr{E}_0$ and excited states ψ_i with energies $\mathscr{E}_i$ where $\omega_i = (\mathscr{E}_i - \mathscr{E}_0)/\hbar$, we can obtain

CHAPTER 13   OPTICAL PROPERTIES OF CRYSTALS   465

$$\varepsilon = 1 + \frac{4\pi e^2 N}{mV} \Sigma_i \frac{f_i}{\omega_i^2 - \omega^2 - i\gamma_i\omega}$$

$$f_i = \frac{2m\omega_i}{e^2\hbar} |\langle \psi_i | e_z | \psi_0 \rangle|^2$$

(13-29)

where f_i is known as the **oscillator strength**, which involves the matrix element of the electric dipole operator (picked arbitrarily in the z-direction) between the ground and excited states. See the Notes for references for the derivation of Eq. 13-29. Equation 13-29 is of the same form as the results in Eq. 13-27c. Here the strength of the "resonance" is determined by the oscillator strength, and the resonant frequencies ω_i correspond to those of allowed transitions.

Part B — Lattice Vibrations

13-8 Introduction

In the last chapter we discussed lattice vibrations and the use of neutron spectroscopy to measure ω vs. k throughout the Brillouin zone. In this section we will discuss the interaction of infrared light with lattice vibrations. The crystal momentum or k-vector of light is very much smaller than the dimension of the Brillouin zone. For visible light the wave vector is $k = 2\pi/\lambda \approx 2\pi/6000 \text{ Å} \approx 10^5 \text{ cm}^{-1}$, while the dimension of the Brillouin zone is $k_{max} = 2\pi/a \approx 2\pi/6 \text{ Å} \approx 10^8 \text{ cm}^{-1}$, a factor 10^3 larger. We remind the reader of some useful formulas for k, ω, and λ.

$$\nu\lambda = \frac{c}{n_1}, \qquad \omega = 2\pi\nu, \qquad k = \frac{2\pi}{\lambda}, \qquad \frac{\omega}{k} = \frac{c}{n_1} \qquad (13\text{-}30)$$

Compared to a typical phonon dispersion curve, as in Chapter 12, the plot of ω vs. k for light is essentially a vertical line starting at the origin. Thus, from conservation of crystal momentum, we expect that light interacts only with phonons that have $k \approx 0$ (i.e., k very small compared to the Brillouin zone edge). We say that infrared and Raman techniques measure the **k = 0** optical modes. Actually, this means that the phonon frequencies for $k \approx 0$ are being measured, but because $d\omega/dk = 0$ at $k = 0$ for the optic modes these various expressions are used. Even then, as we shall see, polaritons modify these concepts, but only at extremely small k ($\approx 100 \text{ cm}^{-1}$).

In Part B we show first that the $k \approx 0$ optical vibrational modes correspond to damped harmonic oscillators, and therefore reflectivity measurements (Section 13-4) can be used to determine the optical mode frequencies (the LO and TO frequencies). Then we discuss other more sophisticated models of the interaction of light with long wavelength phonons ($k \approx 0$). This gives rise to polaritons.

13-9 Long Wavelength Optical Vibrations

13-9a Transverse waves Consider a very simple model of a cubic ionic crystal with two atoms in each primitive unit cell. (The NaCl, CsCl, and ZnS crystal structures have this property.) Let the electromagnetic radiation propagate in the x-direction (direction of **k**) and have its oscillating electric field in the z-direction. Thus, the positive and negative ions move in the z-direction by amounts u_+ and u_-, respectively. Also, we consider only $k \approx 0$.

If each ion has a charge $+q$ or $-q$, with N/V cells per unit volume, then the lattice polarization is $P = (N/V)(+qu_+ - qu_-)$. The use of point charges is a poor approximation for deformable atoms. Thus, a **dynamic apparent charge** Q is defined. (It is sometimes called the **effective charge** or **Born transverse effective charge**.) We show that the resulting equations are equivalent. Expanding P in terms of u_+ and u_- and keeping only the first (i.e., linear) term

$$P = (\partial P/\partial u_+)u_+ + (\partial P/\partial u_-)u_-$$
$$= (N/V)(Qu_+ - Qu_-)$$

where
$$Q \equiv \left(\frac{V}{N}\right)\frac{\partial P}{\partial u_+} = -\left(\frac{V}{N}\right)\frac{\partial P}{\partial u_-} \tag{13-31}$$

$(\partial P/\partial u_+) = -(\partial P/\partial u_-)$ because a uniform translation of both the positive and negative ions, $u_+ = u_-$, must have zero dipole moment. This relation for P has the same form as that for point charges but is theoretically more correct. Apparent charges are discussed shortly.

Let us come back to our major problem and find the equations of motion of the system of positive and negative ions. The electric field exerts a force $\pm QE$ on the ions. In addition, separation of the positive and negative ions $(u_+ - u_-)$ produces a restoring force (as in Eq. 12-2). Thus, the equations of motion are

$$m_+(d^2u_+/dt^2) = -\alpha(u_+ - u_-) + QE$$
$$m_-(d^2u_-/dt^2) = -\alpha(u_- - u_+) - QE \tag{13-32}$$

where α is the force constant and m_+, m_- are the masses of the positive and negative ions. Multiply the first equation by m_-, the second by m_+, subtract the two equations, then divide by $(m_+ + m_-)$, defining a reduced mass $\mu \equiv m_+m_-/(m_+ + m_-)$, and let $r = u_+ - u_-$, the relative displacement of the ions. Thus we obtain Eq. 13-33a

CHAPTER 13   OPTICAL PROPERTIES OF CRYSTALS   467

$$\mu (d^2r/dt^2) + \alpha r = QE \qquad (13\text{-}33a)$$

$$\mu (d^2r/dt^2) + \mu\gamma(dr/dt) + \mu\omega_0^2 r = QE \qquad (13\text{-}33b)$$

Equation 13-33a has the same form as that of a harmonic oscillator. Since lattice vibrations are expected to be damped, we phenomenologically add a damping term and obtain Eq. 13-33b. This is the same expression as Eq. 13-26, the solution of which was discussed extensively. Also we have set the natural frequency to the transverse harmonic frequency, $\omega_0^2 \equiv \alpha/\mu$, which is the expression for the frequency of oscillation of a spring with spring constant α and mass μ, and is also called ω_{TO} (= ω_0). We say transverse frequency as a reminder that the applied electric field (of the light) is transverse to the small but finite wave vector.

The solution of Eq. 13-33b was discussed in Section 13-6 (Eq. 13-27c) and we may immediately write

$$\varepsilon = 1 + \left(\frac{4\pi Q^2 N}{\mu V}\right) \frac{1}{\omega_0^2 - \omega^2 - i\gamma\omega} \qquad (13\text{-}33c)$$

Various aspects of this result are plotted in Fig. 13-3.

13-9b Longitudinal waves One very important point has not been addressed so far. Figure 12-10f shows ω vs. k for KBr, an ionic material with two atoms per primitive cell. The crystal has an LO (**longitudinal optic**) as well as a TO (**transverse optic**) mode. Where is the $k \approx 0$ longitudinal optic mode in our model? Actually, the answer is rather simple. Referring back to Eq. 13-16a, where from $\nabla \cdot D = 0$, we obtain $\varepsilon k \cdot E = 0$. We considered the transverse wave solution with k perpendicular to E, which clearly satisfies this equation and the rest of Eqs. 13-16. However, more generally k (and the displacement of atoms) should be decomposed into parts parallel and perpendicular to E. The parallel or longitudinal part $k_\parallel$ can also satisfy this equation with $k_\parallel \cdot E \neq 0$ for any frequency ω_ℓ provided that

$$\varepsilon(\omega_\ell) = 0 \qquad (13\text{-}34a)$$

that is, the zeros of the dielectric constant (still for $\lambda \gg$ lattice distances). Longitudinal waves propagate when ε is zero.

The concept of longitudinal waves is always a bit troublesome. Well-known systems that have longitudinal waves are plasmas and air (sound waves). Longitudinal waves also play a role in the energy loss of fast, charged particles passing through solids. The result for these longitudinal waves is that the energy loss is proportional to $-\text{Im}(1/\varepsilon)$ (i.e., it peaks whenever $\varepsilon = 0$). Thus, to determine ω_ℓ for crystals we usually plot

$$-\text{Im}(1/\varepsilon) = \frac{\varepsilon_2}{\varepsilon_1^2 + \varepsilon_2^2} = \frac{2n_1 k_2}{n_1^2 + n_2^2} \qquad (13\text{-}34b)$$

For small damping, the reflectivity is almost one between the TO and LO frequencies, as shown in Fig. 13-3c. Between these two frequencies ε_1 is negative, passing from negative to positive at the LO frequency. $\text{Im}(1/\varepsilon)$ is not plotted in Fig. 13-3; its shape is similar to the ε_2 curve but it peaks when $\varepsilon_1 = 0$.

A more detailed derivation of longitudinal waves can be found in Born and Huang, Chapter 2, Sections 7 and 8. They split the relative motion **r** explicitly into parts transverse and parallel to the electric field $\mathbf{r} = \mathbf{r}_t + \mathbf{r}_\ell$ using the condition that $\nabla \cdot \mathbf{r}_t = 0$ and $\nabla \times \mathbf{r}_\ell = 0$.

13-9c Lyddane Sachs Teller (LST) relationship
Since LO and TO modes appear in so many different contexts, it is worthwhile to spend a little more time on them and to develop the useful, well-known LST relation.

It is possible to rewrite the resonance equation Eq. 13-33c by replacing the 1 by ε_∞ as done in Eq. 13-28. The physical reason is simple: ε_∞ summarizes the dielectric response of all the processes that occur at frequencies much higher than those considered here in the lattice vibrational frequency region. The "infinity" subscript is used because the electronic resonances are at very much higher energies than lattice resonance. Second, note that $\Omega^2 \equiv 4\pi Q^2 N/\mu V$ is positive and has the dimensions of squared frequency. Thus,

$$\varepsilon = \varepsilon_\infty + \frac{\Omega^2}{\omega_0^2 - \omega^2 - i\gamma\omega} \qquad (13\text{-}35a)$$

Let $\gamma = 0$ so ε is real. Evaluate $\varepsilon(\omega = 0) \equiv \varepsilon_1(0)$ and also solve for $\varepsilon(\omega_\ell) = 0$:

$$\varepsilon_1(0) = \varepsilon_\infty + \frac{\Omega^2}{\omega_0^2}, \qquad \omega_\ell^2 = \omega_0^2 + \frac{\Omega^2}{\varepsilon_\infty} \qquad (13\text{-}35b)$$

$\varepsilon_1(0)$ is always greater than ε_∞ due to the "strength of the resonance," that is, the pole of ε at ω_0. For strong resonances (due to very large effective charges, high density, small masses, or low values of ω_0 i.e. weak restoring forces), $\varepsilon_1(0)$ can be large. The second equation shows that the LO frequency is always greater than the TO frequency and that the difference is also related to the "strength of the resonance." This is shown in Fig. 12-10f. (The LO-TO splitting can also be determined by realizing that for LO modes, at long wavelength, there is an extra restoring force arising from the macroscopic polarization field that does not effect the transverse modes.) For crystals that have zero

CHAPTER 13   OPTICAL PROPERTIES OF CRYSTALS   469

effective charges, $\varepsilon_1(0) = \varepsilon_\infty$ and $\omega_\ell = \omega_0$. This is the situation in silicon and there is no LO-TO splitting at $k \approx 0$ (Fig. 12-10e).

From Eq. 13-35b, Ω^2 may be eliminated and the Lyddane Sachs Teller equation is obtained

$$\frac{\varepsilon_1(0)}{\varepsilon_\infty} = \frac{\omega_\ell^2}{\omega_0^2} \equiv \frac{\omega_{LO}^2}{\omega_{TO}^2} \tag{13-35d}$$

Remarkably, no microscopic parameters occur in this equation. This is because the differences between the LO and TO frequencies is due to macroscopic polarization fields. Again, we see that if the TO frequency is very low the static dielectric constant is very large. This will turn out to be important in the study of ferroelectric materials.

These results in Eq. 13-35d can be written in two more ways that give insight and also will prove useful later.

$$\varepsilon = \varepsilon_\infty + \frac{[\varepsilon(0) - \varepsilon_\infty]\omega_0^2}{\omega_0^2 - \omega^2} \tag{13-35e}$$

$$\frac{\varepsilon}{\varepsilon_\infty} = \left[\frac{\omega_\ell^2 - \omega^2}{\omega_0^2 - \omega^2}\right] \tag{13-35f}$$

This latter expression shows that the LO mode frequency is given by the zero of ε and that the TO mode is given by the pole of ε.

13-9d Generalizations of LST For ionic crystals with more than two atoms per unit cell there can be more than one set of TO and LO modes. Figure 13-4 shows a unit cell of cubic $BaTiO_3$ and its four different optical modes. In three of these there is an oscillating dipole moment that can couple to the oscillating electric field in the manner that has been discussed, so the same types of equations will arise. In fact, if there are p modes that have oscillating dipole moments, it is straightforward to obtain

$$\varepsilon = \varepsilon_\infty + \sum_{i=1}^{p} \frac{\Omega_i^2}{\omega_{0i}^2 - \omega^2 - i\gamma_i\omega} \tag{13-36a}$$

$$\varepsilon_1(0) = \varepsilon_\infty + \sum_{i=1}^{p} \frac{\Omega_i^2}{\omega_{0i}^2} \tag{13-36b}$$

For $\gamma_i = 0$ the LST and generalization of Eq. 13-35c are

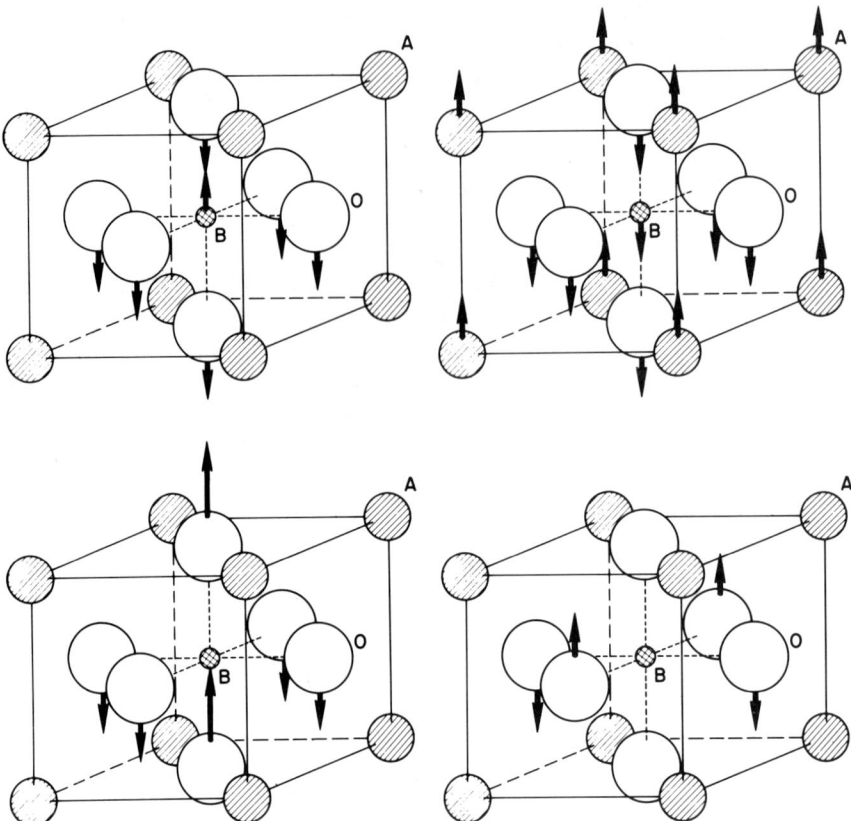

Fig. 13-4 A unit cell of cubic $BaTiO_3$ (i.e., ABO_3 perovskite) showing the atomic displacements of the four different normal modes. The displacements are shown along one of the cubic axes, but occur also along the other two cubic axes. Thus, the vibrations, for each of the four displacements shown, are said to be triply degenerate.

$$\frac{\varepsilon(0)}{\varepsilon_\infty} = \frac{\omega_{\ell 1}^2}{\omega_{01}^2} \cdot \frac{\omega_{\ell 2}^L}{\omega_{02}^2} \ldots = \prod_{i=1}^{p} \frac{\omega_{\ell i}^2}{\omega_{0i}^2} \quad (13\text{-}36c)$$

$$\frac{\varepsilon(\omega)}{\varepsilon_\infty} = \prod_{i=1}^{p} \left[\frac{\omega_{\ell i}^2 - \omega^2}{\omega_{0i}^2 - \omega^2} \right] \quad (13\text{-}36d)$$

These equations can be generalized to crystals with symmetry lower than cubic. For example, for a crystal with orthorhombic symmetry, we can write these four equations of the form of Eqs. 13-36a to 13-36d for each of the three orthogonal directions.

Notice that in Fig. 13-4d the displacements for one normal mode do not produce a time varying dipole moment (i.e., the effective charge is zero), and it does not contribute to ε. Such a mode is infrared inactive, while modes with oscillating dipole moments are called infrared active. We have already noted that the optic mode in silicon is infrared inactive.

13-10 Measurements and Results

13-10a Infrared measurements Determination of the TO (ω_0) and LO (ω_ℓ) modes in ionic crystals is often accomplished by reflection techniques. The measurements are made just as the words imply; the reflectivity, R, of a sample is measured as a function of ω. Normal incidence, from a smooth (compared to the wavelength of the light), thick (compared to the absorption length) sample is most often used and the equations in Sections 13-4 and 13-5 apply.

Figure 13-5a shows some typical results from AlSb a crystal with the cubic zinc blende structure. It is interesting to see how such data are usually analyzed. First, a Kramers-Kronig analysis (Eqs. 13-24) is used to determine the optical constants. This gives ω_0 and ω_ℓ from the peaks in σ_1 and $-\text{Im}(1/\varepsilon)$. The widths can also be obtained from these plots. We might conclude that the plots of these quantities are the end result, but this is not the case. Second, these results are used as starting points for values of ω_0, ω_ℓ, γ, and Ω^2, which are put into a damped harmonic oscillator model, Eq. 13-35a, from which R is calculated. Then some adjustments are made on these input values until a "best" fit is obtained. It is these latter values that are usually quoted as "the" TO and LO phonon frequencies, damping, and oscillator strength, respectively. This indicates how well the damped harmonic oscillator model is trusted. Occasionally these latter fits are not as good as desired. Then, various ad hoc schemes are resorted to, such as coupling various damped harmonic oscillator modes together. Such schemes introduce new parameters that result in a better fit of the data but sometimes result in a loss of the physical understanding.

Figure 13-5b shows reflectivity results for the cubic crystal $SrTiO_3$ (which has the $BaTiO_3$ crystal structure). There are three infrared active modes and the data are fitted with ε from Eq. 13-36a with three modes. In this more complicated crystal the fit is good, although not excellent.

Besides reflectivity measurements, which can give ω_0, ω_ℓ, γ, and Ω^2, the transmission through very thin samples can be measured, yielding ω_0 and γ (Eqs. 13-21 and 13-22). Absorption spectra have been measured and cataloged for a great many organic compounds, and these results are used to identify unknown materials.

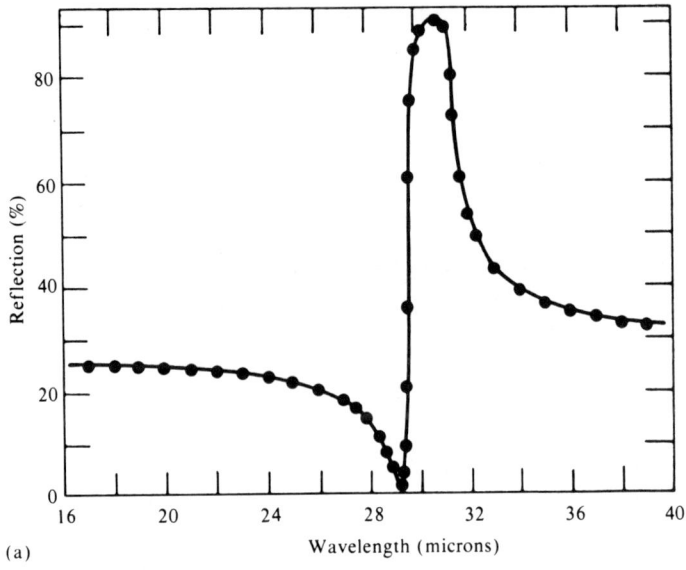

(a)

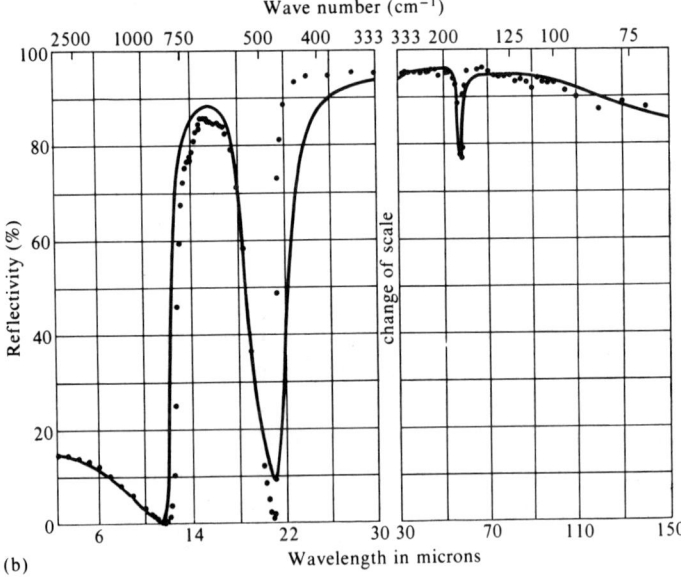

(b)

Fig. 13-5 (a) Infrared reflectivity for AlSb. The points are the measured and the solid curve is the damped harmonic oscillator fit. From W. J. Turner and W. E. Reese, Phys. Rev. **127**, 126 (1962). (b) The same for $SrTiO_3$. From W. G. Spitzer, R. C. Miller, D. Z. Kleinman, and L. E. Howarth, Phys. Rev. **126**, 1710 (1962). See the comment about the ε_∞ effect in Section 13-6. Notice the striking difference between the reflectivity in Fig. 13-3, calculated for $\varepsilon_\infty = 1$ and the results measured here. Clearly $\varepsilon_\infty > 1$.

13-10b Results Infrared reflectivity measurements have been performed on many ionic salts with the simple structures. Table 13-1 (page 448) shows experimental results for the TO and LO modes. Usually, there is good agreement between the rather accurate frequencies obtained from the infrared measurements and those measured by neutron diffraction techniques for $k \approx 0$. There is also usually good agreement between $\varepsilon_1(0)$ calculated via the Lyddane Sachs Teller relation, Eq. 13-35d, and that measured by standard capacitance techniques. We list calculated and experimental room temperature values of $\varepsilon_1(0)$ for some simple materials. (See the Notes for more extensive lists.)

	NaF	NaCl	NaBr	NaI
calc.	5.2	6.04	6.32	7.21
exp.	5.08	5.90	6.27	7.28

The LST relation is used to obtain $\varepsilon_1(0)$ when direct measurements are difficult due to the presence of free carriers, impurities, or ionic conduction. The lead salts (PbSe and PbTe) are an example of such materials. Due to the small band gap of these materials, there are a large number of free carriers, which make capacitance measurements difficult. Reflectivity measurements of PbTe (NaCl structure) give $\omega_0 = 18$ cm^{-1}, $\omega_\ell = 114$ cm^{-1}; using $\varepsilon_\infty = 32$, via the LST the very large value $\varepsilon_1(0) = 1300$ is obtained.

13-10c Raman effect Raman measurements, like infrared measurements, can be used to determine the frequencies of the $k \approx 0$ phonons. As discussed in Section 13-8, the wave vector of the electromagnetic radiation is small, so only $k \approx 0$ phonons have the proper k-value for conservation of crystal momentum.

The Raman scattering process is similar to that of inelastic neutron scattering except for the condition $k \approx 0$. One has incident light, ω_i and k_i, on a sample. The ω_s and k_s of the scattered light is measured, and a phonon with ω and k can be created or destroyed. Conservation of energy and crystal momentum require that

$$\omega_i - \omega_s = \pm \omega \qquad \text{and} \qquad k_i - k_s = \pm k \qquad (13\text{-}37)$$

where the $+$ refers to the absorption (annihilation) of a phonon (anti-Stokes process) and the $-$ sign refers to the emission (creation) of a phonon (Stokes process). Figure 12-13 indicates these inelastic processes. Note that Raman scattering is more complicated than infrared absorption in which one photon is converted directly to a phonon.

Although the conservation equations are the same as for neutron scattering, the Raman scattering mechanism is different from that of infrared absorption or neutron scattering. The mechanism involves the change of electronic polarizability due to atom displacements. We give a very simple classical model to show how the Raman effect arises. Consider the incident light to have a time dependence $E(t)_i = E_i \cos(\omega_i t)$. This will impinge on a crystal and result in a time varying polarization $P(t) = \alpha\, E(t)_i$. Now consider α to be modulated by the time varying lattice vibrations and expand this latter variation in powers of the displacement from equilibrium

$$\alpha(t) = \alpha_0 + \left(\frac{\partial \alpha}{\partial u}\right)_0 u + \left(\frac{\partial^2 \alpha}{\partial u^2}\right)_0 \frac{u^2}{2} + ...$$

$$\equiv \alpha_0 + \alpha_1 u + \alpha_2 u^2 ... \tag{13-38a}$$

The subscript 0 means that the term is evaluated at zero displacement and u is the displacement due to the phonons. The displacement has a time variation, $u = u_0 \cos \omega_0 t$, where ω_0 is the phonon frequency. Then the time varying polarization, $P(t)$, is a source in Maxwell's equations for radiation at the various appropriate frequencies. The α_0 term in Eq. 13-38a yields a polarization that gives rise to **Rayleigh scattering**; this radiation is not frequency shifted from ω_L and is not discussed here. The second term involves α_1 and gives rise to a polarization that is shifted in frequency from ω_i. We have

$$P(t) = (\alpha_1 u_0 E_i)[\cos \omega_i t\,][\cos \omega_0 t\,]$$

$$= (\alpha_1 u_0 E_i/2)[\cos (\omega_i + \omega_0)t + \cos (\omega_i - \omega_0)t\,] \tag{13-38b}$$

Thus, we see how the anti-Stokes and Stokes scattered lines arise from phonons modulating the electronic polarizability. The scattered light intensity is proportional to $[P(t)]^2$ and hence to E_i^2.

Since the frequency of visible light is so much larger than phonon frequencies, we must use very sharply defined incident light in conjunction with a good monochromator in order to observe small shifts. This effect was first observed in the late 1920s by using the sharp, strong lines from a mercury arc lamp as the incident light. However, this field is having a renaissance since laser light has proven to be the perfect tool. Laser radiation is intense, highly monochromatic, unidirectional, and can be highly polarized. The polarization is rather important since a detailed derivation of Raman scattering shows that there are useful selection rules for the various lattice vibrations. For example, LO modes can be measured *directly*.

There is one more interesting aspect of the Raman and infrared selection rules. In crystals that have a center of symmetry, such as the NaCl, CsCl, and diamond structures (but not ZnS), phonons that are infrared active (observable in a one-phonon infrared measurement) cannot be Raman active. This is known as the **exclusion principle**. (This is not related to the Pauli exclusion principle.) However, in

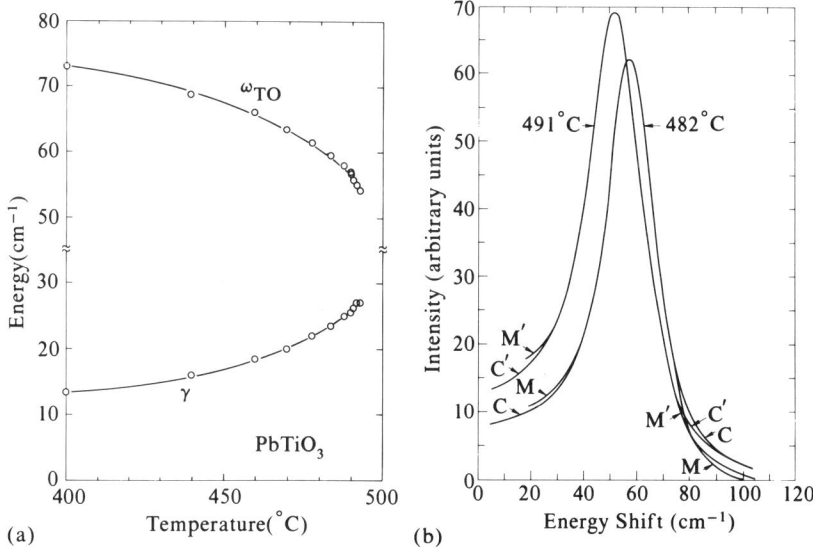

Fig. 13-6 (a) The temperature dependence of the frequency of the lowest TO mode in PbTiO$_3$, and its damping constant, γ. (b) The experimentally measured Raman line, M, and a calculated damped harmonic oscillator fit, C, at two temperatures near T_C. [G. Burns and B. A. Scott, Phys. Rev. B7, 3088 (1973).]

non-centrosymmetric crystals modes may be simultaneously infrared and Raman active.

Figure 13-6 shows data from a mode in the tetragonal, ferroelectric material PbTiO$_3$ (BaTiO$_3$ crystal structure). In the ferroelectric phase this material is non-centrosymmetric, so the mode shown is simultaneously infrared and Raman active. The results shown come from Raman measurements, which are usually more accurate than infrared results. The transverse optic mode ω_{TO} results refer to the mode that has an oscillating dipole moment perpendicular to the c-axis of this tetragonal crystal. The plot shows ω_{T0} and γ from fits of the damped harmonic oscillator model. Figure 13-6b shows the actual measured and calculated line shapes at two temperatures. The fit is excellent, showing the usefulness of the model. The line abruptly disappears at 492°C, because at this temperature the material becomes cubic and the phonon is no longer Raman active.

We make two last points about the experimental observation of Raman lines. First, the ratio of the intensity of the anti-Stokes to Stokes line is given by Eq. 12-27. Thus, even at T = 0°K a Raman line can be observed (with the emission of a phonon). Second, the preceding discussion and Fig. 12-13 apply to a **one-phonon process.** That is, in the scattering only one phonon is emitted or absorbed. A Raman scattering event can occur where two phonons, ω_1 and ω_2, are emitted. The conservation equations, Eq. 13-37, would then be $\omega_i = \omega_s + \omega_1 + \omega_2$ and $\mathbf{k}_i = \mathbf{k}_s + \mathbf{k}_1 + \mathbf{k}_2$. Of course, being a higher order process, the matrix element for a **two-phonon Raman process** is smaller than that of a one-phonon process. However, the k-selection rule now allows a huge number of states to participate in this second-order process. For example, for $\omega_1 = \omega_2$ phonons with equal and opposite crystal momentum can satisfy the selection rule. That is, their k-values can be very large (compared to k_s) and correspond to modes at the Brillouin zone boundary where the density of states is very much larger than at $k \approx 0$. This large density of states makes the two-phonon Raman process observable in many crystals. Similar comments apply to the infrared situation. (See the Notes.)

13-10d Problems The field of Raman and infrared spectroscopy is an active one. All experimentally observed lines cannot be fit by the simple damped harmonic oscillator model. Even when they can, γ obtained from Raman measurements is sometimes different from that determined for the same phonons by infrared results. The fact that such deviations are occasionally found is, of course, not a reason to discard a fine, simple model that is useful in most cases. Rather it prompts us to look at these special situations more carefully.

13-11 Polaritons

13-11a Theory So far in Part B, when discussing transverse optic vibrations, we have not considered the propagation of electromagnetic radiation in the crystals, so have not used Maxwell's equations. At most, in Section 13-9a, the electric field had to be transverse to its propagation direction. This of course leads to the ω_0 being identified with the TO vibration in Eq. 13-33. In this section we consider how light interacts with the mechanical vibrations. The solutions are easy to find since the basic equations have already been obtained. The results are surprising, leading to the notion of polaritons, which, although predicted in 1951 (by Huang), were not found experimentally until 1964.

We did use Maxwell's equations in Section 13-9b in discussing the longitudinal vibrations. There, Maxwell's equations yield $\varepsilon \mathbf{k} \cdot \mathbf{E} = 0$, which implies that modes with $\mathbf{k}$ parallel to $\mathbf{E}$ occur for $\varepsilon = 0$, Eq. 13-34a. However, now we shall consider only transverse vibrations.

For transverse optical waves the same equation, $\varepsilon \mathbf{k} \cdot \mathbf{E} = 0$, is the basis of the solution for the coupling of the TO lattice vibrations to the transverse electromagnetic field. In this paragraph we qualitatively discuss the problem so that it can be appreciated better. Figure 13-7a shows the dispersion curves of a photon, a TO phonon and a LO

CHAPTER 13   OPTICAL PROPERTIES OF CRYSTALS   477

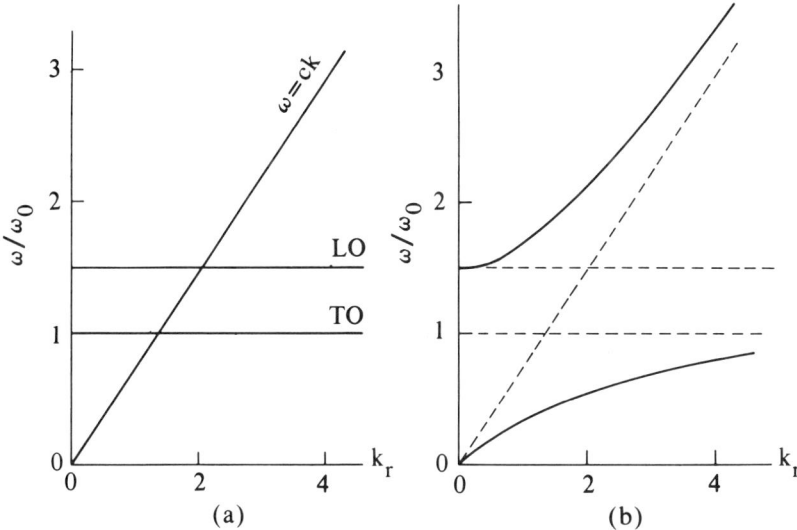

Fig. 13-7 (a) The ω vs. k, in reduced units, for LO, TO, and light in a crystal at very small k for no coupling of the TO modes with the light. [The reduced k is $k_r = k/(\omega_0/c)$.] (b) The same but taking coupling into account. The thick lines are the transverse wave solutions.

phonon for $k \approx 0$. The Brillouin zone boundary is extremely far to the right at $\pi/a \approx 10^8$ cm^{-1}, while this plot ends at about 10^3 cm^{-1}. The curves for the TO and LO modes appear horizontal, as in the figures in Chapter 12, since for $k \approx 0$, $d\omega/dk \approx 0$. On the other hand, ω vs. k for the photon has a slope $\omega/k = c/n_1 = c/\sqrt{\varepsilon_\infty}$, which for $\omega = 100$ cm^{-1} and $\varepsilon_\infty = 4$ gives $k = 1257$ cm^{-1}, a small value indeed. In this region of k, these dispersion curves appear to cross. Thus, for the uncoupled problem, ω vs. k for the transverse vibrations crosses that for the transverse electromagnetic field. However, these two waves are coupled in a fundamental manner (one that cannot be thought of as going to zero, if some coupling parameter goes to zero). (The problem of LO modes interacting with transverse light has already been solved; they do not interact and hence the flat ω_ℓ vs. k curve will persist.)

Thus, the important problem is: how does the transverse electromagnetic radiation interact with the transverse oscillating dipoles in the TO modes? For transverse waves, Maxwell's equations lead to Eq. 13-17, which relates ω to k via the complex dielectric constant. We now study the solutions, taking $\gamma = 0$, which makes ε and therefore k real. For a given k-value, the frequencies are solutions of

$$\omega = c k/[\varepsilon(\omega)]^{1/2} \qquad (13\text{-}39\text{a})$$

We write $\varepsilon(\omega)$ instead of ε as a strong reminder that the dielectric constant is a function of the frequency. In fact, it is given by Eq. 13-35e, which is repeated here for convenience.

$$\varepsilon(\omega) = \varepsilon_\infty + \frac{[\varepsilon(0) - \varepsilon_\infty]\omega_0^2}{\omega_0^2 - \omega^2} \qquad (13\text{-}39b)$$

Figure 13-3 shows a plot of this function with $\gamma \neq 0$. With $\gamma = 0$ it goes to $+\infty$ as ω approaches ω_0 from below and goes to $-\infty$ as ω approaches ω_0 from above. As can be seen from Eq. 13-39a, there are no solutions for real frequencies when $\varepsilon(\omega)$ is negative. When $\varepsilon(\omega)$ is positive, for each value of k there are two values of ω that will solve this equation. Figure 13-7b shows the solutions. The dark solid lines are the solutions for *transverse waves* as a function of k.

There are many remarkable properties associated with these solutions. First, there is a gap in frequency where there are no allowed waves. Incident radiation with these frequencies will be reflected as already shown in Fig. 13-3. (Note that this gap has nothing to do with the translational symmetry of the lattice.) Second, the top of the gap, where transverse waves are again allowed, corresponds exactly to the longitudinal frequency, $\omega = \omega_{LO}$, since it is just at the value of ω where $\varepsilon(\omega) = 0$ that the transverse waves can again propagate. Third, the group velocity for large k and ω is $d\omega/dk = c/\sqrt{\varepsilon_\infty}$, that is, the same as that of light in the solid, as expected. However, at low k and ω, $d\omega/dk = c/\varepsilon(0)^{1/2}$ and, since $\varepsilon(0) > \varepsilon_\infty$, the group velocity is slower. This can be seen in Fig. 13-7b; the lower curve starts with this slope and then bends over to approach ω_0 from below. Thus, we have found the normal modes of this coupled electromagnetic and crystal vibration system. For this very narrow range of k-values, the transverse modes are neither bare photon modes nor bare transverse optical modes but two intimate mixtures of these called **polariton modes**. The quanta in both modes are called **polaritons** (not photons and not transverse optical phonons), and the ω vs. k for the transverse waves, the solid lines in Fig. 13-7b, are usually called **polariton dispersion curves**.

13-11b Experiment Given the rich detail in Fig. 13-7b, it would be nice to observe polaritons directly. Of course, it could be argued that polaritons are being observed in infrared reflectivity measurements (Fig. 13-5). In fact, when electromagnetic waves are propagating in a crystal they propagate only as polaritons. There is nothing else! These arguments are perfectly fine but more believable after polaritons are *directly* observed.

Rather clear and direct experimental confirmation of the polariton character of propagating transverse electromagnetic radiation in crys-

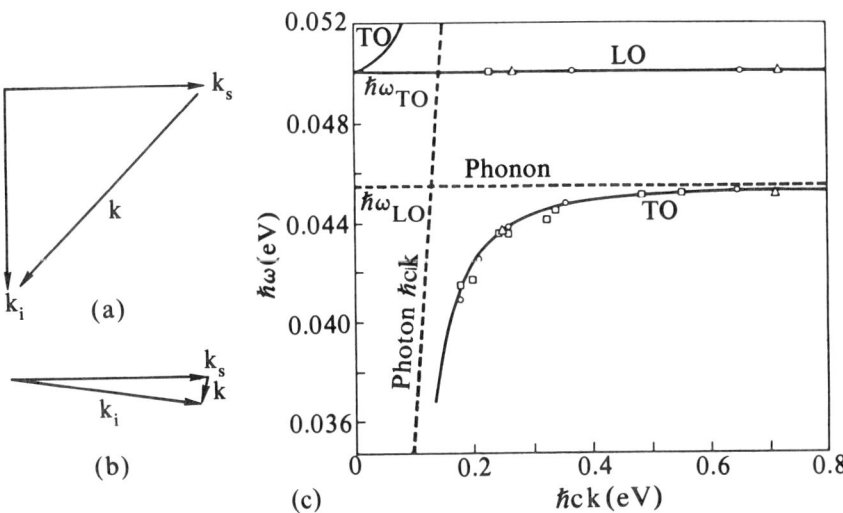

Fig. 13-8 (a) A wave vector diagram for the usual right angle Raman scattering. (b) The same but for forward Raman scattering, showing how small the k value of the phonon can become. (c) A plot of the observed energies and wave vectors of polaritons in GaP. Also shown are the measured LO energies. The solid lines are the theoretical. [C. H. Henry and J. J. Hopfield, Phys. Rev. Letters **15**, 964 (1965).]

tals has been found, and it is indicated in Fig. 13-8. First, the k-vector diagram for right angle Raman scattering is shown in Fig. 13-8a. The emitted phonon wave vector is approximately $\sqrt{2}$ times larger than that of the incident or scattered light. This is the normal arrangement for Raman experiments. For visible light ($\lambda = 6000$ Å) k $= 2\pi/\lambda \approx 10^5$ cm^{-1}, and is out (and off the page) on the flat part of the lower polariton curve in Fig. 13-7b. Thus, the measured frequency is essentially that of the TO phonon. Second, Fig. 13-8b shows that the wave vector of the scattering modes can be significantly smaller in "forward" Raman scattering (i.e., $\mathbf{k}_i$ and $\mathbf{k}_s$ are in practically the same direction). Using forward Raman scattering the bend in the polariton curve can be measured directly. Figure 13-8c shows the first experimental results. These were obtained in GaP, a material with the cubic zinc blende structure. The data points correspond to different (small) angles between $\mathbf{k}_i$ and $\mathbf{k}_s$. The agreement between theory and experiment is excellent.

An interesting sidelight of polariton measurements is that $\varepsilon_1(0)$ can be obtained from the slope of the lower branch of the curve. The polariton ω vs. k curve can be measured down to $\omega \approx 10$ cm^{-1}, or lower, depending on the optical quality of the crystal. Thus $\varepsilon(\omega)$ at 10 cm^{-1} can be obtained. This is a high frequency compared to the normal capacitance bridge measurements. At this high frequency there are essentially no extra contributions due to impurities, rotating dipoles, and so on. For example, by measuring the temperature dependence of $\varepsilon_1(0)$ obtained in this way, in ferroelectric materials and by using the Lyddane Sachs Teller relationship, we can learn a good deal about the modes and the differences between $\varepsilon_1(0)$ and the values measured by capacitance techniques. See the Notes for references.

13-12 A Microscopic Model

We now sketch an atomic model for the long wavelength vibrations. This requires a better treatment of the electric fields that are actually seen by the ions (local field problem) and requires that the electronic polarizabilities be taken into account from the beginning. From this model formulas for ε in which ε_∞, instead of 1, are obtained automatically.

13-12a Local field problem In defining the various optical constants, we have always used the applied electric field E. This is quite reasonable since we want macroscopic constants applicable to volumes that contain many atoms. However, for a microscopic theory we need the electric field seen by an individual electron or at an individual atomic site. This is called the local or effective field, E_{loc}.

It is clear that E_{loc} can be different from the applied electric field E. When E is applied, each ion develops an individual dipole moment, which in turn sets up a dipolar electric field. E_{loc} is E plus the electric field due to the individual dipoles. This latter contribution varies rapidly over atomic dimensions and can be large.

It can be shown by elementary considerations for a cubic structure that E_{loc}, at either a positive or negative ion site that has cubic point symmetry, is given by

$$E_{loc} = E + (4\pi/3)P \qquad (13\text{-}40)$$

(See the Notes for references.) This is what would be expected, namely, the local field is the applied field plus a contribution that is proportional to the size of the induced dipole moments. This same result is valid for a random arrangement of molecules as found in a liquid or a gas. In complicated structures the coefficient may be different from $4\pi/3$ and varies with position within the unit cell. For example, for the oxygen sites in $BaTiO_3$ the factor is approximately eight times larger than $4\pi/3$. (See the Notes.) This leads to an enhancement of certain properties. However, we consider only structures where all the ions are at sites of cubic symmetry so Eq. 13-40 is appropriate.

CHAPTER 13   OPTICAL PROPERTIES OF CRYSTALS   481

13-12b The calculation   Once the effective field is known, the rest of the task is straightforward and we follow Born and Huang.

As in Section 13-9, let $u_\pm$, $\pm Q$, and $a_\pm$ be, respectively, the displacements, effective charges, and atomic polarizabilities of the positive and negative ions. (For example, the calculation applies to crystals with the NaCl and CsCl structures, both have two ions per primitive unit cell.) The first two quantities were considered previously; now we allow for electronic polarizabilities of the ions and we shall use E_{loc}. The effective dipole moments of the two ions are

$$Qu_+ + a_+ E_{loc}$$
$$-Qu_- + a_- E_{loc}$$
(13-41a)

Notice that each ion sees the same local electric field. The macroscopic polarization **P**, the dipole moment per unit volume, is given by the sum of the individual contributions,

$$\mathbf{P} = \frac{1}{V_p}[Q(u_+ - u_-) + (a_+ + a_-)E_{loc}] \tag{13-41b}$$

$$= \frac{1}{1 - \left(\frac{4\pi}{3}\right)\left(\frac{a_+ + a_-}{V_p}\right)}\left[Q\left(\frac{u_+ - u_-}{V_p}\right) + \left(\frac{a_+ + a_-}{V_p}\right)\mathbf{E}\right]$$
(13-41c)

where V_p is the volume of the primitive unit cell and $E_{loc} = E + 4\pi P/3$ is used to eliminate E_{loc} in Eq. 13-41b. Equation 13-41c is the desired relation; it relates **P** to the relative displacements and the *external* field.

Now we would like to find a dynamic equation for the displacements $u_\pm$ comparable to Eq. 13-33b. When the positive and negative ions are displaced relative to each other, there is a force due to the overlap of the ions that keeps them apart. For small displacements the forces are $-\alpha(u_+ - u_-)$ and $\alpha(u_+ - u_-)$ for the positive and negative ions, respectively (i.e., equal and opposite as we expect from Newton's law of reaction). Thus, the equations of motion for the two types of ions are

$$m_+ \ddot{u}_+ = -\alpha(u_+ - u_-) + QE_{loc}$$
$$m_- \ddot{u}_- = -\alpha(u_+ - u_-) - QE_{loc}$$
(13-42a)

This equation, except for E_{loc}, is the same as Eq. 13-32. Using $r \equiv u_+ - u_-$, we arrive at Eq. 13-42b. Equation 13-42c is obtained by using $E_{loc} = E + 4\pi P/3$ and eliminating P via Eq. 13-41c.

$$\mu \ddot{\mathbf{r}} = -\alpha \mathbf{r} + Q\mathbf{E}_{loc} \quad (13\text{-}42\text{b})$$

$$= \left[-\alpha + \frac{\left(\frac{4\pi}{3}\right)(Q^2/V)}{1 - \left(\frac{4\pi}{3}\right)\left(\frac{a_+ + a_-}{V_p}\right)} \right] \mathbf{r} \quad (13\text{-}42\text{c})$$

$$+ \left[\frac{(Q/V^{1/2})}{1 - \left(\frac{4\pi}{3}\right)\left(\frac{a_+ + a_-}{V_p}\right)} \right] \mathbf{E}$$

Notice how much more complicated Eq. 13-42c is than its counterpart, Eq. 13-33a. The differences arise because we now have considered the electronic polarizability, $a_\pm$, and because $\mathscr{E}_{loc}$ couples the ionic displacements and the charge.

Equations 13-41c and 13-42c constitute our final result and a great deal can be learned from them. The coefficient of the $\mathbf{r}$ term in Eq. 13-42c is the transverse optic frequency (as in Eq. 13-33b). The coefficient of $\mathbf{E}$ in Eq. 13-41c is just χ and since $\varepsilon = 1 + 4\pi\chi$ this term is equal to $(\varepsilon_\infty - 1)/4\pi$. Thus,

$$\omega_0^2 = \frac{\alpha}{\mu} - \frac{(4\pi/3)(Q^2/V_p)}{1 - (4\pi/3)(a_+ + a_-)/V_p} \quad (13\text{-}43\text{a})$$

$$\varepsilon_\infty = 1 + \frac{4\pi(a_+ + a_-)/V_p}{1 - (4\pi/3)(a_+ + a_-)/V_p} \quad (13\text{-}43\text{b})$$

The result for ω_0 is used here and that for ε_∞ in the next section.

Compare this frequency, Eq. 13-43a, with the simpler transverse optic frequency, $\omega_0^2 = \alpha/\mu$, calculated before (Eq. 13-33a). The additional mechanisms all tend to *reduce* ω_0, because each one tends to cancel the applied field. The reduction amounts to 20-30% in alkali halide crystals, which is significant.

The $4\pi/3$ value in Eq. 13-43a comes directly from the $4\pi/3$ in the local field in Eq. 13-40. For some other crystal structure this value could be considerably larger. In fact, Slater (see the Notes) has shown that in $BaTiO_3$, some of the ions, that are not at sites with cubic symmetry, see a very much larger local field (eight times larger) than expected from Eq. 13-40. At any rate, the $(a_+ + a_-)/V_p$ term in the denominator is considerably enhanced, increasing the magnitude of the second term on the right of Eq. 13-43a. This results in a very significant decrease in ω_0 from the $(\alpha/\mu)^{1/2}$ value. This can lead to

CHAPTER 13   OPTICAL PROPERTIES OF CRYSTALS   483

Table 13-2 Values of the ratio of the calculated (Eq. 13-45a) to observed compressibility (K_c/K_o) and ratio of the calculated effective charge Q (Eq. 13-45b) to the formal ionic charge Z. (From Born and Huang.)

	K_c/K_o	Q/Z		K_c/K_o	Q/Z
LiF	1.0	0.87	TlCl	0.51	1.08
NaF	0.83	0.93	CuCl	0.85	1.10
NaCl	0.99	0.74	CuBr	0.72	1.00
NaBr	1.13	0.69	MgO	0.47	0.88
KCl	0.96	0.80	CaO	..	0.88
KBr	0.95	0.76	SrO	..	0.58
KI	0.99	0.69	ZnS	0.98	0.48
RbCl	0.89	0.84			
RbBr	0.83	0.82			
RbI	0.66	0.89			
CsCl	0.87	0.84			
CsBr	0.87	0.79			

very small values of TO mode frequencies and, via the Lyddane Sachs Teller relationship, to anomalously large values of $\varepsilon_1(0)$. This idea will be discussed in Chapter 14 on ferroelectrics.

13-12c Szigeti charges   It would be useful to compare theoretical calculations to experimental data. Equation 13-43a contains ω_0 and the electronic polarizabilities, which are measurable, but α, μ, and Q are unknowns. We would like to eliminate them or find an equation with just one unknown so it can be evaluated.

The procedure to do this is straightforward and discussed in detail in Born and Huang as well as by Szigeti (see the Notes). The coefficient of $(\mathbf{u}_+ - \mathbf{u}_-)$ in Eq. 13-41c or the coefficient of $\mathbf{E}$ in Eq. 13-42c is related to the strength of the pole in Eq. 13-39b

$$\frac{[\varepsilon(0) - \varepsilon_\infty]\omega_0^2}{4\pi} = \frac{(Q^2/\mu V_p)}{[1 - (4\pi/3)(a_+ + a_-)/V_p]^2} \quad (13\text{-}44a)$$

$$\frac{\alpha}{\mu} = \omega_0^2 \left(\frac{\varepsilon_1(0) + 2}{\varepsilon_\infty + 2} \right) \quad (13\text{-}44b)$$

The second equation is obtained from Eq. 13-43a by using the first equation to eliminate the polarizabilities. The spring constant α can be calculated as in Section 7-4. The resulting compressibility is

$$\frac{1}{K} = \frac{zd_0^2}{3V_p}\alpha = \frac{z\mu d_0^2}{3V_p}\left(\frac{\varepsilon_1(0) + 2}{\varepsilon_\infty + 2} \right)\omega_0^2 \quad (13\text{-}45a)$$

where the first equality is the same as Eq. 7-7 and the next is obtained by using Eq. 13-44b. In this equation z equals the number of nearest neighbors (six or eight for the NaCl and CsCl structures), and d_0 the distance between the nearest neighbors at equilibrium. This relation, Eq. 13-45a, is one of the **Szigeti equations** and all the quantities in it are measurable. Essentially it relates the TO frequency to the compressibility. Table 13-2 shows the ratios of the calculated values of the compressibility (Eq. 13-45a) divided by the observed values. Results close to one are obtained for most of the alkali halides. For the other materials shown, it could be argued that they have a fair degree of covalent bonding, but actually it is difficult to determine the predominant cause for the deviations.

Another **Szigeti equation** can be found by using Eq. 13-43b to eliminate the electronic polarizabilities from Eq. 13-44a. We obtain

$$\varepsilon_1(0) - \varepsilon_\infty = \left(\frac{\varepsilon_\infty + 2}{3}\right)^2 \frac{4\pi Q^2}{\mu \omega_0^2 V_p} \qquad (13\text{-}45b)$$

From this equation, values of the dynamic charge Q can be determined from experimentally measured quantities. Table 13-2 shows the ratio of Q to the formal ionic charge; the deviations from unity are impressive and the reasons for this deviation are not clear. In the binding energy calculations for ionic crystals (Chapter 7) the full formal ionic charge is used and the agreement with experimental values is good. However, effective charges considerably different from the formal charge must be used to get agreement with these different experimental results. This is disturbing and not totally resolved; it is thought to be associated with the fact that Q in Eq. 13-45b is a dynamic charge since it arises when considering the movement of ions. The charge in the binding energy calculations is a static charge since it arises from ions that are not in motion. Due to overlap effects, when the ions move, their distortions and those of the neighboring ions cause Q to differ from unity. (See the Notes.)

13-13 Clausius-Mossotti (Lorenz-Lorentz) Equations

Equation 13-43b applies for frequencies well above the resonant frequency of the mechanical system (i.e., above $\omega_0 = \omega_{TO}$). It can be rewritten as

$$\frac{\varepsilon_\infty - 1}{\varepsilon_\infty + 2} = \frac{4\pi}{3}\left(\frac{a_+ + a_-}{V_p}\right) = \frac{4\pi}{3} \Sigma_i N_i a_i/V \qquad (13\text{-}46a)$$

OPTICAL PROPERTIES OF CRYSTALS

Table 13-3 The electronic polarizability of many common ions. The Pauling values are from Proc. Roy. Soc. (London) **A114**, 181 (1927); the Tessman, Kahn and Shockley are from Phys. Rev. **92**, 890 (1953). Multiply each value by 10^{-24} cm^3.

				He	Li$^+$	Be^{2+}	B^{3+}	C^{4+}
Pauling				0.201	0.029	0.008	0.003	0.0013
TKS					0.03			
	O^{2-}	F$^-$	Ne	Na$^+$	Mg^{2+}	Al^{3+}	Si^{4+}
Pauling	3.88	1.04	0.390	0.179	0.094	0.052	0.0165
TKS	(2.4)	0.652		0.41			
	S^{2-}	Cl$^-$	A	K$^+$	Ca^{2+}	Sc^{3+}	Ti^{4+}
Pauling	10.2	3.66	1.62	0.83	0.47	0.286	0.185
TKS	(5.6)	2.97		1.33	1.1		(0.19)
	Se^{2-}	Br$^-$	Kr	Rb$^+$	Sr^{2+}	Y^{3+}	Zr^{4+}
Pauling	10.5	4.77	2.46	1.40	0.86	0.55	0.37
TKS	(7.)	4.17	1.981.	6			
	Te^{2-}	I$^-$	Xe	Cs$^+$	Ba^{2+}	La^{3+}	Ce^{4+}
Pauling	14.0	7.10	3.99	2.42	1.55	1.04	0.73
TKS	(9.)	6.44		3.34	2.5		

The second equality is the obvious extension of the first equality, which comes exactly from Eq. 13-43b; in it N_i/V is the number of type i atoms per unit volume, which have polarizabilities a_i. However, there is an important assumption implicit in the second equality: the polarizabilities are additive and are the same in different compounds. (Actually, this assumption is implicit in Eq. 13-41.) Such assumptions are usually fairly good for ionic crystals.

This equation is usually written for Avogadro's number of molecules, N_A, which has a molecular weight, m, and density ρ

$$\frac{m}{\rho} \frac{\varepsilon_\infty - 1}{\varepsilon_\infty + 2} = \frac{4\pi}{3} N_A a_m \qquad (13\text{-}46b)$$

Here a_m is the polarizability of the atoms that make up a molecular formula unit, where additivity is assumed, as stated previously. The quantity in this equation is called the **molar polarizability**. The equation is called the Lorenz-Lorentz equation with $\varepsilon_\infty (= n_1^2)$. However, the equation is more generally applicable. The same result can be derived with $\varepsilon_1(0)$ replacing the ε_∞, in which case it is called the Clausius-Mossotti equation and the frequencies are well below ω_0. It is a convenient result since the quantities on the left side are easily measurable and the result can be derived for any ε just assuming that $E_{loc} = E + 4\pi P/3$. (See the Problems.) Since this value of E_{loc} is appropriate for random systems, the results in Eq. 13-46b are appropriate for liquids and gases. In fact, it is just the molar polarizability that is plotted in Fig. 11-12 for the polar and nonpolar methane compounds; additivity is being assumed.

Tessman, Kahn and Shockley (see the Problems) have looked into this additivity problem for many different ionic crystals where all the ions see cubic symmetry ($E_{loc} = E + 4\pi P/3$ should apply). They, and others, have developed a consistent set of polarizabilities for the various ions (Table 13-3). However they did find that the O^{2-} ion polarizability is dependent on the compound in which it is found.

Part C — Free Carrier Absorption

13-14 Introduction

In Parts C and D optically induced electronic transitions are discussed. In Part C we consider free carrier absorption due to intraband transitions within an unfilled band. For example, in a metal the conduction electrons may half fill a band. Free carrier absorption takes place when photons promote electrons to a higher energy state within the same band. The same statement applies to an n-type semiconductor where there may typically be 10^{13} to 10^{18} cm^{-3} electrons at the bottom of an otherwise empty conduction band. For a p-type semiconductor the valence band is nearly completely filled so there are holes at the top of the band. We speak of a hole transition from the top of the valence band to a deeper state in the band. (Actually within the valence band it is an electron that is excited from deep within the band to a vacant state near the top of the band.)

An easy classical treatment of free carrier absorption agrees with many aspects of the quantum treatment and experiment, so the classical theory is discussed; the mathematics is just a special case of the results in Part B, although the physics is rather different. This simplicity motivates treating free carriers before interband transitions. Note that in Section 9-5 free carrier absorption already was discussed. Here the same considerations apply but we shall go into more detail.

Free carrier electrons move over many interatomic distances. Therefore they respond to the external applied field rather than to any local field. Thus, there are no local field corrections.

Free carrier effects are closely related to plasma effects. A **plasma** is an ionized gas in which the free electrons typically have a high mobility compared to the ions. The different mobility is due to the different masses. Since we treat the optical effects of the free carriers within the Drude free electron model (i.e., a free electron gas) these results are applicable also to a true plasma. In fact, some of the words (i.e., plasma frequency) are used for both the free electron case and a plasma.

In Part D we shall discuss transitions between the various bands (interband transitions) in semiconductors and metals. This is similar to transitions between the sharp energy levels in atoms.

13-15 Oscillator Model

13-15a The model In the classic Drude model electrons are treated as a gas of free particles. They move very rapidly in random directions. An electric field **E** applies a force $-e\mathbf{E}$ and hence a small extra "drift velocity" $\mathbf{v}_d$ is superimposed. (See Section 9-2.) However, the electrons have a mean scattering time, τ, which causes $\mathbf{v}_d$ to relax to zero if **E** is turned off. The crystal plays no direct part in this model, but does cause the electron mass, m, to be replaced with an effective mass m* and it also provides a mechanism for scattering.

We quantify these ideas. To the electron gas, apply an electromagnetic wave with **E** transverse to the propagation direction. This causes an acceleration of the drift velocity.

$$m^* d\mathbf{v}_d/dt = -e\mathbf{E} \qquad (13\text{-}47a)$$

During a small time Δt, an electron's average velocity changes by $\Delta \mathbf{v}_d = (-e\mathbf{E}/m^*)\Delta t$. The probability of a collision is $\Delta t/\tau$. As discussed in Section 9-5a, this is equivalent to a "friction" term, which reduces $\Delta \mathbf{v}_d$ by $\Delta t \mathbf{v}_d/\tau$. Therefore, the equation of motion of $\mathbf{v}_d$ is

$$m^* \frac{d\mathbf{v}_d}{dt} + \frac{m^* \mathbf{v}_d}{\tau} = -e\mathbf{E} \qquad (13\text{-}47b)$$

For example, if a steady field **E** is turned off at $t = 0$, then $\mathbf{v}_d$ returns to zero exponentially $\mathbf{v}_d(t) = \mathbf{v}_d(0) \exp(-t/\tau)$, as expected.

Equation 13-47b can be put in the familiar form of a Lorentz oscillator by using $\mathbf{v}_d = d\mathbf{r}_d/dt$ where $\mathbf{r}_d$ is the mean electron displacement resulting from the applied field.

$$m^* \frac{d^2 \mathbf{r}_d}{dt^2} + \frac{m^*}{\tau} \frac{d\mathbf{r}_d}{dt} = -e\mathbf{E} \qquad (13\text{-}47c)$$

Notice that this equation is of the same form as that of the damped harmonic oscillator, Eq. 13-26, provided that ω_0 is zero and the damping constant $\gamma = \tau^{-1}$. Thus, the solutions for the driven transverse modes are already known and the results for the dielectric constant and the conductivity are

Table 13-4 For various electron densities, values are listed for the angular frequency ω_p, in units of sec^{-1} and cm^{-1}, which we have been using extensively. Also listed are the corresponding plasma wavelengths λ_p. (Given an angular frequency ω_p, an equivalent wavelength always can be defined via, $2\pi\nu = \omega$ and $\nu\lambda = c$ so $\lambda_p = 2\pi c/\omega_p$. Also the wave number unit is just λ_p^{-1}.)

N/V (cm^{-3})		10^{10}	10^{14}	10^{18}	10^{22}
ω_p	sec^{-1}	5.7×10^9	5.7×10^{11}	5.7×10^{13}	5.7×10^{15}
ω_p	(cm^{-1})	3.3×10^{-2}	3.3	3.3×10^2	3.3×10^4
λ_p	(cm)	33	0.33	3.3×10^{-3}	3.3×10^{-5}

$$\varepsilon = 1 - \frac{(4\pi e^2 N/m^*V)}{\omega^2 + i\omega/\tau} = 1 - \frac{\omega_p^2}{\omega^2 + i\omega/\tau} \qquad (13\text{-}47\text{d})$$

where $\omega_p^2 \equiv 4\pi e^2 N/m^*V$

$$\sigma = \frac{\sigma_0}{1 - i\omega\tau} \qquad \text{where } \sigma_0 \equiv \tau e^2 N/m^*V \qquad (13\text{-}47\text{e})$$

ω_p is called the **plasma frequency**. These results are for the free electrons. We could argue that the interband transitions at higher frequencies give a contribution to ε, which would cause the 1 in Eq. 13-47d to be replaced by ε_∞. We shall not do that here; however, the changes in the equations are trivial if an ε_∞ is added. In Eq. 13-47e, notice that the result for the conductivity is just the same as in Eq. 9-13a, and Fig. 9-2 shows what happens for the real and imaginary parts. (Equation 13-47d is the same as Eq. 9-18.)

Once ε is known as a function of frequency, all of the material parameters, such as χ, σ, and so forth, can be calculated. First, however, let us establish a "feel" for the order of magnitudes. Table 13-4 shows values of ω_p (and the corresponding wavelengths) for various values of the charge density, using the electron mass for m^*. Notice that for ordinary metals, where $N/V \approx 10^{22}$ cm^{-3}, the plasma frequency corresponds to 3000 Å or near ultraviolet, while for doped semiconductors the frequencies are in the infrared or even longer wavelengths. On the other hand τ can be evaluated via Eq. 13-47e from values of the dc conductivity if the charge density is known. For the good metals such as the alkali metals and copper, at room temperature, $\tau \simeq 5 \times 10^{-14}$ sec^{-1}. This corresponds to mean free paths of about 500 Å. Thus, τ varies between 5×10^{-14} sec and approximately 5×10^{-13} sec for the types of materials we consider in this chapter. At lower temperatures τ gets larger perhaps by three orders of magnitude in pure materials. So for good metals $\omega_p\tau \gg 1$; thus at the plasma frequency, or what is sometimes called the **plasma edge**, the damp-

CHAPTER 13   OPTICAL PROPERTIES OF CRYSTALS   489

ing is small compared to the frequency. At low temperatures, one can be in the region where $\omega\tau > 1$ even at rather low frequencies.

In **SI units** $\omega_p^2 = e^2N/\varepsilon_0 mV$, so the same result for ε as in Eq. 13-47d is appropriate but with this expression for ω_p.

13-15b   Frequency dependent parameters   Before looking at experimental results, we show the general result using "typical" parameters found in good metals. These results are similar to those in Fig. 13-3 but now $\omega_0 = 0$.

ε from Eq. 13-47d is rewritten here in its real and imaginary parts and the relations (Eq. 13-19a) with the index of refraction n and σ_1 are given. Also, the reflectivity at normal incidence is displayed.

$$\varepsilon_1 = 1 - \frac{\omega_p^2 \tau^2}{1+\omega^2\tau^2} = n_1^2 - n_2^2 \tag{13-47f}$$

$$\varepsilon_2 = \frac{\omega_p^2 \tau}{\omega(1+\omega^2\tau^2)} = 2n_1 n_2 = 4\pi\sigma_1/\omega \tag{13-47g}$$

$$R = \frac{(n_1 - 1)^2 + n_2^2}{(n_1 + 1)^2 + n_2^2} \tag{13-47h}$$

Figure 13-9 shows the results for some of these terms. Notice that at low frequencies the real part of the dielectric constant is negative and the imaginary part is positive. The different frequency regions are considered in the next section in conjunction with experimental data.

For poor metals (often alloys) that have electron densities $<10^{20}$ cm^{-3}, the free electron reflectivity is weak enough so that phonon features also are observed. To analyze such behavior, use is made of a dielectric constant that is the sum of a phonon part (Eq. 13-36a) and a free electron part (Eq. 13-47d). Thus,

$$\varepsilon = \varepsilon_\infty + \sum \frac{\Omega_i^2}{\omega_{0i} - \omega^2 - i\gamma_i \omega} - \frac{\omega_p^2}{\omega^2 + i\omega/\tau} \tag{13-47i}$$

This function is very successful in fitting data; *it is a good example of the utility of the damped harmonic oscillator model in that the total dielectric response is the sum of various different processes that result in a polarization when a field is applied.*

13-16 Experimental Results

13-16a Metals When discussing real systems we must always remember that while these equations apply only to free carrier absorption, other processes take place as well. Particularly at high energies, of the order of ω_p in metals, interband transition can occur. So care must be exercised in looking at actual data. We consider separately the very low frequency region, the mid-region (around ω_p), and the high frequency region (above ω_p).

(1) $\omega \ll \tau^{-1} \ll \omega_p$, so $\omega\tau \ll 1$. In this very low frequency region the preceding equations simplify to

$$\varepsilon_1 \approx -\omega_p^2 \tau^2 \qquad \varepsilon_2 \approx \omega_p^2 \tau/\omega = 4\pi\sigma_0/\omega \qquad (13\text{-}48)$$
$$n_1 \approx n_2 \approx (\varepsilon_2/2)^{1/2} \qquad R = 1 - (2/n_1) = 1 - (2\omega/\pi\sigma_0)^{1/2}$$

While the real part of the dielectric constant is large and negative, the imaginary part is even larger. This cannot be seen in Fig. 13-9 because the frequency is not low enough. Also n_1 is large so the reflectivity is essentially 100%. (This is similar to the phonon case between the TO and LO modes where ε_1 is negative.) However, there is a definite ω (or λ) dependence of R. Figure 13-10 shows that the experimental results are in excellent agreement with what is expected for $(1 - R)^2$ in Eq. 13-48; this equation is sometimes called the **Hagen-Rubens relation** and was found by them in 1903! Notice that they predict that $(1 - R)^2 \sigma_0 = 2\omega/\pi$, where σ_0 is the measured dc conductivity. This expression has no adjustable parameters yet, happily, good agreement with experiment is found for many metals.

The attenuation coefficient, Eq. 13-22, is $\alpha = 2\omega n_2/c$. For reasonable values of ω, the large value of n_2 will cause the electromagnetic wave quickly to be attenuated in a "good" metal. It is customary to talk in terms of a **skin depth** $\delta \equiv 2/\alpha$, which is the distance that the electric field penetrates before being attenuated by e^{-1}. Then $\delta = c/(2\pi\sigma_0\omega)^{1/2}$. This attenuation of the electric field is used in a practical manner to shield various kinds of sensitive equipment from external electromagnetic waves.

The **anomalous skin effect** arises in this frequency region when δ is smaller than the mean free path (mfp) of the electrons. Then the electrons that are excited by the radiation field are not in thermal equilibrium with the crystal because of insufficient collisions. In good metals, at low temperatures when τ gets very long, this results in mfp $\gg \delta$, leading to effects not predicted by the simple theory. A proper theory is complex due to **spatial dispersion** effects (i.e., the dependence of ε on the k-value). The effect arises whenever the relation between **D** and **E** is not exactly a local relation, that is, **D(r)** is

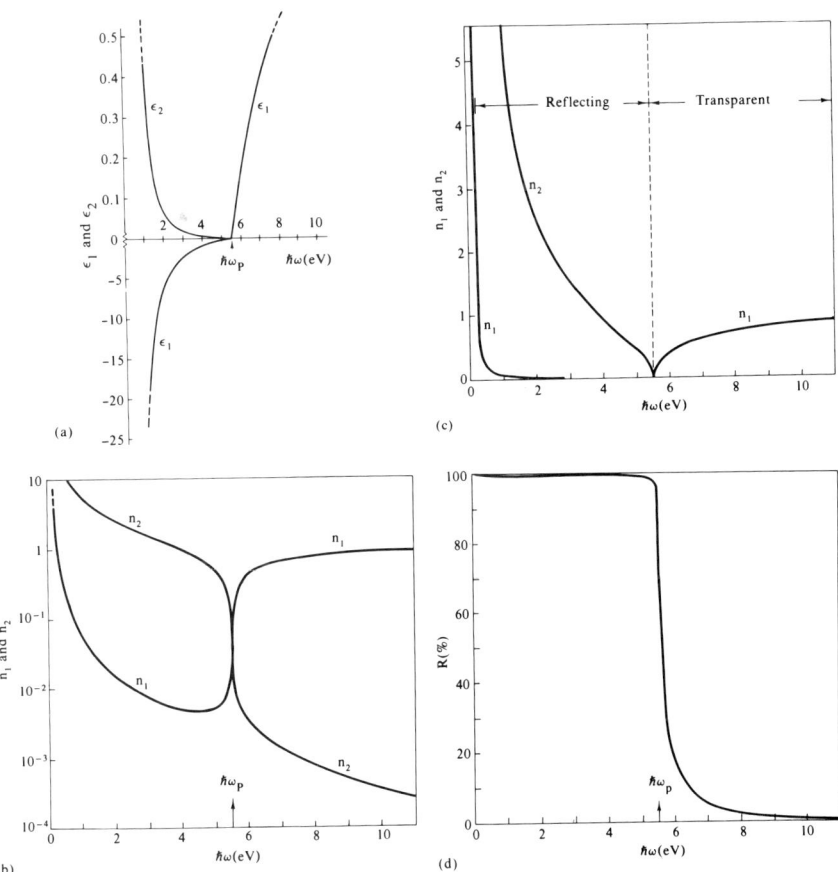

Fig. 13-9 The frequency dependence of the constants for free carriers calculated from the oscillator model. The curves are calculated for $\omega_p^2 = 30$ eV^2, and $\hbar\gamma = 0.02$ eV. [After Wooten referenced in the Notes.] See the comment on the ε_∞ effects in Section 13-6.

not determined solely by **E(r)**. In many crystals the spatial dispersion is small and can be taken care of by the addition of a term proportional to $\nabla \mathbf{E}$ in the expression for **D**. (This leads to optical rotation in, for example, quartz.) However, for metals where the anomalous skin effect is observed, spatial dispersion is extreme, and a proper theory complex.

(2) $\tau^{-1} < \omega < \omega_p$, so $\omega\tau > 1$. The values of the parameters in this region are shown in Fig. 13-9. As can be seen, ε_1 ($\approx -\omega_p^2/\omega^2$) is negative and much larger than ε_2 ($\approx \omega_p^2/\omega^3\tau$). This is the reverse of what occurs in the low frequency region. Thus, in this region $n_2 \gg n_1$. The reflectivity is still very close to 100%.

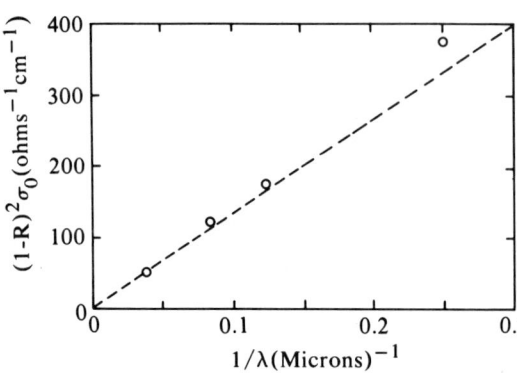

Fig. 13-10 A comparison between the measured and calculated values of $(1 - R)^2 \sigma_0$ of constantan. [From E. Hagen and H. Rubens, Ann Phys. **14**, 986 (1904). See the paper by Givens in the Notes.]

Figure 13-11 shows ε_1, ε_2, and the conductivity of gold in this frequency region. As can be seen the conductivity is decreasing rapidly with increasing frequency, approximately $\propto \omega^{-2}$. However, interband transitions occur at higher energies. This metal, and others such as copper, owe their characteristic colors to the position of the edges of the free electron effects at lower energies and the onset, at higher energies, of interband transitions.

(3) The plasma frequency region At $\omega = \omega_p$ a most remarkable effect can be observed. At this frequency ε_1 changes from negative to positive (ε_2 is negligible) and the metal abruptly becomes transparent. Of course, this can be observed only if interband transitions do not occur in this frequency region. In gold (Fig. 13-11) and in most other metals interband transitions occur in the region. However, in the alkali metals the interband absorptions occur at energies higher than ω_p and this abrupt transparency can be observed! The calculated and experimental values have already been given in Table 9-2 and the agreement between experiment and theory is good. The frequency (or wavelength) at which this occurs is in the ultraviolet and thus to the eye these metals have the characteristic high reflectivity that we associate with metals. In Fig. 13-11a notice that ε_1 comes from $-\infty$ at low frequencies in agreement with Fig. 13-9a. If the restoring force is set to zero, the same result would be seen in Fig. 13-3a.

Above ω_p the properties of the free carriers are trivial but, as has been said, in this spectral region the optical properties are dominated by interband effects, which are discussed in Part D of this chapter. See the Notes for references to review articles that show much more data on metals and that also discuss the anomalous skin effect.

13-16b Semiconductors Semiconductors are interesting materials to examine. Many of them can be grown very pure and then intentionally doped in a controlled manner in order to finely tune their properties. For example, the number of free carriers can be adjusted

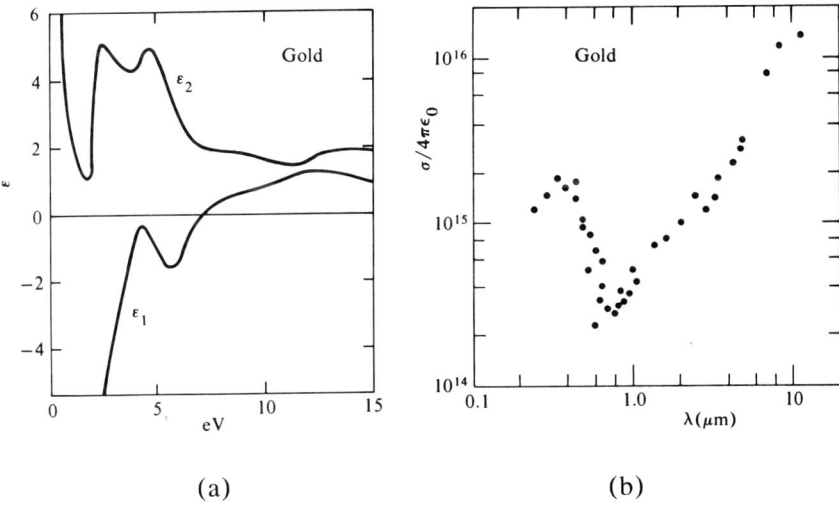

Fig. 13-11 (a) ε_1 and ε_2 for gold. [B. R. Cooper, H. Ehrenreich and H. R. Philipp, Phys. Rev. **138**, A494 (1965).] (b) Dispersion curve for the conductivity of gold [Givens, see the Notes].

as can, to some extent, their relaxation time, τ. The mean free path, ℓ, of electrons is about the same order of magnitude in semiconductors as it is in metals. However, for non-degenerate semiconductors the velocity of the electrons is determined by the electron kinetic energy, $3k_BT/2 \approx 0.04$ eV at room temperature, while for metals it is determined by the Fermi energy, ≈ 5 eV. Therefore, the relaxation time τ ($=\ell/v$ where v is the appropriate velocity) is much larger for electrons in a semiconductor than in a metal, and $\omega\tau \gg 1$ for semiconductors in the infrared where most of the optic measurements are carried out. We shall mention only a few examples of the many types of free carrier experiments that can be performed. See the Notes for references that discuss this active area of research.

A semiconductor like germanium can be grown with either an excess of electrons (n-type) which populate the bottom of the conduction band, or with a deficiency of electrons (p-type) which leaves hole states at the top of the valence band. In either case, the number of free carriers can be controlled and measured.

Figure 13-12a shows some results for free carrier absorption for electrons in n-type Ge and for holes in p-type Ge. The absorption coefficient for electrons is proportional to λ^2 as expected from the preceding equations. In fact, the agreement with experiment is excellent and an effective mass can be obtained from the experimental data. (ω_p^2, Eq. 13-47e, contains the effective mass.) Similar measurements in other materials give information about the frequency dependence of

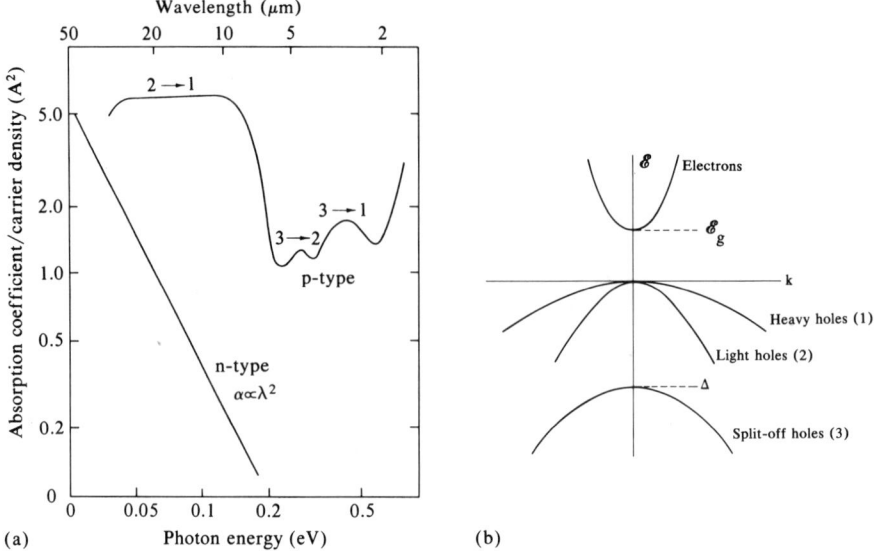

Fig. 13-12 (a) Free carrier absorption in n- and p-type germanium. [W. Kaiser, R. Collins and H. Y. Fan, Phys. Rev. **91**, 1380 (1953).] (b) The details of the valence band in Ge and many III-Vs (see Table 10-3 and Fig. 10-27) showing the light- and heavy-hole bands as well as the split-off hole band. Each band is given a number and, in the p-type material, the transitions for the free carrier absorption are labeled using these numbers.

the scattering process. Until now it has been assumed that τ is a constant. However, more detailed calculations show that for certain types of scattering, for example that from charged impurities, the lifetime of the carriers is dependent on the wavelength of the light used to observe the conductivity. For charged impurity scattering the absorption should be given by $\alpha \propto \lambda^3$. On the other hand, $\alpha \propto \lambda^{2.5}$ is expected for optical mode scattering.

The results for p-type Ge are also shown in Fig. 13-12a and a great deal of structure can be seen. Clearly something more than simple free carrier absorption is being observed. Figure 13-12b shows the complicated top of the valence band in Ge. This is just the same as Fig. 10-19b. The light- and heavy-hole valence bands are shown as well as the split-off spin-orbit band. The transitions that are involved are indicated. These absorptions are really interband absorptions and we show them now only as a reminder of complications that may arise.

Figure 13-13 shows reflectivity spectra for n-type InSb. The density of carriers has been varied; this, of course, changes the plasma edge. The agreement with the equations $\omega_p^2 \propto N/V$ is excellent and the effective mass can be obtained.

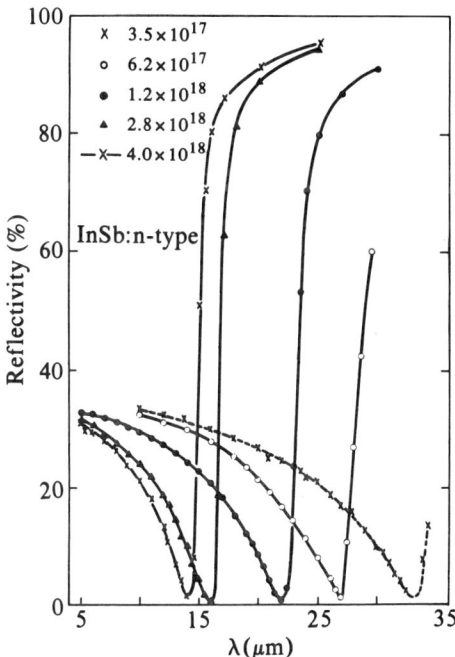

Fig. 13-13 Reflectivity spectra for n-type InSb samples with different numbers of free electrons. [From W. G. Spitzer and H. Y. Fan, Phys. Rev. **106**, 882 (1957).] See the comment about the ε_∞ effect in Section 13-3 and the caption of Fig. 13-5.

In a properly prepared semiconductor sample, the number of free carriers can be electrically varied (via a p-n junction). Then it is possible to modulate ω_p^2 and thus the reflectivity or absorption. This process has found some applications in semiconductor studies.

13-17 Transverse and Longitudinal Free Electron Modes

In our discussion of the interaction of free carriers with light, we obtained the frequency dependence of the macroscopic (long wavelength, small k) parameters and compared the results with experiment. This was done via the simple Lorentz oscillator model in Section 13-15. However, the modes of the free carriers have not been discussed; we will do this here. The procedure is similar to that used for the discussion for the interaction of radiation with lattice vibrations. First, the harmonic oscillator model was used to determine the macroscopic constants. Then Maxwell's equations were used to determine the modes (which in that case, for the transverse modes, were called polaritons).

13-17a Transverse free electron modes The determination of the dispersion curve for the transverse free electron modes proceeds exactly as in the polariton case (Section 13-11). For transverse waves,

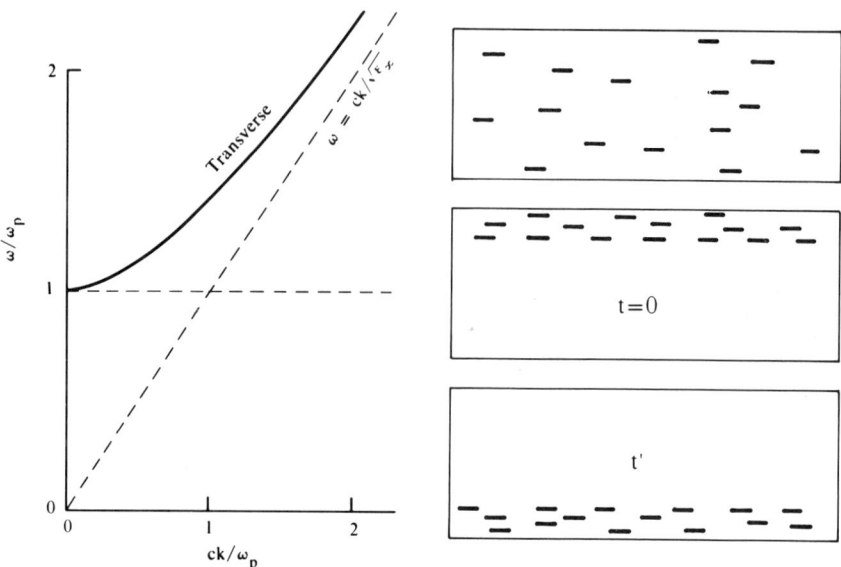

Fig. 13-14 (a) The dispersion curves for electromagnetic waves in a free electron gas. (b) A thin metallic slab showing a longitudinal plasma oscillation. The top part indicates a metal with the electrons randomly distributed in a +ion background (not shown). At t = 0 the electrons are displaced to the top and released. At a time t', corresponding to one-half of the plasma frequency, they are on the other side.

Maxwell's equations give a relation between ω and k in terms of the macroscopic dielectric constant, Eq. 13-17. As for polaritons we assume no loss terms. Then ω vs. k and $\varepsilon(\omega)$ are given by

$$\omega = c\, k/[\varepsilon(\omega)]^{1/2} \tag{13-49a}$$

$$\varepsilon(\omega) = 1 - \omega_p^2/\omega^2 \tag{13-49b}$$

where the second equation is the dielectric constant (Eq. 13-47b). Figure 13-14a shows the dispersion of the transverse modes. There are no propagating modes below ω_p; above ω_p this dispersion relation has a functional form $\omega = (\omega_p^2 + c^2 k^2)^{1/2}$. This result is very similar to the behavior found for polaritons (Fig. 13-7b), if, in that case, ω_0 is set equal to zero.

In summary, transverse electromagnetic waves propagate in a free electron gas (or in a true plasma) with the dispersion relation in Eq. 13-49. The low frequency cutoff for the transverse waves is the plasma frequency $\omega_p = (4\pi e^2 N/mV)^{1/2}$, or $\omega_p' = (4\pi e^2 N/mV\varepsilon_\infty)^{1/2}$ if there is a background dielectric constant, that is, if $\varepsilon(\omega) = \varepsilon_\infty - (\omega_p^2/\omega^2)$.

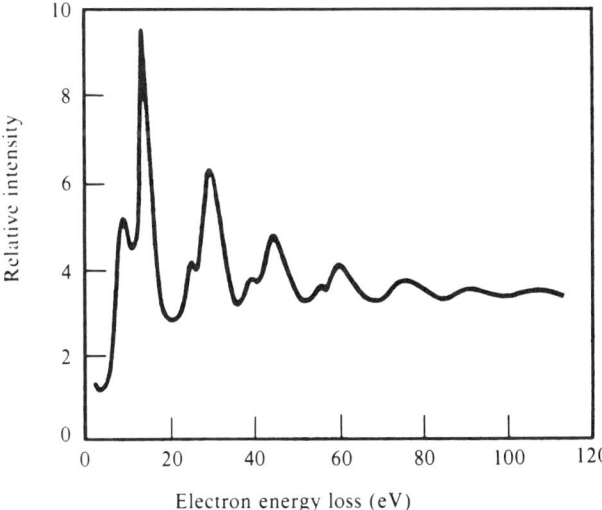

Fig. 13-15 Energy loss measurements for aluminum. [From C. J. Powell and J. B. Swan, Phys. Rev. **115**, 869 (1959).]

13-17b Longitudinal free electron modes What about the longitudinal modes? Exactly as for the longitudinal modes in the lattice vibration case, these modes occur at the zeros of the dielectric constant. This result comes from Eq. 13-16a, $\varepsilon \mathbf{k} \cdot \mathbf{E} = 0$, as discussed in Section 13-9b. Thus, longitudinal modes occur at the plasma frequency ω_p, or ω_p'. This result is shown in Fig. 13-14a as a dashed horizontal line.

Figure 13-14b shows a very simple physical model of a longitudinal wave in a free electron gas (or in a plasma). In this thin metallic slab, all the free charges (the electrons) move by an amount Δ which causes a dipole moment per unit volume or a polarization, $\mathbf{P} = -Ne\Delta/V$. The electric field that occurs is just the depolarizing field $\mathbf{E}_r = -4\pi\mathbf{P}$. (For a longitudinal wave $\mathbf{D} = \mathbf{E} + 4\pi\mathbf{P} = 0$.) Thus, the equation of motion for an electron is

$$m(d^2\Delta/dt^2) = -eE = -4\pi Ne^2\Delta/V$$

or $$d^2\Delta/dt^2 + \omega_p^2 \Delta = 0 \qquad (13\text{-}50)$$

where $$\omega_p = (4\pi e^2 N/V)^{1/2}$$

With zero loss, once such a collective longitudinal wave is set in motion the electrons will oscillate with the plasma frequency.

Quantization of the harmonic Hamiltonian was discussed extensively (Section 12-5) for the phonon case. The point is that whenever a system obeys a simple harmonic oscillator equation, when quantized the energies of the normal modes are equally spaced. Then we can describe the n_kth excited state of the kth normal mode as a state with n_k identical excitations in the mode, each having energy $\hbar\omega_k$ ($\hbar\omega_p$ in this case). Since Eq. 13-50 describes simple harmonic oscillator behavior the system can be quantized. The quantum of plasma oscillation, which describes the degree of excitation of the normal modes of the longitudinal plasma oscillations is called a **plasmon**. The energy of the system is given by $(n + 1/2)\hbar\omega_p$, where n is the number of plasmons.

In practice these longitudinal excitations can be excited and measured by passing fast, charged particles through the solid and measuring their loss of energy. This is called **energy loss measurements.** Figure 13-15 shows some experimental results for a metal obtained by measuring the energy loss of 2 keV electrons reflected from a thin aluminum film. The loss peaks are due to the excitation of multiple bulk plasmons, as well as surface plasmons. Plasmons can also be excited in dielectrics. See the reviews mentioned in the Notes for further information about energy loss spectroscopy.

Part D — Interband Transitions

13-18 Introduction

The last part of this chapter is devoted to transitions between different bands (interband transitions), and only semiconductors are discussed. Optical studies have yielded important information for the understanding of band structures in these materials. However, Fig. 13-16 demonstrates the complexity of the problem. The calculated density of states is shown for the valence and conduction bands of Ge. The width and complexity of the bands shown in this figure compared to the very narrow structure observed for atoms is discouraging. For an intrinsic semiconductor it might be thought that transitions are possible from the entire valence band to the entire empty conduction band, resulting in a very broad spectrum (many eV wide). The optical spectroscopy of energy levels of this type might, thus, look featureless. This is not the case, and it has turned out that interband optical spectroscopy is a very important tool for the study of energy bands. We shall see why.

However, independent of all the rich (to the optimist) or broad (to the pessimist) detail shown in Fig. 13-16, we should not forget the "ground state" or what is usually called the fundamental absorption at $\mathcal{E}_g$. Ground states occupy unique positions in the study of material properties. For example, in the study of the energy levels of the iron series transition metal ions, there is a large amount of optical spectroscopy on the excited energy levels, perhaps up to 5×10^4 cm^{-1}. The ground state and the very first few energy levels, which cover perhaps 1 cm^{-1}, are usually studied by electron spin resonance techniques.

CHAPTER 13   OPTICAL PROPERTIES OF CRYSTALS   499

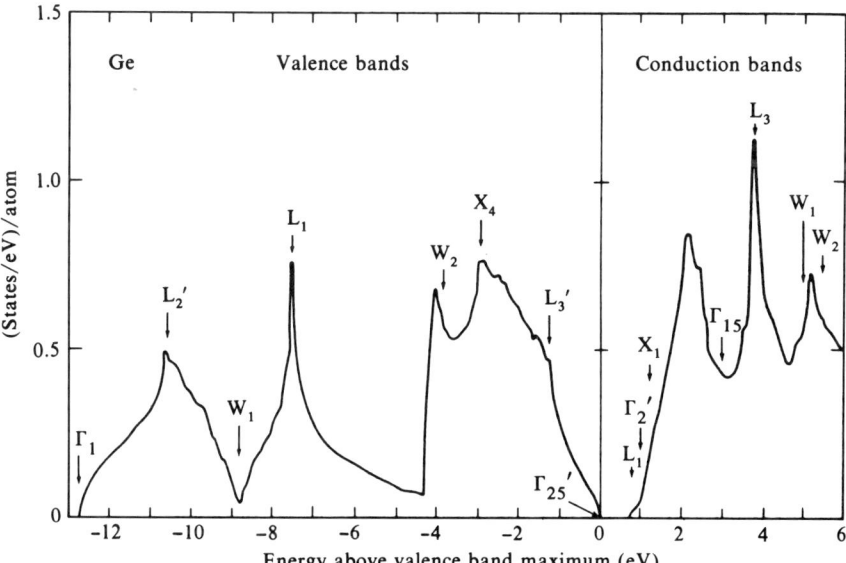

Fig. 13-16 A calculated density of states for the highest lying valence and lowest lying conduction bands in germanium. [From F. Herman, R. L. Kortum, C. D. Kuglin and J. L. Shay, in "II-VI Semiconducting Compounds, 1967 International Conference," D. G. Thomas, Ed. (Benjamin Press, 1967) p. 503.]

Yet these very low energy levels (the ground state) yield as much information as the other 5×10^4 cm^{-1}. The reason is that the quantum states within the first few cm^{-1} are extremely narrow and there are no background effects that overlap with signals from other transitions. Similar effects are found in semiconductors. The initial absorption, near the band gap, $\mathscr{E}_g$, yields a rich amount of detail even though the absorption coefficient may be very weak compared to those at higher energies. There are several reasons for this. First, there is very little background absorption, so the fundamental processes can be studied in detail. Second, the lifetimes of the excited states near $\mathscr{E}_g$ are long, so the effects are not smeared out. Third, usually it is only near the energy gap that phenomena such as excitons, electron hole drops, donor-acceptor pairs, and so on, are seen.

It is for these reasons that the discussion of interband transitions (what is sometimes called the **fundamental absorption**) is divided into two parts. The first is concerned with effects near $\mathscr{E}_g$. The second is concerned with higher lying energy transitions.

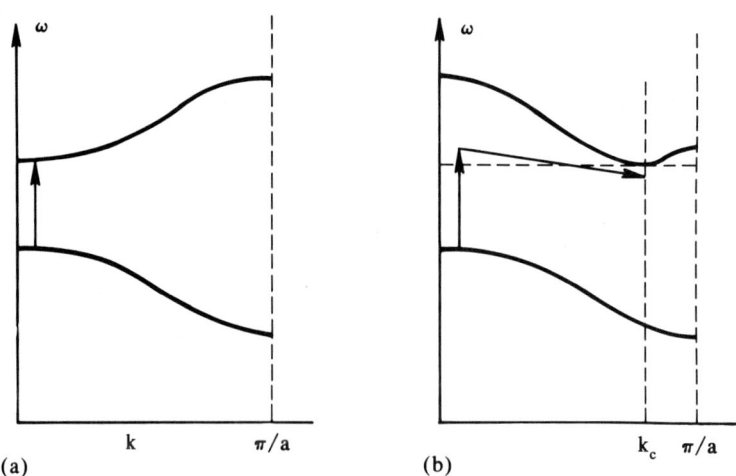

Fig. 13-17 (a) Vertical transitions in a direct band gap material. (b) Phonon assisted transitions in an indirect band gap material where k_c is the wave vector for the minimum of the conduction band.

13-19 Fundamental Absorption near $\mathscr{E}_g$

13-19a Optically induced vertical transitions When the energy of the incident light just exceeds the energy gap $\mathscr{E}_g$, an electron can be excited from the valence to the conduction band. Figure 13-17a shows this process. The momentum of the light, $\hbar k$ $(=\hbar n_1 \omega/c)$, is negligible compared to the momentum of a k-vector at the edge of the Brillouin zone $\hbar\pi/a$; thus the excited electron state must have essentially the same k-value as the initial state. Such transitions are called **vertical transitions**. For this transition a photon is annihilated and an electron and hole are created. (The energy and momentum of a hole are defined in Chapter 10.) The energy and momentum are conserved, and the electron and hole have equal and opposite **k** (neglecting the very small momentum of the light).

Now, consider the fundamental absorption for an indirect band gap semiconductor (i.e., the bottom of the conduction band is at a different **k** than the top of the valence band). Since light has negligible momentum compared to $\hbar\pi/a$, some other process must be available to conserve momentum. Figure 13-17b shows how phonons assist the fundamental absorption. For the electron to be excited from the top of the valence band at $k = 0$ to the bottom of the conduction band at the k_c, a two-step process is required. In the figure the photon is absorbed and a phonon with wave vector k_c is emitted. The energy of the photon is equal to $\mathscr{E}_g$ plus the energy of the phonon. Thus, energy and crystal momentum are conserved. For $T > 0$ a phonon may be annihilated in this process in which case the phonon energy is equal to $\mathscr{E}_g$ minus the energy of the phonon.

It is important to remember that, on the scale of the Brillouin zone, the momentum of light is negligible. With no assistance, light can induce only vertical transitions. Thus, in a direct band gap material the electron makes a vertical transition; but in an indirect gap the electron makes, with phonon assistance, a nonvertical transition.

13-19b Theory We calculate the absorption probabilities for the transitions pictured in Figs. 13-17a and 13-17b. The absorption coefficient depends on the probability per unit time that an electron in the valence band will make a transition to the conduction band. This is a quantum mechanical calculation; classical models, which have been helpful in most of this chapter, are of no use.

Direct (vertical) transitions – From first order time dependent perturbation theory, the probability for a transition from an initial state, i, to a final state, f, in a continuum is given by

$$W_{if} = (2\pi/\hbar) \, |H_{if}|^2 \, \rho(\mathscr{E}_f) \qquad (13\text{-}51a)$$

where H_{if} is the matrix element of the perturbation which connects the i and f states and $\rho(\mathscr{E}_f)$ is the density of final states. The only important process that contributes to the absorption is electric dipole transitions in which case the dipole operator is $\propto \nabla$. Taking the initial and final electron wave functions as Bloch functions of the valence, ψ_v, and conduction, ψ_c, bands we have

$$H_{if} \propto \int \psi_v^* \nabla \psi_c \, d\tau = \int u_v^* e^{-i\mathbf{k}_v \cdot \mathbf{r}} \nabla u_c e^{i\mathbf{k}_c \cdot \mathbf{r}} \, d\tau \qquad (13\text{-}51b)$$

$$= \int u_v^* (\nabla u_c) e^{i(\mathbf{k}_c - \mathbf{k}_v) \cdot \mathbf{r}} d\tau \qquad (13\text{-}51c)$$

$$+ i\mathbf{k}_c \int u_v^* u_c \, e^{i(\mathbf{k}_c - \mathbf{k}_v) \cdot \mathbf{r}} d\tau \qquad (13\text{-}51d)$$

The factor $\exp[i(\mathbf{k}_c - \mathbf{k}_v) \cdot \mathbf{r}]$ oscillates rapidly and causes both integrals in the expressions Eqs. 13-51c and 13-51d to be zero unless $\mathbf{k}_c = \mathbf{k}_v$. Of course this is the vertical selection rule for the conservation of crystal momentum, which is appropriate when only light is involved.

The dipole operator ∇ is an odd operator (has odd parity) so the expression in Eq. 13-51c vanishes unless u_c and u_v have opposite parity. Thus, we use the terminology that transitions are **allowed** if Eq. 13-51c is nonzero and **forbidden** if it is zero. However, the transition is not strictly forbidden since for states that have the same parity the term in Eq. 13-51d can be nonzero leading to weak (k dependent) absorption. For allowed transitions, H_{if} is independent of wave vector. For forbidden ones $H_{if} \propto k$; hence the transition probability $\propto k^2$. Experiment is used to determine which process is important.

For the case pictured in Fig. 13-17a, where the valence band maximum and conduction band minimum occur at the same wave vector (i.e., at **k** = 0), we calculate the energy dependence of the absorption constant, which is proportional to the square of the matrix element in Eq. 13-51b times the density of states $\rho(\mathscr{E}_f)$. Near $\mathscr{E}_g$ the matrix element will have no energy dependence, that is, to first-order u(**r**, **k**) is independent of **k**. Thus, the energy dependence arises only from the density of states, which is easy to determine. The energies of electrons, e, in the conduction band and holes, h, in the valence band are given by Eq. 13-52a, while the energy of the photon required to cause the transition, $\hbar\omega$, is given by Eq. 13-52b, where we have added a k^2 term to represent the parabolic bands away from **k** = 0. (This term in a related context is discussed in Section 13-20b.)

$$\mathscr{E}_e = \mathscr{E}_g + \frac{\hbar^2 k^2}{2m_e} \qquad \mathscr{E}_h = -\frac{\hbar^2 k^2}{2m_h} \qquad (13\text{-}52a)$$

$$\hbar\omega = \mathscr{E}_e - \mathscr{E}_h = \mathscr{E}_g + \frac{\hbar^2 k^2}{2\mu} \qquad \frac{1}{\mu} \equiv \frac{1}{m_e} + \frac{1}{m_h} \qquad (13\text{-}52b)$$

Thus, above the band gap there is a k^2 dependence of the energy which from Eq. 9-30b gives a density of final states $\propto (\hbar\omega - \mathscr{E}_g)^{1/2}$. The direct band allowed absorption, α_{da}, is proportional to $(\hbar\omega - \mathscr{E}_g)^{1/2}$ while the direct band forbidden absorption α_{df} contains an extra k^2 term. Both of these results are written here

$$\alpha_{da} = C_{da} (\hbar\omega - \mathscr{E}_g)^{1/2} \qquad (13\text{-}53a)$$

$$\alpha_{df} = C_{df} (\hbar\omega - \mathscr{E}_g)^{3/2} \qquad (13\text{-}53b)$$

where the C-coefficients involve constants and the matrix elements in Eqs. 13-51c and 13-51d, but have little energy dependence. Clearly there is a large difference between the energy dependences for the allowed and forbidden absorption processes.

The III-V semiconductors such as AlP, GaAs, InSb, AlAs, and others have direct allowed absorption. Many complex oxides such as rutile, SiO_2, Cu_2O, and others have direct forbidden absorption.

Indirect (nonvertical) transitions – Absorption takes place by a two-step process for indirect band gap materials, and second order perturbation theory is required to determine the probability for a transition. Thus, weaker absorption is expected than for the direct band gap case. The two-step process is pictured in Fig. 13-17b. The electron, initially in the top of the valence band at **k** = 0, absorbs a photon and makes a vertical, virtual transition to the intermediate state at **k** = 0. (This energy nonconserving transition is allowed because the lifetime in this state is short enough to be consistent with the uncer-

CHAPTER 13   OPTICAL PROPERTIES OF CRYSTALS   503

tainty principle.) Then, with the emission of a phonon with wave vector $-k_c$, the electron makes a transition to the conduction band minimum. The transition probability depends not only on the density of states and electron-photon matrix elements as would be expected, but also on the electron-phonon matrix elements.

The equations corresponding to those in Eq. 13-53 are not difficult to determine, but we shall leave it as a problem and just state and discuss the results. The absorption constants for the indirect allowed (ia) and indirect forbidden (if) process are

$$\alpha_{ia}^{e} = C_{ia} \frac{(\hbar\omega - \mathcal{E}_g - \hbar\omega_c)^2}{1 - \exp(-\hbar\omega_c/k_B T)} \qquad (13\text{-}54a)$$

$$\alpha_{if}^{e} = C_{if} \frac{(\hbar\omega - \mathcal{E}_g - \hbar\omega_c)^3}{1 - \exp(-\hbar\omega_c/k_B T)} \qquad (13\text{-}54b)$$

where the superscript e corresponds to the emission of a phonon of wave vector k_c and energy $\hbar\omega_c$. Consider just the allowed absorption, Eq. 13-54a, which is observed in the important semiconductors Ge and Si. As expected, at $T = 0°K$ absorption occurs only for energies $\hbar\omega \geq \mathcal{E}_g + \hbar\omega_c$. Also notice the appearance of the Bose-Einstein factor which describes the phonon population; at $T = 0°K$ this is a factor unity but at higher temperatures it is greater than one. In addition to the term in Eq. 13-54a we add a term that corresponds to absorption, α_{ia}^{a}, instead of emission, of a phonon. Then we obtain

$$\alpha_{ia} = \alpha_{ia}^{e} + \alpha_{ia}^{a} = C_{ia}\left[\frac{(\hbar\omega - \mathcal{E}_g - \hbar\omega_c)^2}{1 - \exp(-\hbar\omega_c/k_B T)} + \frac{(\hbar\omega - \mathcal{E}_g + \hbar\omega_c)^2}{\exp(-\hbar\omega_c/k_B T) - 1}\right] \qquad (13\text{-}55)$$

Note that the α_{ia}^{a} contribution can be observed at lower energies than those of the emission term but only for $T > 0°K$.

Figure 13-18 shows a plot of the theoretical results from Eq. 13-55. The energy difference between the phonon emission and absorption part, extrapolated to zero absorption, is just $2\hbar\omega_c$. Also indicated in the figure is the shift of $\mathcal{E}_g$ with temperature and the increase of annihilation term in the equation, which is due to increasing phonon absorption with increasing temperature (i.e., the phonon population increases with T).

There is one further complication. The preceding results assume only one type of phonon is emitted or absorbed in the fundamental optical absorption process. This need not be the case. For example, TA, LA, TO, and LO phonons might be *separately* involved; in fact

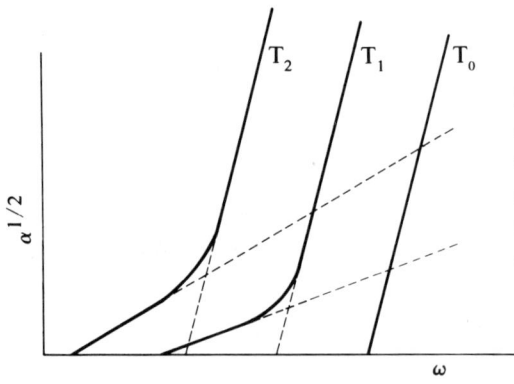

Fig. 13-18 A theoretical plot of $\alpha^{1/2}$ vs. photon energy at several temperatures. T_0 is absolute zero and $T_2 > T_1 > T_0$. Note that the band gap decreases in energy as temperature increases.

this has been observed experimentally.

13-19c Experiment Large sections of books on the optical properties of semiconductors have been devoted to the theory and experiment of the fundamental absorption edge near $\mathscr{E}_g$. (See the Notes.) We mention only a few examples of experimental results to give the reader a "flavor" of the subject.

Figure 13-19 shows some early experimental results for the absorption coefficient of Ge illustrating many aspects of the calculations discussed here. The value of α rises due to the onset of indirect absorption at the lowest energies. At higher energies a sharper rise in α is found when direct (vertical) transitions occur. In Fig. 13-19b the analysis of the 300°K data in different energy regions shows that in the low energy region a plot of $\alpha^{1/2}$ vs. $\hbar\omega$ has a linear region, as expected from Eq. 13-55, and shown in Fig. 13-18. This region corresponds to the indirect band gap that is the smallest gap in Ge. However, at higher energies much stronger direct band gap absorption is observed. Here a plot of α^2 vs. $\hbar\omega$ has a linear region, as expected from Eq. 13-53a. The results in this figure give room temperature values of 0.63 eV and 0.81 eV for the indirect and direct band gaps and ≈ 0.01 eV (= 83 cm^{-1}) for the phonons that assist in the nonvertical transition. We emphasize several points about the data. First, the shifts of $\mathscr{E}_g$ with temperature for both the indirect and direct gaps are apparent and large (Fig. 13-19a). Second, the direct gap absorption is much stronger than the indirect gap absorption as expected.

Figure 13-20a shows more accurate absorption data for germanium in the region of the indirect band gap. The results are similar to the theoretical plots in Fig. 13-18. From this data the energies of the two phonons that help to conserve crystal momentum were determined. These phonons are a TA at 0.008 eV and a LA at 0.027 eV. Subsequent neutron diffraction measurements showed that indeed the TA and LA phonons in the ⟨111⟩ directions have these energies. (The

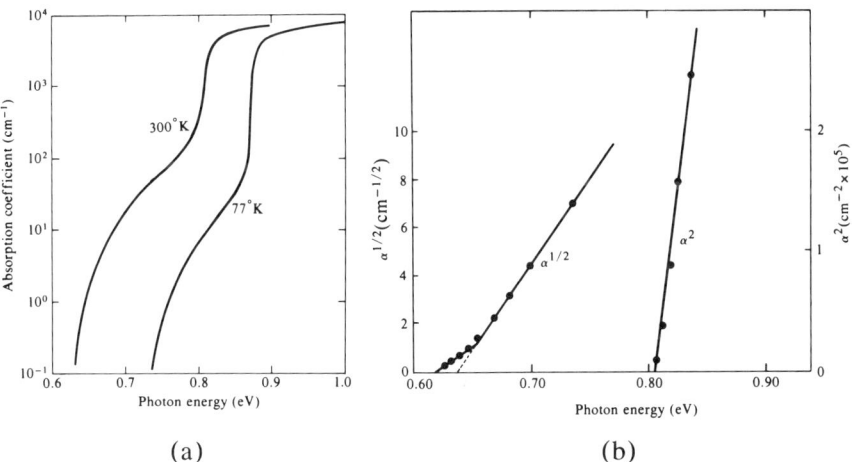

Fig. 13-19 (a) α vs. $\hbar\omega$ for Ge. (b) The analysis of the 300°K experimental data. [From W. C. Dash and R. Newman, Phys. Rev. **99**, 1151 (1955).]

fact that these phonons are found is due to the conduction band minima being at the zone edge in the four $\langle 111 \rangle$ directions, Fig. 10-24.)

GaP is also an indirect band gap semiconductor with conduction band minima in the six $\langle 100 \rangle$ directions near the X-point (Fig. 10-24). Figure 13-20b shows some very high resolution absorption data. Distinct phonon assistance from TA, LA, and TO as well as a LO + TA combination band can be seen in emission (E) at 1.6°K. At higher temperatures absorption (A) of phonons can also be seen.

Figure 13-21a shows α^2 vs. $\hbar\omega$ data for GaAs, which has a direct band gap (Fig. 10-24). The energy dependence of the absorption agrees very well with theory, Eq. 13-53a, and $\mathscr{E}_g = 1.39$ eV is obtained from the data. Actually these data were taken on rather impure (by present-day standards) samples. More recent data on purer samples show another feature in the absorption edge region due to exciton absorption. Excitons are discussed in Section 13-20. However, in Fig. 13-21b absorption at the very highest values (the "top" of the absorption) is shown for another direct band semiconductor, InP. This material clearly shows excitonic absorption and an absorption edge that depends on energy different than expected from the preceding discussion. (See the Notes.)

13-19d Heavily doped materials Photon absorption in heavily doped indirect band gap semiconductors can involve a momentum-conserving process different from that of phonons. Electron-electron or electron-impurity scattering can conserve momentum. These processes are significant because of the large number of impurities and extra electrons (for n-type impurities) that exist in the heavily doped

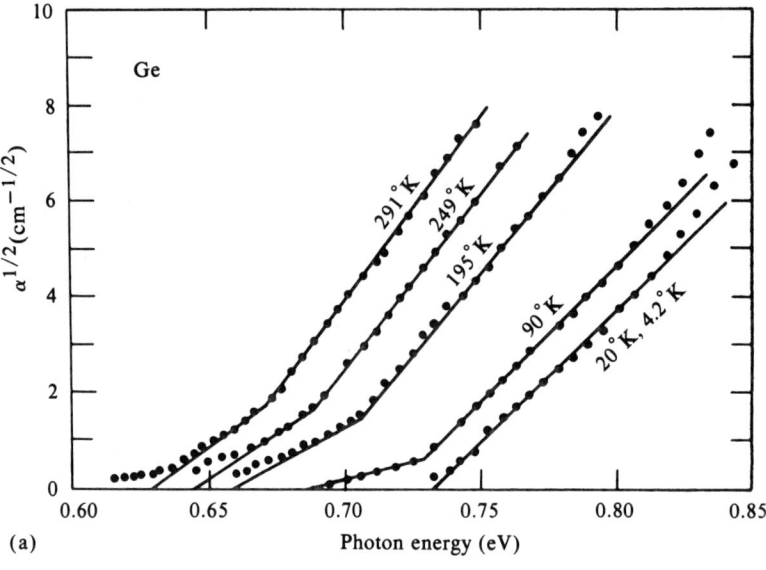

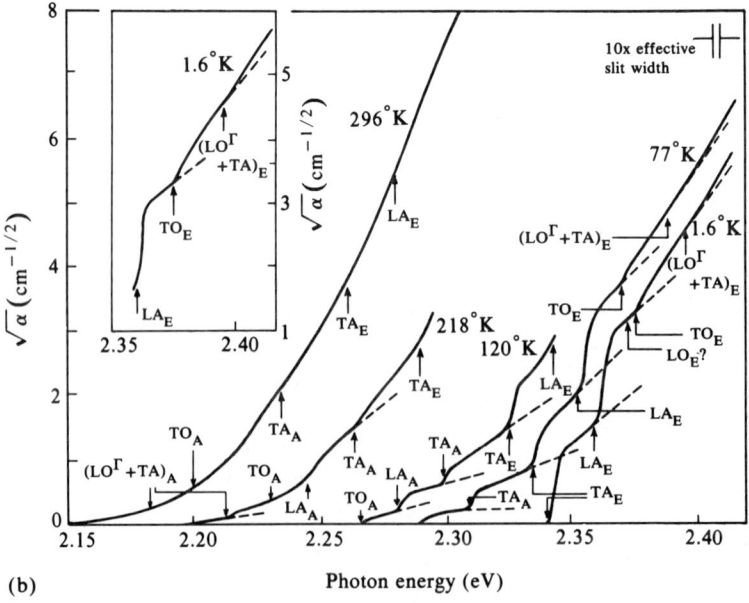

Fig. 13-20 (a) $\alpha^{1/2}$ vs. $\hbar\omega$ for Ge showing details of the absorption results at low absorption constant. [From G. G. MacFarlane and V. Roberts, Phys. Rev. **97**, 1714 (1955).] Their more extensive data can be found in J. Phys. Solids, **8**, 388 (1959). (b) $\alpha^{1/2}$ vs. $\hbar\omega$ for GaP. [From P. J. Dean and D. G. Thomas, Phys. Rev. **150**, 690 (1966).]

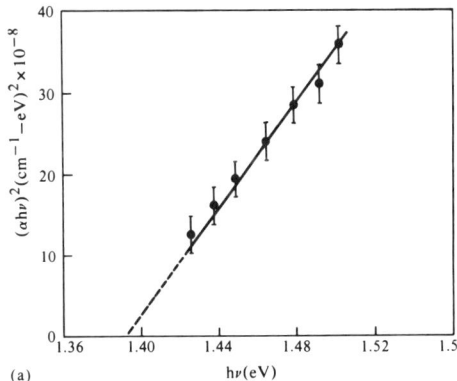

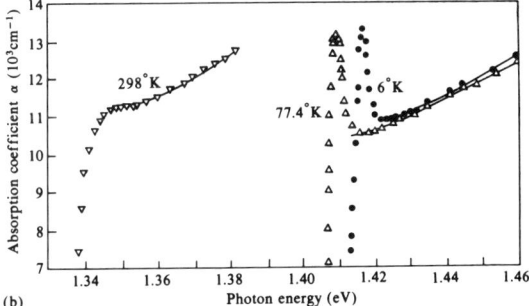

Fig. 13-21 (a) α^2 vs. $\hbar\omega$ for GaAs. [From I. Kudman and T. Seidel, J. Appl. Phys. **33**, 771 (1962)]. (b) The very "top" of the absorption edge in InP. Exciton absorption can be seen. [From W. J. Turner, W. E. Reese and G. D. Pettit, Phys. Rev. **136**, A1467 (1964)].

materials. For such materials the absorption coefficient should be similar to that shown in Eq. 13-55 but without the Bose-Einstein population factor for the phonons. The result is (see the Notes)

$$\alpha \propto N(\hbar\omega - \mathscr{E}_g - \Delta_N)^2 \tag{13-56}$$

where N is the number of scatters and Δ_N is a quantity that describes the fact that for heavily doped (degenerate) semiconductors, the Fermi level moves up, and is in the conduction band (for n-type doping). (This by itself causes an increase in the absorption band gap called the Burstein-Moss shift; see Section 10-15b.) Figure 13-22a shows experimental results for Ge heavily doped with arsenic impurities. In agreement with Eq. 13-56, $d(\alpha^{1/2})/d(\hbar\omega) \propto N^{1/2}$.

Thus, other processes beside phonon emission and absorption can help to conserve crystal momentum; impurities or disorder in crystals often play this role. Disorder destroys the k-selection rule because it destroys translational symmetry, and the entire concept of conservation of crystal momentum arises from translational symmetry. For crystals that contain increasing amounts of disorder, crystal momentum becomes an increasingly meaningless concept. However, it is difficult to predict the quantitative effect of disorder on different properties. As another example of the effects of disorder, we observe that first

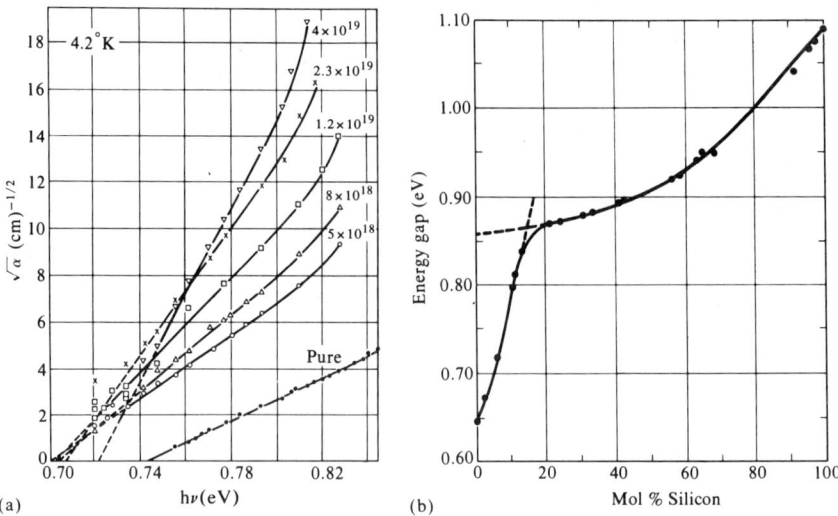

Fig. 13-22 (a) $\alpha^{1/2}$ vs. $\hbar\omega$ for Ge heavily doped with various amounts of arsenic. [From J. I. Pankove and P. Aigrain, Phys. Rev. **126**, 956 (1962).] (b) The composition dependence of $\mathscr{E}_g$ in Ge-Si alloys. [From R. Braunstein, A. R. Moore and F. Herman Phys. Rev. **109**, 695 (1958).]

order infrared and Raman scattering are observed from phonons throughout the Brillouin zone in disordered (amorphous) Ge, while first order scattering in ordered (crystalline) Ge yields only the **k** = 0 phonons.

13-19e Alloys Germanium and silicon are very similar and have the same crystal structure. Thus, we expect, and find, that a continuous solid solution $Ge_{1-x}Si_x$ for all x will be formed (i.e., they are completely miscible in each other). Usually, for such alloys if the value of almost any property is plotted as a function of x, a straight line is obtained. This is called **Vegard's law**. Deviations from this law are usually of considerable interest. That is why the results for the band gap for this alloy system are quite shocking at first glance.

Figure 13-22b shows that the band gap for $Ge_{1-x}Si_x$ is hardly linear with x. Actually, the variation of $\mathscr{E}_g$ is easy to understand. Refer to Figs. 10-27 and 10-28 where by convention the top of the valence band is the zero of energy. For Ge, x = 0, $\mathscr{E}_g$ is determined by the conduction band valleys in the $\langle 111 \rangle$ directions; for Si, the conduction band valleys are in the $\langle 100 \rangle$ directions. Then, as Si is added to Ge the $\langle 111 \rangle$ valleys move to higher energy at a faster rate than the $\langle 100 \rangle$ valleys. At $x \approx 0.15$ they both have the same energy with respect to the valence band, but above this value the conduction band minima are in different directions in k-space.

CHAPTER 13   OPTICAL PROPERTIES OF CRYSTALS   509

The fact that the location of the band gap in k-space changes sometimes leads to interesting effects. For example, the lifetimes of the carriers may be rather different in the different conduction band minima. This would be particularly important for alloys that change from a direct band gap to an indirect band gap material, where the lifetime for electrons is much shorter in the former. $Ga(As_{1-x}P_x)$ is such a system (Fig. 10-27) and interesting things do happen.

13-19f Donors and acceptors (and photoconductivity)   In Section 8-5b the hydrogen-like model of shallow donors and acceptors was discussed. Since the effects of impurities in semiconductors is of technological interest, the ionization energies of many impurities in Si, Ge, GaAs, and related materials have been measured; see the Notes. The hydrogen-like model predicts a series of energy levels (similar to those of the atomic hydrogen series) given by Eq. 8-15c.

Using appropriately doped samples at temperatures low enough that most of the donors are in their ground states, we can observe absorptions due to excitation of the electrons from their n = 1 to higher n-states. Far infrared radiation is used since the separation of the energy levels is ≈20 meV (the ionization energy) and less. The observed lines do not correspond exactly to the simple hydrogen model of Section 8-6b because the conduction band in Ge is multivalued and anisotropic. However, reasonable agreement with experiment is obtained if these factors are taken into account.

Optical detectors can be made employing the impurity energy levels in semiconductors via the process of **photoconductivity**. Consider a doped semiconductor operating at temperatures low enough so that the donors are in states well below their ionization limit ($\mathscr{E}_{bnd}$). Then, if light with energy $\hbar\omega > \mathscr{E}_{bnd}$ impinges on the sample, electrons are excited into the conduction band, greatly enhancing the conductivity of the material. With the proper electrodes such photoconductors can be, and routinely are, used to detect far infrared light. The only inconvenience is that the detector must be cooled to temperatures $k_B T < \mathscr{E}_{bnd}$ which usually means liquid helium temperatures.

13-20 Excitons (Mostly Weakly Bound Excitons)

13-20a Introduction   When an interband transition occurs, an electron is excited from the valence band to a higher conduction band. The electron in the conduction band, and separately the hole in the valence band, moves freely according to the dynamics required of that band (i.e., with the relaxation time, effective mass, etc. appropriate for that band). Such interband transitions occur when the energy of the incoming radiation is greater than or equal to the band gap (i.e.,

$\hbar\omega \geq \mathscr{E}_g$). However, there is a lower energy excited configuration, an exciton state, which is the subject of this section.

An electron can be excited from the valence band to a higher energy state in which the electron is still bound by the Coulomb attraction to the hole that the electron leaves in the valence band. This creates a neutral unit that can move throughout the crystal. This bound electron-hole pair is called an **exciton** and is of considerable interest in semiconductors, wider band gap insulators, and molecular crystals. We discuss primarily the weakly bound excitons found in semiconductors, where a simple exciton model exists.

Excitons can be formed by photon absorption any place in the Brillouin zone where $\partial\mathscr{E}_v/\partial\mathbf{k} = \partial\mathscr{E}_c/\partial\mathbf{k}$ at the same k-value, because at such k-values the group velocities of the electron and hole are equal and the particles may be bound together. However, excitons are most easily observed just below $\mathscr{E}_g$. It is in such spectral regions that there is little background absorption to interfere with the observations.

Usually excitons are observed by one of two methods. (1) Optical absorption just below $\mathscr{E}_g$ can show an exciton line as in Fig. 13-21b. (2) **Recombination radiation** is another experimental tool for observing excitons. In this method electrons are excited to some high energy band by light. These excited electrons (holes) rapidly thermalize, that is, fall into lower energy levels in the same band, usually with the emission of phonons. The electron ends up in the bottom of the conduction band and the hole in the top of the valence band; the difference in energy is $\mathscr{E}_g$ (or slightly less if an exciton is formed). The electron and hole recombine principally by one of two methods. They recombine and emit a photon $\hbar\omega = \mathscr{E}_g$, or they recombine and emit many phonons (assuming $\mathscr{E}_g \gg$ all of the phonon energies). For large band gaps the former process is more probable than the latter because multiphonon emission, being a high order process, is not very probable. Many semiconductors, excited as described, emit radiation, which is known as **recombination radiation**. The radiation can be spectally narrow, intense, and there is usually very little background emission. For these reasons recombination radiation is easy to detect.

In the next several subsections we discuss the character of weakly bound excitons, sometimes called the **Mott-Wannier excitons,** before mentioning tightly bound (Frenkel) excitons.

13-20b Weakly bound excitons When the attraction between an electron and hole is small compared to $\mathscr{E}_g$, the Coulomb potential energy is

$$\text{PE} = -e^2/\varepsilon_\infty r \tag{13-57a}$$

CHAPTER 13   OPTICAL PROPERTIES OF CRYSTALS   511

where r is the distance between the electron and the hole and ε_∞ is the optic dielectric constant for $\hbar\omega \ll \mathscr{E}_g$ but $\hbar\omega \gg$ phonon energies. So ε_∞ is equal to the square of the usual optic index for the material (n_1^2). Why does ε_∞ appear as opposed to $\varepsilon(0)$ or 1? The electron hole pair moves too rapidly for the atom cores to adjust so phonons cannot be involved, hence $\varepsilon(0)$ cannot be used. However, the valence electrons can adjust to this motion so ε_∞ is used. In using this quantity there is also the further assumption that the electron and hole of the exciton are on the average several unit cells apart so that their interaction is screened by the *average macroscopic dielectric constant* ε_∞. This model should be consistent in this respect, that is, the average radius of the exciton must be found to be large compared to the internuclear distances.

The previously mentioned potential energy is the same as we used in Eqs. 8-15 for weakly bound donors in a semiconductor. The physics is quite similar; in both cases the electron and/or hole are weakly bound and the orbit is large so ε_∞ and the electron and hole effective masses can be used. Since the potential energy, Eq. 13-57a, is just that of the hydrogen atom (except for ε_∞) we immediately know the solution. However, while bound impurities are fixed, the electron hole pair can move and there is a kinetic energy term just as in the free electron case. Thus, the discrete energy levels of the exciton are given by the hydrogen-like solution with a k^2 term

$$\mathscr{E}_{ex} = \mathscr{E}_g - \frac{g}{n^2} + \frac{\hbar^2 k^2}{2(m_e + m_h)}$$

$$g = \left(\frac{\mu}{\varepsilon_\infty^2}\right)\left(\frac{e^4}{2\hbar^2}\right) = \left(\frac{\mu}{\mu_H \varepsilon_\infty^2}\right) \mathscr{E}_H$$

(13-57b)

where $\mu = m_e m_h/(m_e + m_h)$ is the reduced mass of the exciton, μ_H is the reduced mass of the hydrogen atom, n is the principal quantum number (equal to 1, 2, ..., ∞), and $\mathscr{E}_H$ (= 13.58 eV) is the ionization energy of hydrogen. Using typical electron and hole effective masses ($m_e^* \approx 0.05m$ and $m_h^* \approx 0.1m$ where m is the mass of an electron) and $\varepsilon_\infty = 16$, appropriate for Ge, we obtain $\mathscr{E}_{ex} = 0.005$ eV for the binding energy of the exciton. Why does $\mathscr{E}_g$ appear in Eq. 13-57b? At the ionization limit of the exciton, the electron and hole are free carriers in their respective bands. We **conventionally** take the top of the valence band as the zero of energy; then for $k = 0$ the bottom of the first conduction band is at $\mathscr{E}_g$ and Fig. 13-23a shows the exciton energy levels in real space at $k = 0$. In k-space, at $k = 0$ these energy values increase parabolically with k as indicated in Eq. 13-57b. This

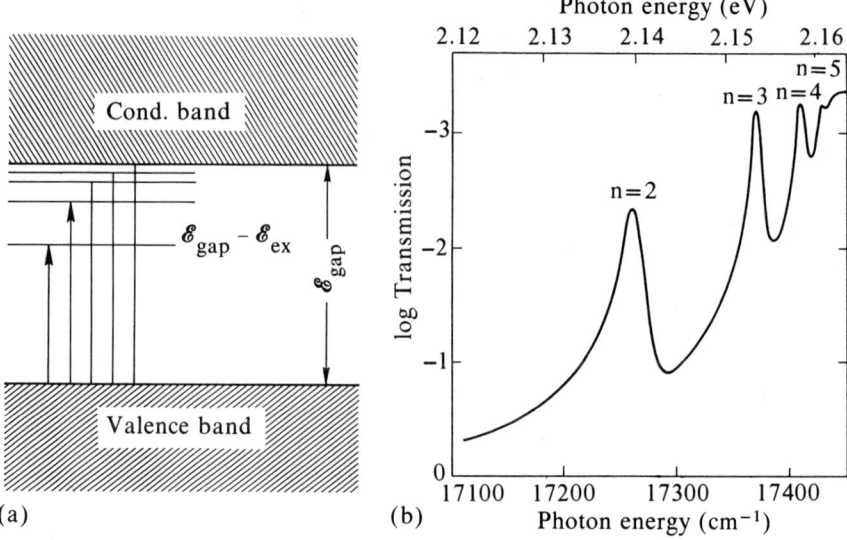

Fig. 13-23 (a) The energy level diagram of an exciton in a direct band gap material. (b) Excitons in Cu_2O. [From P. W. Baumeister, Phys. Rev. **121**, 359 (1961).]

effect and the interactions of the excitons (at $\mathbf{k} \neq 0$) with light are discussed in Section 13-20f.

The Cu_2O, direct forbidden (Eq. 13-53b), exciton absorption spectra are shown in Fig. 13-23b. The data fits (in units of cm^{-1})

$$\mathcal{E} = 17{,}508 \, cm^{-1} - (800/n^2)$$

where 17,508 cm^{-1} (= 2.17 eV) corresponds to the energy gap. Combining the coefficient of the $1/n^2$ term with $\varepsilon_\infty = 10$ and using Eq. 13-57b, we obtain for the reduced mass $\mu \approx 0.7m$.

The results shown in Fig. 13-23b are unusually beautiful. For many semiconductors only a single peak is observed (Fig. 13-21b). However, even though only one line is observed the exciton states do make a sizable contribution to the magnitude of the absorption at and above the absorption edge. Thus, with a theoretical model the binding energy can be estimated. For example, the binding energy for the exciton at the direct gap in Ge is estimated to be 0.0011 eV. See the Notes. Table 13-5 lists exciton binding energy for several materials. Notice the huge variations from several meV in the Si-InP group to much larger values in the alkali halides.

The average radius of the exciton orbital in this hydrogenic model (a screened Coulomb potential energy) is

$$\langle r \rangle = \frac{\mu_H}{\mu} \varepsilon_\infty \left[\frac{a_0 n^2}{Z} \right] \qquad (13\text{-}57c)$$

CHAPTER 13 OPTICAL PROPERTIES OF CRYSTALS 513

Table 13-5 The binding energy of excitons in meV. (1 meV = 10^{-3} eV.)

Si	14.7	CdS	29	KCl	400
Ge	4.1	CdSe	15	KI	480
GaAs	4.2	BaO	56	RbCl	440
GaP	3.5			AgCl	30
InP	4.0			TlCl	11

where a_0 (0.529 × 10^{-8} cm) is the Bohr radius and Z = 1 for hydrogen or an exciton. Thus, the separation of the electron and hole for the n = 1 exciton state in Ge is considerably larger than the nearest neighbor Ge-Ge distance, which means that the weakly bound exciton model is self-consistent (which is also true for Si, Ge, GaAs, GaP, and InP). However, the weakly bound exciton model breaks down for other materials that have much smaller ε_∞ and effective masses closer to the normal electron mass. This leads to the Frenkel model discussed in Section 13-20e.

13-20c Exciton condensation When the density of weakly bound excitons is high, they can condense into a "liquid" and form an **electron-hole drop**. This has been observed in Si and Ge at low temperatures. At high exciton densities mutual interactions become very important. Since the separation between the electron and hole is large (Eq. 13-57c) the densities need not be exceedingly large before these exciton-exciton interactions dominate, leading to a condensed phase. The physics involved is similar to that for the metal-insulator transition, Section 8-6c. However, here the excitons are spatially mobile and can condense in certain regions of the crystal. The donors in the metal-insulator of course are fixed in space.

This condensed phase is just an electron-hole plasma with a binding energy of about 2 meV with respect to the free excitons. The phase occurs for concentrations of electrons (equal to the concentration of holes) of about 2 × 10^{17} cm^{-3}. Figure 13-24 shows the recombination radiation from electron-hole drops. The free exciton (FE) at 714 meV can be seen to be fairly broad due to Doppler broadening. The emission from the electron hole drop (EHD) is much broader. Its width is due to the kinetic energy distribution expected for electrons and holes in a Fermi gas of concentration 2 × 10^{17} cm^{-3}.

13-20d Excitonic complexes Silicon, germanium, and other semiconductors can be grown under extremely well-defined conditions and high purity. These materials have been a source of detailed information about excitons. A number of exciton complexes have been

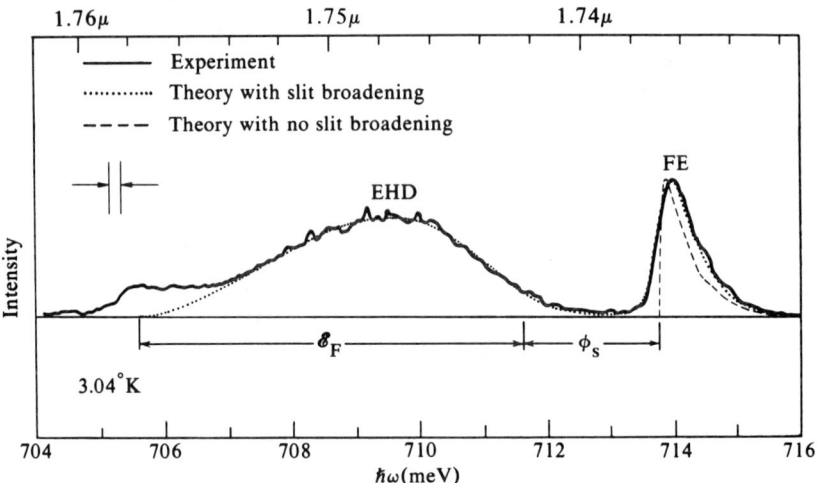

Fig. 13-24 Experimental luminescence spectrum from free (FE) and condensed excitons (EHD) at 3°K. The binding energy of the condensed phase with respect to the free excitons is ϕ_s and the Fermi energy of the condensed phase is $\mathscr{E}_F$. [From T. K. Lo, Solid State Comm. **15**, 1231 (1974).]

observed. For example, a complex of two free electrons and two free holes has been observed. It corresponds to a positronium molecule, and has a lower energy than two free excitons.

An exciton bound to a donor atom and to an acceptor atom also can be observed. Most of these observations are made by studying the recombination radiation or by using electron spin resonance techniques. We mention these complexes only so that the reader might appreciate some of the detailed interactions that can be observed.

13-20e Frenkel excitons So far we have talked about weakly bound excitons where the average distance between the electron and hole is much greater than the nearest neighbor atomic distance. In this section tightly bound excitons are discussed. These are associated with a single atom (or molecule in a molecular crystal), but they are able to travel through the crystal, that is, the exciton can hop from one atom (or molecule) to another. Such an exciton is called a Frenkel exciton. It is similar to an ordinary excited state of the atom (or molecule) except that the excitation can propagate, forming bands. The width of the band is determined by the rate of transfer of the excitation from one atom (or molecule) to the next.

Figure 13-25a shows the absorption spectrum of a thin film of solid krypton. The maximum values of α are very large ($\approx 10^6$ cm^{-1}). The two arrows in the figure correspond to the excitations of *atomic* krypton, that is, excitation of the $4p^6$ to the $4p^5 5s$ level which is split

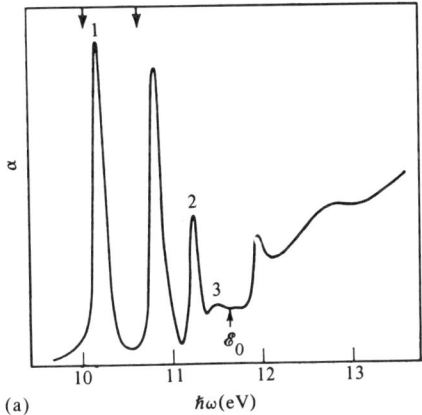

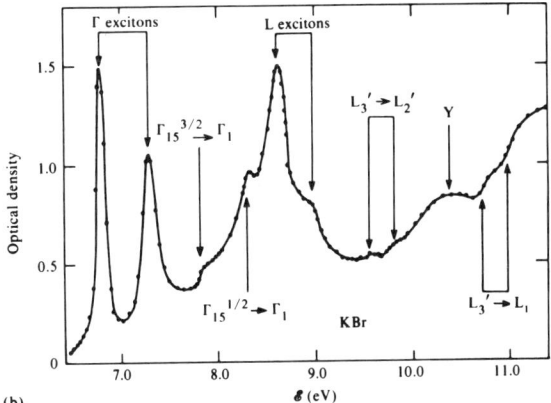

Fig. 13-25 (a) Absorption in a solid film of krypton at 20°K. [From G. Baldini, Phys. Rev. **128**, 1562 (1962).] (b) The absorption in KBr, [from J. E. Eby, J. J. Teegarden, and D. B. Dutton, Phys. Rev. **116**, 1099 (1959). For the labeling see the article by Phillips referenced in the Notes.]

due to spin-orbit coupling. The two lowest and strongest absorption lines have been interpreted as exciton lines, with energies nearly the same as the atomic excitation lines. From the Frenkel model the calculated radius of the lowest energy exciton is ≈ 2 Å.

Figure 13-25b shows the absorption edge for KBr, an alkali halide with the NaCl structure, interpreted in terms of excitons. The lowest energy excitons are expected to be localized on the negative halogen ions because their electronic excitation levels are lower in energy than those of the positive ions. The Br-ion has the same electronic structure as Kr and the lowest energy excitons correspond closely to excitations from the ground $4p^6$ level to the spin-orbit split $4p^5 5s$ level of the free Br^- ion. (This splitting is just like the spin-orbit splitting of the triply degenerate valence band of Ge shown in Fig. 13-12b.) As in the Kr case, the agreement with the free atom values is excellent. Comparison of isoelectronic systems such as Kr and Br^- is one way to help identify excitons. In Fig. 13-25b it is thought that the higher lying excitons have been identified. These are the spin-orbit split excitons in the ⟨111⟩ direction at the zone boundary (the L-point). At this point in reciprocal space $\partial \mathcal{E}/\partial k = 0$ for both the valence and conduction band, so excitons can be formed. The rest of the structure shows typical interband absorption, discussed in Section 13-21.

Molecular crystals readily show Frenkel exciton behavior. Anthracene, $C_{14}H_{10}$, is the most studied of these crystals. Its ground state is an $S = 0$ singlet; there is an $S = 1$ triplet at 1.80 eV, and another singlet at 3.15 eV. Although the singlet to triplet transition is not electric dipole allowed, it is allowed by higher order processes. The $S = 1$ state can be populated by lasers with energies ≈ 1.80 eV. These excitons propagate, and when two collide they combine to form a 3.15 eV exciton where the excess energy is given off by phonons. The singlet states decay rapidly to the singlet ground state by an allowed transition. Using fast pulses of laser light near 1.80 eV and fast detectors at 3.15 eV, both the lifetimes and spatial propagation effects of triplet excitons can be measured. Many other types of exciton studies in molecular crystals have been carried out.

13-20f Polaritons The interaction between photons and phonons, which leads to polaritons, has already been discussed. In the same manner, the coupled photon exciton system has the dispersion curve schematically shown in Fig. 13-26a. The dispersion curve for a free exciton ($\hbar\omega = \mathcal{E}_0 + Ak^2$, Eq. 13-57b) is indicated along with a plot of ω vs. k of light. In the vicinity of the crossing there is a strong interaction between the photon and exciton and a repulsion takes place in exactly the same way as for a photon-phonon interaction, Fig. 13-7. Such effects can be seen experimentally. Figure 13-26b, which is associated with Problem 8, shows similar effects due to coupling between a plasmon and a longitudinal optical phonon.

13-20g Donor-acceptor pairs In this section on excitons we discuss what happens when a <u>bound</u> electron and a <u>bound</u> hole are close enough to interact and form a "pair." The most commonly observed pairs are donor-acceptor pairs in which the electron is bound to a donor and the hole to an acceptor. When the electron (hole) is bound to the donor (acceptor) then the donor (acceptor) is neutral.

When a neutral donor and a neutral acceptor are close enough that their electron and hole orbits overlap slightly, the electron and hole can recombine and emit light. As the energy of this light is generally slightly less than the band gap energy, it is reasonable to think of the neutral donor-acceptor pair as a type of bound exciton. These pairs have the interesting property that the energy of the light emitted varies with the separation r_{DA} between the donor and accep-

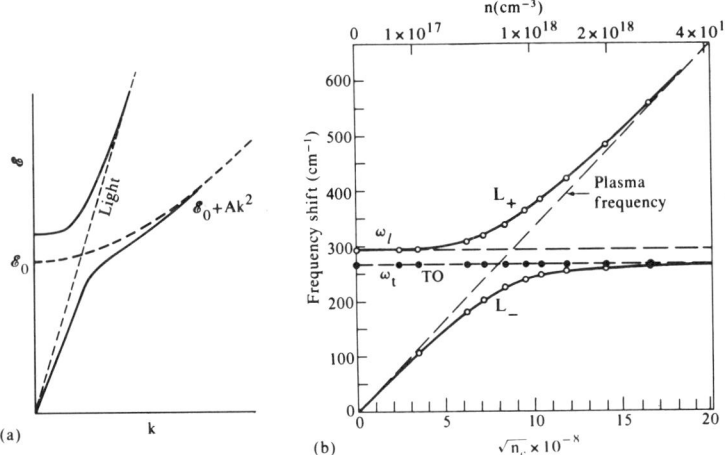

Fig. 13-26 (a) Dispersion curves for photons and excitons. The solid lines show the results when coupling is allowed. (b) Theory and experiment for coupled plasmon-phonon modes in n-type GaAs at 300°K. [From A. Mooradian and G. B. Wright, Phys. Rev. Lett. **16**, 999 (1966); A. Mooradian and A. L. McWhorter, "Proc. Int. Conf. Light Scattering Spectra Solids", (Springer-Verlag, 1969) p. 297.]

tor. They generate a whole spectrum of hundreds of emission lines, one for each possible value of r_{DA}.

The energy available to the emitted photon clearly is the difference between the energies of the state before and after recombination. In the initial state both the donor and acceptor are neutral and there is little interaction between them unless they are quite close. Thus, the energy equals nearly the band gap energy minus the binding energies $\mathcal{E}_D$ and $\mathcal{E}_A$, as is shown in Fig. 13-27a. After the electron and hole have recombined, however, there remain positive and negative ions separated by the distance r_{DA}. Thus, the final state is lowered and the energy of the emitted phonon is raised by the Coulomb interaction energy of these two ions, $\mathcal{E}_C = e^2/(\varepsilon_\infty r_{DA})$. This energy is large, on the order of 100 meV for close pairs but approaches zero as r_{DA} increases. This energy variation is shown in Fig. 13-27b, and two different spectra are shown in Fig. 13-27c. These figures illustrate two important features. (1) The spectra consist of discrete lines. This is because the donors and acceptors are substitutional impurities and occupy sites determined by the crystal structure. (2) The intensities of the lines vary erratically. Their magnitudes are proportional to the number of sites available at each separation and are a property of the crystal structure. It is these variations that enable the identification of the pair separation corresponding to each line. The lines in Fig. 13-27c reflect these relative intensities and the energies

518   CHAPTER 13   OPTICAL PROPERTIES OF CRYSTALS

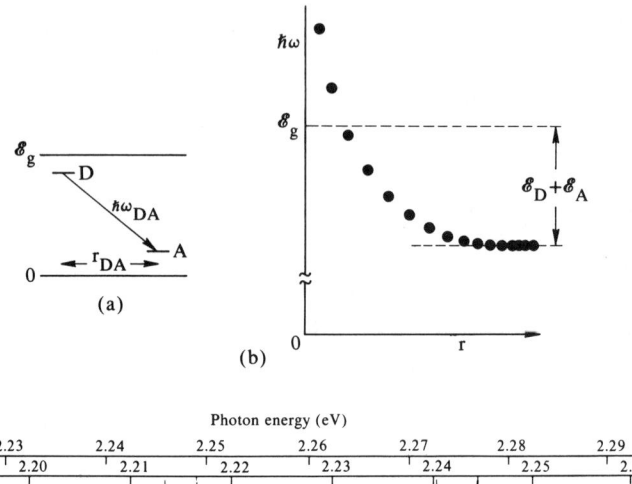

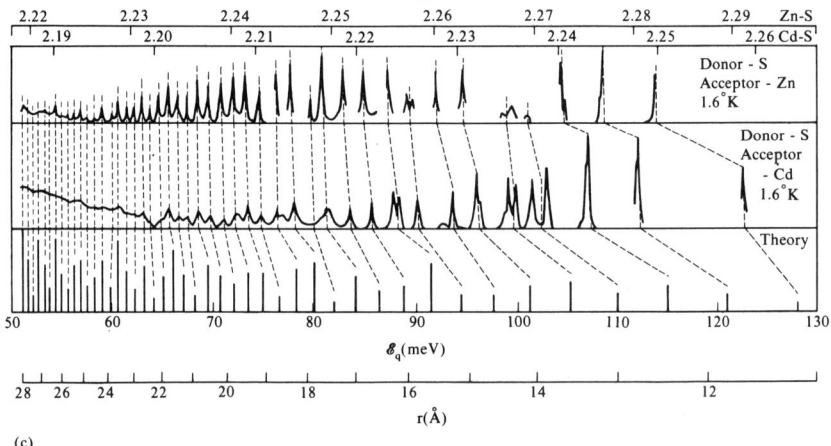

Fig. 13-27 (a) A transition between a donor and an acceptor which are separated by a distance r_{DA}. (b) The effect of the Coulomb interaction on the energy of the emitted photons. (c) A comparison of experiment and theory for two different pair spectra in GeP. The lower scales show r_{DA} and the Coulombic energy E_q derived from r_{DA} [From M. Gershenzon, R. A. Logan, D. F. Nelson, and F. A. Trumbore, "Int. Conf. on Luminescence", ed. G. Szigeti (Hungarian Acad. of Sci., 1968) p. 1737.]

$$\hbar\omega = \mathscr{E}_g - \mathscr{E}_D - \mathscr{E}_A + \mathscr{E}_C \qquad (13\text{-}58)$$

Corrections to these energy values increase as the pair separation decreases and higher order interactions become more important.

The book by Pankove (see the Notes) has more details, shows data analyzed in terms of Eq. 13-58, and gives references to the original papers including studies of different pairs in several materials.

CHAPTER 13   OPTICAL PROPERTIES OF CRYSTALS   519

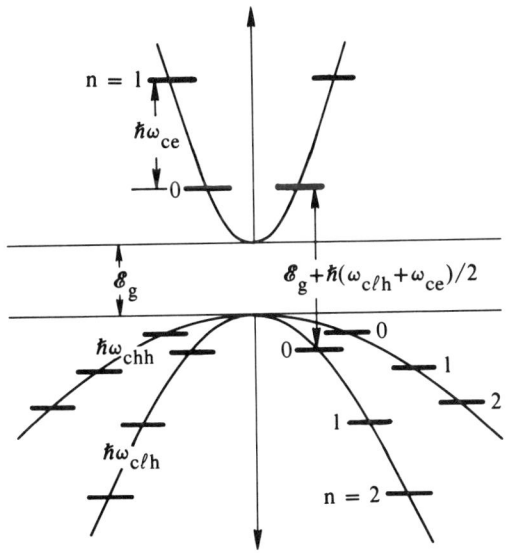

Fig. 13-28 Landau levels for typical semiconductor $k = 0$ bands. Note that the splittings of the Landau levels are larger for the light hole band than for the heave hole band, and largest for the electrons in the conduction band.

13-20h Magnetooptical absorption   The phenomenon of optical absorption involving transitions between different bands (i.e., interband transitions) in an applied magnetic field is termed magnetooptic absorption. We briefly discuss the interband magnetooptic effect in a typical semiconductor. The effect is simple to appreciate, and is capable of yielding the effective masses of the various bands.

Recall from Section 9-16 how, in the presence of strong magnetic fields, the $\mathscr{E}$ vs. k curves are reorganized into Landau levels (in particular see Fig. 9-12a). This reorganization is important whenever the carrier scattering time (τ) is longer than the cyclotron frequency ($\omega_c = eH/m^*c$), that is $\omega_c \tau > 1$. Just as Fig. 9-12a shows the Landau levels for a single, free electron band, Fig. 13-28 shows the Landau levels of the typical $k \approx 0$ bands in the semiconductors discussed here. This figure is essentially the same as Fig. 13-12a but now the Landau levels are shown. In Fig. 13-28 the spin splitting (Eq. 9-67b) is ignored; for qualitative discussions of semiconductors this is usually appropriate since the small effective masses cause the orbital splittings ($\hbar\omega_c$) to be much larger than the spin splittings ($gH\mu_B$). Note how the different effective masses give different orbital splitting.

The magnetooptic interband selection rules essentially are $\Delta n = 0$ (and $\Delta m_s = 0, \pm 1$); thus, optical absorption involves transitions between corresponding Landau levels in different bands. Although typically zero point energies are not observable, in Fig. 10-28 we point out that the sum of the zero point energy in the electron band plus either of the hole band zero point energies is measurable. Experimental absorption, for example, from a full valence band to an empty conduction band follow what would be expected from this figure with the selection rules that have been mentioned. In practice exciton effects are imposed on each of the band-to-band absorptions.

In a similar manner absorption can be seen from donor ground states to the Landau levels in the conduction band. However, in this case, the absorption is at much lower energies than those sketched in Fig. 13-28.

13-21 Fundamental Absorption Above $\mathscr{E}_g$

Since the early 1960s there has been a great deal of experimental and theoretical progress leading to a better understanding of interband transitions. As mentioned (Fig. 13-16), the density of the electron energy levels is complicated and not a very sharp function of energy. However, the situation is not as complex as we might think. First, since the momentum of light is small compared to the wave vectors throughout the Brillouin zone, the strongest optical interband transitions are vertical. Thus, the photon energy for absorption is given by

$$\hbar\omega = \mathscr{E}_c(\mathbf{k}) - \mathscr{E}_v(\mathbf{k}) \tag{13-59a}$$

where $\mathbf{k}$ is the same in the empty conduction and the filled valence bands. Second, the intensity of the absorption is proportional to the number of initial and final states and usually peaks when the conduction and valence bands are parallel in k-space, that is, where for one component of the vector

$$\nabla_\mathbf{k}[\mathscr{E}_c(\mathbf{k}) - \mathscr{E}_v(\mathbf{k})] = 0 \tag{13-59b}$$

Such places in k-space are called **critical points** or **Van Hove singularities**. Much can be learned about these critical points by studying the topology of the band structures, which is independent of the specific material. A simple discussion is in Chapter 3 of Wannier's book or Van Hove's original paper, Phys. Rev. **89**, 1189 (1953).

These two results, vertical transitions and critical points, make optical studies well above $\mathscr{E}_g$ quite useful. Two other tools have made this field more productive. The first is the experimental method of **modulation spectroscopy** or **derivative spectroscopy**. For example, instead of measurements of the intensity $I(\omega)$ vs. ω, $dI(\omega)/d\omega$ vs. ω is measured. This can show the spectral peaks much more clearly because the large, relatively energy independent, background does not appear. Also strain, electric field, temperature, and other externally applied variables can be used to modulate the absorption or reflection. These tend to emphasize various singularities of the density of states. Second, the semiempirical pseudopotential method of calculating energy bands has been very helpful in identifying the critical points and has given stimulus to the experimentalists.

Figure 13-29a shows some experimental and theoretical results for ε_2 of germanium. Notice how sharp the experimental peaks can be. In general, the interband transitions are not appreciably broadened by damping and the line shapes are determined from the density of states. Thus a great deal of information is available. However, good theory is required to get at this information.

Notation – The symbols such as Γ, L, X, Λ, and Σ refer to **special points** or **special lines** in the Brillouin zone, that is, special k-values or entire lines where **k** is special. The word "special" refers to the fact that under some (or all) of the symmetry operations of the point group, these k-values will transform into themselves. When these symbols have subscripts, they refer to the transformation (symmetry) properties of wave functions at these k-values and are convenient labels of the eigenfunctions and eigenvalues (Chapter 10).

Figure 13-29b shows early pseudopotential energy band calculations for Ge along several special lines in the Brillouin zone. For absorption studies of the type that lead to the results in Fig. 13-29a indirect band gaps are of no consequence because only vertical transitions are observed so the emphasis is on direct gaps. The smallest $\hbar\omega$ of any consequence, where Eq. 13-59b is satisfied, is the $\Gamma_{25'} \rightarrow \Gamma_2$ transition at **k** = 0. Notice that the first peak in the calculated results is from the $\Lambda_3 \rightarrow \Lambda_1$ transitions where the valence and conduction bands are parallel, and thus Eq. 13-59b is satisfied. This also occurs for the $\Sigma_4 \rightarrow \Sigma_1$ transitions. Thus, many of the peaks come from transitions along the special lines as well as from the special points in the Brillouin zone.

Figure 13-25b, besides showing excitons in KBr, shows higher interband transitions. For example, $\Gamma_{15}^{3/2} \rightarrow \Gamma_1$ and $\Gamma_{15}^{1/2} \rightarrow \Gamma_1$ are the transitions from the spin-orbit split valence Γ_{15} band (see Fig. 10-24) to the Γ_1 conduction band at **k** = 0. Clearly, good theory is required to interpret these data. This important field is full of rich detail, and the reader is referred to the Notes and Problems.

13-22 Urbach Edge

Introduction – We discussed optical absorption at the fundamental edge in crystals in Section 13-19. The electromagnetic radiation interacts with the electrons and excites an electron from a filled valence band across the fundamental gap into an empty conduction band. For direct transitions the crystal momentum of the electron in the valence and conduction bands is essentially the same, and the absorption constant for allowed transitions is $\alpha_{da} \propto (\hbar\omega - \mathcal{E}_g)^{1/2}$, Eq. 13-53a.

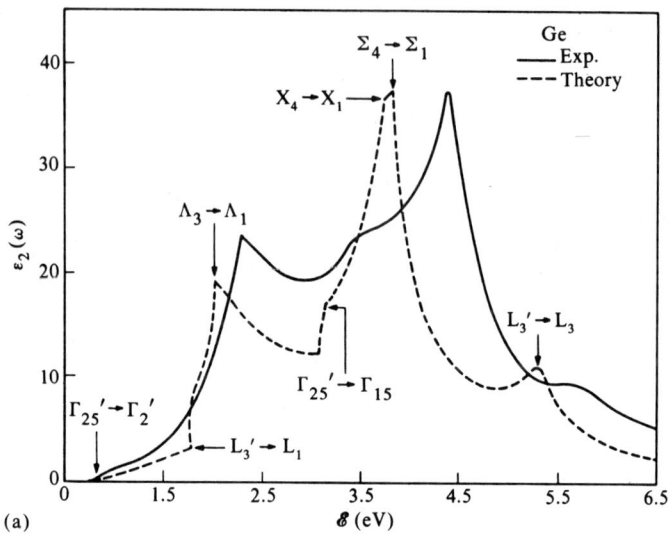

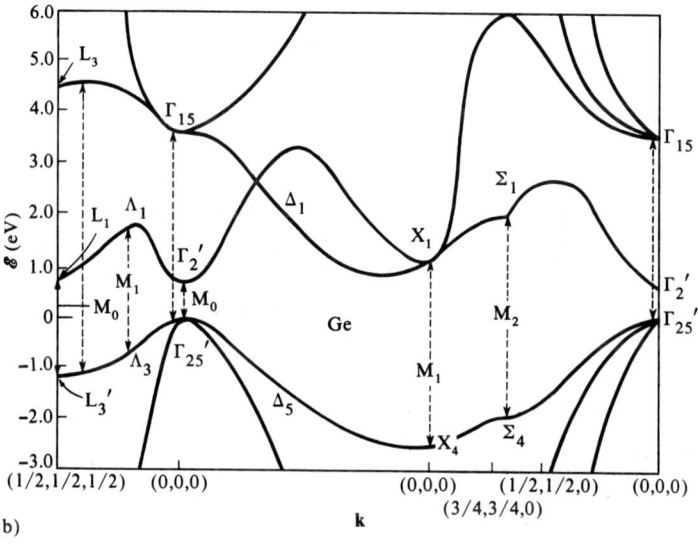

Fig. 13-29 (a) Spectral features of ε_2 for germanium. (b) The calculated pseudopotential energy bands for Ge along some of the principal axes. [From D. Brust, J. C. Phillips, and F. Bassari, Phys. Rev. Lett. **9**, 94 (1962); D. Burst, Phys. Rev. **134**, A 1337 (1964).]

This is a very strong absorption and α_{da} increases by several decades within a few tenths of an electron volt above $\mathscr{E}_g$. For indirect transitions the crystal momentum of the electron must change in a transition from the valence to the conduction band, the change being taken up by phonons. The allowed absorption, given by Eq. 13-55, is not as sharply rising as in the direct transition case.

These simple considerations are altered when the electron-hole interaction is taken directly into account. This interaction leads to a bound state of an electron and hole called an **exciton**. For the direct transition case, absorptions due to hydrogen-like states below $\mathscr{E}_g$ can be observed and the interaction changes the continuum absorption spectrum above $\mathscr{E}_g$ as well. For the indirect transition the electron-hole interaction changes the absorption as well.

Examples of allowed and forbidden, direct and indirect, absorptions have been observed in the many crystalline semiconductors.

Urbach Edge (1953) – There are many materials that are pure but that show an absorption edge different from the types discussed in Section 13-19. They show the so-called Urbach edge with the following empirical form for the absorption constant

$$\alpha = \alpha_0 \exp\left[\gamma(\hbar\omega - \mathscr{E}_0)/k_B T\right] \tag{13-60}$$

where γ is a constant, $\mathscr{E}_0$ is an effective gap, and T is the absolute temperature down to a critical value T_0 and $T = T_0$ for lower temperatures. Materials that exhibit this type of behavior have a more gradual dependence on $\hbar\omega$ than is observed in direct band gap absorption, and the edge becomes broader for temperatures $T > T_0$.

There is no agreement on a single explanation of the Urbach edge in crystals. However, most of the ideas center around internal, random electric fields that either broaden the absorption edge or the exciton states. There is little agreement as to the origin of the internal electric fields.

Amorphous semiconductors – Some properties of amorphous materials have already been discussed in Section 10-20. Amorphous silicon is becoming important technologically, for example, in solar cells. The general field of amorphous semiconductors is attracting increasing research interest. It is unfortunate that the Urbach edge in crystals is not understood because most amorphous semiconductors have this behavior. Figure 13-30 shows the absorption edge for a number of amorphous semiconductors. The Urbach behavior can be seen over a wide range of energy (the straight line portion). In amorphous materials a local electric field may arise from a static spatial variation due to lack of long-range order, variations in the density, or charged defects. However, definitive calculations are difficult to perform because of the lack of knowledge of specific parameters for

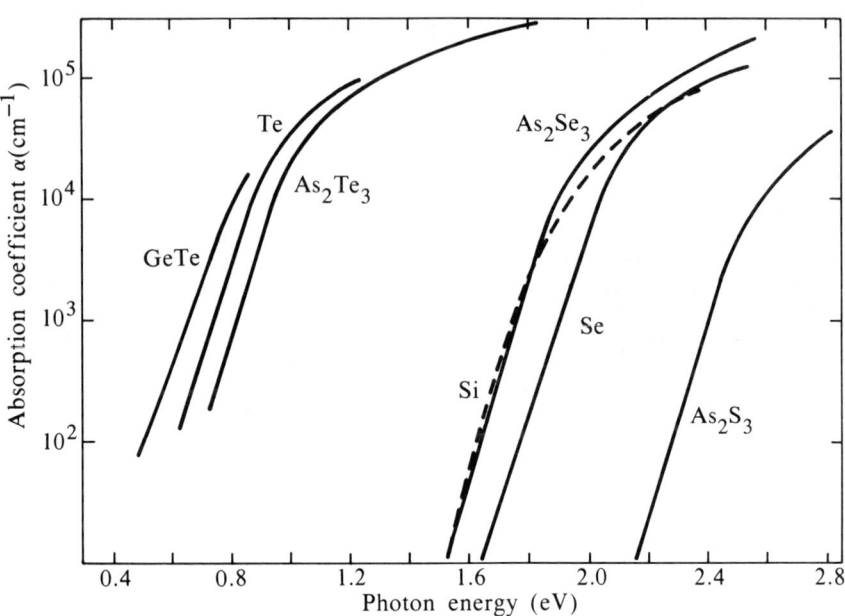

Fig. 13-30 The absorption coefficient for some amorphous semiconductors.

the different systems. The book by Mott and Davis has a good discussion and references to much of the original literature.

Notes

The "little book" by Hodson is a good reference for this chapter in general. It discusses plasma effects in more detail than is done here and covers surface plasmons as well. The book by Wooten is also worthy of reference. Also see F. Stern, Solid State Physics **15**, 299 (1963) and J. D. Axe's chapter on optical properties in "Solid State Chemistry and Physics" Ed. P. F. Weller (Dekker, 1973) Vol. 1, p. 411. "Principles of Optics" by M. Born and E. Wolf (Pergamon Press, 1970) is the classic text on the electromagnetic theory of propagation, interference, and diffraction of light. Optical properties of semiconductors, in particular, are covered in the books listed in the Bibliography under "Oriented to Semiconductors".

Much of the present day research work in the optical properties of solids centers around the use of **synchrotron radiation** to make measurements. It is discussed briefly in Section 17-6a and, due to the huge intensities available ($>10^6$ larger than conventional sources), is causing a revolution in many fields. Several references can be found in the Notes of Chapter 17.

The reader should become familiar with the text by Born and Huang. In particular Chapter II Sections 7 through 9 is a good reference for some of Part B in the present chapter. The papers by Szigeti [Trans. Faraday Soc. **45**, 155 (1949) and Proc. Roy. Soc. (London) **A204**, 51 (1950)] discuss the Szigeti charge.

The book by A. M. Karo and J. R. Hardy on the lattice dynamics of alkali halide crystals is useful as a general reference for lattice dynamic calculations. See also R. P. Lowndes and D. H. Martin, Proc. Roy. Soc. (London) A**308**, 473 (1969), where measurements of ε_∞ and $\varepsilon(0)$ for most of the alkali halides (and the Ag and Tl halides) are reported and discussed in terms of the Szigeti charge. They also present calculations of the second order Raman process. These materials serve as ideal probes for this process since first order (or one phonon) Raman scattering is not allowed by symmetry.

Several review articles on the optical properties of metals are informative. These are M. P. Givens in Solid State Physics, **6**, and M. M. Suffczynski, Phys. Stat. Sol. **4**, 3 (1964). Optical effects in metals near the plasma frequency have proved particularly interesting. For a discussion of spatial dispersion and the appropriate **additional boundary conditions** (**abc**) see F. Forstmann and R. R. Gerhardts, Festkorperprobleme/Advances in Solid State Physics **22**, 291 (1982).

The optical properties of semiconductors in particular are covered in a number of books. For example see: Bube, Pankove, and Moss, Burrell, and Ellis. A very extensive compilation of the band gap parameter of these materials can be found in the 1982 edition of the Landolt-Börnstein tables referenced in the Notes of Chapter 10.

Values for the binding energies for many impurities in Si, Ge, and GaAs are summarized in Sze, p. 21. Sze or Elliott and Gibson present discussions on photoconductivity. The absorption spectrum of As impurities in Ge [H. Y. Fan and P. Fisher, J. Phys. Chem. Solids **8**, 270 (1959)] is a good example of the use of optical techniques to determine fundamental properties of levels in semiconductors.

The book by Mott and Davis discusses many aspects of non-crystalline semiconductors, including the Urbach edge found in many crystals as well as in amorphous materials.

Two reviews of electron energy loss spectroscopy can be found in the "Springer tracts in modern physics" **38**, 84 (1965) and **54**, 77 (1970). The first by H. Raether and the second by J. Daniels, C. V. Festenberg, H. Raether, and K. Zeppenfeld.

For reviews on exciton condensation see Y. E. Pokrovskii, Phys. Stat. Sol. (a) **11**, 385 (1972); C. D. Jeffries, Science **189**, 955 (1975). For exciton complexes see Pankove's book mentioned previously.

Fundamental optical studies above $\mathscr{E}_g$ are reviewed in many places. A good place to start is J. C. Phillips, Solid State Physics **18**, 55 (1966) and M. Cardona, ibid, Supplement **11** (1969). The latter book covers **modulation spectroscopy**. Relatively simple absorption spectra above $\mathcal{E}_g$ are observed in rare gas crystals; this is reviewed in I. Y. Fugol, Adv. in Phys. **27**, 1 (1978).

Problems

1. The average rate of dissipation of energy for an electromagnetic wave is $W = \langle \mathbf{E} \cdot \mathbf{J} \rangle$ where the average is over a complete cycle. Show that $W = (\omega \varepsilon_2/8\pi)E_0^2 = \sigma_1 E_0^2/2 = \sigma_1 E^2$.

2. For the **Lorentz oscillator** show that σ_1 always peaks at ω_0 independently of γ/ω_0. For modest values of γ/ω_0 show that ε_2 peaks at $\omega_0[1 - a(\gamma/\omega_0)^2]$. When does ε_2 peak at $\omega = 0$?

3. Using time dependent perturbation theory, show that the results in Eq. 13-29 can be obtained for allowed electric dipole transitions. (See any quantum mechanics book or the article by Axe.)

4. Show that Eq. 13-44a follows from the earlier equations in the same section. (If you look in the book of Born and Huang, be careful; their relative displacement parameter $\mathbf{w} \equiv \mu^{1/2}(\mathbf{u}_+ - \mathbf{u}_-)/V_p^{1/2}$.)

5. The **Clausius-Mossotti** equation can be derived very simply without all of the apparatus of Section 13-12. $\varepsilon = 1 + 4\pi(P/E)$ and the polarizability of the ith type of atom is defined as $a^i \equiv p^i/E_{loc}^i$ where p^i is the dipole moment. Taking $E_{loc} = E + 4\pi P/3$ for all the atoms, derive the Clausius-Mossotti relation. Using the paper by J. R. Tessman, A. H. Kahn, and W. Shockley, Phys. Rev. **92**, 890 (1953) for a reference, how does the equation compare with experiment?

6. (a) Values for the conductivity are given in (ohms cm)$^{-1}$ in Fig. 13-10 and in sec^{-1} in Fig. 13-11. Find the relation between the two. (b) For the results in Fig. 13-12 for n-type Ge derive an expression for the absorption assuming $\omega\tau \gg 1$ and determine $(N/V)/m$ from the figure. Assuming that the electron effective mass is $1/8$ times the true electron mass what is the density of free carriers? Remember ε_∞ equals 16 for Ge and must be taken into account.

7. (a) Derive the **Hagen-Rubens** relation. (b) For Na at room temperature $\sigma_0 \approx 2 \times 10^{17}$ sec^{-1}. What is the value of τ? Notice that $\sigma_0\tau \gg 1$ and $\sigma_0 \gg \omega$ in the low frequency region. What are the values of ε_1, ε_2, n_1, and R?

CHAPTER 13   OPTICAL PROPERTIES OF CRYSTALS   527

8. Plasmon-longitudinal optic phonon coupling – Figure 13-26b shows the results from an experiment where the plasma frequency in GaAs was varied by using samples with different doping. In this manner the plasma frequency was varied from below to above the phonon frequencies. The results show clearly that the plasmon interacts with the longitudinal phonon and not with the transverse phonon. Show that the appropriate dielectric constant is

$$\varepsilon(\omega) = \varepsilon_\infty + \frac{[\varepsilon(0) - \varepsilon_\infty]\omega_0^2}{\omega_0^2 - \omega^2} - \frac{\omega_p^2}{\omega^2}$$

and solve for the roots of $\varepsilon(\omega) = 0$. Show that these solutions give the results in the figure. Discuss Landau damping of the plasma mode also seen in these experiments. [Besides the original paper, the book by P. M. Platzman and P. A. Wolff, "Waves and Interactions in Solid State Plasmas" (Academic Press, 1973) discusses this work and solid state plasmas in general.]

9. In two or three pages discuss the **pseudopotential method** of calculating the bands in simple solids. (Ashcroft and Mermin is a good place to begin but also see Solid State Phys. **24** for papers by M. L. Cohen, V. Heine, and D. Weaire.)

14

Ferroelectricity and Structural Phase Transitions

14-1 Introduction
14-2 The Free Energy
    a Second-order transitions
    b First-order transition
    c Summary
14-3 Soft Modes
    a Introduction
    b The soft mode idea in ferroelectrics
    c Other structural phase transitions
    d Comparison with experiments
    e The symmetry of the low temperature phases (S, A)
14-4 Microscopic Model of Soft Modes
14-5 Renormalization Group
14-6 Optical Properties of Ferroelectrics (S)
14-7 Other Related Properties
    Notes
    Problems

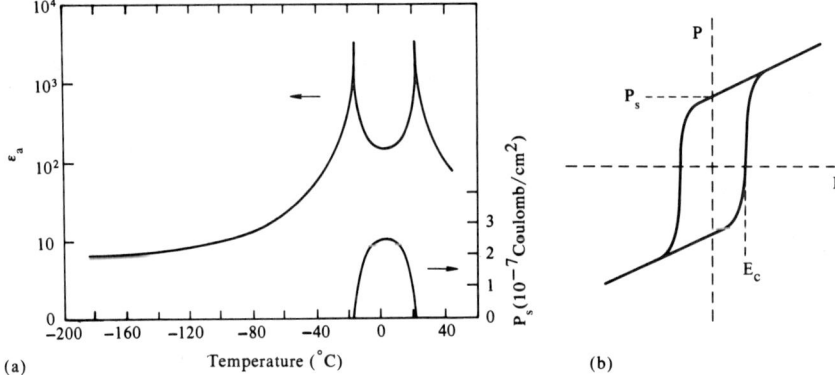

Fig. 14-1 (a) The temperature dependence of the real part of the dielectric constant. Note the log scale. The spontaneous polarization is also shown. (b) Polarization P of Rochelle salt vs. electric field E along the crystallographic a-axis showing a hysteresis loop that is observed in a ferroelectric crystal.

FERROELECTRICITY AND
STRUCTURAL PHASE TRANSITIONS

A modern physicist on examining Galileo's works is surprised to find how little experiment had to do with the establishment of the foundations of mechanics. His principle appeal is to the common sense and 'il lume natural'. He always assumes that the true theory will be found to be a simple and natural one.

Charles S. Pierce, *"The Architecture of Theories"*

The object of this chapter is multifold. First, it introduces the interesting field of ferroelectricity and the more general field of structural phase transitions. Second, a phenomenological free energy is used to describe the ferroelectric phase transition both above and below the transition temperature T_c. The use of such a free energy expression is widespread in solid state physics since it can effectively summarize a large amount of experimental data with very few "parameters" and so serves as the "meeting place" for theories. Last, it discusses the role of lattice dynamics and "soft" modes in causing certain ferroelectric and structural phase transitions. (However, there are many other kinds of phase transitions that are not covered by the "soft" mode theory, such as the changes from the NaCl to CsCl structure transitions, rotations of molecular units such as NO_2 in $NaNO_2$, order-disorder transitions such as in CuAu, and many others.)

14-1 Introduction

The field of ferroelectricity has a definite starting time unlike some other scientific topics. In 1921 the phenomenon was reported for the first time in Rochelle salt, $NaK(C_4H_4O_6) \cdot 4H_2O$, a crystal whose other interesting properties had been studied for some time. Figure 14-1 shows the observations. Two huge dielectric anomalies can be seen. Ordinary materials have $\varepsilon(0)$ values of the order of 10, while for Rochelle salt, values greater than 10^6 have been observed. Between the two temperatures a reversible **spontaneous polarization, P_s**, is observed. In Fig. 14-1b polarization is plotted vs. applied electric field and hysteresis is observed. The definition of a ferroelectric is a crystal in which a *reversible polarization* is observed. (The name arises via the analogy with the ferromagnetic case where we observe a magnetization that is reversible with an applied magnetic field.) Thus, if $E > E_c$ is applied in one direction of the polar axis all the dipoles align in that direction. E_c is defined as the **coercive field**. Setting $E = 0$ will give $P = +P_s$. However, if E is reversed and is $> -E_c$ all the dipoles will reverse (reversible polarization), and then if the electric field is set to zero, we have $P = -P_s$. Thus, the polarization can have one of two values, $\pm |P_s|$, depending on the sign of the electric field when it was shut off. This polarization appears spontaneously between T_c and T_L (the upper and lower transition temperature); hence the full name **reversible spontaneous polarization**.

P_s appears below a transition temperature T_c where the dielectric very sharply peaks (Fig. 14-1a); hence the word spontaneous. Thus, these two effects, a reversible spontaneous polarization P_s, and a peak in the dielectric constant at the transition temperature T_c are the key signatures of a ferroelectric material. We hasten to add that some crystals melt or decompose before T_c is reached. Nevertheless, a reversible polarization can be observed and indeed the crystals are ferroelectric. (And a few ferroelectric crystals show no dielectric peak at T_c for some subtle reasons. $Gd_2(MoO_4)_3$ is one example.)

In Chapter 5 we discussed which of the 32 crystal classes allows pyroelectricity. Since a ferroelectric crystal is a pyroelectric crystal, where the polarization can be reversed, the same symmetry considerations apply; so only in 10 of the 32 crystal systems can be found ferroelectric crystals.

For a long time Rochelle salt (and its deuterated analogue) was the only known ferroelectric materials. It has a complicated, monoclinic crystal structure. Thus, the phenomenon was considered an isolated peculiarity somehow associated with H_2O dipole-dipole forces, and it drew little attention. However, in 1935 ferroelectricity was discovered in KH_2PO_4 and its many **isomorphic compounds** (i.e., K can be replaced by Rb, Cs, or NH_4, and PO_4 can be replaced by AsO_4 and all of the deuterated analogues exist). Above T_c this material has a complicated tetragonal crystal structure. Both Rochelle salt and KH_2PO_4 display large changes in T_c upon deutration (the **isotope effect**). For example, $T_c = 123°K$ for KH_2PO_4 and $213°K$ for KD_2PO_4. Thus, the effect appeared to be associated with hydrogen bonds and still not of general interest.

It was not until 1945 that a simple ferroelectric material was reported, namely $BaTiO_3$; it has no hydrogen bonds and a simple (perovskite) crystal structure, Fig. 3-1c. There are a very large number of isomorphic materials, many of which are ferroelectric or have other structural phase transitions. The order of magnitude of the atomic motions in this crystal can be estimated from the room temperature value of $P_s \sim 8 \times 10^4$ esu cm^{-2}. Since the unit cell size is $(4 \times 10^{-8} \text{ cm})^3$ the dipole moment in one unit cell is $\mu \approx 5 \times 10^{-18}$ esu cm. This should be compared to a movement of the Ba^{2+} and Ti^{4+} ions by an amount 0.1 Å against the oxygen charge, which gives

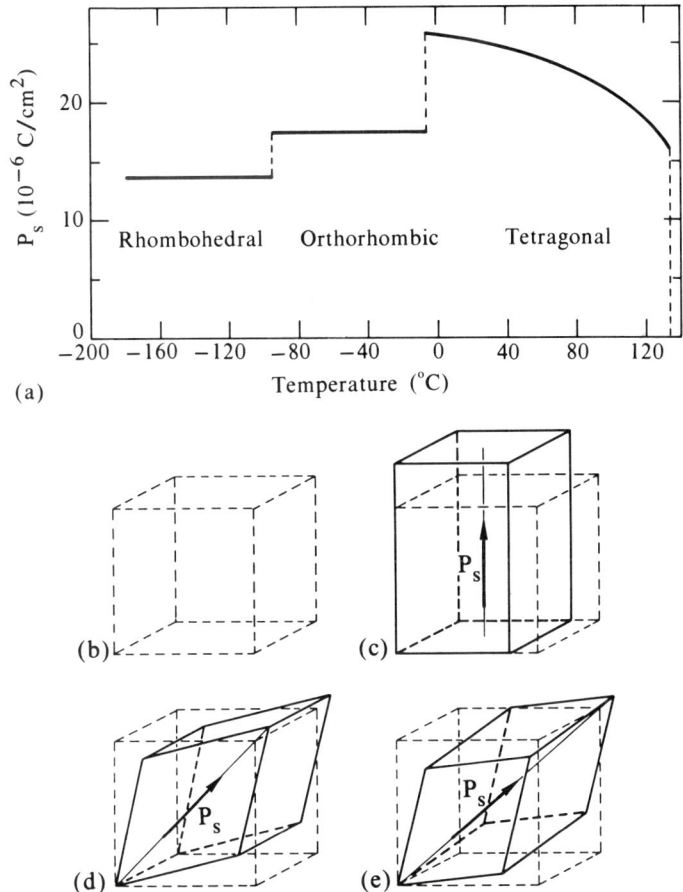

Fig. 14-2 (a) The measured temperature dependence of P_s in $BaTiO_3$. (b) through (e) The direction of **P**$_s$ with respect to the cubic axes in the various phases of $BaTiO_3$.

a dipole moment of $(6)(e)(0.1) \approx 3 \times 10^{-18}$ esu cm. So, contrary to the possible motions of hydrogen bonds, here the atomic displacements are much smaller than the internuclear distance and near T_c, where P_s can approach zero, even smaller displacements are possible.

Above T_c ($\approx 134°C$), in the paraelectric phase, $BaTiO_3$ is cubic. Below T_c the reversible spontaneous polarization develops along one of the six cubic $\langle 100 \rangle$ directions. In most crystals P_s develops along different $\langle 100 \rangle$ directions in different regions of the crystal. Thus, most crystals are "multidomain" below T_c. A **domain** is defined as a region in the crystal where all the dipole moments in the primitive unit cells are aligned parallel. However, if an electric field is applied to the crystal in one of the $\langle 100 \rangle$ directions as it is cooled through T_c, often a single domain crystal results. In $BaTiO_3$ there are other phase transitions at lower temperatures to lower symmetry structures. Below $-5°C$ the crystal becomes orthorhombic and P_s is along one of the original cubic $\langle 110 \rangle$ directions; below $-90°C$ there is yet another phase transition to a rhombohedral structure where P_s is along one of the original cubic $\langle 111 \rangle$ directions. Figure 14-2 shows experimental results for P_s vs. temperature and a sketch indicating the directions of P_s with respect to the original cubic axes. These measurements of P_s, were made by applying metallic contacts to two opposite cubic $\langle 100 \rangle$ faces. Thus, only projections of P_s in the two lower phases are measured and the experimental results should be multiplied by $2^{1/2}$ and $3^{1/2}$ to determine the true P_s.

The observation of multiple phase transitions in some crystals that at high temperatures have structures of high symmetry is common. Often, such crystals are unstable with respect to several types of distortions and as to which occurs first (i.e., at the highest temperature) is just a matter of some small balance of energies. However, the first distortion can be small, not affecting the overall energies in the problem and the other distortions can occur at lower temperatures.

The definition of several more terms will be helpful in our discussions. First, most phase transitions in pure materials can be classified as **first-order** or **second-order phase transitions**. A second-order phase transition is one in which there is no volume change at T_c. Thus, there is a continuous change in the structure across T_c. In a first-order phase transition the change in volume at T_c is discontinuous. For example, Fig. 14-3a shows a diagram of P_s vs. temperature for a first- and second-order phase transition. For the latter P_s goes continuously to zero although dP_s/dT is discontinuous at T_c (and there is a discontinuity in the specific heat at T_c). For a first-order transition, P_s has a discontinuity at T_c (and a latent heat accompanies the transition). The same continuous and discontinuous properties can be seen for the dielectric constant ε in Fig. 14-3b. Thus, the continuity or discontinuity of P_s at T_c can be taken to indicate if the transition is second- or first-order, respectively.

[By dielectric constant, ε, we always mean $\varepsilon_1(0)$, that is, the zero frequency clamped real part of the dielectric constant, for small applied fields **E**. We say "clamped" because for crystals that are piezoelectric there are size and shape dependent mechanical resonances. For 1 mm^3 sizes these resonances usually occur in the region of $10^4 - 10^6$Hz and measurements below these resonances will contain an extra contribution to ε. Measurements at frequencies above these resonances give "clamped" values. We say small **E** because ε can become so huge that nonlinearities in the $P = \varepsilon E$ relation can be seen easily, that is, $P = \varepsilon E + \varepsilon_1 E^2 + ...$ Small **E** is required to measure the "small signal" dielectric constant, ε.]

CHAPTER 14   FERROELECTRICITY   535

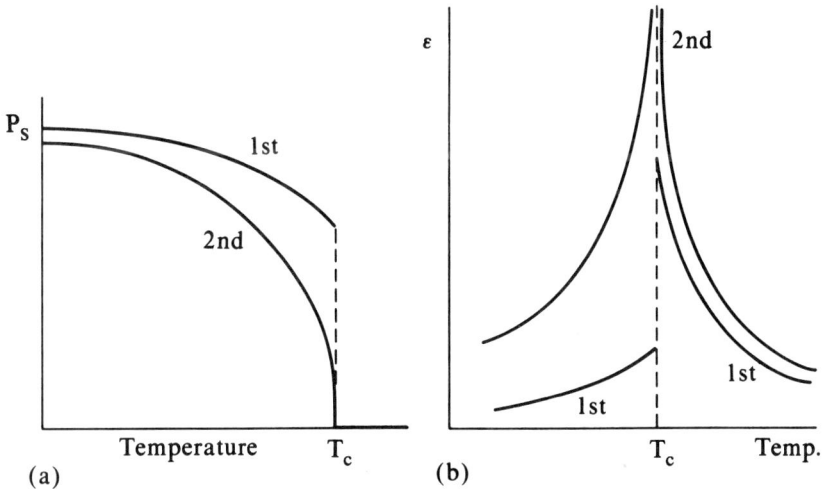

Fig. 14-3 (a) P_s vs. temperature for a first- and second-order ferroelectric phase transition. (b) The same but ε vs. T.

Two useful concepts are displacive and order-disorder phase transitions. These are not exact; they require microscopic knowledge that is often not available or even clear.

A **displacive phase transition** is one in which the displacements of the atoms in the low temperature phase from their positions in the high temperature phase are small compared to the nearest-neighbor internuclear distances. In fact, just below T_c for a second-order displacive phase transition the displacements can be infinitely small. For an **order-disorder transition**, below T_c the displacements are comparable to the internuclear distance. The phase transition in $NaNO_2$, discussed in Section 11-7b, is an example of a ferroelectric order-disorder transition. Above T_c the dipole moments of the NO_2 ions are randomly either pointing up or down parallel to the b-axis. Below T_c essentially all of the dipoles are pointing in the same direction.

In Table 14-1 a few representative ferroelectric materials are listed. (The number of known ferroelectrics is much larger, see the Notes.) For almost all of these materials, other isomorphic ferroelectrics exist. For example, besides $KNbO_3$ there is $NaNbO_3$ as well as crystals where Ta replaces the Nb. GeTe is ferroelectric and has the simple NaCl structure in the paraelectric phase, while the other materials have more complex structures. The list includes materials with first- and second-order phase transitions as well as crystals with displacive and order-disorder phase transitions. The dielectric constant is experimentally seen to vary with temperature T above T_c via a Curie-Weiss law, which determines a **Curie constant** C:

Table 14-1 T_c (in °C) of a few of the more than 1,000 known ferroelectrics. In the case of several phase transitions the higher one is given. (The materials are roughly grouped by structures. See the Notes.)

$BaTiO_3$	120	$KSr_2(NbO_3)_5$	156	$(NH_4)_2SO_4$	−49
$PbTiO_3$	490	$NaBa_2(NbO_3)_5$	560	$(NH_4)_2BeF_4$	−97
$KNbO_3$	435	$Gd_2(MoO_4)_3$	163	K_2SeO_4	−180
$LiNbO_3$	1210	KH_2PO_4	−150	$NaNO_2$	163
$LiTaO_3$	665	KD_2PO_4	−60	KNO_2	47
$PbNb_2O_6$	570	KH_2AsO_4	−174	$GeTe$	354
$PbTa_2O_6$	260	KD_2AsO_4	−114		

$$\varepsilon = \varepsilon(T) = C/(T - T_0) \qquad (14\text{-}1)$$

Notice that T_0 in the equation can be different from T_c. For a second-order phase transition they are equal (i.e., $\varepsilon \to \infty$ at T_c) but for a first-order transition T_0 is the temperature where ε would $\to \infty$ except that this temperature is never reached ($T_0 < T_c$). See Fig. 14-3b. From the value of C we get some idea whether the phase transition is of the order-disorder type or of the displacive type. Most displacive phase transitions have values of $C \approx 10^5\,°K$, while materials with order-disorder phase transitions tend to have values of $C < 10^3\,°K$. Materials like KH_2PO_4 seem to be of an intermediate type.

14-2 The Free Energy

With a thermodynamic theory many of the changes of the macroscopic properties at phase transitions can be interrelated, although such a theory does not give information about the microscopic causes of the transition. A thermodynamic, or phenomenological, theory that is used extensively to describe phase transitions is the **Landau free energy theory**. It is based on the idea of an order parameter. An **order parameter** measures the extent of the departure of the atomic (or electronic) configuration in the less symmetric phase from that in the more symmetric (high temperature) phase. One can say that the appearance of an order parameter at T_c breaks the symmetry of the high temperature phase; the order parameter is zero above T_c and nonzero below T_c. In ferroelectric order-disorder transitions the order parameter would be some measure of the amount of long-range ordering of the permanent dipoles, as was discussed in Chapter 11. For displacive transitions the order parameter could be some measure of the displacement of certain ions from their high temperature equilibrium positions. For most ferroelectrics of either the order-disorder or displacive type, the spontaneous polarization $\mathbf{P}_s$ can be taken as the order parameter. This is very convenient since $\mathbf{P}_s$ is an easily measura-

CHAPTER 14   FERROELECTRICITY   537

ble macroscopic quantity. In other structural phase transitions it is less clear what to choose and each case must be considered individually.

In this section we will consider the simple ferroelectric case with a spontaneous polarization P_s. For example, we have in mind the case of a crystal with the perovskite crystal structure where the transition is between the high temperature centrosymmetric cubic structure and a tetragonal ferroelectric structure. Then P_s is parallel to the tetragonal axis and the Landau free energy expansion is

$$F(T, P) = F(P = 0) + \frac{\alpha'}{2}P^2 + \frac{\beta}{4}P^4 + \frac{\gamma}{6}P^6 + ... \qquad (14\text{-}2)$$

$F(P = 0)$ (i.e., zero polarization) contains a wealth of information but it is associated with the nonferroelectric aspects of the material so it can be ignored here. Only even powers occur in this expansion because we assume that above T_c the crystal has a center of symmetry implying that the free energy only depends on even powers of the polarization.

For a second-order phase change, near T_0 the order parameter P becomes arbitrarily small. Thus, Eq. 14-2 expresses the continuity of the free energy in both states. For most first-order phase changes, for example, where the crystal structure changes from NaCl to CsCl, the free energies are the same in the two states at the transition temperature. However, the free energy in one state *cannot* be expressed as an analytic continuation of that in the other state. Nevertheless, many first-order *ferroelectric* transitions appear to be "very gentle," nearly second-order, (compared to the NaCl to CsCl structural change type), with very small volume changes, and the free energy, Eq. 14-2, is appropriate for both phases. This is an experimental result; namely, the coefficients of the order parameter in Eq. 14-2 are found to be the same in the two phases. Perhaps this should not be a surprise since, at least for displacive ferroelectrics such as $BaTiO_3$, the atomic displacements at the phase transition are small compared to the internuclear distances. Indeed, under modest hydrostatic pressures the phase transition almost is a second-order one. (However, note that fluctuations are being ignored. More will be said about this later.)

For a crystal under zero pressure but in an applied field E, thermodynamics tells us that

$$dF = -S\, dT + E\, dP \qquad \text{or} \qquad E = (\partial F/\partial P)_T \qquad (14\text{-}3)$$

From this we immediately see that the small signal dielectric constant, $P = \varepsilon E$, is given by

$$\frac{1}{\varepsilon} = \alpha' = \frac{T - T_0}{C} \qquad (14\text{-}4)$$

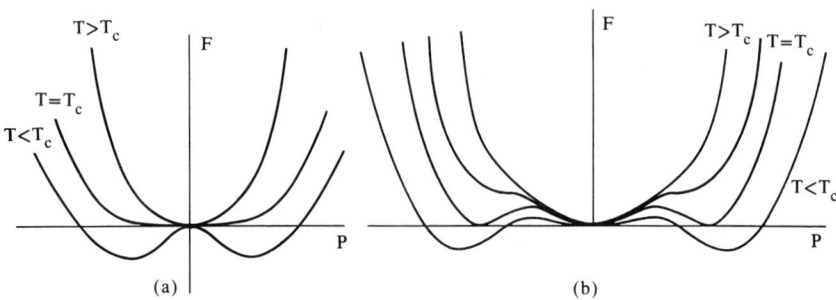

Fig. 14-4 The free energy vs. polarization at several different temperatures: (a) for a second-order phase transition; (b) for a first-order phase transition.

The first equality is obtained from Eq. 14-3 and Eq. 14-2. The second is obtained from experimental results, Eq. 14-1, and since this experimental result generally is found we write

$$\alpha' = \alpha(T - T_0) \qquad (14\text{-}2b)$$

where α is independent of temperature, and equals C^{-1}. This change of sign of α's is really a result of the fact that in the high temperature phase $F(T, P)$ must be a minimum for $P = 0$ and just below the transition (where for a second-order transition the order parameter is finite but very small), $F(T, P)$ must be a minimum for $P \neq 0$. This will be discussed later but now the free energy is rewritten as

$$F(P) = F(0) + \frac{\alpha}{2}(T - T_0)P^2 + \frac{\beta}{4}P^4 + \frac{\gamma}{6}P^6 + \ldots \qquad (14\text{-}2a)$$

The discussion is divided into a section on second-order transitions where β, in Eq. 14-2, must be positive, and first-order transitions where β must be negative, as will be shown.

14-2a Second-order transitions For a second-order phase transition $\varepsilon \to \infty$ at T_c so $T_c = T_0$, that is the transition temperature corresponds to $\alpha' = 0$ in Eq. 14-2 and Eq. 14-4.

In zero applied field, the minima in the free energy determine the thermal equilibrium condition for the polarization, P.

$$(\partial F/\partial P)_T = 0 = \alpha(T - T_0)P + \beta P^3 + \gamma P^5 \qquad (14\text{-}5)$$

$P = 0$ is always a solution of this equation and it obviously applies to high temperature paraelectric phase. For $\alpha(T - T_0)$, β, and γ all positive $P = 0$ is the only solution, which is fine since $T > T_0$. However, if $T < T_0$ and β is positive then there is another solution

CHAPTER 14 FERROELECTRICITY

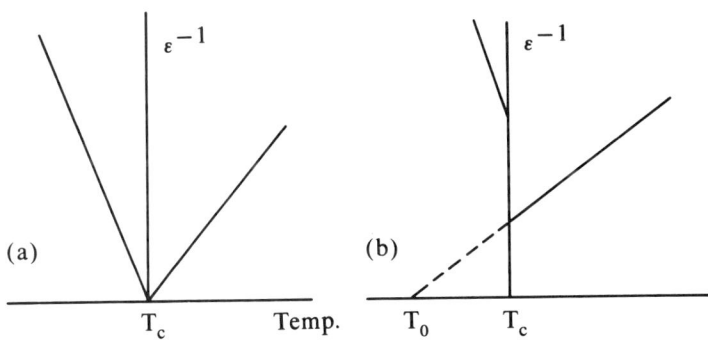

Fig. 14-5 The temperature dependence of ε^{-1} for (a) second-order phase transition; (b) first-order phase transition.

$$P_s = (\alpha/\beta)^{1/2}(T_0 - T)^{1/2} \qquad (14\text{-}6)$$

where we have neglected the γP^6 term (because near T_c it is small compared to the others) and have called this a spontaneous polarization P_s rather than just a polarization because it occurs for zero external applied field. Notice that β is required to be positive and that P_s changes continuously from zero for $T > T_0$ to a finite size for $T < T_0$. This result for P_s is shown in Fig. 14-3a. Figure 14-4a shows the free energy vs. P for T above and below T_0. Notice how the minima moves out from $P = 0$ for $T > T_0$, to finite values of spontaneous polarization below the transition temperature. The value of the coefficient α is obtained from the dielectric behavior above the transition and β via Eq. 14-6 from P_s just below T_c. The coefficient γ could be obtained from corrections to P_s at lower temperature.

We also can obtain another expression that relates to experiments by determining the dielectric constant in the ferroelectric phase just below T_c. Neglecting γP^6 terms again, $E = (\partial F/\partial P)_T$ so

$$E = \alpha(T - T_0)P + \beta P^3 \qquad (14\text{-}7a)$$

and

$$\varepsilon^{-1} = dE/dP = \alpha(T - T_0) + 3\beta P^2 \qquad (14\text{-}7b)$$

$$= 2(T_c - T)/C \qquad (14\text{-}7c)$$

where Eq. 14-7c is obtained by using P_s^2 given by Eq. 14-6. From Eq. 14-7c, $d\varepsilon/dT$ is just half as much below T_c as above. This is what is drawn in Fig. 14-3b for the second-order ε vs. T, but the better way to display this result is to plot ε^{-1} vs. T as in Fig. 14-5a.

To summarize, we have seen how for ferroelectrics the several coefficients in a general free energy expression can be related to experiment and are even overdetermined.

14-2b First-order transition For a first-order phase transition the coefficient β must be negative and γ positive, as will be seen.

Figure 14-4b shows F vs. P for negative β and positive γ. At T = T_c, F = 0 at P = 0, $\pm P_s$, thus there are two stable states for the crystal (i.e., ferroelectric and paraelectric). Just above this temperature only P = 0 gives the absolute minimum in F, and just below this temperature P = $\pm P_s$ gives the absolute minima. Thus, the polarization jumps from zero to P_s at T_c and this is a first-order transition. Notice, even for $T > T_c$ there can be a relative minimum in F vs. P; this will be important later in the discussion of double hysteresis loops. The signs of β and γ mentioned previously are required for the initially downward and then upward curvatures of F vs. P.

Proceeding analytically we write the general equation for P_s from $(\partial F/\partial P)_T = 0$ and the particular equation that at T_c, F = 0 in terms of P_s at T = T_c, that is, $P_s(T_c)$, we see that

$$(\partial F/\partial P)_T = 0 = \alpha(T - T_0) + \beta P^2 + \gamma P^4 \quad (14\text{-}8a)$$

$$F(T_c) = 0 = \frac{\alpha}{2}(T_c - T_0)P_s^2(T_c) + \frac{\beta}{4}P_s^4(T_c) + \frac{\gamma}{6}P_s^6(T_c) \quad (14\text{-}8b)$$

(Here it is necessary to keep terms up to the sixth order to obtain stable solutions.) From these two equations, when the first solved to find $P_s = P_s(T_c)$, we obtain two equations that can be used to evaluate the coefficients in the free energy expansion. These are

$$P_s^2(T_c) = -\frac{3}{4}\left(\frac{\beta}{\gamma}\right); \qquad \alpha(T_c - T_0) = \left(\frac{3}{16}\right)\left(\frac{\beta^2}{\gamma}\right) \quad (14\text{-}8c)$$

Notice that a negative β is required. $\alpha = C^{-1}$ is determined from experiment above T_c. T_c is directly measured. T_0 is readily found from a plot of ε^{-1} vs. temperature, as shown in Fig. 14-5b. Then by measuring the spontaneous polarization at the transition temperature and using Eqs. 14-8c both β and γ can be determined. Further (similar to the second-order case), using Eqs. 14-8c we can show that just below T_c: $\varepsilon^{-1} = 4(T_c - T_0)/C$, which is four times larger than $\varepsilon^{-1} = (T_c - T_0)/C$ just above T_c. These results are shown in Fig. 14-5b. Again, the coefficients in the free energy expansion are experimentally overdetermined.

Actually, expressions for the temperature dependence of P_s and ε below T_c can be obtained, and used to compare to experimental results. The expressions are

$$P_s^2 = (\beta/2\gamma)[1 + (1 - \tau)^{1/2}] \quad (14\text{-}9a)$$

$$\varepsilon^{-1} = (\beta^2/\gamma)(1 - T)^{1/2}[(1 - \tau)^{1/2} + 1] \quad (14\text{-}9b)$$

where $\quad \tau \equiv (4\alpha\gamma/\beta^2)(T - T_0)$

CHAPTER 14   FERROELECTRICITY   541

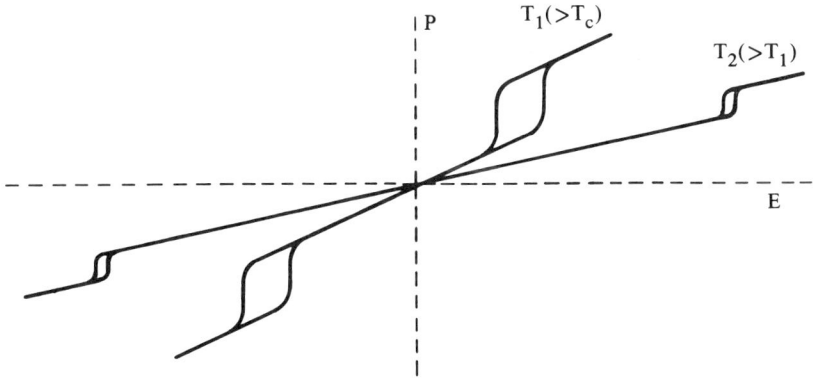

Fig. 14-6 P vs. E above T_c for a first-order phase transition showing double hysteresis loops. Note how the slope of P vs. E, at small E, decreases as the temperature is increased above T_c. This is because the dielectric constant is smaller.

For ferroelectrics with a first-order phase transition an interesting effect can be observed for $T > T_c$. **Double hysteresis loops** can be measured and Fig. 14-6 shows some results. This effect can be understood in terms of the free energy formalism and provides a nice example of its use. When a material is subjected to an electric field then a term $-EP$ must be added to the free energy expression, Eq. 14-2, as known from thermodynamics. In fact, it is just such a term that leads to Eq. 14-3. Then the stable state is found by minimizing this total free energy, including the $-EP$ term. The effect can be pictured easily using Fig. 14-4b where just Eq. 14-2 is drawn. To this figure we want to add a term $-EP$, which is just a straight line, though the origin with a slope of $-E$. Taking E as positive, and drawing such a line we can see that for a first-order transition this line can intersect a relative minima even for $T > T_c$. When this occurs a polarization results as shown in Fig. 14-6. The field required to induce the polarization, E_{ind}, can be calculated by minimizing $F - EP$ with respect to P.

E_{ind} becomes large as $T - T_c$ increases, as would be expected (and for very large $T - T_c$ a double loop cannot be induced because the curvature of F vs. P is such that a stable state does not exist for any value of E). Such loops are not observed in materials with second-order phase transitions, in agreement with what is expected from the free energy with a positive β. See the Problems for further details.

14-2c Summary We have seen how a rather simple free energy expression can be used to summarize a great deal of experimental data. In fact, the coefficients in the expansion are overdetermined. We have also seen that ferroelectric first- and second-order phase transitions are parameterized by the sign of the P^4 term and how the free energy formalism can be used to understand a perhaps unexpected phenomenon such as a double hysteresis loop. In the Notes there are references to other articles that extend this formalism to polarization in more than one direction and thus show how all three phase transitions in $BaTiO_3$ can be accounted for. Additional terms can be added to take into account effects due to stress and other macroscopic variables.

The Notes also contain references to papers that use microscopic models and calculate the coefficients in this free energy expansion, thus completing "the cycle" of experiment to microscopic model. We should make another point about the free energy expansion that has been "slipped in." Equation 14-2 gives the impression of an infinite series of terms that rapidly converges. Actually, for typical values of P_s the third term is as large as the second, which is often larger than the first. There is no apparent convergence. This happens often in solid state physics. In this field the first three terms are *required* to explain the experimental data; less would not do. So only three are *used*. Presumably, the important microscopic thermal averages that determine the α, β, and γ values contain most of the effects of the Hamiltonian. We should also point out that β and γ may be functions of temperature and in fact temperature dependences have been measured. These temperature dependencies simulate the effect of higher order terms in the free energy expansion.

The free energy expansion contains the symmetry of the problem but no microscopic information. It can be used to correlate a variety of different macroscopic experiments in terms of a very few coefficients. This is the strength of the theory. However, little is learned about the microscopic causes of these transitions. In the next section the problem is considered from a more microscopic point of view.

14-3 Soft Modes

14-3a Introduction If a crystal is about to make a phase transition to a structure of lower symmetry how can that be done? First, let us narrow the problem to second-order transitions of the displacive type or the "gentle" first-order ferroelectric transitions where the free energy seems to be continuous across the phase boundary. Thus, we have in mind a low temperature structure that differs from the high temperature by only *very small displacements*. This does not include reconstructive types of phase transitions such as the change from a NaCl to a CsCl structure or even the ferroelectric phase change for $NaNO_2$, for example, which is an order-disorder type because the atomic moments are of the order of the nearest-neighbor distances.

For such small displacements why do the atoms move? The answer to this question is simple: they move because the free energy is lower at the displaced position. Now, come back to the first question. How do they get there? For small displacements there is only one way to describe the motion, namely, normal vibrational modes. Of course, motion and static displacement are not the same thing but the magnitude of the motion is of the same order as the static displacements. Thus, we might suspect a relationship between lattice dynamics and structural phase transitions, in particular the "gentle" displacive ferroelectric transitions such as are found in $BaTiO_3$ and related materials.

14-3b Soft mode idea in ferroelectrics In 1959 Cochran made the connection between lattice dynamics and ferroelectricity. He emphasized that the Lyddane Sachs Teller relationship (Section 13-9c) relates the real part of the clamped static dielectric constant $\varepsilon_1(0)$, conventionally labeled $\varepsilon(0)$, and the optic dielectric constant ε_∞ to the vibrational frequencies, that is

$$\frac{\varepsilon(0)}{\varepsilon_\infty} = \frac{\omega_{LO}^2}{\omega_{TO}^2} \qquad (14\text{-}10)$$

where ω_{LO} and ω_{TO} are the longitudinal and transverse optic modes, respectively (LO and TO modes). Thus, if in the high temperature cubic phase of $BaTiO_3$, for example, when $\varepsilon_1(0)$ gets very large either ω_{LO} must get very large or ω_{TO} very small. We have seen (Section 13-12) that ω_{TO} can be considerably reduced from the value that it would have if only short-range forces were considered (i.e., the electrostatic dipole forces act to reduce ω_{TO}). In a similar manner we can show ω_{LO} is not affected very much. Thus, the increase in the dielectric constant must be caused by a decrease in ω_{TO} as T_c is approached from above. Normally, as temperature decreases, the frequency of most lattice vibrational normal modes increases. However, the TO mode that is causing the phase transition must decrease in frequency as T decreases in order to be consistent with Eq. 14-10. Such vibrational modes are often called **soft modes**. Presumably the phase transition itself is caused by the crystal becoming unstable with respect to displacements of a mode whose frequency is approaching zero.

This soft mode idea was soon verified experimentally by infrared, then neutron and Raman measurements. Notice, via Eq. 14-10 (or Eq. 13-36c) that as long as all the temperature dependence of $\varepsilon(0)$ is associated with one TO mode, its temperature dependence is predicted. By combining the experimental observations in Eq. 14-1 with Eq. 14-10, the behavior in the high temperature phase is

$$\omega_{TO}^2 = K(T - T_0) \qquad (14\text{-}11)$$

where K is a constant determined from the previously mentioned constants and frequencies. This temperature dependence is found experimentally.

Let us examine the soft mode idea more carefully. Figure 13-4 shows a unit cell of an ABO_3 material with the perovskite structure. Consider the lattice vibrational normal modes. The figure shows the displacement for all four of the allowed optic normal modes. All these modes are triply degenerate; the motion along only one axis is shown, and similar displacements along the other two ⟨100⟩ directions are allowed. Three of the four modes are infrared active, which means that there is an oscillating dipole moment, oscillating at the particular TO frequency of the mode, and these modes contribute to $\varepsilon(0)$ as in Eqs. 13-36b, 13-36c, or 14-10. The fourth mode is "silent" (not infrared or Raman active but can be seen with neutrons) and we shall ignore it. The soft mode idea is that the lowest frequency TO mode, which in this case is shown in Fig. 13-4a, is the soft mode. As T_c is approached from above, the frequency of this soft mode decreases with a temperature dependence given by Eq. 14-11, and this determines the temperature dependence of $\varepsilon(0)$ via Eq. 14-10. The decreasing frequency of ω_{TO} is another way to say that the restoring forces for this mode become weaker, causing the vibration to become slower. Finally, the restoring forces become so weak that the high temperature structure distorts into another structure (i.e., there is a phase transition). For a second-order phase transition ω_{TO} actually becomes extremely small. (In such cases, usually some other interesting effects occur as well, perhaps due to coupling to the acoustic modes or anharmonic forces completely dominating the behavior since the harmonic forces became so small.)

One of the outgrowths of the soft mode idea is that the displacements in the low temperature phase are closely related to the soft mode vibrational motion but now "frozen in." Thus, the distortion of the crystal structure found below T_c is just the frozen displacements pictured in Fig. 13-4a with an additional macroscopic strain, in which the crystal elongates along the direction of $\mathbf{P}_s$ and contracts perpendicular to $\mathbf{P}_s$. Since the eigenvectors of the normal mode vibrations can be calculated from a knowledge of the interatomic forces between the ions, this idea can be tested and it is found to be good.

There is another important concept associated with these ideas. Namely, what is the $\mathbf{k}$ value of these soft modes? The answer is $\mathbf{k} \approx 0$. It is only such modes that are involved with the Lyddane Sachs Teller relationship, Eq. 14-10, because this equation relates *macroscopic* quantities, the dielectric constants, to wave motion, and only wave motion with very large wave length (very small k) can directly effect macroscopic quantities. Figure 13-4a shows the normal mode motion in only one unit cell, but if we imagine neighboring unit

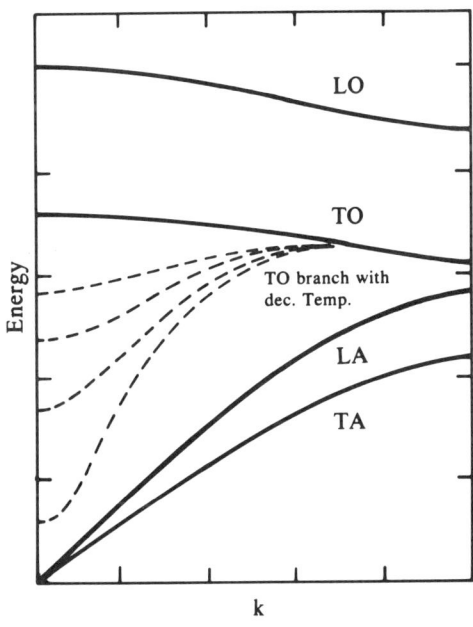

Fig. 14-7 A schematic diagram of $\mathscr{E}$ vs. k for the lowest TO branch in the paraelectric phase as the ferroelectric transition temperature T_c is approached from above. The successively lower dashed curves refer to temperatures closer to T_c.

cells in all directions with the same displacements, then we have $\mathbf{k} \approx 0$. If such motion has a very low frequency and "freezes in" one has the observed displacements in the tetragonal phase. Notice how each unit cell then has a static dipole moment, and they all add together to give a macroscopic polarization. The displacements from the high temperature structure positions are small and in fact a weak electric field can reverse them, changing the sign of P_s.

Figure 14-7 shows the temperature dependence of the lowest TO (soft mode) in a ferroelectric material. As T_c is approached from above, the $\mathbf{k} = 0$ frequency decreases. Since the $\mathbf{k} = 0$ part of the phonon branch is connected to the entire branch the rest of the TO branch tends to move down, as shown.

It also can be appreciated that while a P_s is observed, along the z-axis of this normally cubic crystal, this distortion can have a very small effect on the vibrational motion in the x- and y-directions. That is, the distortion is small so the effects on motions in the two perpendicular directions are relatively small. Thus, TO modes with very small $\mathbf{k}$ (≈ 0) in these two directions can still be soft and at temperatures lower than T_c these modes can cause other transitions. This is just what happens in $BaTiO_3$. At $\approx -5°C$ there is another phase transition due to just such an effect. The displacements are now, let us say, along the y-axis referred to the original cubic structure. The sum of the displacements along the z- and y-axes is a polarization along a $\langle 011 \rangle$ direction of the cubic structure. At still lower temperatures

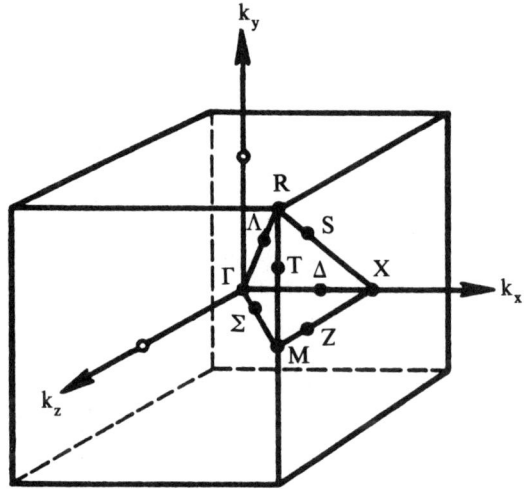

Fig. 14-8 The Brillouin zone of a simple cubic lattice with the labels of the special points and special lines.

displacements along the x-axis occur with a resultant P_s along an original cubic $\langle 111 \rangle$ direction. Thus, the one triply degenerate soft mode in the high temperature cubic phase of $BaTiO_3$ leads, in a very natural way via the soft mode picture, to the three different low temperature ferroelectric phases that are observed. So, interesting and complicated phase transitions and crystal properties can be understood in terms of the lattice dynamics with the simple soft mode concept.

The next question is: what causes these particular modes in these particular crystals to be soft? We defer this question until later, and first continue with the soft mode concept applied to other phase transitions. Again a wealth of new understanding is found.

14-3c Other structural phase transitions The soft mode idea was applied not only to ferroelectrics but to other "gentle" displacive phase transitions and it has met with similar experimental successes.

Reviewing the concept, Fig. 13-4a shows the displacements of the lowest frequency TO mode in an ABO_3 material with a perovskite crystal structure. Not shown, but implied, is that this unit cell should be translated throughout space in the three directions along with the displacements that are shown, resulting in a $k = 0$ mode.

Now consider soft modes for $k \neq 0$. A cubic ABO_3 perovskite crystal structure has a simple cubic lattice and the Brillouin zone is shown in Fig. 14-8. The various letters are the conventional letters for the special points and lines in the Brillouin zone. Figure 14-9a shows two unit cells of an ABO_3 perovskite structure with the B-ions oscillating along the z-axis but in opposite directions in every other unit cell in the x-direction. If these two cells are taken, along with their displacements, and repeated throughout space, we have a wave with $k_x = \pm \pi/a$ (the X-point). If a mode with k at the X-point were to become

CHAPTER 14   FERROELECTRICITY   547

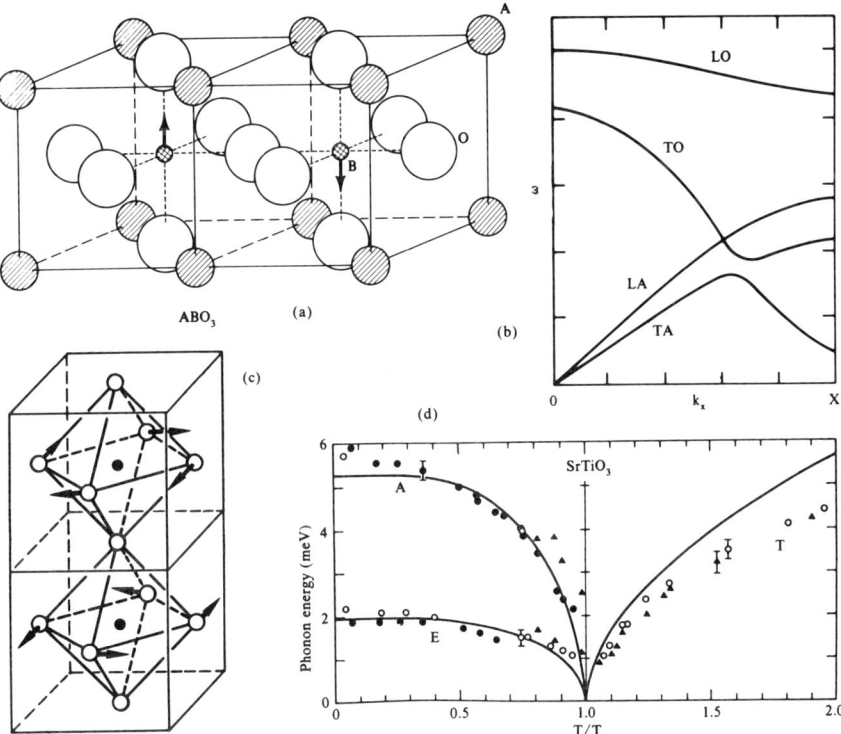

Fig. 14-9 (a) A conceptual diagram showing Ti-ion motion in opposite directions in neighboring cells. (b) A conceptual diagram showing a soft mode at the X-point in the Brillouin zone. Note the no-crossing of the transverse phonon dispersion curves about 2/3 of the way to the X-point. (c) An R-point vibrational mode. (d) The temperature dependence of the soft R-point mode in $SrTiO_3$. Above T_c neutron diffraction is used. Below T_c the soft mode is also measured by Raman techniques (solid dots). The lines are theoretical curves of J. Feder and E. Pytte Phys. Rev. **B1**, 4803 (1970).

soft (i.e., the restoring forces were to decrease as the temperature is lowered) the ω vs. k curves would be as shown in Fig. 14-9b. If a phase transition were to occur due to this soft mode the B-ions in every other high temperature unit cell would be displaced as indicated in Fig. 14-9a. The low temperature structure would have a unit cell length in the X-direction of two times what it was in the cubic phase. Since the direct space unit cell doubles, the Brillouin zone boundary in the k_x-direction halves and the ω vs. k curves fold over (Section 12-3b). Notice that such a crystal is not ferroelectric in the low temperature phase since the dipole moment in the cell on the left in Fig. 14-9a exactly cancels that on the right. Such a crystal is said to be **antiferroelectric** (i.e., neighboring lines of ions are displaced in opposite senses). Table 14-2 lists some antiferroelectric materials along with the transition temperatures. Since this phenomenon results in very few macroscopic changes, sometimes it is difficult to detect. However, there is often a peak or discontinuity in ε vs. T.

Figure 14-9c shows the motion (eigenvector) for one of the triply degenerate modes at $k = [1, 1, 1](\pi/a)$, that is, the Brillouin zone corner or R-point, Fig. 14-8. Notice that only the oxygen ions move for this triply degenerate lattice vibration and motion along the other two axes can be pictured readily. Such a mode is observed to be soft in $SrTiO_3$ and at $-105°C$ it freezes in to produce a tetragonal crystal structure with a doubled unit cell. Figure 14-9d shows experimental measurements of the soft mode frequency vs. temperature above and below the transition temperature. Above $-105°C$ the mode frequency is that of an R-point mode and the measurements were made by neutron scattering. However, below the transition temperature, due to the folding of the Brillouin zone, this soft mode now has $k = 0$ and the results shown are from Raman measurements. (Since the oscillations do not involve an oscillating dipole moment these modes at $k = 0$ are not infrared active.) In the tetragonal low temperature phase the formerly triply degenerate mode breaks up into doubly and singly degenerate modes labeled E and A, respectively.

14-3d Comparison with experiments Some experimental measurements have already been mentioned but here we discuss these and other results more fully.

Figure 14-9d is an excellent example of how the experimental results are in qualitative agreement with the soft mode model. The optic mode at the R-point decreases in frequency as the transition temperature is approached from above. In fact, the inelastic neutron diffraction measurements were made after it was predicted that the soft mode must be at the R-point. Further, the temperature dependence of the soft mode above T_c is given by Eq. 14-11, even though this equation is derived for soft modes in ferroelectrics. Below the transition temperature the modes split into singly (A) and doubly (E) degenerate modes, appear at $k = 0$, and increase in frequency with the E-mode lower. Last, the low temperature structure resembles the frozen in motion (the eigenvector) of the soft mode. All of these results are expected from the soft mode model.

To quantitively predict the dielectric constant, the infrared active soft modes at $k \approx 0$ must be studied. It is only for these that the mode frequencies can be related to $\varepsilon(0)$ via the Lyddane Sachs Teller relationship. When this is done in the high temperature cubic phase of several materials such as $SrTiO_3$, $KTaO_3$, and others, quantitive agreement is found with the prediction in Eqs. 14-10 and 14-11. The agreement can be further tested by a rather nice experiment that we

CHAPTER 14   FERROELECTRICITY   549

Table 14-2 Some antiferroelectric materials and their transition temperatures (°C). It is often difficult to determine if a crystal is truly antiferroelectric so there are many more crystals in which we are not sure. See the Notes.

$NaNbO_3$	354	$NH_4H_2PO_4$	−125
$PbZrO_3$	230	$ND_4D_2PO_4$	−31
Cd_2ScNbO_6	−203	$NH_4H_2AsO_4$	−57
$(NH_4)_2H_3IO_6$	−20	$ND_4D_2AsO_4$	31
$(ND_4)_2D_3IO_6$	−7	$Cu(HCOO)_2 \cdot 4H_2O$	−37
		$Cu(DCOO)_2 \cdot 4D_2O$	−27

now describe. Two materials, $SrTiO_3$ and $KTaO_3$, have very large $\varepsilon(0)$ values at low temperatures, reaching almost 10^4. Neither of these crystals are ferroelectric but they provide good possibilities to test the Lyddane Sachs Teller relationship, which is the cornerstone of the quantitive aspect of the soft mode model. The dielectric constant in these materials at low temperatures is large and nonlinear, that is, ε is a function of the electric field, $\varepsilon(E)$. By applying large electric fields at different temperatures, $\varepsilon(E, T)$ can be measured. Separately, by Raman scattering, ω_{TO} can be measured because the modes became Raman active for modest values of E. The measurements of ω_{TO} is in excellent agreement with the values predicted via the Lyddane Sachs Teller relation where the experimental values of $\varepsilon(E, T)$ are used. Another way to state this result is that the product $[\varepsilon(0)][\omega_{TO}^2]$ = constant, where the constant is independent of E and T, although the quantities on the left side change with E and T. This kind of experiment gives us confidence in the Lyddane Sachs Teller relationship and in the soft mode theory. (See the Notes for references.)

In $BaTiO_3$ the soft mode is highly overdamped (Section 13-6) so is more difficult to measure properly. However, recently it has been measured in the cubic phase by the hyper-Raman technique and found to be in agreement with the measured dielectric constant. The situation in the tetragonal ferroelectric phase is less clear. See the Notes.

14-3e Symmetry of the low temperature phases We briefly mention one aspect of the symmetry changes that may take place at a phase change. (See the Notes for further study.)

As the temperature decreases and T_c is approached from above in a second-order phase change, the structure changes continuously through T_c. However, the symmetry changes discontinuously because at T_c one or more symmetry elements disappear. Since symmetry elements disappear (and no new ones appear), the low temperature structure has a space group that is a subgroup of the high temperature space group. This puts strong restrictions on the low temperature structure. For example, the high temperature structure of $BaTiO_3$ has the space group O_h^1(Pm3m) with 48 point symmetry operations. Below T_c, essentially the Ti^{4+} ion moves along the crystalline +z-axis and the O^{2-} ions move along the −z-axis. This motion eliminates a large number of the symmetry elements and the resultant tetragonal space group, C_{4v}^1(P4mm), is a subgroup of the high temperature cubic space group. Landau's theory shows that only if the space group of the low temperature structure is a subgroup of the space group of the high temperature structure can the transition be a second-order one. Of course, it need not be a second-order one; it can be a first-order one. The point is that it *may* be a second-order transition. The $BaTiO_3$ transition is a first-order one; however under modest hydrostatic pressures it is almost second order. At atmospheric pressures, the perovskite material $K(Nb_{1-x}Ta_x)O_3$ has the same transition and is second order for a range of x.

It is interesting to note that if the displacement of all the Ti^{4+} ions were along one of the $\langle 110 \rangle$ directions then the low temperature space group would be C_{2v}^{11}(Cmm2) or if along one of the $\langle 111 \rangle$ directions it would be C_{3v}^5(R3m). Both of these are actually the space groups of the orthorhombic and rhombohedral phase. However, the phase transition to these states is not from the cubic structure. Rather neither of these structures have a space group that is a subgroup of the structure that is directly above it on a temperature scale so the phase transition may not be second order. This is in agreement with experiment where first-order phase transitions are observed.

14-4 Microscopic Model of Soft Modes

The soft mode idea beautifully gives us an understanding of many types of displacive phase transitions. As the temperature is lowered, a transition occurs because a certain type of lattice vibrational mode of the system becomes soft, that is, its restoring force becomes very small and the crystal is unstable to such a motion. There is a phase transition to a structure whose displacements from the high structure are essentially the eigenvector of the soft mode. The next question to ask is why does a mode in a particular crystal become soft? Why, for example, are there no soft modes in NaCl while GeTe, with the same crystal structure, has such modes? At present the answers to these questions are not totally clear but will be discussed.

In Section 13-12 we discussed, in the harmonic approximation, a microscopic model of an ionic solid with polarizable ions. Rewriting the resulting equation for the relative motion of the positive and negative ions in a transverse electric field, we have

$$\frac{d^2r}{dt^2} + \left[\frac{\alpha}{\mu} - \frac{(4\pi/3)(Q^2/\mu V)}{1 - (4\pi/3)(a_+ + a_-)/V} \right] r = cE \qquad (14\text{-}12)$$

where the symbols are defined in Section 13-12, and c is defined in Eq. 13-42c. The term in the square brackets is just the transverse optic mode (TO) frequency ω_{TO}^2 as discussed in that section; it is made up of short-range part $\alpha_S \equiv \alpha/\mu$ and a long-range part α_L, which is the rest of the expression. Thus, $\omega_{TO}^2 = \alpha_S - \alpha_L$. These two terms are both positive, so the long-range forces act to reduce the frequency that would be obtained if just the α/μ term were considered. Temperature never enters these considerations because everything was done in the harmonic approximation. If we want to allow for temperature effects, that is, allow for anharmonic effects, then one of the results is a damping term, $\gamma \dot{r}$, which we are familiar with, and another is a temperature dependence of the forces. The first term in this temperature expansion should be linear in temperature. For example, in the mid-temperature region $V = V_0(1 + \beta T)$ where β is the volume coefficient of expansion. Then Eq. 14-12 takes the form

$$\ddot{r} + \gamma \dot{r} + (\alpha_S - \alpha_L + \kappa T)r = cE \qquad (14\text{-}13a)$$

$$\omega_{TO}^2 = (\alpha_S - \alpha_L + \kappa T) = \kappa(T - T_0) \qquad (14\text{-}13b)$$

where $T_0 = (\alpha_L - \alpha_S)/\kappa$

The κT term could arise by considering the volume dependence on temperature in Eq. 14-12. However, a much larger effect is found to come from the effective anharmonic restoring forces that arise from the inclusion of anharmonic terms in the potential energy of the modes of vibration. In light of the soft mode model, Eqs. 14-13 contain a great deal of information. First, in writing them this way we are taking $\kappa > 0$ or else soft mode is not obtained. Of course, we should ask why κ is positive in some structures while negative in most others. Second, if we want $T_0 > 0°K$ then the long-range force effect must be larger than the short-range effect in the materials that have a phase transition. So in the harmonic approximation the electrostatic forces, which are pulling the positive and negative charges together are, in the high temperature equilibrium structure, larger than the short-range repulsive forces. Last, once Eq. 14-13b is written (with positive κ and T_0) then we have the form as in Eq. 14-11, and via the Lyddane Sachs Teller equation, we have the Curie-Weiss law Eq. 14-1, and the soft mode picture is understood.

Detailed microscopic theories that attempt to calculate the terms in Eq. 14-13b are complicated. Cowley has included anharmonic interactions between the normal modes of vibration and with few adjustable parameters he is able to fit various measured properties for the ferroelectric transition in $BaTiO_3$ and the structural transition in $SiTiO_3$. From these results he concludes that these types of displacive transitions can be understood in terms of anharmonic interactions between normal modes of vibration. Others claim that in the oxide perovskites anisotropic oxygen ion polarizabilities also contribute to the soft mode behavior. (The oxygen ions sit at a site of tetragonal symmetry so the polarizability is anisotropic.) See the Notes.

In **summary**, the soft mode model is conceptually simple and gives a good physical feeling for a large variety of displacive phase transitions. For some materials, such as $KTaO_3$ and $SrTiO_3$, there is excellent quantitive agreement between the experimental results via the Lyddane Sachs Teller relationship. However, there is still some controversy as to the reason why some modes in certain crystals have this unusual temperature dependence.

14-5 Renormalization Group

The Landau free energy expansion and the soft mode concept provide a simple model for thinking about phase transitions and give us a great deal of quantitative understanding. Nevertheless, these ideas are not completely satisfactory. It might be thought that the expansions used would be most accurate very near the transition temperature, especially for a second-order transition. However, particularly in nonferroelectric structural phase transitions, experimental evidence shows that some of the quantitative predictions of Landau theory are wrong near T_0. Figure 14-10 shows an experimental example of this failure. Landau theory predicts that the order parameter should have a temperature dependence $\propto (T_0 - T)^{1/2}$ while a result closer to $\propto (T_c - T)^{1/3}$ is found experimentally. The understanding of the reasons for the failure of Landau theory near T_0, and the development of the **renormalization group** (**RG**) approach, has been one of the successes of the last decade.

It might be supposed that the errors are inherent in the phenomenological nature of the Landau model, but this is not the case. Let us understand the important physical effects that have been left out of the Landau expansion. The missing element can be described using the soft model concept and Fig. 14-7. While the Landau theory treats the softest, "most critical" mode, it neglects the influence of other nearby modes with slightly different k-values, which while not as soft, are very numerous. Taking into account the collective effect of these "fluctuation" modes is a task of some subtlety, but appropriate techniques have been developed (renormalization group theory). This theory is based on the **notion of universality**; its essence is that in the critical region (very near T_0), physically measurable variables become

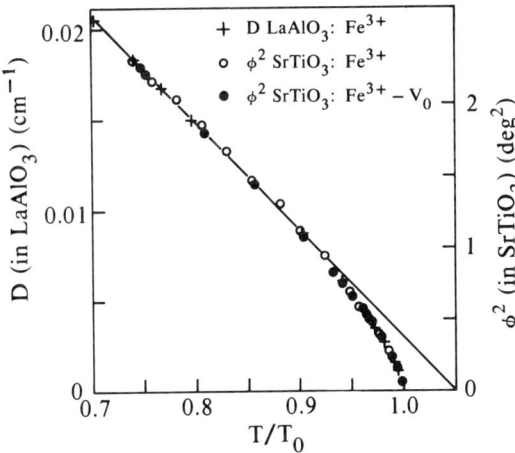

Fig. 14-10 Various experimental quantities that should depend on the square of the order parameter (ϕ) vs. reduced temperatures are plotted. $\phi \propto (T_0 - T)^{1/2}$ is obtained $T/T_0 = 0.7$ to 0.9. However, closer to $T/T_0 = 1$, $\phi \propto (T_0 - T)^{1/3}$ is obtained. (A different plot must be used to show this latter result.) K. A. Müller and W. Berlinger, Phys. Rev. Lett. **26**, 13 (1971).

substantially independent of the details of the particular interatomic interactions. This is because near T_0 the correlation lengths of the fluctuations become longer than the range of the forces between the particles making the critical behavior nearly independent of the details of the particular problem.

Using RG techniques, it is found that critical behavior depends upon the nature and distribution of the "nearly soft" modes. This allows materials to be grouped into a rather small number of **universality classes** which show identical critical behavior. As to which universality class a problem belongs depends on the essential qualitative features of the system. These features include: (1) the dimensionality of the system (perovskite crystals show three dimensional behavior); (2) the degeneracy of the soft mode in the high temperature phase (in the perovskites the soft mode is three-fold degenerate); (3) the range and angular dependence of the interatomic forces (long range Coulomb forces are in a different class from magnetic dipole-dipole forces); (4) the crystal symmetry.

We mention a few examples to give a flavor of the whole subject. It has been shown that uniaxial ferroelectrics should belong to a universality class in which $\varepsilon \propto (T - T_0)^{-1} [\ln(T - T_c)]^{1/3}$, a very weak modification of the Landau prediction $\varepsilon \propto (T - T_0)^{-1}$. However, it should be emphasized that these effects probably will be seen only in second-order ferroelectric phase transition; for first-order phase transitions one can not get close enough to the critical temperature (T_0) before the structure changes.

For nonferroelectric structural phase transitions critical exponents have been calculated which differ substantially from those predicted by Landau theory. For example, it is predicted that the order parameter should have a temperature dependence $P \propto (T_0 - T)^\beta$, where $\beta \approx 0.33$

rather than the Landau theory value of $\beta = 1/2$ (Eq. 14-6). As can be seen in Fig. 14-10, the RG result is very close to the experimental value.

14-6 Optical Properties of Ferroelectrics

One aspect of the optical properties of ferroelectric materials will be discussed, namely, the quadratic electrooptic effect and the biased quadratic electrooptic effect. The former is a very general effect and the latter leads to an important use of these and related materials.

For all materials, even glasses, there is a change of the optic index of refraction n (in Chapter 13 this was called n_1, i.e., the real part) proportional to the square of the polarization. Experimentally, we can apply an external electric field rather than a polarization. However, since the polarization is the more fundamental quantity the tensor is defined with respect to polarization, which always can be related to electric field. In full tensor form this effect is

$$\Delta \left(\frac{1}{n^2}\right)_{jk} = g_{jkmn} P_m P_n \qquad (14\text{-}14)$$

where g is the **quadratic electrooptic** (fourth-rank) **tensor** and all the indices go from 1 to 3 for the x-, y-, and z-directions. Since the effect is symmetric in the jk indices and separately in the mn indices this equation can be written using the contracted notation (Chapter 5) and we specialize it further by allowing the polarization to be along just the 3 or z-direction. Thus,

$$\Delta \left(\frac{1}{n^2}\right)_i = g_{i3} P_3^2 \qquad (14\text{-}15a)$$

$$\Delta n_i = n_i^f - n_i^0 = -(n_i^0)^3 g_{i3} P_3^2 / 2 \qquad (14\text{-}15b)$$

where i goes from 1 to 6. However, considering only cubic materials only n_1, n_2, and n_3 are nonzero. n^0 is the index with zero P_3 and n^f is the final value with the polarization applied. Note that as the polarization varies from zero to P_3 there is a change in index of refraction proportional to the *square* of the polarization.

It is not necessary to apply an electric field to obtain a polarization; a ferroelectric material may be cooled from above T_c to below T_c and a polarization exists. Figure 14-11a shows the results for a material closely related to $BaTiO_3$ except that in the high temperature phase the material is tetragonal, thus $n_3 \neq n_1$. The index of refraction n_3 is along the polarization direction and n_1 which is perpendicular to $P_3 = P_s$; and both g_{33} and g_{13} are positive with $g_{33} \approx 3g_{13}$. From these measurements, which can be done to high accuracy using a

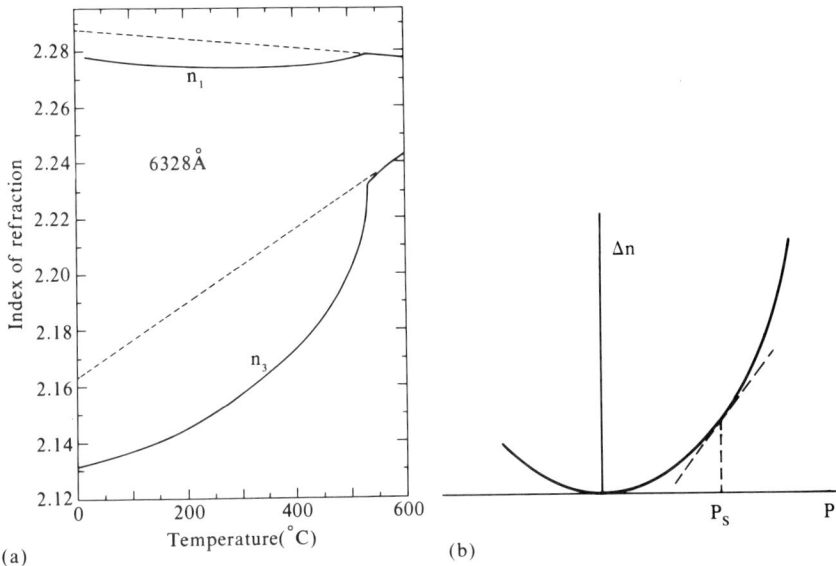

Fig. 14-11 (a) The index of refraction along the ferroelectric axis, n_3, and perpendicular to it, n_1 of a $K_6Li_4(NbO_3)_{10}$ type of material $T_c \approx 525°C$. Since the high temperature phase is tetragonal, rather than cubic, the indices of refraction differ above T_c. However, the deviations from the extrapolated high temperature phase show the P_s^2 dependence just below T_c. From A. W. Smith, et al, J. App. Phys. **42**, 684 (1971). (b) Δn vs. P showing quadratic behavior but also showing how, about a bias point P_s, linear behavior is observed.

prism of the crystal and the minimum deviation technique, values of g can be obtained. The results are $g_{33} = 0.13$ and $g_{13} = 0.03$ in units of $m^4/coul^2$. We can also determine the values of g for $T > T_c$ by applying an electric field and measuring the resultant Δn values. These two methods yield g values in agreement with one another. This experimental result further highlights the fact that the low and high temperature structures of these types of ferroelectrics are very similar and whether P_3 comes from an external field or from a spontaneous polarization the crystal behaves similarly.

The effect we have just discussed is interesting but it does not make ferroelectrics useful. We now extend these ideas and see how the linear electrooptic effect can be understood and the coefficient can be calculated. Consider the polarization in the high temperature centrosymmetric phase to be made up of a large P_s and a small part P_m but both along the 3 or z-axis. These could be obtained by a large low frequency electric field producing P_s and a smaller, higher frequency electric field producing P_m. Since P_m is small it can be related to an external applied electric field, E_3, via the usual (small signal) dielectric constant, ε_3, by $P_m = [(\varepsilon_3 - 1)/4\pi]E_3$ and since $\varepsilon_3 \gg 1$ the one can be dropped. Then substituting $P_3 = P_s + P_m$ in Eq. 14-15a, squaring and dropping the P_m^2 term, we have

$$\Delta\left(\frac{1}{n^2}\right)_i = g_{i3}P_s^2 + \left[\frac{g_{i3}P_s\varepsilon_3}{2\pi}\right]E_3 \quad (14\text{-}16a)$$

$$\Delta n_i^E = -\left(\frac{n_i^0}{2}\right)^3\left[\frac{g_{i3}P_s\varepsilon_3}{2\pi}\right]E_3 \equiv -\left(\frac{n_i^0}{2}\right)^3 r_{i3}\,E_3 \quad (14\text{-}16b)$$

In the second expression only the change in the index associated with the small applied electric field is written. The term in the square brackets defines the **linear electrooptic coefficient**, it is a third-order tensor since it relates a second-order tensor to a first-order tensor. (Remember the first index on r_{i3} goes from 1 to 6 because we are using the contracted notation.) The important physical point is that for small applied field the index of refraction is linear in E_3. Figure 14-11b shows what is happening. From the general, ubiquitous quadratic electrooptic effect $\Delta n \propto P^2$. However, if a large biasing polarization, P_s, is applied, then about this biasing polarization linear effects are seen. Naturally, this large polarization can be taken as the spontaneous polarization in a ferroelectric material. Further, we can measure r_{i3} in the ferroelectric phase and determine if agreement is obtained with the expression in Eq. 14-16b. See the Notes.

Thus, we have seen how the biased quadratic electrooptic effect gives a linear electrooptic effect. Whether the biasing is caused by a large low frequency field or by a spontaneous ferroelectric polarization is of no physical importance.

14-7 Other Related Properties

Most of the discussion in this chapter has been specialized to ferroelectric phase transitions, although many of the ideas are more general. In this last section we mention some related effects.

Pyroelectricity – This topic was discussed, from a symmetry point of view, in Section 5-4a. A crystal is pyroelectric if it possesses a net dipole moment per unit volume. Thus, all ferroelectric crystals are pyroelectric but there are other crystals that are pyroelectric but where the dipole moment is not reversible by an external electric field. Crystals like ZnO and CdS (with the wurtzite structure, Fig. 3-10) are pyroelectric. There are also many ferroelectric crystals that electrically break down at temperatures much lower than T_c, before the polarization reverses. Such crystals can be called pyroelectric; $LiNbO_3$ and $LiTaO_3$ at room temperatures are examples because the barrier between the two states with $\pm P_s$ is larger than the breakdown field.

CHAPTER 14   FERROELECTRICITY

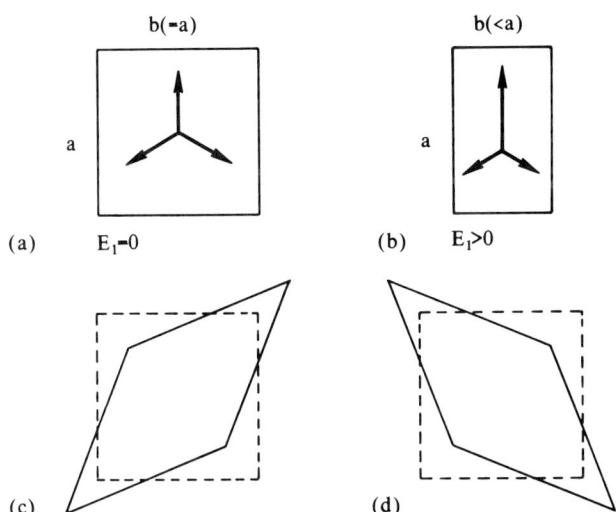

Fig. 14-12 (a) and (b) A schematic diagram showing how the application of an electric field can cause a strain. (c) and (d) A schematic diagram showing the two bistable states of a ferroelastic crystal.

Piezoelectricity — A piezoelectric crystal is one in which there is a linear relation between an applied electric field and a resulting strain. The constant that relates the applied field, E_i, to the strain, e_{jk}, is a third-order tensor, d_{ijk}, and the effects of symmetry on the various components of the tensor were discussed in Section 5-4c. All pyroelectric crystals are piezoelectric but the converse is not true. Figures 14-12a and 14-12b indicate how a unit cell made up of three dipole moments, with zero resultant polarization, can be effected by an electric field. The dipoles in the direction of the field increase while those oppositely directed decrease causing elongation as shown. Crystalline quartz, the most useful piezoelectric material, has point group $D_3(32)$, which because of the twofold axis perpendicular to the principle (threefold) axis cannot be ferroelectric. For quartz the values of d_{ijk} are such that for 10^2 volt/cm the strain $\approx 10^{-8}$; the details depend on the direction of the applied electric field. For a ferroelectric such as $BaTiO_3$ the same field will produce strains about 10^2 times larger.

Of the 32 point groups, 20 are piezoelectric and of these 20, 10 are pyroelectric (sometimes called polar). These latter 10 can also be ferroelectric, as discussed.

Ferroelasticity — A crystal that has two or more stable orientational states and can be switched from one to the other with an external mechanical stress is said to be ferroelastic. An example is shown in Figs. 14-12c and 14-12d, whereby the application of stress can cause the crystal to interchange the a- and b-axes. Also shown is the high temperature reference state. The crystal $Pb_3(PO_4)_3$ is a ferroelastic crystal that goes from the high temperature reference state to the bistable states shown.

$Gd_2(MoO_4)_3$ and KH_2PO_4 are interesting ferroelectrics as well as ferroelastic materials. In their high temperature phase they are tetragonal but below T_c they become orthorhombic, point group C_{2v}(mm2). When observed along the c-axis the crystal is in one of two orientational states as shown in Figs. 14-12c and 14-12d with P_s out of the plane of the paper in Fig. 14-12c and into the plane of the paper in Fig. 14-12d. Since the direction of P_s and the sign of the strain are linearly related to each other, if P_s is reversed so is the strain. Thus, by reversing P_s, the orientation of the crystal is effectively changed by 90°. This effect is called the **longitudinal electrooptic effect,** that is, for light traveling along the c-axis the birefringence is rotated by an electric field along the c-axis.

Antiferroelectricity – We have already defined this term in Section 14-3c and, in Table 14-2, listed some representative antiferroelectrics. Often, but not always, just below the transition temperature the free energy difference between the antiferroelectric (AF) and ferroelectric (F) states is very small and with the application of large electric field one may switch from the AF- to the F-state. For such a situation a double hysterises loop is observed (Fig. 14-6). As the temperature is further reduced from the phase transition, the electric field required to induce the F-state increases.

Other related transitions – The phase transition in $SrTiO_3$ is discussed in Section 14-3c and the soft mode in the high temperature phase and displacements in the low temperature phase are shown in Fig. 14-8c. It is a good example of the "gentle" displacive phase transition that can be understood in terms of the soft mode theory. Soft modes can, in principle, occur throughout the Brillouin zone and phase transitions due to many such modes have been observed.

Nonsoft mode transitions – It should be emphasized that there are very many phase transitions that cannot be understood in terms of soft vibrational modes. The reconstructive phase transition from the NaCl to CsCl structure is one example that has been discussed before. Another example is the transition from the tetrahedrally bonded structures (such as Si or AlSb) to the NaCl or β-tin structures as discussed in Section 8-6c. See the Notes for several references to the very wide field of phase transition in crystals.

Martensitic transformation – This term is used principally by metallurgists to describe a phase transition in a ceramic or metallic alloy that involves a change in crystal structure and which occurs by nucleation and diffusionless growth. The term diffusionless means that the atomic displacements are small compared to interatomic distances. This would apply to the phase transitions in $BaTiO_3$ but not to those in $NaNO_2$. Also it would not apply to order-disorder phase transitions in CuZn alloys, addressed in the problems in Chapter 11, since the atoms must move internuclear distances to get into their ordered state. For references see J. W. Christian, "The Theory of Transformation in Metals and Alloys" (Pergamon Press, 1965); H. Warlimont and L. Delaey, Prog. Materials Sci. **18**, 1 (1979); A. A. Golestaneh, Physics Today **37**, 62 (1984). The latter article discusses how certain martensitic transformations give materials properties that depend on their histories, allowing them to recover earlier shapes after apparent plastic deformation (i.e., **shape-memory phenomena**).

High pressure effects – Under high pressures there can be large variations of T_c and other phase transitions can be induced that are not observed at atmospheric pressure. Routinely achievable high pressures (100 kbar = 10 GPa) are about ten times more effective than temperature changes (of $1000°C$) in producing phase transitions. This is because pressure is a more efficient way to alter the interatomic distances. Pressure has the added advantage in that it can be applied isothermally so the thermal energy of the system is not changed; this can simplify the interpretation of resulting changes of internuclear distance and resulting phase changes.

As a result of the development of the **diamond anvil high pressure cell** a great deal of progress has been made since the early 1970s. This cell is small, relatively inexpensive, portable, and transparent (through the diamond) to x-rays and visible light. Thus, x-ray diffraction, Raman spectroscopy, photoluminescence, and related techniques can be used to measure the properties of crystals. For x-ray measurements, synchrotron radiation is preferable to conventional sources, because of the much higher intensities (Section 17-6a). See E. F. Skelton, Physics Today **37**, 44 (1984); A. Jayaraman, Rev. Mod. Phys. **55**, 65 (1983); G. A. Samara and P. S. Peercy, Solid State Physics **36**, 1 (1981).

Notes

The field of phase transitions is an extremely broad one; we have only touched on a small segment of the phase transitions that occur just in solids. Useful general references are C.N.R. Rao and K. J. Rao, "Phase Transitions in Solids" (McGraw-Hill, 1978), and G. Careri, "Order and Disorder in Matter" (Benjamin/Cummings, 1984). For a review of light scattering studies of structural phase transitions see J. F. Scott, Rev. Mod. Phys. **46**, 83 (1974) and the conference proceeding in Ferroelectrics **52**, 1-263 (1983). See G. Shirane, Rev. Mod. Phys. **46**, 437 (1974), J. D. Axe and R. M. Nicklow, Physics Today, Jan. 1985, page 27, and the Notes in Chapter 12 for reviews of neutron diffraction studies. From the application of **nuclear magnetic**

resonance techniques to study structural phase transitions has been reviewed by A. Rigamonti, Adv. in Phys. **33**, 115 (1984). For reviews of phase transitions under high pressure see the previous paragraph.

For a review of the free energy formalism, as discussed here, see A. F. Devonshire, Advances in Physics **3**, 85 (1954). After W. Cochran proposed the soft mode model in 1959 he wrote two review papers that are both well worth looking at. They are Adv. in Phys. **9**, 387 (1960) and **10**, 401 (1961). For a review of much of the progress in ferroelectrics prior to 1960 including much of the structural work see F. Jona and G. Shirane, "Ferroelectric Crystals" (Pergamon Press, 1962), as well as E. Fatuzzo and W. J. Merz "Ferroelectricity" (North Holland, 1967), and T. Mitsui, I. Tatsuzaki, and Eiji, "An Introduction to the Physics of Ferroelectrics" the Japanese edition was published in 1969 (Gordon and Breach, 1976). The book by R. Blinc and B. Zeks, "Soft Modes in Ferroelectrics and Antiferroelectrics," (North Holland, 1974) is also useful and reviews some of the symmetry considerations appropriate for possible ferroelectric structures. The book by M. E. Lines and A. Glass "Principles and Applications of Ferroelectrics and Related Materials" (Oxford, 1977) summarizes much of the field of ferroelectricity from a soft mode point of view.

For results and discussions on the results of the variation of $\varepsilon(0)$ and the soft mode with applied electric field, such as $\varepsilon(E, T)$ in $SrTiO_3$ and $KTaO_3$, see the works of P. A. Fleury and J. M. Worlock in Phys. Rev. Lett. **18**, 665 (1967); ibid **19**, 1176 (1967); and Phys. Rev. **174**, 613 (1968).

Using **hyper-Raman techniques** modes in centrosymmetric crystals can be observed, by Raman-like techniques, that are not Raman active but are infrared active. For hyper-Raman measurements photomultiplies are used, which are very sensitive to visible light, and laser sources. For infrared experiments there are poorer detectors and much weaker sources. Hyper-Raman experiments on overdamped modes are particularly useful and the results on $BaTiO_3$ and $KNbO_3$ in the cubic phase shows agreement with the soft mode model. See H. Vogt, J. A. Sanjurjo and G. Rossbroich, Phys. Rev. B **26**, 5904 (1982) and H. Vogt et al., Phys. Rev. B, to be published in 1985. In the tetragonal phase the situation is less clear. See G. Burns and E. Burstein, Ferroelectrics **7**, 297 (1974) and Phys. Rev. B **13**, 215 (1976).

The papers by R. A. Cowley that discuss a microscopic model of displacive phase transitions in perovskite crystals are Adv. in Phys. **12**, 421 (1963); Phys. Rev. **134**, A981 (1964); Phil. Mag. **11**, 673 (1965). Also see B. D. Silverman, Phys. Rev. **135** A1596 (1964) for a more transparent calculation in a one-dimensional chain. An anisotropic oxygen polarizability calculation that explains ferroelectricity in perovskite crystals is R. Migoni, H. Bilz, and D. Bäuerle, Phys. Rev. Lett.

37, 1155 (1976). For a review see A. Bussmann-Holder, H. Bilz, and P. Vogl, Springer Tracts in Modern Physics **99** (1983).

The Landau free energy, as discussed here, can be used to predict all three phase transitions in $BaTiO_3$ provided the α, β, and γ terms, Eq. 14-5, are taken as temperature dependent. [A. J. Bell and L. E. Cross, Ferroelectrics **59**, 197 (1984).]

For a discussion of the nonlinear optics of ferroelectric and related materials see Lines and Glass, mentioned previously.

The question of the **central peak** is interesting. The "central peak" is a response (usually observed by neutron or light scattering spectroscopy) observed at $\approx$ zero frequency. It was not expected from the soft mode model, hence caused a great deal of excitement when discovered in 1971. The book by Lines and Glass is a good place to start reading about this topic and see P. A. Fleury and K. Lyons, "Structural Phase Transitions," Vol. 1 Ed. K. A. Müller and H. Thomas (Springer Verlag, 1981) p. 9 and the extensive bibliography in this article. It appears that many of the early experimental observations of very large central peaks between 1971 and about 1978 are actually static in origin, that is, they are triggered by local strains, impurities, or surface effects. However, some much smaller dynamic (finite frequency response) central peaks have been observed.

Topics such as deviations from Landau theory near T_0 can be understood using the **renormalization group** (Section 14-5) approach. This theory has had remarkable success in providing insight and quantitative predictions of critical behavior near the phase transition temperature of most phase transitions. In fact, in 1982 K. G. Wilson received a Nobel Prize for this work. [His Nobel talk is in Rev. Mod. Phys. **55**, 583 (1983).] The theory is difficult and we mention a few references in which the ideas are specialized to structural phase transitions. "Phase Transitions and Critical Phenomena", Ed. C. Domb and M. S. Green, Vol. 6 (Academic Press, 1976); K. A. Müller, "Lecture Notes in Physics," Vol. 104, Ed. C. P. Enz (Springer Verlag, 1979) p. 210; A. D. Bruce and R. A. Cowley, Adv. in Phys. **29**, 1 to 321 (1980).

Many different **pressure units** are used in different fields. The following should help. 1 atmosphere = 760 torr = 760 mm of Hg at $0°C$ = 1.03323 kg/cm^2 = 1.01325 $\times 10^6$ $dynes/cm^2$ = 1.01325 bars. The SI unit of pressure is the pascal defined as 1 Pa = 1 N/m^2. We find that 10 kbar = 10^9 Pa = 1 GPa = 1.4504 $\times 10^5$ psi. Also 133.32 Pa = 1 torr.

Problems

1. Consider a parallel plate condenser made up of a material of thickness d and dielectric constant ε in series with an air gap of thickness δ. Determine the dielectric constant for this composite capacitor. Notice that even if $\delta/d \sim 10^{-4}$ the measured dielectric constant will be much smaller than ε when $\varepsilon \gg 10^4$. This is why care must be taken in measuring materials with very large dielectric constants.

2. For the free energy formulation for a first-order phase transition show that $(\partial F/\partial P)_T = 0$ has real solutions for

$$P^2 = -(\beta/2\gamma)[1 \pm (1-\tau)^{1/2}]$$

where τ is defined in Eq. 14-9. Show that the $+$ sign corresponds to minima in F vs. P while the $-$ sign corresponds to maxima. Show that

$$T_c = T_0 + (3/16)(\beta^2/\alpha\gamma)$$
$$T_u = T_0 + (1/4)(\beta^2/\alpha\gamma)$$

where T_c is the transition temperature and T_u is a temperature above which F only has a minimum at $P = 0$. Also note that ε^{-1} in the ferroelectric phase extropolates to zero at T_u.

3. Critical transitions are transitions in crystals that in the free energy expansion have $\beta = 0$. For such a material determine P_s and ε^{-1} as a function of temperature below T_c. Compare ε^{-1} below and above T_c. How might you hope to make, or get, such a crystal?

4. Write a short, no more than three-page, review of the central peak problem in ferroelectrics.

15

Magnetism

PART A DIAMAGNETISM AND PARAMAGNETISM

15-1 Introduction
15-2 Diamagnetism
15-3 Paramagnetism
    a Paramagnetism of free atoms
    b The quantum theory of paramagnetism
    c Do the free atom equations apply to ions in solids?
    d Magnitude of λ vs. V_c (shielding?)
    e 4d and 5d series
    f Free electron spin paramagnetism

PART B FERROMAGNETISM, ANTIFERROMAGNETISM, AND RELATED TOPICS

15-4 Introduction
15-5 Molecular Field Theory
15-6 The Heisenberg Exchange Interaction
15-7 Magnetic Structures
    a Ferromagnetism
    b Antiferromagnetism
    c Ferrimagnetism
    d Other ordered arrangements
15-8 Special Techniques Used to Study Magnetic Structures
    a Neutron diffraction
    b Magnetic resonance
    c Mössbauer effect

PART C OTHER TOPICS

15-9 Spin Waves (S, A)
    a Molecular field at low temperatures
    b Spin waves
    c Direct experimental observations
    d Quantization of spin waves
    e Antiferromagnetic spin waves
15-10 Anisotropy, Hysteresis, Domains, and Bloch Walls
    a Anisotropy
    b Hysteresis
    c Domains
15-11 Metals and Magnetism (S, A)
15-12 Spin Glasses (S)

Notes
Problems

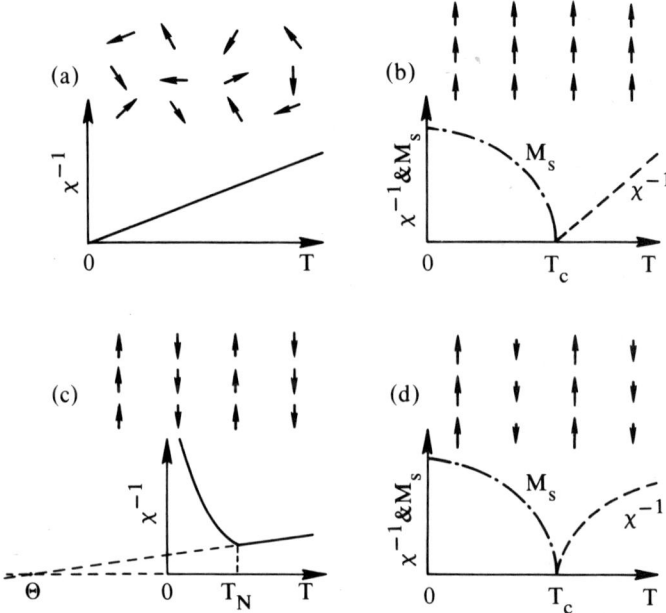

Fig. 15-1 (a) Paramagnetic behavior. (b) Ferromagnetic behavior. (c) Antiferromagnetic behavior. (d) Ferrimagnetic behavior. Each figure shows the low temperature position of the spins. In (a) and (c) χ^{-1} vs. T is shown. In (b) and (d) the same quantity is shown above the transition temperature, but below this temperature the magnetism vs. temperature is plotted.

MAGNETISM

I have so often been through these periods of strain that I have become accustomed to the fact that most of the calamities that we anticipate really never occur.

D. D. Eisenhower, Diary, 1955

In trying to understand physical phenomena, it is often possible to understand some complex observable behavior in terms of an appropriate sum of independent small entities. Yet in many areas of science it is really the cooperative interactions of a large number of entities that is the basic cause of the behavior. In this chapter examples of both types of behavior are seen. Part A is primarily an example of the former behavior and Part B shows the cooperative aspect. In Part C other important aspects of magnetic materials are discussed.

Part A — Diamagnetism and Paramagnetism

15-1 Introduction

Magnetism is one of the oldest branches of solid state science. The Greeks reflected on the strange properties of magnetite, a mineral composed of FeO and Fe_2O_3, found in the province of Magnesia in Asia Minor. In the second century, the magnetic properties of magnetite were put to important use in the form of lodestone (leading stone) for compasses on ships. However, it was not until the sixteenth century that Sir William Gilbert seriously investigated the properties of magnetism. He was one of the early proponents of what is now called the scientific method (theories should be based on experiment, not on verbal arguments). The book by Mattis (see the Notes) has a chapter on the history of magnetism.

Quantum mechanics is required for an understanding of the real physics of most magnetic phenomena. However, some of the basics can be described in classical terms or by treating the constants of a Hamiltonian as parameters (i.e., ignoring the quantum mechanical underpinnings). We take the latter approach to understand some of the phenomena semiclassically and with the use of simple models.

The magnetic response of most magnetic solids involves the orientation of the magnetic dipoles contained in the solid. These magnetic dipoles, or magnetic moments, consist of the electron spin dipole moment and the electron orbital dipole moment. The response to a magnetic field of free atoms and of collections of weakly coupled atoms in a solid will be calculated explicitly. Problems arise when these magnetic moments are strongly coupled, and Fig. 15-1 schematically shows the more important possible results. In the top of each diagram the magnetic dipoles (one for each atom) are shown. The magnetism of the sample, **M**, is the vector sum of all the individual magnetic moments per unit volume. Also plotted is the susceptibility, $\chi \equiv M/H$, defined for small H, where **H** is the applied **macroscopic magnetic field intensity**.

Paramagnetism is shown in Fig. 15-1a. The dipoles point in random directions. Under the influence of an applied field H there will be a slight increase in the number of magnetic moments (or their projections) along +**H** and a slight decrease along −**H**; this will lead to a small **M**. The susceptibility was found by P. Curie (1895) as shown and is known as Curie's law $\chi \propto T^{-1}$. (This is discussed in detail later, as are all the results in Fig. 15-1.) Thermalization of the magnetic moments is the driving force toward keeping the moments random.

Ferromagnetism is shown in Fig. 15-1b below the transition temperature T_c. That is, below T_c there is a spontaneous parallel orientation of the spins. Above T_c the spins are randomly oriented as in Fig. 15-1a. Curie found that $\chi \propto (T - T_0)^{-1}$ where T_0 has a value close to the transition temperature. However, it is not exactly equal to T_c because near T_c critical effects (Section 14-5) become important in magnetic systems, and the Curie law breaks down.

Antiferromagnetism is shown in Fig. 15-1c. Above the transition (Néel) temperature T_N the spins are randomly oriented as in the above cases, but below T_N exactly half of the spins are antiparallel to the other half, resulting in zero net magnetic moment. Well above T_N the susceptibility is Curie-like with $\chi \propto (T + \Theta)^{-1}$, so that an extrapolation of χ^{-1} vanishes at $T = -\Theta$, as shown in Fig. 15-1c.

Ferrimagnetism is shown in Fig. 15-1d. The spins are randomly orientated above T_c but below they are oriented antiparallel. However, unlike the antiferromagnetic case, one set of spins has a larger magnetic moment than the other set, resulting in a net magnetism.

There are other spontaneous spin arrangements, such as canted, helical, conical, and wave-like. However, the arrangements shown in Fig. 15-1 are the most important ones.

For magnetic materials the relationship between the **flux density B** (in units of gauss, abbreviated G), the applied magnetic field **H** (in units of oersteds, abbreviated Oe), and **magnetization M**, given in either unit as 1 G = 1 Oe, is

$$\mathbf{B} = \mathbf{H} + 4\pi\mathbf{M} \qquad (15\text{-}1a)$$

In the linear regime (the diamagnetic and paramagnetic cases) we can think of the applied magnetic field causing the magnetic moments to orient and give a magnetization. Assuming that all of these vectors are

CHAPTER 15   MAGNETISM   567

parallel (i.e., ignoring the tensor properties of the susceptibilities) we define a permeability, μ, a susceptibility, χ, and obtain the relation between the two, via Eq. 15-1a, as

$$\mathbf{B} \equiv \mu \mathbf{H}$$
$$\mathbf{M} \equiv \chi \mathbf{H} \quad (15\text{-}1b)$$
so
$$\mu = 1 + 4\pi\chi$$

In diamagnetic and paramagnetic materials (i.e., dilute magnetic materials) $4\pi M \ll H$ so $\mu \approx 1$ and $B \approx H$. In magnetically ordered materials often $M \gg H$ so μ and χ are very large and, because of the nonlinear, hysteresis behavior between M and H, they depend on H, so great care must be exercised when treating ordered states.

In **SI units** the expressions and units are

$$\mathbf{B} = \mu_0(\mathbf{H} + \mathbf{M})$$
$$\mathbf{B} = \kappa_m \mu_0 \mathbf{H} = \mu \mathbf{H}$$
$$\mathbf{M} = \chi_m \mathbf{H}$$
$$\kappa_m = 1 + \chi_m$$

where μ_0 is the permeability of free space, flux density is given in teslas where 10^4 G = 1 T, and applied magnetic field in amperes-turn per meter where $(4\pi/10^3)$ Oe = 1 A/m.

Diamagnetism is one type of magnetic behavior not discussed yet. It is a small effect but it occurs in all materials, so we evaluate it first.

15-2 Diamagnetism

This effect can be treated in a classical manner. It arises via **Lenz's law**, which says that when a magnetic flux changes in a circuit, a current is induced which opposes the change of flux. Thus, diamagnetism occurs due to a change in the orbital motion of the electrons. It occurs in all forms of matter and it is always a negative contribution to the total magnetic susceptibility.

To calculate the diamagnetism of an atom, think of a charge e, revolving in a circle of radius a, area A, with angular velocity ω (under a central force F). Whether this charge is in an atom or a large circle is irrelevant. The charge passes any point in the circle in a time $2\pi/\omega$ (= $1/\nu$ where $2\pi\nu = \omega$). Thus, the current in the circle (charge/time) and the magnetic moment of this current loop are

$$I = e\omega/2\pi \quad (15\text{-}2a)$$
$$\mu_m = IA/c = e\omega a^2/2c \quad (15\text{-}2b)$$

Now investigate the effect of an external magnetic field on the rotating charge. If a charge e, mass m, and angular velocity ω_0, is

Table 15-1 The molar susceptibilities of some closed shell atoms and ions. Each number must be multiplied by 10^{-6} cm^3/mole. Thus, χ_m must be multiplied by the appropriate moles/cm^3 to obtain a dimensionless susceptibility. (The number of moles/cm^3 is of the order of unity.) (From R. Kubo and T. Nagamiya, Eds. "Solid State Physics", McGraw-Hill, 1969, p. 439.)

Ion	χ_M	Atom	χ_M	Ion	χ_M
		He	-1.9	Li$^+$	-0.7
F$^-$	-9.4	Ne	-7.2	Na$^+$	-6.1
Cl$^-$	-24.2	A	-19.4	K$^+$	-14.6
Br$^-$	-34.5	Kr	-28	Rb$^+$	-22.0
I$^-$	-50.6	Xe	-43	Cs$^+$	-35.1

rotating counterclockwise in a circle, then the centrifugal force is $F = m\omega_0^2 a$. If a magnetic field H pointing toward the reader is applied, then there is an extra force from Lorentz's law (Eq. 9-62). Assume that the charge keeps in the same circle (same radius) and only changes its rotational frequency to compensate for this extra force. Then the additional term is given by $evH/c = e\omega aH/c$. In this situation, the equation of motion is

$$m\omega^2 a = F - (e\omega aH/c) \qquad (15\text{-}2c)$$
$$\omega^2 - \omega_0^2 = -e\omega H/mc \qquad (15\text{-}2d)$$

where the second equation is obtained from the first by using $F = m\omega_0^2 a$. Assume ω is close to ω_0 so $\Delta\omega \equiv \omega - \omega_0$. Thus,

$$\Delta\omega = -eH/2mc \qquad (15\text{-}3)$$

This is known as the **Larmor frequency**. From this relationship, a quantity called the **Bohr magneton**, μ_B, is defined; the energy associated with this frequency is $\Delta\mathscr{E} = \hbar\Delta\omega = -e\hbar H/2mc \equiv -\mu_B H$. The value of μ_B is 9.27×10^{-21} ergs/Oe.

Now consider an atom with closed shells so that for every electron orbit with magnetic moment up, there is another electron orbit with magnetic moment down. With an applied field H there is a small change in the orbital frequency, thus a small induced magnetic moment opposite to that of the field. This is given directly by the substitution of the Larmor frequency into Eq. 15-2b

$$\mu_m = -e^2 a^2 H/4mc^2 \qquad (15\text{-}4a)$$

To apply these results to spherical, closed-shell atoms, first recall that $a^2 = \langle x^2 \rangle + \langle y^2 \rangle$ is the mean square perpendicular distance from the magnetic field axis through the nucleus to the electron. To relate this to the mean square radial distance $\langle r^2 \rangle = \langle x^2 \rangle + \langle y^2 \rangle + \langle z^2 \rangle$ for spherical atoms use $\langle x^2 \rangle = \langle y^2 \rangle = \langle z^2 \rangle$ so that $\langle r^2 \rangle = 3a^2/2$. Second,

CHAPTER 15   MAGNETISM   569

the result in Eq. 15-4a is for one electron so it must be summed for all p electrons in the atom, remembering that $\langle r^2 \rangle$ is different for different electrons; in practice only the electrons in the outer shells have sizable contributions to $\langle r^2 \rangle$. Third, to compare to experiment, μ is calculated not for one atom but for N atoms in a volume V, so $M = \mu_m N/V$ (so N/V is an atomic density here, not an electron density). Last, experimentally a susceptibility $\chi \equiv M/H$ can be measured. Thus, Eq. 15-4a yields

$$\chi \equiv \frac{M}{H} = -\frac{pe^2 N}{6mc^2 V} \langle r^2 \rangle \qquad (15\text{-}4b)$$

[annotation: p — # of electrons in outermost shell]

Notice that this is a negative number so as H increases the induced dipole increases, opposing the external field. This term is known as the **Larmor diamagnetic susceptibility** (or sometimes the Langevin susceptibility). It is rather easy to calculate for atoms since the wave functions for all of the closed-shell atoms or ions have been calculated so $\langle r^2 \rangle = \langle \phi | r^2 | \phi \rangle$ can be evaluated easily. The value of χ is usually given for a mole of atoms, χ_M, and Table 15-1 lists χ_M for a number of closed-shell atoms and ions.

The preceding discussion of diamagnetism is straightforward for closed-shell spherical ions. This behavior in complicated molecules, such as the motion of electrons around a benzene ring, is explained in van Vleck's book (see the Notes). The diamagnetism of free electrons in metals was mentioned briefly in Section 9-12c. Electrons in a metal can execute very large orbits, and the calculations are complex. Last, while all systems have diamagnetism, for most of the magnetic materials discussed in the rest of this chapter there are other positive contributions to the magnetic moment that far outweigh the diamagnetism.

15-3 Paramagnetism

Now that diamagnetism has been discussed, we proceed with the main theme of this chapter: the orientation of magnetic dipoles. The first topic is the paramagnetic behavior (Fig. 15-1a) of free atoms and ions in solids that, from a magnetic point of view, are isolated from each other. A paramagnetic material has a positive contribution to the susceptibility (i.e., the magnetic moment increases with increasing magnetic field). We will find that χ usually depends strongly on the temperature and magnetic field. The materials discussed in this section might be called an **ideal magnetic gas**, which is thought of as a system of noninteracting magnetic atoms. When the interactions are important considerations discussed in Part B must be used.

15-3a Paramagnetism of free atoms   Consider a free atom that contains p electrons in an incomplete shell. The ith electron has an orbital quantum number ℓ_i, with **orbital angular momentum** operator $\hbar\boldsymbol{\ell}_i$ and a spin quantum number s_i, with **spin angular momentum** operator $\hbar\mathbf{s}_i$. The orbital angular momentum operator also can be given in terms of $\hbar\boldsymbol{\ell} = \mathbf{r} \times \mathbf{p}$, where the momentum operator is $\mathbf{p} = (\hbar/i)\nabla$. (Warning: we often say $\boldsymbol{\ell}$ is the "orbital angular momentum"; the units of $\hbar$ understood. The same for $\mathbf{s}$.) The question arises as to how these various angular momenta couple.

Russell-Saunders coupling is found to be the appropriate way to couple the individual angular momenta in all but the very heavy atoms. This occurs because in the lighter atoms the sum of the coulombic interaction between electrons in different orbits, with different ℓ_i's (the orbit-orbit interactions) is stronger than that of the relativistic spin orbit interactions coupling each ℓ_i to the corresponding s_i. Thus, to a good approximation the orbital angular momenta couple to form a resultant orbital vector **L**. Similarly the spins couple, via the intra-atomic spin-spin exchange interaction, and form a resultant spin vector **S** in the following manner:

$$\mathbf{L} = \sum_{i=1}^{p} \boldsymbol{\ell}_i \qquad \mathbf{S} = \sum_{i=1}^{p} \mathbf{s}_i \qquad (15\text{-}5a)$$

The sum is over the p-electrons in incomplete shells. The sum over electrons in complete shells gives zero. Further, **L** and **S** are weakly coupled, via spin-orbit coupling $\lambda\mathbf{L}\cdot\mathbf{S}$, to form a resultant total angular momentum operator $\hbar\mathbf{J}$ with a range of values of the **total angular momentum quantum number**, J, given by

$$J = (L + S), (L + S - 1), \ldots |L - S + 1|, |L - S| \qquad (15\text{-}5b)$$

This method of coupling the various angular momenta, Eqs. 15-5, is called **Russell-Saunders** (or **LS**) coupling; Fig. 15-2a shows the results for a two-electron atom. Since these equations are vector sums, several results are possible; for example, if $s_1 = 1/2$ and $s_2 = 1/2$ then $S = 1$ or 0, and if $\ell_1 = 2$ and $\ell_2 = 2$ (i.e., d-electrons), then $L = 4, 3, 2, 1,$ or 0.

The group of levels indicated in Eq. 15-5b is called a **multiplet**, and the multiplicity of the system is defined as $2S + 1$. Indeed if $L \geq S$ there are $2S + 1$ multiplets but if $L < S$ there are only $2L + 1$ multiplets. The spacing of the multiplet is determined by the **spin-orbit coupling constant** λ defined so that the interaction energy is

$$V_{so} = \lambda\mathbf{L}\cdot\mathbf{S} \qquad (15\text{-}5c)$$

so the Jth energy level is given by

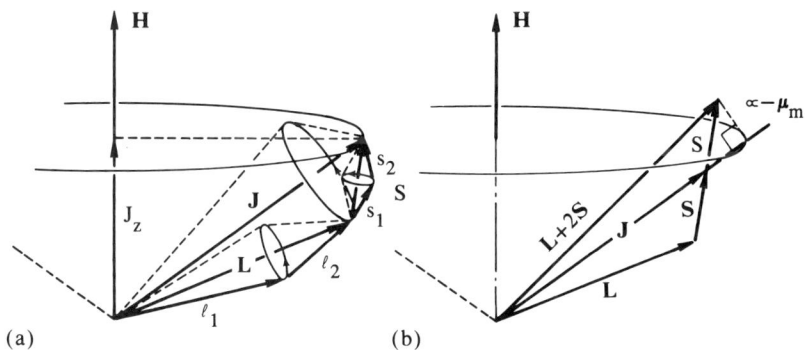

Fig. 15-2 (a) Russell-Saunders coupling for two electrons. (b) A figure to help calculate g_J. Note, $\mu_m = -g_J \mu_B \mathbf{J}$.

$$(\lambda/2)[J(J+1) - L(L+1) - S(S+1)] \tag{15-5d}$$

This is obtained using $\mathbf{J}^2 = \mathbf{L}^2 + \mathbf{S}^2 + 2\mathbf{L}\cdot\mathbf{S}$ and the quantum mechanical result for the eigenvalues of all of these vector operators, namely $\mathbf{J}\cdot\mathbf{J} = J(J+1)$, $\mathbf{L}\cdot\mathbf{L} = L(L+1)$, and $\mathbf{S}\cdot\mathbf{S} = S(S+1)$. First we discuss how to determine the ground state and then the behavior in a magnetic field.

Hund's rules applied to electrons in a given shell characterize the S, L, and J of the ground states of atoms. The rules are

1. The largest S, consistent with the Pauli exclusion principle, will describe the ground state. The reason for this is as follows. If two electrons in an atom are in the same orbital state, they must have opposite spin. However, their coulomb repulsion energy will be significantly lowered if they are in different orbitals. In different orbitals, they can have parallel or antiparallel spins. But the exchange integral for two orthogonal orbitals is always positive, lowering the energy of the parallel spin situation with respect to the antiparallel situation.

2. The maximum L, consistent with S in the first rule, will describe the ground state. This rule is more difficult to explain qualitatively. Quantitive calculations are required. (See the Notes.)

3. $J = |L - S|$ when the electron shell is less than half full and equal to $L + S$ when the shell is more than half full. $J = S$ for a half-filled shell because, as we shall see, $L = 0$ by rule 1.

As an example of the application of these rules consider the one-electron case Ce^{3+} $4f^1$, where $s = 1/2$ and $\ell = 3$ (for an f-electron). Thus for the ground state: $S = 1/2$; $L = 3$; $J = 5/2$. This is usually given the spectroscopic notation $^2F_{5/2}$, which is a rather peculiar but time-honored notation; the upper left, the multiplicity, is

Table 15-2 The ground states, as predicted by Hund's rules, and the spectrographic symbols of partially filled d- and f-shells. The notation ↑ means $m_s = +1/2$ and ↓ means $m_s = -1/2$. On the far right some typical iron series ions and rare earth ions are given that have these ground states.

d-shell ($\ell = 2$)

n	$m_\ell=2$	1	0	−1	−2	S	$L=\|\Sigma m_\ell\|$	J		
1	↑					1/2	2	3/2	$^2D_{3/2}$	Ti^{3+}
2	↑	↑				1	3	2	$^3F_2$	V^{3+}
3	↑	↑	↑			3/2	3	3/2	$^4F_{3/2}$	Cr^{3+}
4	↑	↑	↑	↑		2	2	0	$^5D_0$	Cr^{2+}
5	↑	↑	↑	↑	↑	5/2	0	5/2	$^6S_{5/2}$	Fe^{3+}, Mn^{2+}
6	↑↓	↑	↑	↑	↑	2	2	4	$^6D_4$	Fe^{2+}
7	↑↓	↑↓	↑	↑	↑	3/2	3	9/2	$^4F_{9/2}$	Co^{2+}
8	↑↓	↑↓	↑↓	↑	↑	1	3	4	$^3F_4$	Ni^{2+}
9	↑↓	↑↓	↑↓	↑↓	↑	1/2	2	5/2	$^2D_{5/2}$	Cu^{2+}
10	↑↓	↑↓	↑↓	↑↓	↑↓	0	0	0	$^1S_0$	

f-shell ($\ell = 3$)

n	$m_\ell=3$	2	1	0	−1	−2	−3	S	$L=\|\Sigma m_\ell\|$	J		g_J	
0								0	0	0	$^1S_0$	0	La^{3+}
1	↑							1/2	3	5/2	$^2F_{5/2}$	6/7	Ce^{3+}
2	↑	↑						1	5	4	$^3H_4$	4/5	Pr^{3+}
3	↑	↑	↑					3/2	6	9/2	$^4I_{9/2}$	8/11	Nd^{3+}
4	↑	↑	↑	↑				2	6	4	$^5I_4$	3/5	Pm^{3+}
5	↑	↑	↑	↑	↑			5/2	5	5/2	$^6H_{5/2}$	2/7	Sm^{3+}
6	↑	↑	↑	↑	↑	↑		3	3	0	$^7F_0$	−	Eu^{3+}
7	↑	↑	↑	↑	↑	↑	↑	7/2	0	7/2	$^8S_{7/2}$	2	Gd^{3+}
8	↑↓	↑	↑	↑	↑	↑	↑	3	3	6	$^7F_6$	3/2	Tb^{3+}
9	↑↓	↑↓	↑	↑	↑	↑	↑	5/2	5	15/2	$^6H_{15/2}$	4/3	Dg^{3+}
10	↑↓	↑↓	↑↓	↑	↑	↑	↑	2	6	8	$^5I_8$	5/4	Ho^{3+}
11	↑↓	↑↓	↑↓	↑↓	↑	↑	↑	3/2	6	16/2	$^4I_{15/2}$	6/5	Er^{3+}
12	↑↓	↑↓	↑↓	↑↓	↑↓	↑	↑	1	5	6	$^3H_6$	7/6	Tm^{3+}
13	↑↓	↑↓	↑↓	↑↓	↑↓	↑↓	↑	1/2	3	7/2	$^2F_{7/2}$	8/7	Yb^{3+}
14	↑↓	↑↓	↑↓	↑↓	↑↓	↑↓	↑↓	0	0	0	$^1S_0$	0	Lu^{3+}

the spin degeneracy $2S + 1$, the lower right is the J value, and the letter is a designation of the value of L via

$$L = 0, 1, 2, 3, 4, \ldots$$
$$S, P, D, F, G, \ldots$$

So the symbol is $^{2S+1}L_J$. For the two-electron cases $4f^2(Pr^{3+})$ Hund's rules give: $S = 1$; since the spins are both the same, only the first electron can have $m_\ell = 3$, so the second electron has $m_\ell = 2$, so $L = 5$; $J = 4$. Thus, the state is $^3H_4$. Table 15-2 shows the results for atoms with unfilled f- and d-shells.

The **magnetic dipole moment** of a free atom can be calculated. The potential energy of a magnetic moment in a magnetic field is $-\mu_m \cdot H$. For an orbital magnetic moment $\mu_L = -\mu_B L$ where μ_B is the Bohr magneton. However, for an electron (which has a spin of 1/2) it is found experimentally that the magnetic moment is not $-\mu_B/2$ but rather two times as large, or $-g_e\mu_B/2$ where $g_e = 2$. (This value, $g_e = 2$, is predicted by the Dirac theory of the electron. Actually very accurate measurements give $g_e = 2.0023$, which can be understood via quantum electrodynamic corrections to Dirac's theory, but we use $g_e = 2$.) Thus, the potential energy is $\mu_B(L + 2S)\cdot H$. We also know that when the orbital angular momentum operator, $\hbar L$, and the spin quantum moment operator, $\hbar S$, are coupled, then the total angular momentum operator $\hbar J$ is obtained. Associated with this is the total magnetic moment, proportional to J and is given by $-g_J\mu_B J$. (*Due to the negative charge of the electron, the magnetic moment is in the opposite direction from the angular momentum.*) Since L and separately S precess about J, the term $L + 2S$ also precesses about J. The total magnetic moment of the atom must be parallel to the total angular momentum, $\hbar J$, to give a potential energy of $g_J\mu_B J \cdot H$, where g_J is a number that depends on S, L, and J; it might be written as g(S, L, J) but for historical reasons is written as g_J and is called the **Landé g-factor** or the **spectroscopic splitting factor**. Thus, the magnetic moment of the atom is

$$\mu_m = -g_J\mu_B J = \gamma\hbar J \qquad [= -\mu_B\langle L + 2S\rangle] \qquad (15\text{-}6)$$

where γ, the **gyromagnetic** (or **magnetogyric**) **ratio**, is clearly the ratio of the magnetic moment to the angular momentum, and is included for historical reasons (see Problem 2). The term in the square brackets is only understood in terms of the $-g_J\mu_B J$ term by remembering that $L + 2S$ precesses about J; it is written as a time average.

Figure 15-2b is a picture of the situation described in Eq. 15-6. g_J can be calculated by elementary means as follows:

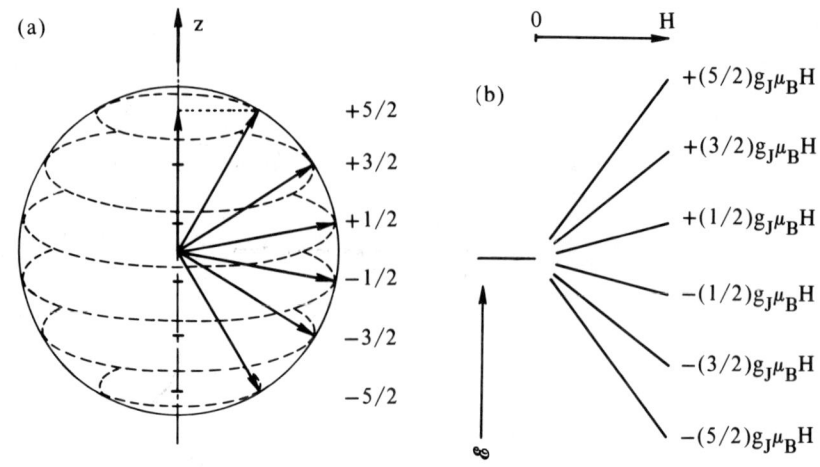

Fig. 15-3 (a) The quantum states for $J = 5/2$ precessing about a magnetic field. (b) The energy vs. magnetic field for these states.

$$(\mathbf{L} + 2\mathbf{S}) \cdot \mathbf{J} = g_J \mathbf{J} \cdot \mathbf{J} = \mathbf{J}^2 + \mathbf{S} \cdot \mathbf{J} \tag{15-7a}$$

yielding
$$g_J = \frac{(3/2)\mathbf{J}^2 + (1/2)\mathbf{S}^2 - (1/2)\mathbf{L}^2}{\mathbf{J}^2} \tag{15-7b}$$

or
$$g_J = \frac{3}{2} + \frac{S(S+1) - L(L+1)}{2J(J+1)} \tag{15-7c}$$

Equation 15-7b is obtained from Eq. 15-7a and Fig. 15-2a by elementary geometry and then Eq. 15-7c is obtained by replacing the square of the angular momentum vector operators by their eigenvalues, $\mathbf{J}^2$ by $J(J + 1)$, and so on. This result (Eq. 15-7c) was used (1923) to explain the multiplet splitting of atomic spectra. Note that when $S = 0$, $L = J$, and $g_J = 1$, while when $L = 0$, $S = J$, so $g_J = 2$.

When an atom with total angular momentum $\hbar \mathbf{J}$ is placed in a magnetic field, the vector $\mathbf{J}$ precesses about the field, Fig. 15-3. The spatial quantization of $\mathbf{J}$ along $\mathbf{H}$ is such that these are $2J + 1$ equally spaced energy levels with a magnetic quantum number m_J taking the values $m_J = J, J - 1, ..., -J + 1, -J$.

Before using the equations derived here, the theory of paramagnetism of ions in solids will be discussed. We will find some unexpected results and see how magnetic moments are oriented in solids.

CHAPTER 15   MAGNETISM   575

15-3b The quantum theory of paramagnetism   Consider the magnetism, M, induced in a solid that has N atoms in a volume V and each atom has a magnetic moment μ_m ($M = \mu_m N/V$). For an applied field of $\approx 10^4$ gauss the $\mu_m \cdot H$ terms $\approx \mu_B 10^4$ gauss $\approx 9.2 \times 10^{-17}$ ergs. If we expect thermal motion to be the principal driving force toward randomness then $\mu_m \cdot H$ should be compared to $k_B T$, which at room temperature $\approx (1.38 \times 10^{-16})(300)$ ergs $\approx 4.1 \times 10^{-14}$ ergs. Thus, at room temperature thermal averaging is much stronger than the magnetic alignment of the dipoles. However, some small alignment should take place and at much lower temperatures, when the energies are comparable, we expect that different effects might occur.

The Langevin (or classical) theory of paramagnetism treats the magnetic dipoles classically and is the same as the treatment of electric dipoles in Section 11-7a. In fact, the original treatment of Langevin was for the magnetic case. Classically the potential energy of a magnetic dipole μ_m in a magnetic field H is $-\mu_m \cdot H$. If thermal agitation is the principal deterrent to alignment, then a calculation of the resultant magnetization **M** for N dipoles in a volume V follows identically as in Eq. 11-36 and 11-37. The result is

$$M = (N/V)\, \mu_m\, [\text{ctnh}(x) - 1/x] \qquad (15\text{-}8a)$$

where $x \equiv \mu_m H/k_B T$ and the term in the square bracket is the Langevin function, Eq. 11-37. In the high temperature, weak field region (i.e., $\mu_m H \ll k_B T$) the susceptibility is

$$\chi \equiv (M/H) = \mu_m^2 N/3 k_B T V \equiv C/T \qquad (15\text{-}8b)$$

Figure 11-13 shows a plot of the Langevin function as a function of x. We shall refer back to this susceptibility when the corresponding quantum mechanical result is calculated. Note the well-known **Curie law** form of χ, namely, it has the form C/T where C is the Curie constant, $C \equiv \mu_m^2 N/3 k_B V$. This law had been experimentally determined (P. Curie, 1895) and Langevin theoretically explained it.

The quantum theory of paramagnetism (Van Vleck) must be used because the magnetic moment of an atom cannot take on any arbitrary angle with the magnetic field, as shown in Fig. 15-3a. The projection of **J** on **H** can assume only a certain finite number of values dictated by the $2J + 1$ values of m_J. Then, the potential energy for a magnetic moment in a magnetic field in the quantum mechanical case is $-\mu_m \cdot H = m_J g_J \mu_B H$. Thus, the statistical mechanics average for the magnetism of a sample with N atoms in a volume V (Eqs. 11-10 and 11-14), summed over the $2J + 1$ values of m_J, is

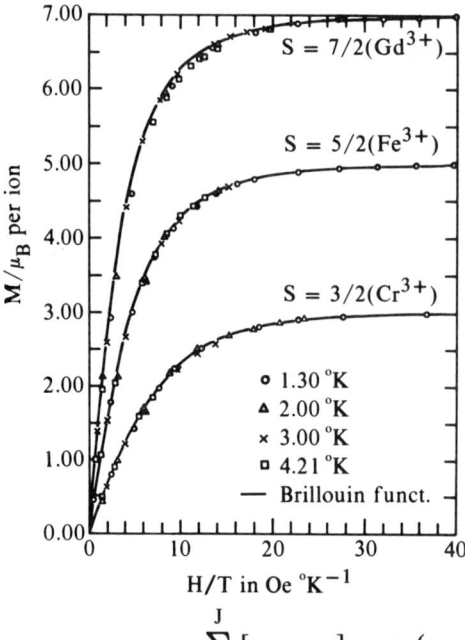

Fig. 15-4 The magnetism vs. H/T for three different parametric crystals. The crystals are gadolinium sulfate octahydrate, ferric ammonium alum, and potassium chromium alum. [W. E. Henry, Phys. Rev. 88, 559 (1952)].

$$M = \frac{N}{V} \frac{\sum_{-J}^{J} [g_J \mu_B m_J] \exp(g_J \mu_B m_J H/k_B T)}{\sum_{-J}^{J} \exp(g_J \mu_B m_J H/k_B T)} \equiv \frac{N}{V} g_J \mu_B J\, B_J(y)$$

$$B_J(y) \equiv \left(\frac{2J+1}{2J}\right)\coth\left(\frac{2J+1}{2J}y\right) - \frac{1}{2J}\coth\left(\frac{y}{2J}\right) \quad (15\text{-}9)$$

where $y \equiv g_J \mu_B J H/k_B T$ and $B_J(y)$ is called the **Brillouin function**. For large J the Brillouin function asymptotically approaches the Langevin function as expected from the correspondence principle; for very low temperatures both functions approach a value of 1. For very small magnetic energy compared to $k_B T$ ($y \ll 1$), then $B_J(y) \approx y(J+1)/3J$. The susceptibility, M/H, is usually considered, which for $y \ll 1$ is

$$\chi = \frac{g_J^2 \mu_B^2 J(J+1) N}{3 k_B T V} \equiv \frac{C}{T} \equiv \frac{\mu_{eff}^2 N}{3 k_B T V} \quad (15\text{-}10)$$

where the Curie constant, C, is defined. This weak field susceptibility goes as T^{-1}, as in the classical case, but now the orbital angular momentum quantum number J is contained in the expression. The **effective magnetic moment**, $\mu_{eff}^2 \equiv g^2 \mu_B^2 J(J+1)$, is defined here. Note that with the definition of μ_{eff} the quantum mechanical result for the susceptibility in Eq. 15-10 has the same form as the classical result in Eq. 15-8b, but would not follow in an obvious manner Eq. 15-8b.

Table 15-3 Values of μ_{eff}/μ_B calculated with g_J defined in Eq. 15-7c for each ground state for the trivalent rare earth ions. The measured values for insulating materials in the vicinity of room temperature are also listed. [Many of the values in this and Table 15-4 are from van Vleck as well as Kubo and Nagamiya; see the Notes.]

Ion	$4f^n$	State	μ_{eff}/μ_B (Calc.)	μ_{eff}/μ_B (Exp.)
La^{3+}	$4f^0$	$^1S_0$	0	0
Ce^{3+}	$4f^1$	$^2F_{5/2}$	2.54	2.4
Pr^{3+}	$4f^2$	$^3H_4$	3.58	3.5
Nd^{3+}	$4f^3$	$^4I_{9/2}$	3.62	3.5
Pm^{3+}	$4f^4$	$^5I_4$	2.68	–
Sm^{3+}	$4f^5$	$^6H_{5/2}$	0.84	1.5
Eu^{3+}	$4f^6$	$^7F_0$	0	3.4
Gd^{3+}	$4f^7$	$^8S_{7/2}$	7.94	8.0
Tb^{3+}	$4f^8$	$^7F_6$	9.72	9.5
Dy^{3+}	$4f^9$	$^6H_{15/2}$	10.63	10.6
Ho^{3+}	$4f^{10}$	$^5I_8$	10.60	10.4
Er^{3+}	$4f^{11}$	$^4I_{15/2}$	9.59	9.5
Tm^{3+}	$4f^{12}$	$^3H_6$	7.57	7.3
Yb^{3+}	$4f^{13}$	$^2F_{7/2}$	4.54	4.5

From this weak field (or high temperature) expansion, Eq. 15-10, agreement between theory and experiment can be tested. However, it is better to compare the measured magnetic moment with experiment not only in the weak field limit but throughout a wide range of H/T. Figure 15-4 shows the experimental results from three different ions with three different spins. In all three cases S = J (Table 15-2); the reason for this choice of ions will become clear in the next section. First, the calculated results (Eq. 15-9) are shown also and the agreement is excellent. Note that at one temperature, results over a wide range of H/T can be obtained by varying the magnetic field. Second, note the linear behavior in the low field case (small H/T) as predicted by Eq. 15-10. Third, $B_J(y) = 1.0$ for $y \gg 1$ so the saturation values of M = M(T = 0°K) per ion are given by $g_J\mu_B J$, which is just the result in the figure (J = S and $g_e = 2$). Last, for these samples, even at the highest values of H/T, where the magnetic moment is saturated, there is no sign of any cooperative effects between the spins (i.e., the ions are behaving like perfectly isolated atoms). The reason for this is that the paramagnetic ions used for these measurements are well isolated from each other, being surrounded by waters of hydration. (There are many materials, discussed in Part B, where the ions are less isolated from each other and cooperative behavior is observed.)

15-3c Do the free atom equations apply to ions in solids? We now apply the ideas developed previously to general classes of ions in solids. The most studied groups of ions in solids are the rare earth ions (4f-electrons) and the iron series ions (3d-electrons) so these are discussed here in turn.

4f-ions – Most rare earth ions are trivalent. Thus, their electronic configuration is written as $[Kr + 4d^{10}]4f^n5s^25p^6$ where n = 0 corresponds to La^{3+}, n = 1 to Ce^{3+}, all the way to n = 13, which corresponds to Yb^{3+} (Lu^{3+} has a full f-shell, i.e., n = 14). The chemical behavior of all of these elements is very similar, as might be expected, since it is determined by the $5s^25p^6$ outer electrons, which are the same for all of these ions. In fact, a strange thing happens as the number of 4f-electrons increases; as one goes from $4f^1$ to $4f^{14}$ the ionic radius decreases. This is known as the **lanthanide contraction**. The $4f^n$ electrons have a much smaller radial extent than the $5s^25p^6$ and thus the 4f-electrons are sometimes called "inner" electrons. It might be expected that in solids these inner 4f-electrons behave similarly to their free atom behavior. Indeed this is found to be the case.

Table 15-3 shows a comparison between μ_{eff} (Eq. 15-10) measured for the rare earth ions in insulating solids, and the calculated value, $g_J\mu_B[J(J + 1)]^{1/2}$, assuming a ground state predicted by Hund's rules, and g_J calculated via Eq. 15-7c. Hund's rules and g_J are formulated for free atoms (or ions) and μ_{eff} is measured via susceptibility measurements. In general the agreement is quite good; the two or three cases showing poor agreement can be directly related to other states, with different J-values, being close to the ground state. By close we mean that the energy separation between the two states is of the order of k_BT. From this agreement of theory and experiment it may be concluded that the rare earth ions in insulators behave very much like free atoms or ions. (The reason for this will be discussed in Section 15-3d.)

Iron series ions – These often are referred to as **transition series ions** or **3d-ions**. In insulating solids their behavior is more complex than that of rare earth ions. Table 15-4 shows the ground state as determined from Hund's rules. For this state J and g_J are known, so $\mu_{eff}/\mu_B = g_J[J(J + 1)]^{1/2}$ can be calculated (Eqs. 15-7c and 15-10). The results, listed under (Calc.-J), are not in good agreement with experiment. Rather, much better agreement is obtained if L = 0 and S = J; calculated values are listed under (Calc.-J = S). In fact, this approach was already used in Fig. 15-4 for Cr^{3+} where S = 3/2 but L = 0 and J = 3/2 was used (g_J = 2) rather than the free atom S = 3/2, L = 3, and J = 3/2 (g_J = 4/10) from Table 15-2. If the latter were used the agreement would be terrible. This phenomenon, L = 0 and J = S, is called **quenching of the orbital angular momentum**. The reason for this effect will be discussed, but briefly, it arises because of

Table 15-4 Calculated values of μ_{eff}/μ_B and representative experimental values for insulating compounds in the vicinity of room temperature.

Ion	$3d^n$	State	μ_{eff}/μ_B (Calc.-J)	μ_{eff}/μ_B (Calc.-J=S)	μ_{eff}/μ_B (Exp.)
Ti^{3+}, V^{4+}	$3d^1$	$^2D_{3/2}$	1.55	1.73	1.8
V^{3+}	$3d^2$	$^3F_2$	1.63	2.83	2.8
Cr^{3+}, V^{2+}	$3d^3$	$^4F_{3/2}$	0.77	3.87	3.8
Mn^{3+}, Cr^{2+}	$3d^4$	$^5D_0$	0	4.90	4.9
Fe^{3+}, Mn^{2+}	$3d^5$	$^6S_{5/2}$	5.92	5.92	5.9
Fe^{2+}	$3d^6$	$^5D_4$	6.70	4.90	5.4
Co^{2+}	$3d^7$	$^4F_{9/2}$	6.63	3.87	4.8
Ni^{2+}	$3d^8$	$^3F_4$	5.59	2.83	3.2
Cu^{2+}	$3d^9$	$^2D_{5/2}$	3.55	1.73	1.9

the existence of an important term in the potential energy for ions in solids that does not exist for free ions. This is the potential energy at the ion site due to the **crystal field** (i.e., due to the rest of the ions in the solid). This crystal field can lift the 2L + 1 degeneracy for a given L. This is called a Stark splitting of the originally degenerate L level. When the Stark splitting is large compared to k_BT and the magnetic energies, then the occupation of the various sublevels cannot be altered by T or H; thus the magnetic moment also cannot be changed by T or H. Quantum mechanically it is found that the resulting components of the orbital angular momentum average to zero, with small deviations observed occasionally.

We should mention that **electron paramagnetic resonance (EPR)** techniques have been used with great success in determining properties of the ground state for the 4f- and 3d-electrons. For example, the g-values of the ground states can be measured directly. Figure 15-3b shows the resultant splitting of the energy levels in a magnetic field. This splitting results from a potential energy = $-\mu_m \cdot H = m_J g_J \mu_B H$ where there are 2J + 1 values of m_J from +J to −J. Using EPR techniques, energy can be absorbed from an external electronic source (usually in a microwave cavity) when the energy of the electromagnetic radiation ($\hbar\omega$) is just equal to the spacing between the energy levels shown in Fig. 15-3b. For $\Delta m_J = \pm 1$ transitions the separation between the energy levels is $g_J \mu_B H$. In practice the microwave frequency is kept constant and the spacing between the energy levels is changed slowly by varying H. When $\hbar\omega = g_J \mu_B H$ the energy lost from the microwave cavity to the sample can be detected. (In practice derivative techniques are used, i.e., H is very slowly increased but also varied at some low ac frequency and the energy loss at this low ac frequency is detected.) In this manner the g-value can be measured very accurately.

Quenching of the orbital angular momentum – To understand better the effects that occur when ions with unpaired spins are incorporated into solids, we must look into the energies due to spin-orbit coupling and the crystal field terms. The latter have been mentioned only in passing but now will be discussed more carefully.

We can appreciate the crystal field problem by looking at a simple model of a crystal. Calculate the potential near the origin of a coordinate system (where eventually an ion with unpaired spins is placed) due to a charge q at $x = -d$, $y = z = 0$. To evaluate this expand $V = q/r$ for small x, y, and z due to this change. Thus, expand $V = q[(d + x)^2 + y^2 + z^2]^{-1/2} = (q/a)[1 - (x/d) + ...]$. Now assume that there are five other charges at $(d, 0, 0)$, $(0, \pm d, 0)$, $(0, 0, \pm d)$. These six charges octahedrally surround the origin and have $O_h(m3m)$ symmetry as six Cl surround Na in the NaCl structure (Fig. 3-5). This is a realistic model of an ionic crystal as discussed in Chapter 7. The potential from the other five charges can also be expanded about the origin. The leading term in the potential is $V = 6q/d$, which is just part of the Madelung energy discussed in Chapter 7; it is large, but being centrosymmetric will not affect the orientation of spins or orbits of a magnetic ion at the origin. Thus, we can neglect it for this discussion and look at the potential of the higher multipole moments of these six charges about the origin. The dipole, quadropole, and other terms cancel. The first nonzero crystal field term is

$$V_c = (35/4)(q/d^5)[x^4 + y^4 + z^4 - (3/5)r^4] \qquad (15\text{-}11a)$$

This expression specifically shows how the crystal field varies in the neighborhood of the origin due to the six charges each at a distance d away. V_c can also be written in terms of spherical harmonics Y_ℓ^n which depend only on the usual θ and ϕ angles.

$$V_c = (7\sqrt{\pi} qr^4/3d^5)\,[Y_0^4 + (5/14)^{1/2}(Y_4^4 + Y_{-4}^4)] \qquad (15\text{-}11b)$$

Although Eq. 15-11a is more straightforwardly obtained, Eq. 15-11b is easier to use with real wave functions since the radial and angular parts of the potential are separated. The potential energy of an electron centered near the origin is $-eV_c$ from which a crystal field

$$\Delta \equiv \langle \phi | -eV_c | \phi \rangle \propto \langle r^4 \rangle / d^5 \qquad (15\text{-}11c)$$

can be calculated. Due to the angular parts, different wave functions will give different values of Δ and hence there will be splitting. However, it is important to see that $\Delta \propto \langle r^4 \rangle / d^5$ where $\langle r^4 \rangle$ is the expectation value of r^4 over the particular wave function. It is from this proportionality that important distinctions between the rare earth and transition series ions can be found.

CHAPTER 15   MAGNETISM

The other term of importance is the spin-orbit coupling $\lambda \mathbf{L} \cdot \mathbf{S}$. This term couples the orbital angular momentum to the spin angular momentum to give the various values of $\mathbf{J}$. The term arises because an electron moving around a nucleus with charge Ze will see a magnetic field due to the charged nucleus. For a one-electron atom the Hamiltonian can be shown to be

$$H_{so} = \frac{1}{2} \frac{Ze^2}{m^2 c^2 r^3} \ell \cdot s \qquad (15\text{-}12a)$$

where the factor $1/2$ is the so-called Thomas factor arising from relativistic effects since the frame of motion of the electron has a constant radial acceleration about the nucleus. For a many-electron atom the spin-orbit coupling is taken as

$$\langle \phi | H_{so} | \phi \rangle = \lambda \mathbf{L} \cdot \mathbf{S}; \qquad \lambda \propto \langle r^{-3} \rangle \qquad (15\text{-}12b)$$

Note that the spin-orbit coupling constant depends on the radial part of the wave function in a much different way than Δ, Eq. 15-11c.

In discussing the rare earth results, the crystal field was never mentioned. Figure 15-5 corrects this and re-emphasizes some points about rare earth ions. As an example consider Tb^{3+}. The lowest state corresponds to $J = 6$ (Tables 15-2 and 15-4) but notice the $J = 5, 4, 3, \ldots 0$ states at higher energies. The $J = 5$ state is about 2200 cm^{-1} above the ground state. Since room temperature, 300°K, corresponds to $(0.6950)\ 300 = 208.5$ cm^{-1}, the idea that only the ground $J = 6$ state is occupied and that the susceptibility comes only from this state is a reasonable approximation. However, notice the thick lines used for the ground and other states. The thickness of the line corresponds to the crystal field splitting of the $J = 6$ state. (The thickness of the various lines are meant to represent the magnitude of the splitting due to terms such as V_c in Eq. 15-11.) As long as the splittings due to the crystal field are small compared to room temperature then the susceptibility as predicted by Eq. 15-10 should be obtained. Thus, we see how the magnitudes of the effects of V_c and λ play important roles in determining the observed behavior.

Now consider the iron series ions and the surprising results of Table 15-4. For the iron series ions it turns out that effects due to V_c are dominant, that is, stronger than those due to the spin-orbit coupling. (The reason for this is discussed in the next section.) Thus, J is no longer a good quantum number; the crystal field interactions (Eq. 15-11) act on the L-states and splits them by amounts much larger than $k_B T$ at room temperature. The entire manifold of states for a given L is no longer available for rotation in a magnetic field as is the case for a free atom. Notice that V_c in Eq. 15-11 has no spin dependent parts, so does not interact with the spins but only with the orbital

582   CHAPTER 15   MAGNETISM

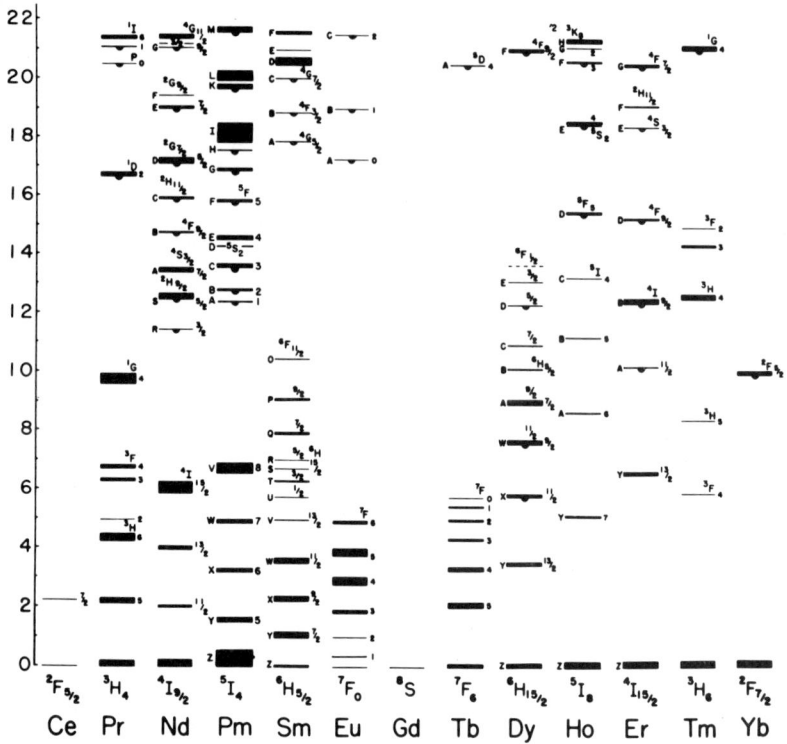

Fig. 15-5 The low lying energy levels of trivalent rare earth ions in $LaCl_3$ (but the levels are representative of these ions in general). The half circles indicate fluorescing levels. (From G. H. Dieke and H. M. Crosswhite.)

part of the wave functions; therefore, S remains a good quantum number. However, when a magnetic field is applied, the g-value is not exactly 2 as expected for a free electron, but due to the $\lambda\mathbf{L}\cdot\mathbf{S}$ term, there is some coupling to the states that have been split by V_c. The g values are calculated to be

$$g \approx 2 + A(\lambda/\Delta) \tag{15-13}$$

where A is a number approximately equal to one, and λ and Δ are in Eqs. 15-11 and 15-12. If the d- or f-shell is less than half full, $\lambda > 0$; $\lambda < 0$ if the shell is more than half full. Thus, g-values slightly larger or smaller than g = 2 can be obtained for the transition series ions.

15-3d Magnitude of λ vs. V_c (shielding?) In the older literature occasionally we see a statement to the effect that 'the rare earth ions in solids behave as if they are free atoms because the 4f-electrons are shielded from the crystal field by the outer $5s^2 5p^6$ electrons.' That the rare earth ions in solids behave as if they are free atoms is true, as discussed. However, attributing this to "shielding" is mistaken. By shielding it is meant that the $5s^2 5p^6$ electrons distort (due to V_c, Eq. 15-11) in such a way so that the V_c seen by the 4f-electrons is significantly smaller than if the $5s^2 5p^6$ electrons did not distort. Calculations of this shielding effect have been performed and it is small (<10%).

Then we may ask why do the rare earth ions behave qualitatively differently from the transition series ions. The reason is straightforward and it can be seen by investigating the several factors that make up the crystal field V_c and the spin-orbit coupling λ. Recall that large λ and small V_c give free atom-like behavior, while large V_c and small λ gives transition series ion behavior.

First, consider V_c. It is essentially made up of $\langle r^4 \rangle$ (the integral is over the 4f- or 3d-electrons) and d^{-5} (distance to the nearest-neighbor ions). The ratio $\langle r^4 \rangle_{3d} / \langle r^4 \rangle_{4f}$ is about a factor 2 because the 3d-electrons have a larger radial extent than the inner 4f-electrons. Also, the entire rare earth ion (naturally including the $5s^2 5p^6$ electrons) is larger than the transition series ions. (See the table of ionic radii, Table 7-2.) So $d_{4f} > d_{3d}$. Since $V_c \propto d^{-5}$ this contribution is sizable and it contributes the same way as the $\langle r^4 \rangle$ term; just this d^{-5} effect makes V_c for transition series ions 3 to 10 times larger than for rare earth ions. (The range, 3 to 10, comes from the different crystal structures.) Thus, two effects cause $(V_c)_{3d}/(V_c)_{4f} \approx$ 6 to 20.

Second, consider λ, which is $\propto \langle r^{-3} \rangle$. Since the 4f-electrons are farther in (closer to the nucleus) than the 3d-electrons, we find that $\lambda_{3d}/\lambda_{4f} \approx 1/2$.

Thus, for straightforward reasons λ/V_c for the rare earth ions is large. And for the same straightforward reasons V_c/λ for transition ions is large. Quantitatively this accounts for the observed effects; see the Notes.

15-3e 4d and 5d series The situation for ions with 4d and 5d valence electrons is more complicated because the spin-orbit and crystal field interactions are of comparable magnitude and larger than $k_B T$ at room temperature. Thus, generalizations are less easy to make.

15-3f Free electron spin paramagnetism A calculation of this effect in metals has already been given in Section 9-12c. The susceptibility was found to be $\chi = 3\mu_B^2 N / 2k_B T_F$, which is temperature independent. A brief discussion of several diamagnetic corrections for metals is also given there.

Part B — Ferromagnetism, Antiferromagnetism, and Related Topics

15-4 Introduction

Now we will emphasize the cooperative aspects of magnetic phenomena. As can be seen in Fig. 15-4, at 1°K the full saturation magnetic moment can be induced by $\approx 3 \times 10^4$ G. At 100°K three million gauss would be required! Yet, there are many ferromagnetic substances above 1000°K, Table 15-5a. Clearly the independent magnetic atom approach of Part A is not always appropriate (Table 15-5). There must be some other, strong quantum mechanical interaction to account for the T_c's. We examine first the phenomenological Weiss theory of ferromagnetism since it is useful and gives an appreciation for the orders of magnitude. Then we will see how the Heisenberg exchange term can give a microscopic picture of the interactions.

15-5 Molecular Field Theory

In 1907 Pierre Weiss first proposed a simple phenomenological theory that explains many of the facts associated with ferromagnetic materials. This theory is called the **molecular field theory** or **Weiss molecular field theory**. The fundamental assumption is that there is an effective field, $\mathbf{B}_E$, acting on the individual magnetic moments that is proportional to the magnetization of the sample $\mathbf{M}$.

$$\mathbf{B}_E = \alpha_E \mathbf{M} \qquad (15\text{-}14a)$$

This is not a normal field that might be used in a $\mathbf{v} \times \mathbf{B}$ expression and it was understood only after the advent of quantum mechanics. The value is $\alpha_E \approx 10^4$ rather than 4π. Since it is not a normal applied field we use the symbol B_E rather than H_E since it is more closely associated with an internal magnetization.

This theory is an example of a **mean field theory,** so named because we assume that all of the spins feel the same, average (mean) field. This offers the simplest approach to many problems and is applied in many other contexts. However, a breakdown in the mean field theory approach should be expected in any situation where the average behavior gives very small or zero average effects, for example, near T_c (i.e., the **critical region**). In such cases correlated fluctuations (spatial and/or time dependent) can be important or even dominant. We shall discuss deviations from mean field behavior for the susceptibility above T_c and in the details of the magnetization at 0°K. (The Landau theory of structural phase transitions, discussed in Chapter 14,

CHAPTER 15   MAGNETISM   585

Table 15-5 (a) Values of the transition temperature and spontaneous magnetism (at 0°K) in gauss for a few ferromagnetic materials. (b) The transition temperatures of a few antiferromagnetic materials. (c) Same as the ferromagnetic only these materials are ferrimagnetics. [Most of the data is from F. Keefer "Handbook der Physik **18** Part 2 (Springer, 1966).]

(a)

Mat.	T_c(°K)	M_s	Mat.	T_c(°K)	M_s
Fe	1043	1752	$CrBr_3$	37	270
Co	1388	1446	Au_2MnAl	200	323
Ni	627	510	Cu_2MnAl	630	726
Gd	293	1980	Cu_2MnIn	500	613
Dy	85	3000	MnAs	318	870
EuO	77	1910	MnBi	670	675
EuS	16.5	1184	$GdCl_3$	2.2	550

(b)

Mat.	T_N(°K)	Mat.	T_N(°K)
MnO	122	$KCoF_3$	125
FeO	198	MnF_2	67.34
CoO	291	FeF_2	78.4
NiO	600	CoF_2	37.7
$RbMnF_3$	54.5	$MnCl_2$	2
$KFeF_3$	115	VS	1040
$KMnF_3$	88.3	Cr	311

(c)

Mat.	T_c(°K)	M_s	Mat.	T_c(°K)	M_s
Fe_3O_4	858	510	$CuFe_2O_4$	728	160
$CoFe_2O_4$	793	475	$MnFe_2O_4$	573	560
$NiFe_2O_4$	858	300	$Y_3Fe_5O_{12}$	560	195

is an example of a mean field theory. Deviations from this theory, and renormalization group, were discussed in Section 14-5.)

The paramagnetic phase — Now consider the high temperature, paramagnetic phase (i.e., $T > T_c$). In this phase we assume that the total field is a sum of B_E and the applied field. These fields produce a magnetization and take the coefficient that relates these two quantities as the paramagnetic susceptibility, Eq. 15-4b, which is called χ_0 in this section. Then $M = \chi_0(H + B_E)$ and using Eq. 15-4a we obtain the normally defined susceptibility, χ, which relates the observed magnetism to the applied field.

$$M = \chi_0(H + B_E) = \chi_0(H + \alpha_E M) \qquad (15\text{-}14b)$$

$$\chi \equiv \frac{M}{H} = \frac{\chi_0}{1 - \chi_0 \alpha_E} \qquad (15\text{-}14c)$$

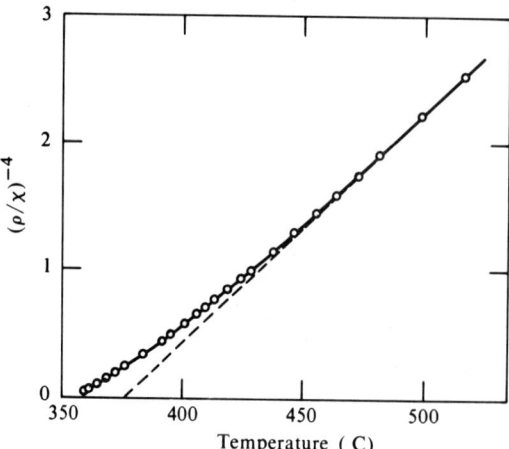

Fig. 15-6 The reciprocal of susceptibility per gram of the ferromagnetic metal nickel. (ρ is the density.) The experimental result (circles) differs from the Curie-Weiss law (dashes) near $T_c = 358°C$. The solid line is due to a more involved analysis. [J. S. Kouvel and M. E. Fisher, Phys. Rev. **136**, A1626 (1964).]

The temperature dependence of χ can be obtained by remembering that χ_0 has a Curie form, $\chi_0 = C/T$, Eqs. 15-8b or 15-10. Thus

$$\chi = \frac{C}{T - C\alpha_E} \equiv \frac{C}{T - T_c} \qquad (15\text{-}14d)$$

This expression for the susceptibility is called the **Curie-Weiss law**. Unlike the Curie law, where the susceptibility diverges at $T = 0°K$, this expression diverges at a temperature T_c. A phase transition is implied (when $\chi \to \infty$) at T_c, where $T_c \equiv C\alpha_E$ and the functional form of χ for $T > T_c$ is obtained. This form is in reasonable agreement with experiment as shown in Fig. 15-6. α_E can be evaluated from Eq. 15-14d by taking approximate values of T_c for iron and Eq. 15-10 for the Curie constant. Approximate values for iron are $T_c \approx 10^3 °K$, $g \approx 2$, $S \approx 1$, giving $\alpha_E \approx 5 \times 10^3$. Using $M \approx 1700$ for Fe we obtain, via Eq. 15-14c, $B_E \approx 10^7$ gauss! This is an astonishingly large field.

If this field were due to a fixed dipole $B_E \approx \mu_a/r_{ab}^3$ where r_{ab} is the distance from μ_a to position b. For typical internuclear distances 10^3 gauss is obtained. Another way to say the same thing is that in terms of an energy $\mu_b B_E/k_B \approx (\mu_a^2/r_{ab}^3)/k_B \approx 10^{-1} °K$. Thus, the magnetic dipole-dipole interaction is much too weak to yield reasonable values for T_c. In fact the origin of these huge exchange fields was not understood until 1929 and the application of quantum mechanics to this problem. (However, for rare earth compounds, which have very low T_c values, these dipole-dipole interactions can be important.)

The ferromagnetic state – Besides making predictions of χ in the paramagnetic state, Weiss' mean field theory can describe the temperature dependence of the spontaneous magnetization below T_c.

From the quantum theory of paramagnetism, Eq. 15-9 describes the dependence of the magnetism on the magnetic field and tempera-

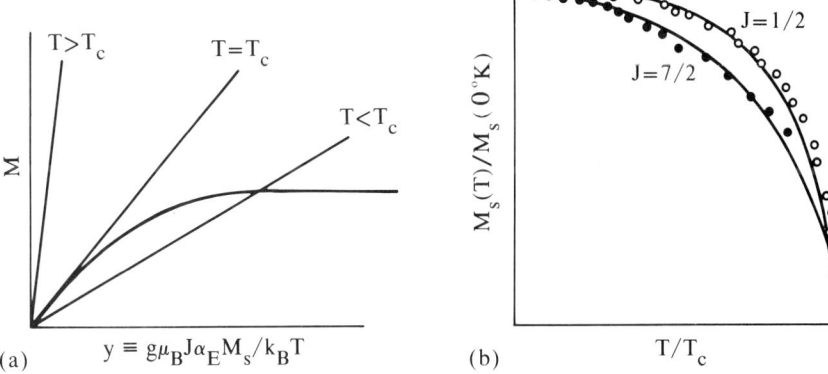

Fig. 15-7 (a) A schematic diagram indicating how the transcendental function Eq. 15-15b is solved. The curve is a plot of M_s vs. y from Eq. 15-15b while the various straight lines are plots of $M_s = (k_B T/g_J \mu_B J \alpha_E) y$ the slope of which decreases as T decreases. As can be seen, below a certain temperature these two plots intersect at a finite M_s. (b) The curves are solutions of M_s vs. T from Eq. 15-15b as described above. The experimental results for $J=7/2$ (solid circles) are for Gd while the $J=1/2$ (open circles) are for Ni, but Co and Fe give very similar results.

ture as $M = (g_J \mu_B J N/V) B_J(g_J \mu_B J H/k_B T)$. For $T < T_c$ we assume that this equation can be used to express the spontaneous magnetism M_s (=M) with no applied field, that is, all of the field comes from B_E in Eq. 15-14a. Thus,

$$M_s = (g_J \mu_B J N/V) B_J(g_J \mu_B J B_E/k_B T) \tag{15-15a}$$

$$= (g_J \mu_B J N/V) B_J(g_J \mu_B J \alpha_E M_s/k_B T) \tag{15-15b}$$

where the second equality is obtained via Eq. 15-14a. This transcendental equation can be solved graphically as shown in Fig. 15-7a. (See Section 11-6a.) For $T \geq T_c$ the only solution is at $M = 0$, but for $T < T_c$ there is also a solution for a finite M. The results are plotted in Fig. 15-7b for two different values of J; reasonable agreement with experiment can be seen. The agreement gives us confidence that this phenomenological mean field contains useful parameters that should be understandable in a microscopic model.

The results for $M_s(T)$ in Fig. 15-7b are similar to the temperature dependence of the long-range order parameter, $\Delta(T)$, shown in Fig. 11-8b. In fact, the magnetism is essentially also a long-range order parameter that describes the ordering of the individual dipole moments. Thus, it is not surprising that there is a heat capacity associated with M(T), $C_V \propto M(dM/dT)$ of the same form as that for $\Delta(T)$, Eq. 11-34, which is shown in Fig. 11-8b. Such $C_V(T)$ are observed but it must be realized that the phonon (and electron if the substance is a metal) heat capacities are also large and have a temperature dependence. Explicit results for $M_s(T)$ are easy to work out for the $J=1/2$ case. This is done in problem 3. In particular, notice that the Curie-Weiss law for the susceptibility is obtained easily.

15-6 The Heisenberg Exchange Interaction

In 1928 Heisenberg applied quantum mechanics to the magnetic problem and showed that the interaction energy between two atoms, i and j, with spins S_i and S_j contains a term

$$V_{ex} = -2J_{ex} S_i \cdot S_j \qquad (15\text{-}16a)$$

The value of J_{ex}, the **exchange coefficient** or **exchange constant**, is strongly related to the overlap of the two atoms and hence to the internuclear distance, that is, $J_{ex} = J_{ex}(r_{ij})$; J_{ex} usually is quoted in units of energy so S is a spin rather than a spin angular momentum $\hbar S$. The origin of V_{ex} does not come from magnetic dipole moments or current loops or the types of magnetic field that go into Maxwell's equations. The origin lies in what are called the quantum mechanical "exchange forces." These are electrostatic in origin, but arise because the distribution of charge (the electron wave functions) is controlled by the antisymmetry principle. This requires that the total many-electron wave function must be antisymmetric with respect to the interchange of two electrons. (Historically the **Pauli exclusion principle** says that no two electrons can have the same set of quantum numbers; for example, two electrons in the same nondegenerate orbital state must have opposite spin quantum numbers. The **antisymmetry principle** is a stronger requirement although the terms are often used interchangeably. For a noninteracting electron system the total wave function can be written as a determinant wave function, a **Slater determinant**, in which case the antisymmetry principle is taken care of automatically.)

If we consider a free electron gas (each electron only sees the mean electronstatic field of all the others) then in calculating the energy of this electron gas using a Slater determinant an exchange term is found

$$V_{ex} = \int d^3 r_1 \int d^3 r_2 \psi_i(r_1) \psi_j(r_1) \frac{e^2}{|r_1 - r_2|} \psi_i(r_2) \psi_j(r_2) \qquad (15\text{-}16b)$$

where $\psi_i(r_1)$ is the (real) wave function for the orbital associated with site i and electron with position coordinate r_1. $\psi_i(r_1)\psi_j(r_1)$ is called the **exchange charge**, and the expression in Eq. 15-16b is the electrostatic self-energy of the "exchange" charge. This exchange also arises

CHAPTER 15   MAGNETISM   589

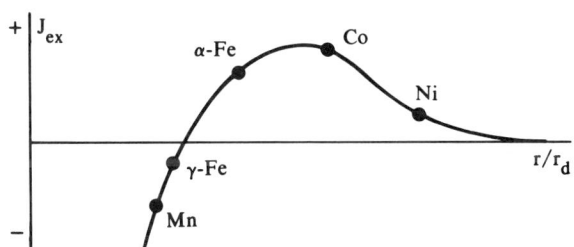

Fig. 15-8 A schematic plot of the exchange constant vs. interatomic distance (r_d is the distance between nearest neighbors). Values for the transition metals are shown.

in atomic and molecular orbital energy calculations. The important point is that this energy does not exist for antiparallel spins. It only exists for electrons with parallel spins and is always positive for orthogonal orbitals; so it lowers the mutual energy of two electrons if their spins are parallel, with respect to their energy with antiparallel spins. Actually, this energy is the explanation of Hund's first rule and is the driving force for band magnetism.

There is an important aspect of Eq. 15-16b that must be considered further. The lowering of the energy of electrons with parallel spins rigorously applies to electrons in orthogonal orbitals. This is fine for the intra-atomic case (orbitals on a single atom). However, for a solid, particularly a d-band metal, the Hamiltonian contains a periodic potential, and the basis functions are on different atomic centers. Typical for magnetic metals, the tight binding approximation is used for the electron wave functions (Section 10-9). Then the inter-atomic exchange integral mainly arises from the overlap of the wave functions of electrons localized on adjacent atomic sites. This leads to exchange energies similar to Eq. 15-16b, but the energy may be positive or negative because of the non-orthogonality of the orbitals on different centers, and it should be sensitive to the internuclear separation between the atoms.

Thus, the exchange energy can be of either sign. Conventionally the signs are picked, in accordance with Eq. 15-16a, so that for the ground state

S_i and S_j parallel, $V_{ex} < 0$, and J_{ex} is positive.

S_i and S_j antiparallel, $V_{ex} > 0$, and J_{ex} is negative.

Realistic calculations of J_{ex} and its distance dependence are difficult to perform. Figure 15-8 is a qualitive picture that shows two things. The first is the general shape of J_{ex} vs. r; as r increases it goes from negative to positive values, as non-orthogonality becomes less important, and then becomes very small. Also shown is a criterion, suggested by Slater, for the occurrence of ferromagnetism in the transition series ions. J_{ex} is plotted vs. the internuclear distance divided by $2r_{ar}$ where

r_{ar} is the atomic radius of the element. While Fe, Co, and Ni satisfy the criterion for ferromagnetic behavior (J_{ex} is positive), Cr and Mn do not even though the latter in particular has a large unpaired spin. However, r_{ab} is larger in MnSb and MnAs, and both of these materials indeed are ferromagnetic (Table 15-5). (Of course, this sort of argument should not be taken very seriously since the Mn-Mn overlap is much different in metallic Mn than in MnAs.)

Approximate values for the exchange constant can be obtained in terms of T_c by considering the central atom as the ith-atom under the influence of an effective field B_E due to an average value of the spins of the z-neighboring atoms $\langle S_j \rangle$. The interaction energy $-\mu_i \cdot B_E$ can be written in terms of the exchange interaction as

$$2\sum_j J_{ex}(r_{ij})\langle S_j \rangle \cdot S_i = -\mu_i \cdot B_E = g\mu_B S_i \cdot B_E \qquad (15\text{-}17a)$$

Then using $M \approx g\mu_B \langle S_j \rangle N/V$ (Eq. 15-9) and $T_c = \alpha_E C$ (Eq. 15-14d) and the value for the Curie constant from Eq. 15-10, we obtain

$$\sum_j J_{ex}(r_{ij}) \approx 3k_B T_c / 2S(S+1) \qquad (15\text{-}17b)$$

assuming that the effects are due to spins only. Taking a nonzero exchange constant only between an atom and its z-nearest neighbors

$$T_c \approx \frac{2zJ_{ex}S(S+1)}{3k_B} \qquad (15\text{-}17c)$$

So for the high T_c values in Table 15-5a, J_{ex} has large values. Another useful relationship is the value of the Weiss inner field at 0°K

$$B_E(0) = \alpha_E M_s(0) = \alpha_E(N/V)g_J\mu_B J \qquad (15\text{-}17d)$$

since the Brillouin function (Eq. 15-9) is one at low temperature.

15-7 Magnetic Structures

It has turned out that neutron diffraction is an excellent technique for determining the magnetic structure of solids. The reasons will be discussed, but now just assume that such structures are known and we shall discuss them. The different values of the sign of J_{ex} in Eq. 15-6 allows us to understand (at least phenomenologically) the wide variety of magnetic structures that are observed.

15-7a Ferromagnetism Table 15-5a lists T_c values for some ferromagnetic materials. For these materials the exchange constant, Eq. 15-6, is positive so all the spins are parallel.

15-7b Antiferromagnetism

In 1932 Néel suggested the possibility that a magnetic material could have neighboring dipoles antiparallel to each other, thus possessing no macroscopic magnetism. Such materials are called **antiferromagnetic**. They can arise if the exchange coefficient, J_{ex} in Eq. 15-16a, is negative since that will cause an antiparallel arrangement of spins to have a lower energy than a parallel arrangement. However, negative J_{ex} is not a sufficient condition for antiferromagnetism since the geometry and distribution of kinds of magnetic ions is important, as we shall see. Antiferromagnetism is discussed via a simple example.

$RbMnF_3$ has the perovskite crystal structure, Fig. 3-1c. Ignore the nonmagnetic ions and focus only on the Mn^{2+} ions; each Mn^{2+} ion is surrounded by six nearest-neighbor Mn^{2+} ions. If there is a negative exchange interaction between nearest neighbors, and if $J_{ex} \approx 0$ between farther neighbors, then at $0°K$ a Mn^{2+} ion with spin up will have six Mn^{2+} ions with spin down and vice versa. Figure 15-9a shows this antiferromagnetic order.

Often the following language is used to describe the magnetic system. Divide the Mn^{2+} ions into an A-sublattice and a B-sublattice so that each A-sublattice ion has six nearest magnetic neighbors all from the B-sublattice and vice versa. Thus, for a negative nearest neighbor J_{ex}, the ions on the A- and B-sublattice will be spontaneously magnetized in opposite directions. This is the condition for antiferromagnetism (Fig. 15-9). As the temperature is increased, the thermal agitation energy begins to become comparable to the exchange energy and eventually the antiferromagnetic ordering disappears. The temperature at which this occurs is called the **Néel temperature** T_N.

The molecular field ideas can be applied to antiferromagnetic materials in a manner similar to what was done for ferromagnetic materials. The effective molecular field on the A-sublattice is $B_A = \alpha_E M_B$ because the A-sublattice will feel the magnetism only from the B-sublattice ($J \approx 0$ for second and farther neighbors). Correspondingly, $B_B = \alpha_E M_A$, so similar to Eq. 15-14b

$$M_A = \chi_{A0}(H + B_A) = \chi_{A0}(H + \alpha_E M_B)$$
$$M_B = \chi_{B0}(H + B_B) = \chi_{B0}(H + \alpha_E M_A) \quad (15\text{-}18)$$

where we have allowed for different paramagnetic susceptibilities for the different sublattices since, in general, the number and types of paramagnetic ions on the two sublattices could be different. Then taking the Curie form for $\chi_{A0} = C_A/T$ and $\chi_{B0} = C_B/T$, substituting into Eqs. 15-18, two simultaneous equations are obtained with M_A and M_B as unknowns. Setting the determinant of these equations to zero gives $T_c = \alpha_E(C_A C_B)^{1/2}$ and solving for M_A and M_B we obtain a susceptibility for $T > T_c$

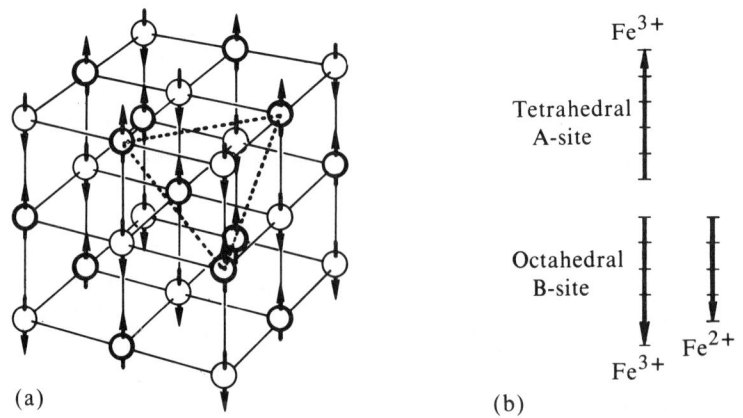

Fig. 15-9 (a) Antiferromagnetic ordering of the Mn^{2+} ions in $RbMnF_3$. The Rb- and F-ions are not shown. In each (111) plane all the Mn-ions have the same spin orientation. (b) A schematic diagram showing the spin directions of the different sites in Fe_3O_4.

$$\chi = \frac{M_A + M_B}{H} = \frac{(C_A + C_B)T - 2\alpha_E C_A C_B}{T^2 - T_c^2} \quad (15\text{-}19a)$$

For $RbMnF_3$ both sublattices are equivalent so $C_A = C_B = C$ and these equations simplify to (still for $T > T_c$)

$$\chi = \frac{2C(T - T_c)}{T^2 - T_c^2} = \frac{2C}{T + T_N} \quad (15\text{-}19b)$$

where $T_N \equiv T_c = \alpha_E C$. For this antiferromagnetic case we have called the transition temperature T_N (the Néel temperature). Notice that the different sign in the denominator of the result for χ in Eq. 15-19b and 15-14c give different types of behavior for χ. In the ferromagnetic case χ approaches infinity as $T \to T_c^+$, Fig. 15-1b, but in the antiferromagnetic case, Eq. 15-19b, χ remains small at T_N, Fig. 15-1c.

In practice for temperatures above the transition temperature (T_N) the susceptibility is better represented by

$$\chi = C/(T + \Theta) \quad (15\text{-}19c)$$

where experimentally it is usually found that $\Theta > T_N$. This expression for the susceptibility says that χ would become infinite (or χ^{-1} would become zero) at $T = -\Theta$. Naturally this negative temperature is never reached. The behavior is shown in Fig. 15-1c. The expression in Eq. 15-19c can be accounted for by more sophisticated models which allow for exchange between more remote neighbors. Usually these models predict $\Theta > T_N$ in agreement with experiment.

It is useful to appreciate that with the simple, intuitive Weiss mean field model we have been able to predict very different behaviors for ferro- and antiferromagnetic susceptibilities (Eqs. 15-14d and 15-19b). Further, we can appreciate qualitatively that for $T < T_N$ the susceptibility should decrease below what would be expected in the paramagnetic case. This is because the spins become "locked" in by the exchange field so that the response to an external applied field, **H**, will be smaller than if they were not locked in place. Furthermore, χ is really a second-order tensor since it relates two vectors, so for a simple antiferromagnetic material such as $RbMnF_3$ it would be possible to consider the anisotropy of χ, that is, $\chi_\parallel$ (parallel to $\mathbf{M}_A$) and $\chi_\perp$ (perpendicular to $\mathbf{M}_A$). Since it is expected to be much easier to tilt the spins slightly by applying **H** perpendicular to $\mathbf{M}_A$ than to reverse some of the spins by applying **H** antiparallel to $\mathbf{M}_A$, we expect $\chi_\perp > \chi_\parallel$. This situation is almost always found in nature and as T approaches zero $\chi_\parallel$ approaches zero.

For the antiferromagnetic compounds of the type discussed here or one such as MnO which has the NaCl structure we must reconsider the origin of the exchange constant in Eq. 15-16a. In both $RbMnF_3$ and MnO each Mn ion's nearest-neighbor Mn ion has an F^- or O^{2-} ion in between. This might be written as Mn-O-Mn. The exchange in such a situation is quite different from the exchange for two nearest-neighbor d-shell ions where the Pauli principle can be applied directly. The antiferromagnetic exchange coupling through an intermediate closed-shell ion is called **super exchange** or **indirect exchange**. This type of exchange is particularly important in insulators. For a simple survey of super exchange see Section 8-5 in Morrish's book or Anderson's article in Vol. 1 of Rado and Suhl's book, both of which are referenced in the Notes.

15-7c Ferrimagnetism In 1948 Néel further considered the situation where the magnetic ions could be divided into two sublattices with unequal magnetic moments. Then even if the exchange interaction is negative there will be a net resultant magnetic moment.

Ferrites – Let us consider the **ferrites** in particular. These crystals have a spontaneous net magnetism at low temperature (Table 15-5c). They are compounds with the formula MFe_2O_4 ($M^{2+}Fe_2^{3+}O_4^{2-}$) where M is a divalent ion (i.e., Fe, Ni, Cu, Co, Zn, or Cd). The crystals have the spinel crystal structure with eight molecules per crystallographic unit cell. The structure contains 16 octahedral and 8 tetrahedral sites for the metallic ions. It turns out that in most cases $8Fe^{3+}$ ions are on the tetrahedral (A) sites and ($8Fe^{3+}$ ions + $8M^{2+}$ ions) are on the octahedral (B) sites.

The most familiar example of the ferrites is magnetite (lodestone) where M = Fe^{2+}, and the formula is sometimes written as $FeO \cdot Fe_2O_3$. If all the spins are parallel, then the saturation dipole moment for each ion is $\mu = g\mu_B J$ (Eq. 15-9). We take g = 2 and J = S so each Fe^{3+} ion (S = 5/2) gives $\mu = 5\mu_B$ and each Fe^{2+} ion (S = 2) gives $4\mu_B$. Thus, there should be a saturation magnetism per formula unit of $2 \times 5\mu_B + 4\mu_B = 14\mu_B$ assuming that all of the spins are parallel. The experimental result is $\approx 4\mu_B$! Néel's model showed that the saturation moment comes from only the Fe^{2+}-ions since the moments of the Fe^{3+}-ions on the tetrahedral and octahedral sites are oppositely directed. Figure 15-9b indicates this effect, and neutron diffraction measurements are in agreement with this model.

The Weiss molecular field model for magnetite has already been worked out for the A- and B-sublattice (Fig. 15-9b), and the result is shown in Eq. 15-19a. The relatively simple result for the susceptibility for the ferro- and antiferromagnetic cases is no longer true for ferrimagnetic materials. Rather a more complicated form for χ is obtained even from this simple model. Figure 15-10 shows an example of the high temperature susceptibility where a more complicated behavior than for the ferro- and antiferromagnetic cases can be seen.

The physical reason for the ferrimagnetic ordering (Fig. 15-9b) is a bit unexpected. It turns out that all of the exchange integrals (J_{AA}, J_{AB}, J_{BB}) are negative but J_{AB} is the strongest interaction so that the spins of the two different kinds of ions on the octahedral sites are parallel to each other because they want to be antiparallel to the ions on the tetrahedral sites. Thus, even though all the exchange integrals are antiferromagnetic (negative) the crystal has a net magnetism (i.e., it is ferrimagnetic). Formerly ferrites had an important use in the memory systems of large computers where desirable magnetic properties can be obtained by a mixture of divalent ions.

Garnets – The garnets are another interesting (and technologically useful) ferrimagnetic material. The general formula is $X_3Fe_5O_{12}$, where X, a trivalent ion, can be a rare earth ion. A useful garnet is $Y_3Fe_5O_{12}$ (YIG) which also is a magnetically convenient reference since Y^{3+} is diamagnetic. There are three Fe^{3+} ions per formula unit on the so-called d-sites, two Fe^{3+} ions on the a-sites, and the three X^{3+} ions are on the c-sites. [The letters a, c, and d refer to positions in the Wyckoff notation of the space group appropriate to these cubic garnets, i.e., $O_h^{10}(Ia3d)$. See Chapter 3.] If the Fe^{3+} ions on the d- and a-sites were aligned parallel, then the saturation moment per formula unit should be $5(g\mu_B S) = 5(2\mu_B 5/2) = 25\mu_B$. However, $\approx 5\mu_B$ is observed and the reason is that the exchange constant is antiferromagnetic (as in the ferrites) and the spins of the Fe^{3+} ions on the d- and a-sites are antiparallel.

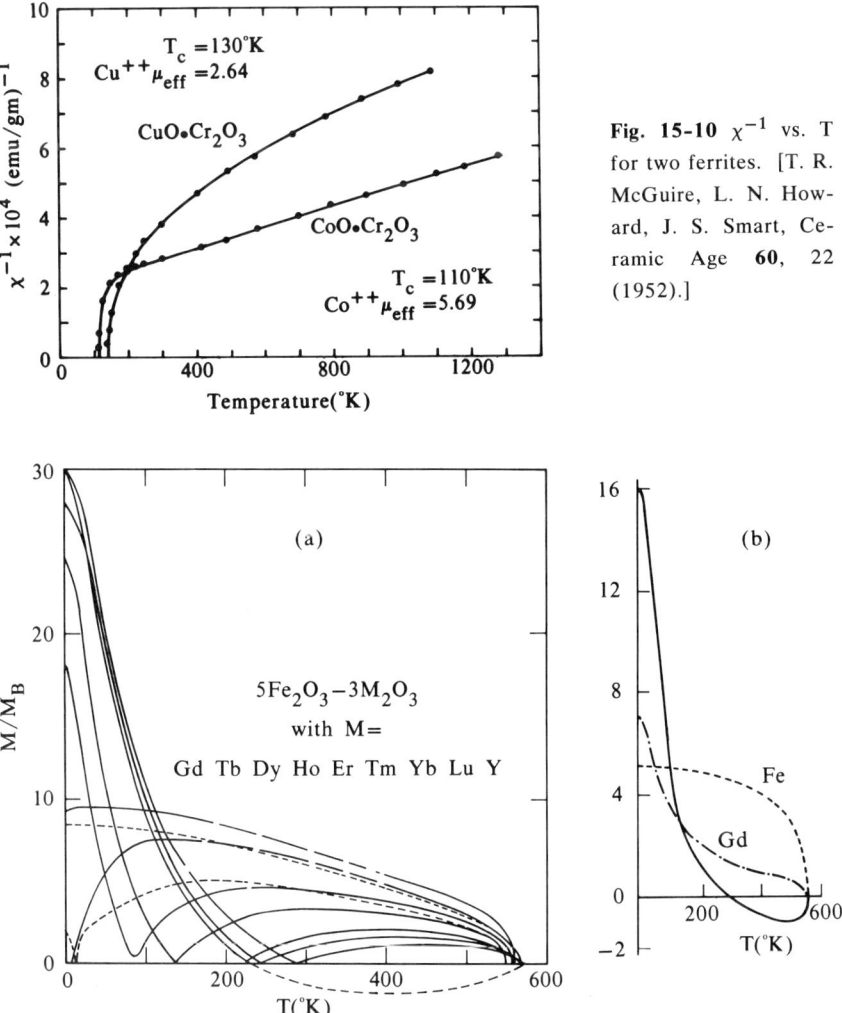

Fig. 15-10 χ^{-1} vs. T for two ferrites. [T. R. McGuire, L. N. Howard, J. S. Smart, Ceramic Age **60**, 22 (1952).]

Fig. 15-11 (a) The temperature dependence of the spontaneous magnetism in garnets. The curves are in the same order as the rare earths listed with the results for Y and Lu showing no compensation points as expected. [F. Bertant and R. Panthenet, Proc. I.E.E.E. Suppl. **B104**, 261 (1957), R. Panthenet, J. Appl. Phys. **29**, 253 (1958).] (b) The same but for only GdIG and showing the individual magnetic contributions [R. Panthenet, Ann. Phys. **3**, 424 (1958).]

For YIG, $T_c = 559°K$ and the magnetism vs. temperature curve displays normal behavior (Fig. 15-1d) and is shown in Fig. 15-11a. Interesting M(T) curves are obtained when X is a rare earth ion, and the results are shown in Fig. 15-11a. The magnetism of the rare earth ion on the c-sites is opposite to that of the (d + a)-sites. Furthermore, the c-d and c-a exchange interactions are weak. Thus, the rare earth ions lose their magnetism at low temperatures. (This is the reason that different Curie constants are needed in Eq. 15-19a.) The temperature dependence of the magnetism of the different sublattices is shown in Fig. 15-11b; $M(T)$ actually changes sign at some temperature, called the **compensation point**, and this accounts for the unusual $M(T)$. (Certain ferrites also can have similar magnetization curves.)

From the small variation of T_c observed in the rare earth (RE) garnets as the rare earth ion is changed (548°K to 570°K where YIG is 560°K), the weak exchange field between the RE and Fe ions can be appreciated. This occurs because magnetic 4f-electrons are inner electrons, that is, they overlap much less with their neighboring ions than the outer 3d-electrons. The c-c exchange is even smaller for the same reason; materials such as $RE_3Al_5O_{12}$ display cooperative behavior only at liquid He temperatures if at all.

15-7d Other ordered arrangements In the most general case $J(r_{ij})$ can vary in sign with distance and, especially in metals, it can be large at distances up to several neighbors. (This is discussed in Part C.) Furthermore, there is no reason that the peaks in $J(r_{ij})$ need occur at the positions of other occupied sites in the material. Figure 15-12 shows some of the unexpected results that are experimentally observed. For the rare earth metal dysprosium there is no resultant magnetic moment but for erbium metal there is a net magnetic moment along the c-axis.

15-8 Some Special Techniques Used to Study Magnetic Structures

We briefly discuss three experimental techniques that are used to study magnetic materials but also are useful in other areas of solid state science. There are certainly many other ways to study magnetic materials. These techniques are discussed just to give the reader a flavor of the possibilities.

15-8a Neutron diffraction Although much had been known about the internal magnetic arrangement of the moments of various ions, the first direct measurement, using neutron diffraction, of spin directions in a magnetic material was not reported until 1949 (Fig. 15-13). This technique is the best way to investigate the internal magnetic structure of a material.

Neutrons can interact with the atoms in crystals in two different ways. First, the neutron interacts with the nuclei in the solid via a nuclear force. When the nuclei are in an ordered array, peaks in the scattering amplitude will occur in certain directions as discussed in

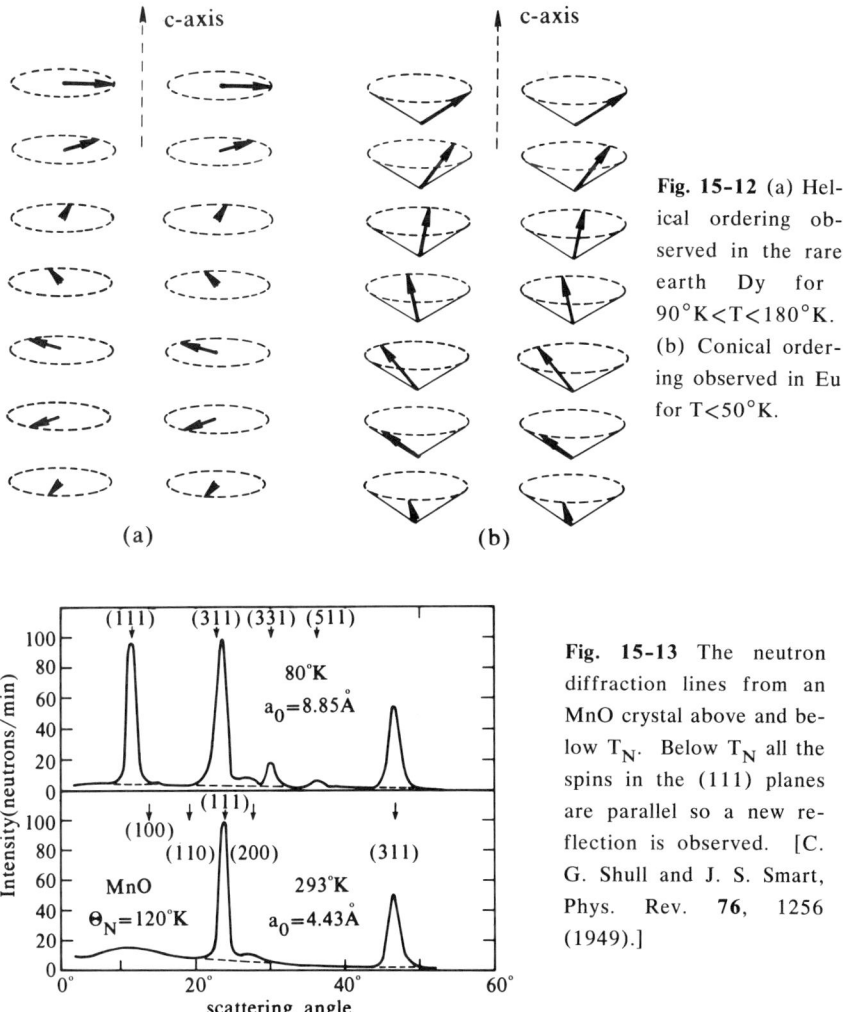

Fig. 15-12 (a) Helical ordering observed in the rare earth Dy for $90°K < T < 180°K$. (b) Conical ordering observed in Eu for $T < 50°K$.

Fig. 15-13 The neutron diffraction lines from an MnO crystal above and below T_N. Below T_N all the spins in the (111) planes are parallel so a new reflection is observed. [C. G. Shull and J. S. Smart, Phys. Rev. **76**, 1256 (1949).]

Chapter 4. Second, and this is the important aspect for this chapter, the neutron has a spin (1/2) and a magnetic moment so it interacts with the magnetic moments of the electron cloud of the atom.

Consider the magnetic ions in Fig. 15-9a. For $T > T_c$ there is no average moment on the magnetic ions so the adjacent (111) planes of Mn-ions are identical and will scatter x-rays or neutrons coherently. However, below T_c adjacent (111) planes contain ions with antiparallel magnetic moments. Thus, the new (111) spacing is twice as large as that above T_N ($\sin \theta$ is twice as small, Eq. 4-3). In fact, a look at Fig. 15-9a will show that the magnetic unit cell ($T < T_N$) is twice the size of the unit cell for $T > T_N$. This is why extra lines, at half angles, are observed below T_N as shown in Fig. 15-13.

The theory of neutron diffraction from magnetic moments shows that the intensity of the lines due to the interaction of the magnetic moment of the neutron with the magnetic moment of the atoms is $\propto M^2$ (i.e., the square of the magnetization). Thus, M^2 can be determined from the temperature dependence of the intensities. The results are in generally good agreement with theory.

15-8b Magnetic resonance For this discussion refer to Fig. 15-3b. The splitting of the $2J + 1$ states for $J = 5/2$ state in a magnetic field is shown. The energy separation between any two levels ($\Delta J_z = \pm 1$) is $g_J \mu_B H$. If using a very sensitive oscillator, with frequency ν_0, then as the magnetic field is increased from zero, there will be some value of magnetic field H_0 such that $h\nu_0 = g_J \mu_B H_0$, and transitions will occur among the quantum states with a net loss of energy from the oscillator. This loss of energy can be detected, and since H_0 can be measured very accurately by other techniques, g-values can be measured. In practice the energy loss from the oscillator is detected by **narrow banding techniques**; besides increasing H slowly, a very small alternating field is superimposed at some low frequency (e.g., 30 cps). The magnetic field is thus driven in and out of resonance at this low frequency, and the loss signal from the oscillator can be detected using narrow band amplifiers tuned to this low frequency.

EPR – These techniques have become quite routine; the measurement of the magnetic resonance of the electrons in their ground states in solids is called **electron paramagnetic resonance** (EPR) or **electron spin resonance** (ESR). There have been many studies of transition metal ions in insulators. For ions with $L = 0$ and $J = S$ the electrons have $g \approx 2$, and if $\nu_0 = 10^{10}$ Hz (x-band) then $H_0 \approx 3600$ gauss.

NMR – Nuclei also posses an angular momentum ($\hbar I$) and a g-factor. The discussion for the nuclear angular momentum is very similar to the preceding for the electron spin or total angular momentum. The Hamiltonian and energy levels are

$$\mathcal{H} = -\mu \cdot H = -g_n \mu_n I \cdot H \quad (15\text{-}20a)$$

$$\mathcal{E}_m = -g_n \mu_n H m_I \quad (15\text{-}20b)$$

where $\quad m_I = I, I - 1, ..., -I$

Thus, there are $2I + 1$ levels, and in a magnetic field they split as shown in Fig. 15-3b. The two distinct differences between EPR and NMR are: first, because of the mass ratio of the proton to the electron the nuclear magneton (μ_n) is 1836 times smaller than μ_B; second, the nuclear g-values (g_n) depend on nucleon-nucleon forces, so it is found that g_n varies between $\approx \pm 10$. These differences cause nuclear magnetic resonance (NMR) experiments to occur at frequencies ≈ 2000 times lower than those of electrons. The frequency range is in the

10^7 Hz range for convenient magnetic fields ($\approx 10^4$ gauss), but the ideas associated with the measurement and detection are similar.

The first NMR measurements in an antiferromagnetic crystal were reported in 1953 on the protons in $CuCl_2 \cdot 2H_2O$ ($T_N \approx 4°K$). The magnetic field seen by the protons consists of the externally applied field plus the internal fields; in fact, it is the latter that are sought after. Then four years later an experiment similar in concept was reported on the fluorine nuclei in the antiferromagnetic MnF_2. Eventually, the NMR of the Mn nuclei also was observed. The fluorine NMR lines are narrow and can be followed right up to $T_N = 67.336°K$. With the possibility of such accurate measurements we can look for critical effects near the phase transition. The F resonance frequency with very high accuracy was found to have a temperature dependence given by $(T_N - T)^{1/3}$.

NMR was not expected to be observed in the ferromagnetic metals (Fe, Co, Ni, etc.) but it was observed in 1959 due to an unexpected enhancement due to domain wall motion. These types of results enable the determination of the magnetic field at the positions of various nuclei in these materials. The theoretical models of such magnetic fields has led to a deeper understanding of internal fields. For a more detailed discussion of NMR see Chikazumi, Chapter 21, or some of the specialized books mentioned in the Notes.

15-8c Mössbauer effect This effect is an interesting tool to use in studying several different aspects of solids including internal magnetic fields; for this reason we mention it here.

Figure 15-14 shows a *nuclear* energy level diagram for a typical Mössbauer transition; the Fe^{57} isotope is used often, but there are many other possibilities. The energy scale in (a) is vastly different from that in (b) through (d). For (a) the lowest energy transition is $\mathscr{E}_0 = 14.4 \times 10^3$ eV corresponding to an excited to ground nuclear state. (Most excited nuclear states actually have higher values of $\mathscr{E}_0$.)

Let us follow the energy level diagram of the nuclear emission (Fig. 15-14a). Co^{57} can be bought as an isotope or it can be produced in situ by irradiating Fe with 4 MeV deuterons. Co^{57} captures an electron (electron capture $\equiv$ E.C.) and decays with a 270-day half-life into an excited Fe^{57} state, as shown in the figure. A further emission of a γ-ray results in the long-lived Fe^{57} under discussion. From the Fe^{57} (137 keV) excited state there is a 9% probability of decay (by γ emission) to the nuclear ground state and a 91% probability to the Fe^{57} (14.4 keV) state. It is this latter excited nuclear state that we concentrate on; it is a very long lived (by nuclear standards) state, having a mean lifetime of 1.4×0^{-7} sec. From the Heisenberg uncertainty principle $\Delta\mathscr{E}\Delta t \sim \hbar$, this corresponds to a line width of $\Gamma = (2\pi \times 1.4 \times 10^{-7} \text{ sec})^{-1} = 1.1 \times 10^6$ c/sec $= 4.6 \times 10^{-9}$ eV $\equiv \Delta\mathscr{E}_0$.

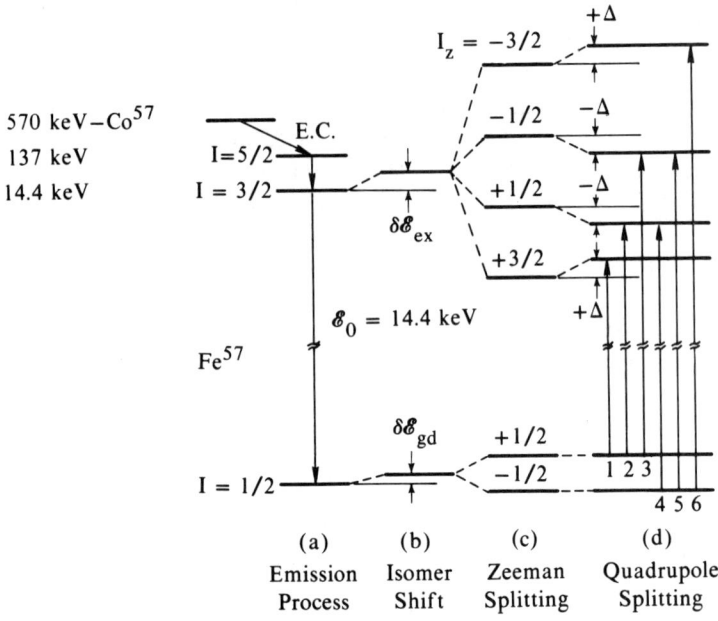

Fig. 15-14 The energy level of Fe^{57}. (a) shows the nuclear level scheme; (b), (c), and (d) show the small perturbations on these levels the scale of which is about 10^{13} smaller than that of (a).

Thus, the transition at 14.4 keV is amazingly sharply defined $\Delta\mathscr{E}_0/\mathscr{E}_0 \approx 3.5 \times 10^{-13}$!

Assume for the moment that the 14.4 keV excited nuclear state can emit its γ-ray without recoil momentum. (Or assume that the nucleus is rigidly clamped to a massive object which can take up the recoil momentum with a correspondingly extremely small velocity. This is what happens in solids.) Then how can the extremely sharp spectrical information contained in the photon source be used? That is, how can an absorber be found with as sharply defined spectrical properties as the emitter? There is only one answer and it is another Fe^{57} nucleus in the ground state; such a nucleus, since it clearly has the correct energy levels, can act as an absorber for these sharply defined photons. In fact, since such an absorber is always in tune with the emitter we can ask how to tune the frequencies one to another, something that is usually required to obtain extended spectral information. Again, the answer is simple: since the photon line is so precisely defined, in frequency space, the Doppler effect (moving the source with respect to the absorber) can be used to tune the absorption in and out of resonance with respect to the source. The cross section of the natural line width of the emission or absorption of γ-ray is given by the Breit-Wigner formula

CHAPTER 15　　MAGNETISM

$$\sigma = \sigma_0 \frac{\Gamma^2/4}{(\mathscr{E} - \mathscr{E}_0)^2 + \Gamma^2/4} \tag{15-21a}$$

where Γ is given above. The relative velocity of the source to absorber in order to scan a full width, 2Γ, of the γ-ray line is given by the Doppler shift

$$v = c(2\Gamma/\mathscr{E}_0) \approx 0.19 \text{ mm/sec} \tag{15-21b}$$

This is a very small, easily achievable velocity.

We can now envision the use of the Mössbauer effect. The Fe^{57} spectrical line of either the source or absorber is perturbed by internal field (magnetic or other kinds) which split the nuclear line similar to the discussion of the NMR experiments. Then by velocity scanning the source with respect to the absorber one can tune through the different lines and thus measure the splittings.

When the nucleus of a free atom emits an x-ray of energy $\mathscr{E}_0$ there is a recoil momentum $p = \mathscr{E}_0/c$ which leads to a recoil energy $\mathscr{E}_0^2/2Mc^2$ where M is the mass of the nucleus. For Fe^{57} the frequency corresponding to this recoil energy is $\approx 8 \times 10^{11}$ c/sec or very much larger than the natural line width of $\approx 10^6$ Hz, which results in a broadening very much larger than its natural line width. This is precisely the situation for free atoms. However, when the emitting Fe^{57} atom is embedded in a crystal, it is found that a certain fraction of the nuclei emit γ-rays with no perceptible recoil. This is known as the Mössbauer effect. Thus a very sharply defined x-ray source ($\Delta\mathscr{E}_0/\mathscr{E}_0 \approx 3.5 \times 10^{-13}$) is available. The fraction of recoilless emissions is related to the stiffness of the springs that hold the atom in place; hence to θ_D the Debye temperature (Chapter 11). For "stiffer" materials (larger θ_D) the fraction of recoilless emissions, f, increases rapidly. For a Debye solid and $T \ll \theta_D$

$$f = \exp(-3\mathscr{E}_0/2k_B\theta_D) \tag{15-21c}$$

Now that the background of the Mössbauer effect has been discussed let us see how it is used to study solids. This gets us to parts b through d of Fig. 15-14; it should be understood that the vertical energy, or frequency, scale that describes the splittings is $\approx 10^{13}$ smaller than that describing nuclear transitions in Fig. 15-14a.

The **isomer shift** is due to the electrostatic interaction of the positively charged nucleus with the electron charge density in the nucleus. In this sense it might be called the electronic monopole interaction since it arises from the first term of an expansion of the type in Eq. 6-4. The quadrupole splitting, Fig. 15-14d, arises from the third term in this expansion. By a straightforward calculation similar to that in Section 6-4, we find that the difference in energy in the excited (ex) and ground (gd) states relative to a point nucleus is

$$\delta\mathcal{E}_{ex} - \delta\mathcal{E}_{gd} = (2\pi/5)Ze^2|\psi(0)|^2(R_{ex}^2 - R_{gd}^2) \quad (15\text{-}22a)$$

where $|\psi(0)|^2$ is the electron charge density at the nucleus and R_{ex} and R_{gd} is the effective nuclear radius in the excited and ground state. Since only s-state electrons have an appreciable probability of being in the nucleus the isomer shift is a measure of the s-electron density. The Mössbauer effect is always measured with the Fe^{57} embedded in a source (s) and absorber (a) so the isomer shift is defined as

$$\delta = \left(\frac{2\pi}{5}\right)Ze^2\left[|\psi_a(0)|^2 - |\psi_s(0)|^2\right](R_{ex}^2 - R_{gd}^2) \quad (15\text{-}22b)$$

where the solid state information is contained in the first square bracket. Characteristic different values for δ are found for both Fe^{2+} and Fe^{3+}. This might seem strange because both of these ions have the same s-electrons; they differ only by a d-electron, which has no direct contribution to $|\psi(0)|^2$. The different isomer shifts for Fe^{2+} and Fe^{3+} arise indirectly. Adding a d-electron reduces the net attractive Coulomb potential for the 3s-electrons, causing their wave function to expand slightly, reducing its charge density at the nucleus. The difference between these two ions corresponds to a velocity shift of the source to absorber of ≈ 0.9 mm/sec, which is considerably larger than the natural width and is easy to measure. There are detailed studies of the isomer shift in materials with electron configurations such as $3d^54s^x$, $3d^64s^x$, $3d^74s^x$, and related materials, where δ can be used to predict x in ionic as well as metallic crystals. Since some aspects of the electronic environment of the ions can be measured, this effect is sometimes called the **chemical shift** as well as the **isomer shift**. The Notes should be consulted for further information.

The **magnetic hyperfine structure** observed in the Mössbauer effect is the most important result for the subject matter of this chapter. Both the expected (14.4 keV) and ground state of Fe^{57} have nuclear spins so their energy levels split in a magnetic field as shown in Fig. 15-14c (and Fig. 15-3). Thus, the Mössbauer effect can be used to measure the internal fields in ferro-, antiferro-, and ferrimagnetic materials. Figure 15-15a shows some very early results of the Mössbauer spectra for metallic iron at various temperatures (a single line source was used). Six lines are observed corresponding to the six transitions shown in Fig. 15-14d which result from the $\Delta m_I = 0, \pm 1$ selection rule. From the temperature dependence of the results, the internal field can be obtained, as shown in Fig. 15-15b.

Measurements similar to these have been performed on hundreds of compounds and alloys. The Mössbauer technique provides a simple yet accurate technique to measure the internal magnetic fields in

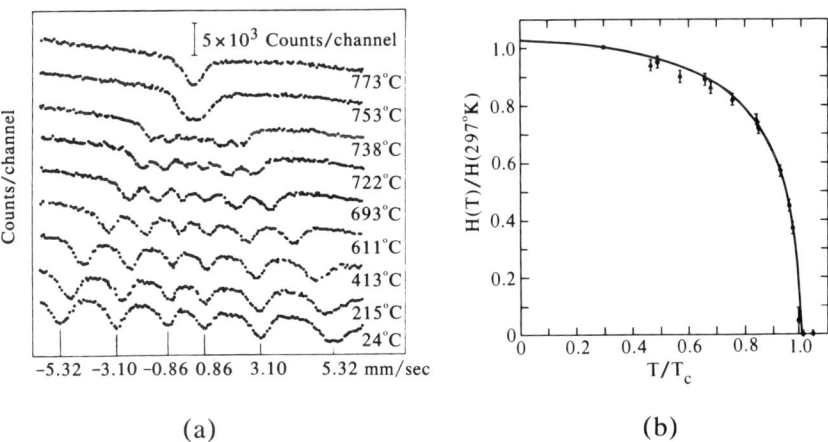

Fig. 15-15 (a) The Mössbauer spectra for a metallic Fe absorber (using a single line source). Positive velocity is the source moving away from the absorber. (b) The relative internal magnetic field at the Fe nuclei as a function of reduced temperature. [D. E. Nagle, H. Frauenfelder, R. D. Taylor, D. R. F. Cochran, and B. T. Matthias, Phys. Rev. Letters **5**, 364 (1960).]

magnetic materials. Besides Fe^{57}, there are about 20 other nuclei that have been used. See the Notes for references.

The **electric quadrupole interaction** is the interaction of the nuclear quadrupole moment, Q, with the gradient of the crystalline electric field at the nucleus. By symmetry Q = 0 for nuclei for I = 0, and 1/2, so this term will affect only the Fe^{57}(14.4 keV) state. Instead of the gradient of the electric field, $\nabla \mathbf{E}$, the interaction is usually written in terms of the potential; we define $q \equiv \partial^2 V/\partial z^2$ and $\eta \equiv [\partial^2 V/\partial x^2 - \partial^2 V/\partial y^2]/q$. Then the eigenvalues are

$$(\mathscr{E}_Q)_{m_I} = \frac{eqQ}{4I(2I-1)}[3m_I^2 - I(I+1)]\left[1 + \frac{\eta^2}{3}\right]^{1/2} \quad (15\text{-}23)$$

For nuclei in cubic symmetry q = 0 so the quadrupole interaction is zero and for nuclei at a position of axial symmetry (tetragonal, hexagonal, and trigonal) $q \neq 0$ but $\eta = 0$. eqQ has been measured by the Mössbauer technique for many different crystalline environments. Using NMR or related techniques eqQ has been measured using many stable nuclei in different crystalline environments. Important information can be obtained about the types of bonding and measurements near a phase transition give the temperature dependence of q. See the Notes for references.

Part C – Other Topics

One of the beautiful aspects of the field of magnetism is the broad diversity of the subject. These are the very important practical applications (transformers, electrical generators, memory cores, bubble devices) as well as a number of different fundamental studies that continually present themselves. While Parts A and B of this chapter focused on the simple basic aspects of magnetism, this part focuses on some of the more subtle aspects.

15-9 Spin Waves

15-9a Molecular field at low temperatures The temperature and magnetic field dependence of the magnetism for a spin-1/2 system (taking $J = S = 1/2$ and $g_J = 2$) is given by the Brillouin function, Eq. 15-9 (see Problem 3). Examine the low temperature behavior of M in the ferromagnetic regime. Using the Weiss theory, assuming that all of the field is an effective field, Eq. 15-14a, we obtain

$$M = \frac{N}{V}\mu_B \tanh\left(\frac{\mu_B \alpha_E M}{k_B T}\right) \tag{15-24a}$$

At very low temperatures $\tanh x \approx 1 - \exp(-2x) + \ldots$ for $x \gg 1$ so $M(0) \approx (N/V)\mu_B$ as expected. Substituting this value of M into the argument of tanh we obtain the low temperature expansion

$$M(T) = M(0)\left[1 - 2\exp(-2\alpha_E \mu_B^2 N/k_B TV) + \ldots\right] \tag{15-24b}$$

We can relate the argument in the exponential to $T_c = \alpha_E \mu_B^2 N/3k_B V$ via Eqs. 15-8b, 15-10, and 15-14d. Thus, for $T \approx T_c/10$ the fractional decrease in the magnetism is found to be $\Delta M/M(0) \approx 4 \times 10^{-9}$. This is about six orders of magnitude smaller than experimentally observed. The experimentally observed temperature dependence is

$$\Delta M/M(0) = AT^{3/2} \tag{15-24c}$$

This behavior is observed in many experimental systems ($A \approx 5 \times 10^{-6}\ °K^{-3/2}$ for the simple metals) and is very different from the molecular field prediction in Eq. 15-24b, which gives exponential behavior. What is wrong? The treatment of the elementary excitations of the spin system is wrong and this is another example of the breakdown of mean field theory (Section 15-5). In this case, we will find, that elementary excitations called spin waves (or magnons) are present.

15-9b Spin waves   Using a simple model F. Bloch (1930) theoretically predicted that spin waves should exist in ferromagnetic materials and obtained a $T^{3/2}$ law for the magnetization at low temperatures. More than 20 years later these waves were observed directly. We sketch a simple one-dimensional calculation to introduce them. The approach is similar to the phonon case in Chapter 12.

Consider N spins on a line with periodic boundary conditions (i.e., in a ring); each spin is coupled to its nearest neighbors by the Heisenberg exchange interaction, Eq. 15-16a. The system has energy

$$U = -2J_{ex} \Sigma_n \mathbf{S}_n \cdot \mathbf{S}_{n+1} \tag{15-25}$$

By performing the sum over the N classical spins we see that the ground state has an energy $U_0 = -2NJ_{ex}S^2$. (Classically $\mathbf{S}_n \cdot \mathbf{S}_{n+1} = S^2$ for spins of angular momentum $\hbar S$ because $\mathbf{S}_n \cdot \mathbf{S}_{n+1} = |\mathbf{S}_n| |\mathbf{S}_{n+1}|$ and $|\mathbf{S}_n| = S$.) In mean field theory the elementary excitations are the reversal (thinking in terms of $S = 1/2$) of one spin, which from Eq. 15-25 has energy $U_1 = U_0 + 8J_{ex}S^2$, but is this the first excited state of this system? Intuitively we sense that since the spins are coupled to each other the reversal of one spin is not a normal mode of the system, because it is a localized excitation. As we shall see, the elementary excitations of the system are delocalized with small deviations spread out over many spins. (The same situation was found for the lattice vibrational case, and is found often in physical systems.)

To get insight into the elementary excitations consider the coupled equations of motion. From classical mechanics we know that the rate of change of angular momentum ($\hbar \mathbf{S}$) of a spin is equal to the torque ($\mu \times \mathbf{B}$) on the spin. Since $\mu = -g\mu_B \mathbf{S}$, Eq. 15-6, and the effective field at the nth site is, via Eqs. 15-25 and 15-17a, given by $\mathbf{B}_n = (-2J_{ex}/g\mu_B)(\mathbf{S}_{n-1} + \mathbf{S}_{n+1})$, assuming nearest-neighbor interactions, then the rate of change of angular momentum is

$$\hbar(d\mathbf{S}_n/dt) = 2J_{ex} \mathbf{S}_n \times (\mathbf{S}_{n-1} + \mathbf{S}_{n+1}) \tag{15-26}$$

This is the basic system of equations that describes the coupling of the various spins. The equation can be linearized by letting

$$\mathbf{S}_n = \mathbf{S}_z + \sigma_n \tag{15-27a}$$

where $\mathbf{S}_z$ is a constant vector parallel to the magnetization and σ_n is a small vector in the x-y plane as in Fig. 15-16a. The idea behind this linearization is that the amplitude of excitation is expected to be small and in this classical approach $(S_n)_z \approx S$, a constant, and products of the small σ_n's are neglected (i.e., the z-component of σ_n would enter squared). Substituting Eq. 15-27a into Eq. 15-26 and retaining only terms that are first order in σ we find that

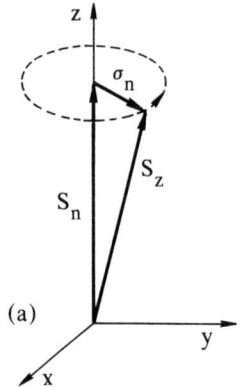
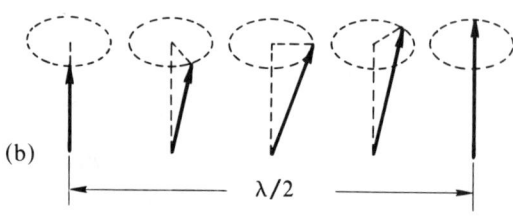

Fig. 15-16 (a) A classical picture of the precession of a spin about the magnetization axis. (b) Spin wave propagation along an axis.

$$\hbar(d\sigma_n/dt) = 2J_{ex}S_z \times (\sigma_{n-1} + \sigma_{n+1}) + 2J\sigma_n \times 2S_z$$
$$= 2J_{ex}S_z \times [\sigma_{n-1} - 2\sigma_n + \sigma_{n+1}] \quad (15\text{-}27b)$$

Again notice the similarity of this equation to Eq. 12-2. In component form Eq. 15-27b is

$$\hbar(d\sigma_n/dt)_x = -2J_{ex}S_z [\sigma_{n-1} - 2\sigma_n + \sigma_{n+1}]_y \quad (15\text{-}27c)$$
$$\hbar(d\sigma_n/dt)_y = +2J_{ex}S_z [\sigma_{n-1} - 2\sigma_n + \sigma_{n+1}]_x \quad (15\text{-}27d)$$

Multiplying Eq. 15-27c by i and subtracting Eq. 15-27d we obtain a single simple equation for the complex variable $\sigma^+ \equiv \sigma_x + i\sigma_y$

$$i\hbar(d\sigma_n^+/dt) = -2J_{ex}S_z [\sigma_{n-1}^+ - 2\sigma_n^+ + \sigma_{n+1}^+] \quad (15\text{-}27e)$$

The only substantial difference between this and Eq. 12-2 is that this equation involves only the first power of the time derivative so only ω will appear in the dispersion relation rather than ω^2 as in Eq. 12-2. Try the same type of propagating wave solution

$$\sigma_n^+ = \sigma^+ \exp i(nka - \omega t) \quad (15\text{-}28a)$$

where a is the spacing between the spins. By substituting this trial solution into Eq. 15-27e we immediately obtain the dispersion relation

$$\hbar\omega = 2J_{ex}S_z [e^{-ika} - 2 + e^{ika}]$$
$$= 4J_{ex}S_z [1 - \cos ka] \quad (15\text{-}28b)$$

and the result, ω vs. k, is similar to that in Fig. 12-2. Of course, the solution is periodic in k with a period of $2\pi/a$, but here, at small ka we have $\omega \approx (2J_{ex}Sa^2/\hbar)k^2$, rather than $\omega \propto k$ as in the lattice vibration case. The time dependent part of the solutions for σ_x and σ_y are

CHAPTER 15   MAGNETISM   607

$$\sigma_x \propto \cos(nka - \omega t), \qquad \sigma_y \propto -\sin(nka - \omega t) \quad (15\text{-}28c)$$

We see that each spin precesses about its z-axis, Fig. 15-16a, and there is a phase change, given by Eq. 15-28a, from spin to spin as shown in Fig. 15-16b.

In summary we have found spin waves where the spins precess about their equilibrium magnetism and the precession is correlated because of the coupling of one to another (by the exchange forces). The spin waves, similar to lattice vibrational waves, are delocalized excitations where the displacements are spread out over many spins. There is an upper cutoff frequency, at $k = \pi/a$ in Eq. 15-28b, at the Brillouin zone boundary. Thus, propagating spin waves are constrained to a band; higher frequency waves are strongly attenuated. At small wave vector $\omega \propto k^2$ for spin waves ($\omega \propto k$ for lattice vibrations) so the group and phase velocities are different. Physically, the small energies of the long wavelength spin waves occur because the spins are almost parallel so the restoring force is small. Mathematically, this enters because the equation of motion involves the first derivative with respect to time, Eq. 15-26 (while for lattice vibrations the second derivative is involved).

15-9c Direct experimental observations   Spin waves can be observed by resonance as well as neutron diffraction techniques.

Demagnetization factor – When a magnetic material is in an applied external field, **H**, magnetic poles are set up on the surface of the specimen. This causes the applied field inside the sample, **H′**, to be different from that outside. For elliptically shaped samples it can be shown that

$$\mathbf{H'} \equiv \mathbf{H} - \mathbf{H}_d = \mathbf{H} - D\mathbf{M} \qquad (15\text{-}29a)$$

where $\mathbf{H}_d$ is the **demagnetization field**, and D is the **demagnetization factor**, which is *shape dependent*. For diamagnetic and paramagnetic materials the difference between **H′** and **H** can be neglected, but demagnetization effects are important for ferro- and ferrimagnetics. In the continuum approximation, the only contribution to $\mathbf{H}_d$ comes from the uncompensated magnetic poles on the surfaces of the sample, and it is with this approach that it is calculated. D is a second-rank tensor; we give values for D along the principal axes of the two usual sample shapes: (a) a sphere; (b) a flat plate in the xy-plane.

$$\begin{aligned}&\text{(a)} \quad D_x = D_y = D_z = 4\pi/3 \\ &\text{(b)} \quad D_z = 4\pi, \qquad D_x = D_y = 0\end{aligned} \qquad (15\text{-}29b)$$

Notice that the sum of the diagonal elements of the D-tensor equals 4π, as it must. In the flat plate geometry, if **M** is along the z-axis then

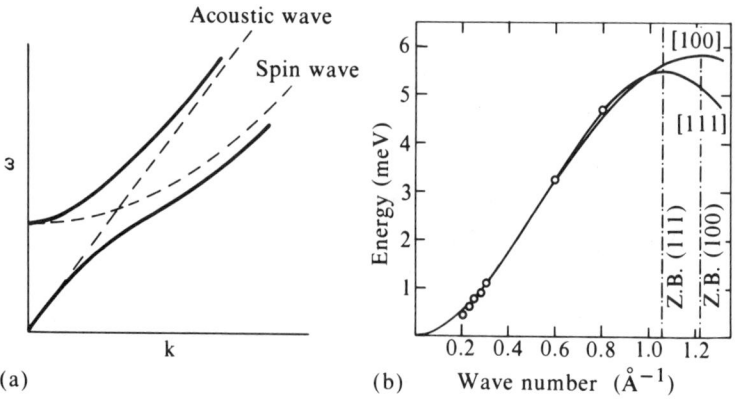

Fig. 15-17 (a) ω vs. k of spin waves and acoustic waves. The dashed and solid lines are the uncoupled and coupled modes respectively. (b) Experimental neutron results for the spin waves in ferromagnetic EuO. The extreme isotropy shows that the Heisenberg exchange expression, Eq. 15-16a, is appropriate for this material. We show ω vs. k for EuO rather than Fe, Co, or Ni because in these ferromagnetic metals one cannot measure the spin wave spectra to the Brillouin zone boundary (Z.B.) because J_{ex} is large so this branch overlaps with the phonon branches. [L. Passell, O. W. Dietrich and J. Als-Nielson, Phys. Rev. B **14**, 4897 (1976).]

$H_d = 4\pi M$, while if **M** is in the plane then $H_d = 0$. An analogous situation arises in the ferroelectric case, with the polarization (**P**) instead of the magnetization, but the consequences usually are small because the applied electric field is small compared to $4\pi P$. With this introduction to the demagnetization factor we can turn our attention to the experimental observation of spin waves.

Spin wave resonances usually are observed in thin magnetic films in a microwave cavity with an external applied field, H_z, normal to the film, which is case (b). Then the spin wave frequency, at small ω and k (long wavelength), is

$$\omega = \gamma(H_z - 4\pi M_z) + (2J_{ex}Sa^2)k_z^2 \quad (15\text{-}30a)$$

$$= \gamma(H_z - 4\pi M_z) + (2J_{ex}Sa^2)(\pi/nL_z)^2 \quad (15\text{-}30b)$$

where γ is the gyromagnetic ratio, Eq. 15-6. The second expression is obtained by realizing that at the surface of the plate the spins are pinned by the surface. Then spin waves are excited every time an integral number of half wavelengths corresponds to the thickness of the film, L_z. The absorption of energy, from a microwave bridge, is what is detected experimentally. By sweeping the magnetic field, we can scan through the absorption conditions and $(2J_{ex}Sa^2)$ can be obtained if L_z is known.

Figure 15-17a illustrates another method of observing spin waves (i.e., by their interaction with acoustic waves). Since at long wavelengths spin waves have $\omega \propto k^2$ while acoustic waves have $\omega \propto k$, the dispersion curves can be made to intersect at a convenient k-value by "biasing" the spin wave curve with an applied field as in Eqs. 15-30. In the region where the two modes cross they couple and repel each other (as in the polariton case in Section 13-11). Using this coupling, the magnetic energy in the spin waves can be converted to acoustic energy. Besides measuring spin waves, this can be used to generate acoustic phonons.

Figure 15-17b shows a complete ω vs. k for spin waves obtained by inelastic neutron diffraction. The technique is the same as described for phonons in Section 12-7. The results are in good agreement with the simple theory as in Eq. 15-28b; the k-dependence at small k is verified. The Notes should be consulted for further reading about the detection of spin waves.

15-9d Quantization of spin waves Any dynamic problem that can be described as a collection of harmonic oscillators can be quantized. This includes phonons (Section 12-5), photons, and the spin waves discussed here. Each normal mode of the spin wave system can be treated approximately as a harmonic oscillator, which is uncoupled from all of the other normal modes, and whose energy is

$$\mathscr{E}_k = (n_k + 1/2)\hbar\omega_k \quad (15\text{-}31a)$$

where $\quad n_k = 0, 1, 2, ...$

and the total energy of the spin wave system is the sum of the individual normal mode energies $\mathscr{E}_{total} = \Sigma_k \mathscr{E}_k$. In analogy with the concept of phonons, introduced as the quanta of excitation of the vibrating crystal, **magnons** are the quanta of excitation of the spin wave system. Thus, a magnon in the kth spin wave mode carries an energy $\hbar\omega_k$. For a state with quantum numbers $\{n_k\}$, we say that there are n_k magnons in the kth spin wave mode. This is possible only because the energy in the kth mode has the form given in Eq. 15-31a, that is, the levels are equally spaced so that the level differences are all the same and can be considered as the energies of identical quanta, the magnons. The magnon of energy $\hbar\omega_k$ can be considered to have a crystal momentum $\hbar k$ which like phonons (Section 12-6) must be conserved in interactions with other quasiparticles and the excitation of one magnon decreases the z-component of the spin of the system by the amount equivalent to the reversal of one spin.

Actually magnons, in contrast to phonons, are not strictly independent, that is, the energy does not precisely have the form given in

Eq. 15-31a. The reason is that while a given atom in a vibrating crystal can receive arbitrary amounts of excitation, a given spin in a magnetic system only has $2S + 1$ states. However, for states with small numbers of magnons the error is very small; multiple excitation of a single spin is improbable because the magnon is an excitation that extends over many unit cells.

Magnons, just as phonons and photons, are bosons (Section 12-5 and problem 7 in Chapter 12) so are described by Bose-Einstein statistics. This is because any number of magnons may be in the same state and the magnons are indistinguishable. That is, if magnons k and k' are created, the resulting state is the same regardless of the order in which they are created.

Then, for magnons at thermal equilibrium the average occupation number $\langle n_k \rangle$ is given by the Bose-Einstein factor

$$\langle n_k \rangle = [\exp(\hbar\omega/k_BT) - 1]^{-1} \qquad (15\text{-}31b)$$

Once the statistics, the ω vs. k, and the degeneracy (which is one for spin waves) are known, then the average thermal energy U of the system can be calculated. Then the specific heat C_V ($= \partial U/\partial T$) as well as the magnetization can be obtained. The general and easy method to do this is outlined in problem 7 in Chapter 12. Neglecting zero point energy

$$U = \sum_k n_k \hbar\omega = \frac{V}{(2\pi)^3} \int d^3k \frac{\hbar\omega}{e^{\hbar\omega/k_BT} - 1} \qquad (15\text{-}32a)$$

where the sum is changed to an integral by remembering that the volume in k-space for each allowed k-value is $(2\pi)^3/V$ (Eq. 9-27b). The integration must be performed over the first Brillouin zone. However, the low temperature results are easy to obtain since the integration can be extended to infinity and the low energy form of Eq. 15-28b can be used to relate ω to k. Thus, spherical symmetry is assumed so $d^3k = (4\pi/3)k^2 dk$ and we must just determine $dk = (dk/d\omega)d\omega$ from the dispersion relation. This is the same problem as for the phonon case, Eq. 11-19, but here $\omega \propto k^2$ while for the phonon case $\omega \propto k$. The solution is discussed in general in problem 7 in Chapter 12. The low temperature result is $U \propto T^{5/2}$. The heat capacity expressed in terms of U is

$$C_V = \frac{\partial}{\partial T} \frac{V}{(2\pi)^3} \int d^3k \frac{\hbar\omega}{e^{\hbar\omega/k_BT} - 1} \qquad (15\text{-}32b)$$

At low temperatures $C_V \propto T^{3/2}$ which is observed experimentally.

For each spin wave that is excited, the total spin is reduced from its saturation value, which is NS, by one unit. Thus, the temperature

CHAPTER 15   MAGNETISM   611

dependence of the magnetization is

$$M(T) = M(0)\left[1 - \frac{1}{NS}\sum_k n_k\right]$$

$$= M(0)\left[1 - \frac{V}{(2\pi)^3 NS}\int \frac{d^3k}{\exp(\hbar\omega/k_B T) - 1}\right] \quad (15\text{-}32c)$$

Again the integration is over the first Brillouin zone but the low temperature result is easy to determine. It is found that at low temperatures the magnetism deviates from the 0°K value as $T^{3/2}$. This is the famous Bloch $T^{3/2}$ law which is in agreement with experiment. At higher temperatures deviations from the $\omega \propto k^2$ dependence (near the Brillouin zone boundary, Fig. 15-17b) become important. Also so many magnons are thermally created that their non-boson character begins to be important; higher order corrections are required.

15-9e Antiferromagnetic spin waves   In a manner analogous to the ferromagnetic spin waves, Section 15-9b, the antiferromagnetic case can be analyzed. If only nearest-neighbor interactions are considered then the classical dispersion relation is

$$(\hbar\omega)^2 = 4|J_{ex}|S[1 - \cos^2 ka] \quad (15\text{-}33a)$$

(For the antiferromagnetic case J is an intrinsically negative number.) This should be compared to the ferromagnetic case in Eq. 15-28b; the results are similar except for the square of the cosine term here. This leads to an important difference at small ka: a linear ω vs. k for antiferromagnetic magnons compared to a quadratic dependence for the ferromagnetic case, that is, at small k

$$\hbar\omega \approx (4|J_{ex}|S)\, ak \qquad \text{antiferromagnetic} \quad (15\text{-}33b)$$

$$\hbar\omega \approx (2J_{ex}S)(ak)^2 \qquad \text{ferromagnetic} \quad (15\text{-}33c)$$

In general these results are in good agreement with experiments.

For the antiferromagnetic case quantization of the classical problem is not as straightforward as for the ferromagnetic case discussed previously. This is because the state with all spins aligned is the ground state for the ferromagnetic problem, but for the antiferromagnetic case the state with the various spins either up or down (for $S = 1/2$) is not the ground state. Nevertheless, the antiferromagnetic magnon theory gives good results in three dimensions, but more care is required for two- and one-dimensional antiferromagnetic systems.

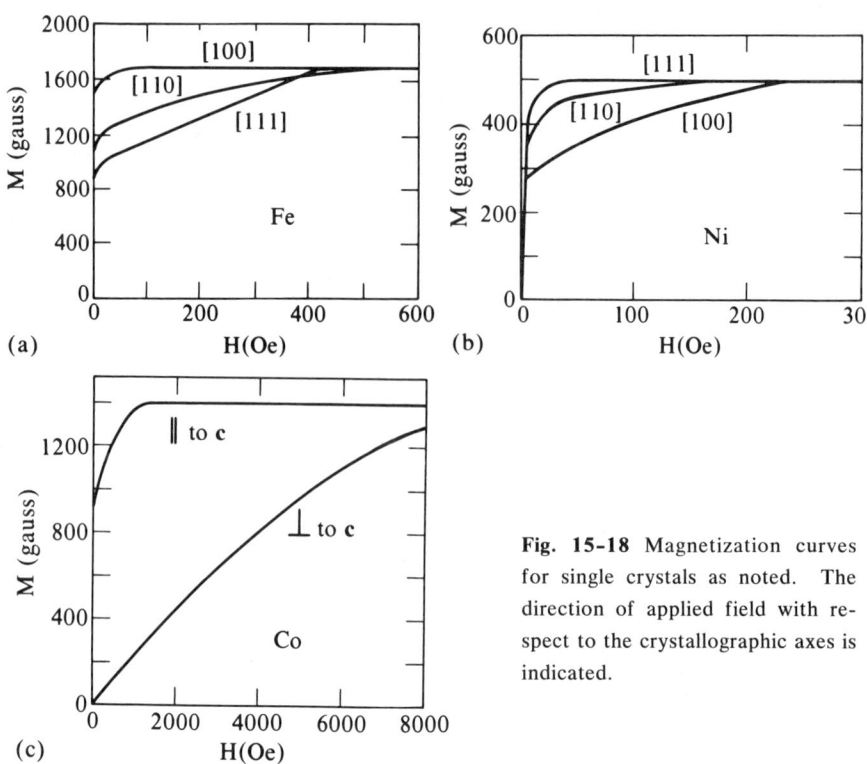

Fig. 15-18 Magnetization curves for single crystals as noted. The direction of applied field with respect to the crystallographic axes is indicated.

15-10 Anisotropy, Hysteresis, Domains, and Bloch Walls

15-10a Anisotropy The exchange Hamiltonian, Eq. 15-16a, is isotropic in the sense that there is no built-in preferred crystallographic direction. Thus, it would be expected that the spontaneous magnetic moments would not have a preferred direction. However, this is not found for real materials, and one reason is that there is an additional one-electron spin-orbital coupling energy, Eq. 15-12b; it is weaker for the d-electron atoms than for f-electron atoms but nonzero for both. This coupling affects terms like the g-values, Eq. 15-13, and leads to an orbital contribution to magnetic moment, which is affected by the crystal field, Section 15-3c. Even when the atomic magnetic moments are located at positions of cubic site symmetry, different directions have different energies, Eq. 15-11. In addition, the straightforward magnetic-dipole interaction energy

$$V_{dd} = \frac{\mu_i \cdot \mu_j}{r^3} - \frac{3(\mu_i \cdot r)(\mu_j \cdot r)}{r^5} \tag{15-34a}$$

between two spin magnetic moments μ_i and μ_j separated by a distance r, also contribute significantly to the directional dependence of the energy. The long-range character of the dipole-dipole interaction generally requires summing contributions from atomic pairs over all distances to estimate the directional dependences properly.

The same fundamental interaction (spin-orbit and dipole-dipole) can, in certain materials, lead to the curious orderings such as those shown in Fig. 15-12 for several rare earth metals (where spin-orbit forces are relatively large).

Thus, it is found that in ferromagnetic materials there is some crystallographic axis along which the magnetic moment tends to align with lower energy than another. For example, for the fcc structures of Fe this axis is a $\langle 100 \rangle$ direction, while for Ni it is a $\langle 111 \rangle$ axis. For Co, which has a hcp structure, the axis is the c-axis. These axes of lowest energy are called **easy axes** or **easy directions**, while the other directions are called **hard directions**. The energy differences associated with the fact that certain axes are easy directions are called **anisotropy energies** or **magnetocrystalline energies**. It is defined as the work required to make the magnetization lie along a certain direction compared to an easy direction. Figure 15-18 shows the magnetization along different axes vs. applied field for Fe, Ni, and Co. The largest anisotropy can be seen for Co where it is relatively difficult to force the magnetization from the c-axis to the basal plane. Equivalently, large magnetic fields are required to achieve saturation magnetization perpendicular to the c-axis compared to fields along the c-axis.

The anisotropy energy density is given by $U_A \approx (1/2)\int \mathbf{H} \cdot d\mathbf{M}$, which can be evaluated from the magnetization curves in Fig. 15-18. U_A can be expressed by a power series in the direction cosines (or sines) of the magnetization vector with the principal axes of the crystal. The expression must be compatible with the various crystal symmetries and, to be useful, the expression should not have too many terms. Provided that the magnetization is independent of rotations by π, U_A must depend on only even powers of the angles. (This follows from the fundamental time-reversal symmetry; reversing the direction of time reverses every spin vector without changing the energy.) For example, for Co, where θ is the angle between $\mathbf{M}$ and the c-axis, this energy can be approximated by two terms in a power series,

$$U_A = K_1 \sin^2\theta + K_2 \sin^4\theta \qquad (15\text{-}34b)$$

where the room temperature values of K_1 and K_2 are $\approx 4.1 \times 10^6$ erg/cm^3 and 1.0×10^6 erg/cm^3, respectively. As can be seen, U_A increases as $\mathbf{M}$ is forced to turn away from the c-axis; the zero of energy is taken as the energy for $\mathbf{M}$ along the easy axis.

In order to describe U_A for cubic materials, the direction cosines α_1, α_2, and α_3 with respect to the cube edges must be used. Again, symmetry requires that the expression must have only even powers of the α_i's, must be invariant to interchanges of the three α_i's, and must

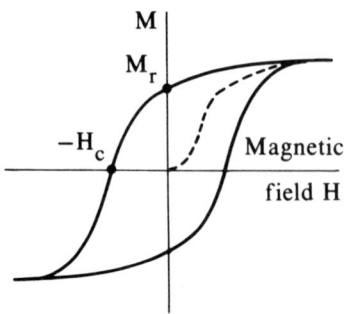

Fig. 15-19 A hysteresis loop also showing the initial magnetization curve (dashed). The critical field and remanent magnetization are defined.

be independent of a change of sign of any one α_i (which excludes terms like $\alpha_1 \alpha_2$). The lowest order term that satisfies these criteria gives no anisotropy since $\alpha_1^2 + \alpha_2^2 + \alpha_3^2 = 1$. However, the anisotropy energy can be represented reasonably well using only fourth- and sixth-degree invariants

$$U_A = K_1[\alpha_1^2\alpha_2^2 + \alpha_2^2\alpha_3^2 + \alpha_3^2\alpha_1^2] + K_2[\alpha_1^2\alpha_2^2\alpha_3^2] \quad (15\text{-}34c)$$

(See the problems as to why certain other terms are not required.) Iron has the $\langle 100 \rangle$ directions as the easy axes so the K_i are positive. At room temperature $K_1 \approx 4.2 \times 10^5$ erg cm^{-3} and $K_2 \approx 1.5 \times 10^5$ erg cm^{-3} decreasing to zero at $T_c = 770°C$. For nickel, where the body diagonals are the easy directions and cube edges are the hard directions, Eq. 15-34b is still appropriate since it is determined by symmetry. However, now K_1 is negative ($\approx -5 \times 10^4$ ergs cm^{-3}). In some cases the K_i can be calculated from band theory.

When the spin-orbit interaction is very small or for crystals with low symmetry there also may be contributions to the anisotropy from dipole-dipole terms. A case in point is the $k_1\sin^2\theta$ in Eq. 15-34b for the hexagonal ferrite $BaFe_{12}O_{19}$. In even lower symmetry situations, terms like $U_A = A\,\alpha_1^2 + B\,\alpha_2^2$ are obtained.

15-10b Hysteresis If the macroscopic magnetization of a crystalline ferri- or ferromagnetic material were measured in zero applied field, most samples would appear to be nonmagnetic. Yet with the application of a rather weak magnetic field the full magnetization could appear, as in Fig. 15-18. Thus, the macroscopic magnetic state of a ferromagnetic material depends on its history.

Figure 15-19 shows an M vs. H hysteresis loop that illustrates some of these facts. The sample starts at $M = 0 = H$ and its initial magnetization follows the dashed curve. The microscopic details of this curve will be discussed soon; for now we can think of this initial curve as essentially the same as one of the curves in Fig. 15-18.

As the magnetic field, H, is increased eventually M saturates. If H is now reduced there is a magnetization, M_r, known as the **remnant magnetization,** at H = 0. To destroy the magnetism completely a negative magnetic field, $-H_c$, called the **coercive field,** is required. Further, larger negative applied magnetic fields will cause the magnetization to saturate but in the opposite direction. The loop shown in the figure is called a **hysteresis loop.** (B vs. H is also often plotted as well as M vs. H. Since for most spontaneous magnetic materials $H \ll 4\pi M$, either plot has the same shape.)

If from the initial magnetization curve the field is reversed when H is much lower than $+H_c$, then at H = 0 a value of $M < M_r$ is obtained. Considering this point, as well as the fact that even after large values of $|H|$ are applied the sample can have $\pm M_r$, it is clear that the magnetism depends on the history of the material.

The magnetic hysteresis loop is extremely important for many applications. The area under the magnetization curve determines the amount of stored or dissipated energy, which for some applications should be maximized, while for others should be minimized. Materials that have small values of H_c with respect to M_r are termed **soft magnetic** materials, while those with large H_c values are called **hard magnetic materials.** The ranges of H_c in commercial materials is typically 2×10^{-3} Oe to 2×10^4 Oe, the technology of making these types of materials being quite advanced. Hard materials are hard to magnetize but once they are magnetized they retain their magnetization near their saturation value. Such materials would be used for permanent magnets. On the other hand, soft materials are used for transformers where it is desirable to minimize the area under the hysteresis loop (which causes heating of the material and is a loss of energy) and obtain a large susceptibility.

15-10c Domains Actual ferromagnetic specimens are composed of small regions within which the magnetization has its remanent value. These regions are called **domains** and the region of transition between domains is called a **domain wall.** The magnetization within different domains points in different directions so the total magnetization of the sample generally is zero if the sample has not been in an applied field, as discussed previously. (For example, for Fe, Fig. 15-18, different domains could have a magnetization along the six $\langle 100 \rangle$ directions.)

The existence of domains was postulated by Weiss in 1907 but not observed until 1931 by Bitter. The **Bitter technique** involves the use of small colloidal particles of a ferromagnetic material, which are attracted to the domain wall by the inhomogeneities in the magnetic field at the walls on the surface of the sample. The particles then can be seen with the aid of a microscope. Domains also can be observed by a Faraday rotation technique in reflection or transmission and now

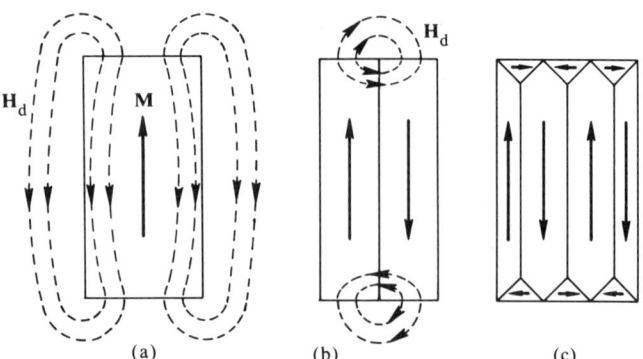

Fig. 15-20 Possible domain arrangements. The one on the right shows closure. The demagnetization field lines (H_d, Eq. 15-29a) eminate from the magnetic poles distributed on the end surfaces of the domains.

are observable using modern-day electronic enhancement techniques with a microscope.

The initial magnetism part of the hysteresis loop shown in Fig. 15-19 can be understood in terms of domain movement. Start with a virgin sample with zero magnetization. At low applied fields domains with magnetization that is favorably oriented with respect to the direction of **H** grow at the expense of unfavorably oriented ones. At very small **H** this is a reversible process while at larger **H**, due to domain wall pinning at imperfection sites, precipitates, and so forth, the process is not reversible. This can lead to hysteresis effects far below M_r. Then for larger **H** the magnetization rotates toward **H**. This initial magnetization curve is not as smooth as drawn in Fig. 15-19; rather small steps in M may be seen as **H** becomes large enough to unpin a domain wall allowing it to move rapidly or, for example, as the magnetization in domain in a cubic crystal reorients from a [100] to a [001] if the field is close to the latter direction. These irregularities in the magnetization curve (called the **Barkhausen effect**) cause noise in magnetic circuits.

Obviously domains are formed because they minimize the energy of the system. On the left of Fig. 15-20 a single domain sample is shown. The self energy of the field in space, known as the **demagnetizing energy** or **magnetostatic energy**, $\approx (1/8\pi)\int H_d^2 d^3r = -(1/2)\int \mathbf{M}\cdot\mathbf{H}_d d^3r$. For this configuration the magnetic energy density is $\approx M_s^2 \approx 10^6$ erg/cm^3. If the crystal had two domains (Fig. 15-20b) then the energy density in space is considerably lowered but there is an energy increase that comes from the energy of the domain wall created between the two domains. However, the energy of the field in space can be reduced to zero by the formation of the small **closure domains** shown in the right. These closure domains totally eliminate the free magnetic poles on the surface and keep the lines of force

CHAPTER 15   MAGNETISM   617

entirely within the material. Simple patterns as shown in Fig. 15-20 are observed particularly in strain-free cubic single crystals. In other crystals the patterns are more complex. Actual domain patterns are determined by minimizing the sums of all of the relevant energies. Besides the magnetocrystalline (magnetic anisotropy energy) and magneto-static energies, discussed previously, two other energies enter.

Magnetostriction refers to the change of dimensions of a specimen when its magnetization is varied. This is a complicated tensor property (see the books by Chikazumi or Morrish referenced in the Notes), but in the simplest experiment, the fractional change in length of a multi-domain sample $(\Delta \ell / \ell) = \lambda M^2$ (i.e., it depends on even powers of the magnetization). More fundamentally, the energy of single domain is studied and its dependence on the direction of **M** and the components of the general strain tensor is measured. This expression is a **magnetoelastic energy** that depends on the interaction between the magnetic anisotropy and the strain.

Bloch walls is the name usually used instead of domain walls. We perform a simple calculation to show that Bloch walls are relatively wide (compared to the interatomic distance). That is, the surface energy of a wall between two antiparallel domains is lower if the reversal of the spin direction is spread out over many spins rather than accomplished by just two neighboring spins. Consider spins labeled 0 to p in the region of a Bloch wall. If all of the spins are parallel and we assume that the exchange force only acts between nearest neighbors then from Eq. 15-16a, the energy of a chain of p nearest-neighbor interactions is

$$U_1 = -2pJ_{ex}S^2 \qquad (15\text{-}35a)$$

where we have adapted a classical picture for the spins. For a wall where the spin reversal is accomplished in one atomic distance, in the y-direction, Fig. 15-21a, the energy is

$$U_2 = -2(p-1)J_{ex}S^2 + 2J_{ex}S^2 = -2pJ_{ex}S^2 + 4J_{ex}S^2 \quad (15\text{-}35b)$$

that is, the energy is larger by $+4J_{ex}S^2$. If the spin reversal is spread uniformly over p spins, as shown in Fig. 15-21b, then the angle between neighboring spins is π/p and the exchange energy is

$$U = -2pJ_{ex}S^2 \cos\frac{\pi}{p} \approx -2pJ_{ex}S^2 \left[1 - \frac{1}{2}\left(\frac{\pi}{p}\right)^2 + ...\right]$$
$$= -2pJ_{ex}S^2 + (\pi^2/p)J_{ex}S^2 + ... \qquad (15\text{-}35c)$$

where we have expanded the cos term assuming small angles between neighboring spins. Notice that by spreading the reversal process over p spins, the excess wall energy is approximately $1/p$ smaller than if the reversal is accomplished in one step, Eq. 15-35b. If only this ex-

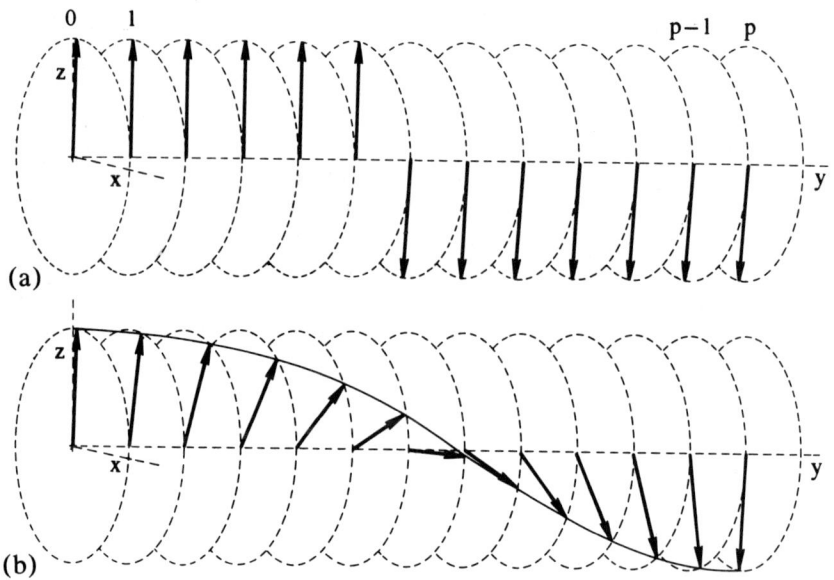

Fig. 15-21 The domain boundary from a microscopic point of view showing (a) an abrupt spin change (b) a gradual spin orientation change.

change interaction were involved, the domain wall would broaden to the size of the sample. However, the previous analysis assumed an isotropic interaction and we have not taken into account the magnetocrystalline energy discussed in Section 15-10a. Although this is a small energy compared to the exchange energy the former will contribute a fixed amount per spin. So as p tends to increase to lower the exchange energy it will eventually be limited by the magnetocrystalline energy effects. In metallic iron the Bloch walls are typically ≈ 300 internuclear distances thick.

The actual domain configuration in a sample depends on the interplay between the magnetocrystalline, magnetostatic, magnetoelastic, and Bloch wall energies. As mentioned, for pure single crystals closure domain patterns, as shown in Fig. 15-20c, indeed are observed. In soft magnetic materials, where high susceptibility is desired, the low anisotropy and low magneto-striction materials are made as pure as possible. The result, in general, is simple domain patterns. However, for hard magnetic materials where large H_c is desired, very finely sintered magnetic powders of high anisotropy are used, often with a precipitated second phase, to suppress domain boundary wall motion; in such materials more complicated domain patterns are observed.

Domains are formed in many types of material besides ferri-, ferro-, and antiferromagnetic materials. A similar situation arises in ferroelectrics (Chapter 14), but for ferroelectrics the forces are different, resulting in domain walls that are much thinner.

CHAPTER 15   MAGNETISM

15-11 Metals and Magnetism

So far in this chapter we have been talking as though the electrons with their magnetic moments are always localized at atomic sites. The underlying quantum mechanics is via the Heisenberg Hamiltonian, which comes from a model for localized electrons. For ionic crystals this is fine but we know that many magnetic crystals are metals, Fe, Co, and Ni being some of the more obvious ferromagnetic metals. Here we make some qualitative remarks about the theory of magnetism in metals and discuss some of the quantitative theories.

The **free electron (Pauli) spin paramagnetism** has already been calculated in Section 9-12c so it will not be discussed here. It only accounts for paramagnetism in a metal and not ferromagnetism.

Itinerant electron picture – The microscopic theories of ferromagnetism were developed soon after the advent of quantum mechanics. The important ingredients of these theories are the antisymmetry (with respect to exchange of any two electrons) of the many-electron wave function and the Coulomb interaction. Explicitly spin-dependent terms in the Hamiltonian are not required. (Some small, spin-dependent terms exist. Of these, the spin-orbit effect does not couple different electrons, and the dipole-dipole interactions are weak. Though important for anisotropy and magneto-striction, they have little effect on the Curie temperature, which is our concern here.) The Coulomb interaction tends to keep the electrons apart and the Pauli and antisymmetry principle correlate the spatial distribution with the relative spin orientations. Thus, the eigenfunctions of the many-electron system really cannot be constructed without referring to all the electrons in the system. (Independent-electron systems cannot give a microscopic model of ferro- or antiferromagnetism.)

Normally, this many-electron system is approached from two limits. One limit is the *localized electron picture* where, in zeroth approximation, each electron is considered to be localized on a single atom. The effects of all of the other atoms and electrons are approximated by a crystal field plus small overlap of the orbitals between neighboring atoms. The degeneracies of the energy levels of the localized electrons are split and levels are shifted due to the crystal field. However, the electrons are constrained to remain localized on a single atom. This approximation is good for ionic crystals and already has been discussed extensively in this chapter.

The other limit is the band approach, usually referred to as the **itinerant electron picture** (itinerant means travelling from place to place). In this approach, each electron, in zeroth approximation, is assumed to be a Bloch wave or in a band state. The effective electron-electron interactions, which are essential in causing ferromagnetism (and/or antiferromagnetism), are of short range and occur within the atoms. However, the band wave functions make it difficult to focus on these correlations.

Instead of covering details of the itinerant electron picture in the theory of magnetism (which is complex), we shall use a much simplified picture due to Stoner (1938). This model gets at the essence of the problem with a minimum of complexities. In this simplified model, the exchange interaction between electrons is treated in a mean field approximation. Since each band electron interacts with every other one (when they happen to be near one another), it makes sense to try to replace the constantly changing field that each electron experiences by a mean field. The band of spin-up electrons can be treated separately from the band with spin-down electrons in a manner similar to the discussion in Section 9-12c, for free electron paramagnetism. Thus, we have a density of states for $m_s = 1/2$ that is different from the density of states for the $m_s = -1/2$ band.

This idea will be applied first to the iron series ions, which form ferromagnetic metals with high Curie temperature. Then the mean field approximation will be discussed; we shall see that other kinds of magnetic and nonmagnetic ordering are possible. Then several aspects of magnetism and related topics of free electron gases will be discussed. Finally, some effects in rare earth metals will be covered; the 4f-electrons in these metals are localized reasonably well.

The **iron group metals** show magnetic moments that are smaller than found in the corresponding insulating salts and correspond to a noninteger number of electrons. For example, at $T \approx 0°K$ the spontaneous magnetism, Eq. 15-15, is $M_s = (N/V)g_J\mu_B J$, since in this limit the Brillouin function ≈ 1. This is the expected result. To compare this to experiment, we list, under "Metal Exp.", measured values of $M(0°K)$ in units of μ_B per atom. Further, classically for J we have $\langle S \rangle + \langle L \rangle$ for each electron associated with each atom. However, the orbital contribution is at most 10% (for Co metal) so, neglecting L, we replace $g_J J$ with $g_e \langle S \rangle$ where $g_e = 2$ and $\langle S \rangle = 1/2$. Thus, the experimental values should directly give the effective number of electrons per atom. Along with the metal experimental values, we list values of S and μ_{eff}/μ_B for the various ions, taking J = S, from Table 15-4. For the ions, $\mu_{eff}/\mu_B = 2[S(S+1)]^{1/2}$ from Eq. 15-10.

	Metal Exp.	Ion	S	$2[S(S+1)]^{1/2}$
Fe	2.22	Fe^{3+}	5/2	5.92
		Fe^{2+}	2	4.90
Co	1.72	Co^{2+}	3/2	3.87
Ni	0.61	Ni^{2+}	1	2.83

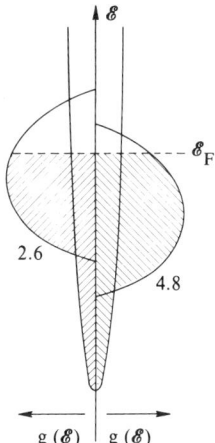

Fig. 15-22 A schematic band diagram of the 3d and 4s energy bands appropriate for Fe. The wide bands are meant to repeat the 4s equally populated because the Weiss field for these electrons is very small; they contain ≈0.3 on each side. The narrower (higher peak density) 3d bands are effected by the inner field and the subband with positive magnetic moment is populated with 4.8 electrons while that with negative magnetic moment has 2.6 electrons. This results in a net of 2.2 electrons, giving the resultant magnetic moment found in metallic iron.

As can be seen, the metal experimental values give an effective number of electrons that are non-integral and quite different from each other. On the other hand, for the ions $g_e S$ (for $g_e = 2$) give values that are in reasonable agreement with μ_{eff}/μ_B; this indicates that for the ions the correct number of magnetic electrons are acounted for. Clearly these metals must be considered from a band point of view.

Figure 15-22 is a simplified band picture of iron using the Stoner approach. The basic understanding is the same as for the insulating materials described in Section 15-5. The Weiss molecular field acts on the 3d electrons and lowers the energy of the spin-up electrons compared to the spin-down electrons. This is the same effect as discussed in Section 9-12c for the free electron spin paramagnetic case. However, for the metals, instead of an external applied magnetic field, it is the Weiss inner field (Eq. 15-14a), due fundamentally to the intra-atomic exchange integral that acts to polarize the electrons in the spin-down band with respect to those in the spin-up band. For pairs of atoms with small overlap the quantum mechanical treatment leading to the Heisenberg Hamiltonian of the localized model (Eq. 15-16a) is complicated but conceptually straightforward. For metals the electron bands make the problem conceptually difficult but a result can be obtained that is similar to a Heisenberg Hamiltonian where J_{ex} is determined by band characteristics. Since each Bloch wave is spread over the entire crystal, two such waves appreciably overlap each other within each atomic cell. Thus, the exchange constant (J_{ex}) is due mostly to the strong intra-atomic overlap rather than the weak inter-atomic overlap that we had in Section 15-6.

In Fig. 15-22 the bands are split into subbands; the electrons with $m_s = 1/2$ are on the left and those with $m_s = -1/2$ are on the right. The 3d electron band is relatively narrow (compared to the 4s band) with a correspondingly more peaked density of states. The reason is that the overlap with neighboring 3d orbitals is much smaller than the 4s-4s overlap. (Thus, the tight bonding approach, discussed in Section 10-9, is appropriate for the 3d electrons.) The Weiss inner field raises the $m_s = 1/2$ bands relative to $m_s = -1/2$ bands. For the 3d bands the inner field is large, shifting one band with respect to the other by large amounts. For iron, this causes the lower energy 3d subband to be populated with 4.8 electrons, while there are only 2.6 electrons in the higher energy 3d subband. For the 4s subbands the Weiss inner field is much smaller (due to the much wider band), so no shift of one subband with respect to the other is shown. At any rate each 4s subband only holds ≈ 0.3 electrons, so the error introduced, by neglecting the shift, is small. The net result of this picture is that there is a net spin of 2.2 electrons per atom in metallic iron.

Thus, proceeding as in Section 9-12c for the free electron spin paramagnetism but using the Weiss inner field concept, $B_E = \alpha_E M$ as in Eq. 15-14a, a spontaneous magnetic moment is obtained when the Fermi energy $\mathscr{E}_F \approx \mu_B B_E$ or in terms of exchange $\mathscr{E}_F \approx 2zJ_{ex}$ where z is the number of nearest neighbors, Eq. 15-17c. (See the book by Morrish, Section 6-7, referenced in the Notes.)

Now we discuss some other theoretical aspects of the theory of magnetism in the presence of free electrons as in a metal.

Free electron gas calculations – Very briefly we now discuss some theoretical results pertaining to ferromagnetic and related behavior in a free electron gas.

Because the full calculation is hopelessly difficult, these calculations usually are done within the **Hartree-Fock approximation**. This implies three things. (1) Each electron sees an average potential due to all of the other electrons. (2) Exchange is taken into account (i.e., the electron wave function has the proper symmetry under the interchange of two electrons and this leads to exchange terms in the energy - without exchange it is called the **Hartree approximation**). (3) The calculation is done self-consistently (i.e., an average potential is assumed, the single-particle wave equation is solved, the resulting eigenfunctions are used to calculate a new average potential, and the cycle is repeated as often as necessary to make the difference between the new and the previous potentials as small as desired).

Ferromagnetism for a free electron gas was first considered by Bloch in 1929. He showed that for low enough electron densities, a ferromagnetic state would satisfy the Hartree-Fock self-consistency condition. To address this problem, it is assumed that that there are more electrons with spin up than with spin down. So all up-spin electrons see one exchange potential and all down-spin electrons see a different exchange potential. Both exchange potentials are spatially uniform and do not flip spin; so it is similar to being in a uniform magnetic field (Fig. 9-8), the exchange field. (Previously in this section, we have called this the Stoner approach.) The magnitude of this field determines the number of up and down spins below the Fermi energy, which in turn determines a new magnetic field. The self-consistent solution with the lowest energy is that in which the old and new fields are the same. These calculations predict a ferromagnetic state (i.e., more up spins than down spins). The state is spatially uniform in both spin and charge (in contrast to the Hartree-Fock solution that will be discussed later). It is now generally believed that for simple metals with a non-degenerate conduction band (e.g., Li and Na), ferromagnetism would never occur even if the band were made very narrow (by imagining the crystal to be uniformly expanded to a lower density). The Hartree-Fock state simply is not a good enough approximation for such a system. In fact, it can be shown for a non-degenerate band in one dimension that the ground state is never ferromagnetic. Ferromagnetism seems to depend on degeneracy, essentially on the Hund's rule interaction between degenerate one-electron states within atoms, that tend to align spins.

The paramagnetic (vanishing Hartree-Fock field) and ferromagnetic states are not the only self-consistent states. One can start with an initial assumption that is different from that used in the last paragraph. An exchange field can be assumed that oscillates periodically in space, either in magnitude or direction. The resulting one-electron states now have spins whose state (or direction) varies with position. Overhauser (1960) considered this possibility and showed that the total energy could be lower, under the same circumstances that were discussed for the ferromagnetic case in the previous paragraph. The Overhauser state has a total spin density that is nonuniform but whose magnitude (and/or direction) varies in space periodically – a **spin density wave (SDW)**. Nevertheless, the charge density is still uniform since the net spin in any direction of bulk sample is zero. This state is sometimes referred to as an **Overhauser antiferromagnetic state**. In general, the periodicities of the SDW are not rational fractions of the periodicities of the underlying crystal structure. For example, Overhauser considered a metal with a spherical Fermi surface for which he found that the wave vector of the SDW, $k_{SDW} = 2\, k_F$. Clearly k_F has no general relationship to the reciprocal lattice vectors.

Whether or not SDW's have been observed in metals with an approximately spherical Fermi surface is not clear; the interpretation of the experimental results is controversial.

Just as the ferromagnetic Hartree-Fock state is favored by narrow bands, SDW states are favored by a Fermi surface that "nests." **Nesting Fermi surfaces** refers to the fact that some portion of a Fermi surface if displayed in k-space by an appropriate vector **q** will nearly coincide with another portion of the Fermi surface. (This cannot happen for a spherical Fermi surface.) The antiferromagnetic metal chromium, which has a complicated Fermi surface, may have a SDW due to a nesting mechanism.

While the Hartree-Fock approximation can lead to both ferromagnetic and antiferromagnetic states, both of which have uniform charge density, it also can lead to states with spatially periodic variations in charge density, which are discussed here for completeness.

Charge density waves (CDW's) also can be predicted using the Hartree-Fock approximation. The simplest of these CDW states is obtained if it is assumed that all of the electrons see the same spin-independent but spatially periodic potential. Then it is found that this assumption can be self-consistent, and that under certain circumstances this self-consistent state will have the lowest energy. The Coulomb energy contribution is increased for such a state, but this is more than compensated for by a decrease in the electron single-particle energies just below the Fermi energy. As for the SDW discussed previously, this CDW state is favored by nesting Fermi surfaces. The resulting CDW periodicities are incommensurate with that of the underlying crystal structures. Usually, the lowering of the single particle energies is enhanced by a distortion of the crystal structure, which can be observed by x-ray or neutron diffraction. The crystal structure distortion energetically stabilizes the CDW. The instability that leads to such a CDW was first discussed in detail by Peierls (and Frölich) and is called a **Peierls instability.**

Last, we note that if the free electron density is very, very low, and the positive charge is uniform in space, then theoretically it can be shown that the electrons will order in a regular periodic structure. This is called **Wigner crystallization** or the formation of a **Wigner lattice** (1938). Again, the periodicity of the periodic electron structure is not simply related to that of the crystal structure. Wigner crystallization has been observed for a two-dimensional gas of electrons above the surface of liquid helium (problem 13 in Chapter 9). We repeat, both Wigner crystallization and CDW's have little to do with magnetism; they simply are competing forms of order and are mentioned here for completeness.

We see that the theoretical possibilities are rich, even for an apparently simple system such as a free electron gas. For further discussion see Chapter 32 of Ashcroft and Mermin.

Other types of exchange — The Heisenberg exchange (Eq. 15-16a) is sometimes called **direct exchange** (or **potential exchange**) because it arises from the direct Coulomb interaction between localized electrons whose orbitals are centered on different sites. However, in metals there are additional types of important exchange mechanisms.

For 4f electrons in rare earth metals the direct overlap of the 4f electron orbits is extremely small because of their small radii; yet some of the metals are magnetic at low temperatures. The mechanism that is used to explain these effects is called **indirect exchange.** The idea is that the spins of the localized 4f electrons, in the partially filled shell, polarize the spins of the surrounding sea of conduction band electrons by direct exchange. (For rare earth metals the 6s electrons have the weakest binding and therefore make up the conduction electrons.) The spin polarized conduction electrons in turn interact with the localized 4f electrons on other atoms. In this manner the spin information from the localized 4f electrons on one atom is transmitted to another atom. It is a weak effect but noticeable because for 4f electrons the direct exchange is even weaker.

This indirect exchange is often known as the **RKKY interaction**; it not only plays an important role in the magnetism of rare earth metals but has important consequences in dilute solid solutions of a magnetic atom in a nonmagnetic metal crystal (i.e. small amounts of Fe in Au). The surprising and interesting aspect of the RKKY interaction is that it has a longer range than just the nearest-neighbor distances and that, although weak at large distances, it oscillates in sign. This long-range and sign oscillation have important consequences in these dilute alloys, including spin-glass behavior discussed in the next section.

The **Kondo effect** also is a consequence of the interaction of localized magnetic moments with the conduction electrons. Experimentally, at low temperatures, a minimum is observed in the resistivity versus temperature in dilute solid solutions of a magnetic ion in a nonmagnetic metal crystal. This result has been shown to be a fundamental many-body effect.

The last exchange, **itinerant exchange,** was discussed previously in terms of the exchange in the iron-series metals. (A very complete discussion of itinerant exchange has been written by C. Herring in the series of books edited by Rado and Suhl referenced in the Notes.)

15-12 Spin Glasses

One of the present thrusts of solid state physics is toward disordered systems. This does not necessarily mean materials that are structurally disordered, which are sometimes referred to as "window" glass as discussed in Section 10-20 (i.e., nothing, neither atoms nor spins has long-range order). Terms such as "xyz-glass" are used to describe systems that have periodic long-range translational symmetry but one, relatively minor, property is random. In the structural phase transition field (Chapter 14) we see terms like dipolar glass, quadrupolar glass and ferroelectric glass. These are found in ordinary crystals but with some impurities. In this section we introduce spin glass. A great deal of effort has been put into these materials.

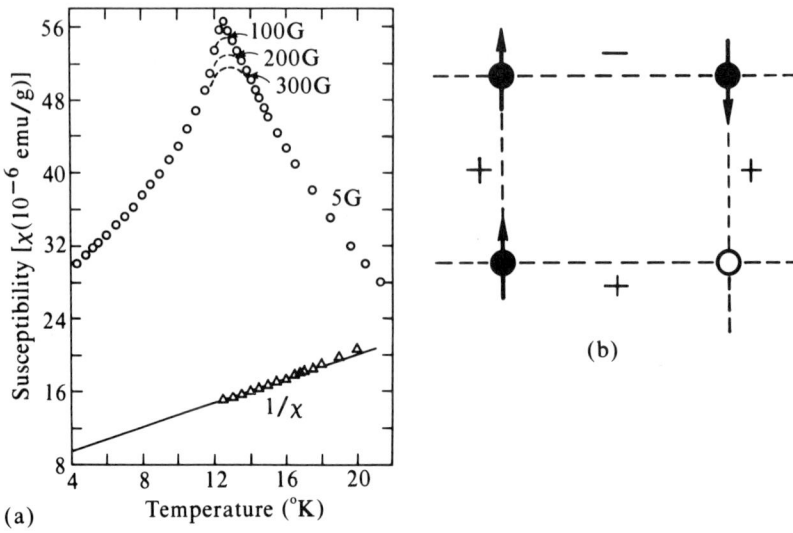

Fig. 15-23 (a) The temperature dependence of the magnetic susceptibility (and its inverse) of Cu containing 1% Mn in different fields as indicated. [V. Cannella, "Amorphous Magnetism" Ed. H. O. Hooper and A. M. deGraff. (Plenum Press, 1973) p. 195.] (b) A simple example of frustration. The + sign refers to ferromagnetic exchange and the − sign to antiferromagnetic exchange. The three aligned spins are shown by solid circles and the frustrated spin is an open circle. The spin on its left wants it to point up, but the spin above it wants to point down.

The first spin glasses were observed (1972) by studying a dilute alloy of iron in crystalline gold. Thus, we are dealing with a normal Au crystal which has an fcc crystal structure. However, this crystal can be grown from a solution containing Fe atoms as well as Au atoms and then x-ray results show that the Fe atoms replace, in a random manner, some of the Au at the normal crystal sites of the fcc crystal structure. Measurements of χ vs. temperature show a peak at a T_c corresponding to some sort of cooperative behavior. For Fe in Au, T_c varies from 8°K for 1% Fe to 218°K for 22% Fe. Figure 15-23a shows the results in a related system Mn in Cu. This is an interesting system since Mn^{2+} is an S-state ion so orbital effects are minimized.

In these spin glasses, there is no ferromagnetic or antiferromagnetic ordering below T_c but the spins of the Fe (or Mn) ions become aligned in definite directions even though the directions appear to be random. We define a spin glass as an interacting magnetic system where the magnetic atoms have positional randomness; it is characterized by a freezing of the spins at a well-defined temperature, T_c, below which there is distinct magnetic behavior without the presence of long-range order. Denote $\langle S_i \rangle$ as a thermal average of a spin at the ith site and $\langle \overline{S} \rangle \equiv N^{-1}\Sigma \langle S_i \rangle$ as the average of $\langle S_i \rangle$ over all N spins. Then a spin glass is characterized as a system which below T_c has $\langle \bar{S} \rangle = 0$, but also $\langle \bar{S} \rangle^2 \equiv N^{-1} \Sigma \langle S_i^2 \rangle \neq 0$. Thus, in a spin glass each spin is frozen in some specific direction but this direction varies throughout the alloy so that there is no macroscopic magnetization.

Qualitatively occurrence of spin glass in dilute metallic alloys can be understood via the RKKY interaction and the concept of "frustration." As mentioned, the sign of the effective field that occurs via the indirect RKKY exchange interaction oscillates with distance, and the wavelength of the oscillation in has little to do with crystal periodicity. Thus, in a random alloy at the site of a given spin, this spin may find that the exchange interaction with some of its near neighbors is positive (ferromagnetic) and with some others it is negative (antiferromagnetic). This given spin cannot satisfy all of these requirements and this is referred to as **frustration**. Figure 15-23b illustrates the idea of frustration, whose consequence is the spatial variability of the frozen spin directions.

As yet the theories of the magnetic spin glasses are not on firm ground. The cusp behavior in the low field susceptibility, as in the figure, is predicted; but a similar predicted cusp in the specific heat is not observed. This has raised the question as to whether this transition is a real equilibrium transition or just the freezing of the spins which have not had enough time to equilibrate.

This spin-glass effect now has been observed in insulating magnetic alloys such as (Eu, Sr)S where a few percent of the nonmagnetic Sr ions are replaced by magnetic Eu ions.

Notes

Most of the solid state books in the bibliography contain one or more chapters on magnetism. Fairly complete books on this field at a complementary level to this book are S. Chikazumi, "Physics of Magnetism" (John Wiley and Sons, 1964); H. H. Morrish, "The Physical Principles of Magnetism" (John Wiley and Sons, 1965). A deeper treatment of the subject can be found in the series of books edited by G. T. Rado and H. Suhl, "Magnetism" (Academic Press, Vol. 1 appeared in 1963). There are about seven books in this series and it is a useful place to begin to look for modern information for most areas in magnetism. Another useful series of books is edited by E. P. Wohlfarth, "Ferromagnetic Materials" (North Holland, Vol. 3 appeared in 1982). The book by C. W. Chen, "Magnetism and Metallurgy of Soft Magnetic Materials" (North Holland, 1977) covers the elementary as well as the more practical aspects of magnetism. For applications of magnetism in information storage see M. H. Kryder and A. B. Bortz, in Physics Today, Dec. 1984.

There are other more specialized books at different levels. Some of them are J. H. van Vleck, "The Theory of Electric and Magnetic Susceptibilities" (Oxford Univ. Press, 1932); D. C. Mattis, "The Theory of Magnetism" (Harper and Row, 1981); R. Kubo and T. Nagamiya, Eds. "Solid State Physics" (McGraw-Hill, 1969) p. 451; J. S. Smart "Effective Field Theories of Magnetism," (W. B. Saunders Co., 1966), as well as many others.

See any quantum mechanics textbook for more detailed explanations of Russell-Saunders coupling, jj-coupling, and Hund's rules.

The problem of "shielding" in rare earth ions is discussed in G. Burns, Phys. Rev. **128**, 2121 (1962). This paper also discusses the various magnitudes of the crystal field and spin-orbit coupling. The shielding is small for the cubic components of the crystal field. The magnitudes of the axial components generally are much smaller than the cubic components. However, for the axial crystal field terms, the shielding can be appreciable.

Charge density waves (and spin density waves) are an important theoretical and experimental topic in the understanding of the ground state of metals. A useful review covering both aspects is A. W. Overhauser, Adv. in Phys. **27**, 343 (1978).

Spin glasses are a lively research topic. They were first observed by V. Cannella and J. A. Mydosh, Phys. Rev. **B6**, 4220 (1972), and the first real theory appeared by S. F. Edwards and P. W. Anderson, J. Phys. F. **5**, 965 (1975). Since the field is moving rapidly it is difficult to suggest a review paper; consult Solid State Physics or a friend.

The Ruderman Kittel Kasuya Yosida (**RKKY**) interaction is sometimes called the RK interaction. The references are M. A. Ruderman and C. Kittel, Phys. Rev. **96**, 99 (1954). T. Kasuya, Progr. Theoret. Phys. (Kyoto) **16**, 45 (1956). K. Yosida, Phys. Rev. **106**, 893 (1957). Also see C. Kittel, Solid State Physics **22**, 1 (1968). For a discussion of the **Kondo effect** see J. Kondo, Solid State Physics **23**, 184 (1969) and H. J. Heeger, Solid State Physics **23**, 248 (1969).

For an introduction to the Mössbauer effect see G. K. Wertheim, "Mössbauer Effect" (Academic Press, 1964); and H. Frauenfelder, "The Mössbauer Effect" (W. A. Benjamin, Inc., 1962). (This volume has reprints.) The proceedings of an early international conference can be found in Rev. Mod. Phys. **36**, 333 (1964).

For a fine overview of "Long range order in solids" including magnetism see R. M. White and T. H. Geballe, Supplement 15 of Solid State Physics (1979).

CHAPTER 15   MAGNETISM   629

Problems

1. **Diamagnetic susceptibility** – For the ground state of atomic hydrogen, show that $\langle r^2 \rangle = 3a_0^2$, and that the molar diamagnetic susceptibility is -2.36×10^{-6} cm^3 mole^{-1}. [$\psi_{1s} = (\pi a_0^3)^{-1/2} \exp(-r/a_0)$, where $a_0 = \hbar^2/me^2$, the Bohr radius.]

2. (a) Show that in the limit of $J \to \infty$ the Brillouin function goes over to the Langevin function. (b) Show that for $J = 1/2$ the expression in Eq. 15-9 reduces to $M = (Ng_J\mu_B/2V) \tanh(g_J\mu_B H/2k_B T)$, and since we expect the $g_J = g_e = 2$ the expression can be simplified.

3. **Ferromagnetic case for $J = 1/2$** – Consider the case of $J = 1/2$. (a) Show that the molecular field theory gives for Eq. 15-15b

$$M_s = (N\mu_b/V) \tanh(\mu_b \alpha_E M_s/k_B T)$$

where $\mu_b \equiv g\mu_B J$. (b) Show that $k_B T_c = (\alpha_E N\mu_b^2/V)$. (c) Further show that near T_c, where M_s is small,

$$\frac{M_S}{(\mu_b N/V)} = 3\left[\frac{T_c}{T} - 1\right]^{1/2}$$

(d) Consider what happens above T_c. In this case

$$M = (\mu_b N/V) \tanh[\mu_b(H + \alpha_E M)/k_B T]$$

but in the paramagnetic range M as well as H is small, so by expanding, determine the Curie-Weiss susceptibility.

4. Consider the molecular field equations for an antiferromagnetic material. Extend Eqs. 15-18 by allowing for the effective field at the A-sublattice from M_A, that is, $M_A = \chi_{A0}(H + \alpha_E M_B + \alpha_E' M_A)$, and so on. Then show that the susceptibility is given by $\chi = C/(T + \Theta)$ where $\Theta/T_N = (\alpha_E + \alpha_E')/(\alpha_E - \alpha_E')$.

5. **Spin wave dispersion** – Expand the spin wave dispersion relation, Eq. 15-28, for small ka and show that $\omega \propto k^2$. Why is this different from the vibrational case? For the spin waves, for small wave vector, the energy is in the form $\mathscr{E} = \hbar^2 k^2/2m^*$. Show that $m^* \approx 100m$ where m is the electron mass.

6. **Spin wave C_V and the Bloch $T^{3/2}$ law** – (a) Evaluate the integrals in Eqs. 15-32 for spin waves at low temperatures and obtain the results discussed in the text. Compare these calculated results with experiments by S. S. Shinozaki, Phys. Rev. **122**, 383 (1961), and F. Holtzberg, et al., J. App. Phys. **35**, 1033 (1964). (b) Show that the ferro-

magnetic spin wave theory predicts a low temperature $T^{3/2}$ magnetization of the form

$$\frac{M(T) - M(0)}{M(0)} = \frac{0.0587}{zS} \left[\frac{k_B T}{2J_{ex}S} \right]^{3/2}$$

where $z = 1$, 2 or 4 for a sc, bcc, or fcc structure. It will be helpful to know the following definite integral

$$\int_0^\infty \frac{x^{1/2} dx}{e^x - 1} = \frac{0.0587}{4\pi^2}$$

7. Two-dimensional spin wave system — By considering Eq. 15-32c, show that spin wave theory predicts that a two-dimensional spin structure will not be ferromagnetic.

8. Anisotropy energy — (a) For cubic symmetry show that the fourth order term $\alpha_1^4 + \alpha_2^4 + \alpha_3^4$, although allowed by symmetry, is not required in Eq. 15-33b because it can be expressed in terms of the fourth order term included in that equation. (b) Approximately evaluate the ratio of K_1/K_2 for Co and for Fe from the data in Fig. 15-18. (Hint: see Kanamori's article in Rado and Suhl's "Magnetism", Vol. 1 referred to in the Notes.)

9. If a ferromagnetic specimen is suspended by a thin cord in a magnetizing coil and the current in the coil reversed, reversing the magnetization of the specimen, then the specimen will rotate. This is called the **Einstein-deHaas gyromagnetic experiment**. Discuss this experiment and show how it can be used to show that ferromagnetism arises mostly from the electron spin rather than from the orbital angular momentum. (Hint: see the early editions of Kittel.)

16

Superconductivity

16-1 Introduction (dc Conductivity)
16-2 The Occurrence of Superconductivity
16-3 Effects that Destroy Superconductivity
16-4 Magnetic Properties
    a Normal metals with $\rho = 0$
    b The Meissner effect
    c The London equation
    d Type I and type II superconductors
16-5 The BCS Theory
    a Introduction
    b The Fröhlich (electron-phonon) interaction
    c Cooper pairs
    d BCS
16-6 BCS Predictions
16-7 BCS Related Measurements
    a Single particle tunneling
    b Isotope effect
    c Optical properties
    d Thermal conductivity
    e Flux quantization
16-8 The Josephson Effect
    a Introduction
    b Phase coherence
    c The dc Josephson effect
    d The ac Josephson effect
    e Quantum interference
Notes
Problems

Table 16-1 Values of T_c for the elements that are superconducting at atmospheric pressure. There are three other phases of Ga that have values 5.9-7.9°K and there are many elements that are superconducting under high pressure. Also listed are a very few of the known compounds, mostly emphasizing those with higher T_c values. $(SN)_x$ is a polymer. Rather complete complications of T_c and H_0 can be found in B. W. Roberts NBS Technical Note 983 (1978); S. V. Vomsovsky, Y. A. Izyumov, and E. Z. Kurmaey, "Superconductivity of Transition Metals" (Springer-Verlag, 1982); CRC Handbook of Chemistry and Physics (1983-1984). The value for Rh is from C. Buchal, F. Pobell, R. M. Mueller, M. Kubota, and J. R. Owers-Bradley, Phys. Rev. Lett. **50**, 64 (1983).

Element	T_c(°K)	Element	T_c(°K)	Compound	T_c(°K)
Nb	9.26	Al	1.17	Nb_3Ge	23
Tc	7.8	Ga	1.08	Nb_3Sn	18
Pb	7.20	Mo	0.915	Nb_3Al	18
Laα(hcp)	4.88	Am	0.85	V_3Si	17
Laβ(fcc)	6.00	Os	0.66	V_3Ga	16.5
V	5.40	Zr	0.61	NbN	15
Ta	4.47	Cd	0.517	La_3In	10
Hg(α)	4.15	Ru	0.49	NbTi	10
Hg(β)	3.95	Ti	0.40	Ti_2Co	3.44
Sn	3.72	Hf	0.128	AuBe	2.64
In	3.41	Ir	0.113	CuS	1.62
Tl	2.38	Lu	0.1	$(SN)_x$	0.26
Re	1.70	Be	0.026	$CeCu_2Si_2$	0.65
Pa	1.4	W	0.0154	UBe_{13}	0.85
Th	1.38	Rh	325×10^{-6}	UPt_3	0.54

SUPERCONDUCTIVITY

It don't make no difference how foolish it is, it's the right way—and it's the regular way. And there ain't no other way, that ever I heard of, and I've read all the books that gives any information about these things.

M. Twain, "The Adventures of Huckleberry Finn"

16-1 Introduction (dc Conductivity)

In 1908, at Leiden, H. Kamerlingh Onnes liquified helium and started the field of low temperature physics. Soon after, in 1911, he observed that the dc resistivity of mercury dropped abruptly to zero at a temperature T_c called the **superconducting transition temperature**. T_c for Hg is $4.15°K$; he found that even the addition of impurities to the samples failed to produce any measurable dc resistance below T_c. Figure 16-1 shows the normally observed low temperature metallic T^5 temperature dependence of the dc resistivity as well as the resistivity of a metal with a superconducting transition.

The transition at T_c can be quite sharp. In very pure Ga it occurs in less than $10^{-5}°K$. On the other hand, in some highly strained alloys the transition region can be broader than $0.1°K$. The sharpness of the transition indicates that the superconducting state is a new state of matter (i.e., a new arrangement of the electron states).

So far we have mentioned only the dc resistance (or dc conductivity). When measured at very high frequencies the resistance does not drop to zero below T_c. Later we discuss what "very high" means.

Is the resistance zero or just very, very small? Below T_c, magnetic induction techniques can be used to start a current circulating in a closed superconducting ring. The current that persists in the ring after the external field variations cease is called a **supercurrent**. The magnetic field generated by the supercurrent can be measured very accurately by nuclear magnetic resonance techniques. From these measurements we conclude that the current would continue for $>10^5$ years. Thus, the resistance is essentially zero.

The magnetic effects associated with superconductivity are equally unusual (and will be discussed later). Thus, perhaps it is not too surprising that it took almost 50 years to obtain a microscopic understanding of this phenomenon. This came in 1957 when Bardeen, Cooper, and Schrieffer showed that electrons with opposite crystal momenta can be attracted to each other (via an electron-lattice interaction), which causes the conduction electrons to form an ordered state with energy lower than that of the uncoupled (free) electrons. This theory, which has met with remarkable success and led to a Nobel

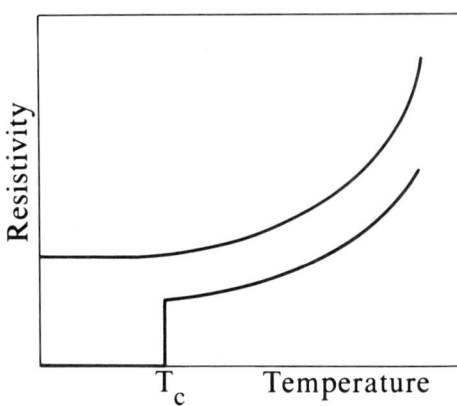

Fig. 16-1 The upper curve shows the behavior of the dc resistance of a normal metal at low temperatures ($\rho = \rho_0 + AT^5$). The lower curve shows the behavior of a superconducting metal (in zero magnetic field).

Prize for these gentlemen in 1972, is universally referred to as the BCS theory. More will be said about it later.

16-2 The Occurrence of Superconductivity

Superconductivity is not a rare phenomenon. Table 16-1 lists the elements that presently are known to be superconductors at atmospheric pressure. T_c ranges from 9 and 7°K for Nb and Pb, to 325 μ°K for Rh. As the few other metallic elements are made free from magnetic impurities and are tested below the millidegree range, the list probably will grow. As we shall see, magnetic impurities strongly suppress superconductivity. There are several important points that we can determine from this table. First, the energies involved with the transition ($\approx k_B T_c$), ranging from 10^{-8} eV to 1 meV, are very small compared to typical values of Fermi energies (≈ 10 eV, Table 9-4). So the fundamental interactions leading to superconductivity must be very weak compared to the usual electronic interactions. Second, the normal state metallic elements with the highest conductivity have not been found to be superconducting. This includes the alkali metals as well as Cu, Ag, and Au. (Actually, cesium is superconducting in a phase observed above 110 kbars, with $T_c = 1.5$°K.) This is now understood theoretically and is discussed later. It turns out that the electron-phonon coupling is weak in these metals and the electron-electron interaction is strong so if a transition occurs it will be at extremely low temperature. Third, the same element that can be obtained in different crystal structures has different values of T_c. Fourth, alloys can be superconducting even when the component elements are not. Fifth, some alloys have a T_c that is more than a factor three higher than the T_c of the elements involved. Figure 16-2

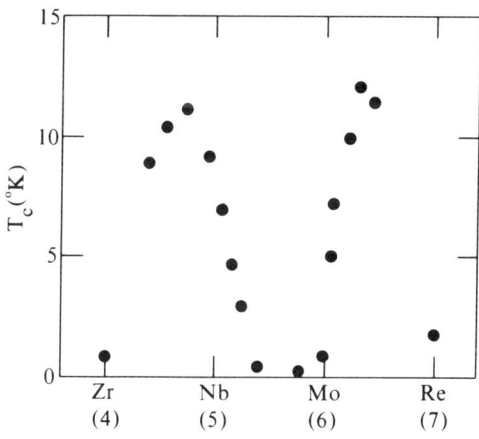

Fig. 16-2 The variation of T_c for alloys from elements of the same row in the periodic table. The numbers are the average number of valence electrons per atom.

shows how quantities such as the average number of valence electrons per atom can be important in determining T_c.

16-3 Effects That Destroy Superconductivity

Magnetic impurities (such as the rare earths or iron-series atoms) have a large effect on reducing T_c. However, in this section we discuss the two most prominent macroscopic ways to destroy the superconducting phase below T_c, namely, applying an external magnetic field or by passing a large current through the superconducting metal. Both of these effects act via magnetic fields and in the next section we will better appreciate the intimate connection between superconductivity and magnetism. The difference between dc and ac conductivity is also mentioned in this section.

Consider a long superconducting wire in an externally magnetic field, H, parallel to the wire. Below T_c the zero resistance state may be destroyed and the material restored to its "normal" state by the application of a sufficiently large magnetic field, the **critical field**, $H_c(T)$. The temperature variation of H_c is shown in Fig. 16-3; $H_c(T_c)$ is of course zero. Below each curve the sample is in the superconducting state, but if a field $H_c(T)$ or larger is applied the sample becomes "normal." The temperature dependence is approximately parabolic

$$H_c = H_0[1 - (T/T_c)^2] \tag{16-1}$$

where H_0 is the value of $H_c(T)$ at 0°K. For some of the high T_c compounds listed in Table 16-1, H_0 is very much larger than for the elements shown in Fig. 16-3.

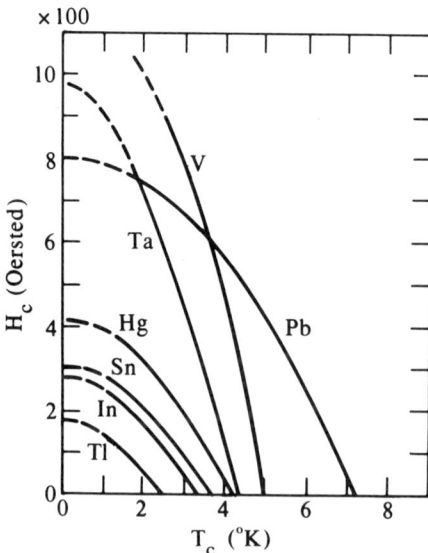

Fig. 16-3 The critical field vs. temperature for several elements. The sample is superconducting below and normal above the curve.

If the superconducting material is supporting a current, the magnetic field generated by the current will also suppress the T_c of the material. Hence there is a **critical current density**, J_c, above which the normal state will be restored. The magnitude of this current and the geometry of the sample determines J_c (it is much larger in a straight wire than in a tightly wound solenoid). If the sample is also in an external magnetic field, J_c will be reduced further. Technologically this is rather important for commercial superconducting magnetics where it is undesirable to have "hot spots" in the coil (places in the wire where the magnetic field is locally higher than in the rest of the wire) or the superconductor will go normal. Some of the very high T_c compounds and alloys listed in Table 16-1 can carry more than 100 amp in a 1 mm diameter wire before going normal, and are used for commercial superconducting magnets.

The BCS theory predicts that for a superconductor there is an energy gap between the electrons in the superconducting state and the electrons in the normal state. That is, in order for the electrons to behave normally they must be excited across this temperature dependent energy gap, of width 2Δ ($\approx 3.5 k_B T_c$). This was appreciated well before BCS; for many years it had been found, that while the dc conductivity goes to zero below T_c, the material behaves normally, above some high ac frequency, which depends on temperature. This ac frequency, of the order $\omega \approx 2\Delta/\hbar$, will be discussed later.

CHAPTER 16 SUPERCONDUCTIVITY

16-4 Magnetic Properties

The magnetic properties of superconductors are unexpected, unusual, and difficult to understand. As we shall see, the behavior is different from that of ordinary metals with perfect conductivity. Yet the magnetic properties are fundamental to superconductivity, and Bardeen has commented that it is perhaps better to think of a superconductor as a material with perfect diamagnetism than with perfect electrical conductivity. In this section, after we review the properties of a normal metal with perfect conductivity, we discuss the Meissner effect, the London equation, and then the difference between type I and type II superconductors.

16-4a Normal metals with $\rho = 0$ Figure 16-4a shows on the left a normal metal in an applied magnetic field. Magnetic flux lines penetrate the sample. Suppose the sample is cooled to below a transition temperature, middle portion of Fig. 16-4a, such that the metal becomes a perfect conductor (i.e., the resistivity, ρ, is zero) but it is not a superconductor. While the magnetic field is still applied, nothing will happen. However, if the external field is now turned off, currents will flow in the surface of the perfect conductor so that the magnetic flux through the sample will not change. This is what is expected and it is shown on the right of Fig. 16-4a.

On the other hand, if the normal metal is cooled in zero applied magnetic field to a temperature such that $\rho = 0$ and then a field is applied, currents will flow in the surface of the metal to keep out the magnetic flux lines. Thus, there will be no magnetic flux in the metal. If the field is reduced to zero the currents will stop and there will be no flux in the sample.

We can see easily why these effects happen in a perfect conductor. In an applied electric field **E**, an electron will be accelerated according to Newton's law

$$m(dv/dt) = -e\mathbf{E} \qquad (16\text{-}2)$$

However, if there is zero resistance (infinite relaxation time as in Section 9-5a) the electron would be accelerated indefinitely, which is impossible so **E** must be 0 inside a metal with zero resistance. (Also macroscopically, from Ohm's law $\mathbf{E} = \rho \mathbf{J}$ and if ρ goes to zero while J is finite then E is zero.) Yet from Maxwell's equations, Eq. 13-14,

$$\nabla \times \mathbf{E} = -\frac{1}{c}\frac{\partial \mathbf{B}}{\partial t} \qquad (16\text{-}3)$$

so **B** cannot change with time in the interior of a perfect conductor.

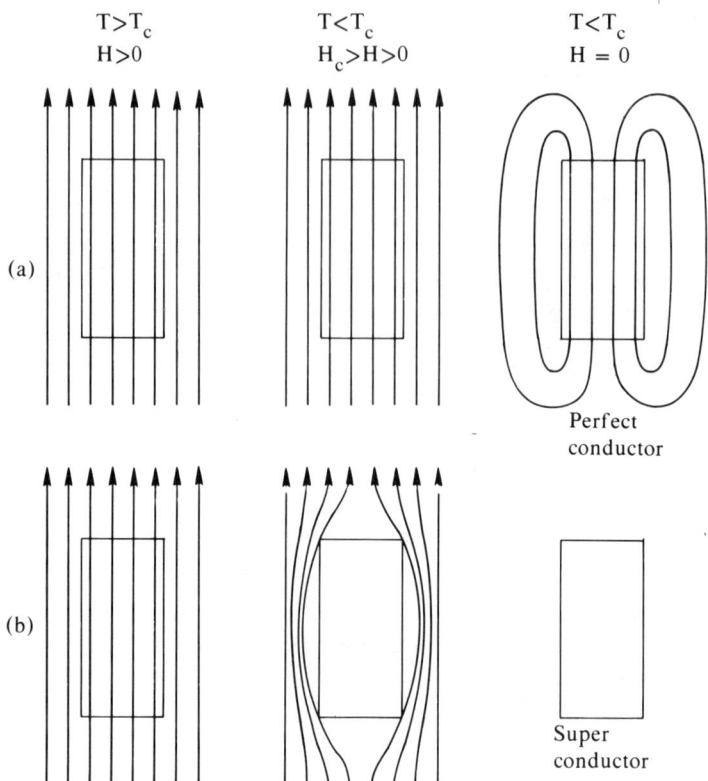

Fig. 16-4 The magnetic field behavior of a "perfect conductor" and a superconductor. (a) A normal metal that has zero resistance below T_c. (b) A metal that is a superconductor below T_c. See the text for a discussion of this figure.

16-4b The Meissner effect A superconductor behaves differently. In 1933 Meissner and Ochsenfeld reported that when a superconductor is cooled below T_c in a magnetic field all of the flux is expelled. [Of course, this occurs only if the applied field is below the critical field $H_c(T)$. With a larger field the metal remains a normal metal.] This is shown in Fig. 16-4b. To expel the flux from the interior of the sample, persistent surface currents are induced. More will be said later about the penetration depth of these currents.

On the other hand, if a magnetic field is applied to a superconductor below T_c, again persistent surface currents are induced to keep the flux lines out. Thus, below T_c the interior of the sample is free from magnetic flux, whether a magnetic field is applied during or after cooling a superconductor.

For the Meissner effect to occur $B = 0$. Thus, from Eqs. 15-1, for the susceptibility of a superconductor, we have **perfect diamagnetism**, $\chi = M/H = -1/4\pi$. ($\chi = -1/\mu_0$ in **SI units**.)

CHAPTER 16   SUPERCONDUCTIVITY   639

16-4c   The London equation   In 1935 the brothers F. and H. London put forth an explanation of the Meissner effect. Their thinking was based on the two-fluid model, which was a model for superfluid helium. The idea is that while N/V is the density of all of the conduction electrons, N_s/V is the density of the superconducting electrons. $N_s(T)$ goes from zero at T_c to $N_s = N$ at $T = 0°K$. The superconducting electrons show zero resistance, while the normal electrons have their usual resistance. Thus, when a transitory electric field is used to induce a current in the sample only the superconducting electrons flow while the normal electrons remain stationary and can be ignored in what follows. Then in Eq. 16-2, $\mathbf{v}$ must be replaced by $\mathbf{v}_s$, which is the drift velocity of the superconducting electrons. The current density $\mathbf{J} = -e\mathbf{v}_s(N_s/V)$ combined with Eq. 16-2 immediately gives

$$\frac{d\mathbf{J}}{dt} = \left(\frac{e^2 N_s}{mV}\right)\mathbf{E} \tag{16-4}$$

Taking the curl of both sides of this equation and substituting the result into Eq. 16-3 gives

$$\frac{\partial}{\partial t}\left[\nabla \times \mathbf{J} + \left(\frac{e^2 N_s}{mcV}\right)\mathbf{B}\right] = 0 \tag{16-5}$$

So far the results are general and this equation applies to any metal with an electron density of N_s/V, so we know the Meissner effect is not accounted for by this equation (as shown previously). The Londons appreciated that while Eq. 16-5 cannot explain the Meissner effect, the effect could be accounted for by restricting this equation and requiring that the expression in the square bracket in Eq. 16-5 be zero, not just its time dependence. Writing this restricted expression and also writing Maxwell's equation, Eq. 13-14, for $\nabla \times \mathbf{B}$ we have

$$\nabla \times \mathbf{J} = -\left(\frac{N_s e^2}{mcV}\right)\mathbf{B} \tag{16-6a}$$

$$\nabla \times \mathbf{B} = (4\pi/c)\mathbf{J} \tag{16-6b}$$

Equation 16-6a is the **London equation**. Taking the curl of both sides of Eq. 16-6b and using Eq. 16-6a, or taking the curl of both sides of Eq. 16-6a and using Eq. 16-6b, we obtain the complementary equations [using the identity $\nabla \times \nabla \times \mathbf{B} = \nabla(\nabla \cdot \mathbf{B}) - \nabla^2 \mathbf{B} = -\nabla^2 \mathbf{B}$ and Eq. 13-14]

$$\nabla^2 \mathbf{B} = \mathbf{B}/\lambda_L^2 \qquad \nabla^2 \mathbf{J} = \mathbf{J}/\lambda_L^2 \tag{16-7a}$$

where $\lambda_L^2 \equiv (mc^2 V/4\pi N_s e^2)$. A solution of Eq. 16-7a is an exponential decay of B from the surface into the bulk of the superconductor as

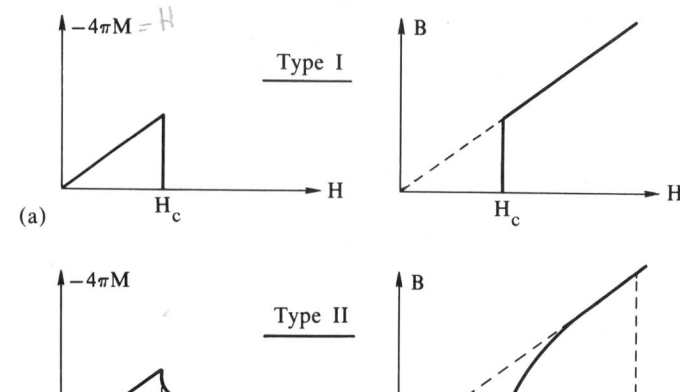

Fig. 16-5 (a) Effects in a type I superconductor. On the left is the magnetism of the sample vs. the applied field, H. On the right the field in the bulk of the sample vs. H is shown. (b) The same but for type II superconductors.

$$B(x) = B(0) \exp(-x/\lambda_L) \qquad (16\text{-}7b)$$

where λ_L is the **London penetration depth** and $B(0)$ is the value at the surface. If N_s is taken as the total density of the conduction electrons then λ_L is 100-1000 Å. The London equation gives a simple picture of what is happening in the Meissner effect. Currents occur within this thin surface layer of a superconductor; these currents screen out the applied magnetic field from the bulk of the superconductor.

Experimentally a magnetic field penetration depth, λ, can be measured and, as might be expected, is temperature dependent. Most experimental results fit a form

$$\left[\frac{\lambda(T)}{\lambda(0)}\right]^2 = \left[1 - \left(\frac{T}{T_c}\right)^4\right]^{-1} = \frac{N}{N_s(T)} \qquad (16\text{-}8)$$

where we assume that $N_s = N$ at $0°K$ and $\lambda(0)$ is the penetration depth at $0°K$. Experimentally we can find values of λ larger than the values of λ_L predicted by Eq. 16-7; furthermore λ depends on impurities and to some extent on **B**. The London equation makes no prediction as to the temperature dependence of λ_L, or really N_s. So the preceding discussion is incomplete and the idea that some electrons can be labeled as superconducting and others as normal cannot be taken too literally. Nevertheless, N_s can be thought of as a parameter that describes the extent of order in the superconducting state. Pippard has given a nonlocal theory of flux penetration that is in better agreement with experiment. See the Notes.

16-4d Type I and type II superconductors Let us consider samples in the form of long cylinders with the applied magnetic field, H, parallel to the cylinder axis except where otherwise described. This shape eliminates demagnetization factors.

Type I — Below the critical field, $H_c(T)$, there are only surface currents and magnetic flux does not penetrate the bulk of the sample, which thus behaves as a perfect diamagnetic material ($\chi = -1/4\pi$). When the applied field exceeds H_c, the entire bulk of the sample becomes normal; the field completely penetrates the sample, the magnetism goes approximately to zero (goes to the value for an ordinary metal which is very small), and the resistance goes from zero to the typical values found for metals at low temperature. Both the magnetization and the field in the sample are shown in Fig. 16-5a for type I superconductors. These are the effects that we have been talking about so far in this chapter. Most elements have type I behavior. Later we note why this is so. (We do not consider the behavior of thin films where the strains can be large.)

Type I materials with other geometries can behave differently. In type I materials we can observe an **intermediate state** (not to be confused with the mixed state found in type II materials). This occurs because of the geometry of the sample. For example, Fig. 16-6a shows the cylinder axis of the sample perpendicular to the applied field, H. Since the sample is a superconductor there is no field inside of it and the flux lines "bunch up" at the top and bottom as shown; at these positions the field is 2H. When H reaches $H_c/2$ this part of the specimen becomes normal while the rest is not normal. Thus, the intermediate state occurs in type I materials and is characterized by regions of macroscopic dimensions ($\sim 10^{-2}$ cm) becoming normal; these regions depend on sample shape and on the direction of H.

Type II — Materials of this type are more complex than those of type I. In type II materials below a lower critical field H_{c1} there is no flux penetration and the behavior is the same as above. See Fig. 16-5b. Above an upper critical field H_{c2} there is complete flux penetration, the material is totally normal, again just as for type I materials above their H_c. However, in between H_{c1} and H_{c2} the material is in a **mixed** (or **vortex**) **state**. This state is characterized by zero resistance but partial flux penetration. The flux penetrates in the form of thin filaments. Within these filaments the field is high and the material is normal ($\rho \neq 0$), but outside these filaments the material is still a superconductor. Thus, around each filament is a screening current that decays in a distance λ as might be given by the London equation. Figure 16-6b shows an array of vortex lines observed in the mixed state of a type II superconductor. The reason that the material still has zero resistance is that all the current flows through the superconducting regions.

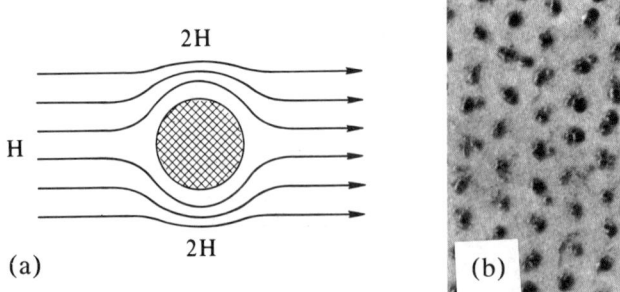

Fig. 16-6 (a) A schematic diagram showing a magnetic field applied perpendicular to a long superconducting wire, shown in cross section. The high field at the top and bottom causes the intermediate state in type I materials. (b) An array of vortex lines observed on the surface of a superconductor. The picture is made possible by decorating the vortex lines with fine ferromagnetic particles. (Courtesy of U. Essmann.)

In some of the high T_c alloys of the type listed in Table 16-1, H_{c2} is higher than 500 kG (50 teslas) at 4.2°K; for type I elements H_c at 0°K (H_0 in Eq. 16-1) is usually less than 1 kG. Commercial superconducting solenoids of course use type II materials and can produce continuous fields of more than 100 kG; these are usually made with a **hard superconductor** (a mechanically worked type II superconductor which thus has a large amount of flux pinning or magnetic hysteresis for similar reasons as discussed in Section 15-10c). H_{c2} for most materials has a temperature dependence approximately similar to that given in Eq. 16-1.

Why the difference? The mechanism is the same for superconductivity in both type I and II materials. The important difference arises from the difference between the mean free path (mfp) of the normal state electrons and the magnetic field penetration depth, λ, Eq. 16-8. If the mfp $> \lambda$ then the superconductor will be type I. Most pure elements satisfy this condition and indeed are type I. However, by alloying certain elements (such as Pb with small amounts of In) the mfp can be reduced strongly, and continuous changes can be observed from type I to type II behavior. The high T_c alloys (Table 16-1) have inherently small mfp even at low temperatures so they are type II superconductors. At the end of the chapter this is discussed in more detail in terms of the Ginzburg-Landau parameter.

16-5 The BCS Theory

In 1957, almost half a century after the discovery of superconductivity, a comprehensive microscopic theory of this phenomenon was proposed. This theory, by Bardeen, Cooper, and Schrieffer and always called BCS has had amazingly good quantitative success. The theory followed the studies by Cooper (1956) of the "Cooper pair" where he found that under certain conditions the ground state energy of a pair of electrons can be lower than that of the two free electrons. The BCS theory is expressed in formalism beyond the level of this book. References are given in the Notes to the original papers as well as to a few books on the subject. As well as possible we shall concentrate on the underlying physical ideas. In Section 16-6 we emphasize the basic predictions of BCS, and in Section 16-7 we discuss experiments related to the predictions of BCS.

A central result of the BCS theory is an energy gap between the electron system in the superconducting ground state and the excited states. Another important result is the dependence of T_c on phonon frequencies, which typically leads to the dependence of T_c on M where M is the atomic mass of the atoms in the crystal (the isotope effect). Actually, before 1957 there were many experiments that indicated an energy gap and showed that T_c depends on M. We shall not discuss these results in historical perspective; rather we summarize them in Section 16-7.

16-5a Introduction We summarize some of the properties that were known in 1957 and what would be expected from the theory of superconductivity. First, zero resistance and a perfect diamagnetic state is observed. Second, the phenomenon is associated with the appearance of a long-range ordering of the conduction band electrons, which leads to an energy gap between the ground and the first excited state. A second-order phase change is associated with the onset of this state. Third, the crystal structure does not show any change at T_c, yet the vibrations of the atoms must play a role in establishing the superconducting state because T_c does depend on M.

There is, of course, a strong long-range Coulomb repulsion between a pair of electrons. However, in a metal the other $\approx 10^{23}$ electrons/cm^3 act to screen this interaction, particularly at large distances. This significantly screened Coulomb interaction is what causes the single electron approximation to be useful (Section 9-14). Nevertheless, even with screening the net interaction between any pair of electrons is repulsion. In order to produce a new, lower energy, ground state what is needed is some sort of weak attractive force between at least some pairs of electrons. It is expected to be a weak interaction because $k_B T_c$ is so small.

16-5b The Fröhlich (electron-phonon) interaction A free electron in a perfect crystal structure propagates freely without attenuation. However, the perfect periodicity is destroyed by thermal vibrations of the atoms. This partial deviation from periodicity causes the

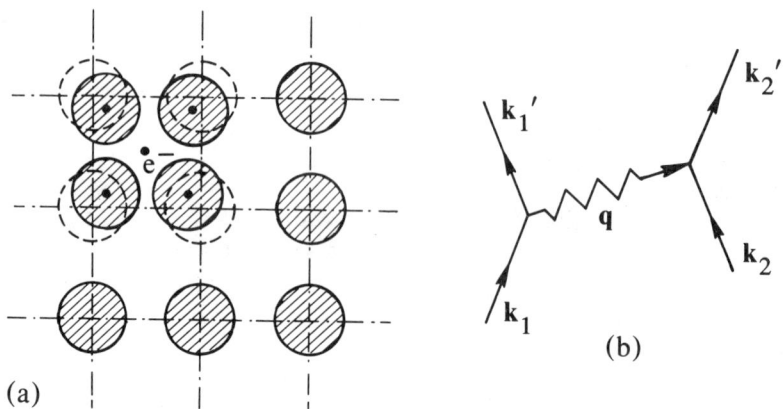

Fig. 16-7 (a) A schematic diagram of an electron polarizing positive ions in its vicinity. (b) An electron-electron interaction transmitted by a phonon of wave vector **q**.

electron wave to have a certain probability of being reflected (or scattered). Since the electron is interacting with the vibrating atoms this is called the **electron-phonon** (or **electron-lattice**) **interaction**. In very pure metals, it is this interaction that determines the resistivity.

In 1950 Fröhlich pointed out that the electron-phonon interaction provides a mechanism for a weak attraction between *two* electrons. Figure 16-7a shows how an electron will attract the positive ions around it. (It is said that the electron polarizes the crystal.) This slightly increases the density of the positive ion cores in the immediate region of the electron. Other electrons in this general vicinity will be drawn toward this region because of the higher density of positive charge. Hence, electrons are attracted to each other via the movement of the ions. Figure 16-7b shows how this interaction is represented (and eventually calculated). An electron of wave vector $\mathbf{k}_1$ emits a phonon of wave vector $\mathbf{q}$ and is scattered into a state with wave vector $\mathbf{k}_1'$ such that $\mathbf{k}_1 = \mathbf{k}_1' + \mathbf{q}$. In the same manner this phonon is absorbed by a second electron which changes its wave vector from $\mathbf{k}_2$ to $\mathbf{k}_2'$. Both processes conserve crystal momentum, hence

$$\mathbf{k}_1 = \mathbf{k}_1' + \mathbf{q}, \qquad \mathbf{k}_2 + \mathbf{q} = \mathbf{k}_2' \qquad (16\text{-}9a)$$
$$\mathbf{k}_1 + \mathbf{k}_2 = \mathbf{k}_1' + \mathbf{k}_2' \qquad (16\text{-}9b)$$

where Eq. 16-9b is just the result of eliminating $\mathbf{q}$ from the first two equations; this equation shows the conservation of crystal momentum between the initial and final electron states as expected. Note that while energy must be conserved between the initial and final states, it need not be conserved between the initial and intermediate state (i.e., the $\mathbf{k}_1$ and $\mathbf{k}_1' + \mathbf{q}$ state), because of the uncertainty principle $\Delta\mathscr{E} \Delta t \approx \hbar$. If the lifetime of the intermediate state is very short there will be a large uncertainty in its energy; thus energy need not be conserved in the emission (or absorption) process. Processes in which energy need not be conserved are called **virtual processes** and their existence depends on there being another electron available to absorb the phonon within a time Δt.

Letting the energy of the first electron before and after scattering be $\mathscr{E}_1$ and $\mathscr{E}_1'$, respectively, detailed calculations show that this electron-phonon interaction produces an attraction between the two electrons only when $\mathscr{E}_1 - \mathscr{E}_1' < \hbar\omega_q$ (the energy of the phonon). In the Debye model the highest density of phonons corresponds to an energy $\hbar\omega_D$, so $\hbar\omega_q \approx \hbar\omega_D$. For most metals this is a small energy (corresponding to $\Theta_D \sim 100°K$) compared to the Fermi energy ($\sim 10^5 °K$). However, superconductivity occurs only at low temperatures so at the Fermi surface all of the important electrons are within a width of $k_B T$ and these can feel this electron-phonon attraction (because $k_B T_c < \hbar\omega_D$ or $T_c < \Theta_D$).

Of course, the screened Coulomb repulsion still exists but this electron-phonon interaction does allow for the *possibility* of a net attractive interaction between electrons at the top of the Fermi sea. Fröhlich argued that this net attractive force might produce a ground state for the electrons that has an energy lower than the filled Fermi sea. This new state would then be separated from normal state electrons by an energy gap. Unfortunately, perturbation theory calculations did not give the desired results.

Nevertheless, from these ideas for an attractive potential, Fröhlich could explain why the best (highest T_c) superconductors were the poorest metals (high resistance in the normal state) while the good metals, such as the alkalies, Cu, Ag, and Au, are the worst superconductors. Namely, while a large electron-phonon interaction would be expected to lead to a high T_c, it also leads to high resistance in the normal state due to greater electron-phonon scattering. Further, since this attractive electron-phonon interaction involves the lattice vibrations of the atoms, it would be expected that T_c depends on the mass of the atoms. To the extent that the important electron energy differences $\propto \hbar\omega_D$, from Eqs. 11-22b and 11-18 we have $\omega_D \propto M^{-1/2}$, so it is expected that $T_c \propto M^{-1/2}$. In fact, experiments on isotopes of Hg also reported in 1950 showed that $T_c \propto M^{-1/2}$. (However, it is now known from experimental results that this exact mass dependence can be different for other elements.)

These experimental measurements showed that somehow lattice vibrations play an important part in establishing the superconducting state. Further, Fröhlich showed that the electron-phonon interaction provides an attractive potential energy which, hopefully, might be larger than the screened Coulomb repulsive energy, and that some important aspects of the superconducting state can be understood.

16-5c Cooper pairs The next critical step towards a microscopic theory of superconductivity was published by Cooper in 1956. Consider $T = 0°K$ so the Fermi sea is filled to energy $\mathscr{E}_F$ and momentum k_F. Add two electrons to this metal; the Pauli principle requires that $k > k_F$ for these two electrons. Assume an attractive interaction. Cooper showed that no matter how weak the attractive interaction between these two electrons, they will form a bound state with a total (kinetic plus potential) energy that is less than $2\mathscr{E}_F$!

To examine some of the concepts consider a simple two-particle product wave function of the individual single electron wave functions. Thus, $\psi(\mathbf{k}_1,\mathbf{k}_2) = \psi(\mathbf{k}_1)\psi(\mathbf{k}_2)$, where we leave out the spatial parts since they are understood (free electrons or Bloch functions) and we ignore the complication of properly antisymmetrizing ψ. Since we have assumed an attractive interaction between this pair of electrons, pictorially speaking we want the electrons to interact, or scatter, many times in the manner illustrated in Fig. 16-7b. After a scattering event the wave function is $\psi(\mathbf{k}_1')\psi(\mathbf{k}_2')$. However, since the total momentum of the two electrons is conserved, Eq. 16-9b, $\mathbf{k}_1 + \mathbf{k}_2 = \mathbf{k}_1' + \mathbf{k}_2' \equiv \overline{\mathbf{b}}$. (Actually, ψ is a stationary state and the matrix elements for scattering must be calculated, but a helpful picture of this idea is the more scattering events for this attractive interaction, the lower the energy of the electron pair.)

Now assume that the attractive interaction is of the electron-phonon type as discussed. We have said that this interaction is important (attractive) only for $\mathscr{E}_1 - \mathscr{E}_1' < \hbar\omega_D$. At the same time $\mathscr{E}_1$ and $\mathscr{E}_1'$ are just above $\mathscr{E}_F$. So we have a picture of two electrons with energies just above the Fermi energy and with this small energy difference (and their momenta are just above k_F), but what about their momentum difference? Since $\mathscr{E} = \hbar^2 k^2/2m$, by differentiation and setting $\Delta\mathscr{E} = \hbar\omega_D$, we find that k_1 and k_1' must lie within a range $\Delta k = m\omega_D/\hbar k_F$ of the Fermi momentum k_F. Further, all of the pairs $\mathbf{k}_1$ and $\mathbf{k}_2$ in the wave function satisfy $\mathbf{k}_1 + \mathbf{k}_2 = \overline{\mathbf{b}}$. The construction shown in Fig. 16-8 is helpful in determining $\overline{\mathbf{b}}$. In this diagram $k_1 \approx k_2 \approx k_F$, Δk is as shown (and is small compared to k_F), and $\mathbf{k}_1 + \mathbf{k}_2 = \overline{\mathbf{b}}$. The number of pairs of electrons with momentum that satisfy all of these conditions is proportional to the crosshatched volume (area in projection). This volume has a sharp maximum when $\overline{\mathbf{b}} = 0$. Thus, the energy will be lowered the most (maximum number of scattering events) by pairing electrons with equal and opposite momenta (i.e., $\mathbf{k}_1 = -\mathbf{k}_2$).

From the spatial part of the wave function detailed quantum mechanical calculations further show that the interaction is largest for electrons with opposite spins. Physically this arises because electrons with opposite spins have a larger probability of being closer to each

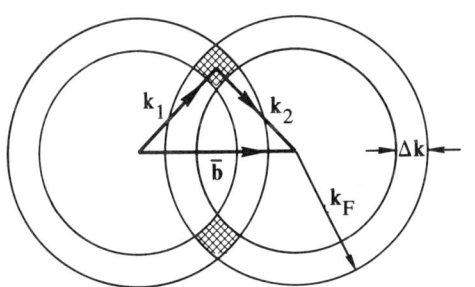

Fig. 16-8 This diagram shows two shells in k-space of radius k_F and thickness Δk. The crosshatched section is a cross section of the ring in which the electrons satisfy the conditions described in the text. The volume of this ring has a very sharp maximum when the spheres are concentric.

other and thus have a stronger electron-phonon interaction (which is assumed to be attractive).

Naturally, Cooper carried out detailed quantum mechanical calculations for this pair of electrons. We summarize the results. For free electrons with Fermi energy and momentum $\mathscr{E}_F$ and k_F at $T = 0°K$, add two electrons with $k > k_F$. As long as there is an attractive potential, a bound pair will form with an energy below $2\mathscr{E}_F$ and whose wave function is of type $\psi(k\uparrow)\psi(-k\downarrow)$, that is, the paired electrons have equal and opposite momentum and spin. This pair is called a **Cooper pair**. An equation, discussed shortly, can be obtained for this binding energy in terms of $\hbar\omega_D$, the density of electron states at the Fermi energy, and matrix element of the attractive electron-phonon interaction, U_0.

Showing that two electrons form a bound pair with a binding energy below $2\mathscr{E}_F$ is a big step toward a theory of superconductivity, but there are $\approx 10^{23}$ conduction electrons per cm^3 so more is required.

16-5d BCS The preceding discussion for a Cooper pair applies to two electrons in the Fermi sea. BCS took the essential steps for a full microscopic theory of superconductivity by constructing a ground state wave function in which all of the electrons form bound pairs. From their model they were able to calculate the equilibrium properties of superconductors and obtain excellent agreement with experiment. The specific predictions are discussed in the next section.

The basic BCS idea is relatively simple, and the original paper (see the summary) is worth reading. We have discussed how the Fermi sea is unstable against the formation of a single Cooper pair provided that the attractive interaction exists. Thus, it might be expected that the electron energy will be lowered further by the condensation of many pairs. This condensation will then continue until the binding energy of an additional pair is zero. To solve this problem BCS assumed that the only interaction between the Cooper pairs comes through the Pauli principle in limiting the states into which the two interacting electrons, which make up the pair, may be scattered. The pair state itself, composed of two fermions with opposite spin, may be regarded as a single entity (i.e., a new quasi-particle with integer spin) that obeys Bose-Einstein statistics. Thus, at $0°K$ all of the Cooper pairs are in the same quantum state with the same energy even though the individual electrons in the pairs are being scattered continually between single electron states having crystal momenta within a range Δk ($= m\omega_D/\hbar k_F$) as described previously.

The mathematics needed to handle this complicated state is not easy and the interested reader is referred to the Notes. We shall proceed directly to the predictions made by BCS.

16-6 BCS Predictions

The 1957 BCS paper accounted well for the equilibrium properties of superconductors. In fact, the paper was aptly summed up by the authors in their concluding section. "This quantitative agreement, as well as the fact that we can account for the main features of superconductivity, is convincing evidence that our model is essentially correct." In this section we discuss the main conclusions of BCS and in the next section a few of the many other experiments related to the energy gap, $\rho = 0$, and so on.

The critical temperature – BCS calculated T_c, the temperature of the transition between the normal and the superconducting state in zero magnetic field. They found that

$$k_B T_c = 1.14 \hbar \omega_q \, e^{-1/g(\mathscr{E}_F)U_0} \qquad (16\text{-}10)$$

$\hbar \omega_q$ is an averaged phonon energy averaged over the phonons that are important for the electron-phonon interaction. It is this interaction that provides the attractive potential that allows Cooper pairs to form. Normally $\hbar \omega_q \approx \hbar \omega_D$, that is, the Debye energy. $g(\mathscr{E}_F)$ is the density of states at the Fermi energy for the normal metal, for a single spin. U_0 is the matrix element of the average attractive electron-phonon interaction. (It is interesting to note that this result is not an analytic function at $U_0 = 0$ so it cannot be obtained by a perturbation expansion even if summed to all orders.)

It is difficult to use Eq. 16-10 to predict T_c accurately because ω_q and U_0 are difficult to calculate precisely. Nevertheless, $\hbar \omega_q \approx \hbar \omega_D = k_B \Theta_D$ and Θ_D is typically $100°K$ so the product $g(\mathscr{E}_F) U_0$ typically ranges from 0.1 to 0.5; the latter figure is for materials with a large electron-phonon interaction. Thus, we see how this exponential behavior accounts for both the low values of T_c (compared to Θ_D) and for the wide range of T_c values. BCS predicts a transition to the superconducting state for any metal in which the attractive electron-phonon interaction is stronger than the screened electron-electron repulsion in some energy range, although the temperature can be extremely low for those with small electron-phonon coupling coefficients. Also, since $\omega_D \propto M^{-1/2}$ the isotopic effect is predicted. However, a more complicated mass dependence is found for materials in which there is strong coupling (of the electrons to the phonons).

The energy gap (2Δ) – Figure 16-9a shows the energy dependence of the density of states for the electrons in a superconductor at T = 0°K. The energy gap is given by 2Δ which at 0°K is written as $2\Delta_0$. BCS found that the energy gap at 0°K is

$$2\Delta_0 = 4\hbar\omega_q \exp{-[g(\mathscr{E}_F)U_0]^{-1}} \qquad (16\text{-}11)$$

The form of this equation is the same as that for T_c, so by taking the ratio we can find a parameter-free result

$$\frac{2\Delta_0}{k_B T_c} = 3.5 \qquad (16\text{-}12a)$$

Here we list values of $2\Delta_0/k_B T_c$ for a number of elemental metals.

Al	3.4	Sn	3.5	Zn	3.2
Cd	3.2	Ta	3.6	Hg(α)	4.6
In	3.6	Tl	3.6	Pb	4.3
Nb	3.8	V	3.4		

The agreement with Eq. 16-12a is good and only for the last two elements, Hg and Pb, is there a distinct difference between the measured and calculated results. These two elements show deviations from the simple BCS theory for many predictions. When the so-called "strong-coupling" BCS theory is used for these materials, better agreement is found for this ratio as well as for the other predictions.

BCS also calculated the temperature dependence of the energy gap, 2Δ. For the ratio Δ/Δ_0 they obtained

$$\Delta/\Delta_0 = \tanh{(T_c\Delta/T\Delta_0)} \qquad (16\text{-}12b)$$

$$\frac{\Delta}{\Delta_0} = 1.74\left(1 - \frac{T}{T_c}\right)^{1/2}, \quad \text{for } T \approx T_c \qquad (16\text{-}12c)$$

Again there is a parameter-free relation between the reduced gap, Δ/Δ_0, vs. the reduced temperature, T/T_c, which is shown in Fig. 16-9b. The experimental data shown in the figure was not available in 1957, but comes from tunnelling experiments to be described shortly. The agreement is very good, as can be seen.

The critical field H_c – BCS also calculated the temperature dependence of H_c and determined the deviations from the simple empirical law given in Eq. 16-1. Figure 16-10 shows the BCS result along with more recent experimental data. Again, even in this very critical manner of displaying the results, the agreement is good except for the strong-coupling superconductors. (In their 1957 paper the experiments

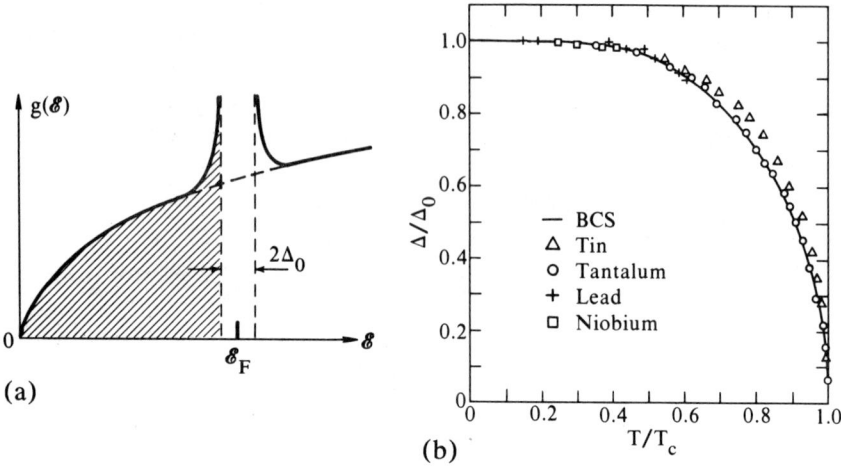

Fig. 16-9 (a) The energy dependence of the density of states for a superconductor. In the normal state $\rho(\mathscr{E})$ is filled up to $\mathscr{E}_F$, but below T_c there is an energy gap and it is shown for the T = 0°K case all of the states below the gap are fully occupied. (b) The reduced energy gap vs. the reduced temperature. The solid line is calculated by BCS and points are measurements by P. Townsend and J. Sutton [Phys. Rev. **128**, 591 (1962)] using electron tunnelling, which is covered later.

were not as accurate and the agreement with experiment was not thought to be good.) At low temperatures ($T \ll T_c$), where the gap has very little temperature dependence (Fig. 16-3), it is found

$$H_c = H_0 [1 - 1.07(T/T_c)^2] \qquad (16\text{-}12\text{d})$$

In their original paper BCS fully developed their theory using only weak coupling between the electrons and phonons. Thus, they have a single average matrix element for the electron-phonon interaction, U_0. However, in some metals this interaction is very strong and it must be considered in more detail. These materials are the so-called **strong-coupling superconductors.** An extension of the BCS theory to the strong-coupling case is required when T_c/Θ_D is relatively large. For strong-coupled superconductors we must take into account the phonon damping rate (lifetime) as this affects the quasiparticle description of the Cooper pairs. Also the retarded nature of the electron-electron interaction, via the phonons, must be considered carefully. Using the strong coupling extension to BCS theory it is found that for Hg and Pb the reduced gap vs. reduced temperature is within a few percent of that found by BCS, that is, in the weak-coupling case, as given in Eq. 16-12b and Fig. 16-9b. However, $2\Delta_0/k_B T_c$ is calculated to be 4.8 for Hg and 4.4 for Pb, in good agreement with the measured values listed previously. The temperature dependence of the critical field is

CHAPTER 16   SUPERCONDUCTIVITY   651

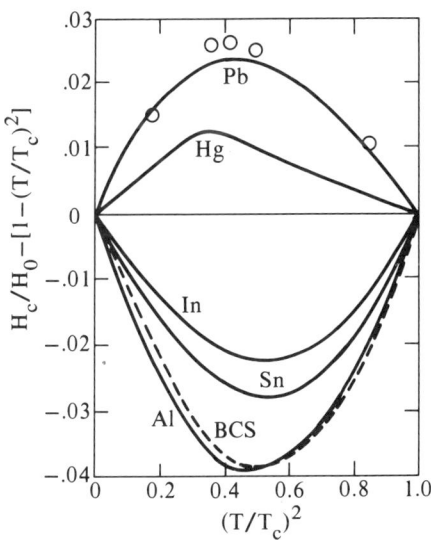

Fig. 16-10 Deviations from the simple magnetic field behavior given in Eq. 16-1. The BCS curve is from their 1957 paper while the other solid lines are from experimental measurements. The open circles are for the numerical strong coupling BCS calculations of J. C. Swihart, D. J. Scalapino, Y. Wada, Phys. Rev. Letters **14**, 106 (1965).

calculated and for Pb is shown in Fig. 16-10, where excellent agreement is seen. (For references see the Notes and the caption of Fig. 16-10.)

Heat capacity − Heat capacity measurements are useful experiments to make on superconductors. It can be measured in the superconducting state and then, with the application of a modest magnetic field, measured in the normal state at the same temperature. (The small fields have no effect on the electronic or lattice contribution to C_V.) Figure 16-11 shows the experimental results for Ga in both the normal and superconducting states. The results can be fit, respectively, by

$$C = 0.596T + 0.0568T^3 \equiv \gamma T + AT^3 \qquad (16\text{-}13a)$$

$$C_{es}/\gamma T_c = 7.46 \exp(-1.39 T_c/T) \qquad (16\text{-}13b)$$

The result in Eq. 16-13a contains the electronic part, γT, and a low temperature lattice Debye term AT^3 as discussed in Chapter 11 (Eq. 9-53 and Eq. 11-25). In Eq. 16-13b, C_{es} applies only to the electronic part of the heat capacity in the superconducting state; the Debye contribution has been subtracted out. When plotted on semilog paper, we find a straight line indicative of the form in Eq. 16-13b. This form is what is expected if there is an energy gap, and the experiment very clearly shows the existence of such a gap. (However, 2Δ is temperature dependent so a detailed comparison between theory and experiment is more involved.)

BCS calculated the difference between the electronic specific heat in the superconducting and normal state. They give at T_c,

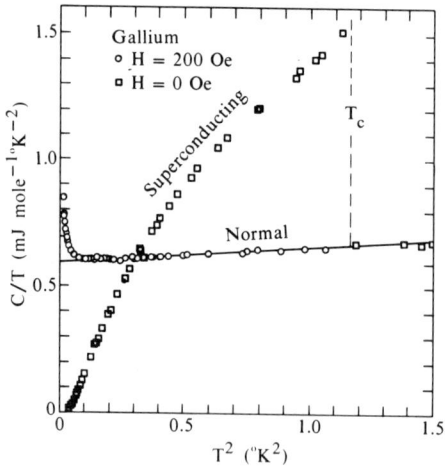

Fig. 16-11 The heat capacity divided by temperature of Ga metal. Zero applied magnetic field is used in the superconducting state but a field of 200 Oe is used to make the material normal. (The sharp rise at the very lowest temperatures is a Schottky anomoly, Section 11-2b, due to the Ga nuclear quadrupole interaction.) From N. E. Phillips, Phys. Rev. **134**, 385 (1964).

$$(C_{es} - \gamma T_c)/\gamma T_c = 1.52 \qquad (16\text{-}13c)$$

This is in very good agreement with the elemental metals that have been measured (see Mersevey and Schwartz referenced in the Notes) except for the strong-coupled superconductors Hg and Pb where 2.4 and 2.7, respectively, are found.

Penetration depth—BCS calculated the magnetic penetration depth and its temperature dependence. The analysis given by London and resulting in a penetration depth given in Eq. 16-7a had been extended by Pippard to more realistic situations. BCS found good agreement with Pippard's results and with the experimental results given in Eq. 16-8. Thus the Meissner effect is understood.

Summary (and $\rho = 0$) — As can be seen BCS gives an excellent account of superconductors. Besides the assumption of an attractive electron-electron interaction (presumably via the electron-phonon interaction), BCS assumed for simplicity that the Fermi surface is spherical and that there is no anisotropy in the matrix elements and in the gap. Also, for simplicity, they assumed a constant matrix element, U_0, for the electron-phonon interaction. This is sometimes called the weak-coupling theory; they discussed the need for strong coupling when T_c/Θ_D is relatively large. They also calculated the excited state, many-particle wave functions.

So far, in this discussion of the BCS theory, we have not mentioned the fact that superconductors have zero resistivity; their name derives from this striking property. Unfortunately, this behavior is difficult to extract from the microscopic theory. At equilibrium $\mathbf{k} = \mathbf{0}$ for each Cooper pair, so there is zero current density; a metal with a current flowing in it is not at equilibrium. In a sense the Meissner effect (which is nicely explained by BCS via the calculation of the penetration depth) implies superconductivity, or persistent currents. In order to expel a macroscopic magnetic field from the bulk of a metal, macroscopic currents must flow in the metal surface; since this situation is independent of time (as long as H_c is not exceeded) the currents must be constant, and hence $\rho = 0$. As discussed in Section 16-4a this is a different situation from what would be observed for a normal metal with $\rho = 0$.

To describe the current carrying states in the bulk of a superconductor it must be realized that in a normal metal or semiconductor the Brillouin zone and its boundaries in reciprocal space are tied to the lattice. However, in the BCS theory the energy gap is not tied to the lattice; rather it occurs at the Fermi surface (even if the surface were to be complicated). For a normal metal Fig. 9-9b shows the result of the application of an electric field. Electrons with k-values on the left can be scattered to *empty* states on the right. These one-electron transitions serve to keep the electrons in equilibrium. However, the situation is different in a superconductor. Suppose, by means of a transient electric field, a situation just like that in Fig. 9-9b is achieved. Since the BCS gap, 2Δ, is carried with the electron states, a member of an electron pair with momentum on the left cannot be scattered to the right side of the diagram; that would require it to be scattered into the gap! The total BCS wave function is made up of identical Cooper pair wave functions which extend over macroscopic distances. The individual pair wave functions cannot be altered without destroying the total paired state. Thus, the usual scattering mechanisms do not provide enough energy to slow down all of the electrons at once. Hence the current persists.

16-7 BCS Related Measurements

Many different experimental probes have been used to study superconductors. However, here we only discuss measurements that can be compared easily and directly with the BCS theory.

16-7a Single particle tunnelling   Tunnelling experiments give direct information about the BCS band gaps, and results from this type of experiment have already been shown in Fig. 16-9b.

When two metals are separated by an insulator, electrons cannot flow from one metal to the other. However, if the insulator is thin (of the order of, or less than, the electron mean face path) then there is a finite probability that an electron from the metal on the right will impinge on the barrier and enter the metal on the left. This is called **tunnelling** and it is an elementary quantum mechanical wave property. However, an electron can tunnel into the metal on the left only if

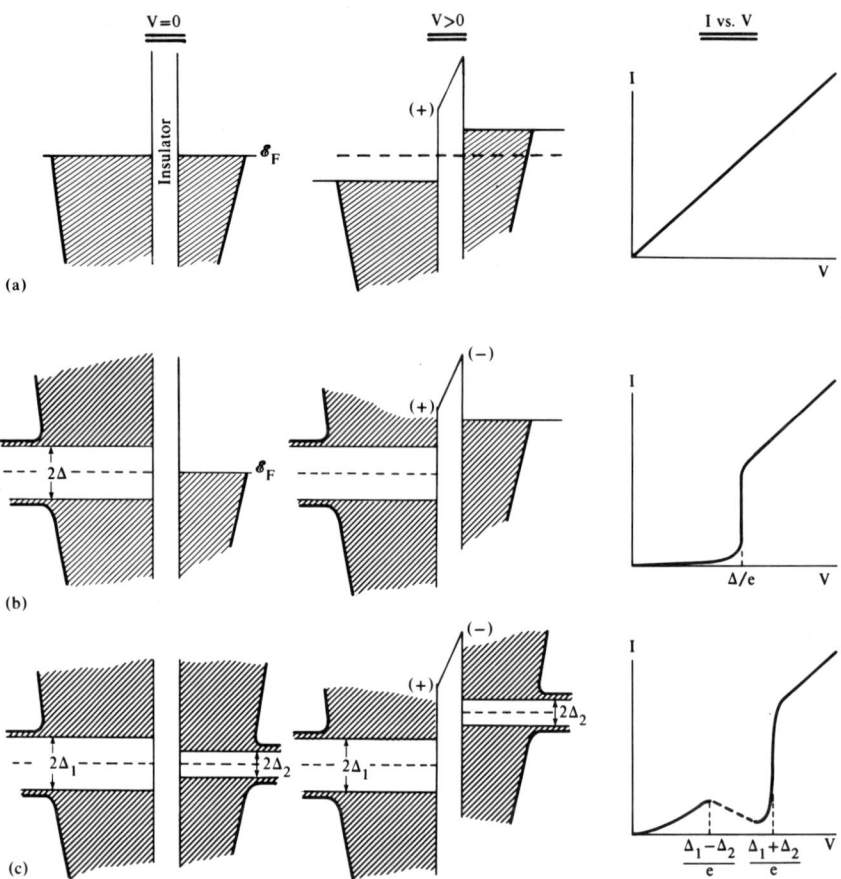

Fig. 16-12 The six density of states vs. energy diagrams on the left are for $T = 0°K$ and the three diagrams on the right are all plots of the current vs. voltage for the various types of metals on the left. (a) This shows two normal metals separated by a thin insulating film with zero applied voltage and then for $V > 0$. (b) The same situation as above but now one of the metals is a superconductor. (c) Now both of the metals are superconductors with different gaps. ($\mathscr{E}_F$ is the Fermi energy.)

empty states are available. Figure 16-12a shows two metals that are separated by a thin insulator with zero applied potential, $V = 0$. (Often the insulator is formed by oxidizing one of the two metals after depositing the metal by evaporation.) If the metals are dissimilar then electrons will flow from the one with the smaller work function until the Fermi levels are equal, as shown. (The equalization of the Fermi levels is discussed in Section 10-18.) Then, at $T = 0°K$, with $V = 0$ there is no current since electrons approaching the barrier from the right side cannot tunnel into the metal on the left. In the middle diagram in Fig. 16-12a, a voltage has been applied between the two metals. Now electrons in the metal on the right can tunnel into the empty electron states on the left and a linear I vs. V curve is expected, and found, as indicated on the far right of Fig. 16-12a.

To describe tunneling of electrons in a superconductor the so-called **semiconductor model** or **semiconductor picture** is used. In this picture Cooper pairs can be treated as (Bogoliubov) quasiparticles. There is a great deal of similarity between the quasiparticles picture in a superconductor and ordinary electrons-hole picture in a semiconductor. The minimum energy needed to create two quasiparticles from a Cooper pair is 2Δ (the gap energy) as in a semiconductor where the gap energy is required to create an electron hole pair. The quasiparticles also can recombine emitting the 2Δ energy in the form of photons or phonons, as do electron hole pairs. On the left of Fig. 16-12b the $T = 0°K$ situation for a metal separated from a superconductor is shown using the semiconductor picture. The energy scale is highly nonlinear. The width of the band is typically 5 eV in metals while the gap, $2\Delta_0 \approx 10^{-4}$ eV. There is a significant bending of the density of states from the parabolic behavior over an energy range $\approx \Delta_0$. As can be seen in Fig. 16-12b, electrons from the metal cannot tunnel into the superconductor because of the superconducting gap and electrons in the superconductor, which is filled up to just below the gap, cannot flow the other way because the states in the metal are occupied. However, if a voltage $eV > \Delta$ is applied as shown, electrons from the metal can tunnel into the empty single particle electron states above the superconducting gap; the resulting I vs. V curve is sketched. This tunnelling experiment is an excellent way to measure the superconducting gap and, as shown in Fig. 16-9b, good agreement with the BCS calculations is found. (The small amount of current for $V < \Delta/e$ arises when $T > 0°K$ because of thermal excitation of the electrons into states above $\mathscr{E}_F$.)

Figure 16-12c shows the semiconductor picture for the slightly more complicated situation when both metals are superconductors with different gaps. At $T = 0°K$ there is a sharp increase in current at $V = (\Delta_1 + \Delta_2)/e$. However, for $T > 0°K$ there is a finite probability of thermally exciting electrons across the smaller gap ($2\Delta_2$ in this case) and they can tunnel into the unoccupied states of the other superconductor at a voltage of $(\Delta_1 - \Delta_2)/e$. This situation is shown to the right in Fig. 16-12c. From these results the gap energies for both superconductors can be obtained.

Tunnelling experiments are an easy and accurate way to determine the superconducting gap. These simple experimental measurements were first used in superconductors by Giaever in 1960. For it he shared the 1973 Nobel Prize with Esaki, for his discovery of the semiconductor tunnel diode, and Josephson, who first theoretically described tunnelling of superconducting pairs as opposed to the single particle tunnelling described here.

16-7b Isotope effect As discussed in Section 16-5, an attractive interaction between two electrons can arise by an electron-phonon interaction. T_c in the original BCS calculation (which is a weak-coupling calculation and is appropriate, when $T_c/\Theta_D \ll 1$) is given by Eq. 16-10. It is natural to assume that the average phonon frequency in this equation is approximately the Debye frequency, ω_D. In this case, $\omega_D \propto M^{-1/2}$ from Eqs. 11-22b and 11-18, so $T_c \propto M^{-1/2}$ is expected, where M is the atomic mass of the atoms in the elemental metal. (Of course, the idea is that the elastic constants, c_{ij}, only depend on the force constants and are independent of the mass.) In fact, this was originally suggested by Fröhlich in 1950, and experiments reported that year using four different isotopes of Hg showed that $T_c \propto M^{-0.504}$. This result strongly indicated the important role that phonons must play in the interaction that causes superconductivity. However, more recent experiments relate T_c to M via

$$T_c \propto M^{-z} \tag{16-14}$$

where experimental values of z vary from 0.6 to 0.0; Ru and Os have experimental values of 0.0 and 0.1, respectively. Values of $z \approx 0.5$ are found for simple metals (i.e., neither transition metals, rare earths, nor actinides) while transition metals tend to have lower values. These results can be understood by considering nonfree electron-like structure of the tightly bound d-states. (See the Notes for references.)

16-7c Optical properties As mentioned, it is the dc resistance of a superconductor that goes to zero. However, when high frequency electric fields are used, the metal behaves normally. This is easy to understand. For example, at $T = 0°K$ when the frequency of the electric field is below that of the energy gap, $\omega < 2\Delta/\hbar$, there is insufficient energy to break up a Cooper pair and the resistance is zero. Thus, electromagnetic radiation in this frequency range will be transmitted through the metal when it is a superconductor. This effect is similar to passing electromagnetic radiation through a semiconductor when $\omega < \mathscr{E}_g/\hbar$. For $\omega > 2\Delta/\hbar$ the metal will behave normally, and the radiation will be reflected. Of course, for $T > 0°K$ the thermal excitation of Cooper pairs into normal, single particle states complicates the quantitative interpretation of the measurements.

Figure 16-13a shows some results of the effect. In the normal state these metals show no transmission of radiation in this frequency range. However, due to the superconducting gap, frequencies $\omega < 2\Delta/\hbar$ are transmitted, and thus the existence of the gap is con-

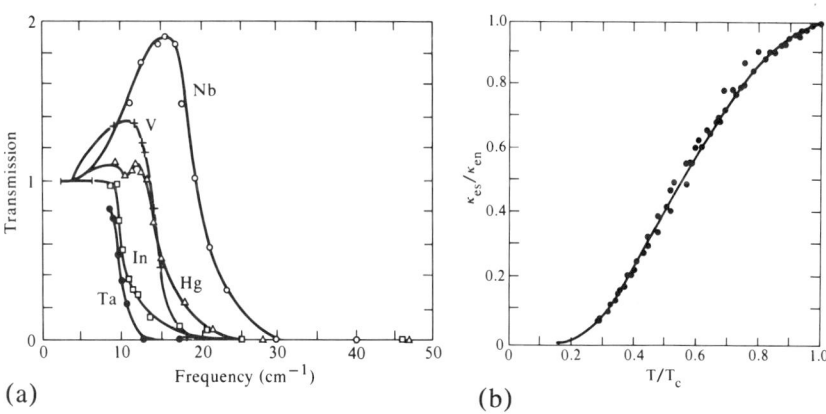

Fig. 16-13 (a) Low temperature ($\approx 1.30°K$) transmission curves for some superconductors. For display purposes the curves are all normalized to give the same value at low frequency. [P. L. Richards and M. Tinkham, Phys. Rev. **119**, 575 (1960)]. The measurements are made by comparing the results in the superconducting phase to the normal phase, the latter obtained with the application of a magnetic field larger than H_c. (b) The ratio of the electronic thermal conductivity in the superconducting to that in the normal state. The curve is calculated from the BCS theory and the dots are measured in Al. [C. B. Satterthwaite, Phys. Rev. **125**, 873 (1962).]

firmed. However, tunnelling experiments have proven to be a simpler and better method of obtaining quantitative values for 2Δ.

16-7d Thermal conductivity For insulators phonons act as the principal transporters of energy (Section 12-8). For metals, the free electrons are much more efficient for transporting heat, at least at low temperatures. Equation 9-8 shows that the free electron thermal conductivity is proportional to the electron density. However, in a superconductor, where at $0°K$ all the electrons are in Cooper pairs, this contribution to the thermal conductivity goes to zero. This is because at $T = 0°K$ all the conduction electrons are paired and in the same (condensed) quantum state, with zero energy and zero entropy. Thus, they cannot transport energy, although they certainly transport electricity. (A Cooper pair behaves like a boson with spin equal to zero so they can all be in the same state, unlike free electrons which are fermions with spin $1/2$ and the Pauli principle only allows two electrons in the same orbital state.) As the temperature is raised above $0°K$ some of the Cooper pairs are thermally excited across the superconducting gap into what are called quasiparticle states. These quasiparticle states represent the excited state of the BCS ground state wave function and the electrons in these states behave quite similarly to the electrons in normal metals and can conduct heat. In Fig. 16-

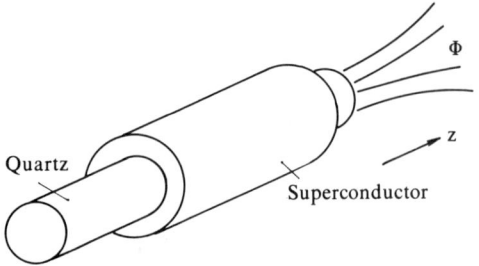

Fig. 16-14 A superconducting ring showing flux lines. The ring might be obtained by evaporating the metal onto a quartz cylinder as indicated but this quartz centerpiece is not necessary. The hole in the superconducting ring can be filled with any nonmetallic material including air.

13b the ratio is given of the electronic thermal conductivity in the superconducting state to that in the normal state. (An external magnetic field serves to make the metal normal.) As can be seen the agreement with what is expected from BCS is excellent.

The thermal conductivity of a superconductor is very low at low temperatures (Fig. 16-13b) yet it can be increased by $>10^2$ if it is driven normally. This property allows these materials to be used as heat switches. A small coil wound around the superconducting wire provides the magnetic field to drive the material normal.

16-7e Flux quantization Figure 16-14 shows a cylindrical ring of superconducting metal. In 1950 London recognized (see his book listed in the Notes) that the magnetic flux, Φ, contained in the space surrounded by the ring, should be quantized. The argument is fairly simple and will be stated in BCS language.

The wave function of a superconducting Cooper pair can be taken as a plane wave of the form $\psi \exp(-i\mathbf{k} \cdot \mathbf{r})$ or $\psi \exp(-i\mathbf{p} \cdot \mathbf{r}/\hbar)$ where $\mathbf{p}$ is the momentum, $\mathbf{p} = \hbar \mathbf{k}$. If a magnetic field is in the z-direction then on the average the pairs can go in a circular path in the superconductor around the hole in the ring. Since we want the wave function to be single valued (i.e., to return to the same value in one complete revolution) we apply the Bohr condition.

$$\oint \mathbf{p} \cdot d\boldsymbol{\ell} = nh \qquad (16\text{-}15a)$$

where the integral is taken over the entire closed path. However, since the particle (i.e., Cooper pair) is in a magnetic field, $\mathbf{p}$ is replaced by $\mathbf{p} + (q/c)\mathbf{A}$ where $\mathbf{A}$ is the vector potential of the magnetic field and q is the charge of the Cooper pair. The path of integration is taken well within the bulk of the superconductor where the magnetic field is zero (i.e., far from the surfaces and hence surface currents). In this case the supercurrent around the loop is zero, that is, $\mathbf{J} = qp N/mV = 0$. Thus, in the presence of a magnetic field Eq. 16-15a is

CHAPTER 16   SUPERCONDUCTIVITY   659

$$nh = (q/c) \oint \mathbf{A} \cdot d\ell = (q/c) \int \nabla \times \mathbf{A} \cdot d\mathbf{S}$$
$$= (q/c) \int \mathbf{B} \cdot d\mathbf{S} = (q/c) \Phi \quad (16\text{-}15b)$$

or $\quad \Phi = n \, (hc/q) \equiv n \, \Phi_0 \quad (16\text{-}15c)$

The second equality in Eq. 16-15b is obtained using Stokes theorem and in the third $B = \nabla \times A$ is used. The integral of B over the area dS bounded by the path through the superconductor is just the definition of the flux Φ as indicated. From these relations we conclude that the flux through the superconducting ring is quantized in units of Φ_0 which is called the **fluxoid** or **flux quantum**. In 1961 this effect was measured directly and $\Phi_0 = 2.0679 \times 10^{-7}$ gauss cm^2 was obtained, which corresponds to $q = -2e$ (rather than $q = -e$, as was originally assumed by London). This is a beautiful direct experimental verification of the idea of Cooper pairs. Also, the experimental results can be used to measure fundamental constants since it just involves hc/2e and can be measured very accurately.

The potent premise that has been assumed but not stated is that in order to use Eq. 16-15a it must be assumed that the coherence of quantum mechanical state extends over the macroscopic dimensions of the order of the circumference of the superconducting ring in Fig. 16-14. In the original 1961 experiments (see the Notes) this corresponds to about 10^{-2} cm, a truly macroscopic distance.

The fluxoid plays a very important role in type II superconductors. In the mixed state these materials have threads of normal material (radius $\approx$ the coherence length) surrounded by circular supercurrents extending out $\approx \lambda$ (the magnetic field penetration depth). This region out to λ encloses one fluxoid as was originally predicted by the Ginzburg-Landau theory in 1950 and has been experimentally confirmed. In a type II superconductor, as the external applied field is increased between H_{c1} and H_{c2}, each vortex still encloses Φ_0 but the density of vortexes increases (Figs. 16-5b and 16-6b).

If a (Dirac) magnetic monopole were to pass through a superconducting cylinder, in the z-direction in Fig. 16-14, then the flux through the cylinder would change by $2\Phi_0$. A single such event of this type was reported in 1982 (see the Notes) and has rekindled interest in magnetic monopoles. This is an excellent example of how different disciplines sometimes come together unexpectedly.

16-8 The Josephson Effect

16-8a Introduction In Section 16-7a single particle electron tunnelling into a superconductor was discussed. These results were first published in 1960 and show very clearly the existence of the BCS gap as well as provide accurate determinations of 2Δ as a function of temperature. Josephson's proposals in 1962 were rather surprising and were greeted initially with skepticism. He proposed that the Cooper pairs could tunnel through a junction and that this tunnelling would give rise to a dc current with zero applied voltage, and an ac current when a dc voltage is applied!

One would think that the probability of tunnelling of a pair of electrons would be very much smaller than the tunnelling of a single electron. The amplitude of the electron wave function on the other side of the barrier is $\propto \exp(-\alpha d) \ll 1$ where d is the thickness of the barrier and α depends on the potential height of the barrier. Then, the probability of single electron tunnelling is $\propto \exp(-2\alpha d)$. Thus, the probability of two electrons tunnelling through the barrier is $\propto \exp(-4\alpha d) \ll \exp(-2\alpha d)$. From this argument it would seem that experimental observation of the Josephson effects would always be masked by single particle tunnelling. However, this argument does not take into account the phase coherence of the pair, which causes the pair current to be proportional to pair amplitude, $\exp(-2\alpha d)$, rather than the square of the pair amplitude. Thus, the same probability factor arises for tunnelling of Cooper pairs as for single particles. Indeed, for sufficiently thin junctions Josephson tunnelling is as easy to observe as single particle tunnelling.

16-8b Phase coherence The travelling, free electron-like wave function is of the form $\psi = A \exp(i\mathbf{k} \cdot \mathbf{r})$ where $\mathbf{k} \cdot \mathbf{r}$ is the phase. When propagating in a normal metal, the electron is scattered by various processes within an average distance given by the electron mean free path (Section 9-4). This causes $\mathbf{k}$ to change in a random manner as the electron traverses the sample. Thus, if the phase is known at position $\mathbf{r}_1$ it will no longer be known at position $\mathbf{r}_2$ if $(\mathbf{r}_2 - \mathbf{r}_1)$ is greater than a mean free path length.

However, if the wave function describes a particle, or collection of particles, which is *not* scattered between $\mathbf{r}_1$ and $\mathbf{r}_2$ then the phase difference is given by $\mathbf{k} \cdot (\mathbf{r}_2 - \mathbf{r}_1)$ and is known for any magnitude of $(\mathbf{r}_2 - \mathbf{r}_1)$ even of truly macroscopic dimensions. The ground state BCS wave function has this property. It is a single, many-body wave function which for N conduction electrons is made up of N/2 (Cooper) pairs of electrons. The spatial range of a given pair is typically 10^3 Å. Thus, within the range of a given pair there are centered millions of other pairs. This interlocking of the pairs is an essential ingredient of the BCS ground state wave function and gives it its stability. In order to scatter this state all the pairs must be broken at the same time; this is very improbable. Hence the phase coherence over macroscopic distances is found.

The temperature dependence of the superconducting properties is determined by an intricate interaction between the fraction of pairs that are thermally dissociated and the remaining paired electrons. As T approaches T_c more pairs become dissociated and finally at T_c all of the pairs are dissociated and the ground state can be given by the usual independent electron ground state wave function.

16-8c The dc Josephson effect Consider two well-separated superconductors M and N. The phases δ_M and δ_N of the ground state wave function in these two materials will be unrelated. Now suppose that these materials are brought closer together. At very, very small separations the Cooper pairs can have a finite probability of tunnelling across the gap so the phases of the wave functions become more and more correlated, and when M and N are joined there is a definite relationship between the phases in these two materials.

However, before actual contact is made, when the metals are ~10Å apart (it could be an air gap but more usually the gap is an oxide of one of the metals), the Cooper pairs from M can tunnel into N and vice versa. (Tunnelling for a pair means that the two electrons maintain their momentum and spin relationship while crossing the gap.) The idea of a Josephson junction is shown in Fig. 16-15a. The two superconductors are separated by an oxide, the configuration being the same as implied in Fig. 16-12. In Figs. 16-15b and 16-15c various kinds of related configurations are shown which are in general called **weak links**. A weak link is a region that has a lower critical current than the superconducting material that adjoins it. As can be seen in Fig. 16-15b, a weak link can be made by just narrowing down a continuous superconducting wire.

For Josephson junctions he showed that a dc tunnelling of pairs can be expected where the current density is given by

$$J = J_c \sin(\delta_M - \delta_N) \equiv J_c \sin \delta \qquad (16\text{-}16)$$

where J_c is the maximum current density that can pass through the junction before driving it normal. (J_c is also proportional to the probability of pair tunnelling across the junction.)

If the tunnel junction is current-biased by a dc current J, then the phase difference between the two wave functions is given by Eq. 16-16, so $\delta = \sin^{-1}(J/J_c)$ provided $J < J_c$. On the other hand, if $J > J_c$ then some of the current must flow via the transport of normal electrons; thus, there must be a voltage drop across the junction.

Figure 16-15d shows the I vs. V characteristic for a tunnel junction; the dc Josephson effect is shown as well as single particle tunnelling (as in Fig. 16-12b).

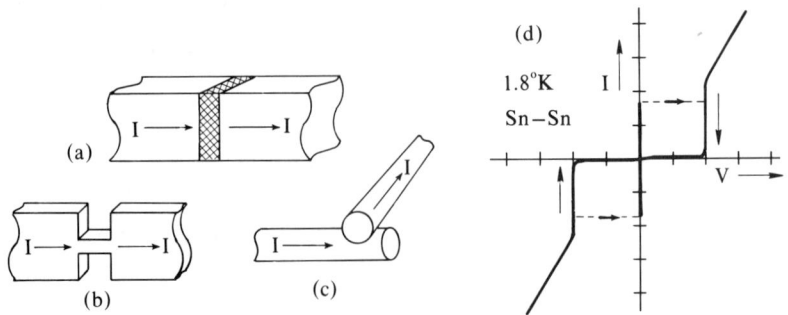

Fig. 16-15 (a) A Josephson junction. (b) and (c) Various kinds of weak links. (d) The I vs. V for a superconducting tunnel junction.

16-8d The ac Josephson effect The time evolution of a quantum state is given by $\exp(i\mathscr{E}t/\hbar)$. The pair tunnelling current arises from a transition from a quantum state on one side of the weak link to a quantum state on the other side but the energy difference is $\mathscr{E} = qV_a = -2eV_a$ for the Cooper pair if a dc voltage V_a is applied. This corresponds to a time evolution of the phase given by

$$\hbar(\partial\delta/\partial t) = -2eV_a \tag{16-17a}$$

For a dc voltage this can be integrated and yields

$$\delta(t) = \delta(0) - (2eV_a/\hbar)t \tag{16-17b}$$

Then the relative phases of the quantum states on both sides of the weak link can be inserted into Eq. 16-16, and defining ω_a, we have

$$J = J_c \sin\left[\delta(0) - \frac{2eV_a}{\hbar}t\right] = J_c \sin[\delta(0) - \omega_a t] \tag{16-17c}$$

Thus, due to an applied *dc voltage* an *ac current* is generated at a frequency given by $2eV_a/\hbar$. This is a rather surprising effect. For a dc voltage of $10\mu V$ an ac frequency of 483.6 MHz is expected.

This effect was observed in 1963. When a Josephson junction was irradiated with microwave radiation, an increase in the dc Josephson current could be observed (due to an absorption of the microwave energy). Other ac Josephson effects include microwave induced steps in the dc I-V characteristics and microwave emission from a Josephson junction when an appropriate voltage is applied. Thus, the quantum states on both sides of the junction behave like a hydrogen atom in the sense that when the energy levels are different (this is done by applying a voltage), absorption and emission of radiation between these states can be observed.

CHAPTER 16   SUPERCONDUCTIVITY   663

Further, if a Josephson junction is dc current-biased with $J > J_c$, as discussed in the previous section, a voltage develops across the junction. This voltage causes the phase δ to vary in time. From Eq. 16-16, this implies that both the fraction of the dc current that flows as pairs and as normal electrons oscillates in time. Thus, a dc current bias in excess of the critical current generates an ac voltage across the junction.

The ac Josephson effect has been used for precise measurements of $e/\hbar$ or as a voltage standard. Typically this is done by irradiating the junction with microwave radiation from a frequency standard (e.g., a Cs atomic clock). However, the effect can be reversed, where an accurate dc voltage is applied and the output radiation can be compared to a frequency standard. Solymar's book, in the Notes, has a simple discussion of this topic.

16-8e Quantum interference   The phase of the ground state wave function, up to an additive constant, is determined throughout any superconducting circuit. Thus, with the appropriate experimental arrangement interference and diffraction effects similar to those observed with light can be demonstrated.

Figure 16-16a is a schematic diagram of a superconducting circuit with two weak links, in which interference effects can be observed similar to that observed in Young's double slit experiment. A current J_{tot} flows in and out at points I and O, respectively. J_{tot} is split into two, passing through the weak links at points A and B as shown. We assume that everywhere the current density is very much less than the critical current; this assures that the only significant phase differences occur at the weak links. The phase change around the closed superconducting circular loop that includes the weak links must be some integral number of 2π, so $\delta_A - \delta_B - 2\pi(\Phi/\Phi_0) = 2\pi n$. If we neglect the self-inductance of the loop then Φ is the externally applied flux. Taking $n = 0$, the phase difference is related to the flux as

$$\delta_A - \delta_B = (2e/\hbar c)\, \Phi \qquad (16\text{-}18a)$$

From the Josephson result, Eq. 16-16, assuming the same J_c for both weak links, and taking the total current as the sum of the currents through A and B

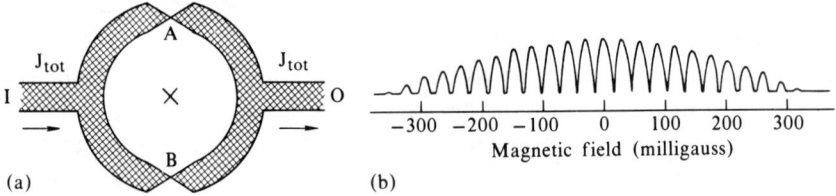

Fig. 16-16 (a) A superconducting loop with two weak links in which interference can be observed. (b) Experimentally observed interference.

$$J_{tot} = J_c \left[\sin \delta_A + \sin \delta_B \right]$$

$$= 2J_c \sin \delta_0 [\cos (\delta_A - \delta_B)/2] \qquad (16\text{-}18b)$$

$$= 2J_c \sin \delta_0 \left[\cos \left(\frac{e\Phi}{\hbar c} \right) \right] = 2J_c \sin \delta_0 \left[\cos \left(\frac{\pi \Phi}{\Phi_0} \right) \right]$$

where $\delta_0 \equiv (\delta_A + \delta_B)/2$ and we used the definition of the flux quantum, Eq. 16-15c. Thus, the total current depends periodically on the flux and typical experimental results are shown in Fig. 16-16b.

With this technique very small magnetic field changes can be measured. Such devices are called SQUIDs (superconducting quantum interference devices). Commercially available SQUIDs can detect a change of $\approx 10^{-5}$ Φ_0 in a 1Hz bandwidth. These devices can be incorporated into voltmeter and ammeter circuits. Besides the dc SQUIDs, of the type shown in Fig. 16-16a, rf SQUIDs also are used; they have only one weak link. SQUIDs are being used to look for magnetic monopoles.

Summary

Many of the major breakthroughs in superconductivity can be summarized by referencing some of the more important papers.

The field was started in 1911 by Kamerlingh Onnes, who discovered that the conductivity of Hg dropped to zero below a critical temperature. In 1933 Meissner and Ochsenfeld discovered that magnetic flux is excluded from the interior of a superconductor. This is the so-called Meissner effect. The flux penetrates only a distance λ. In 1935 F. and H. London developed a theory for the Meissner effect. They showed that the external magnetic field is exponentially screened from the interior of a superconductor within a length λ_L (the magnetic field penetration depth). Pippard, in 1953, reexamined the London results by considering the coherence length, ξ_0, of the amplitude of the superconducting wave function. The London equation is a local equation in that it relates the current density at a point **r** to fields at that point. Pippard considered the nonlocal problem and found that for typical elemental superconductors $\xi_0 \gg \lambda_L(0°K)$. This leads to a prediction that the magnetic field penetrates larger distances than given by λ_L, in agreement with experiment. In the same time period Ginzburg and Landau (GL) proposed phenomenological equations that introduced a complex pseudo-wave function ψ, of the ground state superconducting electrons, as an order parameter. The local density of superconducting electrons is given by $n_s = |\psi(x)|^2$. With these equations they predicted a temperature dependent coherence length $\xi(T)$. For T far below T_c, $\xi(T) = \xi_0$, but near T_c they are different. The **GL parameter** $\kappa \equiv \lambda/\xi$, a dimensionless ratio, is approximately independent of temperature. For typical, pure elemental superconductors $\lambda \approx 500$Å and $\xi \approx 3000$Å so $\kappa \ll 1$.

F. London appreciated the macroscopic extent of the ground state wave function. In 1950 he predicted that the magnetic flux through a superconducting ring would be quantized, with $\Phi = (hc/q)n = n\Phi_0$. In 1961 this was found experimentally to be the case with q = 2e, corresponding to the two electrons in a Cooper pair.

In 1957 Abrikosov, using the GL equations, considered what would happen in the limit where the GL parameter $\kappa > 1$ (i.e., $\xi < \lambda$). In this region, he predicted the very strange (at that time) behavior that he called type II superconductors. He calculated that in the mixed state (between H_{c1} and H_{c2}) the magnetic flux should penetrate in a regular array of flux tubes, each carrying one quantum of flux, Φ_0.

In the meantime a microscopic theory was being pursued by Fröhlich, who in 1950 showed how the electron-phonon interaction could act as a weak attractive force between electrons. He predicted the isotope effect, which was independently measured in the same year. Then in 1956 Cooper showed that for an arbitrarily small attractive interaction, two electrons, with equal and opposite momentum and spin, would form a bound state with a lower energy than if they were not interacting. These electrons are called Cooper pairs. The next year Bardeen, Cooper, and Schrieffer (BCS) published the famous BCS paper that describes calculations of the ground state wave function and equilibrium properties of superconductors.

In 1960 Giaever showed that single particle electron tunnelling could be used to determine the width of the superconducting gap and its temperature dependence. Then in 1962 Josephson theorized that "possible new effects in superconducting tunnelling" could be seen due to the tunnelling of Cooper pairs. This type of tunnelling occurs because of the extremely long-range phase coherence of the superconducting wave function. Josephson predicted at zero voltage, where there is zero single particle tunnelling, that a dc (super)current up to $|J_c|$ could be observed. He also predicted that at finite voltage, besides the usual single particle tunnelling there will also be an ac (super)current of amplitude J_c and frequency $2eV/h$. Both the dc and ac Josephson effects were observed in 1963.

H. Kamerlingh Onnes, Leiden Comm. 120b, 122b, 124c (1911).
W. Meissner and R. Ochsenfeld, Naturwissenschaften **21**, 787 (1933).
F. and H. London, Proc. Roy. Soc. (London) **A149**, 71 (1935).
A. B. Pippard, Proc. Roy. Soc. (London) **A216**, 547 (1953) and T. E. Faber and A. B. Pippard, ibid **A231**, 336 (1955).
V. L. Ginzburg and L. D. Landau, Zh. Eksperim. i. Teor. Fiz. **20**, 1064 (1950).
F. London, "Superfluids" (Wiley, 1950), Vol. 1.
A. A. Abrikosov, Zh. Eksperim. i Teor. Fiz. **32**, 1442 (1957) [Soviet Phys. JETP **5**, 1174 (1957)].
H. Fröhlich, Phys. Rev. **79**, 845 (1950).
L. N. Cooper, Phys. Rev. **104**, 1189 (1956).
J. Bardeen, L. N. Cooper, and J. R. Schrieffer, Phys. Rev. **108**, 1175 (1957).
I. Giaever, Phys. Rev. Letters **5**, 147 and 464 (1960).
B. D. Josephson, Phys. Letters **1**, 251 (1962).

Notes

References, from an historical point of view, can be found in C. J. Gorter, Rev. Mod. Phys. **36**, 1 (1964). There have been many books and conference proceedings published on this active and practical field. The solid state texts in the bibliography discuss superconductivity with different emphasis. We mention some other references.

F. London, Superfluids, Vol. I (Wiley, 1954 and Dover, 1954) is a good text from an historical point of view.

D. Shoenberg, "Superconductivity" (Cambridge University Press, 1952) is a good pre-BCS survey emphasizing experimental results.

C. G. Kuper, "Introduction to the Theory of Superconductivity" (Clarendon Press, 1968). This and Tinkham's book (referenced shortly) deal with the real quantum mechanics of the problem and the BCS solution.

A. C. Rose-Innes and E. H. Rhoderick, "Introduction to Superconductivity" (Pergammon Press, 1968). This very readable treatment is slightly more difficult than the present text.

R. Mersevey and B. B. Schwartz, "Superconductivity," edited by R. D. Parks (Dekker, 1969) Vol. 1. This article is a good source for experimental results. For exampe, the data quoted in the tables within the text in Section 16-6 comes from this source.

L. Solymar, "Superconducting Tunneling and Applications" (Wiley-Interscience, 1972).

M. Tinkham, "Introduction to Superconductivity," (McGraw-Hill, 1975). This is a very useful book.

S. V. Vonsovosky, Y. A. Izyunov, and E. Z. Kurnaev, "Superconductivity of Transition metals," Springer series in Solid-State Sciences **27** (Springer-Verlag, 1982).

For the up-to-date research work, there are the proceedings from the ongoing "International Conferences on Low Temperature Physics." Seventeen conferences have now taken place and they are referred to as LT-17, and so on. LT-15 is published in Journal de Physique **39**, Supp. S, Colloq. 6 (1978). LT-16 and LT-17 are published in Physica (B+C) **107-110** (1982) and **126** (1984).

The paper by J. File and R. G. Mills, Phys. Rev. Letters, **10**, 93 (1963) discusses their nuclear magnetic resonance measurements of the decay time of the supercurrent.

For a discussion of deviations of BCS from the temperature dependence of $H_c(T)$ and a useful introduction to the extension of the BCS theory to **strong-coupling superconductors,** see the reference in Fig. 16-10, the article by D. J. Scalapino in the book edited by R. D. Parks referenced previously, and the book quoted earlier in the Notes. For the isotope effect see J. W. Garland, Jr., Phys. Rev. Letters **11**, 114 (1963).

The **fluxoid** was first measured by Deaver and Fairbank. Their paper [Phys. Rev. Letters 7, 51 (1961)] and the next three papers in that journal make interesting reading. The use of a superconducting loop to detect a **magnetic monopole** was reported by B. Cabrera, Phys. Rev. Letters, 48, 1378 (1982). See Physics Today, June, 1982, p. 17, and "Magnetic Monopoles," Ed. R. A. Carrigan, Jr. and W. P. Trower (Plenum Press, 1983).

The Ginzburg-Landau theory is discussed in many of the books listed here.

The difficulties of observing superconductivity in metallic elements in the $\mu°K$ range, such as Au, probably stems from magnetic impurities in the sub-ppm range. See C. Buchal, R. M. Mueller, F. Pobell, M. Kubota, and H. R. Folle, Solid State Commun. **42**, 43 (1982).

Several **organic metals,** with formulas too complicated to write here, are superconductors. This seems to be a growing field. For a review of it see D. Jérome and H. Schulz, Adv. in Phys. **31**, 299 (1982). Also R. L. Green and P. M. Chaikin, in LT-17.

A fine overview and review of "Long range order in solids" has been written by R. M. White and T. H. Geballe and appears as Supplement 15 of Solid State Physics (1979). The book has extensive coverage of the fields of superconductivity and magnetism but covers other areas as well. The last chapter of this book discusses amorphous superconductors and amorphous magnetic materials. A review of amorphous superconductors can also be found in C. C. Tsuei, in "Superconductor Materials Science" Ed. by S. Foner and B. B. Schwartz (Plenum Publishing Corp., 1981). For a very recent overview of the field see M. R. Beasley and T. H. Geballe, Physics Today, October 1984, p. 60.

In the Notes in Chapter 10, we briefly mentioned **heavy fermion metals** and gave references there. These are metals with a shape peak in the density of states at the Fermi energy, which results in enormous effective masses at $\mathscr{E}_F$ ($\approx 10^2$ to 10^3 m), as can be determined with electronic specific heat measurements, for example. The last three entries in Table 16-1 are heavy fermion metals that are superconducting; their critical fields (Eq. 16-1) typically are in the 10^6O region as opposed to 10^3O for pure metals (Fig. 16-3). Compounds such as $NpBe_{13}$, U_2Zn_{17}, UCd_{11}, and others are heavy fermions metals that order magnetically at low temperatures, while $CeAl_3$, $CeCu_6$ UAl_2, and others are heavy fermion metals that do not show superconductivity or magnetic order at low temperatures. It is fair to say that a proper theoretical understanding of these materials has not, as yet been achieved; there are two key questions to be resolved. First, does the narrow band have its origin in a one-electron effect, the hybridization of the f-band with the d-band? Or is this a many-body effect similar to a Kondo resonance observed for dilute magnetic impurities in metals. Second, is the superconductivity like that of all previously known superconductors?

We have not emphasized this point, but the superconductors discussed in this chapter are sometimes said to have **s-type pairing**. This is because superconductivity comes from the condensation of Cooper pairs of opposite k-values and opposite spin (i.e., a spin of zero). There is some evidence to indicate that the superconductivity in the heavy fermion systems is due to **p-type pairing** of the electrons (i.e., the pair has a total momentum of one). If this is true then the mechanism that causes the pairing may be different than that discussed in this chapter.

Problems

1. A superconducting solenoid is 10 cm long, 2 cm in average diameter and has 10^4 turns of wire. The solenoid is in a liquid helium bath.

CHAPTER 16   SUPERCONDUCTIVITY   669

The current through the wire is slowly increased to 30 A at which time the metal goes normal. Taking the latent heat of vaporization of liquid helium at its boiling point as 2.6 J/cm^3, how much liquid He is evaporated? (Note the large difference with N_2 where the value is 160 J/cm^3.)

2. By calculating the electronic contribution to the specific heat of an intrinsic semiconductor show that exponential results in Eq. 16-13b imply a gap.

3. The BCS gap of lead at 4°K is about 2 meV. If you bias a Pb Josephson junction with a voltage equal to one-half of the gap, what is the frequency of the ac Josephson effect? To two significant figures what is the gap at 4.2°K?

4. A loop of superconducting wire has a diameter of 1 mm and self-inductance $L = 2 \times 10^{-9}$ henrys. The current, I, in the loop can be measured to high accuracy with a nearby SQUID, but ignore the mutual inductance of the SQUID and loop. The earth's magnetic field near the loop is 1 Oe but due to shielding this is reduced by 10^4 at the loop. The flux through the loop is quantized so

$$n\Phi_0 = \Phi_{app} + LI$$

where the applied flux is due to the earth's magnetic field. What is the smallest value of I that will circulate around the loop and in which direction? If, while the loop is superconducting, an external field is applied that reduces Φ_{app} to zero, what is the value of I?

5. Assume a large superconducting solenoid magnetic is below T_c. If a small hearing aid battery is connected across the solenoid what happens? (Assume no resistance of the leads between the battery and the solenoid.) What if a large car battery were used instead? How is the current started in a superconducting magnetic?

6. In no more than four pages discuss the experimental attempts to observe the Dirac magnetic monopole and its significance. Using the "Citation Index" would be helpful.

17

Surface Science

17-1 Introduction — The Need for UHV
17-2 Crystal Shape
17-3 Preparation of Clean Surfaces and LEED
17-4 The Structure of Surfaces
    a Simple surface relaxation
    b Surface structure notation
    c Some examples of reconstruction
17-5 Interaction of Gases with Surfaces
    a Physisorption
    b Chemisorption
    c Co-adsorption
17-6 Surface Related Techniques
    a Synchrotron radiation
    b Ultraviolet reflectivity
    c Low energy electron diffraction (LEED)
    d Photoelectron (or photoemission) spectroscopy
    e Auger process
    f Appearance potential spectroscopy (APS)
    g EXAFS
    h Electron energy loss spectroscopy (EELS)
    i Inverse photoemission (Bremsstrahlung spectroscopy)
17-7 Electronic Surface Structure
    a Surface charge density
    b Work function
    c Charge density effects from chemisorption
Notes
Problems
Appendix to Chapter 17

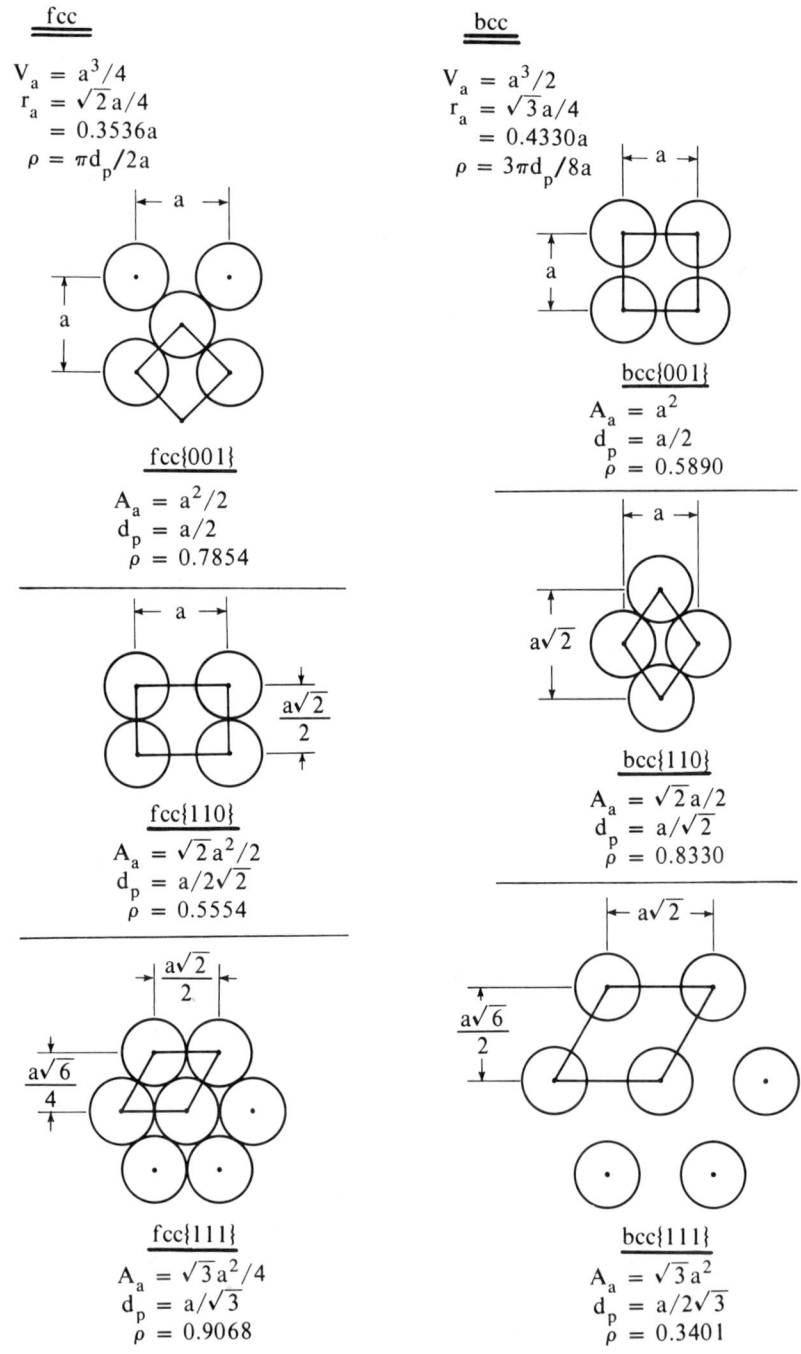

Fig. 17-4 The geometry, to scale, of the common faces of the fcc and bcc structures. The maximum packing of hard spheres is assumed. a = the side of a cubic unit cell, r_a = the atomic radius, d_p = spacing between lattice planes, ρ = surface packing density of the atoms in a lattice plane, A_a = area per atom in a lattice plane. (Courtesy of P. M. Marcus.)

SURFACE SCIENCE

It was like the imminent arrival of Gargantua; preparations had to be made to widen the gutters of Denver and foreshorten certain laws to fit his suffering bulk and bursting ecstasies.

J. Kerouac, "On the Road"

The surfaces of solids have interested scientists for many years. Recent technological advances have made it possible to achieve and maintain atomically clean surfaces, helping surface science to emerge as a major area of solid state physics and chemistry.

Clean surfaces have been achieved by cleaving or sputtering, and then maintained in the clean state with the aid of ultrahigh vacuum (UHV) techniques; this is generally defined as pressures lower than 10^{-10} torr. UHV conditions are required for the preparation and execution of experiments on clean surfaces, as we shall see.

It is not obvious that surfaces of crystals that are as perfect atomically as the bulk can be prepared. Given the extra degree of freedom and the very large gradient of the charge density at a surface, we might expect highly strained or amorphous surface layers. However, perfect crystal-like surfaces can be prepared and studied.

By "surface" we usually mean a region, perhaps 5 Å to 15 Å on both sides of the outermost plane of atoms. By 15 Å into the bulk of metals most electronic properties are totally independent of the presence of the surface. However, if the surface of semiconductors and insulators is not neutral, then effects can be felt at much larger distances. Some of these effects are discussed in Chapters 10 and 18.

The interest in surfaces is rather basic, but the laws of the interactions on surfaces are the same as those appropriate to the bulk; basically they are, Coulomb's law, the wave equation, and the Pauli principle. However, the boundary condition on the equations and the environment of the atoms on the surfaces is very different from that in the bulk. We observe many new phenomena: surface structure relaxation and reconstruction, steps and other defects unique to surfaces, adsorption of impurity atoms sometimes in unusual bonding sites, the formation of bonds between surface atoms that are different from those found in the bulk, two-dimensional phase transitions, and others.

Our curiosity about surfaces is motivated by a desire for a basic understanding as well as a technological interest in catalysis, crystal growth, oxidation, and corrosion, and semiconductor device interfaces. In a sense most of the previously covered topics could be rediscussed, considering surfaces specifically. Only a few general areas are addressed here and even in these areas, we will only scratch the surface.

17-1 Introduction—The Need for UHV

It is easy to have a dirty surface; let us see why by a simple calculation. Let dN_s/dt be the number of particles of a gas striking a surface of 1 cm² per sec. From kinetic theory of gases

$$dN_s/dt = (N/V)\langle v_z \rangle \qquad (17\text{-}1a)$$

where the (N/V) is the number of molecules/cm³ in the gas and $\langle v_z \rangle$ is the mean thermal velocity of the gas molecules in the z-direction (perpendicular to the surface) averaged for all v_z from 0 to ∞. $\langle v_z \rangle$ can be evaluated via Eq. 11-6b using a Maxwell-Boltzmann distribution for the velocities of the gas molecules. The result is

$$\frac{dN_s}{dt} = \left(\frac{N}{V}\right)\left[\frac{k_B T}{2\pi M}\right]^{1/2} \approx 3.5 \times 10^{22} \frac{P}{(MT)^{1/2}} (\text{cm}^{-2} \text{ sec}^{-1}) \qquad (17\text{-}1b)$$

where P is measured in torr. The first equation is obtained via Eq. 11-6b, where M is the molecular weight and T the absolute temperature. The second equation is obtained by using the ideal gas law $PV = Nk_B T$. For typical numbers, take $T = 300°K$, $M = 28$ (nitrogen gas), and assume that monolayer coverage is 3×10^{14} particles/cm². Then from Eq. 17-1b

$$dN_s/dt \approx 10^6 \, P \text{ (monolayers/sec)} \qquad (17\text{-}1c)$$

Thus, for standard high vacuum conditions of 10^{-6} torr a monolayer would build up in one second if each impinging molecule stuck to the surface. The **sticking coefficient**, $S \equiv$ the probability that an impinging molecule becomes adsorbed, is close to one for many surfaces and gases, so pressures very much less than 10^{-6} torr are required. Allowing for a sticking coefficient then the **time of coverage** τ_c to obtain a monolayer is

$$\tau_c = \frac{1}{S(dN_s/dt)} \approx \frac{10^{-6}}{SP} \qquad (17\text{-}1d)$$

In order to have one hour in which to do an experiment, with $S = 1$, the pressure must be $P \approx 10^{-10}$ torr. This pressure range is called **ultrahigh vacuum** or UHV.

We shall not go into the details of UHV systems except to say that such systems can have fairly large experimental chambers of glass and metal. The procedure usually involves baking the entire apparatus at $\approx 150°C$ to remove gases adsorbed on the inner walls of the chamber. Then elaborate procedures in the UHV chamber are required to prepare the crystal surfaces and check that surfaces are, indeed, clean.

Checking for impurities is often done by Auger spectroscopy and the quality of the crystal structure is determined by low energy electron diffraction. These techniques will be discussed. The Notes should be consulted for details of UHV techniques.

17-2 Crystal Shape

One way to appreciate the importance of the surface of a crystal and the broken bonds on the surfaces is to see how these bonds can determine the external morphology. It is, at first, surprising to realize that the shape of a crystal can be determined by the broken bonds on different surface planes. We present a very simple discussion, due to Harrison, that applies particularly to tetrahedral covalent materials that have distinct directed bonds. The model is oversimplified because an ideal surface is assumed and, as will be seen, even clean surfaces are not ideal (they often reconstruct).

In order to check the statements that will be discussed about the bond densities per unit cell it would be helpful to have a three-dimensional "stick and ball" model of a diamond crystal. Figure 17-1a shows a $\{111\}$ surface of Si which has the diamond crystal structure. Look at Fig. 3-7 to be sure of the orientation for various directions and the size of the unit cell. Unit cell refers to the bulk conventional cubic unit cell of length a (containing eight atoms) with the distance between the atoms of $d = \sqrt{3}a/4$. For a $\{111\}$ surface there is one broken bond, for each of the atoms on the surface, Fig. 17-1a. Actually, if the surface of the crystal were just below the top layer shown, there would be three times as many broken bonds, increasing the surface energy by a factor three. Thus, the bonds are broken as shown, and the density of broken bonds is $\sqrt{3}/(4d^2)$ on this surface. Focusing on the $\{100\}$ face we see that each atom would have two broken bonds, so the density is $4/a^2 = 3/(4d^2)$. On a $\{110\}$ face each atom has one broken bond, resulting in a density $3\sqrt{2}/(8d^2)$. The lowest density of broken bonds is on the $\{111\}$ face. Now if a crystal with the diamond structure were cut along these planes the energy required should be proportional to the number of bonds cut. Similarly for crystal growth we expect the $\{111\}$ faces to be the natural growth faces since they have the lowest surface energy per unit area (i.e., they have fewer uncompensated bonds for a given area). There are eight $\{111\}$ planes and a crystal bounded by these can form a regular octahedron. Indeed, crystals with this morphology are found often in nature.

However, given a constant volume of crystal, the total surface area can be reduced by truncating the corners with $\{100\}$ planes as shown in Fig. 17-2. In fact, if the surface energies of both of these

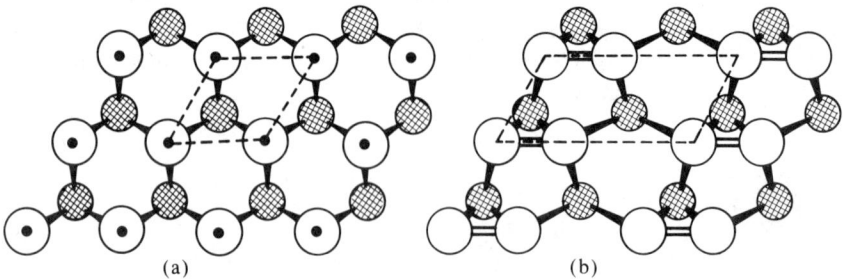

Fig. 17-1 (a) A Si{111} surface that is unreconstructed. The solid black dots represent unpaired, that is, dangling, bonds perpendicular to the surface and the shaded circles are second-layer atoms. (b) The same surface but showing a possible reconstruction that corresponds to a 2 × 1 structure and eliminates the dangling bonds. A structure for Si{111} 2 × 1 that is quite different from that shown here has been proposed [K. Pandey, Phys. Rev. Letters, **47**, 1913 (1982)].

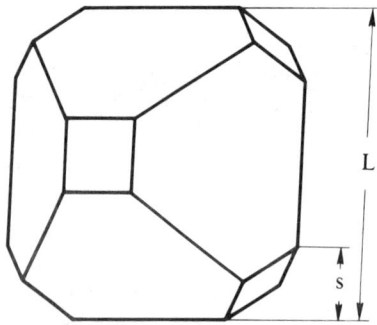

Fig. 17-2 A truncated octahedron where s and L are shown.

surfaces were equal then for a constant volume of crystal s > 0 would reduce the total surface energy. However, the surface energy per unit area is not the same for these two surfaces. At constant volume we can obtain the total surface energy for a crystal with these two kinds of surfaces and then minimize this energy with respect to s/L. Then we obtain s/L = (2/3) $[1 - (\mathscr{E}_{\{100\}}/\sqrt{3}\mathscr{E}_{\{111\}})]$ where the different values of $\mathscr{E}$ refer to the surface energies of the two different planes. Using the values of the energies found previously, from the simple idealized model counting only the unpaired bonds per unit area, this equation gives s/L = 0. However, small corrections to these energies change the result and it is common to find truncated octahedral crystals in nature. Note, from models like this one, the shape of crystals is independent of their size.

Thus, we see that the shape of a crystal can be controlled by a microscopic effect such as the unpaired bonds of surface atoms. However, crystal growth often takes place under non-equilibrium conditions and is dominated by temperature gradients and growth speeds. Also, since many crystal surfaces reconstruct, a calculation of the energies of the various surfaces is more complex than indicated here where only unpaired bonds are counted.

17-3 Preparation of Clean Surfaces and LEED

Preparation — There is no general method for obtaining clean surfaces. We mention some of the most useful procedures.

Usually, the object is to study the properties of certain faces on single crystals. We start with an appropriately oriented crystal in a UHV system. The two most popular ways to obtain clean surfaces are (1) cleaving, which clearly is restricted to certain orientations of a selected group of crystals that cleave easily and (2) noble gas ion sputtering and annealing after orientating, cutting, and polishing the crystal outside the UHV chamber. The latter method is more universally applicable. To proceed, gas ions are produced by electron impact and accelerated toward the sample by $\sim 10^2$ to 10^3 V. The ion current and duration are determined by the type of material and the thickness of the layer to be removed. With this approach the surface impurities and first few layers of the underlying material can be removed. After the bombardment careful annealing of the sample results in recrystallization of the surface. The bombardment and recrystallization may be repeated or mixed with other preparation approaches.

Some of the other preparation approaches are listed here. (3) High temperature annealing — this can be used to evaporate impurities from the surface. (4) Chemical reactions — metal surfaces in particular can be cleaned this way by using hydrogen and/or oxygen. (5) Electric field desorption — this method is usually restricted to sharply pointed samples since the field is largest at such places. (6) Evaporation of thin films — this method provides the growth of an entirely new surface in UHV. (7) Epitaxial growth of one type of compound on top of another one (e.g., Si on Al_2O_3). In recent years this is being done by molecular beams and is called **molecular beam epitaxy (MBE)** and is discussed in Section 18-4. Using this technique we can produce stable and metastable compounds with unusual properties.

Before proceeding, we note that the word "clean" is always connected to the limit of detection. For surface crystallographic studies 1% of a monolayer or less often cannot be detected; so "clean" is fairly dirty by bulk standards. On the other hand, such a level of impurities might be easily detectable if the surface electrical properties of semiconductors are being studied. Usually, to measure impurity content and the crystalline quality of a surface, a combination of Auger electron spectroscopy (or photoelectron spectroscopy) and LEED are used. Both of these approaches are discussed here.

LEED – We now discuss how the geometric arrangements of atoms (the structure) on clean surfaces are studied. There are several techniques but the most widely used for quantitative purposes is **low energy electron diffraction** (LEED). As discussed in Chapter 4, low energy electrons have a wavelength comparable with the internuclear spacings of atoms in crystals. From the de Broglie relation

$$\lambda = h/mv = h/(2m\mathscr{E})^{1/2} \qquad (17\text{-}2a)$$

where m, v, and $\mathscr{E}$ are the mass, velocity, and energy of the electron. From the equation we calculate, for electrons,

$$\lambda(\text{Å}) = [150/\mathscr{E}(\text{eV})]^{1/2} \qquad (17\text{-}2b)$$

meaning that an electron beam with energy ≈ 150 eV will have a wavelength of ≈ 1 Å, which is the correct size to diffract from crystals.

Besides having a suitable wavelength, electrons with this energy penetrate only the first few layers of atoms. This is because of the very strong electron-electron interaction. For example, 100 eV electrons typically penetrate ≈ 5 Å. Thus, these low energy electrons are well suited for surface structural work, and LEED has become an important tool for these studies. (X-rays with this wavelength interact weakly with solids and penetrate much farther, 10^{-3} to 10^{-2} cm.)

The experimental arrangement usually consists of an electron beam produced by the acceleration of electrons from a hot filament to an anode with voltage, V, so $\mathscr{E} = eV$. The anode has a hole in it, allowing electrons to escape. This beam is electrostatically or magnetically focused through a series of grids toward a carefully prepared crystal surface. The backscattered electron beams are concentrated along specific directions because of wave interference (just as in the x-ray case) and appear as spots on a fluorescent screen. Alternately, the backscattered electrons can be detected by a Faraday cup when exact intensity measurements from the various diffracting spots are required for crystallographic structure analysis. (Only a few percent of the incident electrons are elastically scattered. Most of the scattered electrons come out at much lower energy. Thus, it is important to filter out the inelastically scattered electrons. This is done with a series of grids in front of the fluorescent screen.)

The original 1927 work of Davisson and Germer that showed that particles also have wave-like behavior was a LEED experiment. They diffracted electrons from a nickel crystal and analyzed the directions of the diffraction maxima using de Broglie's equation. Besides showing the wave-like behavior of the electrons, they also measured the spacing between atoms on the surface. Later LEED studies showed that ordered surface arrangements of atoms are present and observable in many materials. A related technique, using electrons of about 30 keV but at glancing incidence to the surface, is called **reflection high energy** electron diffraction or RHEED. The mean free path is much greater for electrons of this energy, but, because of the glancing angle, the penetration into the crystal is about the same as in the LEED experiments. (For a 30 keV electron at a glancing angle of 3° the momentum component normal to the surface is about the same as for a 100 eV electron at normal incidence.) RHEED is very sensitive to surface topography and can be used in studies of surface nucleation, faceting of surfaces, and epitaxial growth.

When LEED techniques are used, the periodicities of the surface with respect to the bulk are usually immediately apparent. However, there is more to a structure than simply the size and shape of the unit cell. For classical, bulk x-ray diffraction the intensities of the diffracted beams must be known accurately in order to obtain a complete crystal structure. The situation for LEED is more difficult. The theory for the interaction of low energy electrons with a crystal is reasonably clear. However, the interaction is so strong that multiple scattering occurs. Hence, Fourier transform of the intensities to determine a structure cannot be done (as usually is done for x-rays). Rather, we guess at a trial structure, do a multiple scattering calculation, and see if it is compatible with the LEED intensities. Even if the agreement is good, uniqueness is not assured.

17-4 Structure of Surfaces

17-4a Simple surface relaxation Figure 17-3a shows a cut through a crystal perpendicular to the crystalline surface. The atoms in the bulk have other atoms on all sides that help keep them in place. However, the atoms on the surface have neighbors only to the side and below; above there are no atoms to provide repulsive forces and hence the surface atoms might be expected to displace uniformly away from the bulk as indicated. Table 17-1 shows the results of an early calculation for these displacements in fcc copper using a Morse potential. As can be seen, the displacements are smallest on the highest density plane and they decrease rapidly for the lower layers.

These were interesting results in 1968 but subsequent experiments have shown that most (although not all) metal surfaces contract rather than expand. Figure 17-3b shows that the atoms on the {001} surface of Mo are contracted toward the bulk by 11%. Table 17-2 shows the results for a number of metals and different surfaces. As can be seen, only Al and Pt{111} surfaces show very small expansions while the rest show contraction (up to 15%). The qualitative reason for this contraction is that electronic charge effects are very important and they act as if to smooth out surface roughness (i.e., the charge acts

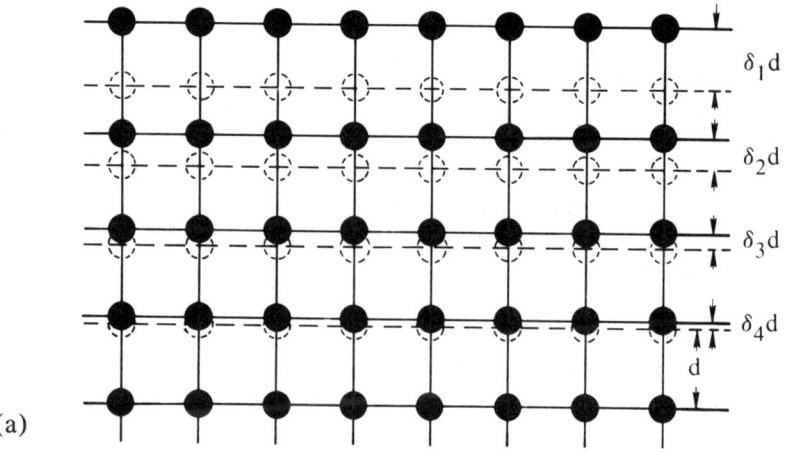

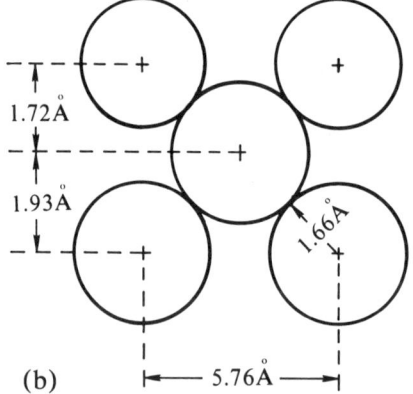

Fig. 17-3 (a) Displacements of atoms from their undisplaced positions (dotted lines) near the surface of a crystal. The displacements are greatly exaggerated. (b) A side view of a {001} surface, which is a {110} plane, of bcc Mo showing the 11% contraction of the surface atoms with respect to the bulk atoms.

Table 17-1 Calculated normal displacements of several layers for several faces for copper surfaces. [From P. Wynblatt and N. A. Gjostein, *Surface Sci.* **12**, 109 (1968).]

Surface	δ_1/d	δ_2/d	δ_3/d	δ_4/d
{100}	0.129	0.033	0.008	0.001
{110}	0.196	0.047	0.019	0.003
{111}	0.055	0.009	0.001	

like an electronic fluid with a surface tension). For references see the article quoted in Table 17-2.

From these results we also learn that surfaces do the unexpected and that very simple thinking can be qualitatively incorrect. Carefully refined LEED analysis has shown, in some cases, successive contraction and expansion of interlayer distances. For example, some careful work on Si{111} surfaces is consistent with a model with contraction

Table 17-2 Surface interlayer distance relaxations for some cubic metals. a = side of the cubic unit cell; d = interlayer distance. [From P. M. Marcus and F. Jona, Applications of Surface Sci., 11, 20 (1982).]

crystal surface	d/a	packing density(ρ)	metal	a(Å)	Δd(Å)	(Δd/d)%
fcc{111}	0.5774	0.9068	Al	4.04	0.05	2.1
			Pt	3.92	0.02	1.0
			Ir	3.83	−0.06	−2.5
			Cu	3.59	−0.10	−4.6
bcc{110}	0.7071	0.8330	Fe	2.86	0.00	0.0
fcc{001}	0.5000	0.7854	Cu	3.59	−0.01	−0.3
bcc{001}	0.5000	0.5890	Fe	2.86	−0.02	−1.4
			W	3.16	−0.09	−5.5
			Mo	3.86	−0.21	−11.0
fcc{110}	0.3536	0.5554	Ni	3.52	−0.06	−5.0
			Ag	4.09	−0.10	−6.6
			Ir	3.83	−0.13	−9.9
			Cu	3.59	−0.13	−10.0
bcc{111}	0.2887	0.3401	Fe	2.86	−0.12	−15.0

of the first layer by 25% but then expansion by 3%, 5%, and 1% of successively deeper layers of atoms.

17-4b Surface structure notation The symmetry of many surfaces is lower than would be expected if the crystal were simply terminated at a plane and all the atoms remained in their original positions (aside from an inward contraction or outward expansion). This rearrangement with lowering of the translational symmetry of the surface layers is called **surface reconstruction**. We develop a nomenclature — the Wood notation — that makes it possible to describe reconstructed and non-reconstructed surfaces.

The surface structure is designated with respect to the bulk unit cell (sometimes called substrate structure). If there is no surface reconstruction then the surface structure is called 1 × 1. Most metal surfaces do not reconstruct. For example, the {110} and {211} surfaces of tungsten do not reconstruct, so we would designate the {110} surface structure of the element W as W{110}1 × 1.

However, reconstruction occurs on the surface of some metals and most semiconductors. Thus, we develop a notation that allows for the designation of reconstructed structures. a_B and b_B are the bulk lattice vectors; then a_s and b_s are the lattice vectors of the surface cell. If

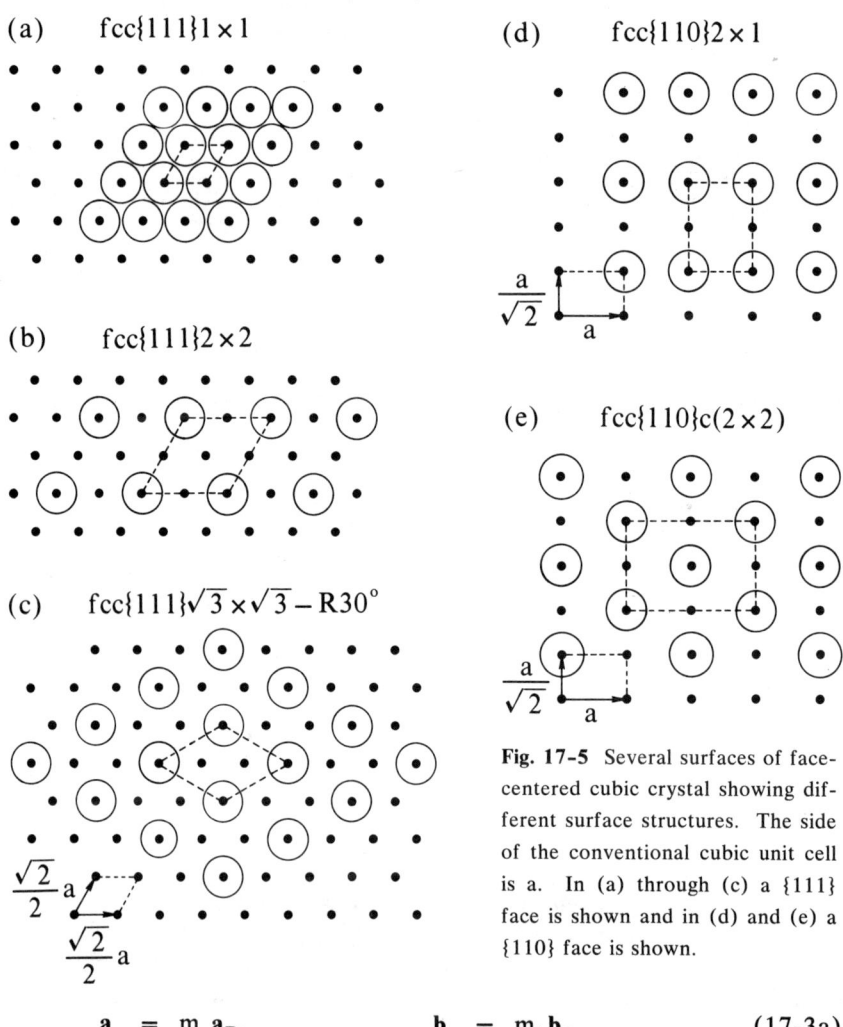

Fig. 17-5 Several surfaces of face-centered cubic crystal showing different surface structures. The side of the conventional cubic unit cell is a. In (a) through (c) a {111} face is shown and in (d) and (e) a {110} face is shown.

$$\mathbf{a}_s = m_a \mathbf{a}_B, \qquad \mathbf{b}_s = m_b \mathbf{b}_B \qquad (17\text{-}3a)$$

then the shorthand notation for the surface unit cell is $m_a \times m_b$ as in the preceding paragraph.

When an adsorbed gas causes a particular reconstruction, the same type of designation is used. For example, when hydrogen is adsorbed on the {211} face of tungsten, the surface unit cell is twice the size in each dimension as the substrate; this is written as W{211}2 × 2 – H. If the surface unit cell is twice as long in one direction but the same length in the other direction as that of the substrate, then it is denoted as 2 × 1. In the case of hydrogen on the Si{111} surface there is no reconstruction; this is denoted as Si{111}1 × 1 – H.

CHAPTER 17 SURFACE SCIENCE   683

Before continuing, we should add some comments for those who actually read the surface science literature. First, when discussing surfaces, often, instead of a lattice sometimes the word **net** and **unit mesh** instead of unit cell is used. Second, if the primitive character of the surface cell is to be emphasized then it could be written W{211}p(2 × 2) − H. Last, a_s and b_s need not be parallel to a_B and b_B so in general

$$a_s = m_{11}a_B + m_{12}b_B$$
$$b_s = m_{21}a_B + m_{22}b_B$$
(17-3b)

We shall talk about cells that have nonzero m_{12} and m_{21} but not go into any detail other than to note that the surface unit cells are rotated with respect to the bulk.

Before showing pictures that display different translational symmetry on some surfaces, Fig. 17-4 (page 672) shows the undistorted geometry of six common faces of the fcc and bcc structures. The diagrams are drawn assuming that the atoms are hard spheres and just touching one another so that the volume of the crystal is filled as compactly as possible (Section 3-8). Then the distance between the touching atoms is $2r_a$ (twice the atom radius), the conventional Bravais unit cell has dimension a, the spacing between the plane shown and the plane directly below is d_p, the **surface packing density** is ρ, and A_a and V_a are the area per atom and volume per atom. The various parts of the figure are drawn to scale using the same atom size (same r_a). Thus, we can see both algebraically and pictorially how the packing densities vary. The fcc{111} plane has the closest packing (Section 3-7), while the bcc{111} plane is a very open structure. Last, the conventional surface primitive unit cell for each face is shown.

In Figs. 17-5a to 17-5c a {111} surface plane of the fcc structure is shown, but each atom is now shrunk down to a solid dot. Surface atoms, shown as open circles, are overlaid, giving various unit cells. Figure 17-5d shows the {110} surface of a fcc structure with a 1 × 2 reconstruction and Fig. 17-5e shows the same surface with a centered 2 × 2 structure, that is, c(2 × 2).

To date most of the LEED studies are on high density, low index crystal surfaces of monatomic (Fig. 17-4) and diatomic solids. Usually, for these surfaces the unit cell is in registry with that of the substrate. For simple rotated surface structures the preceding notation is still applicable. For example, Fig. 17-5c shows a unit cell that is rotated 30° with respect to the substrate; hence − R30°. We shall not discuss more complicated cases.

17-4c Some examples of reconstruction The surfaces of most covalently bonded (rather than metallic) materials reconstruct. Figure 17-1a shows an "ideal" {111} surface of silicon. If this material is

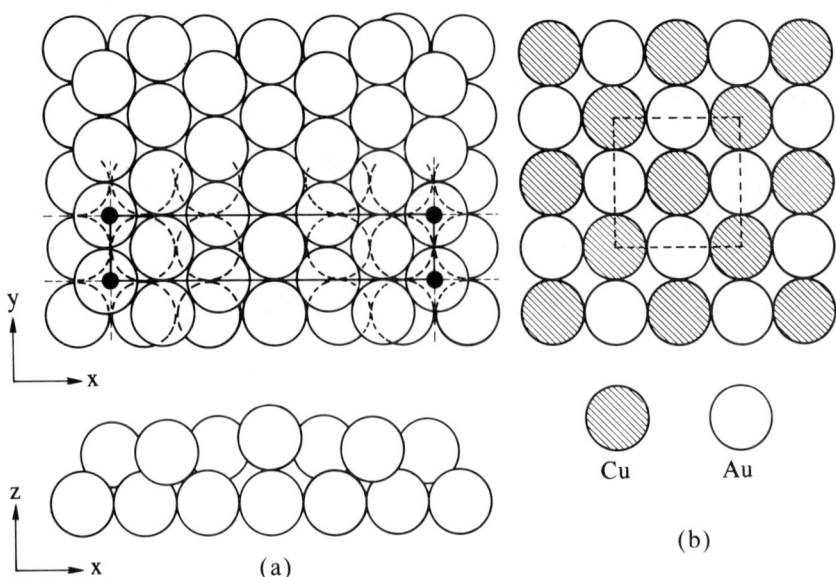

Fig. 17-6 (a) A model for the surface reconstruction on Au{100}. [P. W. Palmberg and T. N. Rhodin, Phys. Rev. **161**, 586 (1967).] (b) A model of the c(2×2) surface of Cu Au on the {100} face of Cu.

properly cleaved at room temperature a reconstructed structure with translational periodicity of a 2 × 1 structure is observed. An oversimplified model for this reconstruction is shown in Fig. 17-1b. Before reconstruction one of the four hybrid orbitals on each Si atom on the surface is directed outward from the surface and is unbonded. These orbitals are called **dangling orbitals** or **dangling bonds**. Recall from Chapter 8 that the energy of a system is lowered when the electrons from neighboring atoms overlap and are in bonding orbitals. If the energy gained by bonding is $\mathscr{E}_b$ then for the unreconstructed {111} face there is a surface energy of $\sqrt{3}\mathscr{E}_b/4d^2$ which is ≈ 7600 erg/cm^2 for silicon. This value is ten times larger then observed. From this fact alone, we expect some sort of reconstruction. The shifts shown in Fig. 17-1b indicate how the dangling bonds on neighboring surface atoms could bond to each other, giving back most of the lost bonding energy. Of course, this movement causes the bonds between the surface and the substrate atoms to form angles rather different from normal bond angles in this tetrahedrally coordinated structure. This interplay of the dangling bond energies and the strain energy of the angles between bonds determines the equilibrium positions.

The surfaces of semiconductors, principally silicon, play a critical role in modern technology. Thus, it is not surprising that there has been a great deal of work on the Si surface and its close relative Ge. Besides the practical importance the problems are intriguing.

An impressively large variety of surface structures are found experimentally on Si and Ge{111} surfaces. For Si: cleavage at room temperature yields a 2 × 1 structure; heating above 400°C causes a transition to a 7 × 7 structure that is stable to at least 870°C and remains when the sample is cooled to room temperature; a 1 × 1 structure can be stabilized at room temperature by small amounts of adsorbed chlorine; the 1 × 1 structure can also be obtained by laser annealing (i.e., using a short burst of very high intensity light) also the 1 × 1 structure is observed at room temperature if the sample is quench-cooled from high temperatures. The complexity of such a large unit cell as the 7 × 7 with 49 atoms is a serious handicap to determining a structural model unambiguously. The recent development of the **scanning tunneling microscope** (see the Notes) will help this and other structures with large complex cells by allowing observations in real – rather than reciprocal – space.

The Ge{111} surface is less completely studied. The results are: cleavage at 4.2°K yields a 1 × 1; cleavage at 77°K yields a 2 × 1; warming the crystals to room temperature yields a c(2 × 8). These results indicate the rich detail that is found in these interesting materials. Understanding of these phases is an active branch of research.

As mentioned, many metal surfaces do not reconstruct but some do, and Fig. 17-6a shows a model that attempts to explain the observed Au{100}5 × 1 structure. The proposal suggests that the reconstructed surface consists of a hexagonal layer arrangement with a small compression in the x-direction and the slight buckling can be seen in the side view. More recent measurements show that the surface may actually be a 5 × 20 structure, indicating the degree of complication that can arise in surface structures.

Figure 17-6b shows a model for a CuAu alloy on a Cu{100} surface. The structure is {100}c(2 × 2), that is, a centered 2 × 2 cell on a {100} surface. The c(2 × 2) pattern is obtained by depositing Au on a Cu{100} surface at room temperature. The model suggests an ordering of the AuCu alloy in the arrangement shown. Only in Cu-Au systems is an ordered two-dimensional alloy observed. For Pd-Ag, Pd-Au, Au-Ag, and Cu-Ag on Cu{100} the LEED patterns do not show ordered arrangements.

These are just a few examples of the wide variety of surface structures observed even among the simple elemental materials. The references in the Notes should be consulted for more examples.

17-5 Interaction of Gases with Surfaces

The interaction of gases with surfaces is a very broad field. On one end, the interactions of atoms with clean well-characterized surfaces are studied; in the other extreme the catalytic processes are studied. In either of these extremes we can study static adsorption behavior or the interaction of a beam of atoms or molecules directed toward, and scattering from, the surface. We discuss the simplest cases: the static behavior of atoms and simple molecules with well-characterized surfaces.

Adsorption is conventionally divided into two types, **chemisorption** and **physisorption**, depending on the strength of the bonding between the molecule and the solid surface. If the bonding is of the van der Waals type, then it is weak and the molecule is said to be physisorbed. On the other hand, chemisorption between molecules and a surface occurs if there is a chemical bond (ionic or covalent) that gives rise to larger bonding energies. Conventionally, if the heat of adsorption (defined in the next paragraph) is less than ≈ 10 kcal/mole then we talk of physisorption; if it is greater than ≈ 10 kcal/mole then chemisorption is said to occur. For O_2 on some metal surfaces, values ≈ 200 kcal/mole are found (23.05 kcal/mole = 1 eV/molecule).

The **heat of adsorption**, ΔH_{ads}, is defined as the average binding energy per molecule between the gas and the surface. Clearly the value of ΔH_{ads} is determined by the depth of the potential energy well created by the interaction. We expect smaller values of ΔH_{ads} for He gas interacting with a metal surface than for atomic H gas, since the former interacts via van der Waals forces while a true chemical bond is formed in the latter case. (Values of heat of adsorption are usually quoted not for one molecule but for Avogadro's number of molecules, so RT will appear in the formulas below rather than $k_B T$.)

After the gas atom is adsorbed on the surface, the ratio of the well depth to the thermal energy $\Delta H_{ads}/RT$, where R is the gas constant, determines the **residence time**, τ, of the gas atom on the surface. τ is usually given by a simple **activation process**

$$\tau = \tau_0 \exp(\Delta H_{ads}/RT) \tag{17-4}$$

where τ_0 is of the order of the time for one oscillation in the potential well, thus $\tau_0 \approx 10^{-12}$ sec. For a weak interaction between the gas and the surface (physisorption) $\tau \approx \tau_0$; for chemisorption $\tau \gg \tau_0$.

17-5a Physisorption Physisorption is defined by convention as $\Delta H_{ads} < 10$ kcal/mole. Assume $\Delta H_{ads} = 2$ kcal/mole and $\tau_0 = 10^{-12}$ sec, then at 300°K, $\tau = 10^{-11}$ sec, but at 100°K, $\tau = 10^{-7}$ sec, so large concentrations of atoms from the gas can be maintained on a

CHAPTER 17   SURFACE SCIENCE   687

surface for a long time by working at low temperatures or high pressures, even in the weakly bonded case.

The **surface concentration**, σ, depends on the number of gas molecules that strike the surface per unit area per second and on the residence time of the molecule on the surface. For small coverage, when many adsorption sites are available, it is given by

$$\sigma = \tau \frac{dN_s}{dt} \approx 3.5 \times 10^{22} \frac{P\,\tau_0}{(MT)^{1/2}} e^{\Delta H_{ads}/RT} \tag{17-5}$$

where Eq. 17-6 has been used. Thus, the coverage may be estimated and it is proportional to the gas pressure as expected.

One of the difficulties in studying physisorption by LEED is that the adsorbate is affected by the LEED electron beam because of the small binding energy. Thus, there are far fewer physisorption studies than chemisorption studies. However, several of the rare gases have been studied on clean metal and graphite surfaces. It has been found that Xe on Pd{100} remains disordered up to high coverages although there is evidence for short-range order. At the highest coverage a close-packed hexagonal structure is found, as would be expected. The Xe-Xe distance is 4.48 Å ($\sigma = 5.8 \times 10^{14}$ cm^{-2}) which is only slightly larger than the 4.37 Å distance found in the bulk Xe. The Xe atoms are not in registry (commensurate) with the surface atoms. However, in other cases when the substrate is chosen so that its surface atoms have almost the proper size, the atoms will have a structure in registry. For example Kr on the basal plane of graphite forms a hexagonal close-packed structure in registry with the substrate structure. On the other hand, Ar and Xe on this same plane form the same structure but it is not commensurate with the carbon atoms.

We are led to conclude that for the physisorption of rare gas atoms the configurations and distances are determined mainly by their mutual interaction potentials and influenced only slightly by the substrate. The distance between the atoms in the close-packed hexagonal layer is very close to what we find in the rare gas solid. These conclusions are consistent with what is expected for atoms that are very weakly bonded to the substrate.

17-5b Chemisorption   There have been many studies of the chemisorption of small molecules (O_2, H_2, D_2, N_2, CO, C_2H_2, etc.) on various metal surfaces. Table 17-3 lists some experimental values of ΔH_{ads}. In a few cases (e.g., CO on Pd) there have been studies of how this quantity varies on different crystalline faces; the differences do not exceed 15%. Thus, the values in Table 17-3 are usually taken as characteristic of the metal in general.

688   CHAPTER 17   SURFACE SCIENCE

Table 17-3 Heats of chemisorption of O_2, H_2, N_2, and CO on several metal surfaces. [From Somorjai's book.]

Gas	Material	ΔH_{ads} (kcal/mole)	Gas	Material	ΔH_{ads} (kcal/mole)
H_2	Ta	45	O_2	W	194
	W	45		Mo	172
	Cr	45		Rh	118
	Mo	40		Pd	67
	Ni	30		Pt	70
	Fe	32			
	Rh	28	CO	Ti	153
	Pd	26		W	82
	Mn	17		Ni	42
				Fe	46
N_2	W	95			
	Ta	140			
	Fe	70			

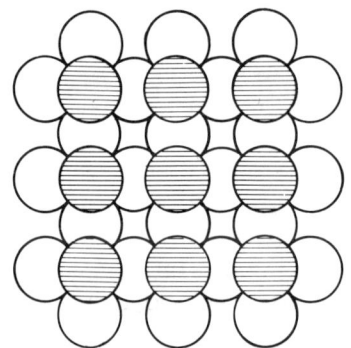

Fig. 17-7 The structure of Cu{001}c(2×2)-Cl where the shaded atoms are the chlorine atoms. [F. Jona, D. Westphal, A. Goldmann, and P. M. Marcus, J. Phys. C. **16**, 3001 (1983)]. The same structure is found for Cl on Ag{001}. (To see easily the basic aspects of the structure turn the figure 45° and find the primitive cell of the fcc surface. The centered cell of the Cl atoms is twice as large.)

The large heats of adsorption generally are associated with high-sticking coefficients, making it easy to obtain large surface coverages, Eq. 17-5. These coverages can be maintained well above room temperature so that high temperature processes can be studied. Since chemical surface reactions usually take place with an adsorbed molecule, (which usually dissociates into atoms) carefully controlled investigations of the properties of chemisorption of molecules should help us to understand catalytic reactions. Typical studies involve structural, thermodynamic, and electrical properties of the chemisorbed layer. We discuss several examples. Somorjai's book (see the Notes) lists well over a hundred examples of small molecules chemisorbed on metal and semiconductor surfaces.

Cl on Cu – Figure 17-7 shows the LEED results for Cl on a {001} face of Cu, which has an fcc structure. (The same results apply to Cl on Ag.) Cl-atoms are on sites of fourfold symmetry and a detailed structural analysis shows that they are on the hollows (as shown) between the surface atoms rather than on top of the surface atoms (which also are sites of fourfold symmetry). As can be seen the structure is Cu{001}c(2 × 2)-Cl.

CO on Ni – In general when CO molecules are adsorbed on a metal surface the bond is between the carbon atom and the metal. The C-O stretching vibration can still be observed with some small interesting frequency shifts, and even the vibration of the entire CO molecule against the metal can be observed. The evidence indicates that the CO molecule is in a fourfold position, perpendicular to the surface with the carbon end closest to the surface.

Ni has a fcc crystal structure and Fig. 17-4 shows the {100} face. Several ordered and disordered structures have been observed for CO on this surface. For a **surface coverage**, Θ, of $\Theta = 1/2$, a c(2 × 2) structure is observed where the CO molecules are located on sites with fourfold symmetry. (Surface coverage is defined with respect to the underlying bulk material.) The fourfold symmetry can be obtained with the CO on top of Ni atoms or in the hollows between the Ni atoms as found for Cl on a Cu{001} surface atom. For this case, experiments indicate that the CO is on top of the Ni atoms.

As more CO is adsorbed on this surface, it is observed that between $\Theta = 0.61$ and 0.69 a hexagonal structure is found. $\Theta = 0.69$ is the maximum density of adsorbed molecules that is observed and it corresponds to a surface density of 1.10×10^{15} cm^{-2}.

H on Ir – On many metal surfaces H_2 dissociates and atomic H will sit at specific sites. The hydrogen atoms themselves are difficult to observe with LEED techniques because of their small electron density. For many covalently bonded solids adsorbed hydrogen prevents reconstruction because it tends to saturate the dangling bonds; it is this uncompensated charge that can be a strong driving force for a reconstruction. However, for some metals hydrogen induces reconstruction.

Ir has a fcc crystal structure and the atomically clean Ir{110} surface shows a 1 × 2 reconstruction. A model for the reconstruction is shown in Fig. 17-8a. It is an interesting reconstruction since the resulting troughs are made up of the densely packed {111} planes inclined to each other as seen in the side view. This is accomplished with every other row of surface atoms missing in the ⟨001⟩ directions. Thus, second as well as third layer Ir atoms are exposed. This model is often called the **missing row model.** In the reconstructed unit cell there are two similar symmetry positions labeled A and B. The former is on top of the rows of Ir atoms and the latter is at a similar site above the third row of Ir atoms. There are also two threefold sites labeled C and D. A number of detailed studies suggest that hydrogen atoms first adsorb on the A and B sites (β_2 state) with a large ΔH_{ads}.

Probable Locations of Hydrogen on Ir{110}1×2

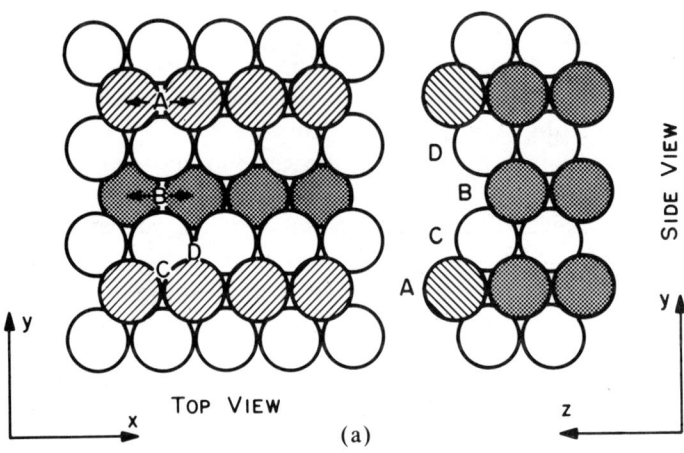

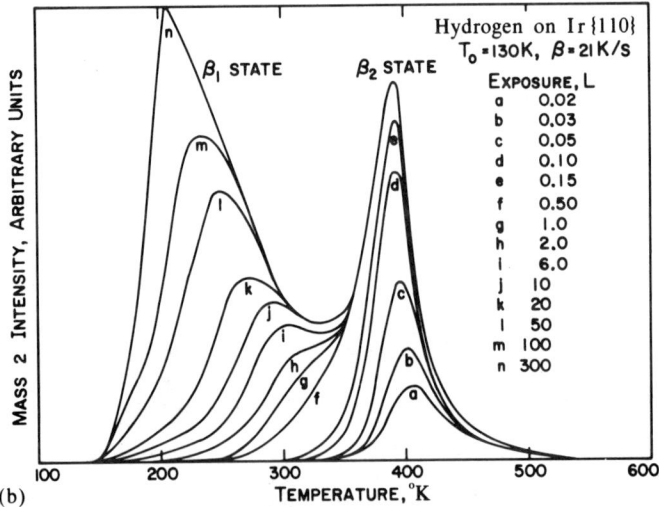

Fig. 17-8 (a) A model of the Ir{110} reconstruction. The letters refer to possible sites for hydrogen. (b) The thermal desorption spectra of hydrogen from Ir{110} for different amounts of gas exposure. The adsorption temperature is 130°K and the heating rate 21°K/sec. The ratio of the β_1 and β_2 states is 2:1 at saturation coverage. [D. E. Ibbotson, T. S. Wittrig and W. H. Weinberg, J. Chem. Phys. **72**, 4885 (1980).]

When these sites are filled, then H atoms adsorb on the C and D sites (β_1 state). Then we expect twice as many H atoms in the β_1 state as in the β_2 state, which can be checked by desorption experiments.

Fig. 17-8b shows a thermal desorption spectrum for many different amounts of gas exposure. Gas exposure is measured in **langmuirs** (L). If a surface is exposed to a pressure of 10^{-6} torr for 1 sec then the exposure is said to be one langmuir (1L $\equiv 10^{-6}$ torr sec). For a sticking coefficient of one, an exposure of 1L leads to approximately one monolayer. A clean Ir{110} surface, maintained at 130°K, is exposed to various amounts of H_2; the lowest amount is 0.02 L. Then, with a mass spectrometer set for mass 2 (i.e., H_2), the desorption of gas is measured for some fixed heating rate, 21°K/sec in this case. As can be seen, for low H_2 gas exposures (i.e., low coverages) the hydrogen is in the β_2 sites. For higher gas exposures hydrogen also goes into the β_1 sites. By means of experiments mentioned shortly, $\Delta H_{ads} \approx 22$ kcal/mole and ≈ 11 kcal/mole for the β_2 and β_1 sites, respectively. From the ratio of the area under the low temperature (β_1) desorption data to the area under the high temperature (β_2) data it is found there are approximately two times as many hydrogen atoms in the β_1 sites as in the β_2 sites. This nicely corresponds to the number of C + D sites and A + B sites, respectively. The rate of desorption dN_d/dt is usually expressed as

$$dN_d/dt = v_d \, \Theta^p \, \exp(-\Delta H_{ads}/RT) \qquad (17\text{-}6)$$

where Θ is the surface coverage and p is an integer that comes from the elementary desorption reaction. The value of p is 2 for this case since $H + H \rightarrow H_2$ and is 1 for Ar desorbing from a surface; v_d is a preexponential factor. From applications of Eq. 17-6, for example, by varying the heating rate, ΔH_{ads} and v_d, both of which vary with Θ to some extent, can be determined experimentally. See the reference quoted in Fig. 17-8b. Some further discussion of this system will be given in the next section on co-adsorption.

17-5c Co-adsorption If two different kinds of molecules, A and B, are adsorbed on the same surface then new phenomena can appear. We can find **cooperative adsorption** where a regular mixed phase will occur. For example, for CO molecules on Pd metal a Pd{110}c(2 × 2) – CO structure is obtained. If a partially CO-covered Pd{110} surface is exposed to H_2 gas, an entirely new 1 × 3 structure, ascribed to a mixed adsorbate complex, is observed.

The other extreme in co-adsorption is **competitive adsorption** where the two different molecules are completely immiscible on the surface. For this condition, when A and B are co-adsorbed onto a surface, they will tend to segregate into domains where there is only A and other domains where there is only B. Both kinds of particles are competing for the same free adsorption sites.

H_2 and CO can co-adsorb on Ir{110}. In this case no changes in the Ir{110}1 × 2 reconstructed surface structure or ordering of the adsorbed layers are observed with LEED. Further, thermal desorption curves, of the type in Fig. 17-8b, show that co-adsorption causes the hydrogen to occupy only the β_1 states. At high CO pressures the hydrogen is completely displaced (if it was adsorbed first) or will not stick (if it is introduced after the CO), that is, the Ir{110} surface is poisoned to hydrogen adsorption. The indications are that for low coverages the CO poisons the β_2 sites for hydrogen by a simple site blocking mechanism and for high coverages the CO excludes hydrogen from the β_1 sites by a CO-hydrogen repulsive force.

As might be appreciated a very large number of different co-adsorption studies can be, and have been, carried out.

17-6 Surface Related Techniques

There are many methods, tools, and specific techniques used in surface studies. Some also are used in other branches of science and some of the surface techniques really give information about the bulk (for example, photoemission provides a way to measure $\mathscr{E}$ vs. k for the valence bands). We briefly discuss a few of the techniques. References containing more details are given in the Notes.

The appendix at the end of this chapter summarizes some of the surface spectroscopies that are discussed here as well as some that are not. There is a terrible tendency for surface scientists to use abbreviations for these techniques and these are given there.

17-6a Synchrotron radiation Synchrotron radiation is electromagnetic radiation that comes out of a synchrotron. This radiation is proving very useful for photoelectron spectroscopy, reflectivity measurements in the ultraviolet region, high resolution lithography (useful in semiconductor device manufacture), and many related fields. Although other optical sources were available before, the use of synchrotron radiation is causing a revolution in many fields. It is very intense, highly polarized, and varies smoothly over a large spectral region. Thus, we briefly discuss this radiation.

In synchrotrons the electrons travel in circular orbits with the acceleration (force) directed toward the center as in Fig. 17-9a. The energy loss of the accelerating electrons is emitted in the form of electromagnetic radiation. When the electrons are traveling at relativistic velocities, $[1 - v^2/c^2] \approx 0$, then the emitted radiation pattern is highly concentrated in the plane of the orbit. The radiation is normally allowed to exit through a slit in the synchrotron wall. Then the radiation is in a narrow cone tangentially directed in the plane of the orbit as in Fig. 17-9a and polarized in the plane. Figure 17-9b shows typical spectral output for three different sources. The number of

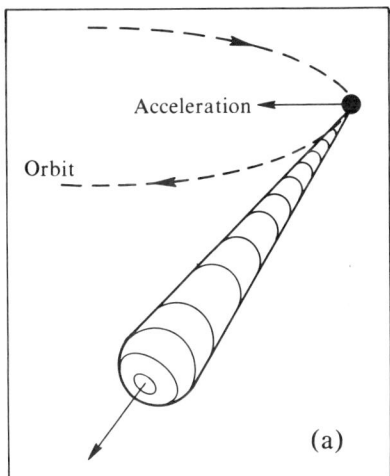
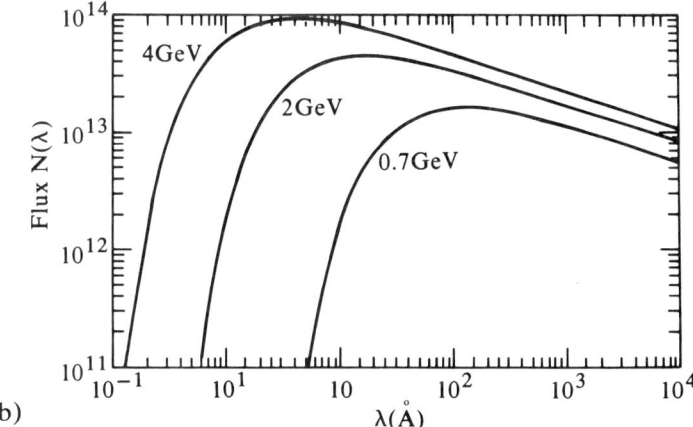

Fig. 17-9 (a) The narrow cone radiation pattern of highly relativistic electrons traveling in a circular orbit. (b) The flux (photons sec^{-1} mrad^{-1} mA^{-1}) for three different storage rings with different energy electrons (in GeV = 10^9 eV) and magnetic fields. The lowest and middle curves correspond approximately to the small and larger Brookhaven storage rings and the higher curve to the Stanford ring as of 1983. (Courtesy of D. Eastman.)

photons available to the experimenter is enormous and the radiation is tunable over an extremely wide spectral region.

The high intensity and continuous tunability can be used in situations where very high resolution (either energy or momentum) is needed. Surfaces contain a relatively small number of atoms, so for many surface experiments the high intensity is needed. Many new synchrotron radiation facilities have been "coming on line" in the early 1980s. Most of the facilities used for solid state studies are called **electron storage rings**. The principle of these devices and radiation emitted is the same as a normal synchrotron, however the purpose of the instrument is different. A classical synchrotron is used for particle physics experiments and the electron beam is built up to its maximum power in (1/60) sec, that is, (line frequency)$^{-1}$, and then "dumped" onto the target. The emitted electromagnetic radiation that is used in solid state experiments is a byproduct of the acceleration process. On the other hand, an electron storage ring is built to keep the electrons in orbit efficiently for a long time (4 to 12 hour half-life). This device is less expensive than a classical synchrotron; however, the principle of the emitted electromagnetic radiation is the same in both cases.

17-6b Ultraviolet reflectivity In Section 13-21 we discussed reflectivity experiments in the ultraviolet region of the spectrum. Many of the early experiments were done using a monochromator to select the appropriate wavelength from the continuum produced by a H_2 gas discharge. It was possible to obtain energies below 11.6 eV, which is the cutoff energy of the LiF windows (this material has the highest energy cutoff). Higher energies and windowless spectrometers are preferable for surface work. For windowless spectrometers resonance lines from rare gas atoms can be used. The most often used resonance lines are HeI (21.2 eV), HeII (40.8 eV), NeI (16.7 eV), and NeII (26.9 eV); I and II refer to neutral and single ionized atoms. Thus, an electron falling from the 2p to the 1s state in He^0 emits a 21.2 eV photon. Since these resonance lines are so intense, a monochromator is not needed. However, the best light source for these experiments is synchrotron radiation. In most systems, the complex dielectric constant obtained from these measurements is that appropriate to the bulk so it is not discussed here.

17-6c Low energy electron diffraction (LEED) LEED was discussed in Section 17-3. It is *the* method for determining surface structures. At typical LEED energies, electrons penetrate only a few layers of the crystal. Hence this technique is particularly suitable for determining the structure of surface layers. The LEED patterns immediately yield the symmetry and lattice parameter of the surface structure. However, because of multiple electron scattering, the diffracted intensities cannot be "inverted" easily to determine the actual surface structure. Rather, we must guess at the structure and see how well the calculated intensities agree with experiment. The guess, or model, is usually based on results in related materials, other data, or "minimum astonishment."

A technique that is related to LEED is the diffraction of He atoms from the surface. The beam is obtained from the expansion of He through a very small nozzle from a high pressure, low temperature container. The He atoms have an energy ≈ 20 meV with a very well-defined velocity distribution ($\delta v/v \sim 1\%$). The wavelength associated with such atoms, ≈ 1 Å, is comparable to lattice dimensions, and the interaction with a solid is very strong so that only the very outermost surface layer is probed. (The LEED electrons penetrate several layers at the surface.) Time of flight techniques are used to detect the scattered He atoms. The elastically scattered He atoms give structural information, and the inelastically scattered He atoms yield information about surface phonons. This technique is just emerging and undoubtedly will play an important role in surface science. For example, surface corrugations show up more clearly with this technique than with LEED.

17-6d Photoelectron (or photoemission) spectroscopy Consider a solid containing electrons bound with various energies $\mathscr{E}_B$ below the Fermi level and electromagnetic radiation of energy $h\nu$. By irradiating the solid with this radiation, electrons, either near the Fermi level or deep in the atomic cores, can be ejected with kinetic energy $\mathscr{E}_{kin}$. Figure 17-10 shows the idea. $\mathscr{E}_{kin}$ is a positive energy measured with respect to the vacuum level and $\mathscr{E}_B$ is conventionally measured with respect to the Fermi level as shown ($\mathscr{E}_B > 0$). In 1905 Einstein wrote a paper on this subject for which he received a Nobel Prize (1921). $\mathscr{E}_{kin}$ is given by

$$\mathscr{E}_{kin} = h\nu - \mathscr{E}_B - \phi \tag{17-7}$$

where ϕ is the **work function** of the crystal (2-5 eV for most clean solids). As indicated in the figure, electrons are emitted from the filled valence bands just below the Fermi level as well as from core states deep in the atoms. However, these various photoelectrons have different kinetic energies. By means of a retarding potential (from a grid between the sample and a detector) the number of electrons with kinetic energies greater than a fixed energy can be determined. To obtain $N(\mathscr{E})$ directly, a hemispherical electron analyzer or a double pass cylindrical analyzer can be used.

Early experiments were performed in the ultraviolet spectral region using a He resonance lamp ($h\nu = 21.2$ eV). Then the field is called **ultraviolet photoemission spectroscopy** (UPS). Even earlier experiments were performed using soft x-rays ($h\nu \approx 1$ keV) in which case the field is called **x-ray photoemission spectroscopy** (XPS) or sometimes **electron spectroscopy for chemical analysis** (ESCA). The latter name derives from the use of the technique for chemical analysis of impurity atoms. Impurity atoms have different $\mathscr{E}_B$ than the normal surface atoms. (Auger electron spectroscopy, to be discussed in the next subsection, is also very good for chemical analysis of the surface.) Now measurements in both UPS and XPS photon energy ranges are often carried out using synchrotron radiation.

UPS – The intensity of the UPS signal depends on the density of states of the electrons from the first few layers of the solid being studied (and the density of final states). This is indicated in Fig. 17-10 (and we mean the UPS signal being emitted at all angles).

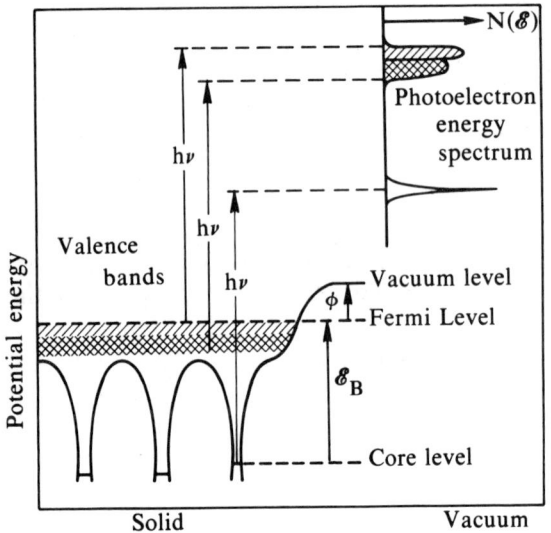

Fig. 17-10 A schematic energy level diagram showing photoemission from the valence bands and a core level in a solid.

Thus, UPS is sensitive to surface geometry and a surface structure can be verified by calculating the density of states for a surface with that structure and comparing it with the observed UPS intensities.

Additional information about the orbital shapes and bonding symmetry can be obtained by measuring the angular distribution of the emitted UPS electrons. These measurements can yield a direct determination of $\mathscr{E}$ vs. $\mathbf{k}$ for the valence (filled) bands in solids. To do this, atomically clean and smooth surfaces are required and the UPS measurements must be done for different values of $h\nu$ and at different emission directions from the sample. Figure 17-11 shows some results for the valence band of GaAs obtained by angle-resolved ultraviolet photoemission spectroscopy experiments. For this crystal the lowest band at about -12 eV (below the zero energy, which is conventionally taken as the top of the valence band) is mostly s-like. The top of the valence band is mostly p-like, with some s-like character, and has a larger $\mathbf{k}$-dependence; the splitting of the triple degeneracy of this band can be seen in the [110] direction and there is, of course, no splitting in the s-like band.

XPS – X-ray photoelectron emission experiments can be performed with conventional x-ray sources or with synchrotron radiation. One important use of XPS is to measure small **chemical shifts** of the atomic core levels. These small shifts of the deep-lying states reflect the different chemical (bonding) environments of the surface atoms compared to the atoms in the bulk. This is due to the fact that the potential energy of the electrons in an atom depend on the environment (bonding) of the atom. For example, shifts of the core levels between the bulk and the surface atoms have been measured on a

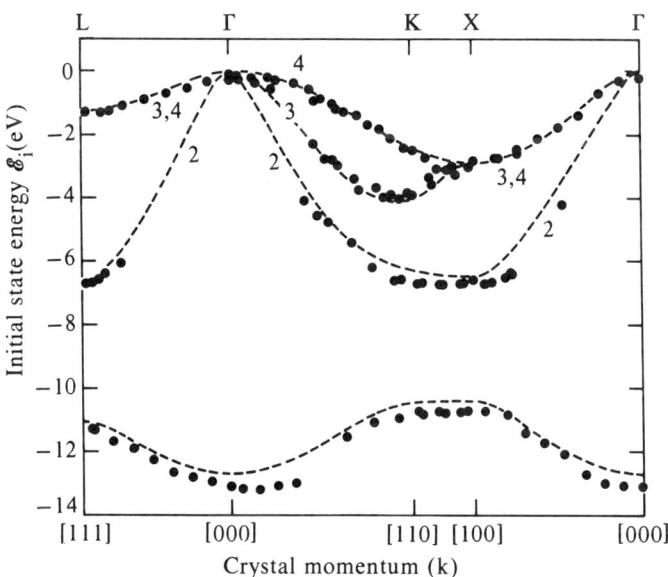

Fig. 17-11 The measured and calculated valence band in GaAs. The points are experimental data from angle resolved UPS and the dashed lines are from a pseudopotential calculation. [T. C. Chiang, J. A. Knapp, M. Aono, and D. E. Eastman, Sol. State Comm. **31**, 917 (1979) and Phys. Rev. **B21**, 3513 (1980).]

{111} surface of the bcc metal tungsten. Core levels at ≈31 eV are measured with a resolution of about 100 meV. The shift between the atoms on the surface and those in the bulk for this core level is about 430 meV. One way to confirm that the extra peaks indeed are due to surface atoms is to observe their sensitivity to changes of coordination number (number of nearest neighbors) which, for surface atoms, can be changed by adding adsorbates. In a study of 5d metals it has been found that the core shift between the bulk and surface atoms, $\delta\mathscr{E}_s$, is proportional to the square root of the number of nearest neighbors, z (i.e., $\delta\mathscr{E}_s \propto \sqrt{z}$). Naturally the coordination number of the surface atoms is smaller than that of the bulk.

As an example of the magnitude of the shifts that can occur, the 1s level in beryllium is found to be shifted by 4.6 eV in BeF_2 and 3.0 eV in BeO with respect to metallic Be.

Figure 17-12 shows that systematic information about electronic structure can be obtained along a row in the Periodic Table. The photoelectron spectra for a series of metals is taken in one row of the Periodic Table. To a first approximation the spectral results are proportional to the density of states of the 4d-band. For Pd the 4d-band is the valence band and it is about 5 eV wide due to the usual $\mathscr{E}$ vs. k

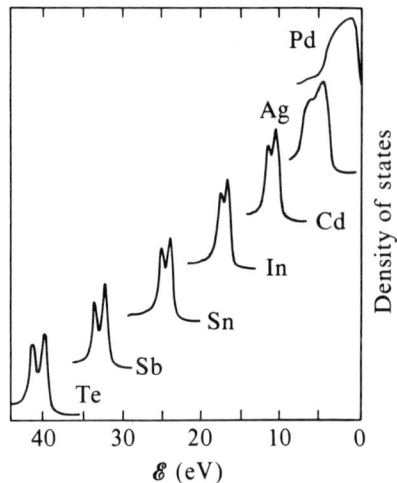

Fig. 17-12 The x-ray photoelectron spectra of the valence band region for palladium through tellurium. [R. A. Pollak, S. Kowalczyk, L. Ley, and D. A. Shirley, Phys. Rev. Lett. **29**, 274 (1972).]

effects in this band. However, notice that with increasing atomic number the 4d structure evolves from d-bands to atomic-like core states. The splitting in the core states is due to the fact that one electron is ejected from the $4d^{10}$ state, leaving one hole, which because of the spin-orbit coupling results in a split peak.

Thus, we see that measurements of core levels are used to determine the environment in which a particular atom is found. Therefore, such measurements can be used to study chemisorption, surface reconstruction, and related phenomena.

17-6e Auger process Detection of Auger (pronounced Ohzhay) electrons is a simple and sensitive way to test for impurities on clean surfaces. LEED and Auger electron spectroscopy (AES) are the two analytic tools used for the everyday characterization of surfaces.

An Auger electron is emitted via the following process. Suppose there is a vacancy in a K electron level (a 1s electron is missing) produced, for example, by bombardment with energetic electrons or photons. This atom can be deexcited in two different ways (Fig. 17-13). A transition can occur from one of the higher lying states, let us say an L state (n = 2 principal quantum state), and fill this vacancy. The difference in energy $\Delta \mathscr{E}$ (= $\mathscr{E}_L - \mathscr{E}_K$) can be carried off by an x-ray with this (characteristic) energy. However, $\Delta \mathscr{E}$ can be carried off in another way and this is called the **Auger process**. An electron (called an **Auger electron**) from a higher state, for example an M state (n = 3 state), can be emitted. The binding energy of this emitted Auger electron is not enough to take up all of $\Delta \mathscr{E}$ so the emitted electron has a characteristic kinetic energy, $\Delta \mathscr{E} - |\mathscr{E}_M|$, *determined by the quantum states of the atom.* Thus, each atom has a characteristic Auger electron spectrum, independent of the way that the electron is

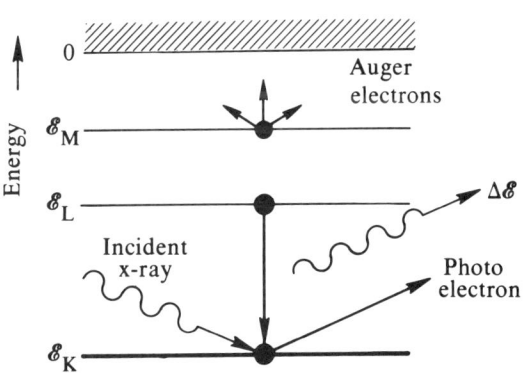

Fig. 17-13 A simplified schematic energy level diagram showing how an atom with an electron missing from its K-level can be deexcited either by the emission of a characteristic x-ray or by an Auger electron.

ejected from the K-state. This description of the Auger process is for gases. In solids the lines are broadened due to valence band widths and the short lifetime of the electron states. Nevertheless, these broadening effects do not reduce the usefulness significantly.

One advantage of the Auger process is that spectra can be measured by LEED electron optics so extra equipment is not needed in the UHV chamber. The initial vacancy in the low lying state is produced by electrons from an electron gun impinging on the surface. With derivative detection, coverages of impurity atoms of 10^{-3} can be detected. (In Auger and any other spectroscopy involving incident electron beams there are space charge build-up problems when insulators are studied. In a UHV system it is a difficult problem but it does not arise when studying metals or most semiconductors.)

17-6f Appearance potential spectroscopy (APS) Like the Auger process, APS is a method of detecting surface impurities using a relatively simple apparatus. An electron beam impinges on the surface to be studied. As the energy of the beam is increased through that of a core electronic level characteristic of the atoms on the surface, this level can ionize; deexcitation of this level produces x-rays. Measurements of the total x-ray yield show modulations as the energy of the beam passes through the excitation thresholds for the various core levels. The sensitivity of APS is increased considerably by the derivative technique. That is, the energy of the incoming electron beam is modulated at some low frequency and the total x-ray yield, usually detected with a photomultiplier, is detected synchronously. This detection scheme severely reduces broad background signals. By using low energy electrons impinging onto the sample at glancing incidence, this technique can be kept surface specific.

17-6g EXAFS As mentioned in Section 17-6a, the use of synchrotron radiation is causing a revolution in many branches of bulk and surface science. The conventional x-ray tube has remained basically the same since 1913. However, the radiation from a synchrotron is about 10^3 brighter than that from an x-ray tube. With the use of a new generation of wiggler and undulator magnet in storage rings, this enhancement factor may be 10^6 or more! This extremely strong and continuous source is being used for measurements that are not possible using a conventional x-ray tube.

Figure 17-14 shows the linear absorption coefficient for Ni_3Fe in the x-ray range. As the energy of the electromagnetic radiation is increased, there is enough energy at 7.1 keV to eject a 1s electron from iron. This process absorbs some photons, causing an increase in the absorption coefficient. Notice that the energy required for this process in nickel is about 1 keV higher, so a source extending over a sizable range of energy is required to make the measurement. However, the important information is contained in the oscillations just above the absorption edge. These oscillations are not noise; they are called **extended x-ray absorption fine structure (EXAFS)**. The amount of x-ray absorption increases when an electron can be removed from an atom more easily, and this latter process depends on the electron final state. When the 1s electron is photoejected from the core of an atom, the yield of the ejected electrons is modulated as a function of incident energy due to interference between outgoing electrons and electrons backscattered from the neighboring atoms. This means that the wave function of the outgoing electron depends on energy so the interference will be constructive or destructive depending on the ratio of the distances to wavelength (the phase shift) of the wave. By varying the energy of the synchrotron radiation we can selectively study a particular element determining the number of neighboring atoms and their distances from the targeted element.

EXAFS is a good tool for use in materials without long-range order because, unlike conventional x-ray diffraction which requires long-range order, the strong interference in EXAFS measurements comes only from near neighbors. Thus, this technique is being applied to study the bulk properties of glasses, amorphous solids, alloys, and materials like superionic conductors. Moreover, the technique is sensitive enough to be applied to surfaces. Important results are emerging from measurements on catalysts. See the references in the Physics Today articles referred to in the Notes. As an example, by measuring the EXAFS of iodine on an Ag{111} surface at low coverage, it is determined that the iodine atoms are in a threefold coordinated site and the I-Ag distance is 2.87 ± 0.03 Å. This distance is 2.5% larger than found in tetrahedrally coordinated bulk AgI. When applied to surface studies the technique is called **SEXAFS**.

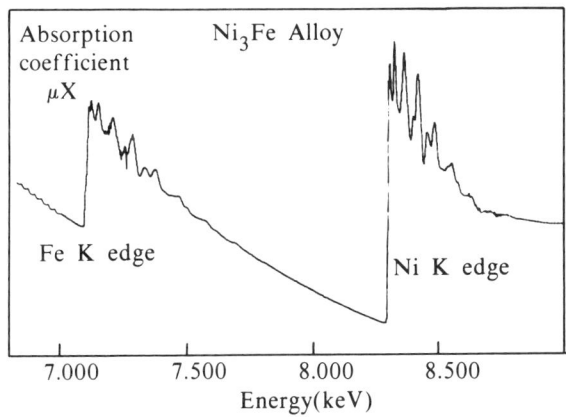

Fig. 17-14 The x-ray absorption for an iron-nickel alloy. (Courtesy of H. Yakel.)

It can be expected that EXAFS and SEXAFS will play an expanding role in obtaining both bulk and surface information.

17-6h Electron energy loss spectroscopy (EELS) Besides elastically scattered electrons that give rise to spots in LEED measurements, there are many processes in which electrons are inelastically scattered. If $\mathscr{E}_p$ is the energy of the incident electron beam and there is an energy loss $\mathscr{E}_\ell$ due to some process in the solid, then for a single loss process there is an electron in the emitted beam with energy $\mathscr{E}_p - \mathscr{E}_\ell$. This emitted beam can be distinguished from Auger electrons by the fact that the latter are observed at *fixed* energies whereas by varying $\mathscr{E}_p$ the feature at $\mathscr{E}_p - \mathscr{E}_\ell$ is shifted.

The EELS technique is being used for many purposes; refer to the references in the Notes. Very sensitive EELS spectrometers with high resolution are used to measure surface vibration frequencies (phonons) and vibrations of adsorbed atoms on the surface. Also, with higher incident energy electrons, EELS is used to measure core level electron energy losses. This yields information about bonding and other characteristics of absorbed atoms.

17-6i Inverse photoemission (Bremsstrahlung spectroscopy) In photoemission spectroscopy (Section 17-6d) a photon impinges on the sample and an electron is emitted. The inverse process consists of an electron impinging on the solid and a photon is emitted. Energy balance (Eq. 17-7) holds except that for inverse photoemission $\mathscr{E}_B$ is negative because the incident electron falls into an unoccupied state *above* the Fermi level. When the energy of the incident electron is of the order of 50 keV this type of radiation is well known. It is the continuum that is emitted from x-ray tubes, Fig. 4-1; electrons that fall to states near the Fermi level emit radiation that is the upper energy of the continuum spectra.

In the last few years low energy electrons (4 to 100 eV) have been used for this type of process. Such low energy electrons only penetrate the first few atomic layers of the sample, opening this technique for surface science studies.

In photoemission spectroscopy the energy and momentum of occupied one-electron states can be measured. Using the inverse process the same information can be obtained but for the *unoccupied* electron states below the vacuum level. Interesting energy levels fall into this category, such as unoccupied broken bonds, empty adsorbate orbitals, and minority spin states in ferromagnets.

17-7 Electronic Surface Structure

The atomic structure of the surface was discussed in Section 17-4. In this section we focus on the theory of the electronic structure of the surface. Broadly speaking we ask: how does the truncation of a bulk crystal affect the electron density distribution and how do adsorbate atoms bind to the surface?

17-7a Surface charge density A simple macroscopic property of a surface is the formation of a **double layer** of charge. Take the surface as the xy-plane so the z-direction is perpendicular to the surface. Then a double layer occurs because the center of gravity of the negative (electron) charge has a different value of z than the center of gravity of the positive (nuclear) charge. However, real surfaces are more complicated; for example, adsorption is strongly site-specific and the bonding energies vary considerably between top atom positions and various bridge and hollow positions for the adatom. This implies important variations of the charge in the xy-plane.

Briefly, we discuss two approaches used for obtaining the surface charge density (and surface potential). One approach takes account of the atomic nature of the surface. A simpler approach uses the so-called **jellium model** in which we assume that the ion cores are smeared out into a uniform density of positive charge. Clearly for the jellium model no xy-dependence is obtained, but we can calculate overall trends more easily, while in the first approach site specific results can be obtained. In both approaches it is important to do self-consistent calculations of electron density.

Atomistic calculation – Figure 17-15a shows the charge density and potential for Na{001} as a function of z, averaged over the xy-plane for simplicity. The calculations are performed by dividing the semi-infinite solid into two regions. The first consists of the vacuum and the first few Na planes in which the electron density is different from the bulk, and the second region is the remainder of the semi-

CHAPTER 17   SURFACE SCIENCE   703

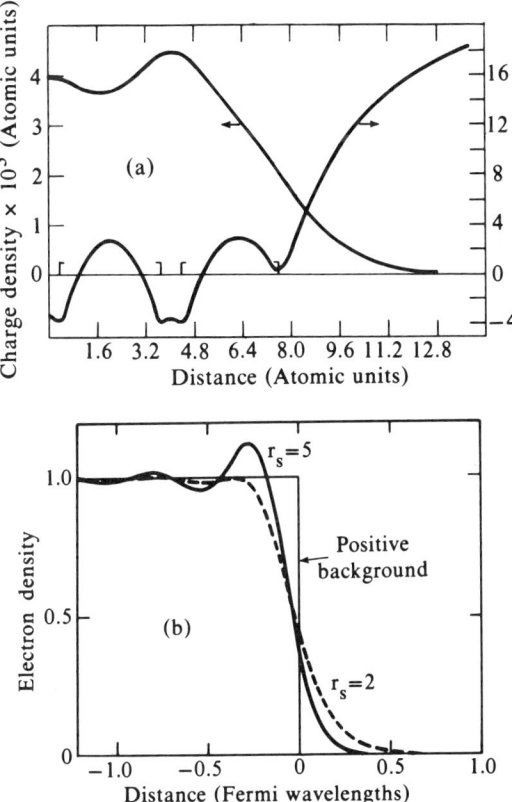

Fig. 17-15 (a) The charge density and potential energy as a function of z. The positions of the ion cores in the last two atomic layers is indicated by brackets. [J. A. Appelbaum and D. R. Hamann, Phys. Rev. B **6**, 2166 (1972).] (b) The electron density near the surface using the jellium model. The results are for two values of r_s. [N. D. Lang and W. Kohn, Phys. Rev. B **1**, 4555 (1970).]

infinite solid in which it is assumed that the bulk properties are known or are calculated separately. The self-consistent solution for the first region shows that the potential in the second plane of atoms is almost equal to the potential in the bulk. The potential results, Fig. 17-15a, show a distinct local minimum near $z = 8a_0$ (where the Bohr radius $a_0 = 0.529$ Å) corresponding to the missing plane of Na atoms. At larger z, the potential asymptotically approaches the vacuum level.

The preceding discussion is for an averaged potential in the xy-plane. However, for this atomistic model the full three-dimensional behavior can be studied. When this is done, the results close to the surface depend strongly on whether one is above a Na-atom or midway between atoms. For example, in a plane 2 Å from the last surface plane of Na-atoms, the potential energy has a 0.4 eV in-plane variation, which is substantial considering that Na is one of the most nearly free electron metals with an almost spherical Fermi surface.

Jellium calculation – In applying the jellium model to surface calculations, the semi-infinite metal structure of positive ions is smeared out into a semi-infinite homogeneous positive background, which has a sharp cut-off at the surface, Fig. 17-15b. Well away from the surface the electron density also is constant and equal to that of the positive background density. However, near the surface the electron density falls off in some fashion that is to be calculated. The only parameter in this model is that of background valence electron density N/V. This quantity is usually expressed in terms of r_s, the equivalent radius for the volume of one conduction electron, Eq. 9-2, $(V/N) = (4\pi/3)r_s^3$. Table 9-1 lists values of r_s for most elemental metals, which vary from $2a_0$ to $6a_0$. Aluminum and potassium correspond roughly to $2a_0$ and $5a_0$ respectively.

The calculated electron density near the surface is shown in Fig. 17-15b for these two values of r_s. The density tails off exponentially in the vacuum and exhibits **Friedel oscillations** in the bulk. These oscillations have a characteristic wavelength of half of the Fermi wavelength ($\lambda_F = 2\pi/k_F$, where values of k_F are in Table 9-4). The amplitude of the oscillation decreases inversely the square of the distance from the surface.

The surface dipole moment (double layer) is such that the resultant negative charge is outside of the surface (i.e., at positive z-values). Thus, the dipole moment points into the metal, causing the potential energy of an electron to be lower in the metal than in the vacuum. This is one of the important factors that keeps the electrons in the metal (i.e., gives rise to ϕ in Fig. 9-1). However, there is a bulk contribution to ϕ that usually is larger than this double layer contribution. In the bulk an electron is, on the average, nearer to the positive nuclei than when it is in the vacuum. Exchange and correlation effects also have some contribution to values of the work function.

17-7b Work function The work function ϕ is defined as the energy necessary to transfer an electron from the Fermi energy to the vacuum level (i.e., to a position outside of the crystal at infinity). See Fig. 9-1. In the preceding paragraph the two mechanisms that give the major contributions to the work function were discussed. ϕ can be measured by several different techniques, which are discussed now.

Thermionic electron emission – At any finite temperature there is a small fraction of electrons that have a thermal energy higher than the vacuum (Figs. 9-1 or 17-16a). The number of such electrons is governed by the Fermi distribution function, the density of states in the metal, and the value of ϕ, Eq. 9-35. The energy term in the Fermi function, $f(\mathscr{E})$, is $\mathscr{E} - \mathscr{E}_F$. For thermionic emission $\mathscr{E} \geq \mathscr{E}_F + \phi$ so the exponential in $f(\mathscr{E})$ is $\gg 1$ since $\phi \gg k_B T$. Thus,

$$f(\mathscr{E}) = \frac{1}{e^{(\mathscr{E}-\mathscr{E}_F)/k_B T} + 1} \approx e^{-(\mathscr{E}-\mathscr{E}_F)/k_B T} \qquad (17\text{-}8)$$

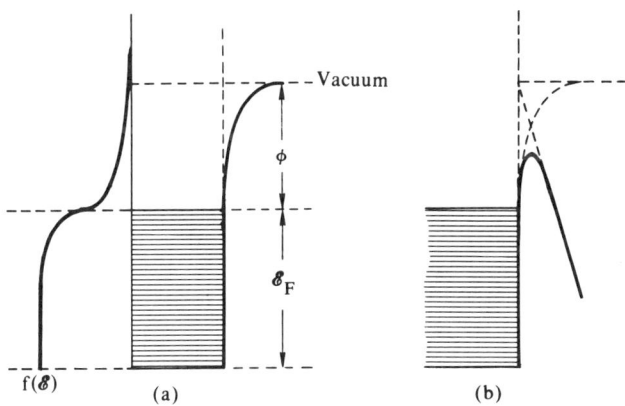

Fig. 17-16 (a) Potential energy diagram helpful to understanding thermionic emission. (b) The same but for field emission.

The density of states in the metal is a relatively weak function of energy while the Fermi function depends sharply on $\mathscr{E}$ and the temperature. Thus, we expect that thermionic electron emission will be dominated by the exponential dependence given in Eq. 17-8.

To collect these electrons we apply a very weak electric field in the z-direction so that as the electrons come out of the metal into the vacuum they are swept away. Then, using the free electron model, the thermionic electron emission current is

$$J = \int e \frac{dN}{V} v_z \qquad (17\text{-}9a)$$

where J is in units of A/cm^2, dN is given by Eq. 9-35, and v_z is the velocity distribution in the z-direction in the metal. For the distribution of electrons in the z-direction we integrate this between $\mathscr{E} = \mathscr{E}_F + \phi$ and ∞. For those in the x- and y-directions we integrate between $-\infty$ and ∞. The result is

$$J = BT^2 e^{-\phi/k_B T} \qquad (17\text{-}9b)$$

This is the **Richardson-Dushman equation** with $B = 4\pi emk_B^2/h^2 = 120$ $A/cm^2 °K^2$. The exponential term dominates the emission so that J increases strongly for small work functions and high temperatures. From experimental data ϕ and B can be determined separately. Values of B range from 60 to 160 $A/cm^2 °K^2$ in reasonable agreement with the calculation; the exponential behavior is always found. (See Somorjai's Table 4-4.)

Photoemission – Another method for determining the work function is the classic Einstein photoemission approach. When the energy ($h\nu$) of incident light is greater than $e\phi$, electrons are emitted from the sample. The current density measured at the collecting electrode is given by the **Fowler equation,**

$$J = CT^2 \, F\,[(h\nu - \phi)/k_B T] \tag{17-10}$$

where C is nearly a constant and F is a tabulated function that is almost exponential. (See Ertl and Küppers referenced in the Notes.) Experimentally $\log(J/T^2)$ vs. $h\nu/k_B T$ is plotted to determine ϕ.

Field emission – If an electric field, E, is applied between the metal sample and an external electrode, the barrier for electron emission from the metal surface is lowered as indicated in Fig. 17-16b. Thus, electrons with lower energies can be emitted from the metal. An exponential dependence of J on E is to be expected because of the exponential behavior of the Fermi function in the high energy range, as is also found for thermionic emission, Eqs. 17-9.

The **Fowler-Nordheim equation** describes this effect.

$$J = \frac{e^3 E^2}{8\pi h \phi C} \, \exp\left[-\frac{4(2m)^{1/2} \phi^{3/2}}{3\hbar e E} C'\right] \tag{17-11}$$

where C and C′ are correction terms ≈1. By plotting $\log(J/E^2)$ vs. $1/E$ a value of ϕ is determined from the slope of the straight line.

Very large electric fields (>10^7 V/cm) are needed to obtain useful current densities. This usually restricts the samples to those with very small radii of curvature (≲1000 Å). The field emission microscope uses this effect and enables us to see many details of the surface of a metal.

One problem with techniques that measure the work function is that we often deal with surfaces that are inhomogeneous either because they are polycrystalline or because of partial surface layer contamination. In such cases the patches of the surface with the lowest work functions will tend to dominate the measured current. Thus, care must be taken in interpreting the data. Accurate values of ϕ can be obtained by contact potential differences measurements and this technique does not have these problems.

Crystal face dependence – If the electron is removed from the material and taken "to infinity," that is, to a distance large compared with the size of the sample, then the work function is independent of the different crystal surfaces. However, in practice, the distance between the sample and the collecting electrode can be small compared with the distances to the edges of the sample. Hence the work function can be different for different faces. (Fringing fields at the sample edges make up the differences in the work function at different faces.) Thus, experimental measurements on single crystals can be used to determine the dependence of ϕ on the crystal face. This is useful

CHAPTER 17   SURFACE SCIENCE   707

Table 17-4 The electron work functions (in eV) for a few elements for polycrystalline samples and specific faces as indicated. [CRC Handbook of Chemistry and Physics (1983-1984)].

Ag	4.26	Au	5.1	Cu	4.65
{100}	4.64	{100}	5.47	{100}	4.59
{110}	4.52	{110}	5.37	{110}	4.48
{111}	4.74	{111}	5.31	{111}	4.94
Ir	5.27	Mo	4.6	W	4.55
{100}	5.67	{100}	4.53	{100}	4.63
{110}	5.42	{110}	4.95	{110}	4.25
{111}	5.76	{111}	4.55	{111}	4.47
{210}	5.00	{112}	4.36	{113}	4.18
		{114}	4.50	{116}	4.30

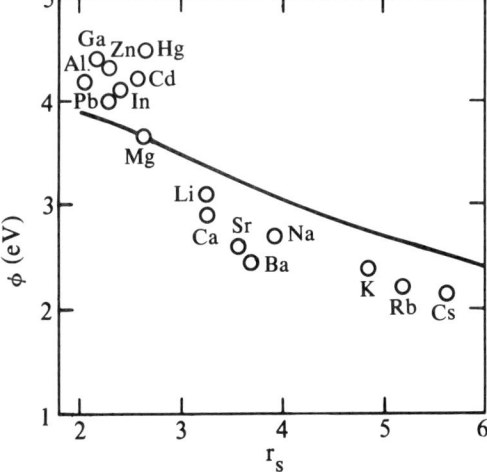

Fig. 17-17 Circles: measured work function for polycrystalline samples. Solid line: ϕ calculated from the jellium model vs. r_s. [N. D. Lang, Sol. State Phys. 28, 225 (1973).]

because this variation can be calculated. Typically, for the elemental metals, the high density faces have the highest work functions.

Table 17-4 lists some values of the work function from different faces for several elements. The variation with crystal face is not strong but also is not negligible. Experimental values of ϕ for different metals are shown in Fig. 17-17; these are polycrystalline average values. The solid line in this figure is the result of a jellium calculation. These calculations give an average value of ϕ since no details of the material are included except the average free electron density. We expect that the model is good for the simple metals but less appropriate for transition metals.

17-7c Charge density effects from chemisorption In this last section we discuss a few aspects of the effects of chemisorption of atoms on the electronic structure of metal surfaces. Two effects are discussed: the change of the work function of a metal with surface coverage of other atoms; and the change of the electronic charge distribution of the adsorbed atom and how it depends on the type of atom. For both effects the results of jellium calculations are used; with this model a broad range of calculations ignoring the details of the surface electronic distribution.

ϕ **dependence on coverage** – Measurements have been made of the change of work function as alkali atoms are chemisorbed on the surface. The results show that ϕ falls with increasing alkali coverage, reaches a minimum, and then increases to the alkali bulk value as the first full layer of chemisorbed atoms is completed.

The qualitative reason for this effect is that alkali atoms lose an electron to the metal, and become positive ions. This charge transfer creates a dipole (double layer) pointing out of the metal, which is opposite to that of the uncovered metal (as mentioned previously). This makes it easier for electrons to leave the metal (i.e., lowers the work function). At higher coverages the dipole-dipole interactions reduce the dipole moment of these polarizable dipoles, which decreases the effectiveness of their reduction of ϕ, and eventually we observe the work function of the chemisorbed material.

The following approach is used to calculate this experimental result using the jellium model. The uniform positive charge background is taken as a step function as in Fig. 17-15b; the adatoms also are represented by a step function slab on the end of the semi-infinite substrate. The width of the slab is taken as the spacing of the most closely packed planes in the bulk alkali and the height of the slab (the surface ionic charge density) is varied to correspond to the alkali coverage. For each value of the slab density a self-consistent calculation is done for the total electron charge distribution; the results generally are similar to those shown in Fig. 17-15b; the details depend on the amount of coverage. Then for each calculation ϕ is obtained just as in the case of the bare surface. The results are shown in Fig. 17-18. This calculation corresponds to an aluminum substrate and is shown for slab thickness corresponding to Na and Cs as indicated. In both cases there is a sharp decrease in ϕ with increasing coverage and a minimum is found in agreement with experiment. Also the magnitude of ϕ at the minimum is in reasonable agreement with experiment and fairly insensitive to the substrate chosen. However, the model is greatly simplified and we should not expect too much accuracy from it. First, ϕ is rather sensitive to the slab thickness (Fig. 17-18) and this is included in a reasonable but arbitrary way. Second, the model uni-

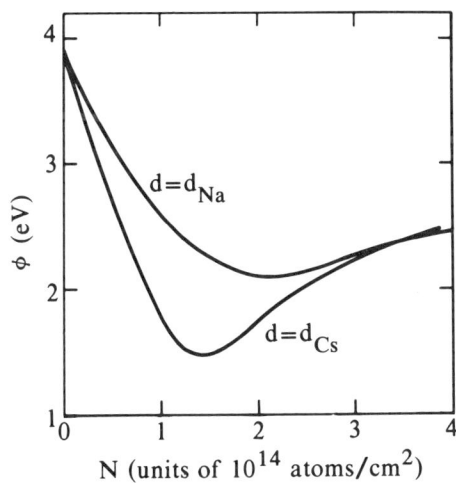

Fig. 17-18 Calculated work function vs. Na and Cs coverage on a jellium metal with $r_s = 2a_0$ (approximates aluminum). [N. D. Lang, Phys. Rev. B **4**, 4234 (1971).]

formly smears the alkali atoms over the substrate surface even when there are only very few adatoms.

Effects for a single atom – Now consider the case of chemisorption of a single atom on a metal surface. Let the atom have a single (valence electron) energy level $\mathscr{E}_a$. If $\mathscr{E}_a$ is higher in energy (closer to the vacuum) than $\mathscr{E}_F$ of the metal, then we expect that the electron from the atom will go into the metal, leaving the atom as a positive ion. If $\mathscr{E}_a$ is lower than $\mathscr{E}_F$, we would expect the atom to acquire an electron from the metal and become a negative ion. If $\mathscr{E}_a \approx \mathscr{E}_F$ then we might expect something like a covalent bond. However, independent of the type of bond, when the sharp atomic energy level interacts with the continuum of states in the metal, the atomic level is broadened.

The preceding statements apply to chemisorbed atoms. It is only in such cases that the interaction between the atom and substrate is strong enough to affect the eigenvalues of the atoms. For physisorption (Section 17-5) the interaction is weak, so the perturbation of the energy levels of physisorbed atoms is extremely small.

One approach to treat the chemisorption of a single atom is to treat the atom as realistically as possible but to use the jellium model for the metal substrate; we discuss these results. Figure 17-19 shows the difference between the electron eigenstate density of the bare metal and that of the metal with the chemisorbed atom. The results for Li, Si, and Cl illustrate the basic three types of chemisorption bond, that is, positive ionic (Li), covalent (Si), and negative ionic (Cl).

For Li the 2s atomic level is about 1.5 eV above $\mathscr{E}_F$ of the metal, with the result that Li loses this valence electron to the metal and become Li$^+$. Remember the results in Fig. 17-19 are not electron

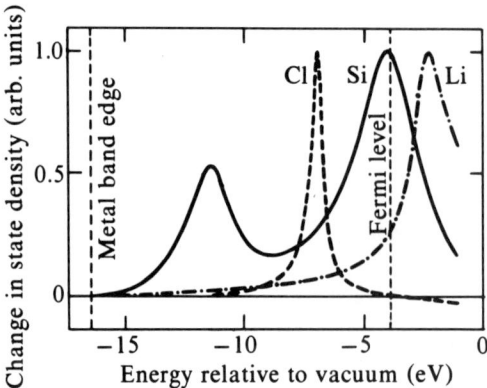

Fig. 17-19 The change in the electron eigenstate density due to chemisorption of a single atom onto a jellium metal with $r_s = 2a_0$. Each curve corresponds to the metal-adatom distance that minimizes the total energy. [N. D. Lang and A. R. Williams, Phys. Rev. B **18**, 616 (1978).]

states; the electrons only fill up the states to $\mathscr{E}_F$. So we must also look at the electron charge density, which indicates the Li$^+$ is the correct picture. For Cl the empty 3p energy level is below $\mathscr{E}_F$ by about 3.5 eV so it acquires electronic charge from the metal to become Cl$^-$. For Si the results are less obvious. The lower broadened level corresponds to the 3s level while the level at $\mathscr{E}_F$ is the 3p level. Calculations of the Si-metal charge density make it is clear that Si is covalently bonded to the metal.

The preceding discussion just "touches the surface" of this important field. The Notes should be consulted for further reading.

Notes

The journal Surface Science is an excellent place to find recent results in this fast-moving field. In particular the proceedings of some of the recent conferences that are published in this journal are in **80** (1979), **89** (1979), **99** (1980), **117** (1982), and **126** (1983). There are a number of books in this field; only a few are listed here:

J. M. Blakely, "Crystal Surfaces" (Pergamon Press, 1973)
J. M. Blakely (Ed.) "Surface Physics of Materials" Vols. 1 and 2 (Academic Press, 1975)
G. Ertl and J. Küppers, "Low Energy Electrons and Surface Chemistry" (Verlag Chemie, 1974)
W. A. Harrison, see the Bibliography, Chapter 10
G. A. Somorjai, "Principles of Surface Chemistry" (Prentice-Hall, 1972)

The preparation of atomically clean surfaces is discussed in R. G. Musket, W. McLean, C. A. Colmenares, D. M. Makowiecki and W. J. Siekhaus, Applications of Surface Science **10**, 143 (1982).

As one example of surface reconstruction of a more complicated crystal see the work on the {110} face of SnO_2 [de Fressart, J. Darville and J. M. Gilles, Solid State Comm. **37**, 13 (1981)] and the many references listed there.

A new device called a **scanning tunneling microscope** has been built and has shown to be useful in surface structure and related studies. It consists of translating a very sharp (≈ 10 Å diameter) metal tip over (~ 10 Å above) a surface. By vertically moving the metal tip, the tunneling current between it and the surface is kept constant. This vertical motion quantitatively reflects the topography of the surface. This technique, introduced in 1982, has the remarkable property of being able to "see" atom positions in real space and is being applied to study surface reconstructions, adatom positions and many related problems. [G. Binnig, H. Rohrer, Ch. Gerber, and E. Weibel, Appl. Phys. Lett. **40**, 178 (1982), Phys. Rev. Lett. **49**, 57 (1982), and ibid. **50**, 120 (1983). Reviews of this technique published by Binnig, Rohrer and their collaborators are: Physica B and C **127**, 37 (1984) and Surface Sci. **144**, 321 (1984).]

For a general account of **synchrotron radiation** see the several articles in the May 1981 issue of Physics Today. Also see H. Winick and S. Doniach, Eds., "Synchrotron Radiation Research" (Plenum Press, 1980); C. Kunz, Ed., "Synchrotron Radiation" (Springer-Verlag, (1979); E. E. Koch, Ed., "Handbook on Synchrotron Radiation" (North Holland, 1983). Further additions to this latter book are planned.

For a recent review of the status of **LEED** for the determination of surface structures see P. M. Marcus and F. Jona, Applied Surface Science **11/12**, 20 (1982). D. Haneman, Adv. in Phys. **31**, 165 (1982) covers the surface structure of semiconductors. A book covering all the aspects of this field is M. Van Hove, C. M. Chan, and W. H. Weinberg, "Low-Energy Electron Diffraction" (Academic Press, 1985). Also see S. Y. Tong, Physics Today **37**, 50 (1984).

The original **He diffraction** work is, J. Estermann and O. Stern, Z. Phys. **61**, 95 (1930). A good review is T. Engel and K. H. Rieder, "Structural Studies of Surfaces," Springer Tracts in Modern Physics **91**, 55 (1982). For a more recent review see I. P. Batra, Surface Sci. **148**, 1 (1984) and the rest of the conference proceedings pp. 1-224.

There are many surface spectroscopic methods other than are discussed in this chapter. For example see H. Ibach and D. L. Mills, "Electron Energy Loss Spectroscopy and Surface Vibrations" (Academic Press, 1982) and J. E. Demuth and Ph. Avouris, Physics Today **36**, 62 (1983). The inelastic scattering of He atoms from solids is described by G. Brusdeylins, R. G. Doak, and J. P. Toennies, Phys. Rev. Letters **44**, 1417 (1980) and **46**, 437 (1981). Angle resolved photoemission studies are discussed by F. J. Himpsel, Adv. in Phys. **32**, 1 (1983).

There are many conferences on various topics in surface science. For example: "Vibrations at Surfaces" Ed. by R. Caudano, J. M. Gilles, and A. A. Lucas (Plenum Press, 1982) or "Vibrations at Surfaces," Ed. C. R. Brundle and H. Morawitz (Elsevier, 1983).

Problems

1. Derive Eq. 17-1b.

2. **Surface packing density** – (a) Determine the surface packing density, ρ, for the fcc structure $\{111\}$ and $\{110\}$ planes. (b) Derive the formulas, given in Fig. 17-4, for the surface packing density in any lattice plane of the fcc and bcc structures. (Hint: Determine the relationship between V_a and A_a, then derive a relation between ρ, A_a, and r_a.)

3. Discuss the use of **Auger spectroscopy** in gases.

4. For the free electron model derive the **Richardson–Dushman equation.** How would band effects change this result?

Appendix to Chapter 17 A list of some techniques used to study surfaces. The procedures are classified according to incident and emitted radiation. A more complete list is in M. J. Higatsberger, Adv. in Electronics and Electron Phys. **56**, 291 (1981) and S. Y. Tong, Physics Today **37**, 50 (1984).

Radiation			
Incident on surface	Emitted from surface	Surface analytical techniques	Abbreviations
Electrons	Electrons	Auger electron spectroscopy	AES
		Electron energy loss spectroscopy	EELS
		High energy electron diffraction	HEED
		Low energy electron diffraction	LEED
		Reflection high energy electron diffraction	RHEED
	Ions	Electron stimulated desorption ion angular distributions	ESDIAD
	Photons	Appearance potential spectroscopy	APS
Ions	Ions	Ion scattering spectroscopy	ISS
		Secondary ion mass spectrometry	SIMS
Photons	Electrons	Electron spectroscopy for chemical analysis	ESCA
		X-Ray photoelectron spectroscopy	XPS
		Ultraviolet photoelectron spectroscopy	UPS
	Neut. Part.	Photostimulated desorption	PSD
Elect. Field	Electrons	Inelastic electron tunneling spectroscopy	IETS
		Field emission microscopy	FEM
	Ions	Field ion microscopy	FIM

18

Artificial Structures

PART A SEMICONDUCTORS

18-1 Introduction
18-2 A Particle in a 1-D Rectangular Well
    a An infinitely deep well
    b A finite well
18-3 3-D Motion with a 1-D Rectangular Well
    a $\mathscr{E}$ vs. k
    b Optical absorption experiments
    c Resonant tunneling experiments
18-4 Experimental Aspects
18-5 Semiconductor Superlattices
    a Introduction
    b Negative differential conductivity
    c Optical absorption
    d Folded acoustic phonons
    e Inelastic light scattering
    f Modulation-doped semiconductor heterojunction superlattices
    g Doping superlattices (nipi structures) (S)
    h The InAs–GaSb superlattice and others (S)
18-6 Inversion Layers
    a Introduction
    b MOSFET
    c Subbands
    d General comments
    e MODFETS

PART B METALS

18-7 Introduction
18-8 Sample Preparation
18-9 Properties of Layered Metal Structures
    a Electrical transport
    b Magnetic properties
    c Superconductors
18-10 Other Artificial Structures (S)

Notes
Problems

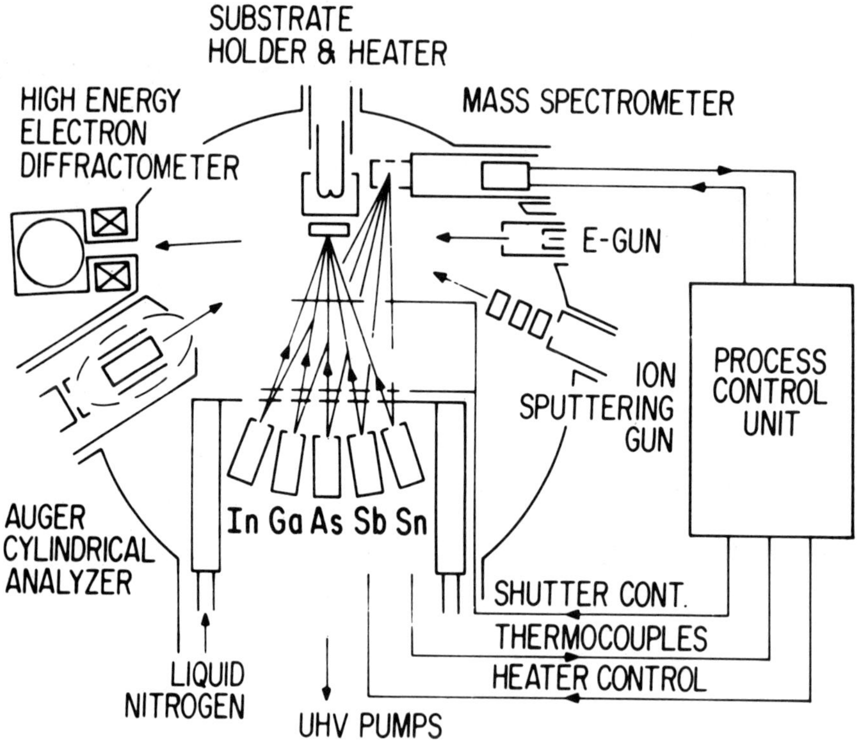

Fig. 18-8 A schematic of a typical molecular beam epitaxy system for the deposition of an InAs-GaSb superlattice. [L. L. Chang and L. Esaki, Surface Science **98**, 70 (1980).]

ARTIFICIAL STRUCTURES

O, wonder!
How many goodly creatures are there here!
How beauteous mankind is!

W. Shakespeare, "The Tempest"

It is now possible to make and study various kinds of "artificial" structures by growing the crystals from two or more layers of different materials. In Part A, artificial semiconductor structures are discussed. The main areas of endeavor are (a) superlattices (two different semiconductors in a periodic layered arrangement); (b) inversion layers (made of a metal-oxide-semiconductor, MOS, or with a heterojunction made from two semiconductors). With these structures we can engage in "band gap engineering," meaning that transport and other properties of the electrons and holes can be altered continuously and independently, leading to interesting physics and new classes of semiconductor devices. In Part B, we will consider layered metal structures. Conductivity, magnetism, and superconductivity effects in these metals are discussed.

As often happens in new fields, the terminology is confusing. The term "superlattice" has been used for a long time by crystallographers to mean an additional periodicity in the system (larger than the primitive unit cell). Semiconductor people tend to reserve this term for systems where the **periodicity is smaller than the electron mean free path**; thus the materials are electronically coupled to each other in a way that affects some standard semiconductor measurement. These materials have been called electronic superlattices to distinguish them from crystallographic superlattices. Periodic arrays of metals are called metallic superlattices or just layered crystals. Others reserve the term superlattice for systems where quantum effects can be seen. For the semiconductor case, we only discuss superlattices that show quantum effects; hence there are no nomenclature problems. For metals we refer to the materials (amorphous or crystalline) as layered structures. For the very important inversion layer case, where quantum effects are observed, the samples simply are called inversion layers.

We make and study the structures discussed here in order to understand systems with dimensions other than three. Various kinds of structures exist in nature with properties that approximate two-dimensional (2-D) or one-dimensional (1-D) structures. In this chapter we shall discuss structures with 2-D behavior but consider only ones that are artificially produced, that is, structures that do not nor-

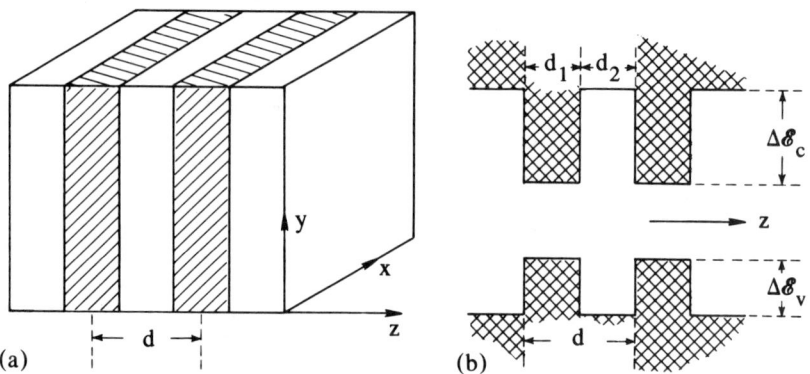

Fig. 18-1 (a) A schematic of a heterostructure superlattice. (b) The energy gap profile for such an artificial structure.

mally exist in nature; man was involved in conceiving as well as in making them, hence the word "artificial" in the chapter title.

Part A – Semiconductors

18-1 Introduction

A good example, and a much-studied artificial structure, is the **semiconductor heterojunction superlattice.** Consider the name of this structure word by word. First, the artificial structure is made from semiconductors. Second, it is formed by the junction of two different semiconductors, for example, InAs with GaSb or GaAs with $(Al_xGa_{1-x})As$. Last, "superlattice" implies that there is an extra periodicity in the material. For the semiconductor case, the structure is made by depositing a thin layer (thickness d_1) of one semiconductor followed by another thin layer of the other semiconductor (thickness d_2), then the first one again, then the second, repeated many times. Thus, there is a new periodicity, $d = d_1 + d_2$, built into the structure. The d-value clearly should be larger than the unit cell size of the individual materials and, to have any overall effect, it must be smaller than the electron mean free path. The period thicknesses are usually between 10 and 500 Å. Figure 18-1a shows the structure. Basically, we are dealing with an artificial structure in which, for a given value of z, there is in the xy-plane essentially an infinite plane of a homogeneous semiconductor, but in the z-direction there are repeated layers of the two semiconductors with this new period.

Why go to all this trouble? The two semiconductors have different band gaps; Fig. 18-1b shows the band gap profile in the z-

CHAPTER 18   ARTIFICIAL STRUCTURES   717

direction. Such a profile can have profound effects on the physical properties of these materials.

More examples of artificial structures could be given, but rather we proceed with detailed studies from this active field of research. We shall cover only the basics of some of the more widely studied phenomena. As in most active research areas, the emphasis shifts rapidly, and the Notes should be consulted for references to review articles and conference proceedings. Because of its relative simplicity in terms of a quantum model, the semiconductor heterostructure superlattice is discussed first.

18-2 A Particle in a 1-D Rectangular Well

18-2a An infinitely deep well A particle of mass m* in an infinitely deep 1-D potential energy well, $V_0 = -\infty$, is an elementary quantum mechanics problem. For a well of length L_z in the z-direction, the wave equation, eigenvalues, and eigenfunctions are

$$-\frac{\hbar^2}{2m^*}\frac{d^2\psi}{dz^2} = \mathcal{E}\psi$$

$$\mathcal{E}_n = \frac{\hbar^2}{2m^*}\left(\frac{n\pi}{L_z}\right)^2 = \left(\frac{\hbar^2\pi^2}{2m^*L_z^2}\right)n^2$$

$$\psi_n = A \sin(n\pi z/L_z) \qquad n = 1, 2, 3, \ldots \qquad (18\text{-}1)$$

The eigenfunctions, which vanish at both ends of the well, are sine waves with energies increasing as n^2. (For a hydrogen-like potential the wave functions are quite different and the energy goes as n^{-2}. So for the 1-D well with infinite depth the energy level spacing increases as n, while for the hydrogen case the spacing decreases at higher n.)

The separations of the energies of the particles in the well go as L_z^{-2} (Eq. 18-1). For the artificial structures discussed here we expect that individual energy levels can be measured, unlike the band case where individual Bloch functions are very closely spaced. Experimental evidence that the energy levels vary as L_z^{-2} indicates that quantum effects are observed; this is called the **quantum size effect**, and the well is a **quantum well**.

Figure 18-2 schematically shows the wave functions and the energy spacings for electrons in the upper quantum well (the conduction band) and holes in the lower quantum well (the valence band). The well in the figure is not infinite but finite, which has serious effects on the wave functions only near the top of the well.

718   CHAPTER 18   ARTIFICIAL STRUCTURES

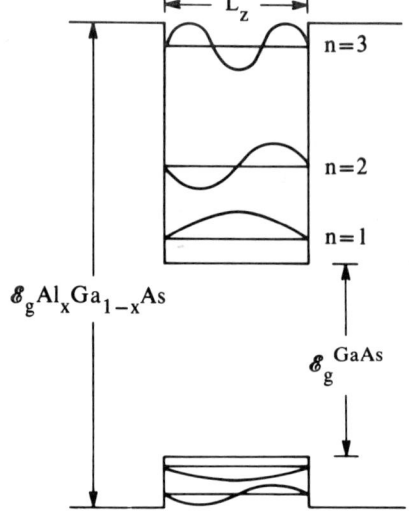

Fig. 18-2 The eigenfunctions of an infinitely deep potential well but shown in two finite wells. The upper well applies to electrons and the lower one to holes.

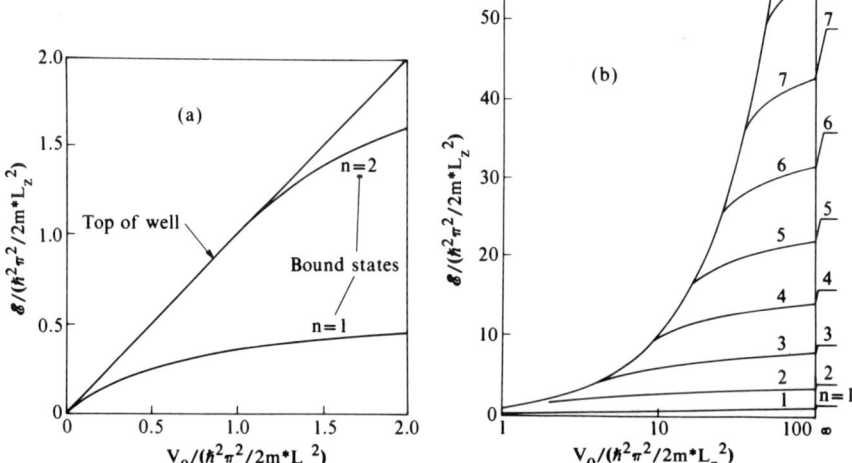

Fig. 18-3 The energy levels of a particle in a potential well of depth V_0. (From R. Dingle; see the Notes.)

18-2b A finite well If the potential energy of the well is not infinite but has a finite depth, V_0, the wave functions are similar to those given in Eq. 18-1 but have decaying exponential into both sides of the well. The solution is an elementary quantum mechanics problem (see the Notes). Figure 18-3 shows $\mathscr{E}_n$ vs. V_0, for the different n values, where both $\mathscr{E}_n$ and V_0 are in reduced units. The results for the infinite quantum well are also indicated. For example, when $V_0/(\hbar^2\pi^2/2m^*L_z^2) \approx 20$ there are five bound states, n = 1, ..., 5; n =

CHAPTER 18 ARTIFICIAL STRUCTURES 719

6 is not bound. The energies of the five bound states are given in the figure. When the particle has an energy greater than V_0, its eigenfunction is that of a continuum state.

18-3 3-D Motion with a 1-D Rectangular Well

18-3a $\mathscr{E}$ vs. k Now that the basic aspects of the 1-D rectangular well are covered, we can move on to a structure closer to the superlattice in Fig. 18-1a. We consider a thin film, that is, very thin in the z-direction but macroscopic size in the x- and y-directions; thus $L_z \ll L_x, L_y$. This film could be a thin film of GaAs in the middle of a very much thicker crystal of $(Al_xGa_{1-x})As$ (i.e., a sandwich).

For a true 3-D case, as discussed in Chapter 9,

$$\mathscr{E} = \frac{\hbar^2}{2m^*}(k_x^2 + k_y^2 + k_z^2) \tag{18-2a}$$

For the thin film, the motion in the x- and y-directions is still free electron-like; thus, k_x and k_y are still good quantum numbers. It is only in the z-direction that there is a sharp quantum well. Thus, $V_0(x, y, z) = V_0(z)$, since the potential energy is independent of x and y. The wave equation can be separated into x-, y-, and z-parts; the first two parts describe free electron behavior and for the z-part we find quantum effects given by Eq. 18-1. Then the energy is

$$\mathscr{E} = \mathscr{E}_n + \left(\frac{\hbar^2}{2m^*}\right)(k_x^2 + k_y^2) \tag{18-2b}$$

where $\mathscr{E}_n$ is given in Eq. 18-1 for the infinitely deep well or can be obtained from Fig. 18-3 for the finite case.

Figure 18-4 shows aspects of Eq. 18-2b. (Naturally, k_x is equivalent to k_y.) Figure 18-4a shows how the parabolic $\mathscr{E}$ vs. **k** is changed because now only certain energy levels are allowed for propagation in the z-direction. Figure 18-4b shows $\mathscr{E}$ vs. k_y, which is just the projection of the curves in Fig. 18-4a onto the $\mathscr{E}$-k_y plane. Thus, each value of $\mathscr{E}_n$ is the bottom (lowest energy) of a two-dimensional continuum called a **subband**. Figure 18-4c shows the density of states vs. energy. We know that for the 2-D free electron case the density of states is constant, independent of energy. (The 2-D density of states is the number of allowed states between energy $\mathscr{E}$ and $\mathscr{E} + d\mathscr{E}$ per unit area. See Problem 10 in Chapter 9.) Thus, the steps in $g(\mathscr{E})$ occur at each allowed value of $\mathscr{E}_n$ in Eq. 18-2b.

720   CHAPTER 18   ARTIFICIAL STRUCTURES

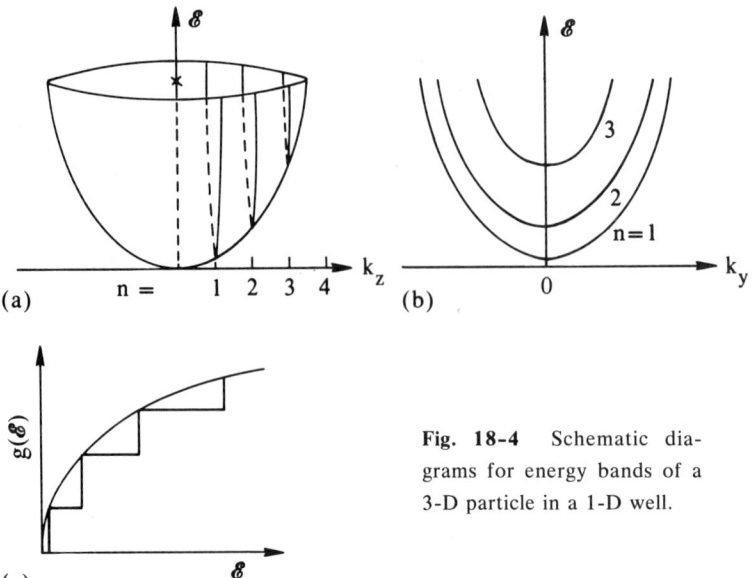

Fig. 18-4 Schematic diagrams for energy bands of a 3-D particle in a 1-D well.

18-3b Optical absorption experiments   The energy levels discussed previously can be observed by optical absorption experiments.

Figure 18-5a shows the energy levels and the experimental results. A thin layer (a well) of GaAs is located in $(Al_xGa_{1-x})As$, where $x \approx 0.25$. The thickness of the GaAs layer is such that only the n = 1 level (Eqs. 18-1 and 18-2b) is bound. Note that absorption no longer starts at an energy corresponding to the energy gap of GaAs but at a somewhat higher energy corresponding to the difference between the lowest energy state of the electrons in the quantum well in the conduction band and the corresponding state of the holes in the valence band. At first thought, just one absorption might be expected to occur from the bound (n = 1) hole to the bound (n = 1) electron. However, as discussed in Section 10-15 and also shown in Fig. 13-12, the valence band in GaAs is more complicated; there are heavy holes (hh) and light holes (ℓh) and each acts as a band unto itself. In Eqs. 18-1 and 18-2 the effective mass is that of m_e, m_{hh}, or $m_{\ell h}$, depending on whether the energy levels in the electron, heavy-hole, or light-hole bands are being considered. Thus, two absorptions are expected and observed. The optical absorption from one well is very weak. In order to enhance the absorption, the experiment is performed (at 2°K) with very many 50 Å GaAs wells separated by thick layers of $(Al_xGa_{1-x})As$, so that tunneling between the wells is negligible. This approach improves the signal compared to what would be observed for just one well.

The credibility of the preceding results is enhanced by the data shown in Fig. 18-5b. In this experiment two wells, each still 50 Å

CHAPTER 18 ARTIFICIAL STRUCTURES 721

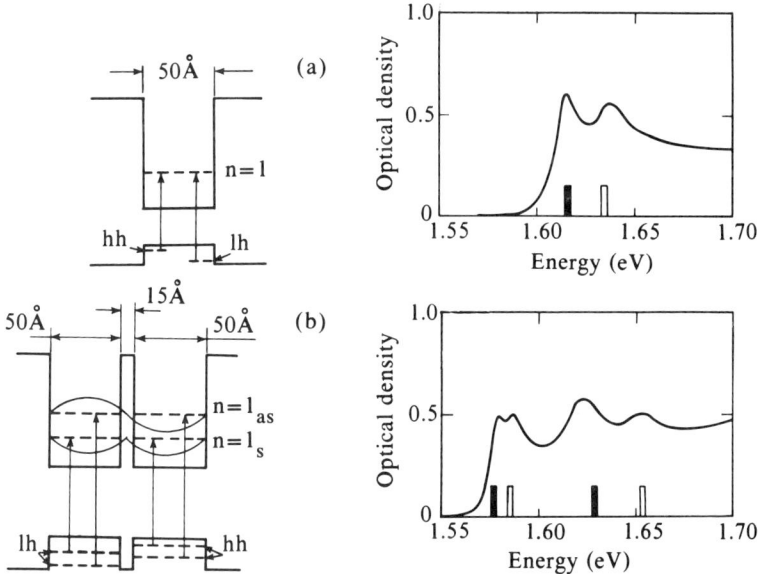

Fig. 18-5 (a) Optical absorption from a single GaAs well. (b) The same but for two coupled GaAs wells. [R. Dingle, A.C. Gossard, and W. Wiegmann, Phys. Rev. Lett. **34**, 1327 (1975).]

wide, were grown only 15 Å apart (i.e., separated by 15 Å of $(Al_xGa_{1-x})As$). Since the eigenfunctions in the two wells are strongly coupled, symmetric and antisymmetric combinations of the single well, n = 1 eigenfunctions occur. These are shown in the figure for the electron states. The coupling occurs for the hole states but it is not shown because of space limitations. Although there are now six coupled-well states within this n = 1 manifold, only four transitions are observed because of selection rules. In addition to the single well Δn = 0 selection rule, an overlap selection rule is also obeyed, which allows only transitions from symmetric to symmetric states as well as from antisymmetric to antisymmetric states. Thus, only four transitions are expected and observed, and the calculations agree with the experiment. (Again, the actual experiment is performed using samples with many pairs of wells isolated from adjacent pairs by distances large compared to the mean free paths.)

Three (and more) coupled-well structures have been measured and the results also are in reasonable agreement with theory. See the reference in the caption of Fig. 18-5b.

Wider 1-D wells – The preceding discussion covers very thin 1-D quantum wells for which only the n = 1 electron and hole levels are bound. What happens if the isolated 1-D well is wider, so that levels with n's higher are bound?

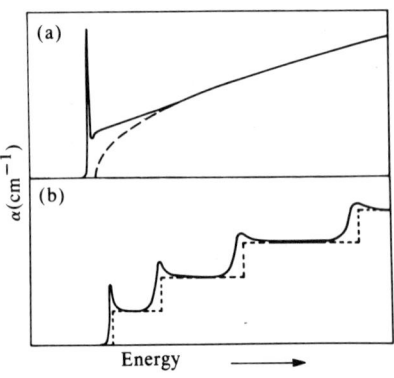
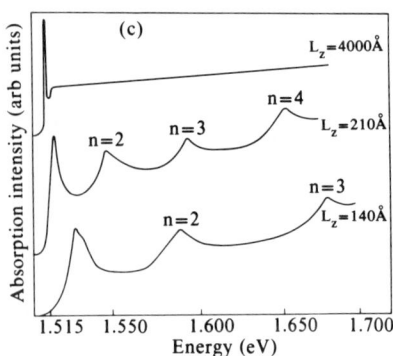

Fig. 18-6 (a) Bulk GaAs band edge absorption including exciton effects. The dashed curve is the density of states. (b) Schematic band edge absorption for a 3-D sample with a 1-D well as discussed in the text. (c) Experimental absorption spectra (2°K) as discussed in the text. [From: R. Dingle, W. Wiegmann, and C. H. Henry, Phys. Rev. Lett. **33**, 827 (1974).]

The dashed line in Fig. 18-6a shows the parabolic density of states in bulk GaAs. The band edge absorption is proportional to $g(\mathscr{E})$ except very near $\mathscr{E}_g$. Near the gap, the Coulomb attraction between the electron and the hole in the valence band leads to exciton formation (Section 13-20); just below and near $\mathscr{E}_g$ the exciton is the dominant cause of absorption. The expected absorption is shown by the solid line in Fig. 18-6a.

Figure 18-6b shows the density of states for a particle that can move in all three directions but with a 1-D well in one direction. The diagram is similar to Fig. 18-4c, however, for the absorption constant we have sketched exciton peaks just below each of the energy levels in $\mathscr{E}_n$. The absorption in the $(Al_xGa_{1-x})As$ layers which separates the 1-D wells of GaAs is not shown because it is at higher energies.

Figure 18-6c shows some experimental results. The upper curve is absorption for a very thick film of GaAs in $(Al_{0.2}Ga_{0.8})As$. The result is essentially indistinguishable from bulk GaAs. However, for $L_z = 210$ Å, electron eigenstates up to $n = 4$ are bound so that absorption to these levels is observed. For thinner films the energy levels separate approximately as L_z^{-2}, illustrating the quantum size effect. Notice, for $L_z = 140$ Å, the $n = 1$ absorption line shows some structure. This is the splitting due to separate absorptions from the heavy- and light-hole bands; for thinner samples it becomes more marked and has been discussed previously and is seen in Fig. 18-5.

The agreement between theory and the experimental results in Fig. 18-6c is very good, as is the agreement with the data in Fig. 18-5. Thus, the basic aspects of these heterostructures are well understood.

CHAPTER 18 ARTIFICIAL STRUCTURES

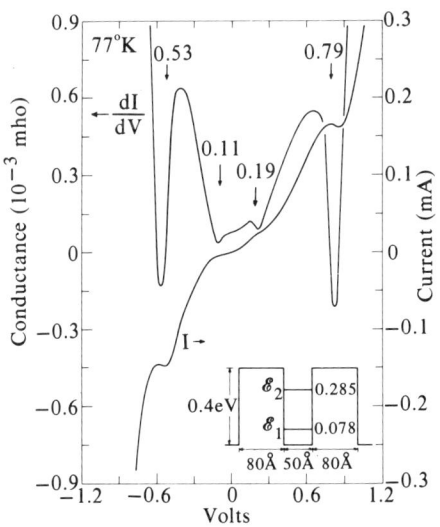

Fig. 18-7 Current and conductance characteristics of a double barrier structure of GaAs between two $Al_{0.7}Ga_{0.3}As$ layers all in a GaAs crystal. [L. L. Chang, L. Esaki, and R. Tsu, Appl. Phys. Letters **24**, 593 (1974).] For more recent measurements see T. C. L. G. Sollner, et al. Appl. Phys. Lett. **43**, 588 (1983).

18-3c Resonant tunneling experiments In a manner similar to the optical experiments, measurements of the current vs. voltage in the appropriate structure show that Eq. 18-2b is valid and that the artificial structures in which quantum effects are observed, indeed, can be produced.

In the insert of Fig. 18-7 a GaAs crystal in which two layers of $(Al_{0.7}Ga_{0.3})As$ serve as barriers for a 50 Å middle layer of GaAs. The energy levels, $\mathscr{E}_n$, for the thin GaAs layer are indicated. To understand what happens, consider electron waves with different energies traveling from the left to the right. If the energy of the incident electrons coincides with the quantum n = 1 and 2 states (at 0.078 and 0.285 eV) indicated in the figure, then the electrons tunnel across the barriers without attenuation. This unit transmissivity arises because, for the electron waves of energies $\mathscr{E}_1$ or $\mathscr{E}_2$, the reflected waves from the first barrier are cancelled by the waves that leave from the well and are going in the same direction; thus only the transmitted waves remain. This is called resonance and it occurs only when the energy of the incident electron wave is the same as a quantum level of the well between the barriers. (This is the same effect as in a Fabry Perot etalon.)

For positive voltage, the current vs. voltage data in the figure shows a local maximum at a voltage approximately equal to two times the values of $\mathscr{E}_n$. The factor two occurs because $\mathscr{E}_n$ is calculated from the origin at the center of the thin GaAs layer, but the voltage drop occurs across the entire double barrier structure. Thus, only one-half of the voltage has dropped between the beginning of the barrier and the origin. The differential conductance is also shown; this displays the energy levels in a much clearer manner. (In calculating the tunneling-voltage characteristic we make the assumption that the momentum, transverse to the z-direction, is conserved at the barriers; this is called **specular tunneling.** Also, it is not a single electron that has an energy = $\mathscr{E}_n$; rather the electron current should peak whenever the Fermi energy in the electrode comes into resonance with a quantum state of the enclosed potential well. This is because it is the electrons at the Fermi energy that dominate the current.)

This experiment demonstrates quantization and its effects on transport. Two other results are also shown. First, as in the optical absorption case, the theory agrees with the experiments, giving us confidence that the behavior of these thin semiconductor layers is understood. Second, in the tunneling experiments, a negative conductivity is observed and this has potential device applications.

By now we should wonder how these very narrow wells with such sharp walls are fabricated. This is the topic of the next section.

18-4 Experimental Aspects

Molecular beam epitaxy (MBE) – The molecular beam epitaxy process refers to a technique in which several atomic (or molecular) beams impinge on a heated substrate material under ultrahigh vacuum conditions (Section 17-1). New layers are single crystalline, in registry with the substrate. (This latter aspect is called **epitaxy.**) Figure 18-8 (page 714) shows a typical MBE system for the fabrication of InAs-GaSb superlattices. A separate source (an **effusion cell**) is used for each constituent and one for Sn, which is used as a dopant if desired. These atomic beams flow from the effusion cells at thermal velocities along the line of sight to the substrate. The temperature of each of these five effusion cells is individually controlled. The flow shutters are turned on and off by a computer whose input comes via a mass spectrometer, which monitors the various flux rates. Growth of the superlattice is typically ≈ 1 to 3 Å/sec. The temperature of the substrate is separately controlled and kept as low as possible, compatible with good epitaxy. The low temperatures minimize diffusion effects, interlayer strain effects when the sample is cooled to room temperature, and results in very abrupt interfaces.

Before evaporation the substrate is first chemically etched, then quickly put in the vacuum chamber, and heated to remove the oxide. Sometimes it is cleaned by ion sputtering (typically Ar^+) and monitored by the Auger analyzer similar to what is done for surface measurements (Section 17-3). The high energy electron diffractometer (HEED—Appendix to Chapter 17) monitors the smoothness of the films and the oxide on the surface. 1-D wells, coupled 1-D wells, and true superlattices made with MBE have very sharply defined interfaces with accurately controllable thicknesses.

The sticking coefficients, S, (Section 17-1) is an important practical concern. For a silicon atom on a silicon substrate it is unity essentially at all temperatures. However, for dopants that we might want to use, S can be considerably less than 1. For example, for Sb on Si the sticking coefficient is ≈ 1 below $550°C$ but falls to $\approx 10^{-3}$ at $1000°C$. If we had to dope Si at $1000°C$ with Sb, the Sb atoms would end up all over the UHV apparatus with disastrous effects on future samples. We mention this so you can appreciate the ultra clean conditions that must be used in this field.

Other techniques, such as vapor phase epitaxy (VPE) may begin to play a role, but for now MBE continues to dominate the field. At present, MBE machines are commercially available.

X-ray – After a superlattice is made there are several simple, standard experimental techniques to check its existence. The most obvious technique is x-ray diffraction from the new periodicity that has been built into the material. From Bragg's law, Eq. 4-3, for d $\approx$ 10^2 Å, the scattering angle $\approx 1°$ for typical x-ray wavelengths, and we can see several orders of diffraction. From detailed considerations of the intensities of the higher order scattered diffraction lines we can estimate roughness of the interfaces.

TEM – Transmission electron microscopy can be used to "see" the different layers in the superlattice since the absorption of the electron waves is different in the two materials. The specimens must be thin since the absorption of the electron waves is so large, but this is always the situation in TEM and the normal experimental methods work well on semiconductors. The TEM results directly yield values for d_1 and d_2 (Fig. 18-1).

We should note that while layered structures can be made (as determined by x-rays and TEM), the true test of the existence of a superlattice, at least as used in the electronics field, is the observation of quantum effects.

Strain layered superlattices – Originally, in making semiconductor heterojunction superlattices, precautions were taken to use two different materials with the same size unit cell, that is, $(\Delta a/a) \lesssim 0.1\%$. It was felt that only by keeping within this close tolerance could a good match be obtained at the boundaries so that the interface would be free of misfit dislocations and other imperfections; poor quality interfaces give rise to charged surface states that adversely affect the electrical properties. However, this close tolerance severely restricts the number of different semiconductors that can be used. It turns out that this restriction is important only for thick layers. For superlattices that are thin, the lattice mismatch can be large. Excellent superlattices can be grown for a $(\Delta a/a)$ mismatch of several percent or larger if the thicknesses, d_1 and d_2, are less than 100 Å. The mismatch is taken up in strain of both of the materials with the strain energy per unit area being less than the energy to form dislocations.

To understand this, take the directions as defined in Fig. 18-1a and assume that the material labeled 1 has a larger unit cell size than that labeled 2. Then for the strain layered superlattice, in the xy-plane the unit cells adjust in size so that they have the same dimension, that is, materials 1 and 2 are compressed and expanded, respectively. To compensate, the materials expand and contract, respectively, in the z-direction; these strains are what would be calculated from the elastic constants, given the strains in the xy-plane.

Thus, from strain layered materials we can obtain high quality superlattice and quantum wells starting with nonlattice matched semiconductors, which allows many more combinations of systems to be studied. The electronic properties are effected by the large strains but we can take advantage of these effects. However, under high power applications these superlattices may tend to degrade.

18-5 Semiconductor Superlattices

The physics behind the ideas of the semiconductor heterojunction superlattice have been discussed as well as the MBE method of preparation. We now discuss a few of the properties and uses of the superlattices themselves to give a flavor of this growing field. The Notes should be consulted for further reading.

18-5a Introduction The thin curve in Fig. 18-9 represents a typical electron $\mathscr{E}$ vs. **k** to $k = \pi/a$ for a crystal with lattice constant a. Near the bottom of the band at $k \approx 0$ (i.e., the conduction band edge) the electron has a positive effective mass, while near the Brillouin zone boundaries ($k \approx \pi/a$) the effective mass is negative. If under the application of an applied electric field an electron could be excited from the conduction band edge into the negative mass region, then negative conductivity might be observed (similar to that shown in Fig. 18-7) or perhaps other high-field phenomena could be found (for example, Gunn oscillations). However, typical conduction band widths are the order of several electron volts and before the electrons gain much energy, they are scattered by electron-phonon or electron-electron interactions. If the band width could be made much narrower, then there is a better chance to observe high-field phenomena.

It was this sort of thinking that led Esaki and Tsu in the 1960s to the idea of the semiconductor heterojunction superlattice. They realized that by building in a much larger periodicity, =d, folding the Brillouin zone (Section 12-3) and forbidden gaps could be obtained. The

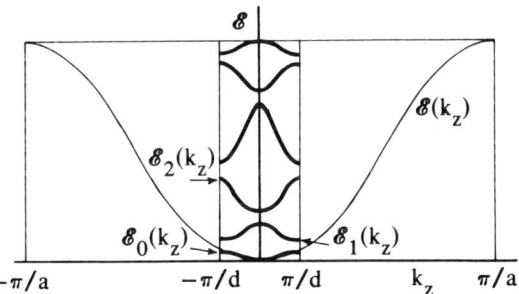

Fig. 18-9 The splitting of a band into subbands by a periodic superlattice potential of period d. There are d/a subbands.

widths of the **minibands** or **subbands** would be very much narrower than the original band. The dark lines in Fig. 18-9 show the splitting of the original band into subbands, necessarily much narrower, and the new gaps also are shown. In normal crystalline semiconductors, the interaction between the free carriers and the longitudinal optic (LO) phonons at frequencies ω_{LO} is a strong relaxation mechanism for the carriers. However, the subband width is typically smaller than $\hbar\omega_{LO}$, thus preventing this particular relaxation process and increasing the possibility of observing high-field phenomena.

The first work in this field was published in 1970; what might be called "early work" refers to work up to about the middle of the 1970s. It was not until about 1978 that a large number of papers started appearing and now the number is increasing exponentially.

18-5b Negative differential conductivity The observation of negative differential conductivity due to tunneling into quantum levels in a simple structure is described in Section 18-3c. Such measurements have been extended to superlattices (i.e., many layers of the two different semiconductors rather than the three in Fig. 18-7). Indeed negative conductivity is observable at modest temperatures (above 77°K). However, usable devices have not resulted as yet.

18-5c Optical absorption Results from optical absorption experiments from single wells and two coupled wells for n = 1 levels are described in Section 18-3b (Fig. 18-5). The results for wider single wells, with more energy levels, are also discussed (Fig. 18-6c). When more coupled wells, or wells with more n-values are measured, optical absorption tends to be continuous and of little quantitative use.

18-5d Folded acoustic phonons Figure 18-9 shows the splitting of an electronic band into subbands due to a folding of the Brillouin zone. This folding also occurs in the phonon dispersion curves in precisely the same way, giving rise to new k ≈ 0 phonons, which are

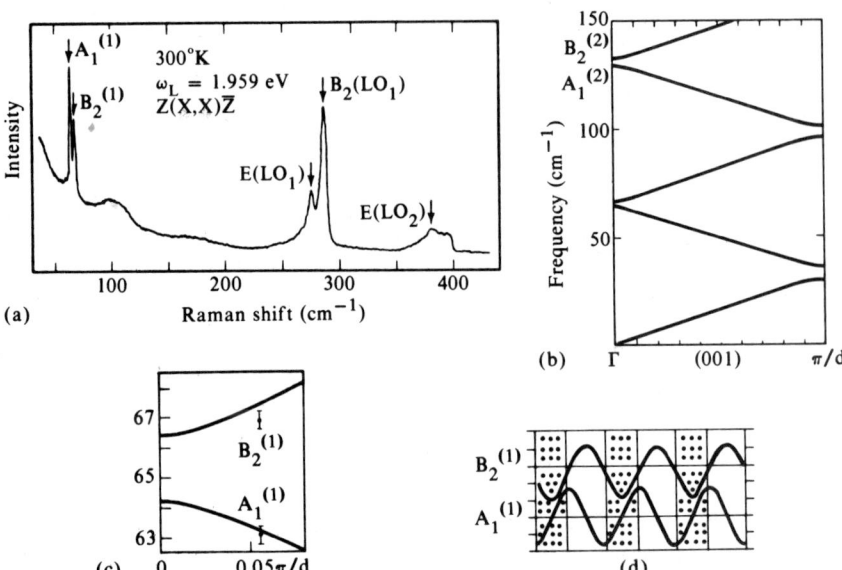

Fig. 18-10 (a) Raman spectra of a superlattice consisting of 1720 periods of 13.6 Å GaAs and 11.4 Å AlAs. The directions are arranged so that only longitudinal phonons are observed. [C. Colvard, R. Merlin, M. V. Klein and A. C. Gossard, Phys. Rev. Letters **45**, 298 (1980).] (b) Dispersion of the LA phonons in the superlattice. (c) A detail of the 65 cm^{-1} region at k ≈ 0. (d) The amplitude of the two observed LA folded phonons. The shaded region is one material and the white region the other material. [There are several articles on this subject in Journal de Physique **45**, Colq. C5 (1984) and see C. Colvard, et al. Phys. Rev. B, **31**, 2080 (1985).]

observable by Raman or infrared techniques.

Figure 18-10a pictures the room temperature, k ≈ 0 Raman spectra of a AlAs-GaAs superlattice showing the effect of the folding of the Brillouin zone. The longitudinal acoustic (LA) subbands are shown in Fig. 18-10b and a detail of the 65 cm^{-1} region is shown in Fig. 18-10c. Both GaAs and AlAs have the zinc blende crystal structure with point group $T_d(\bar{4}3m)$. However, the superlattice with a new periodicity in the z-direction has tetragonal symmetry, with point group $D_{2d}(\bar{4}2m)$. The phonons transform as the different irreducible representations of the D_{2d}, and not the T_d point group, which accounts for the labeling - A_1, B_2, and E.

From Fig. 18-10b, due to the folding, the k ≈ 0, LA lattice vibrations are at ≈130 cm^{-1} and ≈65 cm^{-1}. However, the former are expected to be weaker. The other modes in Fig. 18-10a are longitudinal optic (LO) modes. Finally, the amplitudes (eigenvectors) vs. z of the folded $A_1^{(1)}$ and $B_2^{(1)}$ phonons are shown in Fig. 18-10d. If the superlattice were homogeneous these two vibrations would have the same energy.

18-5e Inelastic light scattering By using the Raman technique (Section 13-10c), we can measure the electronic energy levels in the quantum wells. This is sometimes called **subband** or **sublevel spectroscopy**. As in any inelastic scattering technique, the sample is irradiated with a beam of photons of a known energy, $\mathcal{E}_i$. The sample has certain energy levels whose separation is $\hbar\omega$. Some of the incident particles interact with the sample and emit quanta of energy $\hbar\omega$. Thus, the final energy, $\mathcal{E}_f$, is lower than $\mathcal{E}_i$ by $\hbar\omega$ ($\mathcal{E}_f = \mathcal{E}_i - \hbar\omega$) and the **energy shift** clearly is $\mathcal{E}_i - \mathcal{E}_f = \hbar\omega$. The emission of $\hbar\omega$, which occurs at all temperatures including $0°K$, is called the **Stokes process**. At finite temperatures the energy levels can be populated so that a quantum $\hbar\omega$ can be absorbed by the incident particles resulting in $\mathcal{E}_f = \mathcal{E}_i + \hbar\omega$, which is called the **anti-Stokes process**. When the incident particles are photons, the inelastic scattering is called **Raman scattering**. When this technique is used to study electronic energy levels of the sample, then the process is called **electronic Raman scattering.**

Electronic Raman spectroscopy has proved to be a useful technique for measuring and studying the subband quantum levels. The observed shifts in frequency arise from intersubband excitations, such as between the n = 1 and n = 2 states in a quantum well. Thus, in principle the measurements give information similar to that obtained from the optical absorption experiments described in Section 18-3b. However, the physics of the measurement technique is different. In an optical absorption measurement we excite from the quantum states in the valence band to the quantum states in the conduction band. In an electronic Raman measurements the end result of the measurement is to transfer an electron from an n = 1 to an n = 2 state in the quantum wells in the conduction band. This same effect occurs with infrared measurements.

For these measurements there are important selection rules that distinguish between single particle and collective excitation spectra. These spectra are complicated, but suffice it to say that by varying the polarization of the incoming and outgoing light, we can obtain single particle spectra and separately collective intersubband excitations. The single particle spectrum consists of transitions between, for example, the energy levels in the conduction band quantum well shown in Fig. 18-2. The intersubband excitations are "collective" because the electronic excitations from n = 1 to 2 are all in phase; they set up a polarization field, which increases the excitation energy above the single particle values. Furthermore, in a polar material such as GaAs (as opposed to a nonpolar material such as Si which has a center of symmetry), the charge density excitations couple to the longitudinal optic (LO) phonons; this is the same effect as that shown in Fig. 13-26b and in Problem 8 of that chapter.

There are interesting experimental aspects of this work. In general, the intensity of the spectra is very weak. It is observed only by doing resonance Raman spectroscopy; that is, the exciting laser line is close to a strong, real interband (conduction to valence band) excitation. For the work discussed in this subsection, the laser line energy normally is chosen to be close to the k = 0 split-off valence band to conduction band energy (Fig. 10-19 and Fig. 13-12) which occurs at about 1.90 eV. (In the literature this is usually called the $\mathscr{E}_0 + \Delta_0$ gap, where the 0 implies gaps at k = 0 as opposed to energy gaps at the L, X, or other points. This effect is not observed at the lowest direct gap because too much luminescence occurs at this energy.) The resonance condition is fulfilled for electrons in the conduction band but not for the holes at the valence band edge. Therefore, normally the properties of the electrons are measured. Another interesting observation is that the widths of the observed lines decrease considerably as the mobility of the electrons perpendicular to the z-axis (i.e., within the 2-D plane layer) increases. This is attributed to an increase in the lifetime of electrons due to a decrease in the scattering. The method of increasing the mobility is the subject of the next subsection.

18-5f   Modulation-doped semiconductor heterojunction superlattices The early, unintentionally doped GaAs-(Al_xGa_{1-x})As superlattices were slightly p-type. With uniform Si doping a donor density of n ≈ 10^{18} cm^{-3} is easy to achieve; the electron mobility in the xy-plane, perpendicular to the z-axis, is typically $\mu \approx$ 3000 cm^2 V^{-1} sec^{-1} and independent of temperature. Normally, the dominant sources of carrier scattering are phonons and impurities. Phonons are the more important of the two at high temperature but can be eliminated at low temperatures since the states are not occupied. As we shall discuss, impurity scattering (coulombic scattering from charged doping impurities) can be greatly reduced by modulation doping leading to very large mobilities.

Figure 18-11a shows the energy levels in an undoped superlattice. (In GaAs the charge is not uniformly distributed since the electrons are in quantum states, Eq. 18-1. This results in more electron density in the center of the GaAs slab than near the walls of the well, but for simplicity we ignore this small nonuniformity in Fig. 18-11a.) A uniformly doped superlattice is shown in Fig. 18-11b. The electrons on the donors in the (Al_xGa_{1-x})As layers are ionized and seek their lowest energy, which is at the conduction band edge of the GaAs. Thus, we have positively charged layers of (Al_xGa_{1-x})As followed by negatively charged layers of GaAs. The charge density, ρ, is a function of z only, $\rho(z)$.

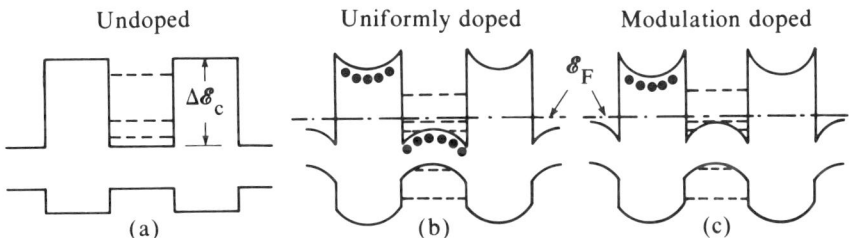

Fig. 18-11 A semiconductor heterojunction superlattice with various methods of doping. [R. Dingle, H. L. Störmer, A. C. Gossard and W. Wiegmann, App. Phys. Lett. **33**, 665 (1978).]

Given a $\rho(z)$, to obtain the potential energy as a function of z, $V_0(z)$, we proceed via Poisson's equation. From Maxwell's equation, Eq. 13-14, $\nabla \cdot \mathbf{D} = 4\pi\rho$, and $\nabla \cdot \mathbf{E} = 4\pi\rho/\varepsilon$, where ε is the dielectric constant of the semiconductor. The relation between the potential, ϕ, and the electric field, $\mathbf{E}$, is $\mathbf{E} = -\nabla\phi$; Poisson's equation thus is obtained, namely $\nabla^2\phi = -4\pi\rho/\varepsilon$. (The calculations in Section 10-17 are totally analogous.) The potential energy is $V_0 = q\phi$ where q is the charge of the particle, in this case it is an electron, so $q = -e$. The potential energy, then must be put into the Schrödinger equation and the charge density obtained is then reinserted into Poisson's equation. This is all done self-consistently (i.e., until there is little difference between the input charge density and the result). Figure 18-11b shows the result. Integrating the linear electric fields, we get parabolic potential energies as shown in Fig. 18-11b. The band bending shown is only schematic. For example, for lightly doped materials this rounding effect is small; also, since the band bending goes as the square of the distance from the interface, the effect is small in thin layers. However, for quantitative results these effects must be considered. The Fermi level $\mathscr{E}_F$ is determined by the number of carriers and is shown for a typical carrier concentration.

Figure 18-11c shows a modulation-doped heterostructure superlattice. The doping is done during the MBE growth (Fig. 18-8) by opening the Sn shutter at the proper times. The result is that the ionized donor impurities are present only in the $(Al_xGa_{1-x})As$ layers while the electrons are in the GaAs layers; the carriers (the electrons) are *spatially separated* from their parent donors. Even in the first such modulated structures the mobility in the x- or y-directions doubled and became greater than 15000 cm^2 V^{-1} sec^{-1} at low temperatures. Clearly the ionized impurity scattering was greatly reduced.

Since the early work in 1978, improvements have been made. Structures are made with an intermediate undoped layer of $(Al_xGa_{1-x})As$ between the doped $(Al_xGa_{1-x})As$ and the GaAs, which further separates the ionized impurities from the carriers. Great care is taken to have sharp and pure interfaces and very pure and dislocation-free GaAs layers. We can obtain an asymptotic expression for the mobility of the carriers in a degenerate 2-D sheet of n carriers/cm^2 separated by a distance d from a 2-D sheet of n_I ions/cm^2. This expression, valid at large d, is

$$\mu \approx 16\sqrt{\pi}ed^3 \; n^{3/2}/\hbar n_I \tag{18-3}$$

In bulk semiconductors μ decreases with increasing n. However, when the carriers and ions are physically separated as described here, μ increases with increasing carrier concentration, n, because the usual scattering from the ionized donors is decreased sharply. Also note that $\mu \propto d^3$. In Section 18-6e, we will come back to structures that use the modulation-doped concept.

18-5g Doping superlattices (nipi structures) In the title of Section 18-5 the word "heterojunction" has been omitted although until now we have been discussing only such materials. The omission is intentional; in this subsection we discuss semiconductor superlattices but not semiconductor heterojunction superlattices. Rather the periodic structure (the superlattice) is obtained by a periodic variation of the doping. That is, the MBE method of crystal growth is used typically for GaAs, but a periodic sequence of n-doped and then p-doped material of thickness d_n and d_p, respectively. There also can be an intrinsic layer between the n- and p-layers. Döhler has named these superlattices **nipi structures.**

There are a few obvious advantages of **doping superlattices** compared to heterojunction **compositional superlattices.** The incorporation of dopants affects only a very small fraction of the atomic sites and is therefore a relatively minor perturbation in an otherwise homogeneous crystal. Also, unlike the heterojunction superlattices, there are no lattice constant mismatches. A doping superlattice can be made from any material provided n- and p-doping is possible. Because of doping limitations, the periods in nipi superlattices cannot be as small as those in a heterojunction superlattice. However, since there is only one kind of semiconductor, liquid phase epitaxy (LPE) or vapor phase epitaxy (VPE) methods of preparation might be usable.

Producing a periodic potential – First, consider how a periodic potential is obtained in a doping superlattice. In Fig. 18-12a the energy gap as a function of z (in real space) is shown. Neutral donors, N_D, and neutral acceptors, N_A, are periodically incorporated (with a period $d = d_n + d_p$) into the crystal. Conceptually, we may think of the processes as follows. Assume that during growth spatial tunneling between the donors and acceptors does not occur. Then the donors and acceptors remain neutral and the band gap $\mathscr{E}_g$ and the conduction

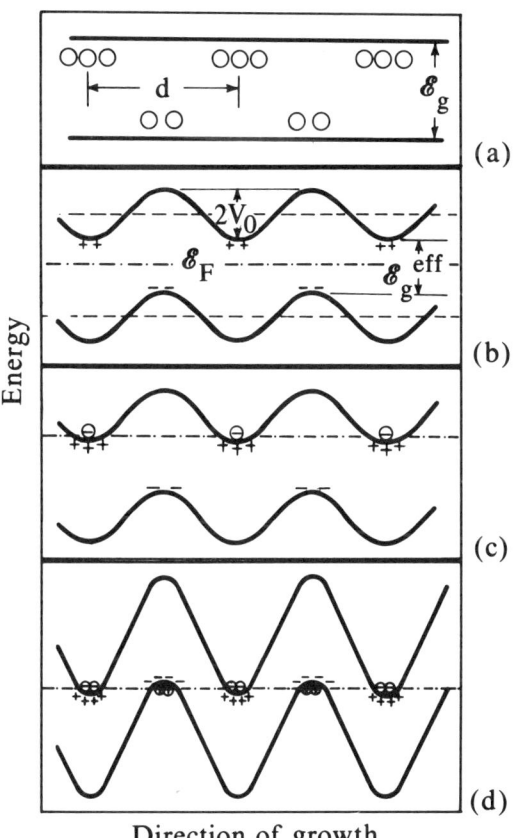

Fig. 18-12 Various aspects of nipi structures. (a) The periodic donor and acceptor doping of a semiconductor using a Maxwell demon to keep them from ionizing. (b) The modulated bands when the donors and acceptors ionize and come to thermal equilibrium. (c) A nipi structure with an excess of donors. (d) A semimetal nipi.

and valence band edges remain as shown, independent of z. However, the donors are thermally ionized, leaving behind positively charged ions, N_D^+, and the electrons lower their energy by attaching themselves to the acceptors, forming negatively charged acceptors, N_A^-. In the same manner as discussed previously, Fig. 18-11c, this periodic space charge, via Poisson's equation, yields a periodic electric field and thereby a periodic electron potential energy, of period d. This results in modulated conduction and valence band edges yielding $\mathscr{E}_c(z)$ and $\mathscr{E}_v(z)$, as shown in Fig. 18-12b. If the doping concentration is not too high and if the number of donors per n-layer ($N_D d_n$) is equal to the number of acceptors per p-layer ($N_A d_p$), then the amplitude of the periodic potential energy, $V_0/2$, can be estimated. The difference in the field in the intrinsic layer can be estimated using as a model a dielectric with positive and negative charge on the opposite surfaces (a double layer). Then $E_i = 2\pi e N_D d_n / \epsilon$, where ϵ is the static dielectric constant of the unmodulated semiconductor, Table 8-3. If the doping layers are thin compared to d, then

$$V_0/2 \approx eE_i d/4 \approx (2\pi e^2 n_D/\epsilon)(dd_n/4) \qquad (18\text{-}4a)$$

If there are no intrinsic layers and the periodic doping is such that $N_D = N_A$ and $d_n = d_p$, then the result is 1/2 of that in Eq. 18-4a or

$$V_0/2 = (2\pi e^2 N_D/\epsilon)(d/4)^2 \qquad (18\text{-}4b)$$

Of course, this periodic potential is superimposed on the normal crystal periodic potential.

Any excess electrons existing or created will go to the lowest energy state in the conduction band. Similarly, holes will "float up" to the highest energy state in the valence band. Figure 18-12c shows an n-type nipi structure that has more donors than acceptors (a p-type material can be made similarly). Figure 18-12d shows the extreme case of a "doping superlattice semimetal" where the bottom of the conduction band has a lower energy than the top of the valence band; however, this overlap occurs in real space rather than in k-space as in an ordinary semimetal (Section 10-20). As can be seen, energy is no longer gained by recombination. In general, an effective band gap of a nipi structure can be defined as

$$\mathcal{E}_g^{\text{eff}} = \mathcal{E}_g - V_0 \qquad (18\text{-}5)$$

From Eqs. 18-4 we can see that V_0 can be increased by varying the period spacing or doping concentration. In practice, doping superlattice semimetals have been made in GaAs, $(Al_xGa_{1-x})As$, PbTe, and other semiconductors.

Quantized energy levels – Qualitatively, the resultant energy profile in Fig. 18-12b resembles the heterojunction superlattice in Fig. 18-1b. The space charge potential quantizes the motion of the carriers in the z-direction, leading to a subband structure as for the compositional superlattice. However, since the shape of the potential energy is different in the doping superlattice case than for the compositional superlattice case, the quantized energy levels are not given by Eq. 18-1. The energy levels at the bottom of the potential well can be approximated by a simple harmonic oscillator. In this approximation the electrons in the conduction band have energies that are given by

$$\mathcal{E}_\nu = \hbar \left(\frac{4\pi N_D e^2}{m_e \epsilon} \right)^{1/2} (\nu + \tfrac{1}{2}); \quad \nu = 0, 1, 2, \ldots \qquad (18\text{-}6)$$

where m_e is the electron effective mass. For holes N_A must be used and either the effective mass of the light holes or heavy holes, depending on which set of energy levels is desired. As can be seen, material parameters such as m^*, N_D, and ϵ affect $\mathcal{E}_\nu$, while for a rectangular well (Eq. 18-1) only the effective mass appears. The energy levels are equally spaced for the doping superlattice case, Eq. 18-6, while for the compositional superlattice, Eq. 18-1, the spacing increases with increasing n. These statements apply to infinitely deep wells. Naturally, in both cases the number of bound energy levels is restricted because of the finite well depths.

To appreciate the order of magnitudes, we note that for GaAs and $N_D = 10^{18}$ cm^{-3}, then $\mathscr{E}_\nu = 40$ $(\nu + \frac{1}{2})$ meV. With this value for N_D and with $d_n = d_p = d/2 = 650$ Å, then $V_0/2$ is about one-half of the band gap of GaAs ($\mathscr{E}_g \approx 1.5$ eV). Thus, the well is deep enough to allow for many quantum states.

Tunability – In many ways a doping superlattice is quite similar to a compositional superlattice. However, a doping superlattice has one general property that is rather different; it has a large range of tunability. That is, for a given sample, certain of the electronic and optical properties can be varied within wide limits by weak excitation, for example, by low intensity optical signals or by small electron and hole injection currents. We discuss this here.

Tunability is a consequence of the origin of the periodic superlattice potential. In the doping superlattices the potential is due to space charges which are variable, while in the heterojunction superlattices the potential results from the different positions of the energies of the bands of the component semiconductors. Thus, for the latter superlattice, changes of the potentials are much more difficult to achieve.

Consider what happens when excess electrons and hole pairs are generated. (The term **excess carriers** refers to a situation where more carriers are present than are in thermal equilibrium.) The electrons will go to their lowest energy state, which is the bottom of the conduction band, and the holes will go to the top of the valence band. Thus, the electrons and holes are *separated physically* by half of a superlattice period. In analogy to the familiar indirect gap in momentum space, a doping superlattice could be called a semiconductor with an "indirect gap in real space." The excess carrier lifetimes in the familiar indirect gap materials (in momentum space) are very much longer ($\sim 10^{-5}$ sec) than those of direct gap materials ($\sim 10^{-9}$ sec) because of the need for phonon emission to conserve crystal momentum (Section 13-19). For the "indirect gap in real space" material the lifetime is determined by tunneling in real space. With the proper design parameters this value can be longer than 10^2 sec! In view of this very long lifetime, large deviations from equilibrium can be obtained and sustained with moderate electron-hole generation or injection rates.

There is a further consequence of the generation of excess electrons and holes. The electrons would go to the bottom of the conduction band and combine, neutralizing a certain fraction of the positively charged donors, and the holes combine with the negatively charged acceptors and neutralize them. Thus, the bands, which started out as in Fig. 18-12b, will change to the configuration shown in Fig. 18-12a (i.e., $\mathscr{E}_g^{\text{eff}}$ increases). This happens because the space charge, which sets up the superlattice periodicity, is neutralized.

A straightforward method of creating electron-hole pairs is to irradiate the sample with light with energy greater than $\mathscr{E}_g$. In bulk GaAs, when the sample is photo-excited above $\mathscr{E}_g$ the recombination of the excess electrons and holes leads to the emission of photons (i.e., photoluminescence at or just below the band gap). The same photoluminescence is observed for the doping superlattices, but the emission occurs at $\mathscr{E}_g^{\text{eff}}$. However, when the rate of generation of electron-hole pairs is increased (by increasing the intensity of the above-band gap light), $\mathscr{E}_g^{\text{eff}}$ increases toward $\mathscr{E}_g$ of bulk GaAs as discussed previously (i.e., going from Fig. 18-12b toward 18-12a). For a semiconductor this energy shift of the photoluminescence is large; experimental results are shown in Fig. 18-13.

Another easy way to modulate $\mathscr{E}_g^{\text{eff}}$ is by selective electron and hole injecting contacts to the n- and p-layers, respectively. At first glance this appears to be a difficult problem, but, with the use of selective injection contacts, it is easy. (See the Notes for references.)

The most important parameter that determines the output light intensity of both the electroluminescence and photoluminescence is the lifetime of the excess carriers. For a small number of excess electron-hole pairs in their lowest energy state, the electrons and holes must spatially tunnel across a distance $\lesssim d$ and must overcome a potential barrier $\approx V_0$. The lifetimes can be very long since they depend exponentially on V_0. However, for higher excitations V_0 decreases toward zero as discussed previously. This increases the tunneling probability, thus reducing the lifetime of the excess carriers and increasing the output intensity; effectively this causes the output intensity to depend very strongly on the position of the peak (i.e., on $\mathscr{E}_g^{\text{eff}}$).

18-5h The InAs-GaSb superlattice and others The InAs-GaSb superlattice is an example of a different type of superlattice. It has peculiar properties due to the relative position of the band edges. Due to the band offsets, the conduction band edge of InAs lies below the upper valence band edge of GaSb. Thus, we have a semimetal where the electrons and holes are separated in real space rather than in k-space as in a normal semimetal. A "real space" semimetal can also be achieved in the nipi superlattice as shown in Fig. 18-12d. A "normal" superlattice, where the electrons and holes are in the same material, is sometimes referred to as a **Type I superlattice**. When the electrons and holes are in the two different materials it is referred to as a **Type II superlattice**. The GaAs-GaP superlattice is another example of a Type II superlattice; it is a superlattice that can be made using the strain layer concept since the lattice constant mismatch is large.

CHAPTER 18 ARTIFICIAL STRUCTURES

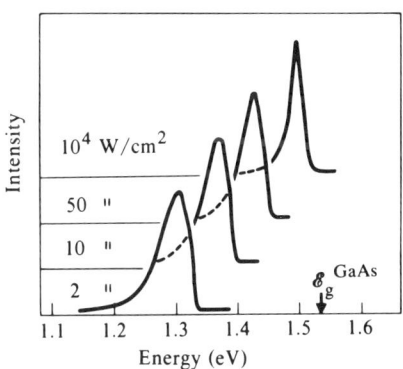

Fig. 18-13 The photoluminescence spectra of a GaAs nipi superlattice at 4°K. The watts/cm² refer to the intensity of the incident light. [G. H. Döhler, H. Künzel, D. Olego, K. Ploog, P. Ruden, H. J. Stolz, and G. Abstreiter, Phys. Rev. Lett. **47**, 864 (1981).]

Also, superlattices of both types with II-VI compounds and PbTe, etc. are being made.

To show the appropriateness of the quote at the beginning of this chapter we touch on one more topic, namely three or more component superlattices. For a three-component system we can have ABCABC..., ABACABAC..., ABCACBABCACB..., and so on. For example, direct band gap semiconductors such as InAs (0.36 eV) and GaSb (0.68 eV) can be combined with a relatively wide, indirect gap semiconductor AlSb (1.6 eV). The AlSb energy gap is much wider than those of the other two materials, making it possible for the AlSb layers to serve as potential barriers for the other two components. With such a three-component superlattice the structure can have an electric field-driven semimetal-semiconductor transition.

The field is just in its infancy. A few of the early review articles are given in the Notes. Due to the rapid pace, conference proceedings should be consulted for more recent developments.

18-6 Inversion Layers

18-6a Introduction An inversion layer is a very narrow layer (10-100 Å) at the surface of a p-type (or n-type) semiconductor where there is a deep enough potential well to contain electrons (or holes) below the Fermi level. For inversion layers the general ideas are similar to those discussed previously for the superlattices or more specifically for a one-dimensional well. However, the wells in inversion layers are not rectangular, but approximately triangular and the original method of obtaining the structure is completely different.

Historically, quantum effects in inversion layers were conceived and observed before superlattices. In 1957 Schrieffer pointed out that electrons confined in the narrow potential well of an inversion layer should show quantum effects. However, it was not until 1966 that quantization of the electrons in an inversion layer on a silicon surface

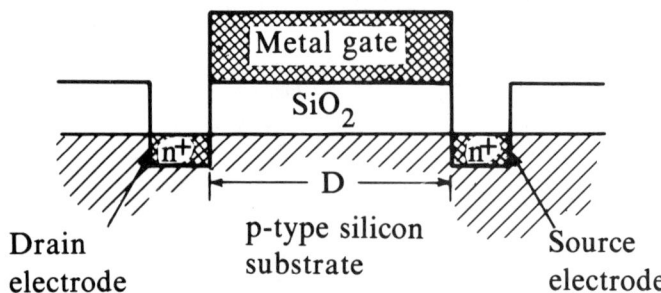

Fig. 18-14 A cross section of a silicon n-channel MOSFET. The current between the grounded source and the drain at a positive potential is controlled by the positive potential on the gate. An ohmic contact to the substrate (not shown) is also used to apply a small potential difference between the substrate and the source.

was observed and calculations of the quantum levels were performed. The calculations involved solving self-consistently the Schrödinger and Poisson equations taking into account the bulk doping and band structure.

Various aspects of inversion layers are discussed later, but first, we discuss the MOSFET structure. This is the structure in which the early work was done and it is still the most important and useful structure. The reason for this discussion is the intimate connection between an inversion layer and the structure. In the last part of this section we will discuss other structures and their technological uses.

18-6b MOSFET We can have an insulating layer between the surface of a semiconductor and a metal, called a **MIS** structure. On silicon, the insulator most often used is its oxide (SiO_2) since a flat, clean Si surface can be made, a controlled amount of oxide can be formed, and then a metal vacuum deposited. This is the silicon example of a general **metal-oxide-semiconductor** (**MOS**) structure. The most important point about the Si-SiO_2 interface is that the density of interface states can be kept small enough so that the potential energy well on the surface can be controlled by an external voltage. If the density of surface states were large the surface potential energy, and Fermi level, would be fixed by these states.

The early work in this field was done on this technologically well-developed metal-SiO_2-Si MOS structure. The technology was developed in the 1960s for the fabrication of field-effect transistors (**FET**) used as amplifying and switching devices for integrated circuits.

Figure 18-14 shows a metal-oxide-semiconductor field-effect transistor (**MOSFET**), which is sometimes called an insulated-gate field-effect transistor (**IGFET**). In the figure, p-type Si is the substrate so, as we shall see, it is an **n-channel** device. The MOSFET is a four-terminal device. Electrons are drawn from the source electrode (a very highly n-doped region called n+), usually at ground potential, along the surface of the p-type silicon to the drain electrode at some positive potential. An **n-channel** is formed at the Si surface by applying a positive potential to the metal gate (which is insulated from the p-type Si by the layer of SiO_2). The fourth terminal is attached to the p-type bulk Si (substrate contact). The general idea, to be discussed in more detail shortly, is that the flow of electrons from the source to the drain, which can take place only in an n-channel along the surface, is strongly controlled by the voltage on the metal gate. Since silicon is used, it is called a Si-MOSFET.

To appreciate some of the many technological problems with a MOSFET, we discuss surface or interface states on the oxide. For a good insulator, a surface charge density of $\gtrsim 10^{13}$ cm^{-2} will prevent control of the surface potential by the gate. This should be compared to the density of atoms on a Si surface $\approx 10^{15}$ cm^{-2}. Surface states at the interface between the oxide and semiconductor, where charges can be trapped, must be kept well below one state per 100 surface atoms. With the present state of the art, Si technology can produce less than 10^{10} states/cm^2. The smoothness of the Si surface, and hence the interface, is another important consideration. Mechanical roughness will scatter the mobile charge in the n-channel, so it must be kept to a minimum (i.e., a smoothness on the 10 Å scale, is required as for superlattices). Naturally, the oxide and the semiconductor must be perfectly bonded; this is no problem for the SiO_2-Si interface. However, minimizing the mechanical roughness and the density of surface states has required considerable effort and is now a highly developed technology. Excess charge within in the oxide must be kept to a minimum; it scatters carriers in the n-channel, limiting the electron mobility at low temperatures. At present, the charge density can be kept below 10^{10} cm^{-2} in good oxide films.

A general discussion of p-n junctions and metal-semiconductor junctions has been given in Sections 10-17 and 10-18, where the effects of forward and reverse bias are discussed. Here, the ideas are reviewed briefly. Referring to Fig. 18-15, consider the effect on the bands of applying a potential to the gate. Figure 18-15a shows the bands in the p-type silicon crystal with no potential applied to the gate and no interface states. The Fermi level is above the acceptor binding energy $\mathscr{E}_A$. In Fig. 18-15b a small negative potential applied to the gate causes an accumulation of holes at the interface, forming an **accumulation layer.** The shape of the band bending is approximately parabolic as in Eq. 10-88. Figure 18-15c shows the result of applying a positive potential to the gate. The bands are bent down, as indicated, depleting the surface of this p-type material of holes, leading to a **depletion layer** of **depletion width** $\approx 10^3$ to 10^4 Å. Figure 18-15d shows (similar to Fig. 10-37) the application of a positive potential to

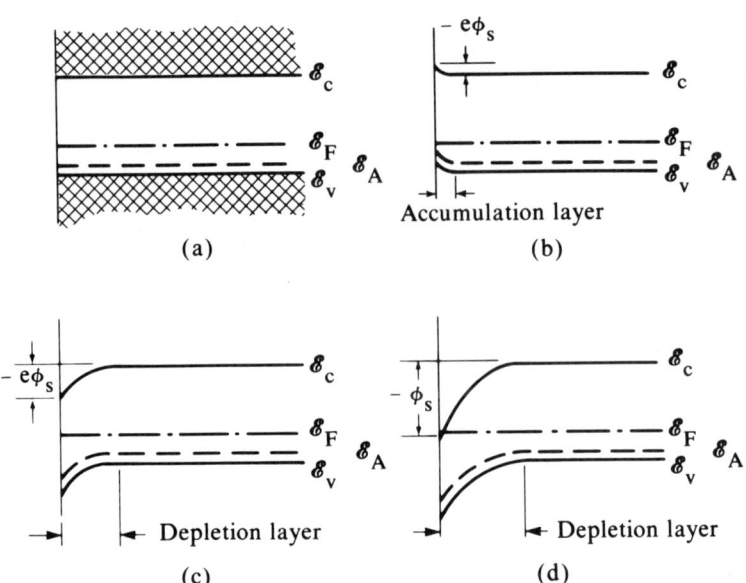

Fig. 18-15 The energy bands of a p-type semiconductor surface for various applied potentials as discussed in the text. $\mathscr{E}_A$ represents the acceptor energy level.

the gate, but the potential here is stronger. It is large enough so the conduction band is bent down below $\mathscr{E}_F$ to form an **inversion layer** (typically of width ≈ 10 to 100 Å). In this inversion layer the minority carriers (electrons in this p-type material) reside. This inversion layer is an **n-channel** where electrons can flow from the source to the drain, depending on the voltages applied. With no potentials applied to the drain (and substrate), electrons can occupy quantum levels in this inversion layer. Naturally, this is possible only if quantum levels exist in the inversion layer, and this is the subject of the next subsection.

(If the semiconductor is n-type instead of p-type and the bands are bent up, as in Fig. 18-15b, then the interface would have fewer electrons and we have a depletion layer. Similarly, if the bands were bent down, as in Fig. 18-15c, then we have an accumulation layer.)

18-6c Subbands For the 1-D rectangular well (Section 18-2) we found that the spacing of the quantum states increased and fewer quantum states are allowed as the wells become narrower and/or shallower. We may inquire as to well depth and width necessary for bound quantum states in the inversion layer case.

Estimation of the ground state — Using the Heisenberg uncertainty principle we can estimate the width and depth required. For the ground state that has a confinement length $\approx \langle z_0 \rangle$, momentum $\langle p_0 \rangle$, and energy $\langle \mathscr{E}_0 \rangle$,

CHAPTER 18   ARTIFICIAL STRUCTURES   741

$$\langle z_0 \rangle \langle p_0 \rangle \approx \hbar \qquad (18\text{-}7)$$

Assume that the conduction band bending in the inversion layer can be approximated by a straight line near the metal-semiconductor interface so that potential varies linearly with z. Then in terms of the surface electric field E and the effective mass for motion perpendicular to the surface (i.e., in the z-direction) m_z^*,

$$\langle \mathscr{E}_0 \rangle \approx eE\langle z_0 \rangle \approx \langle p_0 \rangle^2 / 2m_z^* \qquad (18\text{-}8a)$$

Substituting the two results in Eq. 18-8a into Eq. 18-7, we obtain

$$\langle \mathscr{E}_0 \rangle \approx (\hbar e)^{2/3} E^{2/3} / (2m_z^*)^{1/3} \qquad (18\text{-}8b)$$

For surface fields $\approx 10^5$ V/cm and $m_z^* \approx 0.1 \, m_e$ we find that $\langle \mathscr{E}_0 \rangle \approx$ 30 meV and $\langle z_0 \rangle \approx 30$ Å. These results are of the same order of magnitude as those for the 1-D rectangular case discussed in Sections 18-2 and 18-3. This implies that the surfaces must be smooth to better than 30 Å and the fields must be large enough to bend the conduction band below $\mathscr{E}_F$ by at least ≈ 30 meV in order for the quantum state to be occupied. The surface electric field in the semiconductor, which enters in Eqs. 18-8, can be estimated by assuming that all of the applied gate voltage, V_g, drops across the oxide; then the electric field in the oxide, E_{ox}, is V_g/d_{ox} where d_{ox} is the thickness of the oxide. Taking the normal component of **D** as continuous across the oxide-Si interface, $\epsilon_{ox}E_{ox} = \epsilon E$ where ϵ and E are the dielectric constant and electric field in the silicon. (For SiO_2 $\varepsilon_{ox} = 3.9$ and for Si $\varepsilon = 11.7$, Table 10-1.) With present Si technology the oxide can support as much as $E_{ox} = 10^7$ V/cm so in the Si, E can be as high as 3×10^6 V/cm.

Eigenvalues and eigenfunctions – The triangular approximation for the potential well is useful in solving for the eigenfunctions and eigenvalues of the quantum states in the inversion layer. In this approximation the oxide, z<0, is taken as an infinite barrier and the potential energy is eEz in the semiconductor. For this model the energy levels are

$$\mathscr{E}_\nu \approx \left(\frac{3\pi \hbar e}{2} \right)^{2/3} E^{2/3} \left(\frac{1}{2m_z^*} \right)^{1/3} (\nu + 3/4)^{2/3}$$

where $\nu = 0, 1, 2, ...$ \hfill (18-9)

The exact eigenvalues have $(\nu + 3/4)$ replaced by 0.7587, 1.7540, and 2.7575 for $\nu = 0, 1,$ and 2, respectively. The similarity of the results in Eq. 18-9 with the ground state result in Eq. 18-8b is evident. The eigenfunctions are Airy functions (see the Notes).

Consider the m_z^* term in Eq. 18-9. When discussing superlattices, we considered only direct band gap semiconductors where at **k** = 0

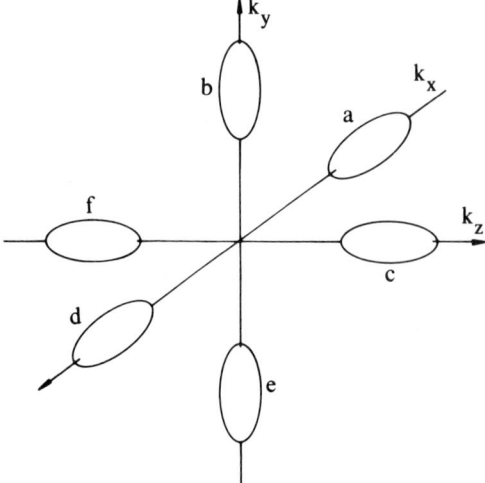

Fig. 18-16 Schematic constant energy surfaces for the conduction band of silicon, showing six conduction-band valleys in the ⟨100⟩ directions of momentum space. The long axis of an ellipsoid corresponds to the longitudinal effective mass of the electrons ($m_L = 0.91 m_L$) while the short axes correspond to the transverse effective mass ($m_T = 0.190 m$).

(the Γ-point) the effective mass of the electrons is isotropic. Thus, for the conduction band there was only one set of subbands; however, due to the structure of the valence band, and the existence of heavy and light holes, there were two separate subband ladders, one for each of the different holes. For the n-channel Si MOSFET we have to deal only with electrons in the conduction band. However, for silicon the conduction band minima are not at $\mathbf{k} = 0$ but at points in k-space 0.85 along the Δ-line to the X-point, that is, along the six ⟨100⟩ directions. The effective mass for the conduction band edge is anisotropic with a heavy, longitudinal mass and a light, transverse mass (Fig. 10-29 and the table at the end of Section 10-15); and there are six equivalent conduction band minima (or **conduction band valleys**) shown in Fig. 18-16. Properly taking the minima into account and using the correct masses is important in calculating the density of states and subband energies. In Section 10-15c the longitudinal and transverse masses are discussed and values are given; Table 10-1 lists the weighted average density of states masses for normal bulk (3-D) material (Section 10-16).

Let us consider the effect of a narrow quantum well in the z-direction with an infinite plane of semiconductor in the x- and y-directions. The energy levels are just like Eq. 18-2,

$$\mathcal{E} = \mathcal{E}_\nu + \frac{\hbar^2}{2m_x^*} k_x^2 + \frac{\hbar^2}{2m_y^*} k_y^2 \tag{18-10}$$

where $\mathcal{E}_\nu$ is the quantized energy values of Eq. 18-9 that arise when considering motion in the z-direction. As in the 3-D rectangular well case (Section 18-3), each $\mathcal{E}_\nu$ is the lowest energy of the νth subband and m_x^* and m_y^* are the appropriate effective masses of the electrons

CHAPTER 18   ARTIFICIAL STRUCTURES   743

in these directions, which are parallel to the semiconductor surface. For the 2-D free electron gas the density of states per unit area is independent of energy and is given by

$$g(\mathcal{E}) = g_s g_\nu (m_x^* m_y^*)^{1/2} / 2\pi\hbar^2 \qquad (18\text{-}11)$$

where g_ν is the conduction band valley degeneracy and g_s (=2) is the spin degeneracy. (See Problem 10 in Chapter 9.)

Now consider the different subband ladders that should be expected. (To do this the usual assumption of neglecting coupling of the electrons in the various conduction band valleys is made.) For this purpose we treat the case studied in the original and many subsequent inversion layer experiments, namely a Si {001} face, which also has the advantage of having the principal axes of the mass tensor orthogonal to one another. Consider the electrons in the c- or f-valleys shown in Fig. 18-16. The effective mass that enters into Eq. 18-9 for $\mathcal{E}_\nu$ is m_L since this is the mass for motion in the z-direction. For these two same valleys $m_x^* = m_y^* = m_T$ enters into Eqs. 18-10 and 18-11 and $g_\nu = 2$. This gives one subband ladder and the lowest energy one since the mass in $\mathcal{E}_\nu$ is the large, longitudinal electron mass. (These are unprimed in Fig. 18-17a.)

The other subband ladder arises when the other four subbands (a, b, d, and e) are considered. For each of these subbands m_T enters into $\mathcal{E}_\nu$ while in the density of states the product $m_T m_L$ enters and $g_\nu = 4$. Since m_T is much smaller than m_L, the quantum energy levels for these four subbands are higher than those arising from the c- and f-subbands considered previously. ($m_T = 0.19\ m_e$ and $m_L = 0.92\ m_e$, Section 10-15). The energy levels are primed in Fig. 18-17a.

Actual values – To solve for the eigenfunctions and eigenvalues we must self-consistently solve Poisson's and Schrödinger's equations. As discussed in Section 18-5f, Poisson's equation is $\nabla^2 \phi = -4\pi\rho/\epsilon$ where ρ is the charge density in the region of interest. From this potential a potential energy is obtained ($V_0 = q\phi$) that is used in the Schrödinger equation to produce eigenfunctions and hence a charge density, which then goes back into Poisson's equation. This procedure is iterated until self-consistent. Figure 18-17a shows typical energy levels for the two subband ladders for typical doping levels, and Fig. 18-17b shows the electron concentration obtained for an inversion layer on two different Si surfaces. The peak electron concentration is only 10 to 20 Å from the interface so the surface must be smooth compared to this distance if we are to see quantum effects. With present Si-MOSFET technology smoothness to ≈3 Å are obtainable.

Subband spectroscopy – For the superlattice case optical experiments at convenient energies close to the band gap can be performed by exciting from quantum states in the valence band to those in the

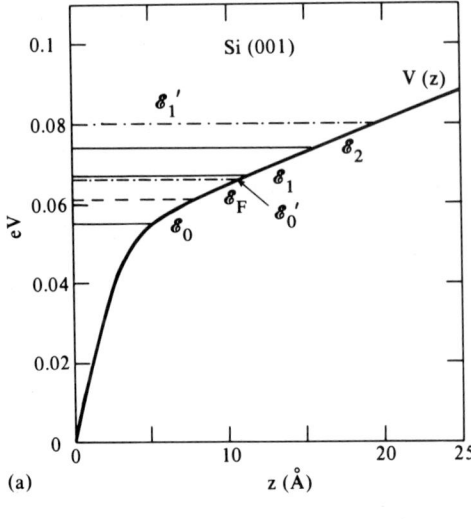

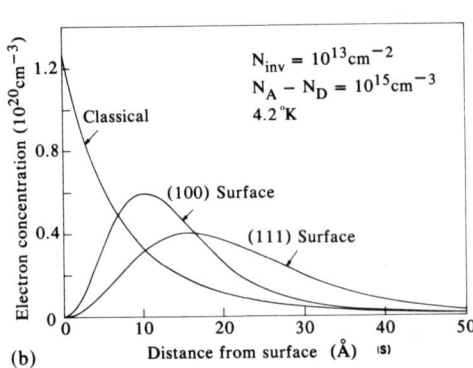

Fig. 18-17 (a) A typical energy level diagram for the two energy ladders as discussed in the text. (The small effects due to the image potential have been ignored.) [F. Stern, private communication.] (b) The charge density vs. z for some typical Si inversion layers for two different surfaces. [F. Stern in "Tenth International Conference of the Physics of Semiconductors" Ed. S. P. Keller, J. C. Hensel, and F. Stern (National Technical Information Service, 1970).]

conduction band. This is not possible for inversion layers because there are quantum levels only in one band, the conduction band for an n-channel device. Far infrared absorption measurements are the easiest to visualize, and have been utilized extensively. With this technique direct absorption of the far infrared radiation can be observed by exciting the electrons between the different quantum levels.

18-6d General comments From a scientific standpoint, an important attribute of the Si-MOSFET and related systems is that, in a single sample, the important parameters can be varied by orders of magnitude. For example, the number of electrons per unit area, $N/A \equiv n_s$, can effectively be varied from 5×10^{10} cm^{-2} to 10^{13} cm^{-2} in a single sample by varying the gate voltage. In normal bulk semiconductor materials, many different samples must be made to obtain the same data base. Let us examine the implications of varying n_s.

We obtain a nonzero value for n_s (for simplicity assume that all of the electrons are in the lowest quantum state) by applying a gate voltage large enough so that the tail of the Fermi distribution function just begins to fill up the first quantum state. For $n_s < 5 \times 10^{10}$ cm^{-2} the electrons tend to be localized at imperfection, surface states, and so on. By increasing the gate voltage n_s can be increased but is limited by voltage breakdown. For a 2-D electron gas $\mathscr{E}_F \propto n_s$ (see Problem 10 in Chapter 9). Thus, for this range of variation of the electron density, $\mathscr{E}_F$ and thus the kinetic energy $\langle KE \rangle$ of the electrons varies by a factor 20, and k_F varies by a factor 7 so that the average distance between electrons varies by a factor $1/7$. The Coulomb or potential energy of this electron system, $\approx e^2/r_0 = e^2 \pi^{1/2} n_s^{1/2}$, has a different dependence on n_s than the $\langle KE \rangle$.

As an example for the significance of these values note that at low n_s the Coulomb interactions may be large compared to the kinetic energy, but at high n_s the situation reverses. Similarly, different types of scattering — surface roughness, oxide charge, or phonon — depend differently on the carrier concentration and thus can be separated from each other. Also at low n_s the electrons may be localized by fluctuations in the potential and effects of varying the Fermi level can be studied in the localization regime. As can be seen, interesting scientific studies can be carried out in these materials. By the clever arrangement of electrodes or other approaches, studies of phenomena in 1-D systems are taking place. See the Notes for references.

18-6e MODFETs So far in Section 18-6 we have been discussing Si-MOSFETs and their use in obtaining inversion layers. Actually, a periodic array of inversion layers made by the modulation-doped technique in semiconductor heterojunction superlattices was discussed in Section 18-5f. As can be seen in Fig. 18-11c, any one of the interfaces is really an inversion layer. One advantage of making and studying inversion layers in semiconductor heterojunctions is that they can be made by the MBE technique with which very flat interfaces are obtainable. In addition, they can be made from III-V materials, for example, GaAs $-$ (Al$_x$Ga$_{1-x}$)As. The electron effective masses for III-V direct band gap materials are small, leading to high mobilities. In fact, electron mobilities $>10^6$ cm^2 V^{-1} sec^{-1} can be obtained in GaAs at 4.2°K for motion parallel to the surface (i.e., in the xy-plane), while values as high as 10^5 can be obtained at 77°K.

Figure 18-18 shows an inversion layer between two semiconductors with a metal gate (forming a Schottky barrier, Section 10-18). Typically, the semiconductor heterojunction is made from systems like GaAs-(Al$_x$Ga$_{1-x}$)As, and the donors are incorporated, during growth, into the wider band gap (Al$_x$Ga$_{1-x}$)As layer. In equilibrium, electrons from the ionized donors reside in the smaller band gap GaAs, as

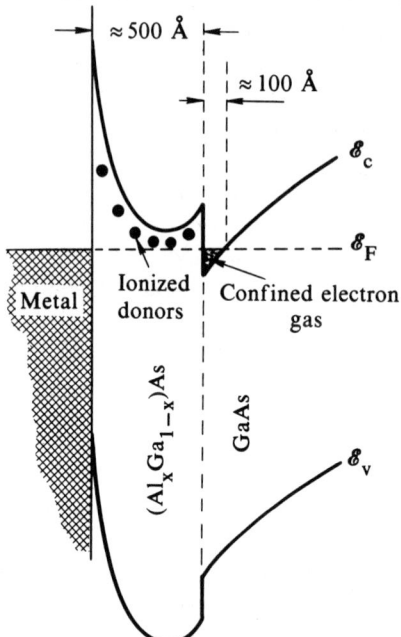

Fig. 18-18 A schematic diagram of a modulation doped semiconductor heterojunction with a metal gate. The energy scale and distances are approximate for x = 0.35, but are only meant to give the reader an idea of the magnitudes. The quantum levels are in the region labeled "confined electron gas". [To obtain the highest mobilities in the confined electron gas, the (Al, Ga) As nearest the heterojunction interface is grown without donors to further reduce the scattering.] (Courtesy of P. M. Mooney.)

indicated. As discussed in Section 18-5f, Poisson's equation and the Schrödinger equation must be solved self-consistently to determine charge distributions. By varying the bias on the metal gate, the density of electrons in the confined electron gas region can be controlled. As indicated in Fig. 18-14, source and drain contacts can be attached to the confined electron gas layer resulting in an FET. These structures go under many names.

HEMT – high electron mobility transistor
MODFET – modulation doped FET
SELFET – selectively doped FET
TEGFET – transferred electron gas FET

There is a great deal of scientific as well as technological interest in these systems. In the z-direction, the electronic states are quantized and the quantum effects can be studied. Also the very large transverse (in the xy-plane) mobilities appear to have important technological aplications, allowing fast FETs to be made.

From the numbers mentioned previously, the transverse mobilities in these MODFETs can be $\approx 10^3$ times larger than electron mobilities in silicon room temperature devices. It might be thought that the speed of operation of such devices should be $\approx 10^3$ times faster. However, this is not so for some physical as well as practical device reasons. These very large mobilities, μ, are small electric field mobilities; for small fields the drift velocity is given by $v_d = \mu E$ (Section 10-16), but for larger fields v_d reaches a maximum of about 2×10^7 cm/sec in most III-V materials. This value occurs because, as the carriers gain sufficient energy they can be transferred to other conduction band minima (e.g., the L-point in the Brillouin zone, Section 10-15).

Consider a device operating at 77°K (perhaps more practical than 4.2°K) where $\mu \approx 10^5$ cm^2/V sec for these MODFETs and $k_B T \approx 6.6$ meV. In order to overcome thermal fluctuation effects, the applied device voltages must be considerably larger than $k_B T/e$, at least 100 mV. These days devices are very small, 1μ ($=10^4$ Å $= 10^{-4}$ cm), so the electric fields in the device are ≈ 100 mV/10^{-4} cm $= 10^3$ V/cm. If the velocity were $=\mu E$, we would have $(10^5$ cm^2 V^{-1} sec^{-1}) $(10^3$ V cm^{-1}) $= 10^8$ cm sec^{-1}, which is well above the saturation velocity. Thus, typical fields in semiconductor devices are of a sufficient magnitude to drive the electrons well above their saturation velocities, indicating that the "effective" mobilities are much smaller than the small field mobilities.

Actually, many semiconductor devices are small enough so that the transport of electrons from the source to the drain is ballistic; this means that the electrons travel along the electric field lines without being scattered (i.e., the scattering time is larger than the transit time). Mobility is linked to average scattering times, so in the ballistic regime it has little meaning. For transport in the ballistic regime it is possible that drift velocities higher than the normal saturated values will occur. This would mean that the device would operate faster. A great deal of effort in the semiconductor field is being put into studying these devices.

Part B – Metals

18-7 Introduction

In this part of the chapter we discuss the layered structures of metals. Typically the measurements are electrical transport measurements, magnetic measurements, and measurements of the superconducting properties of the layered structures. Part B is shorter and more qualitative than the part on semiconductor materials because the state-of-the-art of sample preparation for the metals is behind that found for the semiconductors. This has resulted in less definitive results. Nevertheless, the motivation is similar: to prepare and study materials that exhibit different physical properties than those occurring naturally. Some aspects of this work is quite old, being initiated in 1940 with efforts to produce diffraction gratings for x-rays and later polarizers and monochromators for neutrons.

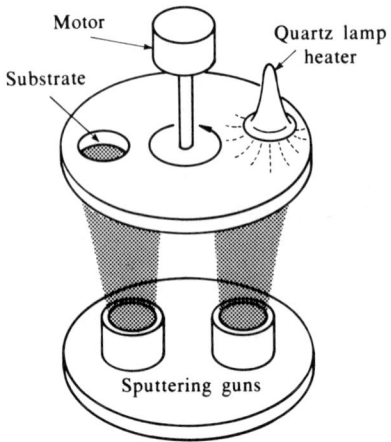

Fig. 18-19 A typical sputtering apparatus for preparing layered metal structures.

18-8 Sample Preparation

Although workers are beginning to use sophisticated growth techniques, at present much of the sample preparation is with apparatus of the type shown in Fig. 18-19. By varying the rotation speed of the substrate and the sputtering rates, various growth rates can be obtained (typically 10 to 200 Å/sec). Some of the disadvantages of this technique are the fast rates are not conducive to single crystal growth; sputtering requires an argon gas pressure of $\approx 10^{-3}$ to 10^{-2} torr, causing some of the gas to become embedded in the samples; the high pressures allow for oxygen contamination, particularly at the interface where it is least desired; during sample preparation there is no continuous in situ surface characterization. The advantage of this preparation technique is that thick samples can be prepared rapidly. Fortunately, many properties of these artificially layered metals are not sensitive to the previously mentioned problems. However, we note that as of the early 1980s people are beginning to study, for example, one, two, and three layers of a single crystal magnetic material deposited on nonmagnetic single crystal metal prepared by UHV surface techniques and other similar systems and approaches. New types of information should come from these structures.

X-ray and TEM techniques are used to determine the thickness of the layers and the crystal properties of the metals just as they are used for the semiconductors. With the rapid growth technique described usually one or both of the materials are disordered. For example, for the Nb/Ge system samples with d_1 and d_2 ranging from 5 to 100 Å have been prepared and for $d_{Nb} > 30$ Å these layers are ordered with a well-defined {110} texture. However, for $d_{Nb} < 30$ Å, the Nb layers are disordered. The absence of x-ray lines from the Ge material indicates that it is amorphous.

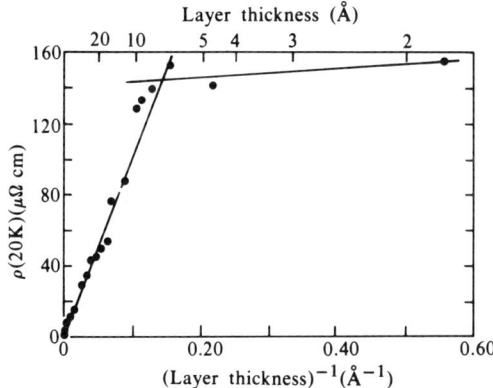

Fig. 18-20 The resistivity vs. inverse layer thickness for a series of Nb/Cu samples. [T. R. Werner, I. Banerjee, Q. S. Yang, C. M. Falco and I. K. Schuller, Phys. Rev. B26, 2224 (1982).]

18-9 Properties of Layered Metal Structures

To give a flavor of the field, we will describe just a few results of the various kinds of measurements taken in studying these materials. The Notes should be consulted for further reading.

18-9a Electrical transport Figure 18-20 is a plot of the 20°K parallel resistivity (in the xy-plane, as in Fig. 18-1) vs. layer thickness for different Nb/Cu artificially layered samples. Below 10 Å layer thickness the resistivity saturates close to 150 $\mu\Omega$ cm, the minimum metallic conductivity value, which is the Ioffe-Regel limit. For samples of intermediate thickness the resistivity depends on d^{-1}, as expected if the electron mean free path is limited by boundary scattering at the Nb-Cu interfaces. For very thick layers the resistivity approaches the value expected for bulk Cu and Nb films in parallel.

18-9b Magnetic properties For the magnetic materials we address the following questions: Can the layering produce enhanced magnetic properties over the homogeneous material? What is the coupling between the magnetic layers? Is it direct exchange, dipole-dipole, or RKKY?

An early magnetic, artificially layered material studied was the Ni/Cu system (1964). It has remained an interesting system and is discussed here.

One experimental technique that is especially useful for measurements in artificially layered materials is ferromagnetic resonance (Section 15-9c) since it is sensitive to the local magnetization, M(z). The original interpretation of the resonance data in the Ni/Cu system was that at temperatures below T_c of Ni the magnetization of Ni atoms within the layers was greater than that of bulk Ni. Needless to say, this result stimulated further work. Self-consistent band structure calculations of the magnetization of three atomic layers of Ni on three atomic layers of Cu in a [111] crystallographic direction have been performed. They indicate that the middle Ni layer has a magnetic moment slightly below the bulk value of 0.54 μ_B, while the two interface Ni layers have a magnetization of only 0.37 μ_B. The experiments have been repeated with samples of various layer thickness, and the macroscopic average magnetization as a function of temperature and magnetic field have been measured. These results, along with neutron measurements, show that the average magnetization ≈ 0.3 μ_B/Ni, in agreement with the calculations.

18-9c Superconductors Of the various artificially layered metallic systems that have been studied, the area of superconductors is the most advanced. The reason for this is that the layering can affect the superconducting properties significantly. Note that in addition to the layered materials discussed here, there are other layered superconducting materials. These include the layered dichalcogenides, their intercalates, and intercalated graphite. The Notes should be consulted for references.

Broadly speaking, there are three basic types of superconducting multilayers that can be prepared. (1) Two different metals, both superconducting, the S/S' system. (2) One of the materials can be a normal metal down to 0°K, the S/N system. (3) Last, the second material can be an insulator, the S/I system. The S/I system is similar to the intercalated materials in which the superconducting layers are coupled to each other by the spatial extent of the BCS wave function. In real structures, when interfaces are not sharp from a compositional point of view, we can have a more complicated system. For example, the S/N can be an S/S'/N system.

For the Nb/Ge, and for most of the systems studied so far, the temperature dependent Ginzburg-Landau coherence length $\xi(T)$ is smaller than the magnetic flux penetration depth λ. (See Chapter 16, Section 16-4d and the Summary.) For the layered materials $\xi(T)$ is anisotropic with the coherence length along the z-axis, $\xi_z(T)$ being the more important length (Fig. 18-1). Since there are several lengths to be considered, various regimes are possible. (1) For example, for the coarsest layering $\lambda \lesssim d \gg \xi_z(0)$ (taking $d_1 \approx d_2$ for convenience) the individual layers retain their bulk properties from the point of view of the order parameter and other microscopic superconducting properties. Such systems are of interest in studying supercurrent flux pinning and critical currents. (2) For very fine layering $d \ll \xi_z(0)$ the system behaves as a bulk anisotropic superconductor. The effects are observed most easily by studying the upper critical field $H_{c2}(T)$. (3)

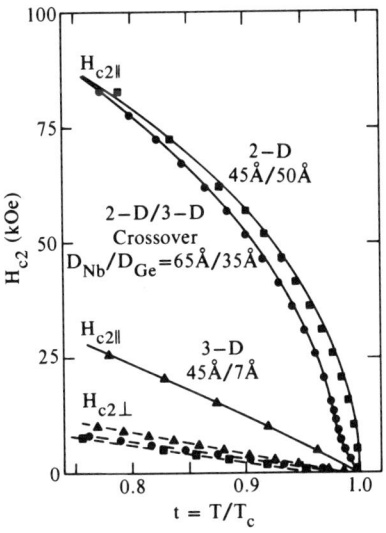

Fig. 18-21 The upper critical field near T_c for Nb/Ge layered samples. The solid lines are theoretical. [S. J. Ruggiery, T. W. Barbee, Jr. and M. R. Beasley, Phys. Rev. B**26**, 4894 (1982).]

When $d \lesssim \xi_z(0)$ the situation is most interesting. The spatial structure of superconducting pair potential, wave function, and ordering parameter are affected directly. Naturally this affects the critical fields, critical currents, and vortex dynamics in a fundamental manner, that is, the material cannot be treated simply as an anisotropic bulk superconductor. After a brief discussion of some of the appropriate theory, we discuss experimental results.

3-D anisotropy – We can extend the Ginzburg-Landau (GL) theory to the case of an anistropic superconductor in three dimensions (3-D). Since superconductors undergo a second-order phase transition (in zero H), then as $T \to T_c$ the coherence length diverges and extends over many superconductor layers. This implies that near T_c, $\xi_z(T)$ can be defined and the anisotropic GL theory is appropriate (neglecting fluctuations). At lower temperatures $\xi_z(T)$ decreases and can become comparable to or smaller than d; in this circumstance the model breaks down and other behavior is expected.

For a 3-D anisotropic superconductor, the GL theory can be used to determine the temperature dependence of the upper critical fields parallel and perpendicular to the z-axis; the results are:

$$H_{c2\perp}(T) = \frac{\Phi_0}{2\pi\xi(0)^2} \frac{T_c - T}{T_c} \qquad (18\text{-}12a)$$

$$H_{c2\parallel}(T) = \frac{\Phi_0}{2\pi\xi(0)\xi_z(0)} \frac{T_c - T}{T_c} \qquad (18\text{-}12b)$$

where Φ_0 is the superconducting flux quantum (Section 16-7e). For both fields there is a linear temperature dependence.

2-D system – For the case where $\xi_z(T) \lesssim d$, a 2-D regime exists and the temperature dependence of $H_{c2\parallel}$ is different while that of $H_{c2\perp}$ remains the same as in Eq. 18-12. Figure 18-21 shows an excellent experimental example of this effect in the Nb/Ge samples (an S/I system). The temperature dependence of H_{c2} is plotted for three different samples where the thicknesses of Nb to Ge were varied. For all three samples $H_{c2\perp}$ shows a linear temperature dependence in agreement with Eq. 18-12a. For the sample with Ge layers 7 Å thick, $H_{c2\parallel}$ also displays a linear temperature dependence. This means that $\xi_z(T)$ remains larger than 7 Å and the material can be treated as a 3-D anisotropic superconductor. However, note the results for the sample where $d_{Ge} = 35$ Å. Near T_c linear behavior is found, but for $T/T_c \approx 0.98$ there is a crossover from 3-D anisotropic behavior to 2-D behavior at lower temperatures. As can be seen, the measurements can be fitted by $H_{c2\parallel}(T) \propto [1 - (T/T_c)]^{1/2}$, in agreement with theory. For the sample with thicker Ge layers, 2-D behavior is observed at all measured temperatures with this same $[1 - (T/T_c)]^{1/2}$ behavior.

The preceding discussion is meant only to show some of the effects that can be observed in the superconducting artificially layered metals. For further reading the Notes should be consulted.

18-10 Other Artificial Structures

So far the topics covered in this chapter have been 2-D layered metals and semiconductors. However, there are other structures, artificially produced, that should be mentioned.

Using a technique developed by Langmuir and Blodgett that depends on the property that oil and water repel each other, large areas of material that is precisely one magnetic atom thick (i.e., a 2-D magnetic layer) can be produced and studied. For example, a Mn^{2+} ion is chemically bonded in the oil and when the film is "pulled" from the water solution a magnetic structure is obtained. As considered in the problems in Chapter 15, simple spin wave theory predicts that a 2-D spin structure will not be ferromagnetic. The experiments seem to differ; see M. Pomerantz, Surface Science **142**, 556 (1984).

Using surface science techniques, the properties of very thin (one, two, and three) atomic layers of magnetic atoms on a single crystal metal surfaces are beginning to be studied. For example, the magnetic properties of a few epitaxial layers of Ni atoms on a Cu (001) metal surface have been studied. This is a new field so very many new results will be appearing. For a review of some of the experimental and theoretical work see G. Bayreuther, J. Magnetism and Magnetic Materials **38**, 273 (1983) and A. J. Freeman, et al., ibid, p. 269.

Very thin layers of Pd, Cr, and V have been prepared epitaxially on Au or Ag and the superconducting as well as magnetic properties have been measured. The variation of the superconducting T_c's with Pd or V layer thickness can be determined and compared to theory. See M. B. Brodsky, Journal de Physique **45**, Supplement #4, Colq. C5, p. C5-349 (1984).

In the previously mentioned conference proceedings, pages C5-1 to C5-554, there are many other articles about various subjects that fall under the title of this chapter.

Notes

For a discussion of the behavior of a particle in a 1-D potential well and the transmission through 1-D potential barriers, see a beginning quantum mechanics books, for example, Chapter 4 of G. Baym, "Lectures on Quantum Mechanics" (Benjamin/Cummings, 1969), Chapter 6 of E. Merzbacher "Quantum Mechanics" (Wiley, 1970), or D. Bohm, "Quantum Theory" (Prentice Hall, 1951) p. 283.

Every two years there is an international conference on the electronic properties of two-dimensional systems. The proceedings have been published in the journal Surface Science **58** (1976), **73** (1978), **98** (1980), **113** (1982), **142** (1984). These and forthcoming proceedings should be consulted for the latest thrusts in this field.

For the semiconductor superlattice field, a few papers of a review nature are (a) R. Dingle, in "Festkörperprobleme: Advances in Solid State Physics", Ed. H. J. Queisser (Vieweg, Braunschweig, 1975), Vol. 15, p. 21; (b) L. Esaki and L. L. Chang, Thin Solid Films **36**, 285 (1976); (c) L. L. Chang and L. Esaki, Surface Science **98**, 70 (1980); (d) G. H. Döhler, Physica Scripta **24**, 430 (1981); (e) G. H. Döhler, Japanese J. of Appl. Phys. **22**, Suppl. 21-1, 29 (1983). The references (a), (b), and (d) are general, while InAs-GaSb is covered in (c) and doping superlattices are covered in (d) and (e).

Thin, apparently high quality, **amorphous semiconductory layered structures** are being produced and studied. See B. Abeles and T. Tiedje, Phys. Rev. Letters **51**, 2003 (1983).

An excellent and complete article reviewing inversion layer has appeared. T. Ando, A. B. Fowler, and F. Stern, Rev. of Mod. Phys. **54**, 437 (1982). Also see F. Koch, Surface Science **80**, 110 (1979).

A good general sourcebook is L. L. Chang and B. C. Giessen, Ed., "Synthetic Modulated Structures" (Academic Press, 1985). Besides covering the semiconductor field there are useful reviews on the less-studied metal and alloy fields. Another book that emphasizes fabrication is L. L. Chang and K. Ploog, Ed., "Molecular Beam Epitaxy and Heterostructures" (Martinus Nijhoff Publishers, 1985).

The **effective masses** used in various formulas for semiconductors with ellipsoidal constant energy surfaces can be subtle. See Section 10-15 and the text by Smith or Kireev. Also for Si and Ge for 3-D see W. F. Brinkman and T. M. Rice, Phys. Rev. B **7**, 1508 (1973) and for 2-D see F. Stern and W. Howard, Phys. Rev. **163**, 816 (1967). For **Strain layered superlattices** see: J. W. Matthews and A. E. Blakeslee, J. Vac. Tech. **14**, 989 (1977) and R. M. Biefeld, G. C. Osbourn, P. L. Gourley and I. J. Fritz, J. Electronic Materials **12**, 903 (1983).

Semiconductor structures are being made and studied in which **1-D conduction** occurs. MOSFETs with extra electrodes have been designed in which an electric field can pinch the 2-D electron gas in an accumulation layer, and a reversible transition from 2-D to 1-D conduction can be observed. See A. B. Fowler, A. Hartstein, and R. A. Webb, Phys. Rev. Letters **48**, 196 (1983), and the first three papers of a conference proceedings published in Science **142** (1984). For a discussion of the observation of the **quantum size effect** and related measurements, in metals see the article by: H. Hoffmann, Festkörperprobleme/Advances in Solid State Physics **22**, 255 (1982).

Quantized Hall effect — This effect was found (1980) at low temperatures and high magnetic fields in two-dimensional semiconductor systems of the type discussed in this chapter. Either Si-MOSFET inversion layers or heterojunction samples can be used. At low temperatures all of the electrons are in the lowest quantized subband (Eqs. 18-2b or 18-10) but of course they move freely in the xy-plane. The magnetic field (H) is applied perpendicular to the xy-plane. Then a given current density is passed through in the x-direction (J_x) and an electric field parallel (E_x) or perpendicular (E_y — the Hall field, Section 9-15) is measured. At high temperatures the **Hall resistivity** $\rho_{xy} \equiv E_y/J_x$ varies linearly with H as

$$\rho_{xy} = H/N_A ec \qquad (A-1)$$

where N_A is the number of carriers per unit area (see Eq. 9-66) and the longitudinal resistivity $\rho_{xx} = E_x/J_x$ is a weak function of H.

However, at sufficiently low temperatures and high magnetic fields, quite different results are found for a 2-D electron gas. As H is varied the Hall voltage develops a series of plateaus. For a given plateau the value of the Hall resistivity is given by

$$\rho_{xy}^{-1} = i(e^2/h) \qquad (A-2)$$

where i is a small integer. (In the plateau region ρ_{xx} decreases by orders of magnitude.) As the temperature is lowered (below 4°K) the plateaus become more distinct enabling Eq. A-2 to be used to determine ratios of the fundamental constants.

This effect can be understood in terms of **Landau levels** which occur when one quantizes free electrons in a magnetic field (Section 9-16). The eigenvalues for motion in the xy-plane are discrete, uniformly spaced simple harmonic oscillator levels

$$\mathscr{E}_n = \hbar\omega_c(n + \tfrac{1}{2}) \tag{A-3}$$

where n = 0, 1, 2, ..., $\omega_c \equiv eH/m^*c$ the cyclotron frequency (Eq. 9-67a), and m* is the appropriate electron effective mass The 2-D geometry is required to eliminate scattering between allowed states in the z-direction. For H = 10^5 kOe (10T) perpendicular to a Si(100) inversion layer, $\hbar\omega_c$ = 5.8 meV (17 meV for a GaAs heterojunction). This is much larger than thermal energies at 1°K ($\approx$0.1 meV). When the Fermi circle in the xy-plane lies between the Landau levels, an integer number of Landau levels are occupied fully and the higher energy ones are empty. The density of states per unit area for each Landau level is given by N_A = eH/hc. For i-levels it is

$$N_A = i(eH/hc) \tag{A-4}$$

Substituting Eq. A-4 into Eq. A-1, the experimental result (Eq. A-2) is obtained. By varying the magnetic field, plateaus in the Hall effect are observed when $\mathscr{E}_F$ is inside regions of localized states in the Landau levels (i.e., between Landau levels, Section 9-16 and Fig. 9-12.)

The quantum Hall effect and the **anomalous** (or **fractional**) **Hall effect** (where i in Eq. A-2 has rational fractional values) are active research fields. The first papers on the quantized and anomalous Hall effects were: K. v. Klitzing, G. Dorda and M. Pepper, Phys. Rev. Letters **45**, 494 (1980); D. C. Tsui, H. L. Stormer and A. C. Gossard, ibid **48**, 1559 (1982). For recent reviews see: K. v. Klitzing, Festkörperprobleme/Advances in Solid State Physics **21**, 1 (1981); B. I. Halperin, Helvetica Physica Acta **56**, 75 (1983); M. E. Cage and S. M. Girvin, Comments Solid State Phys. **11**, 1 (1983).

Problems

1. Band offsets — For the semiconductor heterojunction superlattices the materials have different band gaps ($\Delta\mathscr{E}_g$). Some of $\Delta\mathscr{E}_g$ is taken up as a band offset of the conduction band edges and some as a band offset of the valence band edges (Fig. 18-2). Very recent experiments have found that in the AlAs-GaAs system the band offsets are about equally divided between the conduction and valence bands. Discuss the recent measurements. See R. C. Miller, et al., Phys. Rev. **29**, 3740 and 7085 (1984), W. I. Want, et al., App. Phys. Lett. **45**, 639 (1984), J. Batey, et al., J. App. Phys. **57**, 484 (1985).

2. Do Problem 10 in Chapter 9.
3. Do Problem 12 in Chapter 9.
4. Do Problem 13 in Chapter 9.

APPENDIX - UNITS

It is not enough to develop equations, in the physical sciences one measures or describes quantities in terms of basic **units** that have physical dimensions. Since solid state physics is so diverse, it is not surprising that the units used by the practitioners also are diverse. In this book we use mainly the cgs-Gaussian system of units. This has been the traditional system used in this field. It is a mixed system of units with electrical quantities measured in cgs-electrostatic units (esu) and magnetic quantities measured in cgs-electromagnetic units (emu). The other system of units that is used is the SI system (Système International d'Unités). As mentioned in the preface, most of the equations can be used with either of the two unit systems. However, when the equations would appear differently if SI units were used instead of cgs-Gaussian units, then the equations are also written in SI units, or it is stated explicitly how the conversion to SI units is accomplished.

The **fundamental constants**, on the inside backcover, are listed in both sets of units. The sources are: B. N. Taylor, W. H. Parker, and D. N. Langenberg, Rev. Mod. Phys., **41**, 375 (1969), and E. R. Cohen and B. N. Taylor, Journal of Physical and Chemical Reference Data, **2**(4), 663 (1973). Instead of the official name for charge in the Gaussian system of statcoulombs we, and many others, use esu.

The table of **energy conversions** given in this appendix should help you to convert between the different energy units that are often used. The units of wave number are defined by one over the wave length. So a photon with $\lambda^{-1} = 1$ cm^{-1} has an energy of 0.12398 meV, an equivalent temperature of 1.439°K, and so on; also remember that $\lambda^{-1} = \nu/c$. (Note that wave length is given on the far right, which is $\propto \mathcal{E}^{-1}$ and is given by $\nu\lambda = c$.)

The book by J. D. Jackson, "Classical Electrodynamics" (Wiley, 1975) has an appendix that includes a good discussion of units and dimensions and shows how to convert symbols and equations from one system to the other. Also see R. H. Bube, "Electrons in Solids" (Academic Press, 1981). The **coversion table** given here should be helpful to convert from SI to cgs-Gaussian and vice versa.

Atomic units (a.u.), although not used in this book, are used by some solid state scientists. In this system, $\hbar$, the proton charge (e), and the electron mass (m) all have numerical value of unity. Thus, the unit of length is the Bohr radius (a_0), and the unit of energy is the Hartree ($e^2/a_0 = 27.212$ eV). So in this system

$$1 \text{ a.u. of length} = 0.52918 \times 10^{-8} \text{ cm}$$
$$1 \text{ a.u. of energy} = 27.212 \text{ eV}$$
$$1 \text{ a.u. of time} = 2.41891 \times 10^{-17} \text{ sec}$$

Then in a.u. the velocity of light c = 137.036.

The **various conversion factors** in this appendix lists other terms that might be helpful. Also note the gas constant (R) in various units is given in the Notes of Chapter 6 and various units of pressure are given in the Notes of Chapter 14.

APPENDIX - UNITS

Conversion Table from SI to cgs-Gaussian

Physical Quantity	Symbol	SI		cgs-Gaussian
Length	l	1 meter (m) =	10^2	centemeters (cm)
Mass	m	1 kilogram (kg)	10^3	grams (gm)
Time	t	1 second (sec)	1	second (sec)
Frequency	ν	1 hertz (Hz)	1	hertz (Hz)
Force	F	1 newton	10^5	dynes
Work	W	1 joule	10^7	ergs
Energy	$\mathscr{E}$, U	1 joule	10^7	ergs
Power	P	1 watt	10^7	ergs sec^{-1}
Charge	q	1 coulomb	3×10^9	statcoulombs
Charge density	ρ	1 coul m^{-3}	3×10^3	statcoul cm^{-3}
Current	I	1 ampere (amp)	3×10^9	statamperes
Current density	J	1 amp m^{-2}	3×10^5	statamp cm^{-2}
Electric field	E	1 volt m^{-1}	$1/3 \times 10^{-4}$	statvolt cm^{-1}
Potential	Φ, V	1 volt	1/300	statvolt
Polarization	P	1 coul m^{-2}	3×10^5	dipole moment cm^{-3}
Displacement	D	1 coul m^{-2}	$12\pi \times 10^5$	statvolt cm^{-1} (statcoul cm^{-2})
Conductivity	σ	1 mho m^{-1}	9×10^9	sec^{-1}
Resistance	R	1 ohm	$1/9 \times 10^{-11}$	sec cm^{-1}
Capacitance	C	1 farad	9×10^{11}	cm
Magnetic flux	ϕ	1 weber	10^8	gauss cm^2 or maxwells
Magnetic induction	B	1 tesla	10^4	gauss
Magnetic field	H	1 ampere-turn m^{-1}	$4\pi \times 10^{-3}$	oersted
Magnetizaiton	M	1 ampere m^{-1}	10^{-3}	magnetic moment cm^{-3}
Inductance	L	1 henry	$1/9 \times 10^{-11}$	–

Energy Conversions

	Energy $\mathscr{E}$		Bulk Energy $N_A\mathscr{E}$	Temp. $\mathscr{E}/k$	Freq. $\mathscr{E}/h$	Wave Number $\mathscr{E}/(hc)$	Wave Length $(hc)/\mathscr{E}$
	eV	erg	erg/mol	°K	cycles/sec	cm^{-1}	Å
1 eV	1	1.6021×10^{-12}	9.6487×10^{11}	11605	2.4181×10^{14}	8065.8	12398.5
1 erg	6.2418×10^{11}	1	6.0226×10^{23}	7.244×10^{15}	1.5093×10^{26}	5.0345×10^{15}	1.9863×10^{-8}
1 erg/mol	1.0364×10^{-12}	1.6604×10^{-24}	1	1.203×10^{-8}	250.61	8.3594×10^{-9}	1.1963×10^{16}
1 °K	8.617×10^{-5}	1.381×10^{-16}	8.314×10^{7}	1	2.084×10^{10}	0.6950	1.439×10^{8}
1 cycle/sec	4.1355×10^{-15}	6.6255×10^{-27}	3.9903×10^{-3}	4.799×10^{-11}	1	3.3356×10^{-11}	2.9979×10^{18}
1 cm^{-1}	1.2398×10^{-4}	1.9863×10^{-16}	1.1963×10^{8}	1.439	2.9979×10^{10}	1	10^{8}
1 Å	1.2398×10^{4}	1.9863×10^{-8}	1.1963×10^{16}	1.439×10^{8}	2.9979×10^{18}	10^{8}	1

Various Conversion Factors

length	1 Å = 10^{-8} cm = 10^{-1} nm
	1 μ = 10^{-4} cm = 1 μm
energy	1 erg = 10^{-7} J
	1 cal = 4.184 J
free energy	1 kcal mol^{-1} = 4.184×10^{3} J mol^{-1}
entropy	1 e.u. = 1 cal deg^{-1} mol^{-1} = 4.184 J °K^{-1} mol^{-1}

π = 3.14159
1 radian = 57.296°
e = 2.71828
ln 10 = 2.30259
log$_{10}$e = 0.43429

BIBLIOGRAPHY

Solid State Physics

Altman, S. L. *Band Theory of Metals.* Elmsford, NY: Pergamon Press, 1970.

Ascroft, N. W., and N. D. Mermin. *Solid State Physics.* New York: Holt, Rinehart and Winston, 1976.

Azároff, L. V. *Introduction to Solids.* New York: McGraw-Hill, 1960.

Blakemore, J. S. *Solid State Physics.* Philadelphia: W. B. Saunders, 1969.

Blatt, F. J. *Physics of Electronic Conduction in Solids.* New York: McGraw-Hill, 1968.

Born, M., and K. Huang. *Dynamical Theory of Crystal Lattices.* New York: Oxford University Press, 1954.

Brillouin, L. *Wave Propagation in Periodic Structures.* New York: Dover Publications, 1953.

Brown, F. C. *The Physics of Solids.* Menlo Park, CA: Benjamin-Cummings, 1967.

Bube, R. H. *Electronic Properties of Crystalline Solids.* New York: Academic Press, 1974.

Busch, G., and H. Schade. *Lectures on Solid State Physics.* Elmsford, NY: Pergamon Press, 1978.

Callaway, J. *Energy Band Theory.* New York: Academic Press, 1964.

Dekker, A. J. *Solid State Physics.* Englewood Cliffs, NJ: Prentice-Hall, 1957.

Donovan, B. *Elementary Theory of Metals.* Elmsford, NY: Pergamon Press, 1967.

Elliot, R. J., and A. F. Gibson. *Solid State Physics.* New York: Macmillan, 1974.

Hardy, J. R., and A. M. Karo. *The Lattice Dynamics and Statics of Alkali Halide Crystals.* New York: Plenum Press, 1979.

Kittel, C. *Solid State Physics.* New York: John Wiley and Sons, 1953-1976.

Lovell, M. C.; A. J. Avery; and M. W. Vernon. *Physical Properties of Materials.* New York: Van Nostrand Reinhold, 1976.

Mott, N. F., and E. A. Davis. *Electronic Processes in Non-Crystalline Materials.* Oxford: Clarendon Press, 1979.

Mott, N. F., and H. Jones. *The Theory and Properties of Metals and Alloys.* New York: Dover, 1958.

Peierls, R. E. *Quantum Theory of Solids.* New York: Oxford University Press, 1955.

Seitz, F. *The Modern Theory of Solids.* New York: McGraw-Hill, 1940.

Seitz, F.; D. Turnbull; and H. Ehrenreich. *Solid State Physics, Advances in Research and Applications.* New York: Academic Press, 1955 to 1985. This series is cited as Solid State Physics.

Wannier, G. H. *Elements of Solid State Theory.* New York: Cambridge University Press, 1959.

Weinreich, G. *Solids: Elementary Theory for Advanced Students.* New York: John Wiley and Sons, 1965.

Zhdanov, G. S. *Crystal Physics.* New York: Academic Press, 1965.

Ziman, J. M. *Electrons and Phonons.* New York: Cambridge University Press, 1960.

Ziman, J. M. *Principles of the Theory of Solids.* New York: Cambridge University Press, 1972. References to "Ziman" are to this book.

Oriented to Semiconductors

Dalven, R. *Introduction to Applied Solid State Physics.* New York: Plenum Press, 1980.

Greenaway, D. L., and G. Harbeke. *Optical Properties and Band Structure of Semiconductors.* New York: Pergamon Press, 1968.

Kireev, P. S. *Semiconductor Physics.* Moscow: MIR Publishers, 1978.

Moss, T. S.; G. J. Burrell; and B. Ellis. *Semiconductor Opto-Electronics.* Woburn, MA: Butterworth, 1973.

Pankove, J. I. *Optical Processes in Semiconductors.* Englewood Cliffs, NJ: Prentice-Hall, 1971.

Seeger, K. *Semiconductor Physics.* New York: Springer-Verlag, 1982.

Smith, R. A. *Semiconductors.* New York: Cambridge University Press, 1978.

Streetman, B. G. *Solid State Electronic Devices.* Englewood Cliffs, NJ: Prentice-Hall, 1980.

Sze, S. M. *Physics of Semiconductor Devices.* New York: John Wiley and Sons, 1981.

Optical Effects in Solids

Born, M., and E. Wolf. *Principles of Optics.* New York: Macmillan, 1964.

Fowler, W. B., ed. *Physics of Color Centers.* New York: Academic Press, 1968.

Hodson, J. N. *Optical Absorption and Dispersion in Solids.* New York: Chapman and Hall, 1970.

Wooten, F. *Optical Properties of Solids.* New York: Academic Press, 1972.

Crystal Chemistry, Structures, and Symmetry

Adams, D. M. *Inorganic Solids.* New York: John Wiley and Sons, 1974.

Bhagavantum, S. *Crystal Symmetry and Physical Properties.* New York: Academic Press, 1966.

Burns, G., and A. M. Glazer. *Space Groups for Solid State Scientists.* New York: Academic Press, 1978.

Evans, R. C. *Crystal Chemistry.* New York: Cambridge University Press, 1964.

Galasso, F. S. *Structure and Properties of Inorganic Solids.* New York: Pergamon Press, 1970.

Henry, N. F. M., and K. Lonsdale. *International Tables for X-Ray Crystallography, Vol. 1.* Birmingham, England: Kynoch, 1952, 1965, 1969. This is the book on space groups. For brevity we usually refer to it as the International Tables.

Hirschfelder, J. O.; C. F. Curtiss; and R. B. Bird. *Molecular Theory of Gases and Liquids.* New York: John Wiley and Sons, 1954.

Krebs, H. *Inorganic Crystal Chemistry.* New York: McGraw-Hill, 1968.

Kitaigorodskii, A. I. *Molecular Crystals and Molecules.* New York: Academic Press, 1973.

Kitaigorodskii, A. I. *Organic Chemical Crystallography.* New York: Consultants Bureau, 1961.

Megaw, H. D. *Crystal Structures: A Working Approach.* Philadelphia, PA: W. B. Saunders, 1973.

Nye, J. F. *Physical Properties of Crystals.* New York: Oxford University Press, 1957.

Pauling, L. *The Nature of the Chemical Bond.* Ithaca, NY: Cornell University Press, 1939-1960.

Wooster, W. A. *Tensors and Group Theory for the Physical Properties of Crystals.* New York: Oxford University Press, 1973.

Wykoff, R. W. G. *Crystal Structures,* Vols. **1-5**. New York: John Wiley and Sons, 1963-1968.

Group Theory

Burns, G. *Introduction to Group Theory with Appliations.* New York: Academic Press, 1977.

Cotton, F. A. *Chemical Applications of Group Theory.* New York: John Wiley and Sons, 1971.

Tinkham, M. *Group Theory and Quantum Mechanics.* New York: Mc-Hill, 1964.

INDEX

AB alloys, 402
Absorption above $\mathcal{E}_g$, 520
Absorption near $\mathcal{E}_g$, 500-524
Acceptor, 172
Acceptors, see Semiconductors
Accidental degeneracy, 29
Accumulation layer, 739
Accumulation region, 336
Acoustic branch, 411
Acoustic modes, 411
Activation energy (diffusion), 390
    table of values, 391
Activation process, 686
Additional boundary conditions (abc), 525
Adiabatic approximation, 441
Alloyed junctions, 324
Amorphous metal, 352
Amorphous semiconductors, 343, 349, 523, 525
    layered structures, 753
Anderson localization, 280
Anderson transition, 184
Anharmonic effects, 428, 435, 446
Anisotropic, 87
Anisotropy energy, 613
Anisotropy, 87, 612, 630
    easy directions, 613
    hard directions, 613
Annihilation operators, 330, 346
Anomalous dispersion in solids, 382
Anomalous skin effect, 457, 490
Anti-Stokes scattering, 432, 729
Antibonding, 150
Antiferroelectric, 546, 558
    table of materials, 549
Antiferromagnetism, 566, 585, 591
    tables, 585
Antisymmetry principle, 236, 588
Appearance potential spectroscopy (APS), 699
Attempt frequency, 393
Attenuation coefficient, 456
Auger electron, 698
Auger process, 698
Avalanche breakdown, see p-n junctions
Average thermal energy, 444

Ballistic transport, see Thermal conductivity
Band (one dimensional), 719
Band filling, 350
Band index, 255
Band offsets, 754
Band theory, 241-252
Barkhausen effect, 616
Basis, 37, 51
BCS theory, 642
Biased junction, see p-n junctions
Bitter technique, 615
Bloch condition, 253
Bloch form, 252
Bloch functions, 252, 342, 349
Bloch $T^{3/2}$ law, 629
Bloch walls, 617
Bloch's theorem, 252
Body-centered cubic lattice, 29, 302
Body-centered cubic (bcc) structure, 55
Bohr magneton, 568
Bonding types, 101-128,
    see Covalent bond
    see Hydrogen bond
    see Ionic bond
    see Metals
    see Molecular bond
    repulsive energy, 115
Bonding, 150
Born and von Karman, 399
Born model, 138
Born-Haber cycle, 144
Bose-Einstein factor, 427
Bose-Einstein statistics, 427, 443
Bosons, 443
Bragg planes, 80
Bragg scattering, 75, 251, 260
Bragg's law, 75
Bravais lattices, 24, 29-35
Breakdown diodes, see p-n junctions
Brillouin function, 576, 587
Brillouin zones, 84, 260-273, 302, 304, 350 727
    at a phase change, 416, 441
    $\delta\mathcal{E}/\delta k$ at zone boundaries, 271
Built-in potential, 325
Bulk modulus, 445
Burgers vector, 403
Burstein-Moss effect, 309

c/a ratio, 64
Carrier concentration, 313, see Semiconductors

Cauchy principle value, 459
Cauchy relations, 442
Centering of lattices, 29
Central peak, 561
Chalcopyrite, 441
Characteristic wavelengths, 75
Charge density waves, 624
Chemical shift, 602, 696
Chemisorption, 686-692
Classical statistical Mech., 360
Classification of solids, 247
Clausius-Mossotti equations, 484, 526
Clean Surfaces, 677
Clinographic projections, 61
Close packed structures, 62
Closure domains, 616
Co-adsorption, 691
Coercive field, 532, 615
Cohesive energies of ionic crystals, 138
Cohesive energy, 102, 132
Collisions, 195
Color centers, 396, 400
Compatibility relations, 299
Compensation point, 596
Competitive adsorption, 691
Compositional superlattices, 732
Compressibility, 138, 140
Conduction band edge, 287
Conduction band, 168
Conduction electron density, 188, 194
Conductivity (dc), 196, 223, 633
Conductivity (frequency dependent), 198
Conductivity, 99, 287, 453
Configurational entropy, 375
Conformational effects, 378
Constitutive relations, 451
Contact potential, 325
Contracted notation, 94
Conventions, 284, 286, 292, 314, 325, 329, 511
Cooper pairs, 646
Cooperative adsorption, 691
Coordinates, 44
Coordination number, 52, 65, 133
Correlation energy, 348
Counter doping, 323
Covalent bond, 147-186
  dissimilar atoms, 152
  hybridization, 158
  maximum overlap, 157
  σ-bond, 158
  π-bond, 160
  tetrahedral bonding, 148, 159, 185
Covalent radii, 161, 185
Covering operations, see Symmetry operations

Creation operators, 430, 446
Critical current density, 636
Critical field H_c, 635, 649
Critical points, 423, 520
Critical region, 584
Critical transitions, 562
Crowdion configuration, 395
Crystal field, 579
Crystal momentum, 280, 428
  selection rules, 429
Crystal shape, 675
Crystal structures, 49-70
  close packing, 62
  defect structures, 60
  tables, 50
Crystal systems, 26
Crystallographic point groups, 35
CsCl structure, 52, 372
Cubic close-packed structure, 64
Cuprite structure, 70
Curie constant, 535
Curie law, 575
Curie-Weiss law, 586
Cyclotron frequency, 229, 310
Cyclotron resonance, 231, 310

d-electrons, 578
  3d-ions, 578
  4d-ions, 583
  5d-ions, 583
Dangling bonds, 684
Dangling orbitals, 684
de Broglie law, 73, 204
de Broglie relationship, 204
de Haas-van Alphen (dHvA) technique, 272
Debye frequency, 367
Debye temperature, 367, 401
  table, 354
Debye unit, 402
Debye-Scherrer powder method, 81
Debye's calculation of C_V, see Specific heat
Defect structures, 60
Demagnetization factor, 607
Demagnetization field, 607, 616
Demagnetizing energy, 616
Density of states, $g(\mathcal{E})$, 208, 211
Depletion approximation, 327
Depletion layer, 739
Depletion length, see Depletion width
Depletion region, see Depletion width
Depletion width, 325, 328, 335, 739
Derivative spectroscopy, 520
Diamagnetism (perfect), 638

INDEX 767

Diamagnetism, 567, 638
Diamond anvil cell, 559
Diamond structure, 56, 148
Diatomic chain, 412
Diatomic molecule, 401
Dielectric constant, 180, 453, 484
   of ferroelectrics, 383, 531-546
      Lyddane Sachs Teller relation, 468
   tables, 94, 183
Dielectric polarization, 450
Dielectric response of a quantum system, 461
Dielectric susceptibility, 450, 452
Diffused junctions, 324
Diffusion, 389-396, 400
   attempt frequency, 393
   constant, 389
   Einstein-Nernst equation, 390
   Fick's first law, 389
   Fick's second law, 392
   mechanisms, 393-396
   potential, 325
Diffusivity, 389
Dimensionality effects, 212, 238, 403
Direct exchange, 624
Direct inspection method, 90
Direct lattice, 29, 78, 260
Direct transitions, 500
Directions, 43
Dislocations, 403, 725
   edge, 403
   screw, 403
Disorder, 371-398
Dispersion curve, 407
Dispersion relation, 409, 455
Displacive phase transition, see Phase transition
Domain wall, 615
Domains, 533, 615
Donor-acceptor pairs, 516
Donors, see Semiconductors
Doping superlattices, 732
Double bonding, 160
Double hysteresis loops, see Ferroelectricity
Double layer, 702
Drift velocity, 195
Drude's Assumptions, 195-197
Drude's model, see Metals
Dulong and Petit law, 356, 399
Dynamic apparent charge, 466

Easy axes, 613
Easy directions, 613
Effective density of states, 315

Effective magnetic moment, 576
Effective mass theory, 171
Effective mass, see Semiconductors
Effusion cell, 714
Einstein distribution, 364
Einstein temperature, 364
Einstein-deHaas gyromagnetic experiment, 630
Einstein-Nernst equation, 390
Einstein's model, 362, 399
Elastic coefficients, 95
Electric quadrupole interaction, 603
Electromagnetic waves in solids, see Optical properties of crystals
Electron affinity, 131, 143
Electron charge densities, 177
Electron diffraction, 72
Electron energy loss spectroscopy (EELS), 700
Electron gas, see Metals
Electron heat capacity, see Metals
Electron paramagnetic resonance (EPR), 579, 598
Electron spectroscopy for chemical analysis (ESCA), 695
Electron spin paramagnetism, see Metals
Electron spin resonance (ESR), see Electron paramagnetic resonance
Electron storage rings, 693
Electron-electron collisions, 225
Electron-electron repulsion, 154, 225
Electron-hole drop, 513
Electron-lattice interaction, 643
Electron-phonon interaction, 643
Electronegativity, 175
Electronic Raman scattering, 729
Electronic surface structure, 702-712
Electrooptic effect, 554
   coefficient, 554
   longitudinal, 558
Empty lattice approximation, 293
Enantiomorphic, 5, 48
Energy bands for simple lattice, 350
Energy gap (2Δ), 649
Energy gap, 167
Energy loss measurements, 498
Entropy, 375, 388, 402, 537
Epitaxially growth, 323, 724
ε_∞ effect, 464
Equipartition theorem, 362
Equivalent k-values, 291
Esaki diodes, 332, 351
Essential degeneracy, 292
EXAFS, 700
Exchange charge, 588
Exchange coefficient, 588

Exchange constant, 588
Exchange energy, 589
Excitons, 509-517
　complexes, 513
　conditation, 513
Exclusion principle, 474
Extended x-ray absorption fine structure (EXAFS), 700
Extended zone, 254
Extremal orbits, 273
Extrinsic semiconductors, see Semiconductors
Extrinsic semimetal, see Semimetals

f-electrons, 577, 578
Face-centered lattice, 24, 35, 304
Face-centered cubic (fcc) structure, 56, 62
Fast ion conductors, 60, 389, 400
Fermi energy, 208
　tables, 188, 189
　temperature, 209, 215
　velocity, 209
　wave vector, 209
Fermi gas, 225
Fermi level pinning, 321
Fermi level with doping, 351
Fermi liquid theory, 225, 236, 347
Fermi sphere, 208
Fermi surfaces, 208-211, 266-273
　examples, 266-271
　nesting, 623
Fermi-Dirac statistics, 215, 443
Fermiology, 271
Fermions, 215, 236, 426
Ferrimagnetism, 566
Ferrites, 593
Ferroelasticity, 557
Ferroelectricity, 377, 529-563
　anharmonic model, 551
　displacive, 376, 535
　domain, 523
　free energy, 375, 536, 560
　hysteresis loop, 377, 530, 541
　isotopic effect, 532
　optic properties, 554
　order-disorder, 378, 535
　order parameter, 372, 536
　oxygen polarizability model, 552
　polarization, 382, 531, 538, 540
　pressure effects, 559
　soft modes, 472, 542
　table of materials, 536
Ferromagnetism, 566, 585, 590, 629, also see Magnetism
　metals, 619
　spin waves, 604
　tables, 585
FET, see Field effect transistor
Fick's first law, 389
Fick's second law, 392
Field effect transistor, 738
Field emission, 706
First Brillouin zone, 263
First-order phase transition, see Phase transition
First-order transition, 377, 534, 539
Fluctuating dipole force, 118
Flux density, 566
Flux quantization, 658
Fluxoid, 659, 667
Folding of the Brillouin zone, 416, 441
Forward-biased, 329
Four phonon processes, 430
Fowler equation, 706
Fowler-Nordheim equation, 706
Fröhlich interaction, 643
Free carrier absorption, 486-498
　longtiudinal modes, 497
　metals, 489
　oscillator model, 487
　plasma frequency, 488
　plasmon, 498, 527
　semiconductors, 493
　transverse modes, 495
Free electron properties, see Metals
Free energy, 375, 536, 560
Freeze-out range, 320
Frenkel defect, 386
Frenkel excitons, 514
Friedel oscillations, 704
Frustration, 627
Fundamental absorption, 499

$g(\mathscr{E})$, see Density of states
g-values, 234, 236
Gap, magnitude of, 246, 259
Garnets, 594
General equivalent positions, 8, 41
General points, 8, 41, 291
Generators of a group, 92
Gibbs free energy, 138, 373, 387
Ginzburg Landau theory, 665
　GL parameter, 665
Glasses, 341, 404
Glide planes, 19
Graded junction, 323
Group of Γ (etc.), 290
Group of $\mathbf{k}$, 290
Group velocity, 410, 456

Grown junctions, 323
Gunn Effect, 337
Gyromagnetic ration, 573

Hagen-Rubens relation, 490, 526
Hall effect, 228-231, 236, 351
  angle, 229, 351
  coefficient, 230
  quantum, 754
  two carriers, 351
Hard directions, 613
Hard superconductor, 642
Harmonic oscillator, 361, 461
Hartree approximation, 622
Hartree-Fock approximation, 622
He diffraction, 711
Heat capacity, of electrons, 214
Heat capacity, see Specific heat
Heat of adsorption, 686
Heat pulse experiments, 11
Heavy fermion metals, 348, 668
Heavy-hole band, 306
Heisenberg exchange interaction, 588
Heitler-London approach, 154
Helmholtz free energy, 373, 387
HEMT, 746
Hermann-Mauguin notation, 6
Hexagonal close-packed structure, 62
Holes, 172, 281-285, 306, 312
Holosymmetric point group, 37
Hund's rules, 571
Hybridization, 158, 185, 186
Hydrogen bond, 124
  molecular ion, 150
  molecule, 154
Hyper-Raman techniques, 560
Hysteresis loop, 377, 513, 541, 614

Ice rule, 126
IGFET, 738
Impact ionization, 332
Impurity band conduction, 174, 321
Independent electron approximation, 203, 243, 280
Index of refraction, 183, 456
Indirect band gap, 306
Indirect exchange, 593, 624
Indirect transitions, 502
Infrared measurements, 474
Integrated circuits (IC), 323
Interband transitions, 498, 525
  absorption near $\mathscr{E}_g$, 500-520
  absorption above $\mathscr{E}_g$, 520
  acceptors, 172, 509
  alloys, 508
  direct transitions, 500
  donors, 170, 509
  heavily doped materials, 505
  indirect transition, 502
  magnetooptical, 519
  vertical transitions, 500
Intermediate state, 641
Internal energy, 355
International notation, 5
International Tables for X-ray
  Crystallography, 53
Interstitial diffusion mechanism, 394
Intrinsic range, 319
Intrinsic semiconductors, see
  Semiconductors
Intrinsic semimetals, see Semimetals
Inverse photoemission (Bremsstrahlung)
  spectroscopy, 701
Inversion layers, 737, 740
Ion implantation junctions, 324
Ionic bonding, 129-145
  cohesive enervies, 138, 142
  ionic radii (table), 130
Ionization energy, 131, 143
Iron group metals, 520
Iron series ions, see d-electrons
Irreducible representations, 293
Isomer shift, 601
Isomorphic compounds, 532
Isotope effect, 532, 645, 656
Isotropic, 87
Itinerant electron picture, 619
Itinerant exchange, 625

Jellium model, 702
Josephson junction, 659, 661

k-space space, 79, 207, 210
  volume in, 207
Kondo effect, 625, 628
Kramers-Kronig relations, 458

Landé g-factor, 573
Landau Fermi liquid, see Fermi liquid
  theory
Landau free energy theory, see Free
  energy
Landau levels, 233-235, 272, 754
Langevin function, 381, 575
Langevin theory of paramagnetism, 575
Langmuirs, 691
Lanthanide contraction, 578

Larmor diamagnetic susceptibility, 569
Larmor frequency, 568
Lasers, 333
Lattice complex, see Basis
Lattice vibrations, 361- 371,
      407-446, 466-486
  acoustic, 411, 422
  critical points, 423
  crystal momentum, 428
  diatomic chain, 412, 444
  dispersion curve, 407, 409, 422-424
  frequencies (table), 448
  group velocity, 410
  longitudinal, 408, 480
  molecular case, 417
  monoatomic chain, 408, 443
  normal modes, 408ff, 425
  optic, 413
  optical properties, 465-486
  phase transitions, 416, 441
  phonons, 425
  transverse, 419, 480
Lattice, 25
Laue formulation, 77
Laue method, 82
Law of mass action, 314
Layered metal structures, see Metallic superlattices
Layered structures, see Superlattics
LEED, 678, 711
Lennard-Jones potential, 123, 128
Lenz's law, 567
Light emission, 332
Light emitting diodes (LEDs), 332
Light-hole band, 306
Linear combination of atomic orbitals (LCAO), 277
Linear orthogonal transformations, 89
Liquid crystals, 122, 127
Liquid phase epitaxial (LPE) growth, 323
Local field problem, 480
Localized vibrational modes, 398
London equation, 639
London penetration depth, 640
London-dispersion force, 118
Long-range order parameter, 372, 400, 402, see Ferroelectricity, Magnetism, Superconductivity
Longitudinal free electron modes, 497
Longitudinal modes, 408, 448, 467
Lorentz oscillator, 461, 526
Lorenz number, 198
Lorenz-Lorentz equations, 484, 526
Low energy electron diffraction (LEED), 678
LS coupling, 570

Lyddane Sachs Teller (LST) relationship, 468, 543

$\overline{m}$-space, 209
Mössbauer effect, 599
Madelung constant, 140, 144, 145
Magnetic dipole moment, 573
Magnetic gas, 569
Magnetic hyperfine structure, 602
Magnetic materials (soft), 615
Magnetic materials (hard), 615
Magnetic monopole, 667
Magnetic resonance, 598
Magnetic structures, 590
Magnetism, 563-630
  antiferromagnetism, 566, 585, 591
  diamagnetism, 567, 629
  domains, 615
  ferrimagnetism, 566, 585, 593
  ferrites, 593
  ferromagnetism, 566, 585, 590, 629
  metals, 619
  paramagnetism, 566, 569, 575
  spin waves, 604ff, 629
  structures, 590
  superlattices, 749
  tables, 585
Magnetocrystalline energy, 613
Magnetoelastic energy, 617
Magnetooptical absorption, 519
Magnetoresistance, 228-233
  longitudinal, 231
  transverse, 231
Magnetostatic energy, 616
Magnetostriction, 617
Magnons, 609
Majority carriers, 325
Martensitic transformation, 558
Maximum overlap, 157
Maxwell's equations, 455
Mean field theory, 552, 584
Mean free path, 195, 197
Meissner effect, 638
Melting temperatures, 183
Metal-insulator transition, 174, 184
Metal-nonmetal transition, 174, 184
Metal-oxide-semiconductor, 335, 738
Metal-semiconductor junctions, 334-337
Metallic bond, 190
Metallic superlattices, 747
  electrical peroperties, 749
  magnetic properties, 749
  superconductors, 750
Metals, 187-240, see Superconditivity
$\rho = 0$, 637

INDEX 771

conductivity (ac), 198
conductivity (dc), 99, 196, 223, 749
Drude model, 191-203, 487
electron pressures, 192
electron spin paramagnetism, 221, 58
$\mathscr{E}_F(T)$, 218
free electrons, 187-240
nearly free electrons, 243
optical absorption, 200, 489
properties (tables), 188, 194, 210, 22
specific heat, 219
superlattices, 747-752
thermal properties, 217-223
Miller indices, 44
Minibands, see Subbands
Minority carrier lifetime, 325
Minority carriers, 325
MIS, 738
Missing row model, 689
Mixed state, 641
Mobility edge, 343
Mobility, 288
MODFET, 745
Modulation spectroscopy, 520, 526
Modulation-doped, 730
Molar polarizability, 485
Molecular approximation, 417
Molecular beam epitaxy (MBE), 677, 714, 724
Molecular bond, 118-124
Molecular dissociation, 120
Molecular field theory, 584
Molecular orbital approach, 154
Monatomic chain, 408
  group velocity, 410
Morphotropic transition, 134
MOSFET, 738
Motional narrowing, 393
Mott transition, 184
Mott-Wannier excitons, 510
Multiplet, 570
Multiply primitive unit cells, 26, 33

n-channel, 738
Néel temperature, 591
NaCl structure, 56
Narrow banding techniques, 598
Nearly free electrons, 257, see Metals
Negative differential conductivity, 337, 727
Nesting Fermi surfaces, 623
Net, 683
Neumann's principle, 88
Neutron diffraction, 71, 430, 596
nipi structures, 732

NMR, see Nuclear magnetic resonance
No-crossing rule, 304
Nodal surface, 160
Nondegenerate, see Semiconductors
Nonlinear optics, 96
Nonprimitive unit cells, 26, 33
Nonsymmorphic space group, 38
Nonvertical transitions, see Indirect transitions
Normal modes, 408, 425
Normal processes, 437
Nuclear magnetic resonance, 559, 598
Nuclear quadrupole coupling, 145
Number operator, 446

Ohm's law, 196, 451
Ohmic contacts, 336
One electron energies, 204
  approximation, 347
  wavefunctions, 204
One-phonon process, 475
Optic branch, 413
Optical absorption, 308
Optical polarizability, 454
Optical properties of crystals, 447-527
  dispersion relation, 455, 458
  group velocity, 456
  index of refraction, 456
  infrared, 474
  Kramers-Kronig relations, 458
  lattice vibrations, 465-486
  local field, 480
  macroscopic theory, 450-464
  Maxwell's equation, 455
  phase velocity, 456
  Raman, 474, 479
  reflectivity, 457
Orbital angular momentum, 570
  quenching, 578
Orbital quantum numbers, 206, 570
Order parameter, 536
Order-disorder phase transition, see Phase transition
Organic metal superconductors, 667
Orientational disorder, 371-385
Orthogonal relations, 89
Oscillator strength, 465
Overhauser antiferromagnetic state, 623

p-channel, 740
p-n junctions, 323-337, 351
  avalanche breakdown, 322
  capacitance, 331
  light emission, 332

reverse bias breakdown, 332
see Semiconductors
Zener breakdown, 332

p-type pairing, 668
Packing fraction, 137
Packing of molecules, 121
Paramagnetic phase, 585
Paramagnetism, 566, 569, 575
Partition function, 358
Pauli exclusion principle, 207, 236, 588
Pauling's electronegativity model, 176
Peierls instability, 624
Penetration depth, 652
Penn model, 178
Periodic boundary condition, 206
Periodic potential, 244
Periodic table, 67
Perovskite structure, 52, 470, 532, 536
Phase problem, 347
Phase transition, 371, 531-562
    Brillouin zone effects, 416, 441
    displacive, 376, 535
    first order, 376
    lattice vibrations, 416, 472, 542
    long-range order, see Long-range order
    order-disorder, 376, 400, 535
    second order, 376
    see Ferroelectricity
    see Magnetism
    see Superconductivity
Phase velocity, 456
Phillips' and Van Vechten's method, 178, 184, 186
Phonon mean free path, 434
Phonons, 425-446
Photoconductivity, 170, 509
Photoelectron spectroscopy, 695, 705
Photoemission, see Photoelectron spectroscopy
Physisorption, 686
π-bond, see Covalent bond
Piezoelectricity, 94, 556
Planck's law, 427
Planes, 45
Plasma edge, 488
Plasma frequency, 201, 488
Plasma wavelength, 201
Plasma, 486
Plasmon, see Free carrier absorption
Plastic crystals, 122, 127
Plastic flow, 403
Point group of a space group, 58
Point groups, 9-22, 48

Point Imperfections in Crystals, 385-400
    density of, 387
Point symmetry operation, 1-22
Point symmetry, 12-22
Polaritons, 476-480
Polarizability, 92, 481
Polarization by orientation, 379-385
    ac effects, 382
    gases and liquids, 379
    solids, 381
Polytypes, 64, 442
Pressure effects, 65, 120 134, 175, 559
Pressure units, 561
Primitive lattice translation, 25
Primitive unit cell, 26, 33
Principal axis, 5
Promotional energy, 159
Pyroelectric detectors, 91
Pyroelectricity, 90

Quantum Hall effect, see Hall effect
Quantum interference, 663
Quantum size effect, 717, 754
Quantum theory of paramagnetism, 575
Quantum wells, 717
Quasi-localized state, 398
Quasiparticles, 426

Raman effect, 474, 479, 525, 729
Random close-packed (rcp), 352
Rare earth ions, see f-electrons
Rayleigh scattering, 474
Reciprocal lattice, 78, 83, 84, 260, 350
Reciprocal space, 79
Recombination radiation, 333, 510
Reconstruction (of surfaces), 683
Rectangular well, 719, see Quantum Well
Rectifier, 329
Reduced zone, 253, 261
Reflection high energy electron diffraction (RHEED), 679
Reflectivity at an interface, 457
Relaxation behavior, 199
Relaxation time, 195
Remnant magnetization, 615
Renormalization group, 552, 561
Repeated zone, 254
Repulsive potential energy, 115
Residence time, 686
Resistivity ratio, 342
Resonance energy, 156
Resonance states, 398
Resonant tunneling, 723
Reverse bias breakdown, see p-n junction

INDEX 773

Reversed-biased, 329
Reversible spontaneous polarization, 377, 531
Richardson-Dushman equation, 237, 705, 712
Riemann zeta function, 403
Ring diffusion mechanism, 394
RKKY interaction, 625, 627
Rotating crystal method, 82
Russell-Saunders coupling, 570

s-type pairing, 668
Saturation range, 319
Scanning tunneling microscope, 685, 711
Schoenflies notation, 5
Schottky anomalies, 359
Schottky barriers, 335
Schottky defects, 385
Screw operation, 18
Second-order phase transition, see Phase transition
SEGFET, 746
Seitz notation, 18
Self diffusion, see Diffusion
Semiconductor model, 655
Semiconductor picture, 655
Semiconductor superlattices, 726
Semiconductors, 281-348, 726
    absorption (free carriers), 493
    acceptors, 172, 287
    amorphous, 341, 404, 523, 753
    band structures, 304
    charge density, 177, 285
    conduction band edge, 287
    conduction band, 168, 304
    conductivity, 287
    degenerate, 286
    donors, 170, 287, 730, 745
    doped, 285, 317, 730
    effective mass, 171, 248, 284, 306, 310, 741, 754
    extrinsic, 173, 285, 317
    FETs, see Field effect transistor
    holes, see Holes
    intrinsic, 169, 285, 317
    inversion layers, 737-747
    light emission, 332
    LASERs, 333
    nondegenerate, 286, 315
    optical properties, 486-524
    properties (tables), 119, 172, 183, 242, 308, 312
    quantum wells, 717
    superlattices, 715-737
    tetrahedral bond, see Covalent bond
    valence band edge, 287
    valence band, 268, 304
Semimetals, 339, 349
SEXAFS, 700
Shape-memory phenomena, 558
Shell model, 423, 444
σ-bond
Silent mode, 544
Simple cubic (sc) structure, 52
Site symmetry, 54, 67, 69
Skin depth, 490
Slater determinant, 588
Soft modes, 471, 542, 550
Soft x-rays, 213
Solid electrolytes, 60
Sommerfeld expansion, 216
Sommerfield theory of metals, 203-225
Space charge region, see p-n junctions
Space group, 19, 38, 48, 51, 59
Spatial dispersion, 457, 490
Spatial quantum numbers, 206
Special lines, 291, 521
Special points, 291, 521
Special positions, 41
Specific heat, 355-371
    anharmonic terms, 446
    classical results, 356, 360
    constant pressure, 355
    constant volume, 355
    of a crystal, 361
    Debye model, 365-371, 399
    Debye temperature (table), 354, 401
    diatomic molecule, 401
    Einstein model, 362-365, 399
    of electrons, 219n5
    free particles, 360
    glasses, 404
    harmonic oscillator, 361
    low dimensions, 403
    of spin waves, 610, 629
    of a superconductor, 651
    two-level system, 358, 401, 404
Spectroscopic splitting factor, 573
Specular tunneling, 724
Specularly reflected, 75
Spin angular momentum, 570
Spin degeneracy, 156
Spin density wave (SDW), 623
Spin glasses, 625
Spin wave dispersion, 629
Spin waves, 445, 604-611, 629
    antiferromagnetic, 611
    magons, 609
    observation, 607
    quantization, 609
    resonance, 608

specific heat, 610, 629
Spin-orbit coupling constant, 311, 570
Splat cooling, 343
Standing waves, 205
Step junction, 323
Stereographic projection, 8
Steric forces, 121
Sticking coefficient, 675
Stokes scattering, 432, 729
Strain layered superlattices, 725, 754
Stress strain, 445
Strong-coupling superconductors, 650, 667
Structural phase transitions, 529-562
Subbands, 233, 719, 727, 740
Super exchange, 593
Super ionic conductors, see Fast ion conductor
Superconductivity, 631-669
   BCS, 642
   Cooper pairs, 646
   critical temperature, 632, 633, 648
   energy gap, 649
   GL parameter, 665
   hard material, 642
   intermediate state, 641
   isotope effect, 645, 656
   Josephson effect, 659
   London equation, 639
   magnetic properties, 637
   Meissner effect, 638
   mixed state, 641, 665
   vs. perfect conductivity, 637
   specific heat, 651
   strong coupling, 650, 667
   superlattices, 750
   table of elements, 632
   transition temperature, 632
   tunneling, 653, 659, 665
   two-dimensional, 751
   vortex state, 641, 665
Supercurrent, 633
Superlattices, 715-756
   band offsets, 754
   doping, 732
   phonons, 727
   structures, 441, 715
   subbands, see Subbands
   type I (and II), 736
Surface charge density, 702
Surface concentration, 687
Surface coverage, 689
Surface packing density, 683, 712
Surface reconstruction, 683
Surface relaxation, 679
Surface states of electrons on liquid helium, 238
Surface structure notation, 681
Surface structure, 672, 679
Surfaces, 671-712
   atoms on surfaces, 686
   electronic structure, 702-710
   LEED, 678
   structures, 672, 679
   techniques table, 712
   UHV, 673
Symmetry operations, 1-22
Symmorphic space group, 38
Synchrotron radiation, 524, 692, 711
Szigeti charges, 483, 525
Szigeti equations, 483

TEGFET, 746
TEM, 725
Tensors, 88-99
Tetrahedral bond, 148, 159, 185
Thermal average energy, 361
Thermal conductivity, 433-442
   ballistic transport, 440, 442
   defects, 439
   heat pulse experiments, 440, 442
   superconductors, 657
Thermal effects in solids, 353-404
Thermal expansion, 99, 446
Thermal properties of an electron gas, see Metals
Thermionic electron emission, 704
Thermionic Emission, 237
Three phonon processes, 428
Tight binding approximation, 276-280
Tight binding, 249
Total angular momentum, 570
Transferred electron effect, see Gunn effect
Transition metals, 340
Transition width, see Depletion width
Translational symmetry, 17, 18 51
Transverse free electron modes, 495
Transverse modes, 419, 448
Transverse waves, 419, 466
Triple bonding, 160
Tunnel diodes, 332, 351
Tunnelling (single particle), 653, 665
Tunnelling, 653, 659
Two-level system, 358, 401, 404
Two-phonon process, 476
Type I (II) superconductors, 641
Type I (II) superlattice, 736

Ultrahigh vacuum (UHV), 673
Ultraviolet photoemission spectroscopy (UPS), 695
Ultraviolet transparency of metals, 200
Umklapp processes, 437
Unit cells, 25, 33
Unit mesh, 683
Universality classes, 553
Urbach edge, 521

Vacancy diffusion mechanism, 394
valence band edge, 287
Valence band, 168
Valence bond method, 155
van der Waals equation, 118, 128
Van Hove singularities, 520
vapor phase epitaxial (VPE) growth, 323
Vegard's law, 508
Vertical transition, 308, 501
Vertical transitions, see Direct transitions
Vibrational frequencies table, 448
Vibrational models (localized), 398
Virtual processes, 645
von Laue, see Laue formulation
Vortex state, 641

Wave packet, 433
Wave vector space, 204, see k-space
Weak links, 661
Weiss molecular field, 584
Wiedemann-Franz Law, 197
Wigner crystallization, 239, 624
Wigner lattice, 239, 624
Wigner-Seitz approximation, 273-275
Wigner-Seitz cell, 34
Work function, 193, 695, 704
Wurtzite structure, 66, 69

X-ray diffraction, 71-84, 352, 725
X-ray photoemission spectroscopy (XPS), 695

Young's modulus, 445

Zener breakdown, see p-n junctions
Zero of energy, 306
Zero-point energy, 364, 401, 426, 734
Zinc blend structure, 56